Encyclopaedia of

COMPLEX ANALYSIS

Encyclopaedia of
COMPLEX ANALYSIS

Volume 1

R.K. Pandey

ANMOL PUBLICATIONS PVT. LTD.
NEW DELHI - 110 002 (INDIA)

ANMOL PUBLICATIONS PVT. LTD.
H.O.: 4374/4B, Ansari Road, Daryaganj,
New Delhi-110 002 (India)
Ph.: 23278000, 23261597
B.O.: No. 1015, Ist Main Road, BSK IIIrd Stage
IIIrd Phase, IIIrd Block,
Bangalore - 560 085 (India)
Visit us at: www.anmolpublications.com

Encyclopaedia of Complex Analysis

First Edition, 2009
ISBN 978-81-261-3925-5 (Set)

PRINTED IN INDIA

Printed at Mehra Offset Press, Delhi.

Contents

Preface

The encyclopedia of complex analysis is designed for students in engineering, physics, and mathematics. Without sacrificing ease and clarity of proofs, mathematical preciseness and rigour are stressed. Cross references are used to justify almost every step in each proof.

The authors have made an effort to present some of the deeper and more interesting results, for example, Taylor's theorems, Riemann mapping theorem, Schwarz lemma in few chapters. However, the very basic theory is nevertheless given a thorough treatment so that readers should never feel lost. Each chapter finishes with exercises.

During the preparation of this encyclopedia there is stress on the new topics so as to cover as references for all the universities. This encyclopedia is a condensed correlation of mathematical and engineering concepts. Comments from reader are most invited for better revision of this encyclopedia.

Author

List of Symbols

$\pm$	$\not\subset$	$\subset$	$\subseteq$	$\in$	$\notin$	$\Re$	∂	$\neq$	$\approx$	$\angle$	∇
Σ	Π	Ω	ϕ	α	β	χ	ε	γ	η	λ	μ
ν	$\perp$	π	θ	ρ	σ	τ	$\Re$	ω	ξ	ψ	ζ
Υ		$\exists$	$\int$	$\iint$	$\iiint$	$\times$	$\sqrt{\ }$	$\because$	$\geq$	$\oint$	$\Rightarrow$

Chapter 1

Relations and Functions

INTRODUCTION

The mathematical concept of a function expresses the intuitive idea of deterministic dependence between two quantities, one of which is viewed as primary (the independent variable, argument of the function, or its "input") and the other as secondary (the value of the function, or "output"). A function then is a way to associate a unique output for each input of a specified type, for example, a real number or an element of a given set. This definition covers most elementary functions, maps between algebraic structures, such as groups, and between geometric objects, such as manifolds.

One idea of enormous importance in all of mathematics is composition of functions: if y is a function of x and z is a function of y, then z is a function of x. We may describe it informally by saying that the composite function is obtained by feeding the output of the first function as the input into the second one. This feature of functions distinguishes them from other mathematical constructs, such as numbers or figures, and provides the theory of functions with its most powerful structure.

RELATIONS AND FUNCTIONS

There are many ways to represent a function: by a formula, by a plot or graph, by and algorithm that computes it, by a description of its properties. Sometimes, a function is described through its relationship to other functions (see, for example, inverse function). In applied disciplines functions are frequently specified by their tables of values, or by a formula. Not all ways apply to every possible kind of function, and one has to make a firm distinction between the function itself and multiple ways of presenting or visualizing it.

Relations

After knowing some of the basic elements of set theory (sets), we will move on to the second most elementary concept, the concept of relations and functions.

Let A and B two sets. A relations between A and B is a collection of ordered pairs (a, b)

such that $a \in A$ and $b \in B$. Often we use the notation $a \sim b$ to indicated that a and b are related, rather then the order pair notation (a, b).

Note that this does not mean that *each* element from A needs to be associated with one (or more) elements from B. It is sufficient if *some* associations between elements of A and B are defined. In contrast, there is the definition of a function:

Thus, a set of ordered pairs is called a relation.

For example, (1, 2), (3, 5), (7, 10) … are relations.

The first components in the ordered pairs (x-coordinate) is called the domain, and the second components (y-coordinate of $f(x)$ coordinate) is called the range.

Functions: Domain and Range

Let's return to the subject of domain and range. When functions are first introduced, you will probably have some slightly pathetic "functions" and relations to deal with, being just sets of points. These won't be terribly useful or interesting functions and relations, but your text wants you to get the idea of what the domain and range are.

The first components in the ordered pairs (x-coordinate) is called the domain, and the second components (y-coordinate or $f(x)$ coordinate) is called the range.

Definition: Let A and B be two sets. Since, a function f from A to B is a relation between A and B such that for each $a \in A$ there is one and only one associated $b \in B$. The set A is called the domain of the function, B is called its **range.** Often a function is denoted as $y = f(x)$ or simply $f(x)$ or simply $f(x)$, indicating the relation $\{(x, f(x)\}$.

Definition: Let S and T be sets. A **function** from S into T is a subset F of $S \times T$ such that each element $x \in S$ there is exactly one element $y \in T$ such the $(x, y) \in F$. The set S is called the **domain** of the function, and the set T is called the **codomain** of the function. The subset $\{y \in T \mid (x, y) \in F \text{ for some } x \in S\}$ of the codomain is called the **image** of the function.

Example 1. Let $S = \{1, 2, 3\}$ and $T = \{u, v, w\}$. The subsets

$F_1 = \{(1, u), (2, v), (3, w)\}$ and

$F_2 = \{(1, u), (2, v), (3, w)\}$

Of $S \times T$ both define functions since in both cases each element of S occurs exactly once among the ordered pairs. The subset

$F_3 = \{(1, u), (3, w)\}$

does not define a function with domain S because the element $2 \in S$ does not appear as the first component of any ordered pair.

Note the F_3 is a function if the domain is changed to the set $\{1, 3\}$.

Unlike the conventions used in calculus, the domain and codomain must be specified as well as the "rule of correspondence" (list of pairs) when you are presenting a function.

The subset

$F_4 = \{(1, u), (2, u), (2, v), (3, w)\}$

does not define a function since 2 appears as the first component of two ordered pairs.

When a candidate such as F_4 fails to be a function in this way, we say that it is not "well-defined."

For example, find the range; given the domain {2, 4, 6} of $f(x) = 3x - 1$

$f(2) = 3(2) - 1 = 6 - 1 = 5$

$f(4) = 3(4) - 1 = 12 - 1 = 11$

$f(6) = 3(6) - 1 = 18 - 1 = 17$

the range is {5, 11, 17} and the ordered pairs are (2, 5); (4, 11); (6, 17)

The outcomes of a function (i.e. the elements from the range associated to elements in the domain) do not only depend on the rule of the function (such as x is associated with sin(x) but also on the domain of the function. therefore, we need to specify those outcomes that are possible for a given rule and given domain:

Functions versus Relations

There are different ways of looking at functions. We will consider a few. But first, we need to discuss some terminology.

A "relations" is just a relationship between sets of information. Think of all the people in one of your classes, and think of their heights. The pairing of names and heights is a relation. In relations and functions, the pairs of names and heights are "ordered", which means one comes first and the other comes second. To put it another way, we could set up this pairing so that either you give me a name, and then I give you that person's height, or else you give me a height, and I give you the names of all the people who are that tall. (By the way, the set of all the starting points is called "the domain" and the set of all the ending points is called "the range." The domain is where you start; the range is where you go. The domain is the x' s; the range is the y' s. I'll explain more on the subject of determining domains and ranges later.)

A function is a well-behaved relation. Just as with members of your own family, some members of the relation family are better behaved than other. (This means that, while all functions are relations, since they pair information, not all relations are functions. Functions are a sub-classification of relations.) When we say that a function is "a well-behaved relation", we mean that, given a starting point, we know exactly where to go. That is, given an x, we get only and exactly one y.

Let's return to our relation of your classmates and their heights, and let's suppose that the domain is the set of everybody's heights. And let's suppose that there's a pizza-delivery guy waiting in the hallway. And all the delivery guy knows is that the pizza is for the student in your classroom who is five-foot-five. Now let the guy in. Who does he go to? What if you are five-five? And what if you're out of cash? You see, the relation "height indicates name" is not well-behaved. It is not a function. Given the relationship (x, y) = (five-foot-five person, name), there might be six different possibilities for y = "name". For a relation to be a function, there must be only and exactly one y that corresponds to a given x. Here are some pictures of this:

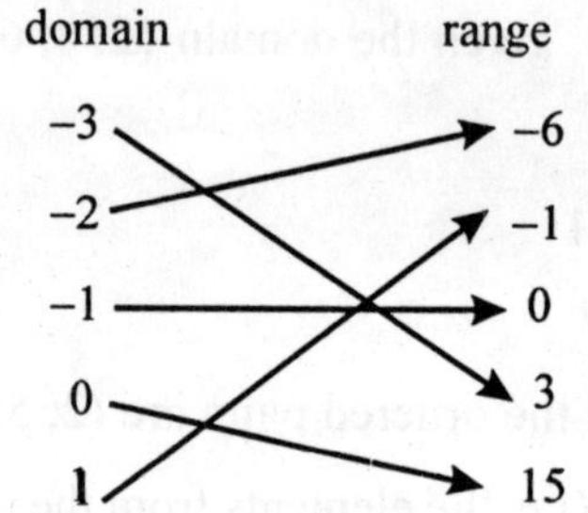

This is a function. You can tell by tracing from each x to each y. There is only one y for each x; there is only one arrow coming from each x.

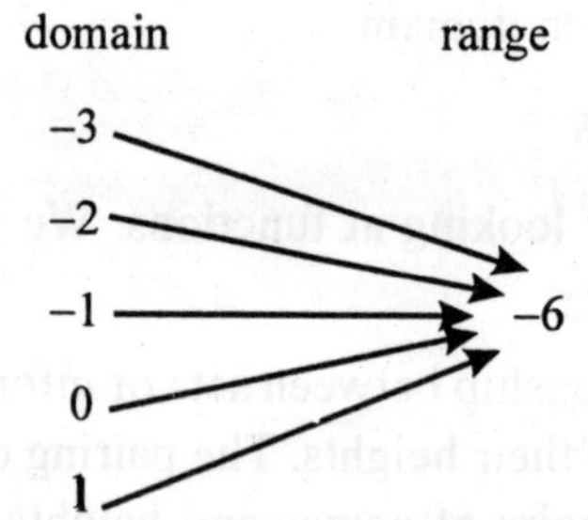

This is a function, There is only one arrow coming from each x; there is only one y for each x. It just so happens that it's always the same y, but it is only one y. So this is a function; it's just an extremely boring function!

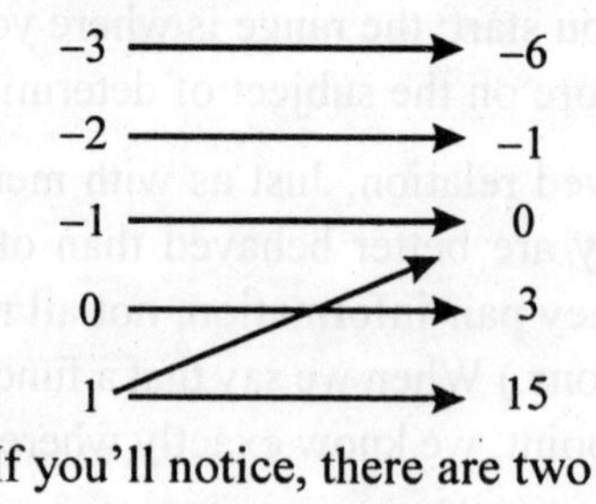

This one is not a function. If you'll notice, there are two arrow coming from the number 1; the number 1 is associated with two different range elements. So this is a relation, but it is not a function.

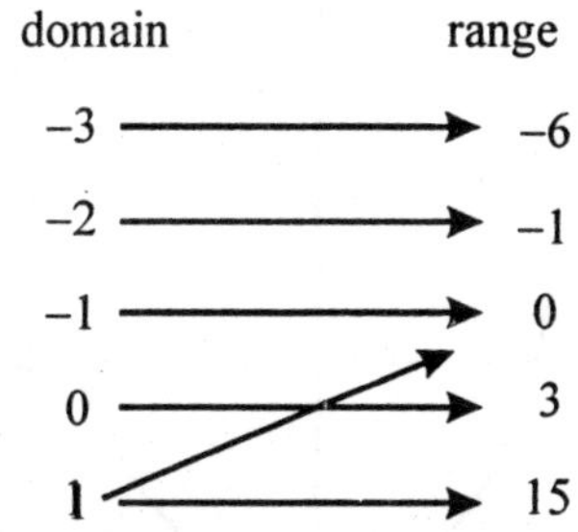

Each element of the domain that has a pair in the range is nicely well-behaved. But what about that 16? It is in the domain, but it has no range element that corresponds to it! This won't work. So then this is not a function.

Vertical Line Test

Given a graph, it can be determined if it is a function or a relation.

In order for the graph to be a function, the vertical line must only intersect the graph at one and only one point.

Looking at this function stuff graphically, what if we had the relation {(2, 3), (2, – 2)}?

We already know that this is not a function, since $x = 2$ goes to both $y = 3$ and $y = –2$.

If we graph this relation, it looks like:

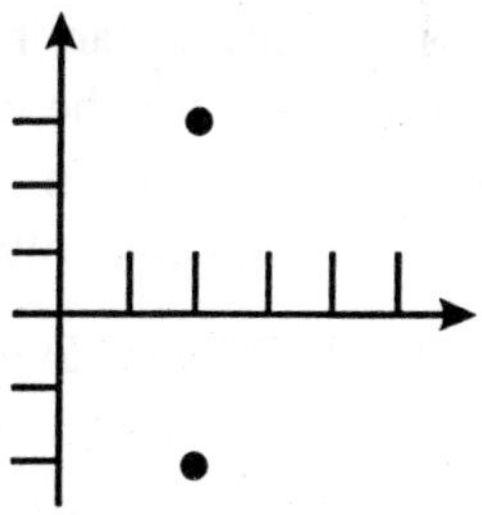

Notice that you can draw a vertical line through the two points, like this:

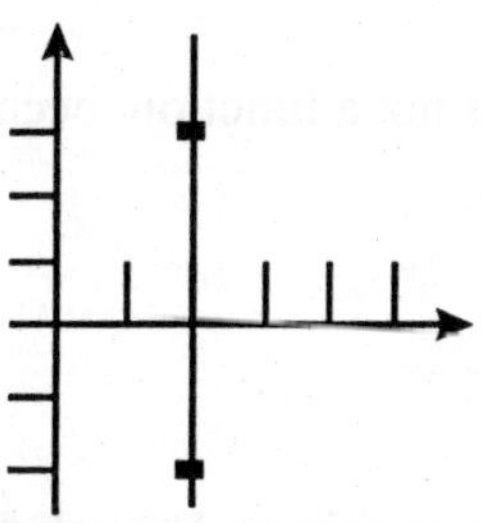

This characteristic of non-functions by I-don't-know-who, and was codified in The Vertical Line Test: Given the graph of a relation, if you can draw a vertical line that crosses the graph in more than one place, then the relation is not a function. Here are a couple examples:

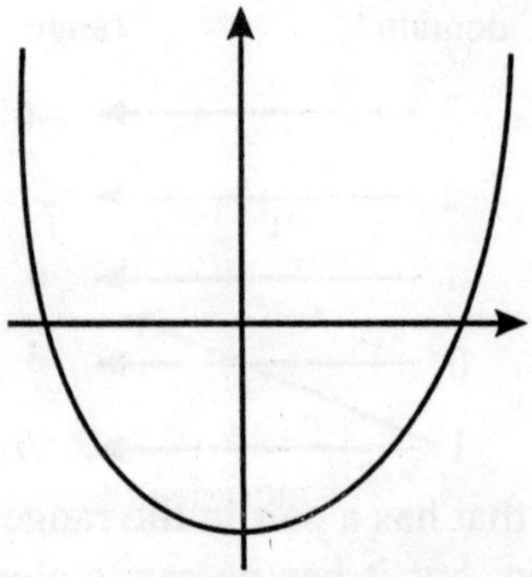

This one is a function. There is no vertical line that will cross this graph twice.

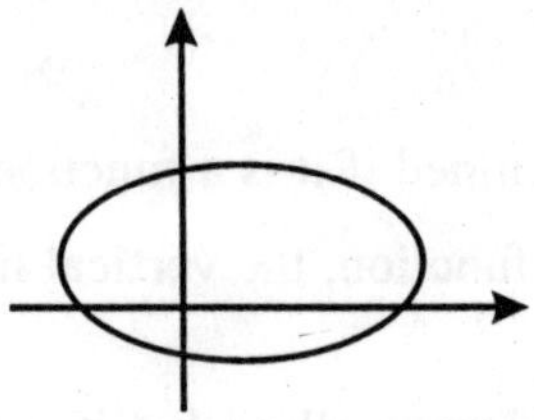

This one is not a function. Any number of vertical lines will intersect this intersect this oval twice. For instance, the y-axis intersects twice.

"Is it a function?" – Quick answer without the graph.

Think of all the graphing that you've done so far. The simplest method is to solve for y. make a T-chart, pick some values for x, solve for the corresponding values of y, plot your points, and connect the dots. Not only is this useful for graphing, but this methodology gives yet another way of identifying functions: If you can solve for "y". then it's a function.

For example, $2y + 3x = 6$ is a function, because you can solve for y:

$$2y + 3x = 6$$

$$2y = -3x + 6$$

$$y = (-3/2)x + 3$$

On the other hand, $y^2 + 3x = 6$ is not a function, because you can not solve for a unique y:

$$y^2 + 3x = 6$$

$$y2 = -3x + 6$$

$$y = \pm\sqrt{-3x+6}$$

This is solved for "y", but it's not unique. Do you take the positive square root,

or the negative? So, in this case, the relation is not a function. (You can also check this by using our first definition from above.

Think of "$x = -1$". Then we get $y2 - 3 = 6$, so $y2 = 9$, and then y can be either –3 or + 3.

That is,

if we did an arrow chart, there would be two arrows coming from $x = -1$.)

Thus, a **function** is a relation were each member of the domain is reserved one and only one member of the range. Functions are named by suing single letters.

For example, $f = \{(x, y) | \; y = x + 4\}$ or $f(x) = x + 4$

The equations above are representations of functions. The second example is the most commonly used representation; it is called the function notation.

The ordered pair for functions is $(x, f(x))$.

Consider the example; $f(x) = x + 4$ at $x = 2$.

$f(2) = 2 + 4 = 6$, thus the ordered pair is (2, 6)

Example 2. Find the domain and range of: $f(x) = 2/(x-3)$

Solution: Since x cannot be zero; the domain is $D = \{x | \; x^1 \; 3\}$ and $f(x)$ can have any value; thus the range is $R = \{f(x) |$ any real number$\}$

Example State the domain and range of the following relation. Is the relation a function?

$$\{(2, -3), (4, 6), (3, -1), (6, 6), (2, 3)\}$$

This list of points, being a relationship between certain x's and certain y's, is a relation. The domain is all the x-values, and the range is all the y-values. you list the values without duplication:

domain: $\{2, 3, 4, 6\}$

range: $\{-3, -1, 3, 6\}$

(It is customary to list these values in numerical order, but it is not required. Sets are called "unordered lists", so you can list the numbers in any order you feel like. Just don't duplicate.)

While this is a relation (because x's and y's are being related to each other), you have two points with the same x-values: (2, –3) and (2, 3). Since $x = 2$ gives you two possible destinations, then this relation is not a function.

Note that all I had to do to check whether the relation was a function was to look for duplicate x- values. If you find a duplicate x-value, then the different y-values mean that you do not have a function.

Example 3. State the domain and range of the following relation. Is the relation a function?

$$\{(-3, 5), (-2, 5), (-1, 5), (0, 5), (1, 5), (2, 5)\}$$

List the x-values for the domain and the y-values for the range:

Domain: $\{-3, -2, -1, 0, 1, 2\}$

range: $\{5\}$

This is another example of a "boring" function (like we saw in the first chart higher up in this page), because all the x-values go to the exact same y-values. But each x-value is different, so, while boring, this relation is indeed a function. (In point of fact, these points lie on the horizontal line $y = 5$.)

There is one other case for finding the domain and range of functions. They will give you a function and ask you to find the domain (and maybe the range, too). I have only ever seen (or can even think of) two things at this stage in your mathematical career that you'll have to check in order to determine the domain of the function they'll give you, and those two things are denominators and square roots.

Example 3. Determine the domain and range of the given function

$$\frac{x^2 + x - 2}{x^2 - x - 2}$$

Solution:

The domain is all the values that x is allowed to take on. The only problem you have with this function is that you have to be careful not to divide by zero. So the only values that x can not take on are those which would cause division by zero. So set the denominator equal to zero and solve; the domain will be everything else.

$$x^2 - x - 2 = 0$$

$$(x - 2)(x + 1) = 0$$

$$x = 2 \text{ or } x = -1$$

Then the domain is "all x not equal to – 1 or 2".

The range is a bit trickier, which is why they may not ask for it. In general, though, they'll want you to graph the function and find the range from the picture. In this case:

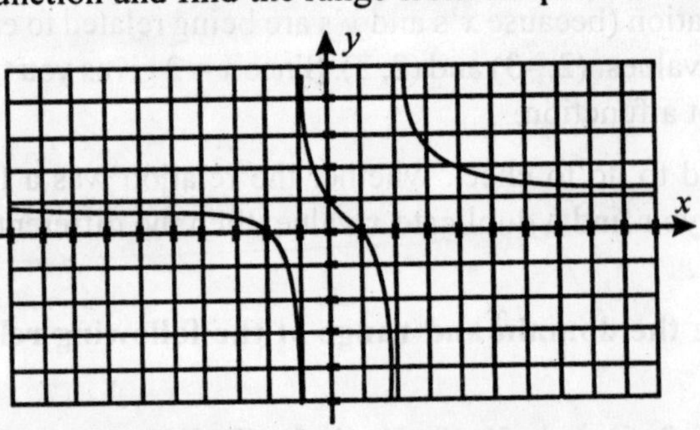

As you can see from the picture, the graph covers all y-values (that is, the graph will go as low as you like, and will also go as high as you like). Since the graph will eventually cover all possible values of y, then the range is "all real numbers".

Example 4. Determine the domain and range of the given function:

$$-\sqrt{-2x+3}$$

The domain is all values that x can take on. The only problem you have with this function is that you cannot have a negative inside the square root. So set the insides greater-than-or-equal-to zero, and solve. This will be the domain:

$-2x + \geq 30$

$-2x \geq -3$

$2x \leq 3$

$x \leq 3/2 = 1.5$

Then the domain is "all $x \leq 3/2$".

The range requires a graph. Be sure to be careful when graphing radicals:

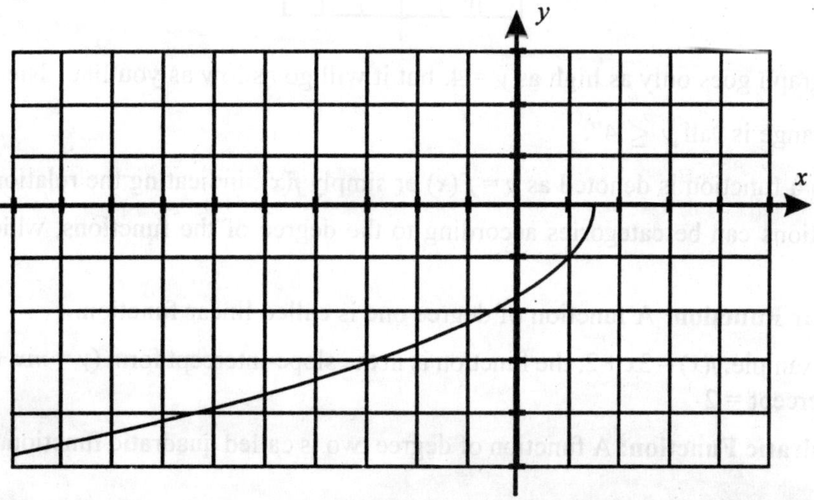

The graph starts at $y = 0$ and goes down from there. While the graph goes down very slowly, you know that, eventually, you can go as low as you like (by picking an x that is sufficiently big). Also, from your experience with graphing, you know that the graph will never start coming back up. Then the range is "$y \leq 0$".

Example 5. Determine the domain and range of the given function:

$$y = -x^4 + 4$$

This is just a garden-variety polynomial. There are no denominators (so no division-by-zero problems) and no radicals (so no square-root-of-a-negative problems). There are no problems with a polynomial. There are no values that you can't plug in for x. When you have a polynomial, the answer is always that the domain is "all x".

The range will vary from polynomial to polynomial, and they probably won't even ask, but, if they do, look at the picture:

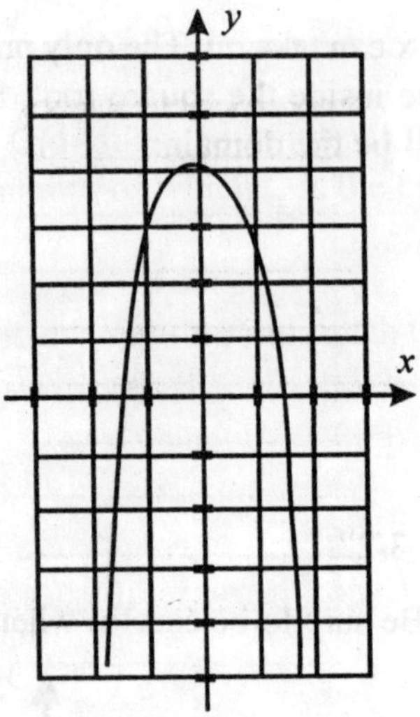

The graph goes only as high as $y = 4$, but it will go as low as you like. Then:

The range is "all $y \leq 4$".

Often a function is denoted as $y = f(x)$ or simply $f(x)$, indicating the relation $\{(x, f(x)\}$.

Functions can be categories according to the degree of the functions, which are listed below:

Linear Function: A function of degree one is called linear function.

For example, $f(x) = 3x + 2$, the function is in the slope-intercept form ($y = mx + b$) m(slope) = 3, y-intercept = 2

Quadratic Function: A function of degree two is called quadratic function. [$f(x) = ax^2 + bx + c$]

The same as graphing quadratic equations. For example, graph: $f(x) = 2x^2 + 8x + 9$

$y = 2x^2 + 8x + 9$ replace $f(x)$ with y

$y - 9 = 2x + 8x$ complete the square

$y - 9 + 4 = 2\ (x^2 + 4x + 4)$

$y - 1 = 2(x + 2)^2$ parabola with vertex at $(-2, 1)$

$y = 2(x + 2)^2 + 1$ solve for y and replace y with $f(x)$

$f(x) = 2(x + 2)^2 + 1$

Cubic Function: A function of degree three is called cubic function. [$f(x) = ax^3 + bx^2 + cx + d$]

Biquadratic Function: A function of degree four is called biquadratic function. [$f(x) = ax^4 + bx^3 + cx^2 + dx + e$]

Graph of Function: To graph functions, just change y into $f(x)$ from graphing equations. Graphing functions is the same as graphing equations with the exception of naming the domain and the range.

Rules for graphing other function:

1. Determine the domain of the graph.
2. Check for symmetry.
   if $f(-x) = f(x)$, then symmetric about the y-axis.
   if $f(-x) = -f(x)$, then symmetric about the origin.
   (functions rule out the possibility that the graph has an x-axis symmetry)
3. Find the x-intercept and y-intercept of the graph
   Evaluate $f(0)$ to find the y-intercept.
   To find the x-intercept, find the value or values that will make $f(x) = 0$
4. Plot some points for the graph and reflect, according to the symmetry test.
   For example, graph: $f(x) = | x |$ the domain is $D = \{x \mid x \text{ is any real number}\}$
   $f(-x) = | -x | = | x | = f(x)$ symmetric about the y-axis.
   $f(0) = | 0 | = 0, f(x) =$ at $x = 0$ $x = 0$ the intercepts are tat the origin.

x	$f(x)$
1	1
–2	2
–1	1
1	2

Examples 6. Let $A = \{1, 2, 3, 4)$, $B = (14, 7, 234)\}$, $C = \{a, b, c\}$, and R = real numbers. Define the following relations:

(*i*) r is the relation between A and B that associates the pairs 1 ~ 234, 2 ~ 7, 3 ~ 14, 4 ~ 234, 2 ~ 234

(*ii*) f is the relation between A and C that relates the pairs $\{(1, c), (2, b), (3, a), (r, b)\}$

(*iii*) g is the relation between A and C consisting of the associations $\{(1, a), (2, a), (3, a)\}$

(*iv*) h is the relation between R and itself consisting of pairs $\{(x, \sin(x))\}$

Solution:

(*i*) The relation r is not a function, because the element 2 from the set A is associated with two elements from B

(*ii*) The relation f is a function, because every element from A has exactly one relation from the set C.

(*iii*) The relation g is not a function, because the element {4} from the domain A has no element associated with it.

(*iv*) The relation h is a function with domain R, because every element $\{x\}$ in R has, exactly one element $\{\sin(x)\}$ associated with it.

The outcomes of a function (i.e. the elements from the range associated to elements in the domain) do not only depend on the rule of the function (such as x is associated with $\sin(x)$) but also on the domain of the function. Therefore, we need to specify those outcomes that are possible for a given rule and a given domain:

Definition: Image and Preimage

Let A and B be two sets and f a function from A to B. Then the image of f is defined as

- image $(f) = \{b \in B : \text{there is an } a \in A \text{ with } f(a) = b\}$.

Let A and B be two sets and f a function from A To B. If C is a subset of the range B then the preimage, or inverse image, of C under the function f is the set defined as

- $f^{-1}(C) = \{x \in A: f(x) \in C\}$

Function can be classified into three groups: those for which every element in the image has one preimage, those for which the range is the same as the image, and those which have both of these properties. Accordingly, we make the following definitions:

As an example, consider the following functions:

Example 7. Let $f(x) = 0$ if x is rational and $f(x) = 1$ if x is irrational. This function is called Dirichlet's Function. The range for f is R

(*i*) Find the image of the domain of the Dirichlet Function when:

(*a*) the domain of f is Q

(*b*) the domain of f is R

(*c*) the domain of f is [0, 1] (the closed interval between 0 and 1]

(*ii*) What is the preimage of R? What is the preimage of [–1/2, 1/2]?

(*iii*) What is the image of [0, 2] and the preimage of [1, 4]?

(*iv*) Find the image and the preimage of [–2, 2].

Solution:

(*i*) (*a*) When the domain is Q, we have that $f(x) = 0$ for any x, because x must be a rational number. Hence, the image of the domain Q is the set consisting of the single element {0}

(*b*) When the domain is R, we have that $f(x)$ could be 0 or 1, because x could be rational or irrational. $f(x)$ can not be any other number. Hence, the image of the domain R is the set consisting of the two element {0, 1}.

(*c*) The interval [0, 1] contains irrationals as well as rational numbers. Therefore, $f(x)$ could be equal to 0 or 1, and the image of [0, 1] under f is the set with the two elements {0, 1}.

(*ii*) The preimage of R is the set of all elements such that $f(x)$ is contained in R. Since R includes the numbers 0 and 1, the preimage of R under f is everything in the domain, i.e. R.

The preimage of [–1/2, 1/2] consists of all elements such that $f(x)$ is contained in [–1/2, 1/2]. This set does not contain 1, so the preimage of that set can not contain any irrational number. hence, the preimage of [–1/2, 1/2] is the set of rational umbers Q.

Let $f(x) - x^2$, with domain and range being R. Then use the graph of the function to determine:

(*iii*) If we draw the graph of the function, and shade the interval [0, 2] on the x axis in red, then its projection onto the y axis will be the image (in green).

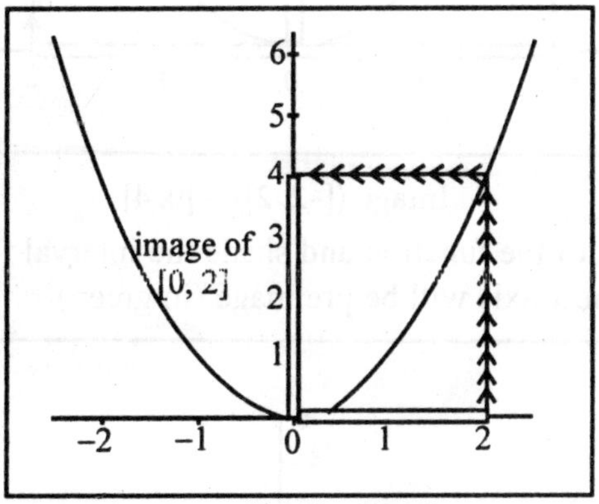

Imag [0, 2] = [0, 4]

If we draw the graph of the function, and shade the interval [1, 4] on the y axis in red, then its projection onto the x axis will be the preimage (in green).

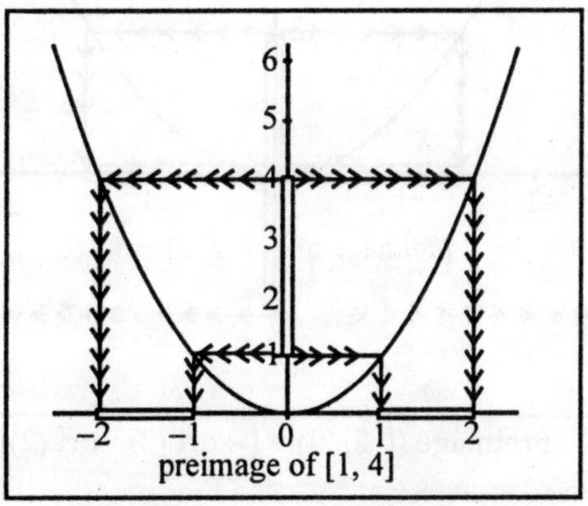

Preimage ([1, 4] = [-2, -1] ∪ [1, 2]

(*iv*) If we draw the graph of the function and shade the interval [–2, 2] on the *x* axis (in red) then its projection onto the *y* axis will be the image (in green).

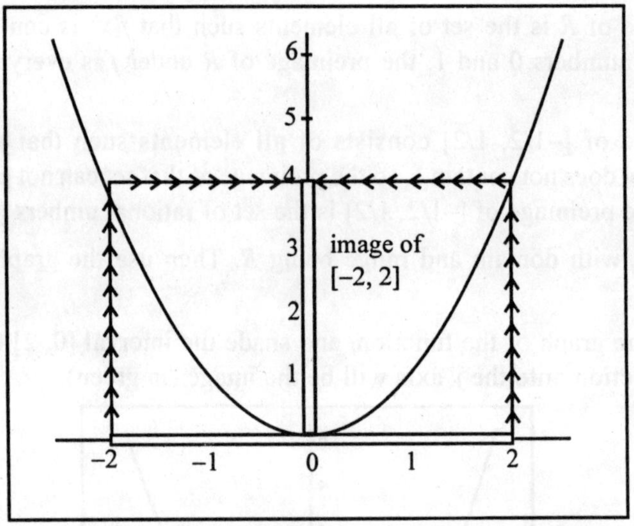

Image ([–2, 2]) = [0, 4]

If we draw the graph of the function and shade the interval [–2, 2] on the *y* axis (in red) then its projection onto the *x* axis will be preimage (in green).

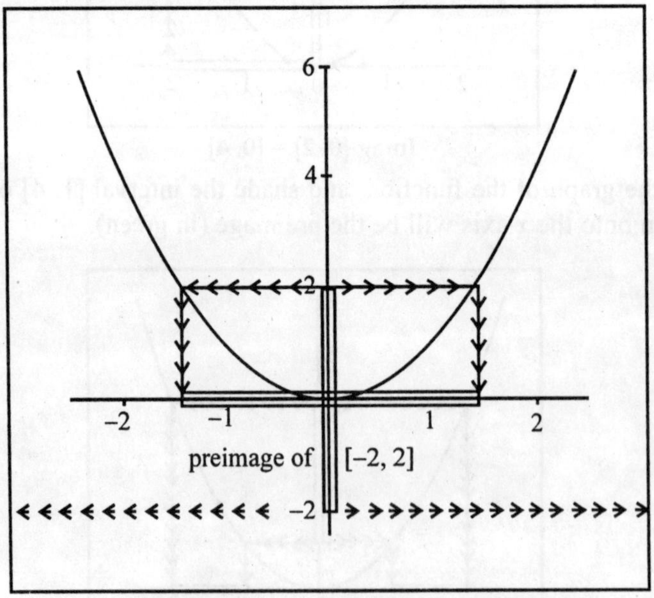

Preimage ([–2, 2]) = [-sqrt (2), sqrt (2)]

Function can be classified into three groups: those for which every element in the image has one preimage, those for which the range is the same as the image, and those which have both of these properties. Accordingly, we make the following definitions:

Types of mapping

Mappings can be classified into a number of different types.

Injective

A function *f* from *A* to *B* is called one to one (or one-one) in whenever *f*(*a*) = *f*(*b*) then a = *b*. Such functions are also called injections.

Thus, the mapping $a : X \rightarrow Y$ is called injective or one-one if for each $x \in X$ there exists at most $y \in Y$ one such that $a(x) = y$. (Note that no such *x* may exist). Another way of saying that *a* is injective is *a*(*x*) implies *y* where *x*. A graphical representation of an injective mapping is show in figure 1.

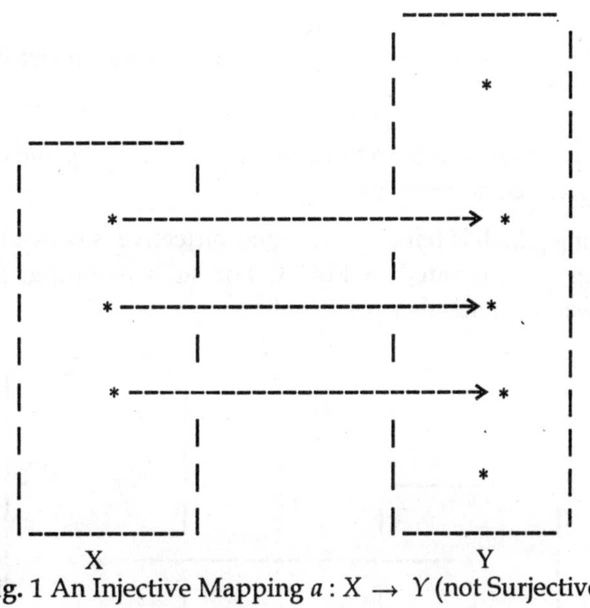

Fig. 1 An Injective Mapping $a : X \rightarrow Y$ (not Surjective)

Surjective

A function *f* from *A* To *B* is called onto if for all *b* in *B* there is an *a* in *A* such that *f*(*a*) = *b*. Such functions are also called surjections.

Thus, the mapping $a : X \rightarrow Y$ is called surjective or onto if for each $x \in X$, there exists at least one $y \in Y$ such that $a(x) = y$. This type of mapping is illustrated in fig. 2.

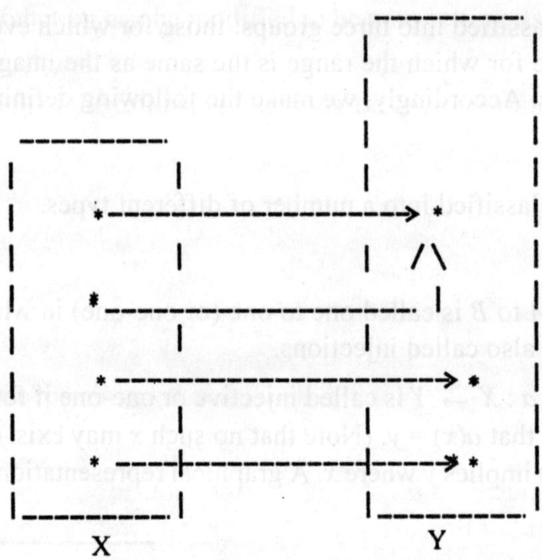

Fig. 2 A Surjective Mapping $a : X \rightarrow Y$ (not Injective)

Bijective

A function *f* from *A* to *B* is called a bijection if it is one to one and onto, i.e. bijections are functions that are injective and surjective.

Thus, the mapping which is both injective and surjective is called bijective or one-to-one and such a mapping is illustrated in Fig. 3. For such mapping, there is a one-to-one correspondence between the members of *X* and *Y*.

Fig. 3 A bijective mapping $a : X \rightarrow Y$ (both injective and surjective)

Horizontal Line Test:

The horizontal line test implies a one to one function, meaning that there is only one value of x associated with each value of $f(x)$.

Example 8. If the graph of a function is known, how can you decide whether a function is one-to-one (injective) or onto (surjective).

Solution

Injection

If a horizontal line intersect the graph of a function in more than one place, then there are two different points a and b for which $f(a) = f(b)$, but $a \# b$. then the function is not onto-to-one.

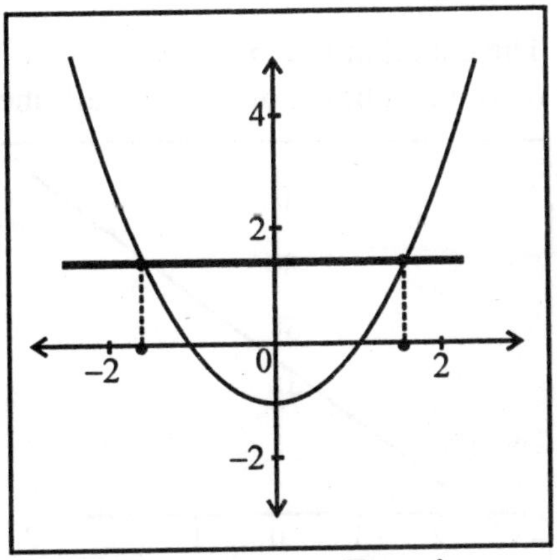

Not a one-to-one Function

Surjection

If you place a light on the left and on the right hand side of the coordinate system, then the shadow of the graph on the y axis is the image of the domain of the function. If that shadow covers the range of the function, then the function is onto.

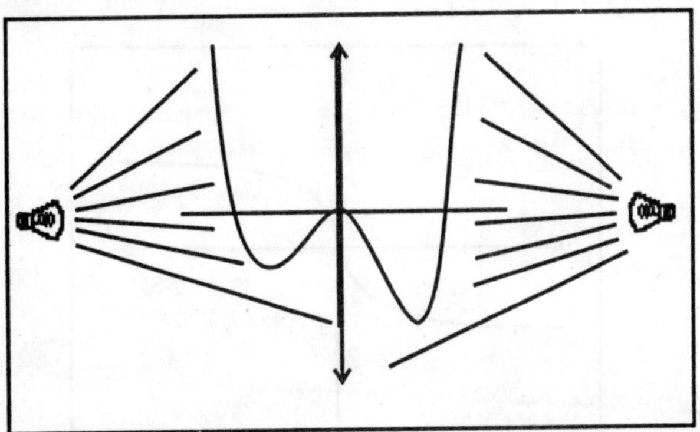

Not an onto function (if range is R)

Note that every function can be modified to be onto setting its range to be the image of its domain

Example 9. Which of the following functions are one-one, onto, or bijections? The domain for all functions is *R*.

(*i*) $f(x) = 2x + 5$

(*ii*) $g(x) = \arctan(x)$

(*iii*) $g(x) = \sin(x)$

(*iv*) $h(x) = 2x^0 + 5x^2 - 7x + 6$

Solution:

(*i*) $f(x) = 2x + 5$ is a linear function

This function is linear. The equation $y = 2x + 5$ has a unique solution for every x, so that the function is one-one and onto, i.e. a bijection. In fact, all linear functions are bijections.

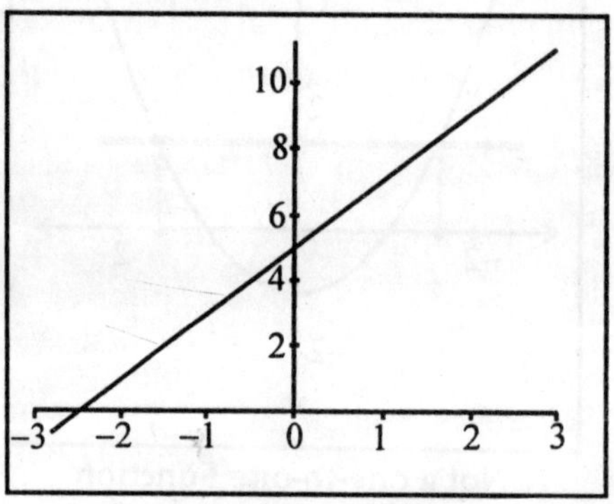

$f(x) = 2x + 5$ is a Bijection.

(*ii*) $g(x)$ = arctan is a inverse Functions

This function is the inverse function of the tangent function. As such, it must be one-to-one. It is onto if the range is $[-\pi/2, \pi/2]$, but not if the range is *R*. In fact, all inverse functions are one-to-one.

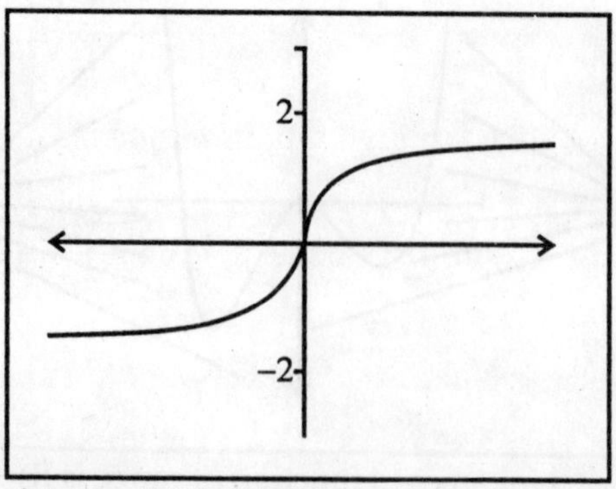

$g(x) = Arctan\ (x)$ is Injective, not Surjective on *R*.

(*iii*) $g(x) = \sin(x)$ is a Periodic Function

Since this function is periodic, it can not be one-to-one. It is onto, if the range is the interval [–1, 1], but not onto if the range is *R*. In fact, every periodic function is not one-to-one.

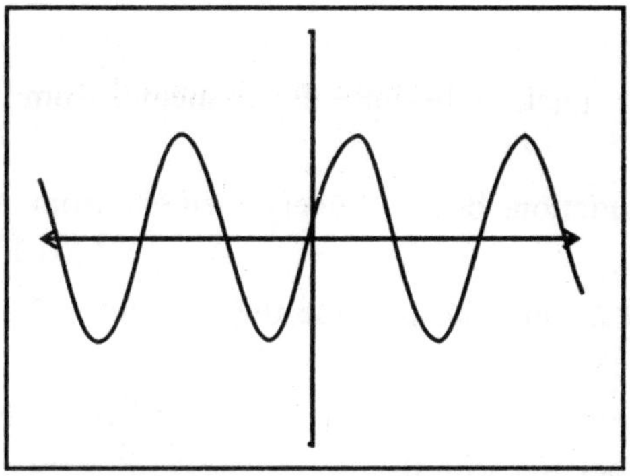

$g(x) = sin(x)$ is Neither one-to-one nor onto R.

(*iv*) $h(x) = 2x3 + 5x2 - 7x + 6$ is a odd Degree Polynomial

This is an odd-degree polynomial. Hence, the limit as *x* approaches plus or minus infinity must be plus or minus infinity, respectively. That means that the function is onto. Since most third degree equations have three zero, this function is probably not one-to-one. A look at the graph confirms this. In fact, every odd-degree polynomial is onto while no even degree polynomial is onto.

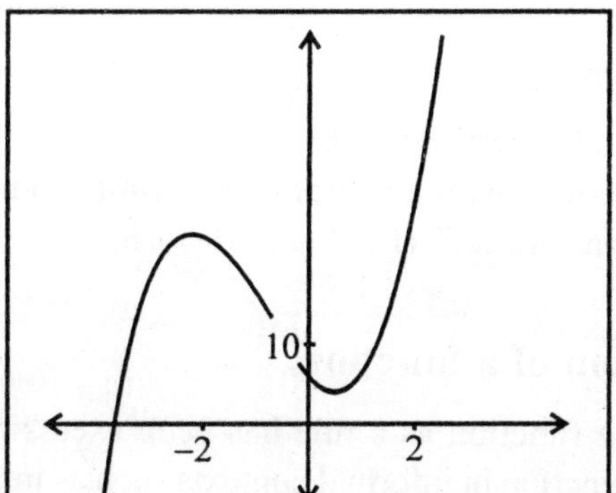

$h(x) = 2x^3 - 7x + 6$ is onto, not one-to-one

After introducing some of the basic elements of set theory (sets), we will move on to the second most elementary concept, the concept of relations and functions.

Examples 10. Let $A = \{1, 2, 3, 4\}$, $B = \{14, 7, 234\}$, $C = \{a, b, c\}$, and R = real numbers.

Define the following relations:

(*i*) *r* is the relation between *A* and *B* that associates the pairs 1 ~ 234 ~ 7, 3 ~ 14, 4 ~ 234, 2 ~ 234 (*ii*) *f* is the relation between *A* and *C* that relates the pairs {(1, *c*), (2, *b*), (3, *a*), (4, *b*)}

(*iii*) *g* is the relation between *a* and *C* consisting of the associations {(1, *a*), (2, *a*), (3, *a*)}

(*iv*) *h* is the relation between *R* and itself consisting of pairs {*x*, sin(*x*)}

Which of those relations are functions?

Solution:

(*i*) The relation *r* is not a function, because the element 2 from the set *A* is associated with two elements from *B*.

(*ii*) The relation *f* is a function, because every element from *A* has exactly one relation from the set *C*.

(*iii*) The relation *g* is not a function, because the element {4} from the domain *A* has no element associated with it.

(*iv*) The relation *h* is a function with domain *R*, because every element {4} from the domain *A* has no element associated with it.

(*v*) The relation *h* is a function with domain *R*, because every element {*x*} in *R* has exactly one element {sin(*x*)} associated with it.

Abstract definition of a function

A fairly precise, but somewhat informal definiton of a function follows. Let A and B be sets, A function from A to B is determined by any association of a unique element of B with each element of A. The set A is called the domain of the function; the set B is called the codomain.

Functions Versus Maps

A map is usually a (total) function as defined here, especially for authors using function as synonym for partial functions with a domain (of definition) not necessarily containing all points of the source or "departure set". (For them, $R \rightarrow R; x \rightarrow 1/x$ is a function with domain $R\backslash\{0\}$.)

Mathematical definition of a function

The informal idea of a function as a rule has been used since ancient times and is still used as the definition of a function in informal contexts such as introductory calculus textbooks. A typical example of this informal definition, is that a function is a rule that assigns to each element of a set *D* a single element of set *C*. This informal definition is sufficient for many purposes, but it relies on the undefined concept of a "rule". In the late 1900s, the question of what constitutes a valid rule defining a function came to the forefront (as described below). The consensus of modern mathematicians is that the word "rule" should be interpreted in the most general sense possible: as an arbitrary binary relation.

Thus it is common in advanced mathematics for an example to formally define a function *f* from a set *D* to a set *C* to be a set *Gf* of ordered pairs (*x*, *y*) in the Cartesian product $D \times C$.

It is required that for each x in D there is at most one pair (x, y) in the set Gf. The "rule" of the function is: given x in D, if there is a pair (x, y) in Gf then $f(x) = y$, and otherwise $f(x)$ is not defined. The set of x for which f is defined is called the domain of f; if the domain of f is all of D then f is called total and the notation $f: D \to C$ is used. The set C is called the codomain of the function; this must be specified because it is not determined by Gf. The set Gf is called the graph of the function.

In most areas of mathematics, the word function is used to mean total function, although non-total functions (called partial functions) are important in functional analysis, mathematical logic, and category theory.

Definition of functions in set theory

Variations of this formal definition are sometimes more convenient for specific disciplines. In some contexts of category theory, even if a function f from a set D is not defined for every element of D, the set D is still called the domain of f. *I* set theory, it is common to identify the function f with its graph Gf; this identification removes the need to specify either D or C in the formal definition.

Total and Partial Mappings

If a mapping $a : X \to Y$ is defined for every $x \in X$, then the mapping a is said to be total. The mappings illustrated in Figs. 2.2, 2.3 and 2.4 are all total (that is every element in X maps to an element in Y). If a mapping a is not defined for all $x \in X$, then the mapping a is said to be partial. Expressed another way, we can say that a mapping a is partial if not all possible inputs have a defined output. An example of a partial mapping is illustrated below in Fig. 2.4.

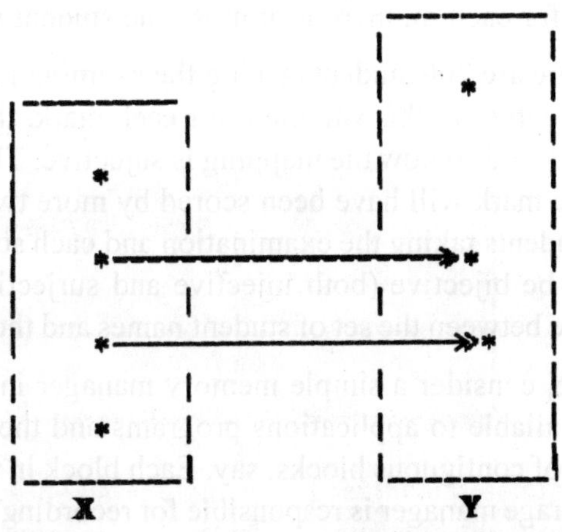

Figure 2.4: Example of a Partial Mapping

This mapping is partial since there are elements in X that are not mapped to a corresponding value in Y.

We should emphasise again that we treat mappings as single-valued, that is for any mapping $a: X \to Y$, we assume that for any $x \in Y$, there is at most one $y \in Y$ such that $y = a(x)$. This is an essential property for computer science applications since, for example, it would be most undesirable for an array element to have to distinct values.

Note also that the terminology is by no means standard in this area. In mathematics, functions (mappings) are by definition, total, that is for each $x \in X$ there is a unique $y \in Y$ such that $y = a(x)$. In computer science applications however, the function (mapping) a from X to Y is usually defined by stating that a maps an element from X onto at most one element of Y. Hence there is no necessity for every member of X to be mapped to a member of Y.

Real life examples of Mappings

In order to capture the various concepts associated with mappings and to summaries, it will be helpful to look briefly at two examples of mappings.

Consider firstly a class of 25 students who sit an examination where marks are awarded in the range) to 100. If we denote the set of student names by S and the set of possible marks by M (so that M will be the set $\{m \mid m \in N, 0 \leq\ m)$, then the "examinations-results-list" constitutes a total mapping since every student has an examination mark and that mark is unique. We can write this mapping as

$$\textit{examination-results-}\ S \rightarrow M$$

If every student whose name appears in S sat the examination, the mapping is total. Note that not every possible examination mark will appear on the results list because there are 101 possible marks but only 25 students. Hence for each mark, it is not the case there is at least one student name, who has attained that mark and so the mapping is not surjective. If it transpired that each of the 25 students achieved a different mark in the examination, then the mapping would be injective (since for each m, there is at most one student with that mark).

Suppose now that there are 200 students taking the examination and that each of the 101 possible marks has been scored. In this situation, for each mark, there is at least one student who has achieved that mark and so now the mapping is sujective. The mapping is not injective however since at least one mark will have been scored by more two or more students. In the (unlikely) event of 101 students taking the examination and each scoring a different mark, the resulting mapping would be bijective (both injective and surjective) and we would have a one-to-one correspondence between the set of student names and the set of examination marks.

As a second example, consider a simple memory manager in an operating system. The heap contains memory available to applications programs and the space in the heap zone is subdivided into a number of contiguous blocks, say. Each block in the zone is either allocated to a user or is free. The storage manager is responsible for recording which blocks are currently owned by which user. If we denote the set of user-names by U and the set of blocks By H (so that H will be the set $\{m \mid m \in N, 0 \leq m \leq h_N\}$, then the storage manager constitutes a mapping from the set of block numbers to the set of users which we can express as

Storage-manager:

Note that this constitutes a partial mapping in the situation where one or more blocks is not owned by any user. Also, since, in general, some users may not own any blocks while other users may own several, then the mapping will be neither surjective nor injective.

Composition of Functions

In mathematics, a composite function, formed by the composition of one function on another, represents the application of the former to the result of the application of the latter to the argument of the composite. The functions $f: X \rightarrow Y$ and $g: Y \rightarrow Z$ can be composed by first applying f to an argument x and then applying g to the result. Thus one obtains a function $g \circ f: X \rightarrow Z$ defined by $(g \circ f)(x) = g(f(x))$ for all x in X. The notation g o f is read as "g circle *f*" or "*g* composed with *f*".

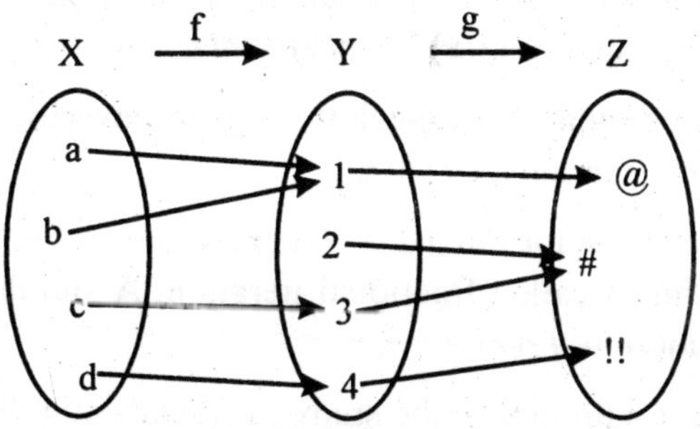

g o f, the Composition of *f* and *g*

The composition of functions is always associative. That is, if *f*, *g*, and h are three functions with suitably chosen domains and codomains, then $f \circ (g \circ h) = (f \circ g) \circ h$. Since there is no distinction between the choices of placement of parentheses, they may be safely left off.

The functions *g* and *f* commute with each other if $g \circ f = f \circ g$. In general, composition of functions will not be commutative. Commutativity is a special property, attained only by particular functions, and often in special circumstances. For example,

$|x| + 3 = |x + 3|$ only when $x \geq 0$. But inverse functions always commute to produce the identity mapping.

Derivatives of compositions involving differentiable functions can be found using the chain rule. "Higher" derivatives of such functions are given by Faa di Bruno's formula. As an example, suppose that an airplane's elevation at time t is given by the function $h(t)$ and that the oxygen concentration at elevation x is given by the function $c(x)$. Then $(c \circ h)(t)$ describes the oxygen concentration around the plane at time t.

Functional powers

If $Y \subseteq X$ then $f: X \rightarrow Y$ may compose with itself; this is sometimes denoted f^2. Thus:

$(f \circ f)(x) = f(f(x)) = f^2(x)$

$(f \circ f \circ)(x) = f(f(f(x))) = f^3(x)$

Repeated composition of a function with itself is called function iteration.

The functional powers f of f n = f n o f = f n + 1 for natural n follow immediately.

By convention, $f(0)$ = idD(f) (the identity map on the domain of f).

If $f : X \to X$ admits an inverse function, negative functional powers f^{-k} ($k > 0$) are defined as the opposite power of the inverse function, $(f^{-1})k$.

Note: If f takes its values in a ring (in particular for real or complex-valued f), there is a risk of confusion, as $f\,n$ could also stand for the n-fold product of f, e.g. $f\,2(x) = f(x)\,.\,f(x)$.

(For trigonometric functions, usually the latter is meant, at least for positive exponents. For example, in trigonometry, this superscript notation represents standard exponentiation when used with trigonometric functions: sin2(x) = sin(x). sin(x)

However, for negative exponents (especially –1), in nevertheless usually refers to the inverse function, e.g., tan–1 = arctan ($\neq$ 1/tan).

In some cases, an expression for f in $g(x) = f\,r(x)$ can be derived from the rule for g given non-integer values of r. This is called fractional iteration. A simple example would be that where f is the successor function, $f\,r(x) = x + r$.

Iterated functions occur naturally in the study of fractals and dynamical systems.

Composition Mapping

Suppose one has two (or more) functions $f : X \to Y, g : X \to X$ having the same domain and range. Then one can form long, potentialy complicated chains of these functions composed together, such as f o f o g o f. Such long chains have the algebraic structure of a maping, sometimes called the composition mapping. In general, composition mappings can have remarkably complicated structure. One particular notable example is the de Rham curve. The set of all functions $f : X \to X$ is called the full transformation semigroup on X.

If the functions are bijective, then the set of all possible combinations of these functions form a group; and one says that the group is generated by these functions.

The set of all bijective functions $f : X \to X$ form a group with respect to the composition operator; this is sometimes called the composition group.

Composition Operator

Given a function g, the composition operator Cg is defined as that operator which maps functions to functions as

$$C_g f = f \circ g$$

Composition operators are studied in the field of operator theory.

Definition. Let $f : S \to T$ and $g : T \to U$ be functions. The composition g o f of f and g is the function from S to U defined by the formula (g o f)(x) = $g(f(x))$ for all $x \in S$.

Proposition. Composition of functions is associative.

Definition. Let f: $S \to T$ be a function then f is said to map S onto T if for each element $y \in T$ there exists an element $x \in S$ with $f(x) = y$.

If $f(x_1) = f(x_2)$ implies $x_1 = x_2$ for all element x_1 $x_2 \in S$, then f is said to be a one-to-one function.

If f is both one-to-one and onto, then it is called a one-to-one correspondence from S to T.

Proposition. Let $f : S \to T$ be a function. Suppose that S and T are finite sets with the same number of elements. Then f is one-to-one if and only if it is onto.

Proposition. Let $f : S \to T$ and $g : T \to U$ be functions.

(*a*) If f and g are one-to-one, the g o f is one-to-one.

(*b*) if f and g are onto, then g o f is onto.

Definition. Let S be a set. The identity function $1_s : S \to S$ is defined by the formula

$$1_s(x) = x \text{ for all } x \in S.$$

If $f : S \to T$ is a function, then a function $g : T \to S$ is called an inverse for f if

$$g \circ f = 1_s \text{ and } f \circ g = 1_T$$

Proposition. Let $f : S \to T$ be a function. If f has an inverse, then it must be one-to-one and onto. Conversely, if f is one-to-one and onto, then it has a unique inverse.

Examples 10. If the graph of a function is known, how can you decide whether a function is one-to-one (injective) or onto (surjective)?

Solution:

Injection

If a horizontal line intersects the graph of a function in more than one place, then there are two different points a and b for which $f(a) = f(b)$, but $a \# b$. Then the function is not one-to-one.

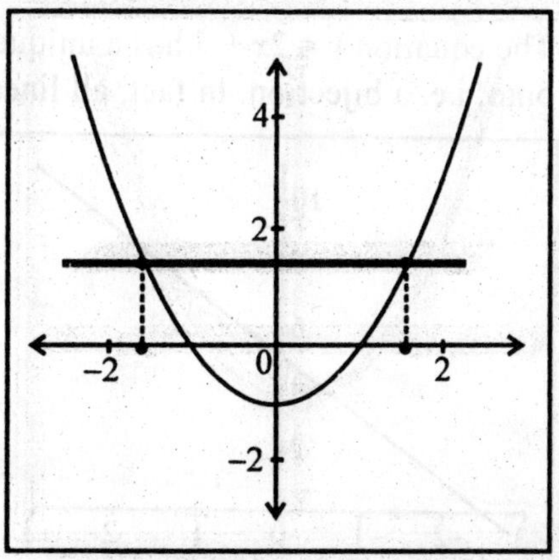

Not a one-to-one Function

Surjection

If you place a light on the left and on the right hand side of the coordinate system, then the shadow of the graph on the y axis is the image of the domain of the function. If that shadow covers the range of the function, then the function is onto.

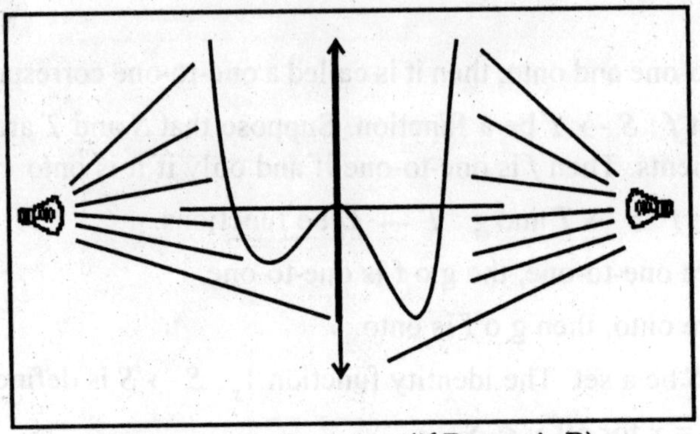

Not an onto Function (if Range is R)

Note that every function can be modified to be onto by setting its range to be the image of its domain.

Example 11. Which of the following functions are one-one, onto, or bijections? The domain for all functions is R.

1. $f(x) = 2x + 5$
2. $g(x) = \arctan(x)$
3. $g(x) = \sin(x)$
4. $h(x) = 2x^3 + 5x^2 - 7x + 6$

Solution:

1. Linear Function

This function is linear. The equation $y = 2x + 5$ has a unique solution for every x, so that the function is one-one and onto, i.e. a bijection. In fact, all linear functions are bijections.

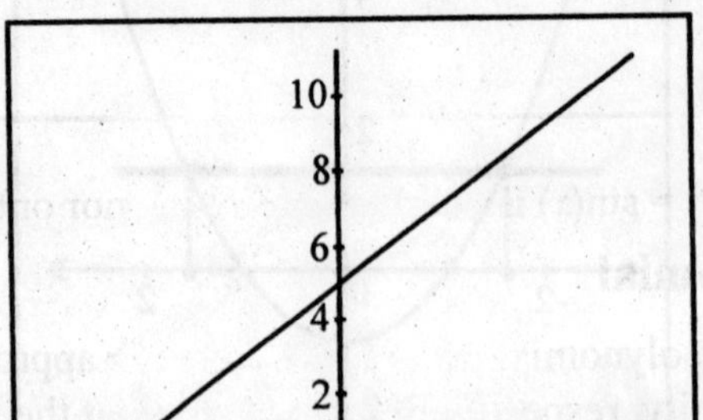

$f(x) = 2x + 5$ is a Bijection.

2. Inverse Functions

This function is the inverse function of the tangent function. A such, it must be one-to-one. It is onto if the range is $[-\pi/2, \pi/2]$, but not if the range is *R*. In fact, all inverse functions are one-to-one.

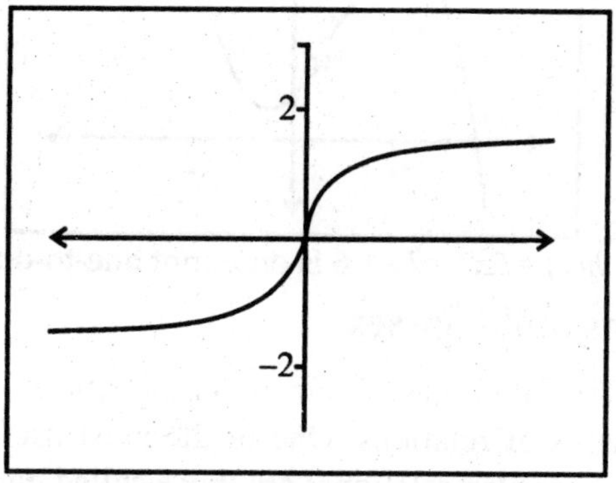

$g(x) = Arctan(x)$ is Injective, not Surjective on *R*.

3. Periodic Function

Since this function is periodic, it can not be one-to-one. It is onto, if the range is the interval [–1, 1], but not onto if the range is *R*. In fact, every periodic function is not one-to-one.

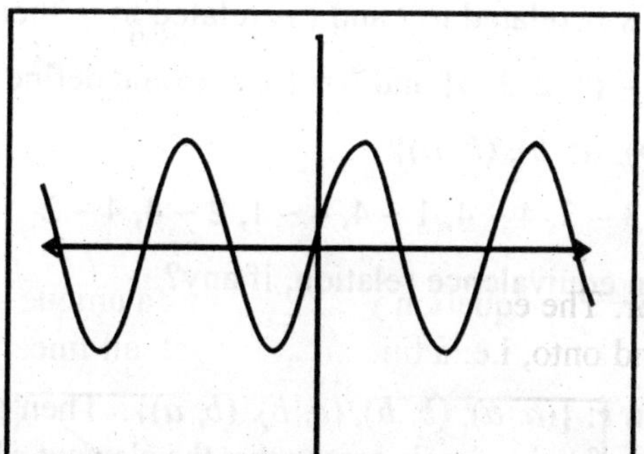

$g(x) = \sin(x)$ is Neither one-to-one nor onto *R*.

4. Odd Degree Polynomial

This is an odd-degree polynomial. Hence, the limit as *x* approaches plus or minus infinity must be plus or minus infinity, respectively. That means that the function is onto. Since most third degree equations have three zero, this function is probably not one-to-one. A look at the graph confirms this. In fact, every odd-degree polynomial is onto while no even degree polynomial is onto.

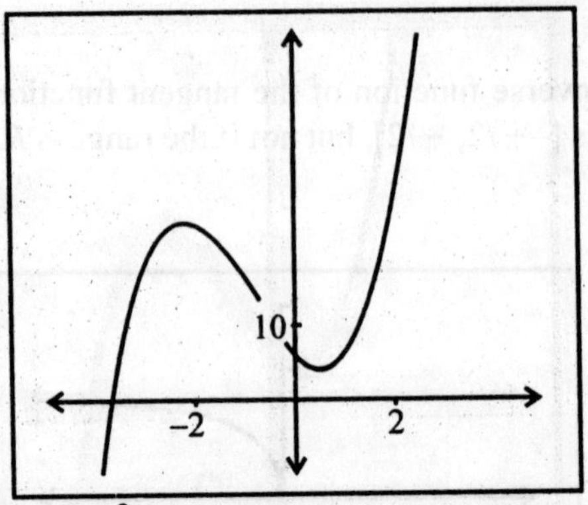

$h(x) = 2x^3 - 7x + 6$ is onto, not one-to-one

Equivalence Relations and Classes

Just as there were different classes of functions (bijections, injections, and surjections), there are also special classes of relations. One of the most useful kind of relation (besides functions, which of course are also relations) are those called equivalence relations.

Definition: Equivalence Relation

- Let S be a set and r a relation between S and itself. We call r and equivalence relation on S if r has the following three properties:
  1. Reflexivity: Every element of S is related to itself
  2. Symmetry: If s is related to t then t is related to s
  3. Transitivity: If s is related to t and t is related to u, then s is related to u.

Examples 12. Let $A = \{1, 2, 3, 4\}$ and $B = \{a, b, c)$ and define the following two relations:

(*i*) $r : \{(a, a), (b, b), (a, b), (b, a)\}$

$s : 1\sim 1, 2 \sim 2, 3 \sim 3, 4 \sim 4, 1 \sim 4, 4 \sim 1, 2 \sim 4, 4 \sim 2$

(*ii*) Which one is an equivalence relation, if any?

Solution:

Consider the relation r: $\{(a, a), (b, b), (a, b), (b, a)\}$. Then this is, technically speaking, not an equivalence relation, for the simple reason that the element c has no association. However, if we restrict our relation to a new domain that does not include the element c, then it might be an equivalence relation. We need to check:

1. **Reflexive:** every element that has a relation does also have a relation with itself
2. **Symmetric:** clearly true
3. **Transitive:** follows from the first two in this case.

- Consider the relation $1\sim 1, 2 \sim 2, 3 \sim 3, 4 \sim 4, 1 \sim 4, 4 \sim 1, 2 \sim 4, 4 \sim 2$. Then this is not an equivalence relation. We need to check:

1. **Reflexive:** every element with a relation is related to itself.
2. **Symmetric:** true as well (3 is only related to 3, so symmetry condition does not apply here)
3. **Not transitive:** $1 \sim 4$ and $4 \sim 2$, but 1 is not related to 2.

At first glance equivalence relations seem to be too abstract to be useful. However, just the opposite is the case. Because they are defined in an abstract fashion, equivalent relations can be utilized in many different situations. In fact, they can be used to define such basic objects as the integers, the rational numbers, and the real numbers.

The main result about an equivalence relation on a set A is that it induces a partition of A into disjoint sets. This property is the one that will allow us to define new mathematical objects based on old ones in the next section.

Theorem: Equivalence Classes

- Let r be an equivalence relation on a set A. Then A can be written as a union of disjoint sets $A = \bigcup_a A_a$ with the following properties:
  1. If a, b are in A then $a \sim b$ if and only if a and b are in the same set A_a
  2. The subsets A_a are non-empty and pairwise disjoint.
- The sets Aa are called equivalence classes.

Proof

We have to decide what the equivalence classes Aa should be: Since by property (1) two elements a and b are supposed to be related if and only if they are in the same class, it seems natural to define:

- $A_a = \{b \in A : b \sim a\}$

To simplify notation a little, we denote this class by $A(a)$ Now we need to check whether this is a good definition, and whether it satisfies the above properties.

Because of reflexivity of the equivalence relation, the class $A(a)$ contains the element a and is therefore not empty.

Next, we will show that two classes that are different can bot have any elements in common, i.e. they are disjoint. Recall that disjoint means that two sets have either an empty intersection or are the same sets. Take two elements a and a' and suppose that $A(a)$ and $A(a')$ have a non-empty intersection; say both classes contain the element c. Then

- $c \sim a$, because c is in $A(a)$
- $a \sim c$, by symmetry and the above line
- $c \sim a$ because c is in $A(a')$
- Therefore, by transitivity, $a \sim a'$

Now take any element b in $A(a')$. Then $b \sim a'$ and $a' \sim a$ so that $b \sim a$ by transitivity, and hence be is contained in $A(a)$. Therefore $A(a')$ is a subset of $A(a)$. The same argument works the other way around as well (by symmetry), showing that $A(a)$ is a subset of $A(a')$. Hence, $A(a) = A(a')$. This shows that two different classes must be disjoint.

Finally, we need to show that all classes $A(a)$ make up the original set A. But that is clear, because every element a in A belongs to the class $A(a)$, hence the union of all classes $A(a)$ must be contained in A, and therefore must be equal to A.

Example 13. Consider the set Z of all integers. Define a relation r by saying that x and y are related if their difference $y = x$ is divisible by 2. Then

1. Check that this relation is an equivalence relation
2. Find the two equivalence classes, and name them appropriately.
3. How would you add these equivalence classes, if at all?

Solution:

1. Equivalence Relation

reflexive:

$x - x$ is equal to zero, which divisible by two. Hence, every element is related to itself.

symmetry:

if $x \sim y$, the $y - x$ is divisible by 2. But then $-(y - x) = x - y$ is divisible by two.

Hence, $y \sim x$

transitivity:

- if $x \sim y$ then $y - x = 2n$ for some integer n
- if $y \sim z$ then $z - y = 2m$ for some integer m

But then $z - x = (2m + y) - (y - 2n) = 2(m + n)$, so that $z - x$ is divisible by 2. In other words, x and z are related.

2. Equivalence Classes

Two elements are in the same equivalence class if and only if they are related. If x and y are in the same class, then $y - x = 2m$ for some integer n.

- If y was even, then $y = 2m$ for some integer m, and $x = 2m - 2n$ must also be even.
- If y was odd, then $y = 2m + 1$ for some integer m, and $x = 2m + 1 - 2n$ must also be

odd

Therefore, there are two equivalence classes, and they are appropriately labeled:

- E = even numbers: $E = [(2)]$ contains all even numbers
- O = odd numbers: $O = [(3)]$ contains all odd numbers.

3. Adding Equivalence Classes

Define $[x] + [y] = [x + y]$. We need to show that this is well- defined, i.e. independent of the particular representative of the equivalence classes of $[x]$ and $[y]$. Take $x \sim x'$ and $y \sim y'$.

- Then $x' - x = 2m$ and $y' - y = 2n$ for some integers n and m.
- Then $(x' + y') - (x + y) = x' - x + y' - y = 2m + 2n = 2(m + n)$

That means $(x' + y')$ and $(x + y)$ are related if $x \sim x'$ and $y \sim y'$. But then

- $[x] + [y] = [x + y] = [x' + y'] = [x'] + [y']$

Hence, addition does not depend on the particular representative from a class, so that addition as defined above is indeed a well-defined operation.

A better would be the following: Defined two classes 0 and 1 by saying:

- all numbers divisible by 2 with no remainder are in class 0
- all numbers divisible by 2 with remainder 1 are in class 1

Then add equivalence classes by adding numbers modulo 2.

Example 14. What kind of equivalence classes do you get when x and y are defined to be related if their difference is divisible by m? How could you add those?

Solution:

- This relation is an equivalence relation (i.e. the three conditions are satisfied)
- There are m equivalence classes:

1. all numbers divisible by m with no remainder are in class 0
2. all numbers divisible by m with remainder 1 are in class 1
3. all numbers divisible by m with remainder 2 are in class 2
4. all numbers divisible by m with remainder m-1 are in class $m - 1$

- Addition can be defined by adding modulo m. That is, if we consider the equivalence classes obtained by dividing the differences by, say, 5, then we have, as an example:

- $[(2)] + [(1)] = [(3)]$
- $[(2)] + [(4)] = [(1)]$
- $[(3)] + [(4)] = [(3)]$
- etc...

Here is another, more complicated example:

Example 15. Consider the set $R \times R \setminus \{(0, 0)\}$ of all points in the plane minus the origin. Define a relation between two points (x, y) and (x', y') by saying that they are related if they are lying on the same straight line passing through the origin. Then, check that this relation is an equivalence relation and find a graphical representation of all equivalence classes by picking an appropriate member for each class.

Definition. Let S be a set. A subset R of $S \times S$ is called an equivalence relation on S if

(*i*) for all $a \in S$, $(a, a) \in R$;

(*ii*) for all $a, b \in S$, if $(a, b) \in R$ then $(b, a) \in R$;

(*iii*) for all $a, b, c \in S$ if $(a, b) R$ and $(b, c) \in R$, then $(a, c) \in R$.

We will write $a \sim b$ to denote the fact that $(a, b) \in R$.

Definition. Let ~be an equivalence relation on the set S. For a given element $a \in S$ we define the equivalence class of a to be the set of all elements of S that are equivalent to a.

We will use the notation $[a]$. In symbols, $[a] = \{x \in S \mid x \sim a\}$.

The notation $S/\sim$ will be used for the collection of all equivalence classes of S under ~.

Example 16. (S/f) 2.2.2. Let $f: S -> T$ ..... be any function. For $x_1, x_2 \in S$ we define $x_1 \sim x_2$ if $f(x_1) = f(x_2)$. Then for all $x_1, x_2, x_3 \in S$ we have

(*i*) $f(x_1) = f(x_1)$;

(*ii*) if $f(x_1) = f(x_2)$, then $f(x_2) = f(x_1)$;

(*iii*) if $f(x_1) = f(x_2) = f(x_3)$, then $f(x_1) = (x_3)$.

This shows that we have defined an equivalence relation on the set S. The proof of this is easy because the equivalence relation is defined in terms of equality of the images $f(x)$, and equality is the most elementary equivalence relation.

The collection of all equivalence classes of S under ~will be denoted by S/f.

Proposition. Let S be a set, and let ~ be an equivalence relation on S. Then each element of S belongs to exactly one of the equivalence classes of S determined by the relation~

Definition. Let S be any set. A family P of subsets of S is called a partition of S if each element of S belongs to exactly one of the members of P.

Proposition. Any partition P of a set S determines an equivalence relation.

Theorem. If $F: S \to T$ is any function, and ~ is the equivalence relation defined on S by letting

$x_1 \sim x_2$ if $f(x_1) = f(x_2)$ for all $x_1, x_2 S$, then there is a one-to-one correspondence between the elements of the image $f(S)$ of S under f and the equivalence classes S/f of the relation ~.

If $f: S \to T$ is a function and y belongs to the image $f(S)$, then the inverse image of y is

$$f^{-1}(y) = \{x \in S \mid f(x) = y\}.$$

The inverse images of elements of $f(S)$ are the equivalence classes in S/f. (Note carefully that we are not implying that f has an inverse function.)

Permutations

Definition. Let S be a set. A function $\sigma : S -> S$ is called a permutation of S if σ is one-to-one and onto.

The set of all permutations of S will be denoted by Sym(S).

The set of all permutations of the set $\{1, 2, ..., n\}$ will be denoted by S_n.

Proposition 2.1.6 shows that the composition of two permutations in Sym(S) is again a permutation. It is obvious that the identity function on S is one-to-one and onto.

Proposition 2.1.8 shows that any permutation in Sym(s) has an inverse function that is also one-to-one and onto. We can summarize these important properties as follows:

(i) If $\sigma, \tau \in$ Sym(S), then $\sigma \tau \in$ Sym(S);

(ii) $1_s \in$ Sym(S);

(iii) if , $\sigma \in$ Sym(S) then $\sigma^{-1} \in$ Sym(S).

Definition. Let S be a set, and let $\sigma \in$ Sym(S). Then is called a cycle of length k if there exist elements $a_1, a_2, ..., a_k \in S$ such that

$\sigma(a_1) = a_2, \sigma(a_2) = a_3 ,....., \sigma(a_{k-1}) = a_k, \sigma(a_k) = a_1$, and

$\sigma(x) = x$ for all other elements $x \in S$ with $x \neq a_1$ for $i = 1, 2, ..., 1, 2, ..., k$.

In this case we write $\sigma = (a_1, a_2, ... a_k)$.

We can also write $\sigma = (a_1, a_2, ... a_k)$ or $= (a_3, a_k, a_1, a_2)$, etc. The notation for a cycle of length k can thus be written in k different ways, depending on the starting point. The notation (1) is used for the identity permutation.

Definition. Let $\sigma = (a_1, a_2, ... a_k)$ and $\tau = (b_1, a_2, ... b_m)$ be cycles in Sym(S), for a set S. Then σ and τ are said to be **disjoint** if $a_i \neq b_j$ for all i, j.

Proposition. Let S be any set. If σ and τ are disjoint cycle in Sym (S), then $\sigma \tau = \tau \sigma$.

Theorem. Every permutation in S_n can be written as a product of disjoint cycles. The cycles that appear in the product are unique.

Definition. Let $\sigma \in S_n$ The least positive integer m such that $\sigma^m = (1)$ is called the order of σ.

Proposition. Let $\sigma \in S_n$ have order m. Then for all integers j, k we have

$$\sigma^j = \sigma^k \text{ if and only if } j \equiv k \pmod{m}.$$

Proposition. Let $\sigma \in S_n$ be written as a product of disjoint cycles. Then the order of is the least common multiple of the lengths of its cycle.

Definition. A cycle (a_1, a_2) of length two is called a transposition.

Proposition. Any permutation in S_n where $n \geq 2$ can be written as a product of transpositions.

Theorem. If a permutation is written as a product of transpositions in two ways, then the number of transpositions is either even in both cases or odd in both cases.

Definition. A permutation σ is called even if it can be written as a product of an even number of transpositions, and odd if it can be written as a product of an odd number of transpositions.

EXERCISE

1. Suppose that the function F defined for all real number t by the formula

   $$F(t) = 7(t+1)+6$$

   Evaluate F at the input values 0, 2 and $x + 2$.

2. Draw the graph of the function $y = f(x) = x^2$ over the interval $[-3,3]$.

3. If $f(x) = \sqrt{x}$ and $g(x) = x+1$, find $(fog)(x), (gof)(x), (fof)(x)$ and $(gog)(x)$.

4. Find the domain and range of each of the following functions:

   (i) $f(x) = 1+x^2$

   (ii) $h(x) = 1-\sqrt{x}$

   (iii) $g(t) = \sqrt{1-\sqrt{1-t^2}}$

   (iv) $h(z) = \dfrac{\sqrt{1-z}}{\sqrt{1+z^2}}$

5. Draw the graph of the function

   $$f(x) = \begin{cases} \lfloor x \rfloor, x \geq 0 \\ \lceil x \rceil, x < 0 \end{cases}$$

   Why $f(x)$ called the intger part of x?

Chapter 2

Derivatives

INTRODUCTION

In the previous chapters we have studied basic tools of complex algebra such set theory, their properties and its application in the study of functions. In this chater we are going to sudy derivatives and their properties, also we study various method of evaluatiing derivatives of functions. Derivatives is another most important tool of complex analysis.

RULES OF DERIVATIVES AND APPLICATION

In higher calculus, it is usually the case that the problems of integration are more important than those of differentiation, although differentiation offers less difficulty than integration. Consequently, the natural method of constructing the integral and differential calculus is first to learn to differentiate the widest possible classes of functions and then, by virtue of the fundamental theorem, make the results thus obtained available for the solution of integration problems. In the following sections, it will be our task to carry out this programme. In this development of the subject, certain rules for differentiation, with the first of which we are already acquainted, will have an important role.

Rules for Derivatives

We assume that in the interval under consideration the functions $f(x)$ and $g(x)$ are differentiable; our rules then are:

Rule 1: Multiplication by a constant.

If c is a constant and $\phi(x) = cf(x)$, then $\phi(x)$ is differentiable and

$$\phi'(x) = cf'(x)$$

This follows immediately from the relation

$$\frac{\phi(x+h)-\phi(x)}{h} = c\frac{f(x+h)-f(x)}{h}$$

if we take the limit $h \} 0$.

Rule 2: Derivative of a sum.

If $\phi(x) = f(x) + g(x)$, then $\phi(x)$ is differentiable, and

$$\phi'(x) = f'(x)+g'(x)$$

that is, the processes of differentiation and addition are interchangeable. The same holds for the sum of any finite number n of terms

$$\phi(x) = \sum_{\nu-1}^{n} f_{\nu}(x),$$

for which we obtain

$$\phi'(x) = \sum_{\nu-1}^{n} f'_{\nu}(x).$$

We may pass over the proof, which, by 2.3, is fairly obvious.

Rule 3: Derivative of a product.

If $\phi(x) = f(x)g(x)$, then $\phi(x)$ is differentiable and

$$\phi'(x) = f(x)+g'(x)+g(x)+f'(x)$$

The proof follows from the equation

$$\frac{\phi(x+h)-\phi(x)}{h} = \frac{f(x+h)g(x+h)-f(x)g(x)}{h}$$

$$= \frac{f(x+h)g(x+h)-f(x+h)g(x)+f(x+h)g(x)-f(x)g(x)}{h}$$

$$= f(x+h)\frac{g(x+h)-g(x)}{h}+g(x)\frac{f(x+h)-f(x)}{h}$$

The passage to the limit $h \}$ 0 can here be carried out directly and yields the result shown.

This formula takes a still more elegant form, if we divide throughout by $\phi(x) = f(x)g(x)$. We then obtain

$$\frac{\phi'(x)}{\phi(x)} = \frac{f'(x)}{f(x)} + \frac{g'(x)}{g(x)}$$

where we, of course, assume that $\phi(x)$ is nowhere equal to zero.

By repeated application of this product formula, we obtain, by induction, for the derivative of a product of n factors an expression consisting of n terms, each of which consists of the derivative of one factor multiplied by all the other factors of the original product. In symbols, this is

$$\phi'(x) = \frac{d}{dx}\{f_1(x)f_2(x).....f_n(x)\}$$

$$= f_1'(x)f_2(x)....f_n(x) + f_1(x)f_2'(x)f_3(x)....f_n(x)$$

$$+ .... + f_1(x)f_2(x).....f_n'(x)$$

$$= \sum_{\nu-1}^{n} f_\nu'(x)\frac{\phi(x)}{f_\nu(x)},$$

or on division by $\phi(x) = f_1(x)f_2(x)...f_n(x)$, assuming again that $\phi(x)$ does not vanish anywhere,

$$\frac{\phi'(x)}{\phi(x)} = \frac{f_1'(x)}{\phi(x)} + \frac{f_2'(x)}{\phi(x)} + ... + \frac{f_n'(x)}{f_n(x)}$$

$$= \sum_{\nu=1}^{n} \frac{f_\nu'(x)}{f_\nu(x)}$$

Rule 4: Derivative of a quotient.

We have for a quotient

$$\phi(x) = \frac{f(x)}{g(x)}$$

the rule: The function $\phi(x)$ is differentiable at every point at which $g(x)$ does not vanish and

$$\phi'(x) = \frac{g(x)f'(x) - g'(x)f(x)}{\{g(x)\}^2}$$

If $\phi(x) < \;\; 0$, this can be rewritten

$$\frac{\phi'(x)}{\phi(x)} = \frac{f'(x)}{f(x)} - \frac{g'(x)}{g(x)}.$$

If we accept the differentiability of $\phi(x)$ as a hypothesis, we can apply the product rule to $f(x)=\phi(x)g(x)$ and conclude that

$$f'(x) = \phi(x)g'(x) + g(x)\phi'(x)$$

By substituting $f(x)/g(x)$ for $\phi(x)$ on the right hand side and solving for $\phi'(x)$, we obtain the above rule. In order to prove the differentiability of $\phi(x)$ as well as the rule, we use the method: We write

$$\frac{\phi(x+h)-\phi(x)}{h} = \frac{\dfrac{f(x+h)}{g(x+h)} - \dfrac{f(x)}{g(x)}}{h}$$

$$= \frac{g(x)\dfrac{f(x+h)-f(x)}{h} - \dfrac{g(x+h)-g(x)}{h}f(x)}{g(x)g(x+h)}$$

If we now let h tend to 0, we arrive at the result stated, because, by assumption, the two terms obtained by performing the division on the right hand side have definite limits, which are

$$\frac{g(x)f'(x)}{\{g(x)\}^2} \quad \text{and} \quad \frac{g'(x)f(x)}{\{g(x)\}^2},$$

respectively. This proves immediately the existence of the limit on the left hand side as well as the differentiation formula.

Differentiation of the Rational Functions: To begin with, we shall again deduce the differentiation formula

$$\frac{d}{dx}x^n = nx^{n-1}$$

for every positive integer n, basing the proof on the rule for differentiation of a product. x and hence obtain

$$\frac{d}{dx}x^n = 1.x^{n-1} + 1.x^{n-1} + ... + 1.x^{n-1}$$

$$= nx^{n-1}.$$

If we use this formula and the first rule of differentiation, the second derivative of the function x^n becomes:

$$\frac{d^2}{dx^2}x^n = n(n-1)x^{n-2}$$

Continuing this process, we obtain

$$\frac{d^3}{dx^3}x^n = n(n-1)(n-2)x^{n-3}$$

$$\frac{d^3}{dx^3}x^n = 1.2...n = n!$$

It is clear from the last of these formulae that the $(n + 1)$-th derivative of x^n vanishes everywhere.

By virtue of our first two rules, a knowledge of the differentiation of powers enables us at once to differentiate any polynomial

$$y = a_0 + a_1x + a_2x + ... + a_nx^n.$$

We have simply

$$y' = a_1 + 2a_2x + 3a_3x^2 + ... + na_nx^{n-1}$$

and

$$y'' = 2a_2 + 3.2a_3x + 4.3a_4x^2 + ... + n(n-1)a_nx^{n-2}$$

and so on.

The differentiation of any rational function now follows with the help of the quotient rule. In particular, we shall again derive the differentiation formula for the function x^n, where $n = -m$ is a negative integer. The application of the quotient rule together with the fact that the derivative of a constant is equal to zero yields

$$\frac{d}{dx}\left(\frac{1}{x^m}\right) = -\frac{mx^{m-1}}{x^{2m}}$$

$$= -\frac{m}{x^{m+1}}$$

or, if we take $m = -n$,

$$\frac{d}{dx}x^n = nx^{n-1}$$

which agrees formally with the result for positive values of n and with the results given earlier.

DERIVATIVE OF TRIGONOMETRIC FUNCTIONS

In the previous section we have learnt to, how to diffrentiate a function. In this section we are going to study the derivatives of trignometric functions.

For the trigonometric functions sin x and cos x, we have already obtained the derivative formula of sin x and cos x such as

$$\frac{d}{dx}\sin x = \cos x$$

and

$$\frac{d}{dx}\cos x = -\sin x$$

The quotient rule now enables us to differentiate the functions

$$y = \tan x$$

$$= \frac{\sin x}{\cos x} \text{ and}$$

$$y = \cot x$$

$$= \frac{\cos x}{\sin x}$$

According to the rule, the derivative of the first of these functions is

$$\frac{dy}{dx} = y' = \frac{\cos^2 x + \sin^2 x}{\cos^2 x}$$

$$= \frac{1}{\cos^2 x}$$

and we obtain the result

$$\frac{d}{dx}\tan x = \frac{1}{\cos^2 x}$$

$$= \sec^2 x$$

$$= 1 + \tan^2 x$$

Similarly, we obtain

$$\frac{d}{dx}\cot x = -\frac{1}{\sin^2 x}$$

$$= -\cos ec^2 x$$

$$= -\left(1+\cot^2 x\right).$$

Derivatives of a Function of Several Variables

In the previous section we have studied the derivative of the fucntions of single variable. Now we are going to discuss the derivatives of a Function of Several Variable.

Definition. Geometrical Representation: If we assign in a function of several variables definite numerical values to all but one of the variables and allow only that one variable, say x, as shown in figure1 to vary, the function becomes a function of one variable. For

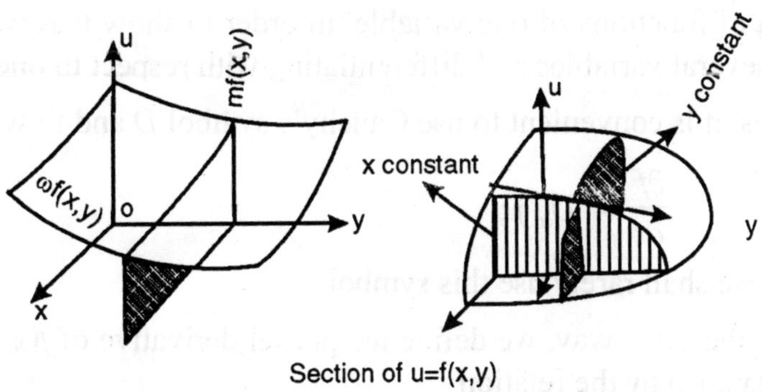

Fig. 1 Derivatives of Functions of Variables.

example, consider a function $u = f(x,y)$ of the two variables x and y and give y the definite fixed value $y = y_0 = c$. The function $u = f(x, y_0)$ of the single variable x, which is thus formed, may be simply represented geometrically by letting the surface $u = f(x, y)$ be cut by the plane $y = y_0$ (Figs. 5 and 6). The curve of intersection thus formed in the plane is represented by the equation $u=f(x,y_0)$. If we differentiate this function in the usual way at the point $x = x_0$ (assuming that the derivative exists), we obtain the partial derivative of $f(x,y)$ with respect to x at the point (x_0, y_0). According to the ordinary definition of the derivative, this is the limit

$$\lim_{h\to 0} \frac{f(x_0+h, y_0)-f(x_0, y_0)}{h}$$

If (x_0, y_0) is a point on the boundary of the region of definition, we make the restriction that in the passage to the limit the point $(x + h, y_0)$ must always remain in the region.

Geometrically speaking, this partial derivative denotes the tangent of the angle between a parallel to the *x*-axis and the tangent line to the curve $u = (x,y_0)$. It is therefore the slope of the surface $u = f(x,y)$ in the direction of the *x*-axis.

In order to represent these partial derivatives, several different notations are in use of which we mention:

$$\lim_{h\to 0} \frac{f(x_0+h,y_0)-f(x_0,y_0)}{h} = f_\infty(x_0,y_0) = u_\infty(x_0,y_0)$$

If we wish to emphasize -that the partial derivative is the limit of a difference quotient, we denote it by

$$\frac{\partial f}{\partial x} or \frac{\partial}{\partial x} f .$$

We use here a special curved letter ∂ instead of the ordinary *d* used in the differentiation of functions of one variable, in order to show that we are dealing with a function of several variables and differentiating with respect to one of them.

Sometimes. it is convenient to use Cauchy's symbol *D* and to write

$$\frac{\partial f}{\partial x} = D_\infty f;$$

however, we shall rarely use this symbol.

In exactly the same way, we define the partial derivative of $f(x, y)$ with respect to *y* at the point (x_0, y_0) by the relation

$$\lim_{h\to 0} \frac{f(x_0,y_0+k)-f(x_0,y_0)}{k} = f_v(x_0,y_0) = \frac{\partial f}{\partial y} = D_v f(x_0,y_0).$$

It represents the slope of the curve of intersection of the surface $u = f(x, y)$ with the plane $x = x_0$, perpendicular to the *x*-axis.

Now, let us think of the point (x_0, y_0), hitherto considered to be fixed, as variable and accordingly omit the suffices. In other word, we think of the differentiation as being carried out at any point (x, y) of the region of definition of $f(x, y)$. Then the two derivatives are themselves functions of *x* and *y*:

$$u_x(x,y) = f_x(x,y) = \frac{\partial f(x,y)}{\partial x} \text{ and } u_y(x,y) = f_y(x,y) = \frac{\partial f(x,y)}{\partial y}$$

Similarly, we make the definition for any number (*n*) of independent variables:

$$\frac{\partial f(x_1,x_2,...,x_n)}{\partial x_1} = \lim_{h\to 0} \frac{f(x_1+h,x_2...,x_n)-f(x_1+x_2...,x_n)}{h}$$

$$= fx_1(x_1,x_2,....,x_n) = Dx_1 f(x_1,x_2,...,x_n),$$

it being assumed that the limit exists.

Of course, we can also form higher partial derivatives of $f(x, y)$ by again differentiating the partial, first order derivatives $f_x(x,y)$ and $f_v(x,y)$ with respect to one of the variables, and repeating this process. We indicate the order of differentiation by the order of the suffixes or by the order of the symbols ∂x and ∂y in the denominator from the right to the left hand side, and use for the second partial derivatives the:

On the other hand, in Continental usage, $\frac{\partial}{\partial x}\left(\frac{\partial f}{\partial y}\right)$ becomes $\frac{\partial^2 f}{\partial y \partial x}$

We likewise denote the third partial derivatives by

$$\frac{\partial}{\partial x}\left(\frac{\partial^2 f}{\partial x^2}\right) = \frac{\partial^2 f}{\partial x^3} = f_{xxx},$$

$$\frac{\partial}{\partial y}\left(\frac{\partial^2 f}{\partial x^2}\right) = \frac{\partial^3 f}{\partial y dx^2} = f_{yxx},$$

$$\frac{\partial}{\partial x}\left(\frac{\partial^2 f}{\partial x \partial y}\right) = \frac{\partial^2 f}{\partial x^2 \partial y} = f_{xxy}, \text{ etc.}$$

and, in general, the n-th derivatives

$$\frac{\partial}{\partial x}\left(\frac{\partial^{n-1} f}{\partial x^{n-1}}\right) = \frac{\partial^n f}{\partial x^n} = fx^n$$

$$\frac{\partial}{\partial y}\left(\frac{\partial^{n-1} f}{\partial x^{n-1}}\right) = \frac{\partial^n f}{\partial y \partial x^{n-1}} = fyx^{n-1}, \&c.$$

Finally, we shall study a few examples of the actual calculation of partial derivatives. According to the definition, all the independent variables are to be kept constant except the one with respect to which we are differentiating. We therefore have merely to regard the other variables as constants and carry out out the differentiation according to the rules by which we differentiate functions of a single independent variable.

For example, we have:

Function $f(x,y)=xy$

first derivatives, $f_x = y, f_v = x$

second derivatives $f_{xx}=0, f_{xy}=f_{yx}=1, f_{yy}=0$

Function $f(x,y)=\sqrt{(x^2+y^2)}$

first derivatives, $f_x=\dfrac{x}{\sqrt{(x^2+y^2)}}, f_v=\dfrac{y}{\sqrt{x^2+y^2}}\cdot r$

(Thus for the radius vector $r=\sqrt{(x^2+y^2)}$ from the origin to the point (x, y), the partial derivatives with respect to x and y are given by the cosine, $\cos\varphi = x/r$, and the sine, $\sin\varphi = y/r$ of the angle φ, which the radius vector makes with the positive direction of the x-axis.)

Second derivatives

$$f_{xx}=\frac{\sqrt{(x^2+y^2)-\dfrac{x^3}{\sqrt{(x^2+y^2)}}}}{x^2+y^2}=\frac{y^2}{\sqrt{(x^2+y^2)^2}}=\frac{\sin^2\varphi}{r}$$

$$f_{xy}=f_{yx}=-\frac{xy}{\sqrt{(x^2+y^2)^2}}=-\frac{\sin\varphi\cos\varphi}{r}$$

$$f_{yy}=\sqrt{(x^2+y^2)-\frac{y^2}{\sqrt{(x^2+y^2)}}}=\frac{x^2}{\sqrt{(x^2+y^2)^2}}=\frac{\cos^2\varphi}{r}$$

Reciprocal of the radius vector in three dimensions;

$$f(x,y,z)=\frac{1}{\sqrt{(x^2+y^2+z^2)}}=\frac{1}{r}$$

first derivatives,

$$f_x = \frac{-x}{\sqrt{\left(x^2+y^2+z^2\right)^2}} = -\frac{x}{r^2}$$

$$f_y = \frac{-x}{\sqrt{\left(x^2+y^2+z^2\right)^2}} = -\frac{y}{r^2}$$

$$f_z = \frac{-x}{\sqrt{\left(x^2+y^2+z^2\right)^2}} = -\frac{z}{r^2}$$

second derivatives,

$$f_{xx} = -\frac{1}{r^2}+\frac{3x^2}{r^2}, f_{yy} = -\frac{1}{r^2}+\frac{3y^2}{r^2}, f_{zz} = -\frac{1}{r^2}+\frac{3z^2}{r^2}$$

$$f_{xy} = f_{yx} = \frac{3xy}{r^2}, f_{yz} = f_{zy} = \frac{3yz}{r^2}, f_{zx} = f_{zx} = \frac{3zx}{r^2}$$

We see from this that there holds for the function $f = \frac{1}{\sqrt{\left(x^2+y^2+z^2\right)}}$ the equation

$$f_{xx}+f_{yy}+f_{zz} = -\frac{3}{r^2}+\frac{3\left(x^2+y^2+z^2\right)}{r^2} = 0,$$

for all values of *x, y, z* except 0,0,0; as we may say, the equation

$$f_{xx}+f_{yy}+f_{zz} = 0$$

is satisfied identically in x,y,z by the function f(x,y,z) = 1/*r*.

Functions $f(x,y) = \frac{1}{\sqrt{y}}e^{-(x-a)^2/4y}$

first derivatives,

$$f_x = \frac{1}{\sqrt{y}}\frac{-(x-a)}{2y}e^{-(x-a)^2/4y}$$

$$f_y = \left(\frac{-1}{2y^{3/2}} + \frac{(x-a)^2}{4y^{3/2}}\right)e^{-(x-a)^2/4y}$$

second derivatives,

$$f_{xx} = \left(\frac{-1}{2y^{3/2}} + \frac{(x-a)^2}{4y^{3/2}}\right)e^{-(x-a)^2/4y}$$

The equation

$$f_{xx} - f_y = 0$$

is therefore satisfied identically in x and y.

Just as in the case of one independent variable, the possession of derivatives is a special property of a function. (Note that the expression differentiable implies more than that the partial derivatives with respect to x and y exist (cf. Vol. II!) All the same, this property is possessed by all functions of practical importance except perhaps at isolated exceptional points.

In contrast to functions of one variable, the possession of derivatives does not imply the continuity of the function. This is clearly shown by the example $u = \dfrac{2xy}{x^2 + u^2}$ already considered. because the partial derivatives exist for it everywhere and yet the function is discontinuous at the origin. But, as is stated by the following theorem, the possession of bounded derivatives does imply continuity:

If a function $f(x,y)$ has partial derivatives f_x and f_y everywhere in a region R, and these derivatives satisfy everywhere the in equalities $|f_x(x,y)| < M, |f_x(x,y)| < M$,

where M is independent of x and y, then $f(x, y)$ is continuous everywhere in R.

In particular, if f_x and f_y are continuous, they are necessarily bounded, so that $f(x,y)$ is also continuous.

In other words, it made no difference whether we differentiated first with respect to x and then with respect to y or *vice versa*. This is no accidental occurrence. In fact, we have the theorem:

If the mixed partial derivatives f_{xy} and f_{yx} of a function $f(x,y)$ are continuous in a region R, then the equation

$$f_{yx} = f_{xy}$$

holds everywhere in the interior of this region, i.e., the order of differentiation with respect to x and y is immaterial.

Applying this theorem to f_x and f_y, then to f_{xx}, f_{xy}, f_{yy}, etc., we find that

$$f_{xxy} = f_{xyx} = f_{yxx},$$
$$f_{xyy} = f_{xyx} = f_{yyx},$$
$$f_{xxyy} = f_{xyxy} = f_{xyyx} = f_{yxxy} = f_{yxyx} = f_{yyxx}, \ \& \ c.,$$

and, in general, we have the result:

In repeated differentiation of a function of two variables, the order of differentiation can be changed arbitrarily, provided only that the derivatives in question are continuous functions.

Chain Rule and Differentiation of Inverse Functions

Functions of Functions (Compound Functions):. It often happens that a function u of the independent variables x, y is stated in the form

$$u = f(\xi, \eta, ...),$$

where the arguments ξ, η, ... of the function f are themselves functions of x and y:

$$\xi = \phi(x, y), \ \eta = \psi(x, y), ...$$

We then say that

$$u = f(\xi, \eta, ...), f = (\phi(x, y), ... \psi(x, y), ...) = F(x, y)$$

is given as a compound function of x and y.

For example, the function

$$u = e^{x^3} y(x+y)^3$$

may be written as a compound function by means of the relations

$$u = e^{\xi}\eta^3 = f(\xi,\eta); \ \xi = x^2y, \ \eta = x + y.$$

Similarly, the function

$$u = \log(x + 1) \ . \ \text{arc cos} \ \sqrt{(4 - x^2 - y^2)}$$

can be expressed in the form

$$u = \eta \arccos\xi = f(\xi,\eta), \ \xi = \sqrt{(4 - x^2 - y^2)}, \eta = \log(x+1).$$

In order to make this concept more precise, we assume to begin with that the functions

$\xi = \phi(x, y)$, $\eta = \psi(x, y)$, ... are defined in a certain region R of the independent variables x, y. Then there corresponds to every point (x,y) of R a point $(\xi,\eta, ..)$. As the point (x,y) ranges over R, the point $(\xi,\eta,\ldots)$ will range over a certain set of values. We assume that the point $(\xi,\eta, \ldots)$ always lies within a region S in which $f(\xi,\eta, \ldots)$ is defined. The function

$$u = f(x, y), \psi(x, y), \ldots) = F(x, y)$$

is then defined in the region R.

Referring to our examples, we find in the first one of them that ξ and η are defined for every x, y and $f(\xi,\eta)$ is defined for every ξ,η, so that our region R can be taken to be the entire xy-plane. However, in the second example, the region S is restricted by the inequality $|\xi| \leq 1$, since for $|\xi| > 1$ the function

arcos ξ is undefined. Secondly, the region R is restricted by the inequalities $x+1>0$ and $x + y \geq 4$, since for other values both ξ and η are not defined. Thirdly, the region R must be further limited by the inequality $3 \geq x + y$ in order that the point with co-ordinates ξ,η shall fall into S, i.e., the restriction $|\xi| \leq 1$ implies that $x + y \geq 3$. Hence R consists of the part of the ring $3 \geq x+y \geq 4$ lying to the right of the line $x = -1$.

The following theorem on compound functions is an immediate consequence of the definitions:

If the function $u = f(\xi,\eta, \ldots)$ is continuous in S and the functions $\xi = \phi(x, y)$, $\eta = \psi(x, y)$), are continuous in R, then the compound function $u = F(x, y)$ is continuous in R.$\xi = \phi(x, y)$

The reader should be able to prove this statement

The Chain Rule : We now turn our attention to compound functions of the type $u = f(\xi,\eta, \ldots)$, where $\xi,\eta, \ldots$ depend on the single variable x:

$$\xi = \phi(x), \eta = \psi(x), \ldots.$$

For such functions, we have the important theorem known as the chain rule:

If the function $u = f(\xi,\eta, \ldots)$ has continuous partial derivatives of the first order in S and the functions $\xi = \phi(x)$, $\eta = \psi(x)$, ... have continuous first derivatives in the interval R, $a \geq x \geq b$, then $u = f(\phi(x), \psi(x), \ldots) = Fxi)$ has a continuous derivative in R and

$$F'(x) = f_\xi \phi'(x) + f_\eta \eta'(x) + \ldots.$$

The right-hand side of this equation is an abbreviation for

$$f_\xi \{\phi(x), \psi(x), \ldots\} \phi'(x) + \ldots$$

In order to simplify the notation, we shall assume that f is a function of the three arguments ξ, η, ζ. We shall denote by x_0 an arbitrary fixed point of the interval $a \rbrace x \rbrace b$, by ξ_0, η_0, ζ_0 the corresponding values $\xi_0 = \phi(x_0)$, $\eta_0 = \psi(x_0)$, $\xi_0 = \chi(x_0)$ and by ξ, η, ζ the values $\phi(x)$, $\psi(x)$, $\chi(x)$, corresponding to a variable point $x = x_0 + h$. We first write down the identity

$$F(x) - F(x_0) = f(\xi,\eta,\zeta) - f(\xi_0,\eta_0,\zeta_0)$$

$$= f(\xi,\eta,\zeta) - f(\xi_0,\eta_0,\zeta_0) + \{f(\xi_0,\eta,\zeta) - f(\xi_0,\eta_0,\zeta)\}$$

$$\{f(\xi_0,\eta_0,\zeta) - f(\xi_0,\eta_0,\zeta_0)\}$$

In each bracket on the right hand side, we observe that only one of the independent variables changes its value. Hence we can apply to each bracket the mean value theorem for functions of a single variable and obtain

$$F(x) - F(x_0) = (\xi - \xi_0) f_\xi(\bar{\xi},\eta,\xi) + (\eta - \eta_0) f_\eta(\xi_0,\bar{\eta},\zeta)$$

$$+(\xi - \xi_0) f_\zeta(\xi_0,\eta_0,\bar{\zeta}),$$

where $\bar{\xi}$ lies between x and x_0, $\bar{\eta}$ between h_0 and h, and $\bar{\zeta}$ between z_0 and z. Moreover, by the mean value theorem, we have

$$\xi - \xi_0 = \phi(x) - \phi(x_0) = (x - x_0)\phi'(x_1),$$

$$\eta - \eta_0 = \psi(x) - \psi(x_0) = (x - x_0)\psi'(x_2),$$

$$\zeta - \zeta_0 = \chi(x) - \chi(x_0) = (x - x_0)\chi'(x_3),$$

where all x_1, x_2, and x_3 lie between x_0 and x. Substituting these values in the last equation and dividing by $x - x_0$, we have

$$\frac{F(x) - F(x_0)}{x - x_0} = f_\xi(\bar{\xi},\eta,\zeta)\phi'(x_1) + f_\eta(\xi_0,\bar{\eta},\zeta)\psi'(x_2) + f_\zeta(\xi_0,\eta_0,\bar{\zeta})\chi'(x_3)$$

We now let x tend to x_0. Owing to the continuity of $\phi(x)$, $\psi(x)$, $\chi(x)$, the quantities ξ,η,ζ tend to ξ_0, η_0, ζ_0, respectively, and, *a fortiori*, so do $\bar{\xi}$, $\bar{\eta}$, $\bar{\zeta}$. Also x_1, x_2 and x_3 tend to x_0. Since all the functions on the right hand side are continuous, we have

$$\lim_{n \to x_0} \frac{F(x) - F(x_0)}{x - x_0} = F'(x_0)$$

$$= f_\xi(\xi_0,\eta_0,\zeta_0)\phi'(x_0) + f_\eta(\xi_0,\eta_0,\zeta_0)\psi'(x_0) + f_\zeta(\xi_0,\eta_0,\zeta_0)\chi'(x_0),$$

thus establishing the formula for $F'(x)$.

The continuity of $F'(x)$ follows immediately from the formula, since, by assumption, ϕ', ψ', χ', are continuous f_ξ, f_η and f_ζ are are continuous functions of continuous functions.

This theorem may be extended to compound functions of two or more variables, as follows:

If the funding $u = f(\xi, \eta, ...)$ has continuous partial derivatives of the first order in the region S and the functions $\xi = \phi(x,y)$, $\eta = \psi(x,y)$, x,y have continuous partial derivatives of the first order in R, then $u = F(x, y) = f\{\phi(x,y), \psi(x,y), ...$ has continuous partial derivatives of the first order in R and their derivatives are given by

$$F_x = f_\xi \phi_x + f_\eta \psi_x + ...,$$

$$F_y = f_\xi \phi_y + f_\eta \psi_y + ...$$

These formulae are often written in the abbreviated form

$$u_x = u_\xi \xi_x + u_\eta \psi_x + ...,$$

$$u_y = u_\xi \xi_y + u_\eta \eta_y + ...$$

In order to derive them, we temporarily introduce the notation $g(x) = \phi(x,y_0)$, $h(x) = \psi(x,y_0)$, ..., where y_0 is a fixed value of y. By the definition of the partial derivatives, it follows that $g'(x) = \phi_x(x,y_0)$, $h'(x) = \psi_x(x,y_0)$, ... Similarly, if we writs $H(x) = F(x, y_0)$, we have $H'(x) = F_x(x, y_0)$. We now apply the theorem just proved to the function $u = H(x) = f(\xi, \eta, ...) = f\{g(x), h(x), ...\}$ and obtain

$$H'(x_0) = f_\xi g'(x_0) + f_\eta h'(x_0) + ...$$

Returning to the original symbols, we have

$$F_x(x_0, y_0) = f_\xi \phi_x(x_0, y_0) + f_\eta \psi_x(x_0, y_0) + ...$$

The other formula is proved in a similar manner.

If we wish to calculate higher order derivatives, we need only differentiate again the right hand side of our formulae with respect to x and y, regarding f_ξ, f_η, ... as compound functions. Thus, for u $= f(\xi, \eta) = f\{\phi(x,y), \psi(x,y)\}$, we have

$$u_{xx} = f_{\xi\xi}\phi_x^2 + 2f_{\xi\eta}\phi_x\psi_x + f_{\eta\eta}\psi_x^2 + f_\xi\phi_{xx} + f_\eta\psi_{xx},$$

$$u_{xx} = f_{\xi\xi}\phi_x\phi_y + f_{\xi\eta}(\phi_x\psi_y + \phi_y\psi_x) + f_{\eta\eta}\psi_x\psi_y + f_\xi\phi_{xy} + f_\eta\psi_{xy},$$

$$u_{yy} = f_{\xi\xi}\phi_y^2 + 2f_{\xi\eta}\phi_y\psi_y + f_{\eta\eta}\psi_y^2 + f_\xi\phi_{yy} + f_\eta\psi_{yy}.$$

Change of the Independent Variables : A particularly important type of compound function occurs in the process of changing the independent variables. For example, let $u = f(\xi,\eta)$ be a function of ξ and η, which we interpret as rectangular co-ordinates in the $\xi\eta$-plane. If we rotate the axes in the $\xi\eta$-planes through an angle θ, we obtain a new system of co-ordinates x,y, related to the co-ordinates (ξ,η) by the equations:

$$\xi = x\cos\theta - yx\sin\theta,\ \eta = x\sin\theta + y\cos\theta, \text{ or}$$

$$x = \xi\cos\theta + \eta\sin\theta,\ y = x\sin\theta + y\cos\theta, \text{ or}$$

The function $u = f(\xi,\eta)$ can then be expressed as a function of the new variables x, y:

$$u = f(\xi,\eta) = f(x, y).$$

The chain rule yields then immediately:

$$u_x = u_\xi\cos\theta + u_\eta\sin\theta,\ u_v = -u_\xi\sin\theta + u_\eta\cos\theta$$

Thus, the partial derivatives are transformed by the same formulae as the independent variables. This is true for rotation of the axes in space also. Another important type of change of co-ordinates is the change from rectangular co-ordinates x, y to polar co-ordinates r, θ. This is done by means of the equations

$$x = r\cos\theta,$$

$$y = r\sin\theta,$$

$$r = \sqrt{(x^2+y^2)}$$

$$\theta = \arctan\frac{y}{x}$$

We then find that for an arbitrary function $u = f(x,x)$ with continuous partial derivatives of the first order we have

$$u = f(x, y) = f(r\cos\theta, r\sin\theta) = F(r, \theta),$$

$$u_x = u_r r_x + u_\theta\theta_x = u_r\frac{x}{r} - u_\theta\frac{y}{r^2} = u_r\cos\theta - u_\theta\frac{\sin\theta}{r},$$

$$u_y = u_r r_y + u_\theta\theta_y = u_r\frac{y}{r} - u_\theta\frac{x}{r^2} = u_r\sin\theta - u_\theta\frac{\cos\theta}{r}.$$

Whence we obtain the often useful equation

$$u_x^2 + u_r^2 = u_r^2 + \frac{1}{r^2}u_\theta^2$$

In general, let us consider a pair of functions $\xi = \phi(x, y)$, $\eta = \psi(x, y)$ which are continuous and have continuous derivatives in a region R of the xy-plane. These equations assign to each point (x, y) in R a point $\xi = \phi(x, y)$, $\eta = \psi(x, y)$ in the $\xi\eta$-plane. As (x, y) ranges over R, the corresponding point (ξ,η) will range over some set of values S in the $\xi\eta$-plane. Naturally, it is possible that several distinct points (x, y) will give the same values for (ξ,η), so that there corre-spond to several points (x, y) only one point (ξ,η). We shall assume that this is not s, but instead that there corresponds to one point $Q(\xi,\eta)$ in S exactly one point $P(x,y)$ in R. We may therefore look at the correspondence from either point of view - saying that Q corresponds to P or that P corresponds to Q. The latter point of view can be expressed as follows: There corresponds to each point (ξ,η) in S one x and one y, namely, the co-ordinates of P, or, in , equations, there are two functions $x = g(\xi,\eta)$, $y = h(\xi,\eta)$, defined in S, which represent the corresponding inverse to $\xi = \phi(x,y)$, $\eta = \psi(x,y)$.

It happens often that the functions $g(\xi,\eta)$, $h(\xi,\eta)$ are by no means easy to calculate, even when they do exist, whence we shall now discover how to obtain the partial derivatives g_ξ, g_η, h_ξ, h_η directly from the partial derivatives ϕ_x, ϕ_y, ψ_x, ψ_y without at all calculating g and h themselves. For this purpose, we observe that, if we choose any point $Q(\xi, \eta)$, find the corresponding point $P\{g(\xi,\eta),h(\xi,\eta)\}$ in R and then find the point in S corresponding to P, which is $\phi\{g(\xi,\eta),h(\xi,\eta)\}$, $\psi\{g(\xi,\eta),h(\xi,\eta)\}$, we have simply returned to the point Q. In other words, the equations $\xi = \phi\{g(\xi,\eta),h(\xi,\eta)\}$, $\eta = \psi\{g(\xi,\eta),h(\xi,\eta)\}$ are identities in ξ and η. We now differentiate both sides of both equations with respect to ξ and η and find

$$1 = \phi_x g_\xi + \phi_y h_\xi,\ 0 = \phi_x g_\eta + \phi_y h_\eta,$$

$$0 = \psi_x g_\xi + \psi_y h_\xi,\ 1 = \psi_x g_\eta + \psi_y h_\eta,$$

If an equation expresses an identical relationship, differentiation with respect to any independent variable in it yields an identity, as follows immediately from the definition.

By solving these equations, we find that

$$g_\xi = \frac{\psi_y}{D}, g_n = -\frac{\phi_y}{D}, h_\xi = -\frac{\psi_x}{D}, h_\eta = \frac{\phi_x}{D}$$

or

$$g_\xi = \frac{\eta_y}{D}, x_n = -\frac{\xi_y}{D}, y_\xi = -\frac{\eta_x}{D}, y_\eta = \frac{\xi_x}{D}$$

where by D is the determinant

$$D=\xi_x\eta_y-\xi_y\eta_x=\begin{vmatrix}\frac{\partial\xi}{\partial x} & \frac{\partial\xi}{\partial y}\\ \frac{\partial\eta}{\partial x} & \frac{\partial\eta}{\partial y}\end{vmatrix}$$

which we assume to be non-zero.

This determinant D, called the functional determinant or Jacobian of (ξ,η) with respect to (x,y), occurs so frequently that one uses frequently the special symbol

$$D=\frac{\partial(\xi,\eta)}{\partial(x,y)}$$

Implicit Functions

In our study of functions of several variables, we have as yet had no analogue to the inverse function. We can regard the inverse function of $y=f(x)$ to be the function obtained by solving the equation $y - f(x) = 0$ for x. In this section, we shall seek more generally to solve equations $F(x, y) = 0$ for x or y and to discuss functions of several variables in a corresponding way.

Even in elementary analytical geometry, curves are frequently represented not by equations $y=f(x)$ or $x=\phi(y)$, but by an equation involving x and y in the form $F(x, y) = 0$. For example, we have the circle $x + y - 1 = 0$, the ellipse $x/a + y/b - 1 = 0$ and. the lemniscate $(x + y) - 2a(x - y) = 0$. In order to obtain y as a function of x or x as a function of y, we must solve the equation for y or for x. We then say that the function $y=f(x)$ or $x=\phi(y)$ found in this way is defined implicitly by the equation $F(x, y) = 0$, and that the solution of this equation yields the function explicitly. In the examples cited and in many others, the solution can be carried out and the solutions stated explicitly in terms of the elementary functions. In other cases, the solution can be obtained in terms of an infinite series or other limiting processes, i.e., we can approximate to the solution $y = f(x)$ or $x = \phi(y)$ as closely as we please.

However, for many purposes, it is more convenient to base the discussion on the implicit definition $F(x, y) = 0$, instead of resorting to an exact or approximate solution of the equation.

The idea that every function $F(x, y)$ yields a function $y = f(x)$ or $x = \phi(y)$ given implicitly by means of the equation $F(x, y) = 0$ is erroneous. On the contrary, it is easy to give examples of functions $F(x, y)$ which, when equated to zero, permit no solution in terms of functions of one variable. Thus, for example, the equation $x + y = 0$ is satisfied by the single pair of values x = 0, y = 0 only, while the equation x + y + 1= 0 is not satisfied by any real values. It is therefore necessary to investigate this matter more closely, in order to find out whether an equation $F(x, y) = 0$ can actually be solved.

Geometrical Interpretation of Implicit Functions: In order to discuss this problem geometrically, we represent the function $u = F(x, y)$ by a surface in three-dimensional space. Finding values (x, y) which satisfy the equation $F(x, y) = 0$ is the same thing as finding values (x, y) which satisfy two equations $F(x, y) = u$, $u = 0$; in other words, we wish to find the intersection of the surface $u = F(x, y)$ and the plane $u = 0$ - the *xy*-plane. We then suppose that we have a definite point (x_0,y_0) which satisfies the equation $F(x_0, y_0) = 0$), i.e., at (x_0, y_0), the surface $u = F(x, y)$ has a point in common with the plane $u = 0$. (If there does not exist such a point, there is no intersection and the equation $F(x, y) = 0$ cannot be solved.) If the tangent plane to the surface $u = F(x, y)$ at the point (x_0,y_0) is not horizontal, it cuts the plane $u = 0$ in a single straight line. Intuition then tells us that the surface $u = F(x, y)$, lying near the tangent plane, likewise cuts the plane $u = 0$ in a single well defined curve. How far this curve extends does not at present concern us. The tangent plane will be horizontal, if both the curves $u = F(x_0, y)$ and $u = F(x, y_0)$ have horizontal tangent lines at (x_0, y_0), i.e., if $F_x(x_0, y_0) = 0$ and $F_y(x_0, y_0)) = 0$. Thus, if either $F_x(x_0, y_0) \neq 0$ or $F_y(x_0, y_0) \neq 0$, the tangent plane is not horizontal, and, as we ave just seen, we may expect that a solution in the form $y = f(x)$ or $x = \phi(y)$ hwill exist.

On the other hand, if both $F_x(x_0, y_0)$ and $F_y(x_0, y_0)$ have the value 0, we readily see that there is no guarantee that a solution is possible.

For example, for $F = 1 - \sqrt{(1 - x^2 - y^2)}$, the corresponding spherical surface is $u = 1 - \sqrt{(1 - x^2 - y^2)}$, has the point (0, 0) in connection with the *xy*-plane. The partial derivatives $F_x(0,0)$ and $F_y(0,0)$ are both zero; we find that no point other than (0,0) satisfies the equation $F = 0$. For the function $F(x,y)=xy$, we find that $F(0, 0) = 0$, while $F_x(0, 0) = F_y(0, 0) = 0$. Here all the points on the *x*-axis and all the points on the *y*-axis satisfy the equation $F(x,y)=0$; in the neighbourhood of the origin, we have no unique solution $x=\phi(y)$ or $y=f(x)$. Thus, we see that when $F_x(x_0,y_0) = F_y(x_0,y_0) = 0$, we cannot be sure that a solution exists.

Accordingly, we return to the case in which one of the partial derivatives—say, $F_y(x_0,y_0)$ to be specific—is not zero, the graphical suggestion that a smooth surface should be cut by a non-tangent plane in a smooth curve leads us to expect that one has the theorem:

If the function $F(x,y)$ has continuous derivatives F_x and F_y, and if at the point (x_0,y_0) holds the equation $F(x_0,y_0) = 0$, while $F_y(x_0,y_0)$ is not zero, then we can mark off about the point (x_0,y_0) a rectangle $x_1 < x < x_2$, $y_1 < y < y_2$ such that for every x in the interval $x_1 < x < x_2$ the equation $F(x, y) = 0$ determines just value $y=f(x)$, lying in the interval $y_1 < y < y_2$. This function $y = f(x)$ satisfies the equation $y=f(x_0)$, and the equation $F\{x, f(x)\} = 0$ is satisfied for every x in the interval. Moreover, the function $y = f(x)$ is continuous and has a continuous derivative.

Assuming it to be true, we can add the following:

The derivative of the function $y = f(x)$ is given by the equation

$$y' \quad = \quad f'(x) = -\frac{F_x}{F_y}$$

This follows immediately by using the chain rule. In fact,

$$\frac{d}{dx}F\{x, f(x)\} = F_x\frac{dx}{dx} + F_y\frac{dx}{dx} = F_x + F_y f'$$

However, since $F\{x,f(x)\}$ identically zero, its derivative is also zero, whence $F_x + F_y f' = 0$, and the formula is established.

If we regard the right hand side of the formula as a compound function of x and differentiate according to the chain rule, replacing y' by $-F_x/F_y$, we have

$$y'' = -\frac{F_y\left(F_{xx} + F_{yx}y'\right) - F_x\left(F_{xy} + F_{yy}y'\right)}{F_y^2}$$

$$= -\frac{F_{xx}F_y^2 - 2F_{xy}F_xF_y + F_{yy}F_x^2}{F_y^3}$$

Continuing the process, we may calculate y^{iii}, ^{iv}, etc.

By using this formula, we can usually find the derivative of a function given in implicit form much more easily than by solving first and then differentiating.

For example, for the circle

$$F(x, y) = x^2 + y^2 - 1 = 0$$

we have

$$y' = -\frac{F_x}{F_y} = -\frac{x}{y}$$

This is easily verified. In fact, solving the equation of the circle for y, we obtain two solutions, namely, $y=\sqrt{(1-x^2)}$ and $y=-\sqrt{(1-x^2)}$ giving the upper and lower semi-circles, respectively. For the upper part,

$$y'=\frac{-x}{\sqrt{(1-x^2)}}\text{, for the lower }\quad y'=\frac{+x}{\sqrt{(1-x^2)}}$$

so that in either case

$$y'=-\frac{x}{y}$$

Another example is $F(x,y) = e^{x+y} + y - x = 0$. We find $F_x = 0$, while $F_y = 2$. Thus, the equation has a solution $y = f(x)$; but its actual explicit calculation is not simple. Nevertheless, we have

$$y' = -\frac{F_x}{F_y} = -\frac{e^{x+y}-1}{e^{x+y}+1}$$

In order that the function $f(x)$ may have a maximum or minimum, we must have $y' = 0$, i.e., $e^{x+y}-1=0$ whence $y = -x$. Substitution of $y = -x$ into the equation $F(x, y) = 0$ yields $1 - 2x = 0$, whence $x = 0, y = -0$. If we calculate

$f''(x)$ for $x = 0$, we find it to be negative so is the maximum of y.

An extension of this theorem for implicit functions to functions of a larger number of independent variables readily suggests itself:

Let $F(x, y, z, u)$ be a continuous function of the independent variables x, y, z, u with continuous partial derivatives F_x, F_y, F_z, F_u. For the system of values $(x_0, y_0, , z_0, u_0)$, let $F(x_0,y_0,...,z_0,u_0)=0$ and $F_u(x_0,y_0,..,z_0,u_0)\}0$. Then we can mark off an interval $u_1 \}u\} u_2$ about u_0 and a region R containing $(x_0,y_0,..,z_0)$ such that for every (x, y, z) in R the equation $F(x,y,\ldots,z,u)= 0$ is satisfied by just one value of a in the interval $u_1 \} u \} u_2$. This value of n, which $u =$ we denote by $u = f(x, y, .. , z)$ is a continuous function of $x, y, \ldots , z$ and possesses continuous partial derivatives $f_x, f_y, \ldots , f_z$ and

$$u_0 = f(x_0, y_0,\ldots, z_0)$$

The derivatives of f are given by the equations

$$F_x + F_u f_x = 0$$
$$F_y + F_u f_v = 0$$

....

$$F_x + F_u f_z = 0$$

For the proof of the existence and continuity of u. The formulae for f_{xy}, etc., follow immediately from the chain rule.

Incidentally, the concept of an implicit function enables us to give a general definition of the term algebraic function.

We say that $u = f(x, y, z)$ is an algebraic function of the independent variables $x,y,,z$ if u can be defined implicitly by an equation $F(x,y,,z,u)=0$, where F *is* a polynomial in $x,y,,z,u$, that is, if u satisfies an algebraic equation. Functions which do not satisfy any algebraic equation are called transcendental equations.

EXERCISE

1. What role does the derivative play in defining slopes, tangents and rate of change?

2. Could the unit step function $U(x)=\begin{cases}0, x<0\\ 1, x\geq 0\end{cases}$ possibly be the derivatives of some other function on $[-1,1]$? Explain.

3. Find the derivatives of the following functions:

   (i) $f(x) = x^6 + 4\sqrt{x} - \dfrac{1}{1+x^2}$

   (ii) $g(x) = 2\tan x^3 + 5\sec\sqrt{x} - \dfrac{\cos x}{1+\sin x^2}$

4. Find $\dfrac{d^2y}{dx^2}$ by implicit diffrentiation:

   (i) $x^3 + y^3 = 1$

   (ii) $y^2 = 1 - \dfrac{2}{x}$

5. If $x^{\frac{1}{3}} + y^{\frac{1}{3}} = 4$, find $\dfrac{d^2y}{dx^2}$ at the point (5, 5).

Chapter 3

Applications of Derivatives

INTRODUCTION

In the previous chapter we have studied the derivatives and evaluation of derivatives of a functions using various method. Also we have studied the derivatives of functions of several variable. In thsi chapter we are going to study the application of derivatives. Derivatives play an important role in the field of Mathematics, Physics, Chemistry and other subject where there is a need of basic tool as calculus.

TANGENT AND NORMAL

We have already studied tangent and normal to the given curve in analytical geometry. Now we are going to study tangent and normal to the given curve in calculus, i.e., with the help of derivatives.

Equation of tangent to a Curve

Let $y = f(x)$ be a given continous curve as shown in the figure given below.

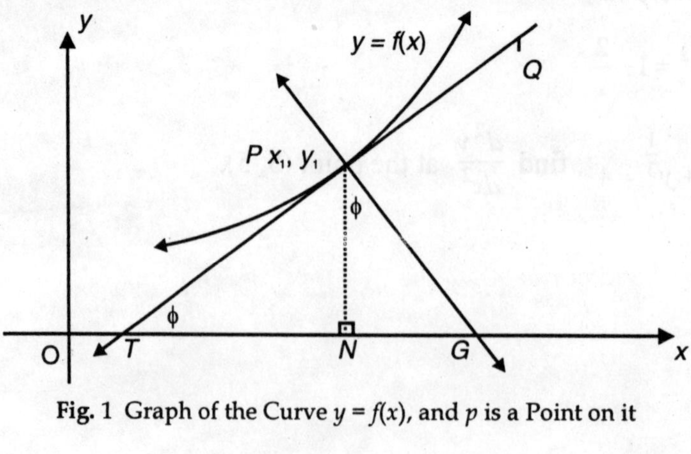

Fig. 1 Graph of the Curve $y = f(x)$, and p is a Point on it

The figure 1 shows the graph of the curve $y = f(x)$, and p is a point on it. TPQ is the tangent to the curve at P. This is inclined at an angle ϕ to the positive of x-axis.

Then as seen previously, $\tan\phi = \left(\frac{dy}{dx}\right)_P$ represents the slope of the tangent line TPQ to the

curve $y = f(x)$ at the point P.

Thus, the equation of the tangent line to the curve at,

$$P(x_1, y_1) \text{ is } y - y_1 = \left(\frac{dy}{dx}\right)_P (x - x_1)$$

The Normal Line is defined as the line which is perpendicular to the tangent at the point of its contact to the curve. Therefore the slope of this normal line at point

$$P(x_1, y_1) \text{ is } -\frac{1}{\left(\frac{dy}{dx}\right)_P} \text{ or } \frac{-1}{f'(x)}$$

Therefore, the equation of the normal line at P to the curve $y = f(x)$ will be

$$y - y_1 = \frac{1}{\left(\frac{dy}{dx}\right)_P}(x - x_1)$$

Angle Between Two Curves

Angle between two curves is the angle between two tangents lines drawn to the two curves at their point of intersection. If m_1 and m_2 are two slopes of these two tangents and θ is the acute angle between them then we have

$$\tan\theta = \left|\frac{m_1 - m_2}{1 + m_1 \cdot m_2}\right|$$

Let C_1 and C_2 be two curves intersecting at a point P. Then these curves are said to intersect 'orthogonally' at P. If tangents at P to C_1 and C_2 are at right angles i.e. $m_1 m_2 = -1$

Definition

Let the tangent and normal at $P(x_1, y_1)$ to the curve $y = f(x)$ meet x - axis at T and G respectively and Let PN be perpendicular to the x – axis at N. Then

1.PT is called the "tangent length" at P

2.PG is called the "normal length" at P

3. The segment TN is known as the "sub-tangent" at P and

4. The segment NG is known as the "sub-normal" at P

Then we have

1. $PT = \frac{y_1}{y'}\sqrt{1+(y')^2}$ 2. $PG = y_1\sqrt{1+(y')^2}$

3. $NT = \frac{y_1}{y'}$ and 4. $NG = y_1 y'$

Example 1. Find the equation of the tangent and normal to the curve $y = 3x^2 - 4x - 2$ at the point (1, –3)

Solution: $y = 3x^2 - 4x - 2 \rightarrow$ The curve.

By Differentiating w. r. to x, we get

$$\frac{dy}{dx} = 6x - 4$$

Let $$P = (1, -3)$$

Then $$\left(\frac{dy}{dx}\right)_{P(1,-3)} = 6(1) - 4 = 2$$

$\therefore$ Equation of the tangent at point P (1, –3) is

$$y - (-3) = 2(x - 1)$$

i.e. $$y + 3 = 2x - 2$$

i.e. $$y = 2x - 5$$

Next, slope of the normal at

$$P = -\frac{1}{\left(\frac{dy}{dx}\right)_p}$$

$$= -\frac{1}{2}$$

$\therefore$ Equation of the normal at P (1, –3) is

$$y-(-3) = -\frac{1}{2}(x-1)$$

$$\text{i.e. } 2(y + 3) = -(x - 1)$$

$$\text{i.e. } x + 2y + 5 = 0$$

Example 2. Find the equation of the tangent and normal line to the curve $y = \sqrt{x^2+8}$ at the point (–1, 3)

Solution: $y = \sqrt{x^2+8} \rightarrow$ The curve and P = (1,3)

Differentiating w. r. to x we get

$$\frac{dy}{dx} = \frac{2x}{2\sqrt{x^2+8}} = \frac{x}{\sqrt{x^2+8}}$$

$$\therefore \left(\frac{du}{dx}\right)_p = \frac{1}{\sqrt{1+8}} = \frac{1}{3}$$

∴ Equation of the tangent line at P is,

$$y - y_1 = \left(\frac{dy}{dx}\right) p(x - x_1)$$

$$\text{i.e. } y - 3 = \frac{1}{3}(x - -1)$$

$$\therefore 3(y - 3) = x - 1$$

$$\text{i.e. } x - 3y + 8 = 0$$

$$\text{Also, slope of the normal is} = -\frac{1}{\left(\frac{dy}{dx}\right)_p}$$

$$= -\frac{1}{1/3}$$

$$= -3$$

Example 3. Find the co-ordinates of the point of contact of the tangent line to the curve $y = x$ log x, which is inclined at an angle of 450 with the x-axis

Solution: Let $P(x, y)$ be the point of contact

The curve is $y = x \log x$

$$\therefore \quad \frac{dy}{dx} = x\left[\frac{1}{x}\right] + \log x[1]$$

$$\therefore \quad \left(\frac{dy}{dx}\right)_p = 1 + \log x$$

∴ Slope of the tangent at P = 1 + log x

Since this tangent has inclination of 45° with x-axis, its slope is

$$\therefore \quad \tan 45° = 1$$

$$\therefore \quad 1 + \log x = 1$$

$$\therefore \quad \log x = 0 \quad \Rightarrow x = 1$$

But P lies on the curve

$y = x \log x$

$\therefore y = 1 \log 1 = 0$

∴ P has co-ordinates = (1,0)

Example 4. Find the angle between the two curves $2y^2 = x^3$ and $y^2 = 32x$ at their point of intersection in the Ist quadrant

Solution: $2y^2 = x^3$ and

$y^2 = 32x$

Solving equations, we get

$2\,(32x) = x^3$

$\therefore x^3 - 64x = 0$

$\therefore x(x^2 - 64) = 0$

$\therefore x = 0$ and $x = \pm 8$

As the point of intersection is in the I[st] quadrant it must be +ve and ≠ 0

∴ x = 8

Putting in $y^2 = 32x$

we get $y^2 = 32(8)$

$\therefore y = 256$

$\therefore y = \pm 8$

Accepting $y = 16$, we have the point of intersection is $P = (8,16)$ which lies in the Ist quadrant

Now Differentiating equations w. r. to x, we get

$$4y\frac{dy}{dx} = 3x^2$$

$$\therefore \frac{dy}{dx} = \frac{3x^2}{4y}$$

$$\therefore \left(\frac{dy}{dx}\right)_p = \frac{3(8)^2}{4\times16} = 3$$

$\therefore$ slope of tangent PT_1 $(m_1) = 3$

$$\text{Again } 2y\frac{dy}{dx} = 32$$

$$\therefore \left(\frac{dy}{dx}\right)_p = \frac{16}{16} = m_1$$

$$= 1$$

$\therefore$ slope of tangent PT2 $(m_2) = 1$

Let θ be the acute angle between curves (1) and (2) at *P*.

$$\text{Then } \tan\theta = \left|\frac{m_1 - m_2}{1 + m_1 \cdot m_2}\right|$$

$$= \left|\frac{3-1}{1+3(1)}\right|$$

$$= \frac{2}{4}$$

$$= \frac{1}{2}$$

$$\therefore \theta = \tan^{-1}(1/2)$$

Example 5. Prove that the curves $y^2 = 16x$ and $2x^2 + y^2 = 4$ cut each other orthogonally.

Solution: Let P (x_1, y_1) be a point of intersection of the curves

$y^2 = 16x$ and

$2x^2 + y^2 = 4$

$\therefore$ $y_1^2 = 16x_1$ and

$2x_1^2 + y_1^2 = 4$

Let m_1 and m_2 be the two slopes of the two tangents to the curve at P respectively.

By Differentiating w. r. to x , the equation, we get

$$2y\frac{dy}{dx} = 16$$

$$\therefore \frac{dy}{dx} = \frac{8}{y}$$

$$\therefore m_1 = \left(\frac{dy}{dx}\right)_{p(x_1,y_1)}$$

$$= \frac{8}{y_1}$$

By Differentiating w. r. to x , the equation, we get

$$4x+2y\frac{dy}{dx} = 0$$

$$\therefore \frac{dy}{dx} = -\frac{2x}{y}$$

$$\therefore m_2 = \left(\frac{dy}{dx}\right)_{p(x_1,y_1)}$$

$$= \frac{-2\times 1}{y_1} \quad ...(6)$$

from equations we have

$$m_1 \times m_2 = \frac{-16\times 1}{y_1^2}$$

$$= \frac{-16\times 1}{16\times 1}$$

$$= -1 \text{ by } (3)$$

Thus the curves intersect orthogonally at P

Interpretation of the Sign of Derivative

We will now cover the meaning of the sign of the derivative and see what inference can be drawn from it and also study its applications to some practical problems from onwards. We will assume that the functions that we consider are continuous and differentiable at all points and all intervals in questions unless otherwise stated. Further the numbers, denoted by h or Δ, will always be taken as positive.

Locally Increasing or Decreasing Functions

1. Definition of a function increasing at a point :

   There exists a number shown in figure 2, such that $f(c-h) < f(c) < f(c)+h)$ for all h satisfying $0 < h < \Delta$ then the function 'f' is said to increasing at $x = c$.

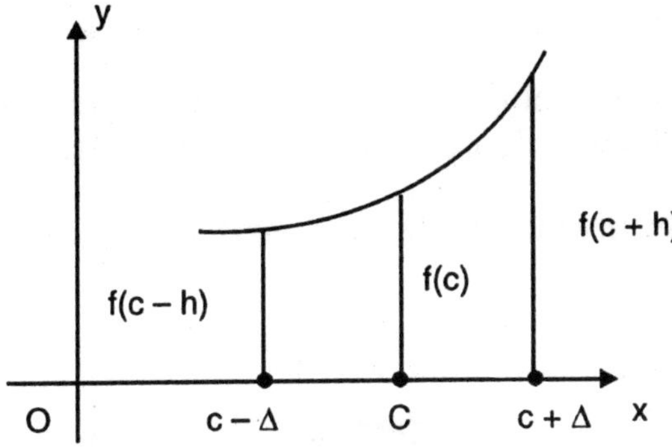

Fig. 2 Function Increasing at a Point

2. Definition of a function decreasing at a point:

   There exists a number Δ shown in figure 3, such that $f(c-h) > f(c) > f(c+h)$ for all h satisfying $0 < h < \Delta$ then the function 'f' is said to be decreasing at $x = c$ The careful study of these definitions, clearly states that the function $y = f(x)$ is increasing at x = c if $\left(\frac{dy}{dx}\right)_{x=c} > 0$ i.e. '$f'(c) > 0$ and $y = f(x)$ is decreasing at x = c if $\left(\frac{dy}{dx}\right)_{x=c} < 0$ i.e. '$f'(c) < 0$. Thus if a function is increasing at a point, then the derivative is positive at that point and if a function is decreasing at a point, then the derivative is negative at that point. Also it is clear that a function $f(x)$ is increasing in an interval (a, b) if $f'(x) > 0$, at every $x \in (a, b)$. Similarly a function $f(x)$ is decreasing in an interval (a, b) if $f'(x) < 0$, at every $x \in (a, b)$.

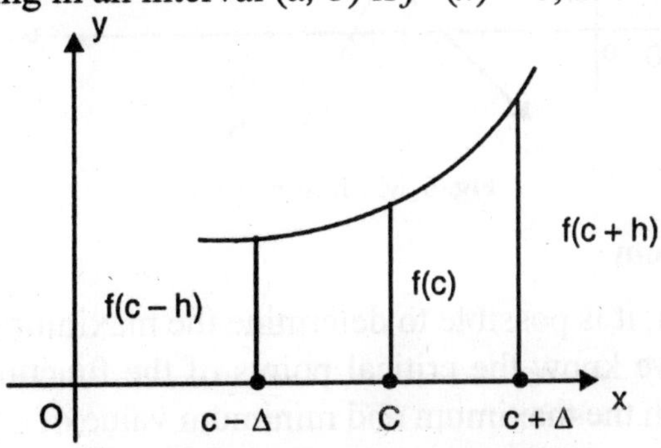

Fig. 3 Function Decreasing at a Point

Critical Points

Points on the graph of a function at which the derivative is zero or does not exist at all are called critical points i.e. The point $(x, f(x))$ is a critical point of $f(x)$, if x is in the domain of 'f' and either $f'(x) = 0$ or DNE. Geometrically speaking, the tangent at such a point to the curve representing the function $y = f(x)$ is either horizontal, vertical or DNE.

Turning Points

There are two types of turning points. They are separately called maximum points and minimum points shown in figure 4 and 5. At such points the curves turns. At such points the tangent to the curve is parallel to x - axis and so, at these points $\frac{dy}{dx} = 0$

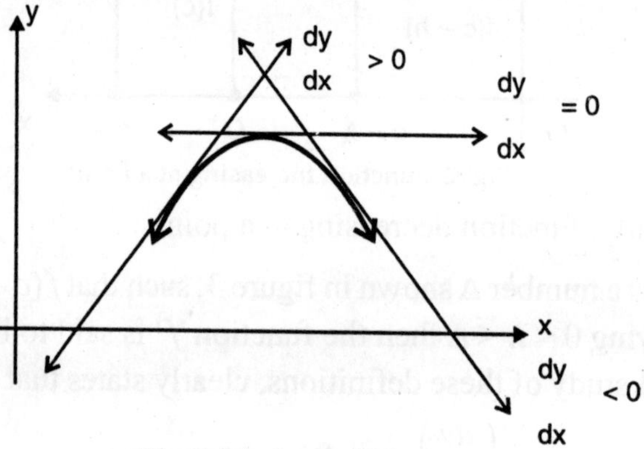

Fig. 4 Maximum Point

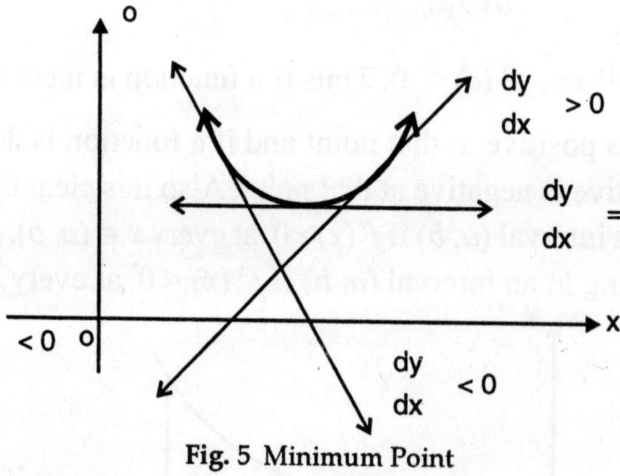

Fig. 5 Minimum Point

Extreme Value Theorem

In a certain interval, it is possible to determine the maximum and minimum values of the function when we know the critical points of the function. The extreme value theorem guarantees both the maximum and minimum value of a function under certain conditions.

It is stated as:

If a function $f(x)$ is continuous on a closed interval $[a, b]$, then $f(x)$ has both maximum and minimum value on $[a, b]$

Procedure

I. First establish the continuity of $f(x)$ on $[a, b]$

II. Determine all critical points of $f(x)$ in $[a, b]$

III. Evaluate the function $f(x)$ at these critical points and at the extremities of $[a, b]$

IV. The largest value is the maximum value and the smallest value is the minimum value of $f(x)$ on $[a, b]$

Example 6. I. If $f'(x) > 0$ for all x in the interval I, then 'f' is an increasing in I and

II. If $f'(x) < 0$ for all x in I then 'f' is decreasing function in I. Is the converse of these two rules true?

Justify your answer with a suitable example.

Solution: Consider the curve $y = x^3$ as shown in the figure 6.

x	–4	–3	–2	–1	0	1	2	3	4
$y = x^3$	–64	–27	–8	–1	0	1	8	27	64

From the graph the curve is increasing at $x = 0$. But $f'(0) = 0$ i.e. $f'(0)$ is not positive Hence the converse of (1) is not true. In a similar way by considering $y = x^3$ at $x = 0$, we can show that the curve is decreasing at $x = 0$ but $f'(0) = 0$. Hence the curve decreasing at $x = 0$ but $f'(0)$ is not negative i.e. converse of (II) is also true.

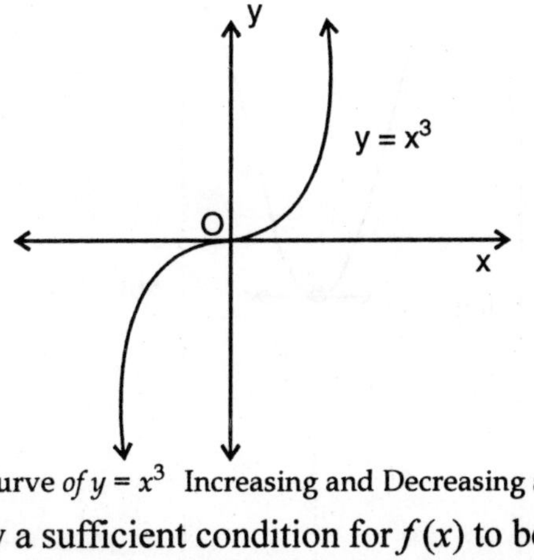

Fig. 6 Curve of $y = x^3$ Increasing and Decreasing at $x = 0$.

Thus $f'(c) > 0$ is only a sufficient condition for $f(x)$ to be increasing at x = c and $f'(c) < 0$ is only a sufficient condition for $f(x)$ to be decreasing at $x = c$ but it is not a necessary condition.

Example 7. Find the values of x for which the function $f(x) = x^3 - 12x + 5$ is (I) decreasing and (II) increasing.

Solution: $f(x) = x^3 - 12x + 5$

$\therefore f'(x) = 3x^2 - 12$

$= 3(x^2 - 4)$

Now 'f' is decreasing if $f'(x) < 0$

i.e. $3(x^2 - 4) < 0$

i.e. $(x^2 - 4) < 0$

i.e. $x^2 < 4$

i.e. $-2 < x < 2$

$\therefore f(x)$ decreases in $(-2, 2)$

Similarly 'f' increases if $f'(x) > 0$

i.e. $3(x^2 - 4) > 0$

i.e. $x^2 > 0$

i.e. $x < -2$ and $x > 2$

The Mean-value Theorem

For the theoretical purpose, the mean-value theorem is one of the most useful tools. The special and easy case of mean value theorem is the 'Rolle's' theorem, which will serve as a lemma for our main results.

Rolle's Theorem : If 'f' is continuous on $[a, b]$ and differentiable in (a, b) and if, further $f(b) = f(a)$, then their exists at least one point $x = c$ in (a, b) such that $f'(c) = 0$

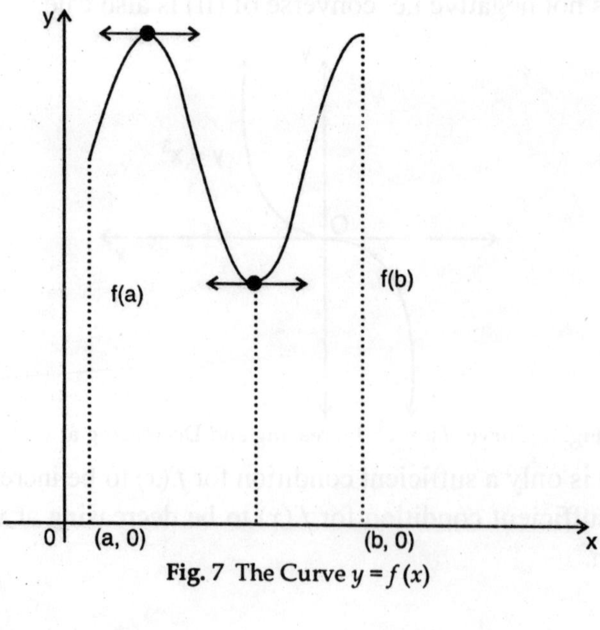

Fig. 7 The Curve $y = f(x)$

In this theorem, $y = f(x)$ is continuous on $[a, b]$ and differentiable in (a, b). Also $f(b) = f(a)$. Hence the curve $y = f(x)$ can be drawn as in the figure 7. The result $f'(c) = 0$ means the tangent at $x = c$ is parallel to x - axis. Hence $y = f(x)$ is continuous on $[a,b]$, differentiable in (a, b) and $f(b) = f(a)$, then there exists at least one point on this curve at which the tangent is parallel to x - axis.

Lagrange's Mean Value Theorem

It is the generalization of the Rolle's theorem. Let $f(x)$ be any continuous function on $[a, b]$ and differentiable in (a, b), then there exists at least one value of $x = c \in (a, b)$ such that

$$\frac{f(b)-f(a)}{b-a} = f'(c)$$

i.e. $f(b) - f(a) = (b - a) f'(c)$

For, consider a function $\phi(x) = f(x) + Ax$

Where A is constant to be determined such that $\phi(a) = \phi(b)$

$\therefore f(a) + Aa = f(b) + Ab$

$$\therefore \quad A = -\frac{f(b)-f(a)}{b-a}$$

Now $\phi(x)$ satisfies all conditions of the Rolle's theorem. Hence there exists at least one

$x = c$ in (a, b) such that $\phi'(c) = 0$

$\therefore f'(c) + A = 0$

$\therefore \quad A = -f'(c)$, we get

$$\therefore -f'(c) = \frac{f(b)-f(a)}{b-a} \text{ or } f(b) = (b-a)f'(c)$$

Geometrically: If the graph of $y = f(x)$ is continuous as shown in figure 8, be drawn between $x = a$ and $x = b$ then $A(a, f(a))$ and $B(b, f(b))$.

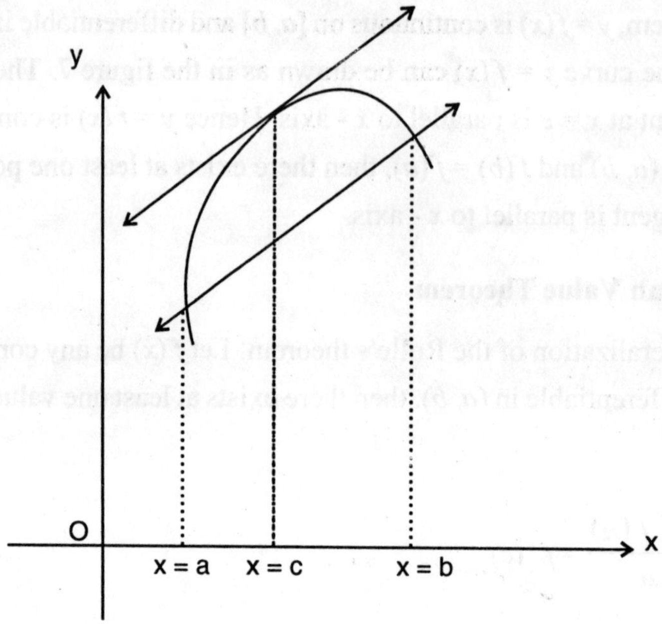

Fig. 8 Graph of $y = f(x)$

Slope of the chord $AB = \dfrac{f(b)-f(a)}{b-a}$ by L.M.V. , this is equal to the slope of the tangent to $y = f(x)$ at $x = c$ i.e. $f'(c)$.

Hence the theorem states that there exists at least one tangent at $x = c$ which is parallel to the chord AB.

First Derivative Test For Local Extrema

If the derivative of a function changes its sign while passing through a critical point along a given curve (i.e. around the critical point), the function possesses a Local (relative) Extrema at that point.

1. If the derivative changes its sign from positive (increasing function) to negative (decreasing function), the function has a Local maxima at that critical point
2. If the derivative changes its sign from negative (decreasing function) to positive (increasing function), the function has a Local minima at that critical point.

Technique

1. Find $f'(x)$
2. Find roots of $f'(x) = 0$
   i.e. critical points of $f(x)$

3. Let $x = c$ be a critical point of $f(x)$ find $f'(c-h)$ and $f'(c+h)$, $h > 0$ (however small)
   a. If $f'(c-h)$, $h > 0$ and $f'(c+h) < 0$
      Then 'f' has a Local maximum at $x = c$
      i.e. $f(c) =$ Local maximum
   b. If $f'(c+h) < 0$ and $f'(c+h) > 0$
      Then 'f' has a Local minimum at $x = c$
      i.e. f (c) = Local minimum
4. Use the same procedure for the other roots.

Second Derivative Test For Local Extrema

Maximum Points

Consider the point P. The gradient at P is zero for the curve $y = f(x)$ shown in figure 9. The gradient is positive for all points to the immediate left of P and negative for all points to the immediate right of P.

Thus near P $\frac{dy}{dx}$ is changing its sign from positive, through zero to negative values, therefore at P, $\frac{d}{dx}\left(\frac{dy}{dx}\right)$ i.e. $\frac{d^2y}{dx^2}$ is negative.

Therefore, for maximum point, $\frac{dy}{dx} = 0$ and $\frac{d^2y}{dx^2} < 0$

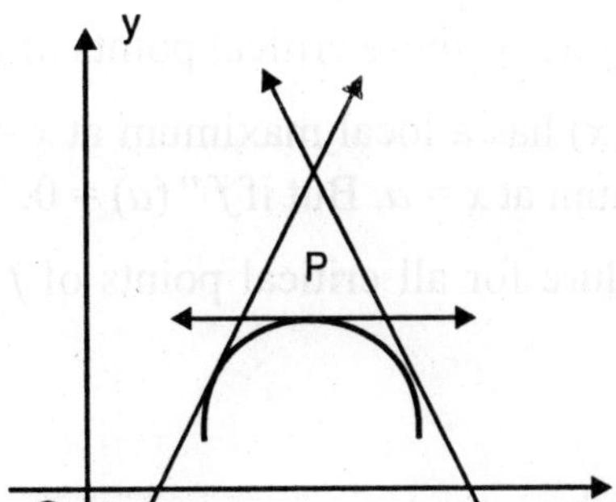

Fig. 9 Curve $y = f(x)$

Minimum Points

Consider the point P. The gradient there is zero. For all points to the immediate right of P, the gradient is positive while for all points to the immediate left of P, the gradient is negative. Thus near P, $\frac{dy}{dx}$ changes from negative, through zero to positive values as shown in figure 10.

Therefore, at $P \frac{d}{dx}\left(\frac{dy}{dx}\right) = \frac{d^2y}{dx^2}$ is positive.

Therefore,for a minimum point $\frac{dy}{dx} = 0$ and $\frac{d^2y}{dx^2} > 0$.

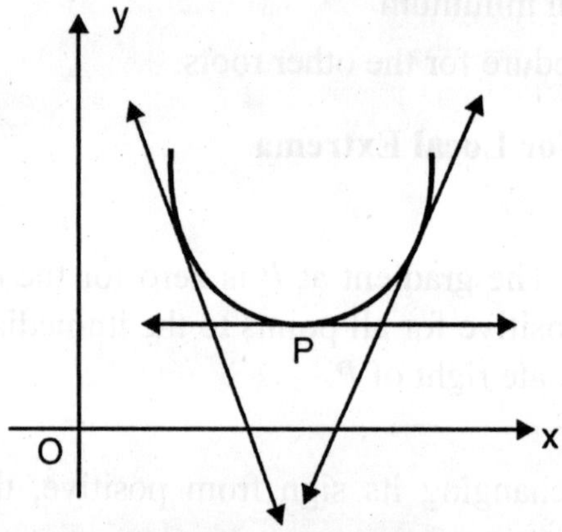

Fig. 10 Curve $y = f(x)$

Technique

1. Find $f'(x)$ and $f''(x)$.
2. Find the roots (i.e. critical points) of $f'(x) = 0$.

   Let $x = a$, $x = b$, $x = c$ etc be these critical points of $f(x)$.
3. If $f''(a) < 0$, then $f(x)$ has a local maximum at $x = a$ and if $f''(a) > 0$, then $f(x)$ has a local minimum at $x = a$. But if $f''(a) = 0$. Then the test fails.
4. Use the same procedure for all critical points of $f(x)$ as it has been given in (3).

Drawbacks

(I)

1. $f'(x) = 0$ and $f''(x) = 0$
2. $f'(x) = 0$ and $f''(x)$ DNE
3. $f'(x)$ DNE

Under any of these conditions, the first derivative test would have to be used instead the second derivative test.

II. In case of some functions, the second derivatives are very difficult and tedious to calculate. Thus relent back to the first derivative test to find the Local extrema.

Stationary Points

1. The condition that $f'(x)$ i.e. $= 0$ at $x = c$, for $f(x)$ should be maximum or minimum is necessary but it is not sufficient.

   Consider $f(x) = \quad x^3$

   Now $\frac{dy}{dx} = 3x^2$

   $\frac{dy}{dx} = 0$ at $x = 0$

   Also, $\frac{d^2y}{dx^2} = 6x$ then at $x = 0$, $\frac{d^2y}{dx^2} = 0$

   $y = x^3$

   Hence $\frac{d^2y}{dx^2}$ at $x = 0$ is neither negative nor positive

   hence at $x = c$ is neither a maximum nor a minimum.

2. The maxima and minima of $f(x)$ are also known as 'turning values' of f(x) since $\frac{dy}{dx}$ changes its sign from positive to negative or from negative to positive. They are also known as "extrema or extreme values."

3. The values of the function for $\frac{dy}{dx} = 0$ are known as *critical values* or *stationary values* since $f(x)$ is neither increasing nor decreasing at these values (points).

   It is a value or point at which the gradient $\frac{dy}{dx}$ is zero.

Note that a stationary point is not necessarily a turning point. It is clear from the above example given. At (0, 0) = 0 and the x – axis is the tangent there. However, (0, 0) $\frac{dy}{dx} = 0$ is not a turning point because $f(x)$ continues to increase when passing through it. The origin is stationary since the curve $y = f(x)$ is momentarily stationary there.

Concavity and Points of Inflection

The second derivative of a function may be also be used to determine the general shape of $y = f(x)$ on a selected interval.

A point of *inflection* on a curve is a point where the curve changes its direction of curvatureas shown in figure 11. You will notice that at *P* the curve changes from being concave upward to

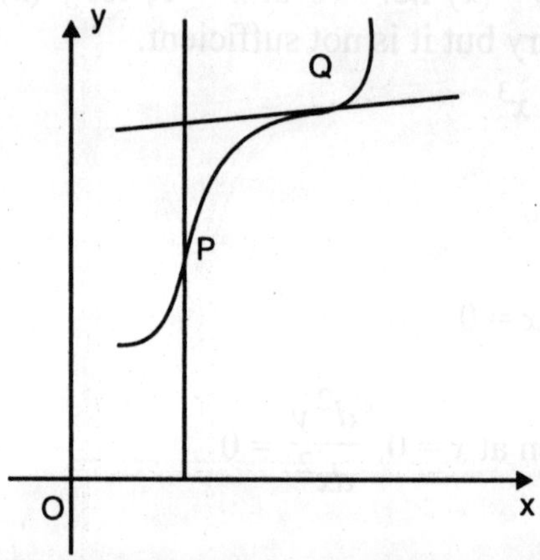

Fig. 11 Tangent to a Curve at a Point of Inflection Crosses the Curve.

being concave downwards. At *Q*, it changes from concave downwards to being concave upwards. Since $f''(x)$ i.e. $\frac{d^2y}{dx^2}$ is the rate of change of the gradient $\left(\frac{dy}{dx}\right)$, a graph is concave upwards where $f''(x) > 0$ and concave downwards where $f''(x) < 0$.

Now at a point of inflection, the gradient stops increasing and begins decreasing, or it stops decreasing and begins increasing. Thus a point of inflection is one at which $\frac{dy}{dx}$ is either a maximum or a minimum. Thus the condition for a point of inflection is that $\frac{d^2y}{dx^2} = 0$ at the point and it changes sign in passing through it i.e. $\frac{d^2y}{dx^2} = 0$ and $\frac{d^3y}{dx^3} \neq 0$. Note that, the tangent to a curve at a point of inflection crosses the curve, or more accurately, it cuts the curve at three coincident points.

Rate Measure (Distance, Velocity and Acceleration)

If $y = s(t)$ represents the position function, then $y = s'(t)$ represents the instantaneous $\lim_{\Delta t \to 0} \frac{\Delta s}{\Delta t} = \frac{ds}{dt} = v$ velocity i.e. and $a = v'(t) = s''$ (t) represents the instantaneous acceleration at time t i.e. $a = \lim_{\Delta t \to 0} \frac{\Delta v}{\Delta t} = \frac{dv}{dt}$ or $\frac{d^2 s}{dt^2}$.

Note that $a = \frac{dv}{dt} = \frac{dv}{ds} \cdot \frac{ds}{dt} = v \cdot \frac{dv}{ds}$

1. If $v > 0$ (i.e. positive velocity) indicates that the position is increasing with increasing time while $v < 0$ (i.e. negative velocity) indicates the decreasing position as time decreases.
2. If $v = 0$ implies that distance 's' remains constant on the given interval of time.
3. If $a > 0$ implies that velocity is increasing with respect to time and $a < 0$ implies that the velocity is decreasing with respect to time.
4. If v = constant on an interval of time, then $a = 0$ on that interval.

Related Rates

We have seen that $\frac{ds}{dt}$ is the derivative of $s(t)$ w. r. to t and can be interpreted as the velocity. This physical interpretation can be extended further. If (x, y) are the co-ordinates of a moving particle then both x and y are functions of 't'.

$\frac{dx}{dt}$, then, is the rate of change of x w. r. to 't' and $\frac{dy}{dt}$ is the rate of change of y w. r. to 't'

$\frac{dx}{dt}$ and $\frac{dy}{dt}$ since both x and y are related by a function $y = f(x)$, $\frac{dx}{dt}$ and $\frac{dy}{dt}$ are also related. If one of is known the other can be found using the relation between these rates.

Problem of this type are known as problem of related rates.

Technique :

$y = f(x)$ where both x and y are dependent on time 't'

Hence

$$\frac{dy}{dt} = \frac{d}{dx} f(x) \cdot \frac{dx}{dt}$$

i.e. $$\frac{dy}{dx} = f'(x) \frac{dx}{dt}$$

If one of $\frac{dx}{dt}$ and $\frac{dy}{dt}$ is known the other can be found out using the above relation.

Differentials: Approximation and Errors

Recall the definition of the derivative of a function $y = f(x)$

$$f'(x) = \lim_{\Delta x \to 0} \frac{f(x+\Delta x) - f(x)}{\Delta x}$$

This represents the slope of tangent to the curve $y = f(x)$ at some point $(x, f(x))$

Now if Δx is very small ($\Delta x \neq 0$), then the slope of tangent is approximately the same as the slope of the secant through $(x, f(x))$

$$\text{i.e. } f'(x) \approx \frac{f(x+\Delta x) - f(x)}{\Delta x}$$

$$\therefore \ f'(x)\Delta x \approx f(x+\Delta x) - f(x)$$

Now the differential of the independent variable x is written as dx and it is the same as Δx

$$\therefore \quad dx = \Delta x, \ \Delta x \neq 0$$

Hence, $f'(x) \cdot dx \approx f(x + \Delta x) - f(x)$ ... (1)

Similarly $dy = \Delta y$ for the variable y, $\Delta y \neq 0$.

we also have that $\Delta y = f(x + \Delta x) - f(x)$

$$\therefore \quad dy = f(x + Dx) - f(x) \ ...(2)$$

Therefore from relations (1) and (2) we have $dy = f'(x)\, dx = \Delta y$

Thus, when the change in x ($\Delta x = dx$) is relatively very small, then the differential of y (dy) is approximately equal to the exact change in y (Δy) i.e. The smaller the change in x, the closer dy to Δy. This fact enables us to approximate the functional value close to $f(x)$

An Alternate Definition for Approximation is as:

If a and $a+h$ belongs to the domain of a differentiable function 'f ', then the approximate value of $f(a+h)$ will be $f(a+h) \approx f(a) + hf'(a), f'(a) \neq 0$. By using this, approximate problems can be solved easily.

DIFFERENTIATION OF FUNCTIONS OF SEVERAL VARIABLES

We shall mainly be concerned with differentiation and integration of functions of more than one variable. We describe

- *How* each process can be done;
- *Why* it is interesting, in terms of applications; and
- How to interpret the process geometrically.

Functions of Several Variables

Last year, you did a significant amount of work studying functions, typically written as $y = f(x)$, which represented the variation that occurred in some (dependent) variable y, as another (independent) variable, x, changed. For example you might have been interested in the height y after a given time x, or the area y, enclosed by a rectangle with sides x and $10 - x$. Once the function was known, the usual rules of calculus could be applied, and results such as the time when the particle hits the ground, or the maximum possible area of the rectangle, could be calculated. In the earlier part of the course, we have extended this work by taking a more rigorous approach to a lot of the same ideas.

We are going to do the same thing now for functions of several variables. For example the height y of a particle may depend on the position x and the time t, so we have $y = f(x, t)$; the volume V of a cylinder depends on the radius r of the base and its height h, and indeed, as you know, $V = r^2h$; or the pressure of a gas may depend on its volume V and temperature T, so $P = P(V, T)$. Note the trick I have just used; it is often convenient to use P both for the (defendant) variable, and for the function itself: we don't always need separate symbols as in the $y = f(x)$ example.

When studying the real world, it is unusual to have functions which depend solely on a single variable. Of course the single variable situation is a little simpler to study, which is why we started with it last year. And just as last year, we shall usually have a "standard" function name; instead of $y = f(x)$, we often work with $z = f(x, y)$, since most of the extra complications occur when we have two, rather than one (independent) variable, and we don't need to consider more general cases like $w = f(x, y, z)$, or even $y = f(x_1, x_2, \ldots, x_n)$.

Graphing functions of Several Variables

One way we tried to understand the function $y = f(x)$ was by drawing its graph. We then used such a graph to pick out points such as the local minimum at $x = 3/2$, and to see how we could get the same result using calculus as shown in figure 12..

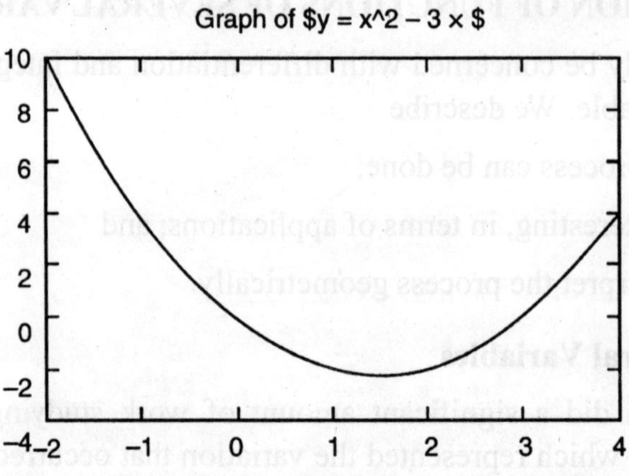

Fig. 12 Graph of a Simple Function of One Variable

Working with two or more independent variables is more complicated, but the ideas are familiar. To plot $z = f(x, y)$ we think of z as the height of the function f at the point (x, y), and then try to sketch the resulting surface in three dimensions. So we represent a function as a surface rather than a curve.

Example 8. Sketch the surface given by $z = 2 - x/2 - 2y/3$

Solution: We know the surface will be a plane, because z is a linear function of x and y. Thus it is enough to plot three points that the plane passes through shown in figure 13.

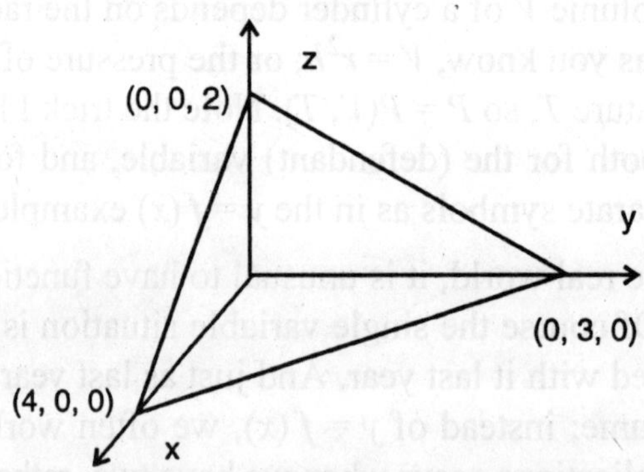

Fig. 13 Sketching a Function of Two Variables

Of course it is easy to sketch something as simple as a plane. There are graphical difficulties when dealing with more complicated functions, which make sketching and visualisation rather harder than for functions of one variable. And if there are three or more independent variables, there is really no good way of visualising the behaviour of the function directly. But for just two independent variables, there are some tricks.

Example 9. Sketch the surface given by $z = x^2 - y^2$.

Solution: We can represent the surface directly by drawing it as shownin figure 14.

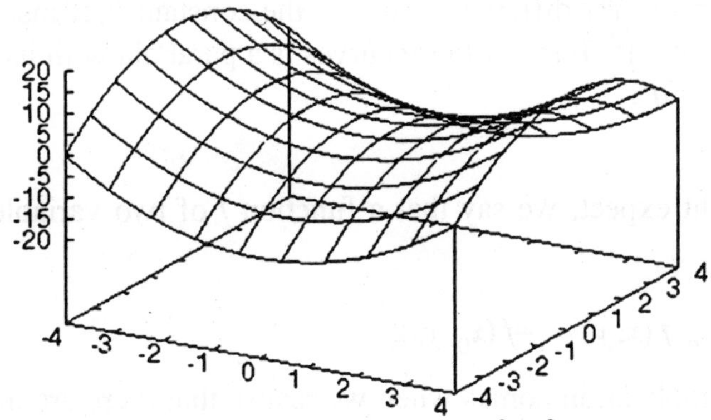

Fig. 14 Surface Plot of $z = x^2 - y^2$.

Such a representation is easy to create using suitable software and Fig. shows the resulting surface. We now describe how to looking at similar examples without such a program. One approach is to draw a contour map of the surface, and then use the usual tricks to visualise the seurface.

For the surface $z = x^2 - y^2$, the points where $z = 0$ lie on $x^2 = y^2$, so form the lines $y = x$ and $y = - x$. We can continue in this way, and look at the points where $z = 1$; so $x^2 - y^2 = 1$. This is one of the hyperbolae indeed, fixing z at different values shows the contours (lines of constant height or z value) are all the same shape, but with different constants. We this get the alternative representation as a contour map shown in figure 15.

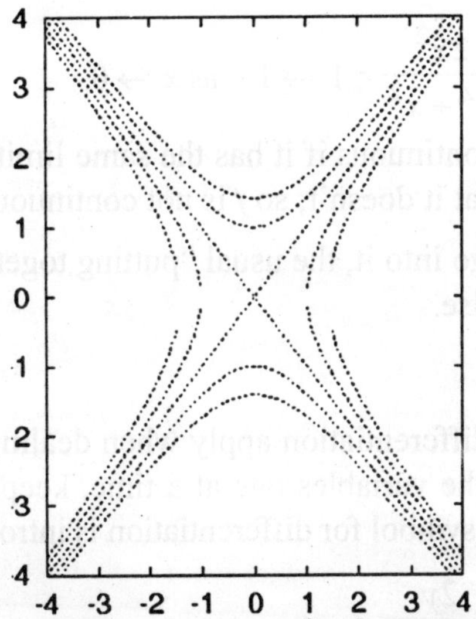

Fig. 15 Contour Plot of the Surface $z = x^2 - y^2$. The Missing Points Near the x-axis are an Artifact of the Plotting Program.

A final way to confirm that you have the right view of the surface is to section it in different planes. So far we have looked at the intersection of the planes $z = k$ with the surface $z = x^2 - y^2$ for different values of the constant k. If instead we fix x, at the value a, then $z = a^2 - y^2$. Each of these curves is a parabola with its vertex upwards, at the point $y = 0, z = a^2$.

Continuity

As you might expect, we say that a function f of two variables is continuous at (x_0, y_0) if

$$\lim_{x \to x_0 \to \infty} f(x, y) \quad = f(x_0, y_0).$$

The only complication comes when we realise that there are many different ways if which $x \to x_0$ and $y \to y_0$. We illustrate with a simple example.

Example 10. Investigate the continuity of $f(x, y) = \dfrac{2xy}{x^2 + y^2}$ at the point (0, 0).

Solution: Consider first the case when $x \to 0$ along the x-axis, so that throughout the process, $y = 0$. We have

$$f(x, 0) = \frac{2x \cdot 0}{x^2 + y^2} = 0 \to 0 \quad \text{as } x \to 0.$$

Next consider the case when $x \to 0$ and $y \to 0$ on the line $y = x$, so we are looking at the special case when $x = y$. We have

$$f(x, x) = \frac{2x^2}{x^2 + x^2} = 1 \to 1 \quad \text{as } x \to 0.$$

Of course f is only continuous if it has the same limit however $x \to 0$ and $y \to 0$, and we have now seen that it doesn't; so f is not continuous at (0, 0).

Although we won't go into it, the usual "putting together" theorems show that f is continuous everywhere else.

Partial Differentiation

The usual rules for differentiation apply when dealing with several variables, but we now require to treat the variables one at a time, keeping the others constant. It is for this reason that a new symbol for differentiation is introduced. Consider the function

$$f(x, y) = \frac{2y}{y + \cos x}$$

We can consider y fixed, and so treat it as a constant, to get a partial derivative

$$\frac{\partial f}{\partial x} = \frac{2y\sin x}{(2y+\cos x)^2}$$

where we have differentiated with respect to x as usual. Or we can treat x as a constant, and differentiate with respect to y, to get

$$\frac{\partial f}{\partial y} = \frac{(2y\cos x)\cdot 2 - 2y\cdot 2}{(2y+\cos x)^2} = \frac{2\cos x}{(2y+\cos x)^2}$$

Although a partial derivative is itself a function of several variables, we often want to evaluate it at some fixed point, such as (x_0, y_0). We thus often write the partial derivative as

$$\frac{\partial f}{\partial y}(x_0, y_0).$$

There are a number of different notations in use to try to help understanding in different situations. All of the following mean the same thing:

$$\frac{\partial f}{\partial y}(x_0, y_0), \quad f_1(x_0, y_0), \quad f_x(x_0, y_0) \text{ and } D_1F(x_0, y_0).$$

Note also that there is a simple definition of the derivative in terms of a Newton quotient:-

$$\frac{\partial f}{\partial y} = \lim_{\delta x \to \infty} \frac{f(x_0+\delta x, y_0) - f(x_0, y_0)}{\delta x}$$

provided of course that the limit exists.

Example 11. Let $z = \sin(x/y)$. Compute $x\,\frac{\partial z}{\partial x} + y\,\frac{\partial z}{\partial y}$.

Solution: Treating first y and then x as constants, we have

$$\frac{\partial z}{\partial x} = \frac{1}{y}\cos\left(\frac{x}{y}\right) \text{ and } \frac{\partial z}{\partial y} = \frac{-x}{y^2}\cos\left(\frac{x}{y}\right).$$

thus

$$x\,\frac{\partial z}{\partial x} + y\,\frac{\partial z}{\partial y} = \frac{x}{y}\cos\left(\frac{x}{y}\right) - \frac{x}{y}\cos\left(\frac{x}{y}\right) = 0.$$

Theorem Assume that f and all its partial derivatives f_x and f_y are continuous, and that $x = x(t)$ and $y = y(t)$ are themselves differentiable functions of t. Let

$$F(t) = f(x(t), y(t)).$$

Then F is differentiable and

$$\frac{dF}{dt} = \frac{\partial f}{\partial x}\frac{dx}{dt} + \frac{\partial f}{\partial y}\frac{dy}{dt}.$$

Proof: Write $x = x(t)$, $x_0 = x(t_0)$ etc. Then we calculate the Newton quotient for F.

$$\begin{aligned} F(t) - F(t_0) &= f(x, y) - f(x_0, y_0) \\ &= f(x, y) - f(x_0, y) + f(x_0, y) - f(x_0, y_0) \\ &= \frac{\partial f}{\partial x}(\xi, y)(x - x_0) + \frac{\partial f}{\partial x}(x_0, \eta)(y - y_0) \end{aligned}$$

Here we have used the Mean Value Theorem to write

$$f(x, y) - f(x_0, y) = \frac{\partial f}{\partial x}(\xi, y)(x - x_0)$$

for some point ξ between x and x_0, and have argued similarly for the other part. Note that ξ, pronounced "Xi" is the Greek letter "x"; in the same way η, pronounced "Eta" is the Greek letter "y". Thus

$$\frac{F(t) - F(t_0)}{t - t_0} = \frac{\partial f}{\partial x}(\xi, y)\frac{x - x_0}{t - t_0} + \frac{\partial f}{\partial y}(x_0, \frac{y - y_0}{t - t_0})$$

Now let tt_0, and note that in this case, xx_0 and yy_0; and since ξ and η are trapped between x and x_0, and y and y_0 respectively, then also $\xi \to x_0$ and $\eta \to y_0$. The result then follows from the continuity of the partial derivatives.

Example 12. Let $f(x, y) = xy$, and let $x = \cos t$, $y = \sin t$.Compute $\frac{\partial f}{\partial t}$ when $t = \pi/$ 2.

Solution:

From the chain rule,

$$\frac{\partial f}{\partial t} = \frac{\partial f}{\partial x}\frac{dx}{dt} + \frac{\partial f}{\partial y}\frac{dy}{dt} = -y(t)\sin t + x(t)\cos t = -1.\sin(\pi/2) = -1.$$

The chain rule easily extends to the several variable case; only the notation is complicated. We state a typical example

Proposition 8: 10 Let $x = x(u, v)$, $y = y(u, v)$ and $z = z(u, v)$, and let f be a function defined on a subset $U \in R^3$, and suppose that all the partial derivatives of f are continuous. Write

$$F(u, v) = f(x(u, v), y(u, v), z(u, v)).$$

Then

$$\frac{\partial F}{\partial u} = \frac{\partial f}{\partial x}\frac{dx}{du} + \frac{\partial f}{\partial y}\frac{dy}{du} + \frac{\partial f}{\partial z}\frac{\partial z}{\partial u}$$

and

$$\frac{\partial F}{\partial v} = \frac{\partial f}{\partial x}\frac{dx}{dv} + \frac{\partial f}{\partial y}\frac{dy}{dv} + \frac{\partial f}{\partial z}\frac{\partial z}{\partial v}$$

The introduction of the domain of f above, simply to show it was a function of three variables is clumsy. We often do it more quickly by saying

Let $f(x, y, z)$ have continuous partial derivatives

This has the advantage that you are reminded of the names of the variables on which f acts, although strictly speaking, these names are not bound to the corresponding places. This is an example where we adopt the notation which is very common in engineering maths. But note the confusion if you ever want to talk about the value $f(y, z, x)$, perhaps to define a new function $g(x, y, z)$.

Example 13. Assume that $f(u, v, w)$ has continuous partial derivatives, and that

$$u = x - y,\ v = y - z\ \ w = z - x$$

Let

$$F(x, y, z) = f(u(x, y, z), v(x, y, z), w(x, y, z))$$

Show that

$$\frac{\partial F}{\partial x} + \frac{\partial F}{\partial x} + \frac{\partial F}{\partial z} = 0$$

Solution: We apply the chain rule, noting first that from the change of variable formulae, we have

$$\frac{\partial u}{\partial x} = 1,$$

$$\frac{\partial u}{\partial y} = -1,$$

$$\frac{\partial u}{\partial z} = 0,$$

$$\frac{\partial v}{\partial x} = 0,\ \frac{\partial w}{\partial x} = -1,$$

$$\frac{\partial v}{\partial y} = 1,\ \frac{\partial w}{\partial y} = 0,$$

$$\frac{\partial v}{\partial z} = -1, \frac{\partial w}{\partial z} = 1,$$

Then

$$\frac{\partial F}{\partial x} = \frac{\partial f}{\partial u}.1 + \frac{\partial f}{\partial v}.0 + \frac{\partial f}{\partial w}.-1 = \frac{\partial f}{\partial u} - \frac{\partial f}{\partial w}$$
$$\frac{\partial F}{\partial y} = \frac{\partial f}{\partial u}.-1 + \frac{\partial f}{\partial v}.1 + \frac{\partial f}{\partial w}.0 = -\frac{\partial f}{\partial u} + \frac{\partial f}{\partial v}$$
$$\frac{\partial F}{\partial z} = \frac{\partial f}{\partial u}.0 + \frac{\partial f}{\partial v}.-1 + \frac{\partial f}{\partial w}.1 = -\frac{\partial f}{\partial v} + \frac{\partial f}{\partial w}$$

Adding then gives the result claimed.

Higher Derivatives

Note that a partial derivative is itself a function of two variables, and so further partial derivatives can be calculated. We write

$$\frac{\partial}{\partial x}\frac{\partial f}{\partial x} = \frac{\partial^2 f}{\partial x^2}, \frac{\partial}{\partial x}\frac{\partial f}{\partial y} = \frac{\partial^2 f}{\partial x \partial y}, \frac{\partial}{\partial y}\frac{\partial f}{\partial x} = \frac{\partial^2 f}{\partial y \partial x}, \frac{\partial}{\partial y}\frac{\partial f}{\partial y} = \frac{\partial^2 f}{\partial y^2}.$$

This notation generalises to more than two variables, and to more than two derivatives in the way you would expect. There is a complication that does not occur when dealing with functions of a single variable; there are *four* derivatives of second order, as follows:

$$\frac{\partial^2 f}{\partial x^2}, \frac{\partial^2 f}{\partial x \partial y}, \frac{\partial^2 f}{\partial y \partial x} \text{ and } \frac{\partial^2 f}{\partial y^2}$$

Fortunately, when f has mild restrictions, the order in which the differentiation is done doesn't matter.

Proposition : Assume that all second order derivatives of f exist and are continuous. Then the mixed second order partial derivatives of f are equal. i.e.

$$\frac{\partial^2 f}{\partial x \partial y} = \frac{\partial^2 f}{\partial y \partial x}.$$

Example 14. Suppose that $f(x, y)$ is written in terms of u and v where $x = u + \log v$ and $y = u - \log v$. Show that, with the usual convention,

$$\frac{\partial^2 f}{\partial u^2} = \frac{\partial^2 f}{\partial x^2} + 2\frac{\partial^2 f}{\partial x \partial y} + \frac{\partial^2 f}{\partial y^2}$$

and

$$v^2\frac{\partial^2 f}{\partial v^2} = \frac{\partial f}{\partial y} - \frac{\partial f}{\partial x} + \frac{\partial^2 f}{\partial x^2} + 2\frac{\partial^2 f}{\partial x \partial y} + \frac{\partial^2 f}{\partial y^2}$$

You may assume that all second order derivatives of f exist and are continuous.

Solution: Using the chain rule, we have

$$\frac{\partial f}{\partial u} = \frac{\partial f}{\partial x}\frac{\partial x}{\partial u} + \frac{\partial f}{\partial y}\frac{\partial y}{\partial u} = \frac{\partial f}{\partial x} + \frac{\partial f}{\partial y}$$

and

$$\frac{\partial f}{\partial v} = \frac{\partial f}{\partial x}\frac{\partial x}{\partial v} + \frac{\partial f}{\partial y}\frac{\partial y}{\partial v} = \frac{1}{v}\frac{\partial f}{\partial x} - \frac{1}{v}\frac{\partial f}{\partial y}$$

Thus using both these and their operator form, we have

$$\frac{\partial^2 f}{\partial u^2} = \frac{\partial}{\partial u}\left(\frac{\partial f}{\partial x} + \frac{\partial f}{\partial y}\right) = \frac{\partial}{\partial x} + \frac{\partial}{\partial y}\left(\frac{\partial f}{\partial x} + \frac{\partial f}{\partial y}\right)$$

$$= \frac{\partial^2 f}{\partial x^2} + \frac{\partial^2 f}{\partial x \partial y} + \frac{\partial^2 f}{\partial y \partial y} + \frac{\partial^2 f}{\partial y^2},$$

While differentiating with respect to \$v\$, we have

$$\frac{\partial^2 f}{\partial v^2} = \frac{\partial}{\partial v}\left(\frac{1}{v}\frac{\partial f}{\partial x} - \frac{1}{v}\frac{\partial f}{\partial y}\right)$$

$$= \frac{1}{v^2}\frac{\partial f}{\partial x} + \frac{1}{v}\frac{\partial}{\partial v}\left(\frac{\partial f}{\partial x}\right) + \frac{1}{v^2}\frac{\partial f}{\partial y} - \frac{1}{v}\frac{\partial}{\partial v}\left(\frac{\partial f}{\partial y}\right)$$

$$= -\frac{1}{v^2}\frac{\partial f}{\partial x} + \frac{1}{v}\left(\frac{1}{v}\frac{\partial^2 f}{\partial x^2} - \frac{1}{v}\frac{\partial^2 f}{\partial x \partial y}\right) + \frac{1}{v^2}\frac{\partial f}{\partial y} - \frac{1}{v}\left(\frac{1}{v}\frac{\partial^2 f}{\partial x \partial y} - \frac{1}{v}\frac{\partial^2 f}{\partial y^2}\right)$$

$$= \frac{1}{v^2}\left(\frac{\partial f}{\partial y} - \frac{\partial f}{\partial x} + \frac{\partial^2 f}{\partial x^2} - 2\frac{\partial^2 f}{\partial x \partial y} + \frac{\partial^2 f}{\partial y^2}\right)$$

SOLVING EQUATIONS BY SUBSTITUTION

One of the main interests in partial differentiation is because it enables us to write down how we expect the natural world to behave. We move away from 1-variable results as soon as we have properties which depend on e.g. at least one space variable, together with time. We illustrate with just one example, designed to whet the appetite for the whole subject of mathematical physics.

Assume the displacement of a length of string at time t from its rest position is described by the function $f(x, t)$. The laws of physics describe how the string behaves when released in this position, and allowed to move under the elastic forces of the string; the function f satisfies the *wave equation* shown in figure 16.

$$\frac{\partial^2 f}{\partial t^2} = c^2 \frac{\partial^2 f}{\partial x^2}.$$

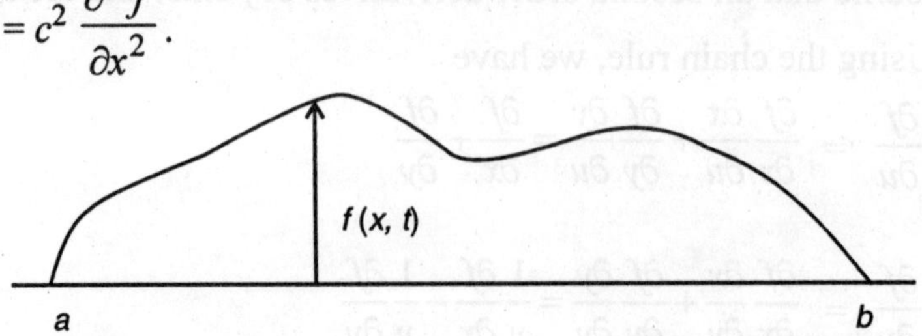

Fig. 16 A String Displaced from the Equilibrium Position

Example 15. Solve the equation $\frac{\partial^2 F}{\partial u \partial v} = 0$.

Solution: Such a function is easy to integrate, because the two variables appear independently. So $\frac{\partial F}{\partial v} = g_1(v)$, where g_1 is an arbitrary (differentiable) function. since when differentiated with respect to u we are given that $\frac{\partial^2 F}{\partial u \partial v} = 0$. Thus we can integrate with respect to v to get

$$F(u, v) = \int g_1(v)dv + h(u) = g(v) + h(u),$$

where h is also an arbitrary (differentiable) function.

Example 16. Rewrite the wave equation using co-ordinates $u = x - ct$ and $v = x + ct$.

Solution: Write $f(x, t) = F(u, v)$ and now in principle confuse F with f, so we can tell them apart only by the names of their arguments. In practice we use different symbols to help the learning process; but note that in a practical case, all the F's that appear below, would normally be written as f's By the chain rule

$$\frac{\partial}{\partial x} = \left(\frac{\partial}{\partial u}\right).1 + \left(\frac{\partial}{\partial v}\right).1$$

and

$$\frac{\partial}{\partial t} = \left(\frac{\partial}{\partial u}\right).(-c) + \left(\frac{\partial}{\partial v}\right).c.$$

differentiating again, and using the operator form of the chain rule as well,

$$\frac{\partial^2 f}{\partial t^2} = \left(c\frac{\partial}{\partial v} - c\frac{\partial}{\partial u}\right)\left(c\frac{\partial F}{\partial v} - c\frac{\partial F}{\partial u}\right)$$

$$= c^2\frac{\partial^2 F}{\partial v^2}-c^2\frac{\partial^2 F}{\partial u\partial v}-c^2\frac{\partial^2 F}{\partial u\partial u}+c^2\frac{\partial^2 F}{\partial u^2}$$

$$= c^2\left(\frac{\partial^2 F}{\partial v^2}+\frac{\partial^2 F}{\partial u^2}\right)-2c^2\frac{\partial^2 F}{\partial u\partial v}$$

and similarly $$\frac{\partial^2 f}{\partial x^2} = \left(\frac{\partial^2 F}{\partial v^2}+\frac{\partial^2 F}{\partial u^2}\right)+2\frac{\partial^2 F}{\partial u\partial v}$$

Substituting in the wave equation, we thus get

$$4c^2\frac{\partial^2 F}{\partial u\partial v} = 0,$$

an equation which we have already solved. Thus solutions to the wave equation are of the form $f(u) + g(v)$ for any (suitably differentiable) functions f and g. For cxample we may have $\sin(x - ct)$. Note that this is not just any function; it must be constant when $x = ct$.

Tangent Planes

Consider the surface $F(x, y, z) = c$, perhaps as $z = f(x, y)$, and suppose that f and F have continuous partial derivatives. Suppose now we have a smooth curve on the surface, say $\phi(t) = (x(t), y(t), z(t))$. Then since the curve lies in the surface, we have

$$F(x(t), y(t), z(t)) = c,$$

and so, applying the chain rule, we have

$$\frac{\partial F}{dt} = \frac{\partial F}{dx}\frac{dx}{dt}+\frac{\partial F}{dy}\frac{dy}{dt}+\frac{\partial F}{dz}\frac{dz}{dt} = 0$$

or, writing this in terms of vectors, we have

$$\nabla F.\mathrm{v}(t) = \left(\frac{\partial F}{\partial x},\frac{\partial F}{\partial z},\frac{\partial F}{\partial z}\right).\left(\frac{dx}{dt},\frac{dy}{dt},\frac{dz}{dt}\right) = 0$$

Since the RH vector is the velocity of a point on the curve, which lies on the surface, we see that the left hand vector must be the normal to the curve.

Note that we have defined the gradient vector ∇F associated with the function F by

$$\nabla F = \left(\frac{\partial F}{\partial x},\frac{\partial F}{\partial z},\frac{\partial F}{\partial z}\right)$$

Theorem The tangent to the surface $F(x, y, z) = c$ at the point (x_0, y_0, z_0) is given by

$$\frac{dF}{\partial x}(x-x_0)+\frac{dF}{\partial y}(y-y_0)+\frac{dF}{\partial z}(z-z_0)=0$$

Proof: This is a simple example of the use of vector geometry. Given that (x_0, y_0, z_0) lies on the surface, and so in the tangent, then for any other point (x, y, z) in the tangent plane, the vector $(x - x_0, y - y_0, z - z_0)$ must lie in the tangent plane, and so must be normal to the normal to the curve (i.e. to ∇F). Thus $(x - x_0, y - y_0, z - z_0)$ and ∇F are perpendicular, and that requirement is the equation which gives the tangent plane. □

Example 17. Find the equation of the tangent plane to the surface $F(x, y, z) = x^2 + y^2 + z - 9 = 0$ at the point $P = (1, 2, 4)$.

Solution: We have $\nabla F|_{(1, 2, 4)} = (2, 4, 1)$, and the equation of the tangent plane is $2(x - 1) + 4(y - 2) + (z - 4) = 0$.

Linearisation and Differentials

We obtained a geometrical view of the function $f(x, y)$ by considering the surface $z = f(x, y)$, or $F(x, y, z) = z - f(x, y) = 0$. Note that the tangent plane to this surface at the point $(x_0, y_0, f(x_0, y_0))$ lies close to the surface itself. Just as in one variable, we used the tangent line to approximate the graph of a function, so we shall use the tangent plane to approximate the surface defined by a function of two variables.

The equation of our tangent plane is

$$\frac{\partial F}{\partial x}(x-x_0)+\frac{\partial F}{\partial y}(y-y_0)+\frac{\partial F}{\partial z}(z-f(x_0,y_0))=0,$$

and writing the derivatives in terms of f, we have

$$\frac{\partial f}{\partial x}(x_0,y_0)(x-x_0)+\frac{\partial f}{\partial y}(x_0,y_0)(y-y_0)+(-1)(z-f(x_0,y_0))=0,$$

or, writing in terms of z, the height of the tangent plane above the ground plane,

$$z=f(x_0,y_0)+\frac{\partial f}{\partial x}(x_0,y_0)(x-x_0)+\frac{\partial f}{\partial y}(x_0,y_0)(y-y_0).$$

Our assumption that the tangent plane lies close to the surface is that $z f(x, y)$, or that

$$f(x,y) \approx f(x_0,y_0)+\frac{\partial f}{\partial x}(x_0,y_0)(x-x_0)+\frac{\partial f}{\partial y}(x_0,y_0)(y-y_0).$$

We call the right hand side the linear approximation to f at (x_0, y_0).

We can rewrite this with $h = x - x_0$ and $k = y - y_0$, to get

$$f(x,y) - f(x_0,y_0) \approx h\frac{\partial f}{\partial x}\bigg|(x_0,y_0) + k\frac{\partial f}{\partial y}\bigg|(x_0,y_0),$$

or
$$df \approx h\left.\frac{\partial f}{\partial x}\right|_{(x_0,y_0)} + k\left.\frac{\partial f}{\partial y}\right|_{(x_0,y_0)}$$

This has applications; we can use it to see how a function changes when its independent variables are subjected to small changes.

Example 18. A cylindrical oil tank is 25 m high and has a radius of 5 m. How sensitive is the volume of the tank to small variations in the radius and height.

Solution: Let V be the volume of the cylindrical tank of height h and radius r. Then $V = r^2H$, and so

$$\begin{aligned} dV &= \left.\frac{\partial f}{\partial r}\right|_{(x_0,h_0)} dr + \left.\frac{\partial f}{\partial h}\right|_{(x_0,h_0)} dh, \\ &= 250\,\pi dr + 25\,\pi dh. \end{aligned}$$

Thus the volume is 10 times as sensitive to errors in measuring r as it is to measuring h.

Example 19. A cone is measured. The radius has a measurement error of 3%, and the height an error of 2%. What is the error in measuring the volume?

Solution: The volume V of a cone is given by $V = \pi r^2 h/3$, where r is the radius of the cone, and h is the height. Thus

$$\begin{aligned} dV &= \frac{2}{3}\pi\, rh\, dr + \frac{1}{3}\pi\, r^2\, dh \\ &= 2.(0.03).\,\frac{\pi r^2 h}{3} + \frac{\pi r^2 h}{3}.(0.02) = V(0.06 + 0.02) \end{aligned}$$

Thus there is an 8% error in measuring the volume.

Exercise: The volume of a cylindrical oil tank is to be calculated from measured values of r and h. What is the percentage error in the volume, if r is measured with an accuracy of 2%, and h measured with an accuracy of 0.5%.

Implicit Functions of Three Variables

Finally in this section we discuss another application of the chain rule. Assume we have variables x, y and z, related by the equation $F(x, y, z) = 0$. Then assuming that $F_z \neq 0$, there is a version of the implicit function theorem which means we can *in principle*, write $z = z(x, y)$, so we can "solve for z". Doing this gives

$F(x, y, z(x, y)) = 0.$

Now differentiate both sides partially with respect to x. We get

$$0 = \frac{\partial}{\partial x} F(x, y, z(x, y)) = \frac{\partial F}{\partial x}\frac{\partial x}{\partial x} + \frac{\partial F}{\partial y}\frac{\partial y}{\partial x} + \frac{\partial F}{\partial z}\frac{\partial z}{\partial x},$$

and so since $\frac{\partial}{\partial x} \neq 0,$

$$0 = \frac{\partial F}{\partial x} + \frac{\partial F}{\partial z}\frac{\partial z}{\partial x} \text{ and } \frac{\partial z}{\partial x} = -\frac{\frac{\partial F}{\partial x}}{\frac{\partial F}{\partial z}}$$

Now assume in the same way that $F_x \neq 0$ and $F_y \neq 0$, so we can get two more relations like this, with $x = x(y, z)$ and $y = y(z, x)$. Then form three such equations,

$$\frac{\partial z}{\partial x}\cdot\frac{\partial x}{\partial y}\cdot\frac{\partial y}{\partial z} = -\left(\frac{\frac{\partial F}{\partial x}}{\frac{\partial F}{\partial z}}\right)\left(\frac{\frac{\partial F}{\partial y}}{\frac{\partial F}{\partial x}}\right)\left(\frac{\frac{\partial F}{\partial z}}{\frac{\partial F}{\partial y}}\right) = -1!$$

Maxima and Minima

As in one variable calculations, one use for derivatives in several variables is in calculating maxima and minima. Again as for one variable, we shall rely on the theorem that if f is continuous on a closed bounded subset of R^2, then it has a global maximum and a global minimum. And again as before, we note that these must occur *either* at a local maximum or minimum, or else on the boundary of the region. Of course in R, the boundary of the region usually consisted of a pair of end points, while in R^2, the situation is more complicated. However, the principle remains the same. And we can test for local maxima and minima in the same way as for one variable.

Definition : Say that $f(x, y)$ has a critical point at (a, b) if and only if

$$\frac{\partial f}{\partial x}(a, b) = \frac{\partial f}{\partial y}(a, b) = 0.$$

It is clear by comparison with the single variable result, that a necessary condition that f have a local extremum at (a, b) is that it have a critical point there, although that is not a sufficient condition. We refer to this as the first derivative test.

We can get more information by looking at the second derivative. Recall that we gave a number of different notations for partial derivatives, and in what follows we use f_x rather than the more cumbersome $\frac{\partial f}{\partial x}$ etc. This idea extends to higher derivatives; we shall use

$$f_{xx} \text{ instead of } \frac{\partial^2 f}{\partial x^2}, \text{ and } f_{xy} \text{ instead of } \frac{\partial^2 f}{\partial x \partial y} \text{ etc.}$$

Theorem (Second Derivative Test) Assume that (a, b) is a critical point for f. Then

- If, at (a, b), we have $f_{xx} < 0$ and $f_{xx}f_{yy} - f^2{}_{xy} > 0$, then f has a local maximum at (a, b).
- If, at (a, b), we have $f_{xx} > 0$ and $f_{xx}f_{yy} - f^2{}_{xy} > 0$, then f has a local minimum at (a, b).
- If, at (a, b), we have $f_{xx}f_{yy} - f^2{}_{xy} < 0$, then f has a saddle point at (a, b).

The test is inconclusive at (a, b) if $f_{xx}f_{yy} - f^2{}_{xy} = 0$, and the investigation has to be continued some other way.

Note that the discriminant is easily remembered as

$$\Delta = \begin{vmatrix} f_{xx} & f_{xy} \\ f_{yx} & f_{yy} \end{vmatrix} = f_{xx}f_{yy} - f^2{}_{xy}$$

A number of very simple examples can help to remember this. After all, the result of the test should work on things where we can do the calculation anyway!

Example 20. Show that $f(x, y) = x^2 + y^2$ has a minimum at $(0, 0)$.

Of course we know it has a global minimum there, but here goes with the test:

Solution: We have $f_x = 2x$; $f_y = 2y$, so $f_x = f_y$ precisely when $x = y = 0$, and this is the only critical point. We have $f_{xx} = f_{yy} = 2$; $f_{xy} = 0$, so $\Delta = f_{xx}f_{yy} - f^2{}_{xy} = 4 > 0$ and there is a local minimum at $(0, 0)$.

Example 21: Let $f(x, y) = xy$. Show there is a unique critical point, which is a saddle point

Proof: We give an indication of how the theorem can be derived — or if necessary how it can be remembered. We start with the two dimensional version of Taylor's theorem, see section 5.6. We have

$$f(a+h, b+k) \sim f(a, b) + h\frac{\partial f}{\partial x}(a, b) + k\frac{\partial f}{\partial y}(a, b)$$
$$\frac{1}{2}\left(h^2\frac{\partial^2 f}{\partial x^2} + 2kh\frac{\partial^2 f}{\partial x \partial y} + k^2\frac{\partial^2 f}{\partial y^2}\right)$$

where we have actually taken an expansion to second order and assumed the corresponding remainder is small.

We are looking at a critical point, so for any pair (h, k), we have $h\frac{\partial f}{\partial x}(a, b) + k\frac{\partial f}{\partial y}(a, b) = 0$ and everything hinges on the behaviour of the second order terms. It is thus enough to study the behaviour of the quadratic $Ah^2 + 2Bhk + Ck^2$, where we have written

$$A = \frac{\partial^2 f}{\partial x^2},$$

$$B = \frac{\partial^2 f}{\partial x \partial y},$$

and $$C = \frac{\partial^2 f}{\partial y^2}.$$

Assuming that $A \neq 0$ we can write

$$Ah^2 + 2Bhk + Ck^2 = A\left(h + \frac{Bk}{A}\right)^2 + \left(C - \frac{B^2}{A}\right)k^2$$

$$= A\left(h + \frac{Bk}{A}\right)^2 + \left(\frac{\Delta}{A}\right)k^2$$

where we write $\Delta = CA - B^2$ for the discriminant. We have thus expressed the quadratic as the sum of two squares. It is thus clear that

- If $A < 0$ and $\Delta > 0$ we have a local maximum;
- If $A > 0$ and $\Delta > 0$ we have a local minimum; and
- If $\Delta < 0$ then the coefficients of the two squared terms have opposite signs, so by going out in two different directions, the quadratic may be made either to increase or to decrease.

Note also that we could have completed the square in the same way, but starting from the k term, rather than the h term; so the result could just as easily be stated in terms of C instead of A.

Example 22: Let $f(x, y) = 2x^3 - 6x^2 - 3y^2 - 6xy$. Find and classify the critical points of f. By considering $f(x, 0)$, or otherwise, show that f does not achieve a global maximum.

Solution: We have $f_x = 6x^2 - 12x - 6y$ and $f_y = -6y - 6x$. Thus critical points occur when $y = -x$ and $x^2 - x = 0$, and so at (0, 0) and (1, –1). Differentiating again, $f_{xx} = 12x - 12$, $f_{yy} = -6$ and $f_{xy} = -6$. Thus the discriminant is $\Delta = -6.(12x - 12) - 36$. When $x = 0$, $\Delta = 36 > 0$ and since $f_{xx} = -12$, we have a local maximum at (0, 0). When $x = 1$, $\Delta = -36 < 0$, so there is a saddle at (1, – 1).

To see there is no global maximum, note that $f(x, 0) = 2x^3(1 - 3/x) \to \infty$ as $x \to \infty$, since $x^3 \to \infty$ as $x \to \infty$.

Example 23. An open-topped rectangular tank is to be constructed so that the sum of the height and the perimeter of the base is 30 metres. Find the dimensions which maximise the surface area of the tank. What is the maximum value of the surface area? [You may assume that the maximum exists, and that the corresponding dimensions of the tank are strictly positive.

Solution: Let the dimensions of the box be as shown in figure 17.

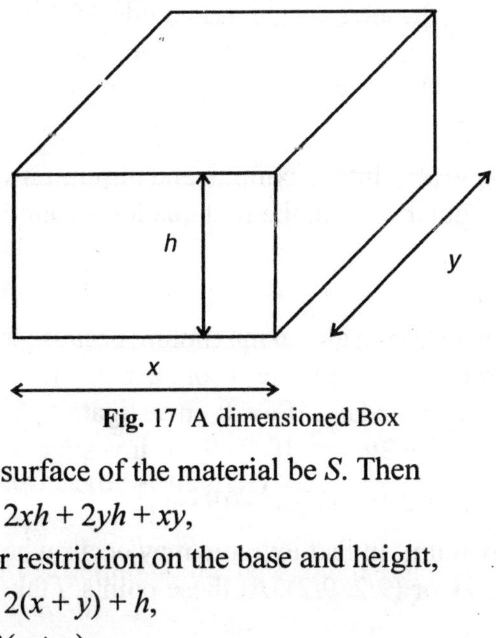

Fig. 17 A dimensioned Box

Let the area of the surface of the material be S. Then

$$S = 2xh + 2yh + xy,$$

and since, from our restriction on the base and height,

$$30 = 2(x + y) + h,$$

we have $h = 30 - 2(x + y)$.

Substituting, we have

$$S = 2(x + y)(30 - 2(x + y)) + xy = 60(x + y) - 4(x + y)^2 + xy,$$

and for physical reasons, S is defined for $x \geq 0$, $y \geq 0$ and $x + y \leq 15$.

A global maximum (which we are given exists) can only occur on the boundary of the domain of definition of S, or at a critical point, when $\frac{\partial S}{\partial x} = \frac{\partial S}{\partial y} = 0$. On the boundary of the domain of definition of S, we have $x = 0$ or $y = 0$ or $x + y = 15$, in which case $h = 0$. We are given that we may ignore these cases. Now

$$S = -4x^2 - 4y^2 - 7xy + 60x + 60y, \quad \text{so}$$

$$\frac{\partial S}{\partial x} = -8x - 7y + 60 = 0,$$

$$\frac{\partial S}{\partial y} = -8y - 7x + 60 = 0.$$

Subtracting gives $x = y$ and so $15x = 60$, or $x = y = 4$. Thus $h = 14$ and the surface area is $S = 16(-4 - 4 - 7 + 15 + 15) = 240$ square metres. Since we are given that a maximum exists, this must be it. [If both sides of the surface are counted, the area is doubled, but the critical proportions are still the same.

Sometimes a function necessarily has an absolute maximum and absolute minimum — in the following case because we have a continuous function defined on a closed bounded subset of ^{2}, and so the analogue of 4.35 holds. In this case exactly as in the one variable case, we need only search the boundary (using ad - hoc methods, which in fact reduce to 1-variable methods) and the critical points in the interior, using our ability to find local maxima.

Example 24. Find the absolute maximum and minimum values of $f(x, y) = 2 + 2x + 2y - x^2 - y^2$ on the triangular plate in the first quadrant bounded by the lines $x = 0$, $y = 0$ and $y = 9 - x$.

Solution: We know there is a global maximum, because the function is continuous on a closed bounded subset of R^2. Thus the absolute max will occur either in the interior, at a critical point, or on the boundary. If $y = 0$, investigate $f(x, 0) = 2 + 2x - x^2$, while if $x = 0$, investigate $f(0, y) = 2 + 2y - y^2$. If $y = 9 - x$, investigate

$$f(x, 9 - x) = 2 + 2x + 2(9 - x) - x^2 - (9 - x)^2$$

for an absolute maximum. In fact extreme may occur when $(x, y) = (0, 1)$ or $(1, 0)$ or $(0, 0)$ or $(9, 0)$ or $(0, 9)$, or $(9/2, 9/2)$. At these points, f takes the values -41/2, 2, 3, - 61.

Next we seek critical points in the interior of the plate,

$f_x = 2 - 2x = 0$ and $f_y = 2 - 2y = 0$.

so $(x, y) = (1, 1)$ and $f(1, 1) = 4$, so this must be the global maximum. Can check also using the second derivative test, that it is a *local* maximum.

Maxima and Minima

Now that we have attained a certain mastery of the problem of differentiating the elementary functions and the functions compounded from them, we are in a position to make a variety of applications. We shall consider here the simplest of these applications - the theory of maxima and minima of a function - in conjunction with a geometrical discussion of the second derivative and then, in the next section, we shall again take up the general theory.

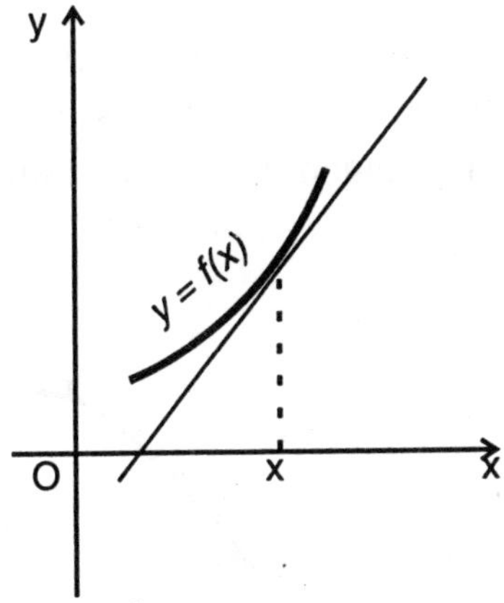

Fig. 18 $f''(x) > 0$

Convexity or Concavity of Curves : By definition, the derivative $df(x)/dx$ of a function $f(x)$ gives the slope of the curve $y = f(x)$. This slope itself can be represented by a curve $y = df(x)/dx$ - the derived curve of the given curve. The slope of this last curve will be given by $df'(x)/dx = df(x)/dx = f''(x)$ - the second derivative of $f(x)$ – and so on. If the second derivative $f''(x)$ is positive at a point x - so that owing to continuity (which we here assume) it is positive in a certain neighbourhood of the point x - then the derivative $f'(x)$ must increase as it passes this point in the direction of increasing values of x. Hence, the curve $y = f(x)$ turns its convex side towards the direction of decreasing values of y. The opposite is true if $f'(x)$ is negative. Hence, in the first case, the curve in the neighbourhood of the point lies above the tangent, in the second case below the tangent.

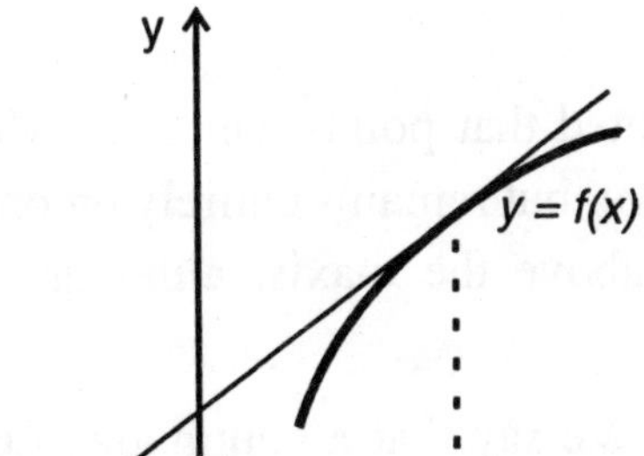

Fig. 19 $f''(x) < 0$

Special consideration is required only in the case of points where $f''(x) = 0$. As a rule, on passing through such a point, the second derivative $f''(x)$ will change its sign. Such a point will then be a point of transition between the two cases indicated above; that is, the tangent will on one side be above the curve and on the other side below it, so that, besides touching the curve, it will also cross it. This is a point of inflection of the curve, and the corresponding tangent is called an inflectional tangent as shown in figure 20.

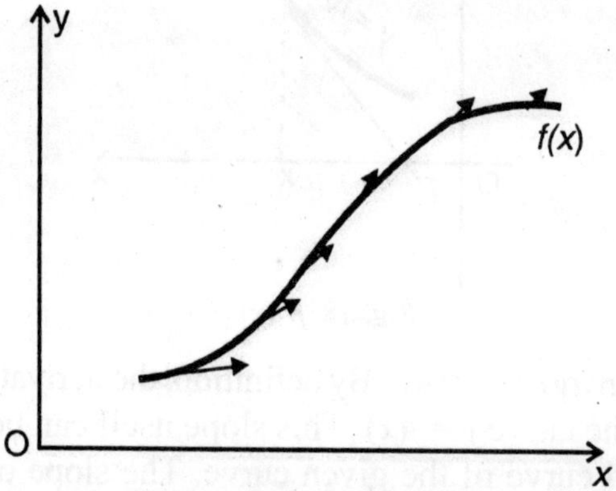

Fig. 20 Point of Inflection

The simplest example is given by the function $y = a$, the cubical parabola, for which the x-axis itself is an inflectional tangent at the point $x = 0$. Another example is given by the function $f(x) = \sin x$, for which

$$f'(x) = d(\sin x)/dx = \cos x \text{ and } f''(x) = d(\sin x)/dx = -\sin x$$

Consequently, $f'(0) = 1$ and $f''(0) = 0$; since the sign of $f''(x)$ changes at $x = 0$, the sine curve has at the origin an inflectional tangent, inclined at an angle of 45 to the x-axis.

However, it must be noted that points can exist where $f''(x) = 0$, although the tangent does not cut the curve, but remains entirely on one side of it. For example, the curve $y = x^4$ lies entirely above the x-axis, although the second derivative $f''(x)$ vanishes for $x = 0$.

Maxima and Minima : We say that a continuous function or a curve $f(x)$ has a maximum (minimum) at a point ξ, if in at least some neighbourhood of the point $x = \xi$ the values of the function $f(x)$ for $x \rangle \xi$ are all less than $f(\xi)$ (greater than $f(\xi)$).We mean by a neighbourhood of a point an interval $\alpha \rangle x \rangle \beta$ which contains the point ξ in its interior. Geometrically speaking, such maxima and minima are the wave-crests and wave-troughs of the curve, respectively. As shown in the figure 21 the value of the maximum at the point P_5 may very well be less than the value of the minimum at another point P_2; thus, on account of the restriction to a certain neighbourhood, the concept of maximum and minimum is always relative **to some extent.**

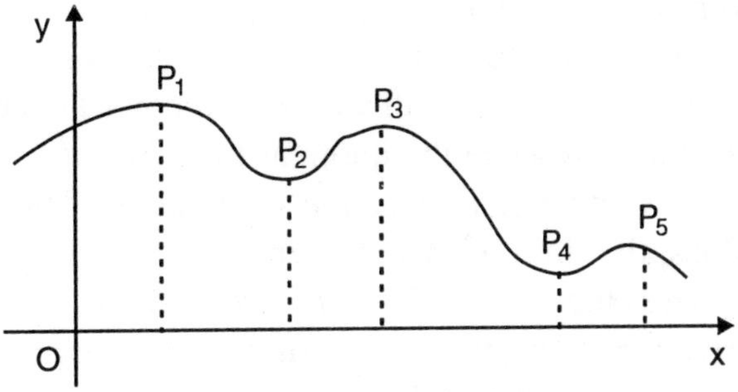

Fig. 21 Maxima and Minima

If we wish to focus on the actual greatest or least value of a function, we must employ special means for deciding how this value is to be selected from among the maxima or minima.

Our objective at present is to find the (relative) maxima or minima, or, to use a word that covers both maxima and minima, the relative extreme values (extrema) of a given function or curve. This problem, which is *very* frequently encountered in geometry, mechanics, and physics and which occurs in many other applications, was in the Seventeenth Century one of the principal incentives for the development of the differential and integral calculus.

The expressions turning value, turning point, are also used. On the other hand, the terms stationary value, stationary point, include inflections as well as maxima and minima.

We see at once that, if a function is assumed to be differentiable, the tangent to its curve at an extreme value ξ must be horizontal. Hence, the condition

$$f'(\xi) = 0$$

is a necessary condition for an extreme value; by solving this equation for the unknown ξ, we obtain the points at which an extreme value may possibly occur. Our condition, however, is by no means a sufficient condition for an extreme value; there may be points at which the derivative vanishes, i.e., at which the tangent is horizontal, although the curve has there neither a maximum nor a minimum. This occurs, if at the given point the curve has a horizontal inflectional tangent cutting it, as in the above example of the function $y = x\acute{y}$ at the point $x = 0$.

However, if we have found a point at which $f'(x)$ vanishes, we may immediately conclude that the function has a maximum at that point, if $f''(\xi) < 0$, a minimum, if f

$''(\xi) > 0$. In fact, in the first case, the curve in the neighbourhood of this point lies completely below the tangent, in the second case, completely above the tangent.

Instead of basing the deduction of our necessary condition on intuition, we could, of course, have given an easy proof by purely analytical methods (cf. the exactly analogous considerations for Rolle's theorem. If the function $f(x)$ has a maximum at the point ξ, then, for all sufficiently small values of h other than 0, the expression $f(\xi) - f(\xi + h)$ must be positive, whence the quotient $[f(\xi+h) - f(\xi)]/h$ will be positive or negative, according to whether h is negative or positive. Thus, if h tends to zero through negative values, the limit of this quotient cannot be negative, while if h tends to zero through positive values, the limit cannot be positive. However, since we have assumed that the derivative exists, these two limits must be equal to each other and, in fact, to $f'(\xi)$, which therefore can only have the value zero; we must have $f'(\xi) = 0$. A similar proof holds for the case of a minimum.

We can also formulate and prove analytically conditions, which are necessary and sufficient for the occurrence of a *maximum* or a minimum, without involving the second derivative. We assume that the function $f(x)$ is continuous and has a continuous derivative $f'(x)$ which vanishes only at a finite number of points.

Then $f(x)$ has a maximum or a minimum at the point $x = \xi$, if and only if the derivative $f'(x)$ changes sign on passing through this point; in particular, the function has a minimum, if the derivative is negative to the left of ξ and positive to the right of it, while, in the opposite case, it has a maximum.

We prove this by using the mean value theorem. First, we observe that there exist to the left and right of ξ intervals $\xi_1 < x < \xi$ and $\xi < x < \xi_2$ (extending to the nearest points at which $f'(x) = 0$) at the each of which $f'(x)$ has only one sign. If the signs of $f'(x)$ in these two intervals differ, then $f(\xi+h) - f(\xi) = hf'(\xi+\theta h)$ has the same sign for all numerically small values of h, whether h is positive or negative, so that $f(\xi)$ is an extreme value. If $f(x)$ has the same sign in both intervals, then $hf'(\xi+\theta h)$ changes sign with h so that $f(\xi + \theta h)$ is greater than $f(\xi)$ on one side and less than $f(\xi)$ on the other side, whence there is no extreme value. Our theorem has thus been proved.

At the same time, we see that the value $f(\xi)$ is a greatest or least value of the function in every interval, containing the point ξ, in which the only change of sign of $f'(x)$ occurs at ξ itself.

The mean value theorem, on which this proof is based, can still be used even if $f(x)$ is not differentiable at an end-point of the interval to which it is applied, provided that $f(x)$ is differentiable at all the other points of the interval; for example, the above proof still holds if $f'(x)$ does not exist at $x = \xi$. This leads us to the following more general result: If the function $f(x)$ is continuous in an interval containing the point ξ and everywhere in this interval, with the possible exception of ξ itself, has a derivative $f'(x)$ which vanishes at not more than a finite number of points, then $f(x)$ has an extreme value at the point $x = \xi$ if, and only if, the point ξ separates two intervals in which $f'(x)$ has different signs. For example, the function $y = |x|$ has a minimum at $x =$ 0, since $y' > 0$ for $x > 0$ and $y' < 0$ for $x < 0$. The function $y = \sqrt[3]{x^2}$ likewise has a minimum at the point x=0, even though its derivative $2x^{-1/2}/3$ is infinite there.

In addition, we make the following remark on the general theory of maxima and minima: The finding of maxima and minima is not directly equivalent to the finding of the greatest and least values of a function in a closed interval. In the case of a monotonic function, these greatest and least values will be assumed at the ends of the interval and are therefore not maxima and minima in our sense; in fact, this latter concept refers to a complete neighbourhood of the location in question. For example, the function $f(x) = x_0$ } x } 1 in the interval assumes its greatest value at the point $x =$ 1, and its least value at $x = 0$, and a corresponding statement holds for every monotonic function. The function y=artan x with the derivative $1/(1 + x)$ is monotonic for –} } x } } and has in that open interval neither a maximum nor a minimum nor a greatest or a least value.

If, after finding the zeros of $f'(x)$, we wish to make sure that we have thereby found the points at which the function has its extreme values, we can often employ the criterion:

A point ξ , at which $f'(x)$ vanishes, yields the least or greatest value of the function $f(x)$ in an entire interval, if throughout that interval $f''(x) > 0$ or $f''(x) < 0$, respectively.

In fact, if both, ξ and $\xi + h$, belong to the interval, by the mean value theorem

Hence, at the point $x = \xi + h$, the derivative $f'(x)$ has the same sign as h or the opposite sign, according to whether $f''(x) > 0$ $f''(x) < 0$; the statement then follows from the remark following the theorem above.

Examples of Maxima and Minima

1. Of all rectangles of given area, find that with the least perimeter.

Let a be the area of the rectangle and x the length of one side (we must consider here x as ranging over the interval $0 < x <$}), then the length of the other side is $a$ýÿ/x and half theperimeter is given by

$$f(x) = x + \frac{a^2}{x},$$

We have

$$f'(x) = 1 - \frac{a^2}{x^2}, f''(x) = \frac{2a^2}{x^3}.$$

The equation f '(ξ) = 0 has the single positive root $\xi = a$. For this value, $f''(x)$ is positive (as it is for any positive value of x), whence it yields the required least value and we obtain the very plausible result that of all rectangles of given area the square has the smallest perimeter.

2. Of all triangles with given base and area, find that with the least perimeter.

In order to solve this problem, we place the x-axis along the given base AB and the central point of AB as the origin. If C is the vertex of the triangle, h its altitude (which is fixed) and (x,h) are the co-ordinates of the vertex, then the sum of the two sides AC and BC of the triangle, which are to be determined, will be given by

$$f(x) = \sqrt{\{(x+a)^2 + h^2\}} + \sqrt{\{(x-a)^2 + h^2\}},$$

where $2a$ is the length of the base, whence

$$f'(x) = \frac{x+a}{\sqrt{\{(x+a)^2 + h^2\}}} + \frac{x-a}{\sqrt{\{(x-a)^2 + h^2\}}},$$

$$f''(x) = \frac{-(x+a)^2}{\sqrt{\{(x+a)^2 + h^2\}^2}} + \frac{1}{\sqrt{\{(x+a)^2 + h^2\}}} + \frac{-(x-a)^2}{\sqrt{\{(x-a)^2 + h^2\}^2}} + \frac{1}{\sqrt{\{(x-a)^2 + h^2\}}}$$

$$= \frac{h^2}{\sqrt{\{(x+a)^2 + h^2\}^3}} + \frac{h^2}{\sqrt{\{(x-a)^2 + h^2\}^3}}$$

We see at once (1) that $f'(0)$ vanishes, (2) that $f''(x)$ is always positive, whence at $x = 0$ there is a least value. In fact, since $f''(x) > 0$, the first derivative $f'(x)$ always increases and therefore cannot be equal to zero at any other point, so that the point $x = 0$ must really yield the least value of $f(x)$. Hence, this least value is given by the isosceles triangle.

Similarly, we find that of all triangles with given perimeter and given base the isosceles triangle has the greatest area.

3. Find a point on a given straight line such that the sum of its distances from two given fixed points is a minimum.

Let there be given a straight line and two fixed points *A* and *B* on the same side of the line. We wish to find a point *P* on the straight line such that the distance *PA* + *PB* has the least possible value.

We take the x-axis as the given line and use the notation of Fig.. Then the distance in question is given by

$$f(x) = \sqrt{(x^2+h^2)}+\sqrt{\{(x-a)^2+h_1^2\}},$$

and we obtain

$$f'(x) = \frac{x}{\sqrt{(x^2+h^2)}}+\frac{x-a}{\sqrt{(x-a)^2+h_1^2}},$$

$$f''(x) = \frac{-x^2}{\sqrt{(x^2+h^2)^3}}+\frac{1}{\sqrt{(x^2-h^2)}}+\frac{-(x-a)^2}{\sqrt{\{(x-a)^2+h_1^2\}^3}}$$

$$+\frac{1}{\sqrt{\{(x-a)^2+h_1^2\}}}$$

$$=\frac{h^2}{\sqrt{(x^2+h^2)^3}}+\frac{h_1^2}{\sqrt{\{(x-a)^2+h_1^2\}^3}}$$

Hence. the equation $f'(\xi) = 0$ yields

$$\frac{\xi}{\sqrt{(\xi^2+h^2)}} = \frac{a-\xi}{\sqrt{\{(\xi-a)^2+h_1^2\}}} \text{ or } \cos\alpha = \cos\alpha$$

which means that the two lines *PA* and *PB* must form equal angles with the given line. The positive sign of $f''(x)$ shows us that we really have a least value.

The solution of this problem is closely linked to the optical law of reflection as shownin figure 22. By an important principle of optics known as Fermat's principle of least time, the path of a light ray is determined by the property that the time taken by the light to go from a point *A* to at point *B* under known conditions must be the least possible. If the condition is imposed that a ray of light shall on its way from *A* to *B* pass through some point on a given straight line (say, on a mirror), we see that the shortest time will be taken along the ray for which the angle of incidence is equal to the angle of reflection.

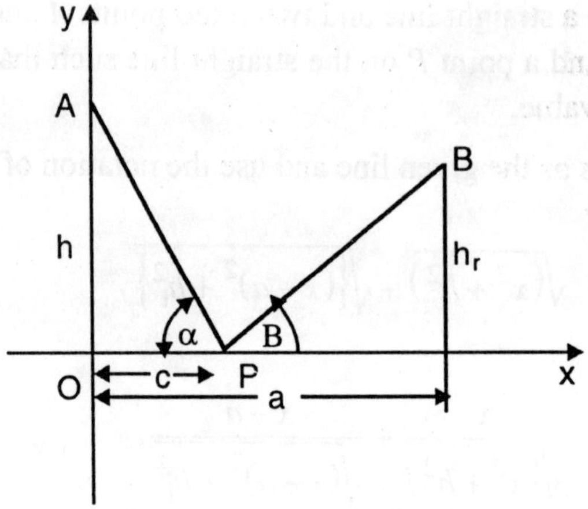

Fig. 22 Law of Reflection

The Law of refraction

Let there be given two points *A* and *B* on opposite sides of the *x*-axis. Which path from *A* to *B* corresponds to the shortest possible time, if the velocity on one side of the *x*-axis is c_1 and on the other side c_2?

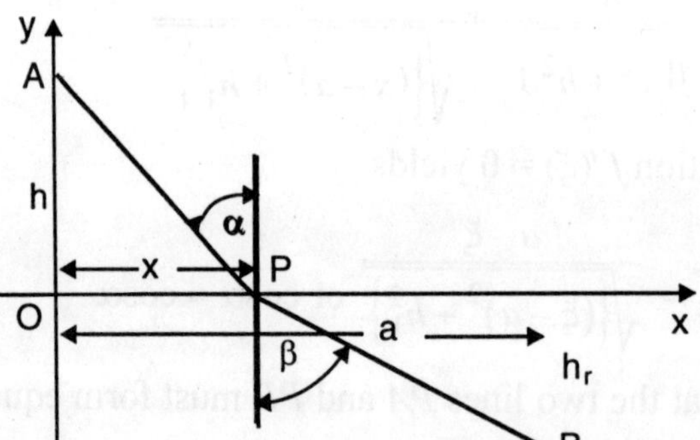

Fig. 23 Law of Refraction

It is clear that the shortest path must lie along two portions of straight lines meeting each other at a point *P* on the *x*-axis. Using the notation of Figure 23, we obtain the two expressions

$$\sqrt{(h^2+x^2)} \text{ and } \sqrt{\left(h_1^2+(a-x)^2\right)}$$

for the lengths *PA*, *PB*, respectively, and we find the time of passage along this path by dividing the lengths of the two segments by the corresponding velocities and adding the results. This yields for the time taken

$$f(x) = \frac{1}{c_1}\sqrt{(h^2+x^2)} + \frac{1}{c_2}\sqrt{\left\{h_1^2+(a-x)^2\right\}}$$

By differentiation, we obtain

$$f'(x) = \frac{1}{c_1}\frac{x}{\sqrt{(h^2+x^2)}} - \frac{1}{c_2}\frac{a-x}{\sqrt{\left\{h_1^2+(a-x)^2\right\}}},$$

$$f''(x) = \frac{1}{c_1}\frac{h^2}{\sqrt{(h^2+x^2)^3}} - \frac{1}{c_2}\frac{h_1^2}{\sqrt{\left\{h_1^2+(a-x)^2\right\}^3}},$$

As we readily see from the figure, the equation $f'(x) = 0$, i.e.,

$$\frac{1}{c_1}\frac{x}{\sqrt{h^2+x^2}} = \frac{1}{c_2}\frac{a-x}{\left\{h_1^2+(a-x)^2\right\}},$$

is equivalent to the condition

$$\frac{1}{c_1}\sin\alpha = \frac{1}{c_2}\sin\beta, \text{or} \frac{\sin\alpha}{\sin\beta} = \frac{c_1}{c_2}$$

We leave it to the reader to prove that there is only one point which satisfies this condition and that this point actually yields the required least value. The physical meaning of our example is again given by the optical principle of least time. A ray of light travelling between two points describes the path of shortest time. If c_1 and c_2 q are the velocities of light on either side of the boundary of two optical media, the path of the light will be that given by our result, which accordingly yields Snell's law of refraction.

MOTION, VELOCITY AND ACCELERATION

We have developed the essential theory for defining and analyzing the mathematical function. The derivative of a function was defined as the instantaneous rate of change of a function at any given point or moment. Geometrically the derivative was simply the slope of the tangent to the graph at a particular point.

As a student you are probably still a little confused by the concept of instantaneous rate of change. To strengthen your understanding of the derivative, let us scientifically analyze a physical phenomena and see how exactly the derivative is defined for the situation. The situation we will study is that of the dynamic motion of a body.

Motion is characterized by a changing distance between a reference point and the object itself as shown in figure 24. Distance is measured in the dimension of length with units of meters, miles, or feet. We know how to measure distance but how do we measure a changing distance? Whenever a dimension changes with respect to itself it constitutes an action that defines time. Thus a changing distance has to be measured relative to a change in time.

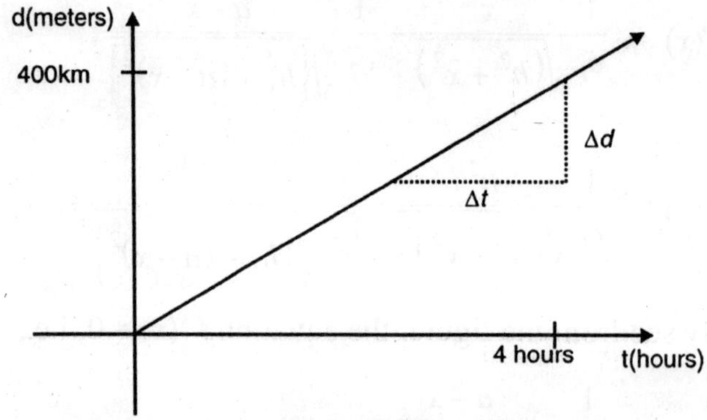

Fig. 24. Changing Distance between a Reference Point

We can define velocity to be a measure of how fast an object in motion moves. Thus velocity is the rate at which the distance is changing relative to time. A simpler definition is that velocity is the distance covered per unit time. If an object moves 100 m in one second, then its velocity for that interval is $\frac{100m}{1s} = 100m/s$. The units of velocity are meters per second or change in distance per change in unit time. Consider a for a sports car moving at a constant velocity of 100 *km/hr*. If we plot graph distance covered from a reference point, we get.

The graph is a linearly increasing straight line, where the distance covered increases directly with time. The steepness or rate of change of this graph is by definition the change in distance over the change in time $\frac{\Delta s}{\Delta t}$ where is an abbreviation for distance. Since motion is characterized by a change in position relative to time, then velocity is only defined over that definite time interval. If an object moves from a to b, the velocity of the object over this interval is this interval distance covered, $s_b - s_a$ divided by the time taken to cover this distance, $t_b - t_a$ Thus velocity is the rate of change of distance covered with respect to time.

$$\frac{s_b - s_a}{t_b - t_a} = \frac{\Delta s}{\Delta t} = \text{velocity}$$

For example, the car above takes one hour to go from a station 400 kilometers down the highway to another station 500 kilometers down the same highway. The velocity of the car is then $\frac{\Delta d}{\Delta t}\frac{500-400}{1}$=100km/hr

Since the graph of the car's distance covered with respect to time is a straight line, it has a constant rate of change, such that $\frac{\Delta d}{\Delta t} = v$, where v is a constant velocity as shown in figure 25. This tells us that the velocity over the entire journey is a constant or 100km/hr. The graph of this velocity function is just a straight line, or is the same at any time t.

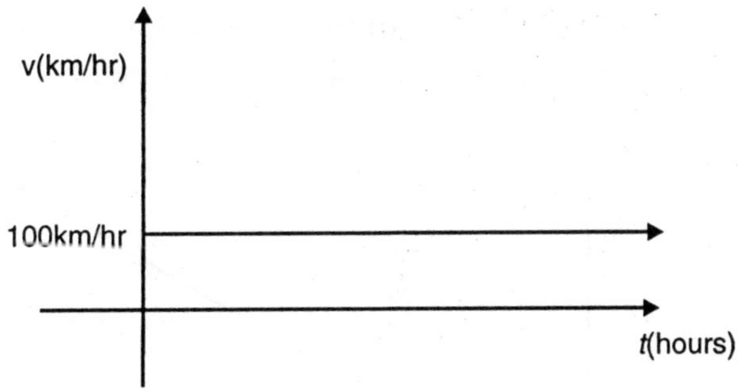

Fig. 25 Graph of the Car's Distance Covered with Respect to Time is a Straight Line

This confirms logic since throughout the journey the speedometer reads a constant 100 km./hr. Therefore, distance covered is directly related to the elapsed time, in each hour it will cover a 100 more kilometers.

Now let us define acceleration. Acceleration is the rate of change of velocity with respect to time. Acceleration is to velocity as what velocity is to distance. The concept of acceleration refers to a changing velocity per unit time. The units of acceleration are $\frac{m/s}{s}=\frac{m}{s^2}$

For example if the acceleration of a Ferrari is 4 km/hr per second, all this means is that each second the velocity increases or changes by 4 km/hr. If the velocity at $t = 3$ seconds is 48 km/hr, then the velocity at $t = 6$ seconds will be:

$$48\text{km/hr} + \left(4\frac{km}{hr}\Big/\sec\right).(3\sec) = 60\text{km/hr}$$

Lets see how this applies to a car moving with constant acceleration. Since the acceleration is constant, the graph is a horizontal line, where the acceleration of the car is at any time t is the same.

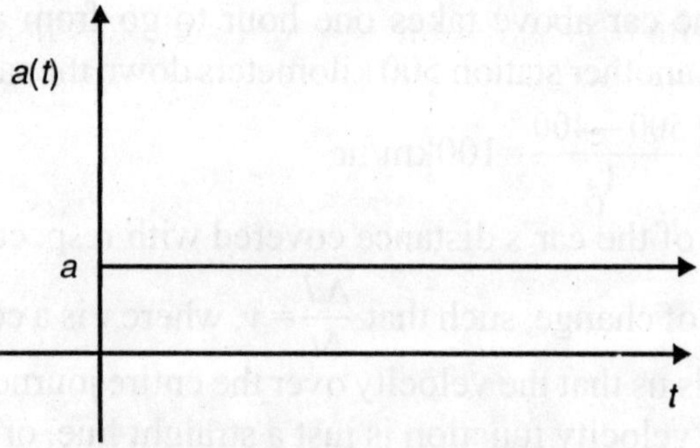

Fig. 26 Graph of Velocity as a Function of Time

The graph of velocity as a function of time, $v(t)$ will increase directly with time for a car moving with constant accelerationas shown in figure 26.

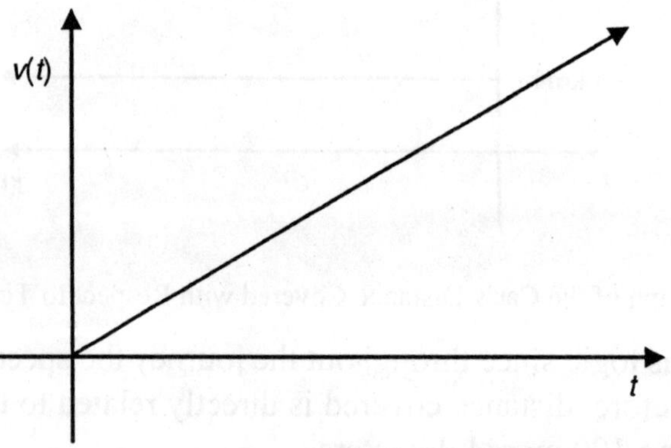

Fig. 27 Rate of Change of Velocity with Respect to Time

Acceleration is therefore the rate of change of velocity with respect to time or $\frac{\Delta v}{\Delta t}$ as shown in figure 27. An acceleration of 4 m/s/s means that every second the velocity increases by 4 m/s.

The concept of acceleration and velocity are fairly obvious to understand when dealing with constant accelerations and velocities. We now need to define a more precise way of explaining velocities as the derivative of the position or distance function with respect to time and acceleration as the derivative of the velocity function with respect to time.

INSTANTANEOUS VELOCITY AND ACCELERATION

When an object's distance changes with time, its velocity is the rate at which the distance is changing with respect to time, while its acceleration is the rate at which the velocity is changing with respect to time. As our time interval goes to zero, the velocity and acceleration of an object take on instantaneous values at a certain moment. These instantaneous rate of changes represent the derivatives with respect to time.

To understand how the derivative relates to a moving object, consider a Porche that accelerates *from rest* at a constant rate of 15 km/hr/s from a starting point d=0. It continues at this acceleration until its velocity is 160 km/hr after which it stops accelerating and maintains its velocity.

If we freeze the moment when 4 seconds have past then its speedometer will read a velocity of exactly 60 km/hr *at that instant only*. However at *t*= *4.1* seconds the velocity will be slightly higher since the car is accelerating. This is why we use Calculus to analyze how these accelerations give rise to a changing velocity that results in a changing distance that is covered shown in figure 28.

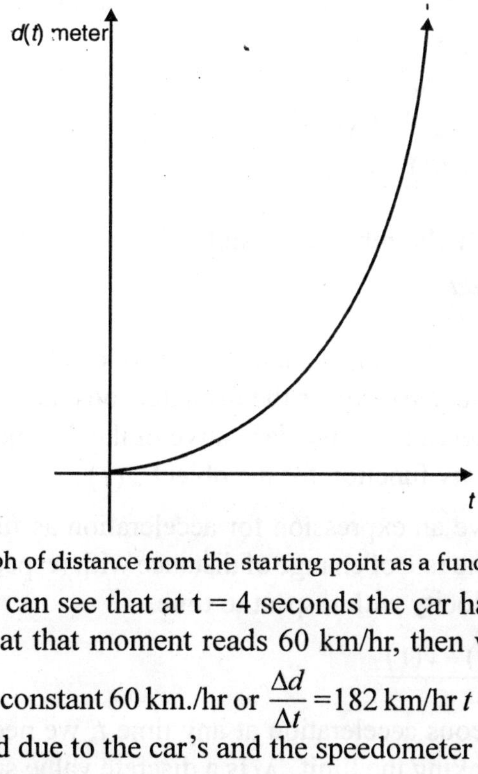

Fig. 28 Graph of distance from the starting point as a function of time.

From the graph we can see that at t = 4 seconds the car has covered 2 kilometers. Since the speedometer at that moment reads 60 km/hr, then we can say that at t = 4 seconds, its velocity is a constant 60 km./hr or $\frac{\Delta d}{\Delta t}$ =182 km/hr t = 4.1 seconds; however, the velocity has changed due to the car's and the speedometer now reads, 61 km/hr..

We can assume that from t = 4.0 seconds to t = 4.1 seconds the velocity of the car is a constant 60 km/hr. By definition velocity is the distance covered divide by the time taken to cover the distance or $\frac{\Delta d}{\Delta t}$ Thus, the distance covered by the car in this small 5time interval, divided by the time, .1 seconds, will give us 60 km/hr.

Since the velocity of the car is increasing, due to its constant rate of acceleration, the velocity of the Porche at any instant, t, will be whatever the speedometer reads at that moment.

If we were given the relationship for distance covered as a function of time t, then velocity of the car at any time t can be found by calculating the distance covered $d(t+\Delta t)-d(t)$ over a time interval, Δt:

$$v(t) = \frac{d(t+\Delta t)-d(t)}{\Delta t}$$

Since the car is accelerating, its velocity is not constant over the interval Δt. We can assume the velocity is constant over an infinitely small time interval, Δt

Therefore we have to take the limit as Δt goes to zero to find the instantaneous rate of change of distance with respect to time. The instantaneous rate of change of distance will correspond exactly to what the speedometer reads at time t.

$$v(t) = \lim_{M \to 0} \frac{d(t+\Delta t)-d(t)}{\Delta t}$$

From the definition of the derivative:

$$d'(t) = \lim_{\Delta t \to 0} \frac{d(t+\Delta t)-d(t)}{\Delta t}$$

This leads to the extremely important result:

$$v(t) = d'(t) = \frac{dd}{dt}$$

The velocity at any time t is the instantaneous rate of change of the distance function at a time t. By definition the derivative is the instantaneous rate of change of a function over an infinitely small interval. thus the derivative of the distance function, $d'(t)$ with respect to time is the velocity function for the object $v(t)$

We now need to derive an expression for acceleration as function of time. In the same way that velocity is the rate of change of distance with respect to time, acceleration is the rate of change of velocity with respect to time.

$$a(t) = \frac{v(t+\Delta t)-v(t)}{\Delta t}$$

To find the instantaneous acceleration at any time *t*, we need to take the limit as Δt goes to zero. Without taking the limit, Δt is a discrete value such that the calculated acceleration is the average acceleration is for that interval. By taking the limit as $\Delta t \to 0$, we are assuming the acceleration is constant over that time interval.

$$a(t) = \lim_{\Delta x \to 0} \frac{v(t+\Delta t)-v(t)}{\Delta t}$$

$$a(t) = v'(t)$$

This proves that acceleration is the derivative of the velocity function with respect to time. Since velocity is the first derivative of the distance function with respect to time, the acceleration function is the second derivative of the distance function. In other words the acceleration function is obtained by differentiating the distance function twice. Our results can be summarized as follows:

$$v(t) = d'(t)$$
$$a(t) = v'(t) = d''(t)$$
$$a(t) = d''(t)$$

DYNAMICS

FORCES IN MOTIONS

An object in motion is characterized by a changing position as a function of time. The derivative of the distance function functions with respect to time, gives us the velocity of the object at as a function of time. Furthermore the derivative of the velocity function, gives us the acceleration of the object as a function of time. But what causes an object to move? To understand particle dynamics we need to first understand the concept of force.

Newton's first law of motion states that an object in motion will remain in motion until acted on by a *force*. This observation is one of most important ones ever made as it offers a way of defining what a force is. According to this law, an object traveling in space at 100,000 km/hr will remain at that velocity forever provided no force acts on it.

For example, if you were to throw a ball in space it would forever continue in the same direction along with the same velocity with which it left your hand. Here, *space* refers to anywhere that is free from the influence of any gravitational force, electro-magnetic force, air-resistance or any other forces. Once set in motion an object will continue with that same velocity forever.

Since no force is required to keep an object in motion, then a force can be defined as that which changes the velocity of the object. Thus force is a measure of a resistance to a change in motion. Since motion is characterized by a constant velocity, then a change in motion results in a change in velocity. By definition, a change in velocity is an acceleration.

This simple, yet profound conclusion tells us that forces are defined by accelerations. The force required to accelerate an object is proportional to the magnitude of the acceleration. The mass of the object is also a factor since the greater the mass, the greater its resistance to motion. Observation shows that the resistance to a change in motion is directly dependent on the amount of matter being accelerated.

We can define *a Newton* as the force equipped accelerate a body of unit mass, 1 *kg*, a^m/s^2 Therefore to accelerate a body of mass, *m*, the required force would be *m* times *a*.

Force = mass × acceleration

$F(m, a) = ma$

This is read as, the force required to accelerate a body is directly related to its mass and the magnitude of the acceleration of the mass. The important concept to understand is that forces are defined as accelerations or changes in velocity. It requires no force to keep a body in motion Once in motion it will remain in motion. A force is required only to change its velocity or accelerate it. Thus force is a quantifiable measurement of a mass's resistance to a change in motion. If an object had no resistance to a change in motion then there would be no such thing as force!

This might seem to contradict reason. One can better understand this by considering an airplane flying in space, where space is some imaginary place that contains no matter or force fields inside it. If its four engines produce an acceleration $5m/s^2$ and the mass of the plane is 8×10^6 kg, then the thrust or force acting on the plane is $(5)\cdot(8\times10^6)=4\times10^7$. If we assume an inexhaustible and weightless fuel source then theoretically the engines will push the plane forward with a constant force of 4 million newtons.

Now since the plane is flying in an imaginary space under a constant force, it is free to accelerate forever. Remember forces are defined as accelerations and not velocities. The plane will accelerate at a constant acceleration of $5m/s^2$. This means the planes velocity would increases and increase at the rate of $5\dfrac{m/s}{s}$. The graph of its velocity as a function of time would be a linearly increasing function:

$v(t) = 5.t$

The derivative of the velocity function is the acceleration function:

$$a(t) = \frac{dv}{dt} = 5$$

The fundamental concept to understand here is that a force is required only to change an objects velocity. A change in velocity is by definition an acceleration. Therefore forces are required only to accelerate an object. A constant force acting on a body will accelerate the body with a constant acceleration, which means the body's velocity will increase and increase forever, all due to a constant force. Furthermore the greater the mass, the greater its resistance to a change in velocity. Thus the force required to accelerate a mass is directly proportional to its mass also.

UNDERSTANDING FREE FALL MOTION

Having laid down the conceptual basis of what velocity, acceleration and forces are, we can now study the motion of free falling bodies on earth. An object falls to earth because of the gravitational force of attraction that the earth experiences for the object. What then is gravity? From Newton's law of gravitation, the force attracting two bodies is given by:

$$F_{\text{gravitational}} = G.\frac{m_1 \cdot m_2}{d^2} \text{ (G is the gravitational constant)}$$

To derive this, think of a unit mass of 1 kg separated a distance from a larger mass, *M*. The gravitational force is the force of attraction the larger mass expresses for the unit mass and vice-versa. Observation confirms that the gravitational force is proportional to mass the larger body and decreases with the square of the distance separating them. The reason it is distance squared ad not just directly related to the distance is because masses are 3-dimensional. In 3-dimensional space, properties are related to the projected areas as opposed to 2-dimensional geometry that are dependent on the length only. Therefore, the gravitational force between the two masses is:

$$F_{\text{gravitational}} = c.\frac{M \cdot 1}{d^2} \quad (c \text{ is some constant})$$

If the unit mass was replaced by a mass *m*, the force of attraction would be *m* times the amount it was with the unit mass.

$$F_{\text{gravitational}} = G.\frac{M_1 \cdot m_2}{d^2}$$

G replaces "*c*" as a gravitational constant determined from experiments. The gravitational force of earth acting on a body of mass located *near the surface of the earth* is then:

$$F_{\text{gravity}} = G.\frac{M_{earth} \cdot m_1}{d^2}$$

Since most of our free falling bodies occur near the surface of the earth, we can take *d*, to be the radius of the earth. Substituting the known values in along with the value for *G* reduces the equation to:

$$F_{\text{gravity}} = 9.8.m_1$$

Since force equal *ma*, we have:

$$\cancel{m_1} \cdot a = 9.8.\cancel{m_1}$$

$$a = 9.8$$

This important result tells us that the acceleration of a body of any mass is 9.8 meters per second per second near the surface of the earth. The gravitational force acting on a body near the surface of the earth would be its mass, *m*, times the constant acceleration, 9.8. For those who have a ground to hold them up this does not mean much but for a free-falling body in air its acceleration as it falls toward the earth will be a constant $9.8.m/s^2$, regardless of its mass. A body of twice the mass will be pulled in by twice the force, but the acceleration *due to the force of gravity* remains the same.

This may sound a bit confusing but just remember any body will fall to the earth with a constant acceleration, independent of its mass. In terms of particle motion, mass means nothing for a falling body! While the gravitational force increases with mass, the acceleration remains the same.

We can now write the acceleration function for a falling body near the earth's surface as:

$$a(t) = 9.8m/s^2 \text{ (constant)}$$

Since acceleration as a function of time is by definition the derivative of the velocity function with respect to time then what we have is the same as:

$$a(t) = \frac{dv}{dt} = v'(t)$$

$$\therefore\ v'(t) = 9.8$$

The derivative of the velocity function (acceleration) is 9.8. Since we know that the derivative of any function $c.\ x^n$ is $c.\ nx^{n-1}$ then we can easily find the velocity function since we know what its derivative is.

The process of finding a function, given its derivative is known as anti-differentiation. In this case 9.8 can also be written as $9.8.t^{\circ}(t^{\circ} = 1)$. We see that:

$$(n - 1) = 0$$

Therefore n equals 1. Consequently the anti-derivative of $9.8.t^{\circ}$ is $9.8.t$. This should hopefully be obvious since the derivative of $v(t) = 9.8.t$ is just $v'(t) = 9.8$. We now know that velocity as a function of time is:

$$v(t) = 9.8t$$

whose derivative with respect to time is:

$$v'(t) = a(t) = 9.8$$

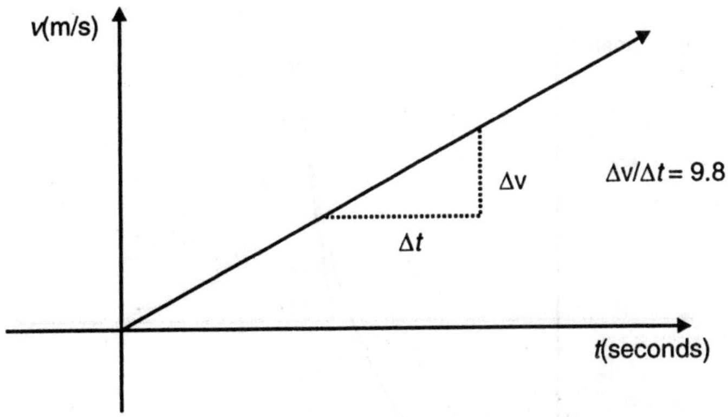

Fig. 29 The Graph of the Velocity Function is a Linearly Increasing Function with Constant Rate of Change or Slope $9.8m/s^2$.

This graph of the velocity function gives us the objects velocity as any time t, assuming that air-resistance is negligible. For example at $t = 10$ seconds, the object's velocity is:

$$v(10) - 9.8.10 = 98m/s$$

At t = 94 seconds its velocity is:

$$v(94) = 9.8.94 = 910m/s$$

Or almost 30,000 km/ hr. Due to air-resistance no object reaches such high velocities. Remember the greater the height it is dropped from the more time it has to increase its velocity or accelerate before it slams into the earth.

Now how do we find the distance function or the distance covered from the point of dropping the object. From the definition of velocity we know that:

$$v(t) = \frac{ds}{dt} = s'(t)$$

$$\therefore s'(t) = 9.8t$$

Since s'(t) = 9.8.t, we have n=1 so its anti-derivative will be $9.8.\frac{t^2}{2}$ where the derivative of $c.\frac{t^2}{2}$ is c.t = 9.8t. The solution is then:

$$d(t) = 9.8.\frac{t^2}{2} \text{ (or } 4.9.t^2)$$

The derivative of this function is the velocity function or:

$$v(t) = 9.8.t$$

We can graph the distance function $d(t) = 9.8.\frac{t^2}{2}$ as shown in figure 30.

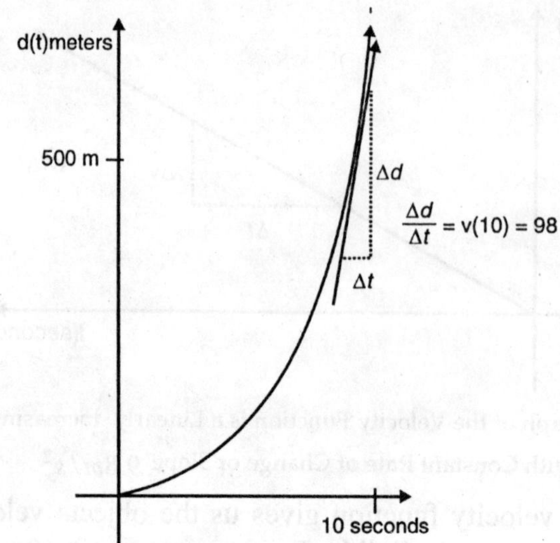

Fig. 30 Graph of vertical distance traveled

Clearly as time, *t*, increases, the rate at which distance is being covered is very great . For example between $t = 0\ s$ to $t = 5s$, the object has covered totally.

$$d(5) - d(0) = 9.8.\frac{5^2}{2} - 0 = 122.5 \text{ meters}$$

Or the object has covered 112.5 meters in the first five seconds of its free-fall. However from t = 20 s to t = 25 s, the object has covered:

$$d(25) - d(20) = 9.8.\frac{25^2}{2} - 9.8.\frac{20^2}{2} = 1102.5 \text{ meters}$$

The object has covered more than a kilometer during this five second interval!! This should make sense because initially the body's velocity is small and thus does not cover much distance over a time interval Δt;. However, after some time its velocity has increased (look at the graph of the velocity function graph), such that over a same interval Δt, the object covers a greater distance. Remember constant acceleration means that the velocity is increasing linearly with time and distance increases with the half square of time.

INITIAL CONDITIONS OF MOTION

The concepts of forces, accelerations, velocities and distance are not limited to free-fall motion. Constant accelerations exist in many other physical phenomena's. Before ending we need to understand how we can modify our equations of motion to be consistent with any initial conditions that may exist. For example, a ball may be dropped with an initial velocity or a car may accelerate from a certain distance from a starting point.

If an object was moving at constant velocity, v_c, its velocity function would be:

$v(t) = v_c$(constant at any time t)

From the definition of the derivative, the derivative of a constant function of the form $f(x) = c$ is zero because:

$$\lim_{\Delta x \to 0} \frac{f(x+\Delta x) - f(x)}{\Delta x} = \frac{c-c}{\Delta x} = \frac{0}{\Delta x} = 0$$

Therefore the derivative of the velocity function is:

$v'(t) = a(t) = 0$, v_c is a constant

Looking back at our free-falling body, we know that its acceleration was $9.8m/s^2$. This can now be written as:

$v'(t) = a(t) = 98 + 0$

The anti-derivative of this acceleration function is then:

$v(t) = 9.8.t + v_i$

Remember that the anti-derivative of zero is a constant. At $t = 0$ we have the initial condition:

$v(0) = 0 + v_i$

$v(0) = v_i$

We call , $v_i v_i$ or the velocity of the object at t = 0. It represents a situation that may exist when the initial condition is zero. When anti-differentiating we need to remember to add a constant along to reflect the initial conditions that may exist in the situation. For example if a car is traveling down the highway at 82 mph , suddenly sees a cop, and then steps on the pedal, accelerating the cart at 3 mph/s, then its velocity at any time t, were t is measured as soon as he steps on the gas is:

$v(t) = 82 + 3t$

We know that the velocity function of a free-falling body is:

$v(t) = 9.8.t + v_i + 0$

The anti-derivative of this function gives us the distance covered as a function of time:

$$d(t) = 9.8.\frac{t^2}{2} + v_i.t + d_i$$

d_i is the objects initial position at $t = 0$. Our result can be generalized for initial acceleration (a_i) (v_i)or distance as follows:

$$a(t) = a + 0 + 0$$

$$v(t) = a.t + v_i + 0$$

$$d(t) = a.\frac{t^2}{2} + v_i.t + d_i$$

EXERCISE

1. Find the absolute maximum and absolute minum values of the function $f(x) = x^2$ on [–2, 1].
2. Find the absolute extrema values of the function $h(z) = 8z - z^2$ on $[-2,1]$.
3. Find the absolute extrema values of the function $f(x) = x^{2/3}$ on $[-2,3]$.
4. Give an example of a function defined on [0, 1] that has neither a local maximum nor a local minimum value at 0.
5. Show that $|\sin b - \sin a| \le |b - a|$ for any real number *a* and *b*.
6. For what values of *a*, *m* and *b* does the function

$$f(x) = \begin{cases} 3, & x = 0 \\ -x^2 + 3x + a, & 0 < x < 1 \\ mx + b, & 1 \le x \le 2 \end{cases}$$

satisfy the hypotheses of the Mean Value Theorem on the interval [0, 2]?

7. Let *F*(*x*, *t*) = log(2*x* + 2*ct*) for *x* > - *ct*, where *c* is a fixed constant. Show that

$$\frac{\partial^2 F}{\partial t^2} - c^2 \frac{\partial^2 F}{\partial t^2} = 0.$$

Chapter 4

Basic Concept of Complex Number

INTRODUCTION

In this chapter we firstly studied abot real and complex number number system , after that we deeply focused on complex number and their application. Complex analysis is the study of complex numbers together with their derivatives, manipulation, and other properties. Complex analysis is an extremely powerful tool with an unexpectedly large number of practical applications to the solution of physical problems. They also have applications in a variety of areas, ranging from fluid flow, to electric circuits, to the mysterious quantum world. Generally, they are described as belonging to the area of mathematics known as complex analysis.

The key result in complex analysis is the Cauchy integral theorem, which is the reason that single-variable complex analysis has so many nice results. A fundamental result of complex analysis is the Cauchy-Riemann equations, which give the conditions a function must satisfy in order for a complex generalization of the derivative, the so-called complex derivative, to exist. When the complex derivative is defined "everywhere," the function is said to be analytic.

NATURAL NUMBERS

The basic root of all the number system is natural number. Here we are going to study about natural numbers and the properties related to it.

The natural numbers

1, 2, 3, ...

are used for counting, and for labeling the terms of an infinite sequence. They obey the following laws.

Laws of Algebra

Associative law of addition:

$$(a + b) + c = a + (b + c).$$

Commutative law of addition:

$$a + b = b + a.$$

Cancellation law of addition:

If $a + c = b + c$, then $a = b$.

Distributive law with multiplication on the left:

$$a.(b + c) = a.b + a.c.$$

Distributive law with multiplication on the right:

$$(a + b).c = a.c + b.c.$$

Associative law of multiplication:

$$(a.b).c = a.(b.c).$$

Commutative law of multiplication:

$$a.b = b.a.$$

Cancellation law of multiplication:

If $a.c = b.c$, then $a = b$.

Laws of Order

Trichotomy law:

For two given natural numbers, a and b, exactly one of the following relations is true:

$$a < b,\ a = b,\ a > b.$$

Transitive law:

If $a < b$ and $b < c$, then $a < c$.

Order and addition:

If $a < b$, then $a + c < b + c$.

Order and multiplication:

If $a < b$, then $a.c < b.c$.

Other Number Systems

It is not always possible to subtract one natural number from another. For example, there is no natural number n such that $8 + n = 5$. In other words we cannot subtract 8 from 5. Therefore the natural numbers are not a group under addition.

In order to get a group under addition we introduce the number zero and negative whole numbers to make the system of **integers**:

$$\ldots -3, -2, -1, 0, 1, 2, 3, \ldots$$

Also, it is not always possible to divide one natural number by another. For example, there is no natural number n such that $5 \times n = 8$. In other words we cannot divide 8 by 5. Therefore the natural numbers are not a group under multiplication.

In order to get a group under multiplication we introduce the fractions to make the system of positive rational numbers:

1/1, 2/1, 3/1 ...

1/2, 2/2, 3/2 ...

1/3, 2/3, 3/3 ...

... ...

In order to get a number system that is a group under addition and also a group under multiplication, we introduce the negative fractions to make the rational numbers:

... – 2/1, – 1/1, 0/1, 1/1, 2/1, ...

... – 2/2, – 1/2, 0/2, 1/2, 2/2, ...

... – 2/3, – 1/3, 0/3, 1/3, 2/3, ...

... ...

The system of rational numbers is sufficient for all practical arithmetical calculations. But it is not sufficient for mathematical analysis, because there are theorctical problems that cannot be solved using rational numbers. Here are two examples:

- The equation $x^2 = 2$ cannot be solved in the rational number system.
- The equation $x^2 = -1$ cannot be solved in the rational number system.

To solve these problems we need real numbers and complex numbers

Representation of Real Numbers

The real number system consists of whole numbers, fractions, and irrational numbers. These numbers may be positive, zero, or negative. Real numbers may be represented by infinite decimals, for example:

11 = 11.00000000...
2/11 = 0.18181818...
sqrt 2 = 1.41421356...
pi = 3.14159265...

In practical calculations we use finite decimals as approximations to real numbers.

A real number may be represented by a point on the real line (see Fig. 1).

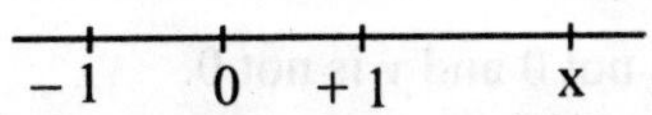

Fig. 1. The real number x on the real line.

Laws of Algebra

The real numbers are a commutative group under addition:

Associative law of addition:

$$x + (y + z) = (x + y) + z.$$

Commutative law of addition:

$$x + y = y + x.$$

Identity law of addition:

There is a unique zero 0 such that, for all x:

$$x + 0 = x.$$

Inverse law of addition:

Every real number x has a unique negative $-x$ such that:

$$x + (-x) = 0.$$

In the real number system multiplication is distributive over addition:

$$x.(y + z) = x.y + x.z.$$

The non-zero real numbers are a commutative group under multiplication:

Associative law of multiplication:

$$x.(y.z) = (x.y).z.$$

Commutative law of multiplication:

$$x.y = y.x.$$

Identity law of multiplication:

There is a unique unit 1 such that, for all x:

$$x.1 = x.$$

Inverse law of multiplication:

Every non-zero real number x has a unique reciprocal $1/x$ such that:

$$x.(1/x) = 1.$$

These laws of algebra show that the real number system is a field. Important laws derived from the laws of a field are as follows:

- For all $x : x.0 = 0$.
- If $x.y = 0$, then $x = 0$ or $y = 0$.
- If $x.y$ is not 0, then x is not 0 and y is not 0.

Laws of Order

The real numbers are simply ordered. They obey the following laws of order:

Trichotomy law:

For two given real numbers, x and y, exactly one of the following relations is true:

$$x < y,\ x = y,\ x > y.$$

Transitive law:

If $x < y$ and $y < z$, then $x < z$.

Order and addition:

If $x < y$, then $x + z < y + z$.

Order and multiplication:

If $x < y$ and $z > 0$, then $x.z < y.z$.

Important laws derived from the laws of order are as follows:

- If $u < v$ and $x < y$, then $u + x < v + y$.
- If $x > 0$ and $y > 0$, then $x.y > 0$.
- If $x < 0$ and $y > 0$, then $x.y < 0$.
- If $x < 0$ and $y < 0$, then $x.y > 0$.

Roots of Polynomials

It follows from the laws of order that there is no real number x such that $x^2 < 0$. For example, there is no real number x such that $x^2 = -1$. More generally, there is not always a real number x which satisfies a given polynomial equation:

$$x^n + a_1x^{n-1} + \ldots + a_{n-1}x + a_n = 0.$$

Absolute Values

The **absolute value** $|x|$ of a real number x is defined as follows:

If $x > 0$, then $|x| = x$,

If $x = 0$, then $|x| = 0$,

If $x < 0$, then $|x| = -x$.

Important laws with absolute values are as follows:

- If $|x| = 0$, then $x = 0$.
- $|x| = |-x|$.
- $|x| - |y| \leq |x + y| \leq |x| + |y|$.
- $|x| - |y| \leq ||x| - |y|| \leq |x - y| \leq |x| + |y|$.
- $|x.y| = |x|.|y|$.
- $|x + y|^2 \leq |x|^2 + 2.|x|.|y| + |y|^2$.
- $|x + y|^2 + |x - y|^2 = 2.|x|^2 + 2.|y|^2$.

Note: The following relations are not always true:

- $|x - y|^2 \leq |x|^2 - 2.|x|.|y| + |y|^2$. [*generally false*]
- $|x + y|^2 \leq |x|^2 + |y|^2$. [*generally false*]

Sequences

A sequence is an infinitely long list of numbers:

$$x_1, x_2, x_3, \ldots$$

These numbers are called the **terms** of the sequence. The sequence may also be written: x_k, where $k = 1, 2, 3, \ldots$

When the difference between the terms of a sequence and a number *u* approaches zero as *k* increases we call *u* a **limit** of the sequence, and we write:

$$\lim x_k = u.$$

The exact definition of the limit is as follows:

The number *u* is a limit of the sequence x_k if and only if, for all *epsilon* > 0, there exists n such that $k > n$ implies $|x_k - u| <$ *epsilon*.

When a sequence has a limit *u* we also say the sequence converges to *u*.

Important properties of limits are as follows:

A limit (when it exists) is unique:

If $\lim x_k = u$, and $\lim x_k = v$, then $u = v$.

Limits and sums:

If $\lim x_k = u$ and $\lim y_k = v$, then $\lim (x_k + y_k) = u + v$.

Limits and products.

If $\lim x_k = u$ and $\lim y_k = v$, then $\lim (x_k.y_k) = u.v$.

A sequence x_k which has a limit satisfies the Cauchy condition:

For all *epsilon* > 0, there exists n such that $j > n$ and $k > n$ imply $|x_k - x_j| <$ *epsilon*. (In other words: if the terms of a sequence approach a limit, then the terms of the sequence approach each other.)

Completeness

A sequence which satisfies the Cauchy condition is called a Cauchy sequence. Therefore, every sequence which has a limit is a Cauchy sequence. In the real number system the converse is also true:

Every Cauchy sequence has a limit.

This property of the real number system is called completeness.

We may regard the infinite decimals which represent real numbers as special Cauchy sequences. For example, the infinite decimal

$$\text{sqrt } 2 = 1.41421356...$$

may be regarded as the Cauchy sequence

1, 1.4, 1.41, 1.414, 1.4142, 1.41421, 1.424213, 1.4142135, 1.41421356, ...

This represents the real number sqrt 2, because sqrt 2 is the limit of the sequence.

Bounded Sequences

A sequence x_k is called bounded when there exist real numbers a and b such that, for all k, $a \le x_k \le b$.

The following theorems show the relationship between bounded sequences and Cauchy sequences:

1. Every Cauchy sequence is bounded.
2. The Bolzano-Weierstrass Theorem: Every bounded sequence has a Cauchy subsequence.

   Since a sequence of real numbers is a Cauchy sequence if and only if it has a limit, it follows from the above theorems that:
3. If a sequence has a limit, then it is bounded.
4. Every bounded sequence has a subsequence with a limit.

Series

A series is an infinite list of numbers added together:

$$x_1 + x_2 + x_3 + ...$$

A series is summable when the sequence of partial sums

$$x_1, x_1 + x_2, x_1 + x_2 + x_3, ...$$

has a finite limit s. This limit is the sum of the series, and we write:

$$x_1 + x_2 + x_3 + ... = s.$$

We also say the series converges to s. A series which is not summable is said to diverge.

If the series $x_1 + x_2 + x_3 + ...$ is summable, then the sequence $x_1, x_2, x_3, ...$ converges to zero. But the converse is false. For example, lim(1, 1/2, 1/3, 1/4, ...) = 0, but the series 1 + 1/2 + 1/3 + 1/4 + ... diverges.

A series $x_1 + x_2 + x_3 + ...$ is absolutely summable, or absolutely convergent, when the series

$$|x_1| + |x_2| + |x_3| + ...$$

is summable.

A series $x_1 + x_2 + x_3 + ...$ converges to a unique sum independent of the arrangement of the terms if and only if it is absolutely summable.

Example

The following series are all rearrangements of the same series:

- 1 – 1/2 + 1/3 – 1/4 + 1/5 – 1/6 + 1/7 – 1/8 ... = 0.693147...
- 1 + 1/3 – 1/2 + 1/5 + 1/7 – 1/4 + ... = 1.03972...
- 1 – 1/2 + 1/3 + 1/5 – 1/4 + 1/7 + 1/9 + 1/11 – 1/6 + ... diverges.

These series are not absolutely summable because 1 + 1/2 + 1/3 + 1/4 + ... diverges.

REAL VARIABLES

Continuity

A many-one relation between real numbers is called a real-valued function of a real variable. We write: $y = f(x)$.

A function f is continuous at u when, for all sequences x_k which have the limit u, the sequence $f(x_k)$ has the limit $f(u)$.

An equivalent definition of continuity is as follows: A function f is continuous at u when, for all *epsilon* > 0, there exists *delta* > 0 such that $| x - u |$ < *delta* implies $| f(x) - f(u) |$ < *epsilon*.

Examples

1. $f(x) = x^2$ is continuous everywhere.
2. $f(x) = | x |$ is continuous everywhere.
3. $f(x) = 1/x$ is continuous everywhere except at $x = 0$. At $x = 0$ the function has an infinite discontinuity.
4. $f(x) = 1/(x^2 - 1)$ is continuous everywhere except at $x = \pm 1$. At $x = -1$ and $x = +1$ the function has infinite discontinuities.
5. $f(x) = x/| x |$ is continuous everywhere except at $x = 0$. At $x = 0$ the function has a finite jump discontinuity from $f(x) = -1$ where $x < 0$ to $f(x) = +1$ where $x > 0$.

Algebraic Properties of Continuous Functions

If $f(x)$ and $g(x)$ are both continuous at u, then:

1. $f(x) + g(x)$ is continuous at u.
2. $f(x).g(x)$ is continuous at u.

Continuous Functions in $[a, b]$

The closed bounded interval $[a, b]$ is the set of all real numbers x such that $a \leq x \leq b$. Important properties of continuous functions in closed bounded intervals are given in the following theorems. The open interval (a, b) is the set of all real numbers x such that $a < x < b$. Open intervals are also important in real variable analysis.

THEOREM. If $f(x)$ is continuous at every point in $[a, b]$, then $f(x)$ is bounded in $[a, b]$.

PROOF. Suppose $f(x)$ is unbounded in $[a, b]$. Then for all k there exists x_k in $[a, b]$ such that $|f(x_k)| > k$. By the properties of bounded sequences, the sequence $x_1, x_2, x_3, \ldots$ has a subsequence $x_1', x_2', x_3', \ldots$ which has a limit x' in $[a, b]$. Since the sequence $|f(x_1')|, |f(x_2')|, |f(x_3')|, \ldots$ is unbounded, it cannot have a finite limit at x'. Therefore there exists x' in $[a, b]$ where $f(x)$ is not continuous.

THEOREM. A function which is continuous at every point in $[a, b]$ has a maximum in $[a, b]$.

PROOF. By the previous theorem, if $f(x)$ is continuous at every point in $[a, b]$, then $f(x)$ is bounded in $[a, b]$. Let $M = \sup f(x)$ in $[a, b]$. Then for all k there exists x_k in $[a, b]$ such that $M - f(x_k) < 1/k$. Therefore $\lim f(x_k) = M$. The sequence $x_1, x_2, x_3, \ldots$ has a subsequence x_1', $x_2', x_3', \ldots$ which has a limit x' in $[a, b]$. Then $\lim f(x_k') = M$. Since $f(x)$ is continuous, we also have $\lim f(x_k') = f(x')$. Therefore $f(x') = M$. In other words $f(x)$ has a maximum at x' in $[a, b]$.

THEOREM. A function which is continuous at every point in $[a, b]$ has a minimum in $[a, b]$.

The proof is similar to the proof of the above theorem

The Intermediate Value Theorem. If $f(x)$ is continuous in $[a, b]$, and if y' is any number between $f(a)$ and $f(b)$, then there exists x' in (a, b) such that $f(x') = y'$.

PROOF. Let E be the set of points in $[a, b]$ where $f(x) < y'$, and let $x' = \sup E$. Then there is a sequence $u_1, u_2, \ldots$ of points in E which has the limit x'. Since $f(x)$ is continuous, we have $\lim f(u_k) = f(x')$, where $f(x') \leq y'$. Let F be the set of points in $[a, b]$ where $f(x) \geq y'$. Then there is a sequence $v_1, v_2, \ldots$ of points in F which has the limit x'. Since $f(x)$ is continuous, we have $\lim f(v_k) = f(x')$, where $f(x') \geq y'$. Therefore $f(x') = y'$.

Simple Approximations to Continuous Functions

Suppose $f(x)$ is continuous in $[a, b]$. Then $f(x)$ is uniformly continuous in $[a, b]$. In other words, for all *epsilon* > 0 there exists *delta* > 0 such that $|x_1 - x_2| <$ *delta* implies $|f(x_1) - f(x_2)| <$ *epsilon* in $[a, b]$. Choose *epsilon* to give the accuracy you want. Make a partition of $[a, b]$ by points $x_0 = a < x_1 < x_2 < \ldots x_{n-1} < x_n = b$ such that for all k you have $x_k - x_{k-1} <$ *delta*. Then you can make the following simple approximations to $f(x)$ with accuracy *epsilon*.

The step function

This function (shown in Fig. 1) is useful for approximate numerical integration. It is defined as follows: $g(x) = f[(½).(x_{k-1} + x_k)]$, where x is in (x_{k-1}, x_k).

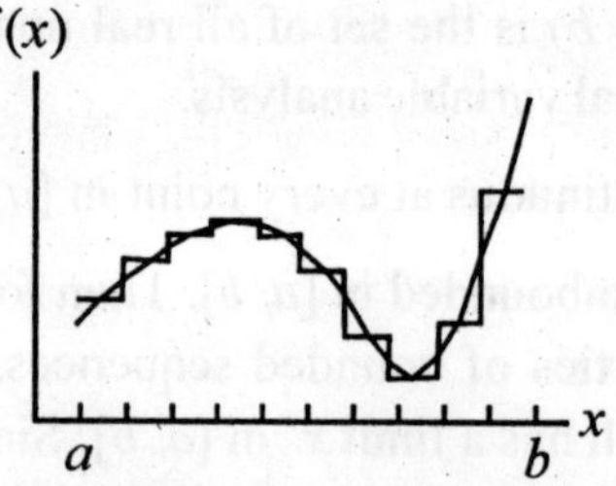

Fig. 1. The step function approximation.

The piecewise linear function

This function (shown in Fig. 2) is useful for approximate numerical differentiation. It is defined as follows: $h(x) = f(x_{k-1}) + m_k.(x - x_{k-1})$, where $m_k = [f(x_k) - f(x_{k-1})] / (x_k - x_{k-1})$ and x is in $[x_{k-1}, x_k]$.

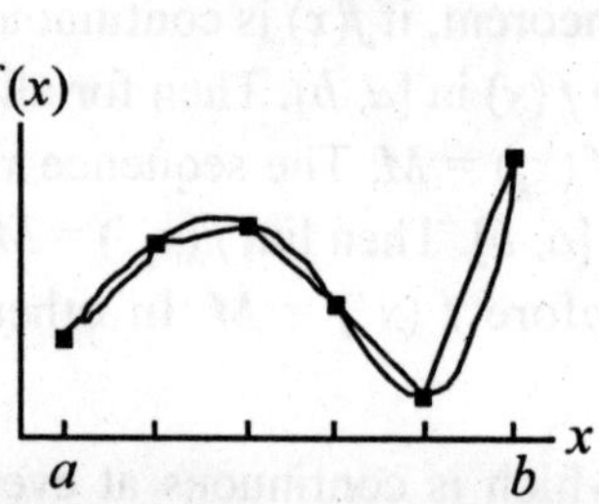

Fig. 2. The piecewise linear function approximation.

Derivatives

Let $f(x)$ be a given function. Suppose that, for every sequence x_k such that $\lim x_k = x$,

$$f'(x) = \lim [f(x) - f(x_k)] / [x - x_k].$$

Then the function $f'(x)$ is called the **(first) derivative** of $f(x)$.

The **second derivative** of $f(x)$ is the derivative of the first derivative:

$$f''(x) = \lim [f'(x) - f'(x_k)] / [x - x_k].$$

The **rules of algebra for calculating derivatives** are as follows.

If $f(x) = u(x) + v(x)$, then $f'(x) = u'(x) + v'(x)$.

If $f(x) = u(x) - v(x)$, then $f'(x) = u'(x) - v'(x)$.

If $f(x) = u(x).v(x)$, then $f'(x) = u'(x).v(x) + u(x).v'(x)$.

If $f(x) = u(x)/v(x)$, then $f'(x) = [u'(x).v(x) - u(x).v'(x)] / [v(x)]^2$.

Using these rules we can differentiate polynomials as follows:

If $p(x) = a_0.x^n + a_1.x^{n-1} + ... + a_{n-1}.x + a_n$, then $p'(x) = n.a_0.x^{n-1} + (n-1).a_1.x^{n-2} + ... + 2.a_{n-2}.x + a_{n-1}$.

Using these rules we can also differentiate

$f(x) = p(x) / q(x)$, where $p(x)$ and $q(x)$ are polynomials.

Elementary Special Functions

The elementary special functions are continuous differentiable functions *defined by the following properties*:

The exponential:

$$exp\ (0) = 1,\ exp'\ (x) = exp\ (x).$$

The logarithm:

The inverse of the exponential: $exp\ (\log(x)) = x$, where $x > 0$, $\log(1) = 0$, $\log'(x) = 1/x$.

The sine:

$$\sin\ (0) = 0,\ \sin'\ (0) = 1,\ \sin''\ (x) = -\sin\ (x).$$

The cosine:

$$\cos\ (0) = 1,\ \cos'\ (0) = 0,\ \cos''\ (x) = -\cos\ (x).$$

Differentials

In the equation

$$y = f(x),$$

x is the independent variable, f is the function, and y is the dependent variable. Let dx be an *independent parameter*, which is called the differential of x. Then dy, which is called the differential of y, is defined by

$$dy = f'\ (x).dx.$$

Note that dy depends on *two independent quantities*: the variable x and the parameter dx.

Suppose that f is a function of n independent variables $x_1, \ldots, x_n$:

$$y = f(x_1, \ldots, x_n).$$

Let $dx_1, \ldots, dx_n$ be the n independent differentials of $x_1, \ldots, x_n$. Then the differential of y is defined by

$$dy = f_1(x_1, \ldots, x_n).dx_1 + \ldots + f_n(x_1, \ldots, x_n).dx_n,$$

where $f_1, \ldots, f_n$ are the *partial derivatives* of f with respect to $x_1, \ldots, x_n$. Note that dy depends on *2n independent quantities*: the n variables $x_1, \ldots, x_n$ and the n parameters $dx_1, \ldots, dx_n$.

Let $y = f(x)$. Then the second differential of y is a new dependent variable defined by

$$d^2y = d(dy) = [f'\ (x).dx]'.dx = [(f'\ (x))'.dx + (f'\ (x)).(dx)'].dx.$$

Since $(f'(x))' = f''(x)$, and $(dx)' = 0$, because dx is a parameter independent of x, we have

$$d^2y = f''\ (x).(dx)^2.$$

The Chain Rule

Suppose $y = f(x)$ and $x = g(u)$. Then $y = f[g(u)]$. Differentiating we have

$$dy = f'(x).dx = f'[g(u)].d[g(u)] = f'[g(u)].g'(u).du.$$

This formula for dy in terms of u and du (instead of x and dx) is called the chain rule.

Implicit Functions

Suppose the relation between n variables is given by

$$f(x_1, \ldots x_n) = 0.$$

Note that we do not distinguish between dependent and independent variables; all the variables have the same status. Differentiating, we get

$$f_1(x_1, \ldots, x_n).dx_1 + \ldots + f_n(x_1, \ldots, x_n).dx_n = 0.$$

The rate of change of x_i with respect to x_j can be calculated from this equation by making all the differentials zero except dx_i and dx_j. The rate of change is then given by:

$$dx_i/dx_j = -f_j(x_1, \ldots, x_n) / f_i(x_1, \ldots, x_n).$$

Integration

Divide the interval $[a, b]$ into k subintervals of length $h_k = (b-a)/k$, and let $x_1, \ldots, x_n$ be the mid points of the subintervals. Let $f(x)$ be a piecewise continuous function defined in $[a, b]$. Then the Riemann integral of $f(x)$ from $x = a$ to $x = b$ is

$$\int_a^b f(x)dx = \lim [f(x_1) + \ldots + f(x_k)].h_k, \text{ where } k = 1, 2, \ldots$$

Note: The differential dx is not used in the calculation of the integral; it only shows that x is the variable of integration.

The integral has the following elementary properties:

(i) $\int_a^b f(x)dx = -\int_b^a f(x)dx$

(ii) $\int_a^b [f(x) + g(x)]dx = \int_b^a f(x)dx + \int_b^a g(x)dx$

(iii) $\int_b^a cf(x)dx = c\int_b^a f(x)dx$

, where c is a constant.

(iv) If $f(x)$ is even, in other words $f(-x) = f(x)$, then

$\int_{-a}^a f(x)dx = 2\int_0^a f(x)dx$

(v) If $f(x)$ is odd, in other words $f(-x) = -f(x)$, then

$\int_{-a}^a f(x)dx = 0.$

The Fundamental Theorem of Calculus

Let $F(x)$ be a function such that $F'(x) = f(x)$. Then

$$\int_a^b f(x)dx = F(b) - F(a)$$

Simplified Proof. We shall use the same notation as in the definition of the Riemann integral. Let $[a, b]$ be divided into k equal subintervals. Then $b = a + k.h_k$, where h_k is the length of each subinterval. Let x_k be the mid point of each subinterval.

First, let $F(a)$ be given, and let $F(a + h_k) = F(a) + f(x_1).h_k$. Then $f(x_1) = [F(a + h_k) - F(a)] / h_k = [F(x_1 + (½).h_k) - F(x_1 - (½).h_k)] / h_k$.

Next, for $j = 2, \ldots k - 1$, let $F(a + j.h_k) = F(a + (j - 1).h_k) + f(x_j).h_k$. Then $f(x_j) = [F(a + j.h_k) - F(a + (j - 1).h_k)] / h_k = [F(x_j + (½).h_k) - F(x_j - (½).h_k)] / h_k$.

Finally, let $F(b) = F(b - h_k) + f(x_k).h_k$. Then $f(x_k) = [F(b) - F(b - h_k)] / h_k = [F(x_k + (½).h_k) - F(x_k - (½).h_k)] / h_k$, and $F(b) - F(a) = [f(x_1) + \ldots + f(x_k)].h_k$.

As k increases we get in the limit: $F'(x) = f(x)$, and $F(b) - F(a) = \int_a^b f(x)dx$.

It may be difficult or impossible to find a formula for $F(x)$ when $f(x)$ is given. Known formulas are tabulated in many mathematical handbooks. If there is no formula, the integral must be evaluated numerically.

MORE PROPERTIES OF INTEGRALS

Integrals of Products

There is no simple formula for the integral of a product, but the following equation (integration by parts) is often useful:

Integral (a, b): $f'(x).g(x).dx$ + Integral (a, b): $f(x).g'(x).dx = f(b).g(b) - f(a).g(a)$.

Change of Variable

Let $x = g(u)$. Then $u = g^{-1}(x)$, and we have:

Integral (a, b): $f(x).dx$ = Integral$(g^{-1}(a), g^{-1}(b))$: $f[g(u)].g'(u).du$.

Differentiation with Respect to a Parameter

Suppose the function $f(x, s)$ to be integrated and the endpoints $a(s)$, $b(s)$ of the integral contain a parameter s. Let $I(s)$ = Integral $(a(s), b(s))$: $f(x, s).dx$. Then the derivative of the integral with respect to s is:

$I'(s)$ = Integral $(a(s), b(s))$: $f_2(x, s).dx + f(b, s).b'(s) - f(a, s).a'(s)$,

where $f_2(x, s)$ is the partial derivative of $f(x, s)$ with respect to s.

The complex numbers are the field $\mathbb{C}$ of numbers of the form $x + iy$, where x and y are real numbers and i is the imaginary unit equal to the square root of -1.

COMPLEX NUMBERS

Complex numbers are useful abstract quantities that can be used in calculations and result in physically meaningful solutions. However, recognition of this fact is one that took a long time for mathematicians to accept.

THE ORIGIN OF COMPLEX NUMBERS C

Complex analysis can roughly be thought of as the subject that applies the theory of calculus to imaginary numbers. But what exactly are imaginary numbers? Usually, students learn about them in high school with introductory remarks from their teachers along the following lines: "We can't take the square root of a negative number. But let's pretend we can and begin by using the symbol $i = \sqrt{-1}$."

If $i = \sqrt{-1}$, it stands to reason that $i^2 = -1$.

Representation of Complex Numbers

A complex number z has the form

$$z = x + i.y,$$

where $i^2 = -1$. The real number x is called the real part of z, and the real number y is called the imaginary part of z. We write:

$$x = \text{Re}(z),\ y = \text{Im}(z).$$

The complex number z may also be written in the form

$$z = r(\cos\theta + i\sin\theta),$$

where $r^2 = x^2 + y^2$, $\cos\theta = x/r$, and $\sin\theta = y/r$. The real number r is called the absolute value of z, and the real number θ is called the argument of z. We write:

$$r = |z|,\ \theta = \arg(z).$$

A complex number may be represented by a point in the complex plane as shown in the figure given below.

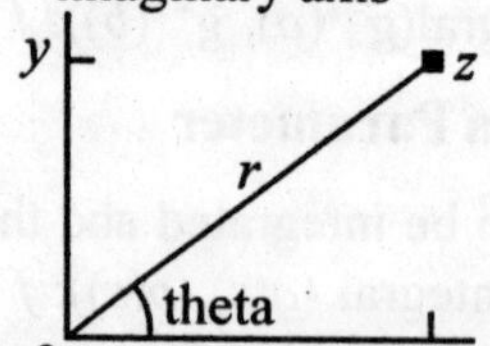

Fig. The complex number z in the complex plane.

ALGEBRA OF COMPLEX NUMBERS

We have seen that complex numbers came to be viewed as ordered pairs of real numbers. That is, a complex number z is defined to be

$$z = (x, y),$$

where x and y are both real numbers.

The reason we say ordered pair is because we are thinking of a point in the plane. The point (2, 3), for example, is not the same as (3, 2). The order in which we write x and y in the equation makes a difference. Clearly, then, two complex numbers are equal if and only if their x coordinates are and their y coordinates are equal. In other words,

$(x, y) = (u, v)$ iff $x = u$ and $y = v$.

A meaningful number system requires a method for combining ordered pairs. The definition of algebraic operations must be consistent so that the sum, difference, product, and quotient of any two ordered pairs will again be an ordered pair. The key to defining how these numbers should be manipulated is to follow Gauss's lead and equate (x, y) with $x + iy$. Then, if $z_1 = (x_1, y_1)$ and $z_2 = (x_2, y_2)$ are arbitrary complex numbers, we have

$$\begin{aligned} z_1 + z_2 &= (x_1, y_1) + (x_2, y_2) \\ &= (x_1 + iy_1) + (x_2 + iy_2) \\ &= (x_1 + x_2) + i\,(y_1 + y_2) \\ &= (x_1 + x_2, y_1 + y_2) \end{aligned}$$

Thus, the following definitions should make sense.

Definition 1, (Addition of complex numbers)

Two complex numbers $z = x + iy$ and $z' = x' + iy'$ are added together component-wise,

$$z + z = (x + x') + i\,(y + y').$$

In component form,

$$(x, y) + (x', y') = (x + x', y + y')$$

$$z_1 + z_2 = (x_1 + iy_1) + (x_2 + iy_2) = (x_1 + x_2, y_1 + y_2).$$

$$z_1 = x_1 + iy_1,\ z_2 = x_2 + iy_2$$

$$z_1 + z_2 = x_1 + x_2 + i\,y_1 + i\,y_2)$$

$$z_1 + z_2 = x_1 + x_2 + i\,(y_1 + y_2) \qquad ...(1)$$

Definition 2, (Subtraction of complex numbers)

Formula (1-9) $z_1 - z_2 = (x_1 + i\,y_1) - (x_2 + i\,y_2) = (x_1 - x_2, y_1 - y_2)$.

$$z_1 = x_1 + i\,y_1, \qquad z_2 = x_2 + i\,y_2$$

$$z_1 - z_2 = x_1 - x_2 + i\,y_1 - i\,y_2$$

$$z_1 - z_2 = x_1 - x_2 + i\,(y_1 - y_2) \qquad ...(2)$$

Example 1. Given $z_1 = 3 + 7\,i$ and $z_2 = 5 - 6\,i$.

(a) Find $z_1 + z_2$ and (b) $z_1 - z_2$.

$$z_1 + z_2 = (3, 7) + (5, -6) = (3 + 5, 7 - 6) = (8, 1) \text{ and}$$
$$z_1 + z_2 = (3, 7) - (5, -6) = (3 - 5, 7 + 6) = (-2, 13).$$

We can also use the notation $z_1 = 3 + 7\,i$ and $z_2 = 5 - 6\,i$:

$$z_1 + z_2 = 3 + 7\,i + 5 - 6\,i = 8 + i$$

and $$z_1 + z_2 = 3 + 7\,i - (5 - 6\,i) = -2 + 13\,i.$$

Definition 3 (Complex Multiplication)

Two complex numbers $x = a + i\,b$ and $c + i\,d$ are multiplied as follows:

$$x\,y = (a + i\,b)\,(c + i\,d) = ac + i\,b\,c + i\,a\,d - bd$$
$$= (ac - bd) + i\,(ad + bc).$$

In component form,

$$(x, y)\,(x', y') = (x\,x' - y\,y', x\,y' + y\,x')$$

The special case of a complex number multiplied by a scalar is then given by

$$a\,(x, y) = (a, 0)\,(x, y) = (a\,x, a\,y).$$

Surprisingly, complex multiplication can be carried out using only three real multiplications, ac, bd, and $(a + b)\,(c + d)$ as

$$R\,[(a + i\,b)\,(c + i\,d)] = a\,c - b\,d$$
$$\text{I}\,[(a + i\,b)\,(c + i\,d)] = (a + b)\,(c + d) - ac - bd.$$

Complex multiplication has a special meaning for elliptic curves.

$$z_1 z_2 = (x_1, y_2)\,(x_2, y_2) = (x_1 x_2 - y_1 y_2, x_1 y_2 + x_2 y_1). \quad ...(3)$$

Example 2. Given $z_1 = 3 + 7\,i$ and $z_2 = 5 - 6\,i$. Find $z_1\,z_2$.

Solution.

Ist Method

$$z_1\,z_2 = (3, 7)\,(5, -6)$$
$$= (3 * 5 - 7\,(-6), 3\,(-6) + 5 * 7)$$
$$= (15 + 42, -18 + 35) = (57, 17)$$

IInd Method

$$z_1 = 3 + 7\,i \text{ and } z_2 = 5 - 6\,i:$$
$$z_1\,z_2 = (3, 7)\,(5, -6) = (3 + 7\,i)\,(5 - 6\,i)$$
$$= 15 - 18\,i + 35\,i - 42\,i^2$$

$$= 15 - 42\,(-1) + (-18 + 35)\,i$$
$$= 57 + 17\,i$$
$$= (57, 17)$$

Of course, it makes sense that the answer came out as we expected because we used the notation $x + iy$ as motivation for our definition in the first place.

To motivate our definition for division, we will proceed along the same lines as we did for multiplication, assuming $z_2 \neq 0$:

Complex Division

$$\frac{z_1}{z_2} = \frac{(x_1, y_1)}{(x_2, y_2)} = \frac{x_1 + i\,y_1}{x_2 + i\,y_2}$$

$$\frac{z_1}{z_2} = \frac{(x_1, i\,y_1)\,(x_2 - i\,y_2)}{(x_2, i\,y_2)\,(x_2 - i\,y_2)}$$

$$= \frac{x_1, x_2 + y_1 y_2 + i\,(-x_1 y_2 + x_2 y_1)}{x_2^2 + y_2^2}$$

$$= \frac{x_1 x_2 + y_1 y_2}{x_2^2 + y_2^2} + i\,\frac{-x_1 y_2 + x_2 y_1}{x_2^2 + y_2^2}$$

$$= \left(\frac{x_1 x_2 + y_1 y_2}{x_2^2 + y_2^2}, \frac{-x_1 y_2 + x_2 y_1}{x_2^2 + y_2^2}\right) \qquad ...(4)$$

Definition 4. (Complex Division)

The division of two complex numbers can be accomplished by multiplying the numerator and denominator by the complex conjugate of the denominator, for example, with $z_1 = a + b\,i$ and $z_2 = c + d\,i$, $z = z_1/z_2$ is given by

$$z = \frac{a + bi}{c + di} = \frac{(a + bi)\overline{c + di}}{(c + di)\,\overline{c + di}}$$

$$= \frac{(a + bi)(c - di)}{(c + di)\,(c - di)} = \frac{(ac + bd) + i(bc - ad)}{c^2 + d^2},$$

where $\bar{z}$ denotes the complex conjugate. In component notation with $(x, y) = x + i\,y$,

$$\frac{(a + b)}{(c + d)} = \left(\frac{ac + bd}{c^2 + d^2}, \frac{bc - ad}{c^2 + d^2}\right)$$

$$\frac{z_1}{z_2} = \frac{(x_1, y_1)}{(x_2, y_2)} = \left(\frac{x_1 x_2 + y_1 y_2}{x_2^2 + y_2^2}, \frac{-x_1 d_2 + x_2 y_1}{x_2^2 + y_2^2}\right), \text{ for } z_2 \neq 0.$$

Example 3. Given $z_1 = 3 + 7$ II and $z_2 = 5 - 6$ II. Find $\frac{z_1}{z_2}$.

$$\frac{z_1}{z_2} = \frac{(3,7)}{(5,-6)} = \left(\frac{15-42}{25+36}, \frac{18+35}{25+36}\right) = \left(-\frac{27}{61}, \frac{53}{61}\right)$$

As with the example for multiplication, we also get this answer if we use the notation x + IIY :

$$\frac{z_1}{z_2} = \frac{(3,7)}{(5,-6)} = \left(\frac{3+7\text{II}}{5-6\text{II}}\right)$$

$$= \frac{(3+7\pi)\,(5+6\pi)}{(5-6\pi)\,(5+6\pi)} = \frac{15+18\pi+35\pi+42\pi^2}{25+30\pi-30\pi-36\pi^2}$$

$$= \frac{15-42+(18+35)\pi}{(25+3)} = -\frac{27}{61} + \frac{53}{61}\pi$$

$$= \left(-\frac{27}{61}, \frac{53}{61}\right)$$

To perform operations on complex numbers, most mathematicians would use the notation $x + i\,y$ and engage in algebraic manipulations, as we did here, rather than apply the complicated-looking definitions we gave for those operations on ordered pairs.

It turns out that our algebraic definitions give complex numbers all the properties we normally ascribe to the real number system. Taken together, they describe what algebraists call a field. In formal terms, a field is a set (in this case, the complex numbers) together with two binary operations (in this case, addition and multiplication) having the following properties.

Laws Associated with Addition with Complex Number

Here we are going to discuss the laws associated with complex numbers.

1. Commutative Law

$$z_1 + z_2 = z_2 + z_1.$$

2. Associative Law

$$z_1 + (z_2 + z_3) = (z_1 + z_2) + z_3.$$

3. Additive Identity.

There is a complex number ω such that

$z + \omega = z$ for all complex numbers z.

The number ω is obviously the ordered pair (0, 0).

4. Additive Inverses.

Given any complex number z, there is a complex number n (depending on z) with the property that

$$z + n = (0, 0).$$

Obviously, if z = (x, y) = $x + i\,y$, the number n will be $n = (-x, -y) = -x - i\,y = -z$.

Laws Associated with Multiplication with Complex Number

1. Commutative Law

$$z_1z_2 = z_2z_1.$$

2. Associative Law

$$z_1\,(z_2z_3) = (z_1z_2)\,z_3.$$

3. Multiplicative Identity

There is a complex number ξ such that

ξ z = z for all complex numbers z.

As one might expect, it turns out that (1, 0) is the unique complex number ξ with this property.

4. Multiplicative Inverses

Given any number z other than the number (0, 0), there is a complex number (depending on z) which we shall denote by z^{-1} with the property that

$z\,z^{-1} = (1, 0) = 1.$

Based on our definition for division, it seems reasonable that the number z^{-1} would be

$$z^{-1} = \frac{(1,0)}{z} = \frac{1}{x+iy}$$

$$= \frac{x-iy}{(x+iy)(x-iy)} = \frac{x-iy}{x^2+y^2}$$

$$= \frac{x}{x^2+y^2} + i\frac{-y}{x^2+y^2}$$

$$= \left(\frac{x}{x^2+y^2}, \frac{-y}{x^2+y^2}\right)$$

5. The Distributive Law.

$$z_1(z_2 + z_3) = z_1(z_2 + z_3).$$

Definition 5. (Real Part of z) The real part of z = x + $\dot{\pi}$ y, denoted by Re(z), is the real number x.

Definition 6. (Imaginary Part of z) The imaginary part of z = x + $\dot{\pi}$ y, denoted by Im(z), is the real number y.

Definition 7. (Conjugate of z) The complex conjugate of a complex number z ≡ a + bi is defined to be

$$\bar{z} \equiv a - bi. \quad ...(5)$$

The conjugate matrix of a matrix A = (a_{ij}) is the matrix obtained by replacing each element a_{ij} with its complex conjugate, $\bar{A} = (\bar{a}_{ij})$.

By definition, the complex conjugate satisfies

$$\bar{\bar{z}} = z. \quad ...(6)$$

The complex conjugate is distributive under complex addition,

$$\overline{z_1 + z_2} = \bar{z}_1 + \bar{z}_2, \quad ...(7)$$

since

$$\begin{aligned}\overline{(a_1 + ib_1) + (a_2 + ib_2} &= \overline{(a_1 + a_2) + i(b_1 + b_2)} \\ &= (a_1 + a_2) - i(b_1 + b_2) \\ &= (a_1 - ib_1) + (a_2 - ia_2) \\ &= \overline{a_1 + ib_1} + \overline{a_1 + ib_2}, \quad ...(8)\end{aligned}$$

and distributive over complex multiplication,

$$\overline{z_1 z_2} = \bar{z}_1\, \bar{z}_2, \quad ...(9)$$

since

$$\begin{aligned}\overline{(a_1 + b_1 i)(a_2 + b_2 i)} &= \overline{(a_1 a_2 - b_1 b_2 + i(a_1 b_2 + a_2 b_1)} \\ &= (a_1 a_2 - b_1 b_2) - i(a_1 b_2 + a_2 b_1) \\ &= (a_1 - ib_1)(a_2 - ib_2) \\ &= \overline{a_1 + ib_1}\ \overline{a_2 + ib_2}. \quad ...(10)\end{aligned}$$

The conjugate of $z = x + i\,y$, denoted by $\bar{z}$, is the complex number (x, –y) = x – $\bar{z}$ y.

The following theorem gives some important facts relating to these operations. You will be asked for a proof in the exercises

Theorem Suppose z_1, z_2 and z_3 are arbitrary complex numbers. Then if $z_2 \neq 0$

$$\overline{\overline{z}} = z$$

$$\overline{z_1 + z_2} = \overline{z_1} + \overline{z_2}$$

$$\overline{z_1 z_2} = \overline{z_1}\ \overline{z_2}$$

$$\overline{\left(\frac{z_1}{z_2}\right)} = \frac{\overline{z_1}}{\overline{z_2}} \text{ if } z_2 \neq 0$$

$$\text{Im}[z] = \frac{z - \overline{z}}{2i}$$

$$\text{Re}[i\,z] = -\text{Im}[z]$$

$$\text{Im}\,[i\,z] = \text{Re}[z]$$

EXERECISE

1. Find each of the following:
   (a) $(1 + 2i)(3 + 4i)$
   (b) $(1 - 2i)^2/(1 - i)$
   (c) $\{1/(1 + i) + 1/(1 - 2i)\}$
2. Find all the roots of the equation $(z^8 - 1)(z^3 + 8) = 0$.
3. Suppose that $z = x + iy$, where $x, y \in R$. Express each of the following in terms of x and y:

   (a) $\left|\frac{z+1}{z-1}\right|$ (b) $\left|\frac{z+i}{1-iz}\right|$
4. For each of the following, compute all the values and plot them on the plane:

   (a) $(1+i)^{\frac{1}{2}}$ (b) $(-4)^{\frac{3}{4}}$ (c) $(1-i)^{\frac{3}{8}}$
5. Solve the equation $z^2 + z + 1 = 0$ for $z = (x, y)$ by writing $(x, y)\,(x, y) + (x, y) + (1, 0) = (0, 0)$ and then solving a pair of simultaneous equations in x and y.

Chapter 5

Basic Concept of Real and Complex Analysis

INTRODUCTION

In the previous chapter we have already studied the basic tools of complex variable. In this chapter we are going to discuss the basic concepts of real and complex analysis. Firstly we are going to learn some advanced topic of real analysis such as accumulation point, sequence of point, point of convergence. After that we studied the basic topic of complex analysis such as properties of complex number and little concept of De movires theorem.

Principle of the Point of Accumulation

In the rigorous discussion of the fundamentals of analysis, the leading part is played by Weierstrass' principle of the point of accumulation.

The principle follows:

If infinitely many numbers are given in a finite interval, these numbers possess at least one point of accumulation, that is, there is at least one point x such that in every interval about the point x, however small, there lie infinitely many of the given numbers.

Proof of Weierstrass' principle of the point of accumulation

In order to prove this principle arithmetically, we assume to begin with that the given interval is the interval [0, 1]. We now subdivide this interval into ten equal parts by means of the points 0.1, 0.2,..., 0.9. At least one of these subintervals must contain infinitely many points. Let the interval beginning with the number $0.a_1$ be that interval (or one of those intervals if there are several). We now subdivide this interval into ten parts by means of the points of subdivision $0.a_1 1$, $0.a_1 2$,..., $0.a_1 9$. Again, it is true that at least one of these subintervals must contain infinitely many points; let it be the subinterval beginning with the number $0.a_1a_2$. We again subdivide into ten parts, note that one of these parts must contain infinitely many points and continue the process. We thus arrive at a sequence of digits $a_1, a_2, a_3, \ldots$, each having one of the values 0, 1, 2, ... , 9. We now consider the decimal

$$\xi = 0.a_1a_2a_3 \ldots$$

It is clear that this is a point of accumulation of our set of numbers. In fact, every interval, no matter how small, in the interior of which the point ξ lies, contains the subintervals of our system of subdivision from a certain degree of fineness onward, and these subintervals contain infinitely many numbers of the set.

If the interval under consideration instead of being the interval from 0 to 1 is, say, the interval from a to $a + A$, nothing essential in the above argument is changed. The point of accumulation is then represented simply by a number of the form

$$a + h < 0.a_1a_2a_3 \ldots$$

Limits of Sequences

These considerations throw new light upon the concept of the limit of an infinite sequence of numbers $a_1, a_2, a_3, \ldots, a_n, \ldots$. We first consider the exceptional case in which infinitely many numbers of the sequence are equal to one another and extend our definition by applying the name point of accumulation also to this point (or these points). If there are infinitely many different numbers in the sequence and if we assume that the numbers a_n of this sequence are bounded, i.e., that there is a number M such that $| a_n | < M$ for all values of n, the numbers of the sequence form an infinite set of numbers in a finite interval, since they all lie between - M and M. Hence they must possess at least one point of accumulation (ξ).

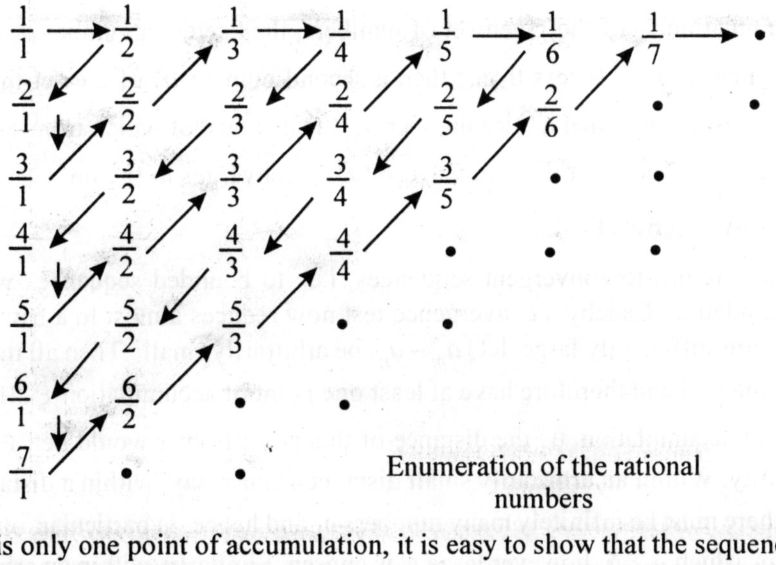

Enumeration of the rational numbers

If there is only one point of accumulation, it is easy to show that the sequence converges and that its limit is the number ξ. In fact, let us mark off any small interval about the point ξ. If infinitely many points of the sequence were outside this interval, they would have a limit point other than ξ, contrary to the hypothesis, whence only a finite number of the numbers of the sequence are exterior to the interval and, by definition, the sequence approaches ξ. On the other hand, if there are several points of accumulation, the sequence approaches no limit. The existence of a limit and the uniqueness of the point of accumulation of a bounded sequence of numbers are therefore equivalent ideas.

The case of the non-existence of a limit is to be regarded as the rule rather than the exception. For example, the sequence with the terms

$$a_{2n} = \frac{1}{n}, a_{2n-1} = 1 - , (n = 1, 2, \ldots)$$

has the two points of accumulation 0 and 1.

The aggregate of the positive rational numbers may be regarded as a sequence of numbers, in which the ordering by magnitude is, of course, completely destroyed. We arrive most easily at such an arrangement in a sequence by first writing down the rational numbers as shown and then running through this array as shown by the arrows, disregarding those numbers which have already been encountered (such as 2/4). The system of rational numbers obviously has all rational and irrational points as points of accumulation. It therefore forms a simple example of a sequence with an infinite number of points of accumulation.

By means of the concept of convergence, we can state the principle of the point of accumulation in a remarkable form which is often convenient for applications.

From every bounded infinite set of numbers, it is possible to choose an infinite sequence $a_1, a_2, a_3, \ldots$ which converges to a definite limit ξ. For this purpose, we have only to take a point of accumulation ξ of the given set of numbers, then select a number a_1 of the set the distance of which from ξ is less than , then a second number a_2 of the set the distance of which from x is less than , then a third number a_3, the distance of which from ξ is less than , and so on. We see at once that this sequence actually converges to the limit ξ.

Cauchy's Convergence Test

Let us now return to convergent sequences, i.e., to bounded sequences with only one point of accumulation. Cauchy's convergence test now reduces almost to a triviality. In fact, when m and n are sufficiently large, let $| a_m - a_n |$ be arbitrarily small . Then all the numbers a_n lie in a finite interval and therefore have at least one point of accumulation ξ. If there were a second point of accumulation η, the distance of this point from x would be $| \xi - \eta | = a$, a positive quantity. Within an arbitrarily small distance from x, say, within a distance less than $a/3$ from ξ, there must be infinitely many numbers a_n and hence, in particular, infinitely many numbers a_n for which $n > N$, however large N is chosen. Similarly, within an arbitrarily small distance from the point h, say, within a distance less than $a/3$ from η, there are infinitely many numbers a_m of the sequence; in particular, infinitely many numbers a_m, for which $m > N$. For these values a_n and a_m, it is true that $| a_m - a_n | > a/3$, and this relation is incompatible with the hypothesis that for sufficiently large values of N the difference $| a_n - a_m |$ is arbitrarily small provided that n and m are both greater than N. Hence there are not two distinct points of accumulation and Cauchy's test has been proved.

The Existence of Limits of Bounded Monotonic Sequences

It is equally easy to see that a bounded monotonic increasing or monotonic decreasing sequence of numbers must possess a limit. In fact, let the sequence be monotonic increasing and ξ be a point of accumulation of the sequence; such a point of accumulation must certainly exist. Then x must be greater than any number of the sequence. For if a number a_l of the sequence were equal to or greater than ξ, every number a_n for which $n > l + 1$ would satisfy the inequality $a_n > a_{l+1} > a_l$? ξ. Hence all numbers of the sequence, except at most the first (l + 1), would lie outside the interval of length $2(a_{l+1} - \xi)$ the mid-point of which is at the point ξ. However, this contradicts the assumption that ξ is a point of accumulation. Hence no numbers of the sequence, and, a no points of accumulation lie above x. Thus, if another point of accumulation h exists, we must have $\eta < \xi$. But if we repeat the above argument with η in place of ξ, we obtain $\xi < \eta$, which is a contradiction. Hence only one point of accumulation can exist and the convergence is proved. Naturally, an argument exactly analogous to this one applies to monotonic decreasing sequences.

As in 1.6.4, we can extend our statements about monotonic sequences by including the limiting case in which successive numbers of the sequence are equal, to one another. It is in this case better to speak of monotonic non-decreasing and monotonic non-increasing sequences, respectively. The theorem about the existence of a limit remains valid for such sequences.

Upper and Lower Points of Accumulation

Here we are going to discuss about upper and lower bounds and their properties.

Upper and Lower Bounds of a Set of Numbers

In the construction of subintervals, which led us to a point of accumulation ξ, we must at each step choose a subinterval containing infinitely many points of the set. Had we always chosen the last subinterval which contained an infinite number of points, we should have been led to a certain definite point of accumulation β. This point of accumulation β is called the upper point of accumulation or the upper limit of the set of numbers and is denoted by $\overline{\lim}$. It is that point of accumulation of the sequence which lies furthest to the right, i.e., it is quite possible that an infinite number of points of the sequence lie above β, but no matter how small the positive number ε may be, there are not an infinite number above $\beta + \varepsilon$.

If we had always chosen in the construction of the first of the intervals containing an infinite number of points of the set, we should again have arrived at a certain definite point of accumulation ξ. This point ξ is called the lower point of accumulation or lower limit of the set and is denoted by $\underline{\lim}$.

There may be infinitely many numbers of the set below α, but, no matter how small is the positive number *e*, there are only a finite number below $\alpha - \varepsilon$. The proofs of this facts is left for the students.

Neither the upper limit β nor the lower limit α need belong to the set. For example, for the set of numbers a_{2n} = l/*n*, a_{2n-1} = 2 –, these limits are α = 0 and β = 2, respectively, but the numbers 0 and 2 do not themselves occur in the set.

There is in this example no number of the set above β = 2. In this case, we say that β = 2 is also the upper bound M of the set, according to the following definition: M is called the least upper bound or simply the upper bound of a set of numbers, if: (1) there is no number of the set greater than M, but (2) there is a number of the set greater than M – ε for every positive number ε. The upper bound may coincide with the upper bound, as in the example above. But the set $a_n = 1 + \frac{1}{n}$ (n = 1, 2,...) shows that this is not necessarily the case. Here M = 2 and β = 1.

Every bounded set of numbers has a least upper bound. In fact, let β be the upper limit of the set. Either there are no numbers of the set larger than β or there are such numbers. In the first case, β is the least upper bound, since no numbers are above β and there are numbers arbitrarily close to β below it. In the second case, let α be a number of the set greater than β. There are only a finite number of numbers of the set equal to or greater than α, since otherwise there would be a point of accumulation above β, which is impossible. We therefore need only choose the greatest of these numbers; it will be the upper bound of the set.

We see that in any case M > β and we recognize the fact: If the upper bound of a set does not coincide with the upper limit, it must belong to the set and is an isolated point of the set.

Corresponding statements hold for the lower bound m; it is always equal to or less than a, and if m and a do not coincide, m belongs to the set and is an isolated point of the set.

Theorems on Continuous Functions

Greatest and Least Values of Continuous functions

A bounded, infinite set of numbers must possess a least upper bound M and a largest lower bound m. However, as we have seen, these numbers M and m do not necessarily belong to the set; as we say: The set does not necessarily have a greatest or a least value.

In view of this fact, the following theorem on continuous functions is by no means as obvious as it appears to be to simple intuition: Every function *f*(*x*), which is continuous in a closed interval $a \leq x \leq b$, assumes at least once a greatest and a least value or, as we say, it possesses a greatest and a least value.

This may easily be proved as follows: The values assumed by the continuous function *f*(*x*) in the interval *a* ? *x* ? *b* form a bounded set of numbers and therefore possess a least upper bound *M*. Otherwise, a sequence of numbers ξ_1,, ξ_2, ... , ξ_n, ... in our interval would exist for which *f*(ξ) increases beyond all bounds. This sequence would have at least one point of accumulation $\bar{\xi}$ in the interval. In that case, arbitrarily near to $\bar{\xi}$, there would always be numbers ξ_n of our sequence for which the expression $|f(\bar{\xi})-f(\xi_n|$ exceeds 1 (and, in fact, is arbitrarily large), that is, the function would be discontinuous at the point $\bar{\xi}$. Thus, a least upper bound M exists and hence either there is a point ξ such that $f(\xi) = M$, which would prove the statement, or there is a sequence of numbers x_1, x_2, ??? , x_n, ??? in the interval for which

$$\lim_{n\to\infty} f(x_n) = M.$$

According to the formulated principle of the point of accumulation, we can select a subsequence of the numbers ξ_n,, which converges to a limit ξ. Let us call this subsequence $\xi_1, \xi_2, \ldots, \xi_n$, so that

$$\lim_{n\to\infty} \xi_n = \xi$$

It is then certain that

$$\lim_{n\to\infty} f(\xi_n) = M.$$

Ott the other hand, the function has been assumed to be continuous in the interval and, in particular, at ξ, whence

$$\lim_{n\to\infty} f(\xi_n) = f(\xi).$$

Hence $f(\xi) = M$. The value M is therefore assumed by the function at a definite point ξ in the interior or on the boundary of the interval, as stated. An exactly similar discussion applies to the least value.

In general, the theorem about the greatest and least values of continuous functions does not remain true unless we expressly assume the interval to be closed, that is, unless we make the hypothesis of continuity also refer to the end-points. For example, the function $y = 1/x$ is continuous in the open interval $0 < x < \xi$. However, it has no greatest value, but has arbitrarily large values near $x = 0$. Similarly, it has no least value, but comes arbitrarily near 0 for sufficiently large values of x without ever assuming the value 0.

The Uniformity of Continuity

As we have already seen and as we shall further see, the continuity of a function $f(x)$ in a closed interval a ? x ? b leaves room for a variety of possibilities, which do not suggest themselves intuitively. For this reason, we shall give logically rigorous proofs of certain consequences of the idea of continuity which, from a na point of view, seem to be quite obvious. The definition of continuity simply states that there follows from the relation

$$\lim_{n\to\infty} \xi_n = \xi \;\Rightarrow\; \lim_{n\to\infty} f(\xi_n) = f(\xi).$$

We can also express this as follows: There correspond to every $\varepsilon > 0$ for each point ξ a number $\delta > 0$ such that $|f(x) - f(\xi)| < \varepsilon$ whenever $|x - \xi| < \delta$, provided that all the numbers x considered lie in the interval $a < x < b$.

For example, in the case of the function $y = cx$ (where $c > 0$), such a number δ is given by the relation $\delta = \varepsilon | c |$. For the function $y = f(x)$, we can find such a number as follows: We assume that $a = 0$ and $b = 1$, and ask ourselves how near to ξ the number x must lie in order that the expression $| x^2 - \xi^2 |$ may be less than ε. For this purpose, we write

$$|x^2 - \xi^2| = |x - \xi||x + \xi| \leq |x - \xi|(1 + \xi).$$

Hence, if we choose $\delta < \varepsilon/(1 + \xi)$, we can be sure that $| x^2 - \xi^2 | < \varepsilon$. We see in this example that the number δ found in this way depends not only on ε, but also on the point of the interval at which we are investigating the continuity of the function. However, if we give up the attempt to make the best possible choice of δ for each ξ, we can eliminate this dependence of δ on ξ. In fact, we need only replace ξ on the right hand side by the number 1 and thus obtain for δ the expression $\varepsilon/2$, which is smaller than the previous expression for δ, but serves equally well for all points ξ.

There arises now the question whether something similar does not hold for every function which is continuous in a closed interval, that is, we enquire whether it may not be possible to determine for each e a $d = d(e)$ which depends only on e and not on x such that the inequality

$$|f(x) - f(\xi)| < \in$$

holds, provided $| x - \xi | < \delta$ for all values of ξ at the same time (or, better expressed, uniformly with respect to ξ). As a matter of fact, this is possible merely as a consequence of the general definition of continuity without any additional assumptions. This fact, which first attracted attention late in the Nineteenth Century, is called the Theorem of the Uniform Continuity of Continuous Functions.

We shall prove this theorem indirectly, that is, we shall show that the assumption that a function $f(x)$ exists, which in a closed interval $a \leq x \leq b$ is continuous and yet not uniformly continuous, leads us to a contradiction. Uniform continuity means that, if we wish to make the difference $| f(u) - f(v) |$ less than an arbitrarily chosen positive number ε, the numbers u and v being chosen in the closed interval $a \leq x \leq b$, we need only choose u and v near enough to one another, namely, at a distance apart which is less than $\delta = \delta(\varepsilon)$; it is immaterial where in the interval the pair of numbers u, v is chosen. Now, if $f(x)$ were not uniformly continuous, there would exist a positive (perhaps very small) number a with the property: There corresponds to every number δ_n of an arbitrary sequence $\delta_1, \delta_2, \ldots$ of positive numbers tending to zero a pair of values u_n, v_n of the interval for which $| u_n - v_n | < \delta_n$ and $| f(u_n) - f(v_n) | > \alpha$. According to the principle of the point of accumulation, the numbers u_n must have a point of accumulation ξ and the numbers v_n must have the same point of accumulation. If we select an arbitrarily small interval $|x - \xi| < \delta$ about this point ξ, an infinite number of the pairs u_n, v_n will lie in this interval. But this contradicts the assumed continuity of $f(x)$ at the point ξ; in fact, by Cauchy's convergence test, this requires that the points x_2 and x_1 are near enough to ξ.

$$|f(x) - f(x_2)| < a.$$

The uniformity of the continuity is thus proved.

The theorem of uniformity of continuity does not hold for intervals which are not closed.

- Otherwise the point of accumulation need not belong to the interval.

For example, the function l/x is continuous in the half-open interval $0 < x < 1$, but it is not uniformly continuous, because no matter how small the length $\delta(< l)$ of an interval is chosen, the function will take values differing by any fixed number, say 1, in the interval, if only the interval lies near enough to the origin, say, $\delta/2 < x < 3\delta/2$. Of course, the non-uniformity of continuity is due to the fact that in the closed interval $0 \leq x \leq 1$ the function possesses at the origin a discontinuity.

Intermediate Value Theorem

A function $f(x)$, continuous in a closed interval $a \leq x \leq b$, which is negative for $x = a$ and positive for $x = b$ (or conversely), assumes the value 0 at least once in the interval.

Geometrically speaking, this theorem is trivial, since it merely states that a curve, which begins below the x-axis and ends above it, must cut the axis somewhere in between. Analytically speaking, the theorem is very easily proved. There are in the interval an infinite number of points for which $f(x) < 0$; in fact, due to the continuity of the function, this is true for an entire interval beginning at the point a. The set consisting of those points x for which $f(x) < 0$ has a least upper bound x which is greater than a. Since there are points in every neighbourhood of x for which $f(x) < 0$, we must have $f(x) < 0$ (whence, in particular, x ? b). However, it is impossible that $f(\xi) < 0$, (whence, in particular, $\xi < \beta$). However, it is impossible that $f(\xi) < 0$, because then $f(\xi)$ would be negative in a sufficiently small neighbourhood of ξ, including values of $x > \xi$, in contradiction to the assumption that ξ is the upper bound of the values of ξ for which $f(x) < 0$. Hence $f(\xi) = 0$ and our assertion is proved.

A slight generalization of our theorem is: If we assume that $f(a) = \alpha$ and $f(b) = \beta$ and if μ is any value between α and β, the continuous function $f(x)$ assumes the value $f(\mu)$ at least once in the interval, because the continuous function

$$\phi(x) = f(x) - \mu$$

will have different signs at the two ends of the interval and will therefore assume the value 0 somewhere inside it.

The Inverse of a Continuous Monotonic Function

If the continuous function $y = f(x)$ is monotonic in the interval $a < x < b$, it will assume each value between $f(a)$ and $f(b)$ once and only once; hence, if y describes the closed interval between the values $\alpha = f(a)$ and $\beta = f(b)$, there will correspond to each value of y exactly one value of x. We can therefore think of x as a single-valued function of y in this interval, i.e., the function $y = f(x)$ has a unique inverse. We assert that this inverse function $x = \phi(y)$ is also a continuous, monotonic function of y, as y varies within the interval between α and β.

The monotonic character of the inverse function $x = \phi(y)$ is obvious. In order to prove its continuity, we observe that it follows from the monotonic character of $f(x)$ that

$$|f(x_2) - f(x_1)| = |y_2 - y_1| > 0,$$

provided that x_1 and x_2 are distinct numbers of the interval. If h is a positive number less than $b - a$, the function

$$|f(x + h) - f(x)|$$

is continuous in the closed interval $a ? x ? b - h$, whence it has a least value

$$|(\xi + h) - f(\xi)| = a(h),$$

which by our preceding remark is not zero.* We conclude from this that, if x_1 and x_2 are two points in the interval for which $|x_1 - x_2| < h$, then

$$|f(x_1) - f(x_2)| \geq a(h).$$

However, this implies the continuity of the inverse function, because, if $|y_1 - y_2|$ falls below the positive number $\alpha(h)$, then we must have $|x_1 - x_2| < h$, whence, if a positive number e is given, we need only choose δ equal to $\alpha(\varepsilon)$ in order to ensure that for all values y for which $|y_1 - y_2| < \delta$ it is also true that $|\phi(y_1) - \phi(y_2)| < \varepsilon$.

* On account of the continuity of $f(x)$, naturally, $\alpha(h)$ itself tends to 0 with h.

Hence we have established the theorem: If the function $y = f(x)$ is continuous and monotonic in the interval $a < x < b$ and $f(a) = \alpha$, $f(b) = \beta$, then it has a single-valued inverse function $x = \phi(y)$, $\alpha < y < \beta$ and this inverse function is also continuous and monotonic.

Further Theorems on Continuous Functions

We leave it to the reader to prove the following almost trivial fact: A continuous function of a continuous function is itself continuous, i.e., if $\phi(x)$ is a function, continuous in the interval $a < x < b$, its functional values lie in the interval $\alpha < \phi < \beta$ and, in addition, if $f(\phi)$ is a continuous function of ϕ in this last interval, then $f(\phi(x))$ is a continuous function of x in the interval $a < x < b$. It has already been mentioned that the sum, difference and product of continuous functions are themselves continuous and that the quotient of continuous functions is continuous, provided that the denominator remains different from zero.

EXERCISES

1. Give the upper and lower bounds and upper and lower limits for the following sequences and state which belong to the sequence:

   (a) $\dfrac{6^n}{n!}, n = 1, 2, \ldots.$

   (b) $0, \dfrac{(-1)^n}{n!}, n = 1, 2, \ldots.$

   (c) $\dfrac{(-1)^n}{n} + \dfrac{n}{2n+1}, n = 1, 2, \ldots.$

   (d) $1 + \dfrac{(-1)^n}{n} + \dfrac{(-1)^n n}{2n+1}, n = 1, 2, \ldots.$

   (e) $\dfrac{1}{m^2} + \dfrac{1}{n^2}, m, n = 1, 2, \ldots.$

2. Prove that if $f(x)$ is continuous for $a \leq x \leq b$, then there exists for every $\varepsilon > 0$ a polygonal function $\phi(x)$ (i.e., a continuous function the graph of which consists of a finite number of rectilinear segments meeting at corners} such that $| f(x) - \phi(x) | < \varepsilon$ for every x in the interval.

3. Prove that every polygonal function $\phi(x)$ can be represented by a sum

   $\varphi(x) = a + bx + \sum c_i | x - x_i |,$

   where the x_i are the abscissa of the corners.

   Find a formula of this kind for the function $f(x)$, defined *by* the equations:

   $f(x) = 2x - 1 (0 \leqq x \leqq 2),$

   $f(x) = 5 - x (2 \leqq x \leqq 3),$

   $f(x) = x - 1 (3 \leqq x \leqq 5),$

   $f(x) = 4 (5 \leqq x \leqq 7).$

4. Find $\delta(\varepsilon)$ for the following functions such that $(f(x_1) - f(x_2) | < \delta(\varepsilon)$:

   (a) $f(x) = 2x^3, -1 \leqq x \leqq 1.$

   (b) $f(x) = x^n, -a \leqq x \leqq a.$

   *(c) $f(x) = \sqrt{1 - x^2} \; -1 \leqq x \leqq 1.$

5. The function $y = \sin(1/x)$ has a discontinuity in the interval. $0 < x < 1$. Prove that it is not uniformly continuous in that open interval.
6. A function $f(x)$ is defined for all values of x as follows:

   $f(x) = 0$ for all irrational values of x,

   $f(x) = 1/q$ for all rational $x = p/q$,

   where p/q is a fraction in its lowest terms (thus, for $x = 16/29, f(x) = 1/29$).

   Prove that $f(x)$ is continuous for all irrational and discontinuous for all rational values of

Polar Co-ordinates

Functions can be represented functions geometrically by means of curves. However, it is useful to recall that analytical geometry follows the reverse procedure; it begins with a curve given by some geometrical property and represents this curve by a function, for example, by a function which expresses one of the coordinates of a point of the curve in terms of the other coordinate. This point of view naturally leads us to consider, apart from rectangular coordinates, other systems of coordinates which may be better suited for the representation of curves which are given geometrically,

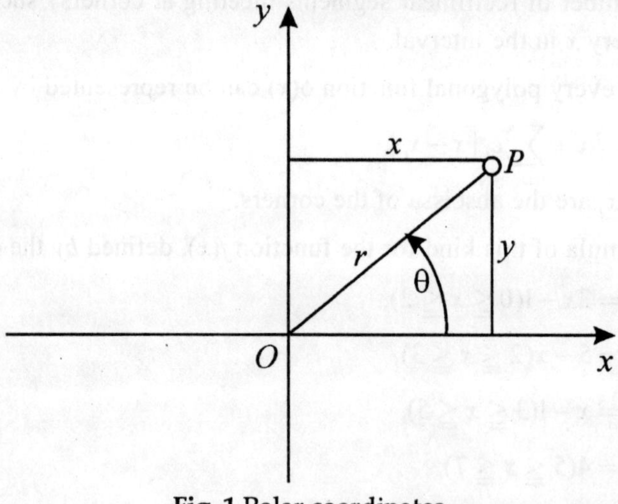

Fig. 1 Polar coordinates

The most important example is that of the polar coordinates r, θ, connected with the rectangular co-ordinates x, y of a point P by the equations

$$x = r\cos\theta, y = r\sin\theta, r^2 = x^2 + y^2, \tan\theta = \frac{y}{x},$$

and the geometrical interpretation of which follows from Figure 1.

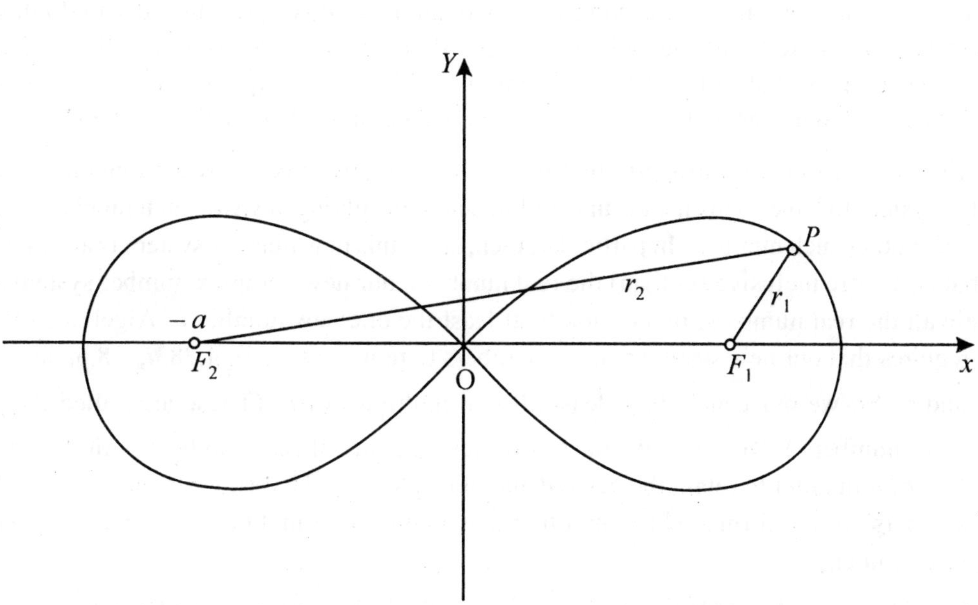

Fig. 2 Graph

Consider as an example the lemniscate. It is geometrically defined as the locus of all points for which the product of the distances r_1 and r_2 from the fixed points P_1 and P_2 with the rectangular co-ordinates $x = a$, $y = 0$ and $x = -a$, $y = 0$, respectively, has the constant value a (Fig. 2). Since

$$r_1^2 = (x - a)^2 + y^2, \; r_2^2 = (x + a)^2 + y^2,$$

a simple calculation yields the equation of the lemniscate in the form

$$(x^2 + y^2)^2 - 2a^2 (x^2 - y^2) = 0.$$

If we now introduce polar coordinates, we obtain

$$r^4 - 2a^2r^2 (\cos^2\theta - \sin^2\theta) = 0;$$

now, division by r and use of a simple trigonometric formula yields

$$r^2 = 2a^2 \cos 2\theta.$$

Thus, we see that the equation of the lemniscate in polar coordinates is simpler than in rectangular coordinates.

Complex Numbers

Does the process of creation of new "numbers" end with the real numbers? In spite of creating these successively larger number systems in order to have solutions to more and more equations, there are still equations that have no solution, the simplest being the innocent-looking $x^2 + 1 = 0$. Rather than simply declaring such equations to be unsolvable, why not extend the system of real numbers to include solutions to equations of this type? In order to do this, let us postulate the existence of such a larger system of numbers, which we will call the *complex numbers*, and agree that in this larger system, a solution to the equation $x^2 + 1 = 0$ will be denoted by the letter i. Must there be any other numbers in such a number system?

Algebraic closure is a property that we would certainly expect any set that calls itself a number system to have. This means that adding and multiplying any two such numbers should result in another such number. In particular, then, since this new number system is an extension of (that is, a more inclusive set than) the real numbers, our new complex number system must contain all the real numbers, in addition to at least the one new number i. Algebraic closure then requires that our new system contain numbers to represent $2i$, $3i$, $4.783i$, $-8.6i$, and even $\sqrt{2}i$ and πi. So we must include at least all real multiples of i. (These are called the *pure imaginary* numbers.) But then we must also include sums of real numbers with the various multiples of i, in order to retain the desired algebraic closure. That is, we must include at least the "numbers" of the form . Must any other new numbers be included in our new system of complex numbers? .

Consider a simple quadratic equation $x^2 + 1 = 0$. There is no real number which satisfies this equation. So there was a need to find a system which could answer to this problem. Euler used the symbol 'i' to denote $\sqrt{-1}$ to solve the above equation.

Complex number system consists of the set of all ordered pairs of real numbers (a, b) denoted by $a + ib$, where $i = \sqrt{-1}$.

Complex number

Imaginary Number

Square root of a negative number is known as an imaginary number. $\sqrt{-a}$, $a > 0$ is an imaginary number.

or

A number whose square is negative is known as an imaginary number.

$$\sqrt{-1}, \sqrt{-2}, \sqrt{-3}, \ldots$$

The symbol i,

$$i = \sqrt{-1}$$

We write,

$$\sqrt{-3} = \sqrt{3}\sqrt{-1} = \sqrt{3}i = i\sqrt{3}$$

$$\sqrt{-4} = 2i = i2$$

Powers of i,

$$i = \sqrt{-1}$$

$$i^2 = -1$$

$$i^3 = (i)^2\, i = -i$$

$$i^4 = (-1)^2 = 1$$

Let $p > 0$ be a positive integer such that $p > 4$. Let q be the quotient and r be the remainder, when p is divided by 4.

i.e., $p = 4q + r$ where $0 \leq r < 4$.

Hence $i^p = i^{4q+r} = i^{4q}.i^r = (i^4)^q.i^r = 1^q.i^r = i^r \quad (\because\ i^4 = 1)$

Complex Numbers

If x and y are real numbers, then $x + iy$ is called a complex number.

x is called the real part and y is called the imaginary part.

The complex number $x + iy$ is also written as an ordered pair (x, y) and is denoted by z.

i.e., $z = x + iy$

The positive value $\sqrt{x^2 + y^2}$ is called the modulus of Z and is denoted by $|Z|$.

i.e., $Z = x + iy$, then $|Z| = \sqrt{x^2 + y^2}$

In $Z = x + iy$, if $y = 0$, then Z is purely real and if $x = 0$, then Z is imaginary.

Example: $2 + 4i$, $6 - 5i$, $3 + \sqrt{2i}, \sqrt{3} - 2i$ are complex numbers.

Note: Set of real numbers is a proper subset of the set of Complex numbers.

Equality of Complex numbers

Two complex numbers are equal iff their corresponding real parts and imaginary parts are separately equal.

i.e., $a + ib = x + iy \Leftrightarrow a = x$ and $b = y$

Sum of two Complex numbers

If $Z_1 = a + ib$ and $Z_2 = x + iy$, then we define their sum as

$$Z_1 + Z_2 = (a + ib) + (x + iy)$$

$= (a + x) + i(b + y)$ which is a complex number.

Negative of a Complex number

If $Z = a + ib$, then Z is called the negative of Z.

$$-Z = -(a + ib)$$
$$= -a - ib$$

Additive Identity of the Complex Number

The complex number $0 + i0$ is the additive identity for the set of complex numbers.

$0 + i0$ is called the additive identity for the complex number.

$$Z + (0 + i\,0) = a + i\,b + (0 + i\,0)$$
$$= (a + 0) + i(b + 0)$$
$$= a + i\,b$$

Additive Inverse of a Complex Number

Let $Z = a + i\,b$ and $Z' = x + iy$ be the additive inverse of Z, then

$$(a + ib) + (x + iy) = 0 + i0$$
$$(a + x) + i(b + y) = 0 + i0$$
$$a + x = 0 \text{ and } b + iy = 0$$

$\Rightarrow x = -a$ and $y = -b$

$\therefore$ Additive inverse of $a + ib$ is $-a - ib$.

Product of Two Complex Numbers

Let $Z_1 = a + ib$ and $Z_2 = c + id$, then

$$Z_1Z_2 = (a + ib)(c + id)$$
$$= ac + iad + ibc + i^2db$$
$$= ac + i(ad + bc) - bd \quad (\because i^2 = -1)$$

= (ac – bd) + i(ad + bc) which is a

Multiplicative Identity of Complex Numbers

Let $Z = a + i\,b$ and $Z' = x + iy$, then

$$(a + ib)(x + iy) = a + ib$$
$$(ax - by) + i(ay + bx) = a + ib$$
$$ax - by = a \quad \text{... (i)}$$

and
$$ay + bx = b \quad \text{... (ii)}$$

Solving (*i*) and (*ii*), we have

$$x = 1, y = 0$$

Multiplicative identity is $1 + i0$.

Conjugate Complex Numbers

If $Z = a + ib$ is a complex number, then $a - ib$ is called the complex conjugate of $a + ib$ and the conjugate is denoted by $\overline{Z}$.

Remark:

The sum and product of two conjugate numbers is always real.

For $Z = a + ib$, $\overline{Z} = a - ib$

$$Z + \overline{Z} = (a + ib) + (a - ib)$$

$$= 2a + i(0) = 2a$$

$$Z.\overline{Z} = (a + ib)(a - ib)$$

$$= a^2 - (ib)^2$$

$$= a^2 - (-1)b^2 = a^2 + b^2$$

Hence $Z + \overline{Z}$ and $Z.\overline{Z}$ are both real.

Quotient of two non-zero Complex numbers

If $Z_1 = a + ib$ and $Z_2 = c + id$ are two complex numbers, then $\frac{Z_1}{Z_2}$, quotient is defined as

$$\frac{Z_1}{Z_2} = \frac{a+ib}{c+id} = \frac{a+ib}{(c+id)} \cdot \frac{c-id}{(c-id)}$$

$$\frac{(ac+bd)+i(bc-ad)}{c^2+d^2}$$

$$\left(\frac{ac+bd}{c^2+d^2}\right)+i\left(\frac{bc-ad}{c^2+d^2}\right)$$

If $Z_1 = 0 + i0$, then quotient is also $0 + i0$ if $Z_2 \neq 0$.

Reciprocal of a non-zero complex number or multiplicative inverse of a non-zero complex number

Let $Z = a + ib$, then $\frac{1}{Z} = \frac{1}{a+ib}$

$$\frac{1}{Z} = \frac{1}{a+ib}\frac{(a-ib)}{(a-ib)}$$

$$= \frac{a - ib}{a^2 - b^2}$$

$$= \left(\frac{a}{a^2 + b^2} - \frac{ib}{a^2 + b^2} \right)$$

Properties of Complex numbers

If Z, Z_1 and Z_2 are complex numbers, then

(i) $|Z| = |\overline{Z}|$

$$Z = x + iy \Rightarrow |Z| = \sqrt{x^2 + y^2}$$

$$\overline{Z} = x - iy \Rightarrow |\overline{Z}| = \sqrt{x^2 + (-y)^2} = \sqrt{x^2 + y^2}$$

(ii) $Z.\overline{Z} = |Z|^2$

$$(x + iy)(x - iy) = x^2 + y^2$$

$$= \left[\sqrt{x^2 + y^2} \right]^2$$

$$= |Z|^2$$

(iii) $|Z_1.Z_2| = |Z_1|.|Z_2|$

Let $Z = a + ib$, $Z_2 = c + id$

so that $|Z_1| = \sqrt{a^2 + b^2}, |Z_2| = \sqrt{c^2 + d^2}$

$$Z_1.Z_2 = (a + ib)(c + id)$$

$$= (ac - bd) + i(bc + ad)$$

$$|Z_1.Z_2| = \sqrt{(ac - bd)^2 + (bc + ad)^2}$$

$$= \sqrt{a^2c^2 + b^2d^2 + b^2c^2 + a^2d^2}$$

$$= \sqrt{a^2(c^2 + d^2) + b^2(c^2 + d^2)}$$

$$= \sqrt{a^2 + b^2}\sqrt{c^2 + d^2}$$

$$= |Z_1|.|Z_2|$$

(iv) $|kZ| = |k||Z| \forall k \in R$

Let $Z = a + ib$

$$k|Z| = k(a + ib) = ka + ikb$$

$$|kZ| = \sqrt{k^2a^2 + k^2b^2}$$

$$= |k|\sqrt{a^2 + b^2}$$

$$= |k||Z|$$

(v) $Z + \overline{Z} = 2R_e(Z), Z - \overline{Z} = a - ib.$

Let Z = a + ib then $\overline{Z}$ = a – ib.

$$Z + \overline{Z} = (a + ib) + (a - ib) Z - \overline{Z} = (a + ib) - (a - ib)$$

$$= 2a + i0 = (a - a) + i(b + b)$$

$$= 2a = 0 + i(2b)$$

$$= 2R_e(Z) = 2I_m(Z)$$

(vi) $|R_e(Z)| \le |Z|, |I_m(\overline{Z})| \le |Z|$

Let Z = a + ib, then $|Z| = \sqrt{a^2 + b^2}$

$$R_e(Z) = a$$

$a \le \sqrt{a^2 + b^2}$ ($\because$ Both terms in the radical sign are positive)

Hence $R_e(Z) \le |Z|$

$$I_m(Z) = b$$

$$b \le \sqrt{a^2 + b^2}$$

$$I_m(Z) \le |Z|$$

(vii) $|Z_1 + Z_2|^2 = +|Z_1|^2 + |Z_2|^2 + 2R_e(Z_1.\overline{Z}_2)$

LHS = $(Z_1 + Z_2)(\overline{Z_1 + Z_2})$

$$= (Z_1 + Z_2)(\overline{Z}_1 + \overline{Z}_2)$$

$$= Z_1\overline{Z}_1 + Z_1\overline{Z}_2 + Z_2\overline{Z}_1 + Z_2\overline{Z}_2$$

$$= Z_1\overline{Z}_1 + Z_2\overline{Z}_2 + Z_1\overline{Z}_2 + Z_2\overline{Z}_1$$

$$= |Z_1|^2 + |Z_2|^2 + 2R_e(Z_1.\overline{Z}_2)$$

= RHS

(viii) $|Z_1 + Z_2|^2 + |Z_1 - Z_2|^2 = 2(|Z_1|^2 + |Z_2|^2)$

$$|Z_1 + Z_2|^2 + |Z_1 - Z_2|^2$$

$$= |Z_1|^2 + |Z_2|^2 + 2R_e(Z_1.\overline{Z}_2) + |Z_1|^2 + |Z_2|^2 + 2R_e(Z_1(-\overline{Z}_2))$$

$$= 2|Z_1|^2 + 2|Z_2|^2 + 2R_e(Z_1.\overline{Z}_2) - 2R_e(Z_1\overline{Z}_2)$$

$$= 2|Z_1|^2 + 2|Z_2|^2$$

(ix) $\overline{(\overline{Z})} = Z$

Let $Z = a + ib$, then

$$\overline{Z} = \overline{(a - ib)} = (a + ib) = Z$$

(x) $(Z - \overline{Z})$ is zero or imaginary.

Let Z = a + ib then $\overline{Z} = a - ib$

$$Z - \overline{Z} = (a + ib) - \overline{(a - ib)}$$

$$= (a + ib) - a - ib)$$

$$= 2ib$$

which is zero when $b = 0$ and purely imaginary when $b \neq 0$.

(xi) $\overline{Z_1 + Z_2} = (\overline{Z}_1 + \overline{Z}_2)$

Let $Z_1 = a + ib$, $Z_2 = c + id$

So that $\overline{Z_1} = a - ib$ and $\overline{Z_2} = c - id$

$$Z_1 + Z_2 = (a + c) + i(b + d)$$

$$\overline{Z_1 + Z_2} = (a + c) - i(b + d)$$

$$= (a - ib) + (c - id)$$

$$= \overline{Z_1} + \overline{Z_2}$$

(xii) $\overline{Z_1.Z_2} = \overline{Z}_1.\overline{Z}_2$

Let $Z_1 = a + ib$, $Z_2 = c + id$

So that $\overline{Z}_1 = a - ib$ and $\overline{Z}_2 = c - id$

$$Z_1.Z_2 = (ac - bd) + i\,(ad + bc)$$

$$\overline{Z_1.Z_2} = (ac - bd) - i(ad + bc)$$

$$= c(a - ib) - id(a - ib)$$

$$= (a - ib)\,(c - id)$$

$$= \overline{Z}_1.\overline{Z}_2$$

(xiii) $|Z_1 + Z_2| \leq |Z_1| + |Z_2|$

$$= |Z_1 + Z_2|^2 = (Z_2 + Z_2)(\overline{Z_1 + Z_2}) \quad (\because |Z|^2 = Z.\overline{Z})$$

$$= (Z_1 + Z_2)(\overline{Z}_1 + \overline{Z}_2) \quad (\because \overline{Z_1 + Z_2} = \overline{Z}_1 + \overline{Z}_2)$$

$$= Z_1.\overline{Z}_1 + Z_1.\overline{Z}_2 + Z_2.\overline{Z}_1 + Z_2.\overline{Z}_2$$

$$= |Z_1|^2 + |Z_2|^2 + Z_1.\overline{Z}_2 + (\overline{Z_1\overline{Z}_2})$$

$$= |Z_1|^2 + |Z_2|^2 + 2R_e(Z_1.\overline{Z}_2) \quad (\because Z + \overline{Z} = 2R_e(Z))$$

$$\leq |Z_1|^2 + |Z_2|^2 + 2|Z_1.\overline{Z}_2| \quad (\because R_e(Z) \leq |Z|)$$

$$= |Z_1|^2 + |Z_2|^2 + 2|Z_1|.|\overline{Z}_2|$$

$$= |Z_1|^2 + |Z_2|^2 + 2|Z_1|.|\overline{Z}_2| \quad (\because |\overline{Z}_2| = |Z_2|)$$

$$= [|Z_1| + |Z_2|]^2$$

$\therefore$
$$|Z_1 + Z_2|^2 \leq [|Z_1| + |Z_2|]^2$$

$$|Z_1 + Z_2| \leq |Z_1| + |Z_2|$$

(xiv) $|Z_1 - Z_2| \geq |Z_1| - |Z_2|$

$$|Z_1| = |Z_1 - Z_2 + Z_2| \leq |Z_1 - Z_2| + |Z_2|$$

$$|Z_1| - |Z_2| \leq |Z_1| - |Z_2|$$

$$|Z_1 - Z_2| \geq |Z_1| - |Z_2|$$

Graphical Representation of Complex Numbers

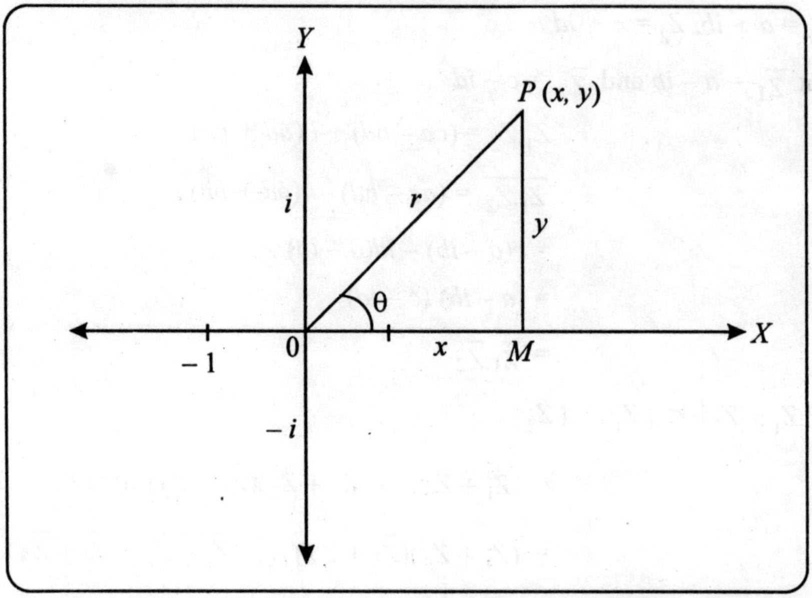

The complex number $Z = x + iy$ may be represented graphically by the point P whose rectangular co-ordinates are (x, y). Thus each point in the plane is associated with a complex number. In the figure, P defines $Z = x + iy$. It is customary to choose x-axis as real axis and y-axis as imaginary axis.

Draw PM perpendicular to x-axis.

In ΔPOM,

$$P\hat{O}M = \theta, \ OM = x, \ PM = y, \ OP = r$$

$$r = \sqrt{x^2 + y^2}, x = r\cos\theta, y = r\sin\theta, \tan\theta = \frac{y}{x}$$

Here the positive value of r is called the modulus of Z or absolute value of Z. ? is called the amplitude or argument of Z.

$$\therefore \quad r = |Z| = |x + iy| = \sqrt{x^2 + y^2}$$

$$\theta = ampZ = \arg Z = \tan^{-1}\frac{y}{x}$$

Geometrical Representation of a Complex Number, Argand Diagram

Since every complex number $z = x + iy$ is an order pair of real numbers (x, y), it can therefore be represented by a point $P(x, y)$ in the xy plane by taking the real part along the x-axis and the imaginary part along the y-axis.

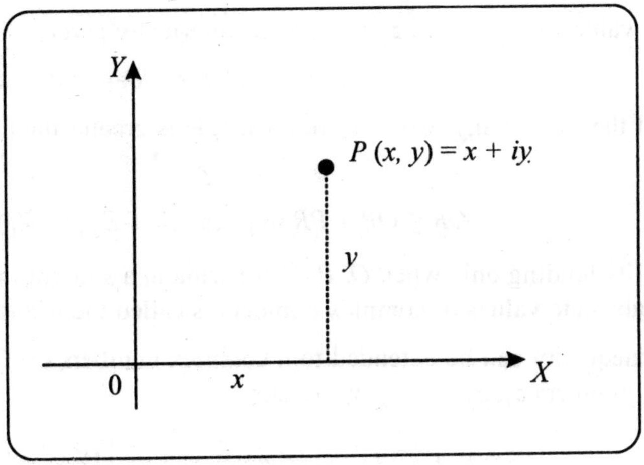

This representation of a complex number in a plane is called the argand diagram.

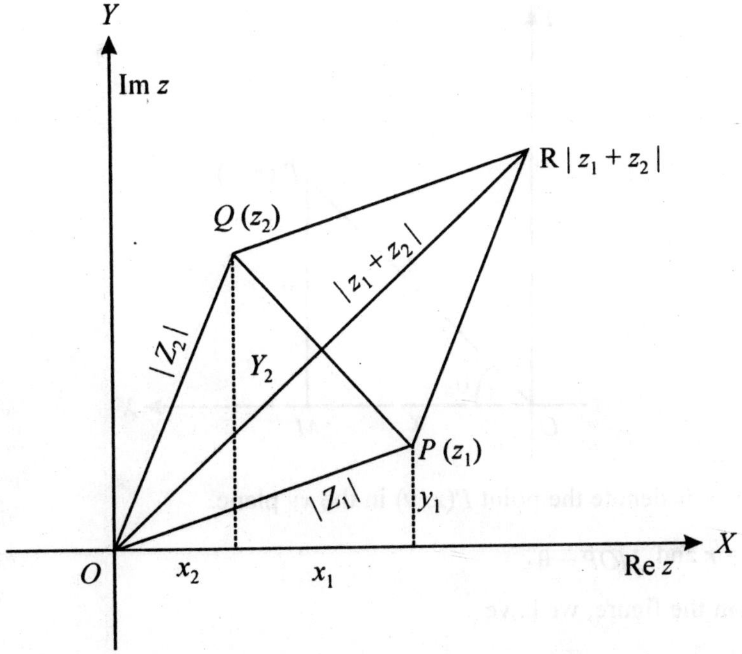

Let points P and Q in the complex plane represent two complex numbers $z_1 = x_1 + iy_1$ and $z_2 = x_2 + iy_2$, respectively, as shown in figure.

Join the origin O with the points P and Q and complete the parallelogram $OPRQ$.

It is clear from the figure that the coordinates of R are

$(x_1 + x_2, y_1 + y_2)$ and it represents the complex number

$(x_1 + x_2) + i(y_1 + y_2)$, i.e., $z_1 + z_2$.

The absolute values of z_1, z_2 and $z_1 + z_2$ are geometrically given by

$$|Z_1| = OP, |Z_2| = OQ = PR \text{ and } |Z_1 + Z_2| OR.$$

We know that the sum of any two sides of a triangle is greater than the third. Hence, in ?ORP, we have

$$OR \leq OP + PR \text{ implies } |Z_1 + Z_2| \leq |Z_1| + |Z_2|$$

with the equality holding only when *O, P, Q* are lying in a straight line. That is why this inequality for the absolute values of complex numbers is called the triangle inequality.

The triangle inequality can be extended to n complex numbers by finite induction, i.e., for any *n* complex numbers $z_1, z_2 \ldots\ldots, z_n$, we obtain

$$|Z_1 + Z_2 + \ldots + Z_n| \leq |Z_1| + |Z_2| + \ldots + |Z_n|$$

Polar form of a Complex Number

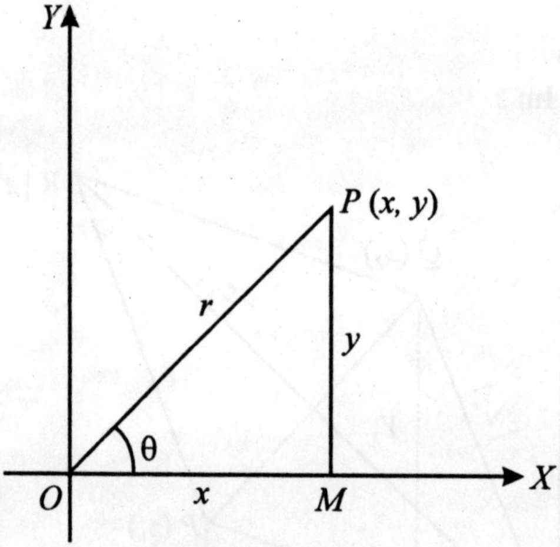

Let $z = x + iy$ denote the point $P(x, y)$ in the *xy* plane.

Let $OP = r$ and $M\hat{O}P = \theta$.

Then from the figure, we have

$$\cos\theta = \frac{x}{r} \text{ and } \sin\theta = \frac{y}{r}$$

$$\therefore z = x + iy = r\cos\theta + ir\sin\theta$$

$$z = r(\cos\theta + i\sin\theta)$$

$$r^2 = x^2 + y^2 \Rightarrow r = \sqrt{x^2 + y^2}$$

$z = r\{\cos\theta + i\sin\theta\}$ is called the polar form of the complex number.

$x + iy$ (which is in the cartesian form) where $r = \sqrt{x^2 + y^2}$ is called the modulus of the complex number z denoted by $|z|$ and $= \tan^{-1} y/x$ is called the amplitude or argument of the complex number z denoted by amp (z) or arg(z).

The value of θ is such that $-\pi < \theta \le \pi$, is called the principal value of the amplitude.

The general value of the amplitude is $\theta + 2n\pi$, where n is positive or negative integer and is the principal value of the amplitude.

To find the principal value of (amplitude) from the equation $\cos\theta = x/r$ and $\sin\theta = y/r$, the following table is useful.

$\cos\theta$	$\sin\theta$	θ
+ ve	+ ve	α (say)
+ ve	– ve	$-\alpha$ (say)
– ve	+ ve	$(\pi - \alpha)$
– ve	– ve	$-(\pi - \alpha)$

De Moivre's Theorem

(i) Let $Z_1 = (\cos\theta_1 + i\sin\theta_1)$

$$Z_2 = \cos\theta_2 + i\sin\theta_2$$

$$Z_1.Z_2 = (\cos\theta_1 + i\sin\theta_1)(\cos\theta_2 + i\sin\theta_2)$$

$$= \cos\theta_1\cos\theta_2 + i(\cos\theta_1\sin\theta_2 + \sin\theta_1\cos\theta_2) + i^2\sin\theta_1\sin\theta_2$$

$$= \cos\theta_1\cos\theta_2 - \sin\theta_1\sin\theta_2 + i(\sin\theta_1\cos\theta_2 + \cos\theta_1\sin\theta_2)$$

$$= \cos(\theta_1 + \theta_2) + i\sin(\theta_1 + \theta_2)$$

(ii) Let $Z_1 = (\cos\theta_1 + i\sin\theta_1)$

$$Z_2 = (\cos\theta_2 - i\sin\theta_2)$$

$$Z_1.Z_2 = (\cos\theta_1 + i\sin\theta_1)(\cos\theta_2 - i\sin\theta_2)$$

$$= (\cos\theta_1 \cos\theta_2 + \sin\theta_1 \sin\theta_2) + i(\sin\theta_1 \cos\sin\theta_2 - \cos\theta_1 \sin\theta_2)$$

$$= \cos(\theta_1 - \theta_2) + i\sin(\theta_1 - \theta_2)$$

(iii) Let $Z_1 = (\cos\theta_1 - i\sin\theta_1)$

$$Z_1.Z_2 = \cos\theta_1 \cos\theta_2 - \sin\theta_1 \sin\theta_2 - i(\sin\theta_1 \cos\theta_2 + \cos\theta_1 \sin\theta_2)$$

$$= \cos(\theta_1 + \theta_2) - i\sin(\theta_1 + \theta_2)$$

Statement of DeMoivre's Theorem

If '*n*' be any rational number, positive or negative, then

$$[\cos\theta + i\sin\theta]^n = \cos n\theta + i\sin n\theta$$

Proof:

Part I: When '*n*' is any integer, positive or negative then

$$[\cos\theta + i\sin\theta]^n = \cos n\theta + i\sin n\theta$$

Part II: When '*n*' is a fraction.

Part I:

Case 1:

When *n* is a positive integer.

By actual multiplication, we have

$$(\cos\theta_1 + i\sin\theta_1)(\cos\theta_2 + i\sin\theta_2) = \cos\overline{\theta_1 + \theta_2} + i\sin\overline{\theta_1 + \theta_2}$$

Again $(\cos\theta_1 + i\sin\theta_1)(\cos\theta_2 + i\sin\theta_2)(\cos\theta_3 + i\sin\theta_3)$

$$= [\cos(\theta_1 + \theta_2) + i\sin(\theta_1 + \theta_2)][\cos\theta_3 + i\sin\theta_3]$$

$$= \cos(\theta_1 + \theta_2 + \theta_3) + i\sin(\theta_1 + \theta_2 + \theta_3)$$

Similarly by the method of induction,

$$(\cos\theta_1 + i\sin\theta_1)(\cos\theta_2 + i\sin\theta_2) \ldots\ldots (\cos\theta_n + i\sin\theta_n)$$

$$= \cos(\theta_1 + \theta_2 + \theta_3 + \ldots + \theta_n) + i\sin(\theta_1 + \theta_2 + \theta_3 + \ldots + \theta_n)$$

In this result, if we put $\theta_1 = \theta_2 = \theta_3 = \ldots. = \theta_n = \theta$,

then we obtain, $(\cos\theta + i\sin\theta)^n = \cos n\theta + i\sin n\theta$

Case 2:

When 'n' is a negative integer.

Let $n = -m$ and $m > 0$

Now $(\cos\theta + i\sin\theta)^n = (\cos\theta + i\sin\theta)^{-m}$

$$= \frac{1}{(\cos\theta + i\sin\theta)^m}$$

$$= \frac{1}{\cos m\theta + i\sin m\theta} \text{ (by case 1)}$$

$$= \frac{\cos m\theta - i\sin m\theta}{(\cos m\theta - i\sin m\theta)(\cos m\theta - i\sin m\theta)}$$

$$= \frac{\cos m\theta - i\sin m\theta}{\cos^2 m\theta + \sin^2 m\theta}$$

$$= \cos m\theta - i\sin m\theta$$

$$= \cos(-m\theta) + i\sin(-m\theta)$$

$$= \cos n\theta + i\sin n\theta$$

Part II:

Case 3:

When 'n' is a fraction.

Let $n = \frac{p}{q}$

where 'p' is any positive integer and 'q' is any integer.

From Part I,

$$(\cos\theta + i\sin\theta)^p = (\cos p\theta + i\sin p\theta)$$

$$= \cos\frac{p}{q}.q\theta + i\sin\frac{p}{q}.q\theta$$

$$= \left[\cos\frac{p}{q}\theta + i\sin\frac{q}{p}\theta\right]^q$$

Now taking q^{th} root of both sides,

$$(\cos\theta + i\sin\theta)^{\frac{p}{q}} \text{ has one of its value } = \cos\frac{p}{q}\theta + i\sin\frac{p}{q}\theta$$

Now writing $\frac{p}{q} = n$,

$$= (\cos\theta + i\sin\theta)^n = \cos n\theta + i\sin n\theta$$

Hence the theorem.

Corollary 1

$$(\cos\theta + i\sin\theta)^{-m} = \cos(-m\theta) + i\sin(-m\theta)$$

$$= \cos m(-\theta) + i\sin m(-\theta)$$

$$= [\cos(-\theta) + i\sin(-\theta)]^m$$

$$= (\cos\theta - i\sin\theta)^m$$

Corollary 2

$$(\cos\theta + i\sin\theta)^{mn} = \cos mn\theta + i\sin mn\theta$$

$$= (\cos m\theta + i\sin m\theta)^n$$

$$= (\cos n\theta + i\sin n\theta)^m$$

Corollary 3

$$\frac{1}{\cos\theta + i\sin\theta} = (\cos\theta + i\sin\theta)^{-1} = \cos\theta - i\sin\theta$$

Find the qth Roots of a Complex Number

One of the most important applications of De Moivre's theorem is to find the qth roots of a complex number.

Let $z = x + iy$ be a complex number.

Let $z = r\ \{\cos\theta + i\sin\theta\}$ be its polar form.

And $z^{\frac{1}{q}}$ be the qth root of z.

We have $z = r\,\{\cos(2n\pi + \theta) + i \sin(2n\pi + \theta)\}$

[2n? + ? is the general amplitude]

$$\therefore z^{\frac{1}{q}} = [r\{\cos(2n\pi+\theta)+i\sin(2n\pi+\theta)\}]^{\frac{1}{q}}$$

$$r^{\frac{1}{q}}\left\{\cos\left(\frac{2n\pi+\theta}{q}\right)+i\sin\left(\frac{2n\pi+\theta}{q}\right)\right\}$$ (using De Moivre's theorem

where $n = 0, 1, 2, 3, \ldots\ldots, q - 1, \ldots$

Since we are finding the qth roots of a complex number, $z^{1/q}$ should satisfy the equation $x^q = z$ which is an algebraic equation of degree q and should have exactly q roots.

$$\therefore z^{\frac{1}{q}} = r^{\frac{1}{q}}\left\{\cos\left(\frac{2n\pi+\theta}{q}\right)+i\sin\left(\frac{2n\pi+\theta}{q}\right)\right\}$$

where $n = 0,1,2,3,\ldots\ldots, q - 1$

By giving values for $n = 0, 1, 2, 3, \ldots\ldots, q - 1$, we get q distinct q^{th} roots of z.

Cube Roots of Unity

Let $z^3 = 1 = 1\{\cos 0 + i \sin 0\}$

$$z^3 = 1\{\cos(2n\pi+0)+i\sin(2n\pi+0)\}$$

where 2n? + 0 is the general amplitude.

$$z = 1^{\frac{1}{3}}\{\cos 2n\pi + i\sin 2n\pi\}^{\frac{1}{3}}$$

$$z = \cos\frac{2n\pi}{3} + i\sin\frac{2n\pi}{3},$$ where n = 0, 1, 2 ...

when $n = 0$, $z = \cos 0 + i \sin 0 = 1$

$$n = 1, z = \cos\frac{2\pi}{3} + i\sin\frac{2\pi}{3} = \frac{-1}{2} + i\frac{\sqrt{3}}{2}$$

$$n = 2, z = \cos\frac{4\pi}{3} + i\sin\frac{4\pi}{3} = -\frac{1}{2} - i\frac{\sqrt{3}}{2}$$

$\therefore$ The cube roots of unity are $1, \frac{-1}{2} + i\frac{\sqrt{3}}{2}, \frac{-1}{2} - \frac{i\sqrt{3}}{2}$ which are usually denoted by 1, ?

Note 1:

$$\omega^3 = 1, (\omega^2)^3 = 1$$

Note 2:

$$1 = \omega^3 = \omega^6 = \omega^9 = \omega^{12} = \ldots$$

$$\omega = \omega^4 = \omega^7 = \omega^{10} = \omega^{13} = \ldots$$

$$\omega^2 = \omega^5 = \omega^8 = \omega^{11} = \omega^{14} = \ldots$$

Note 3:

We have $\omega^3 = 1$

$$\Rightarrow \quad (\omega - 1)(\omega^2 + \omega + 1) = 0$$

$$\Rightarrow \quad \omega^2 + \omega + 1 = 0$$

Note 4:

$$\frac{1}{\omega} = \omega^2 \text{ and } \frac{1}{\omega^2} = \omega$$

Fourth roots of unity

Let $z^4 = 1 = 1\{\cos 0 + i \sin 0\}$

Replacing the amplitude 0 by general amplitude 2*n*? + 0, we get

$$z^4 = \cos 2n\pi + i \sin 2n\pi$$

$$z = (\cos 2n\pi + i \sin 2n\pi)^{\frac{1}{4}}$$

$$z = \cos\frac{n\pi}{2} + i \sin\frac{n\pi}{2}, n = 0, 1, 2, 3 \ldots$$

when $n = 0$, $z = \cos 0 + i \sin 0 = 1$

$$n = 1, z = \cos\frac{\pi}{2} + i \sin\frac{\pi}{2} = i$$

$$n = 2, z = \cos\pi + i \sin\pi = -1$$

$$n = 3, z = \cos\frac{3\pi}{2} + i \sin\frac{3\pi}{2} = -i$$

$\therefore$ The fourth roots of unity are $\pm 1, \pm i$.

Note:

$$1 = 1^4 = i^8 = i^{12} = \ldots$$

$$i = i^5 = i^9 = i^3 = \ldots$$

$$-1 = i^2 = i^6 = i^{10} = \ldots$$

$$-i = i^3 = i^7 = i^{11} = \ldots$$

Argand diagram of the qth roots of a Complex number

All the q-th roots of z lie on a circle centred at the origin O and having radius equal to the real, positive q^{th} root of r. One of them has amplitude $\dfrac{\theta}{q}$ and others are uniformly spaced around the circle separated from others by an angle $\dfrac{2\pi}{q}$.

Representation of cube roots of unity

radius = 1

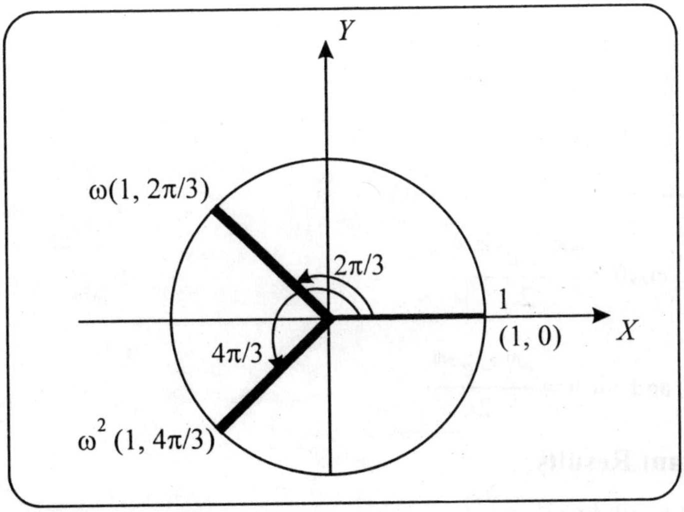

Representation of Fourth Roots of Unity

radius = 1

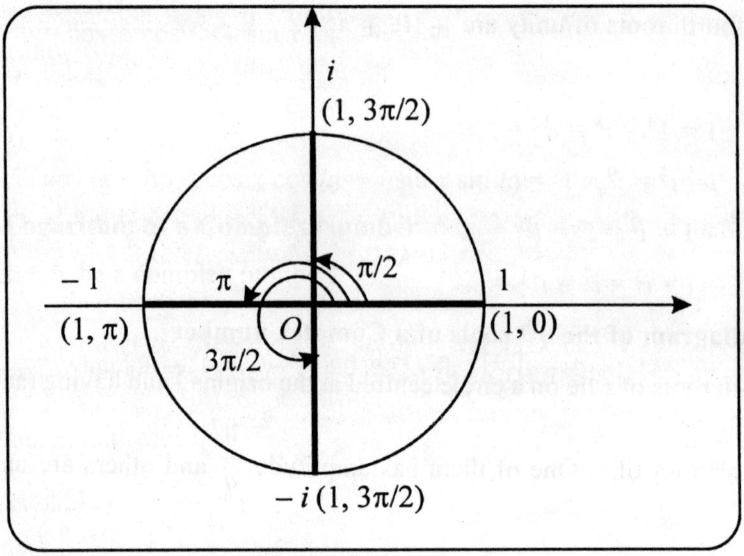

Exponential form of a Complex Number

If $z = x + iy$

Then $e^z = e^{x+iy}$

$= e^x e^{iy}$

$e^z = e^x \{\cos y + i \sin y\}$ is called the exponential form of the complex number.

We have,

$$e^{i\theta} = \cos\theta + i\sin\theta$$

and $e^{-i\theta} = \cos\theta - i\sin\theta$

$\therefore$ We get,

$$\cos\theta = \frac{\theta^{i\theta} + e^{-i\theta}}{2}$$

$$\text{and } \sin\theta = \frac{e^{i\theta} - e^{-i\theta}}{2i}$$

Some Important Results

(i) $\cos(ix) = \cos hx$

Proof:

We have $\cos\theta = \dfrac{e^{i\theta} + e^{-i\theta}}{2}$

Replacing θ by *ix*,

$$\cos(ix) = \frac{e^{i(ix)} + e^{-i(ix)}}{2}$$

$$= \frac{e^{-x} + e^{x}}{2}$$

$$= \cos hx$$

(ii) $\sin(ix) = i \sin hx$

Proof:

We have,

$$\sin\theta = \frac{e^{i\theta} - e^{-i\theta}}{2i}$$

Replacing θ by *ix*,

$$\sin(ix) = \frac{e^{i(ix)} - e^{-i(ix)}}{2i}$$

$$= \frac{1}{i}\left(\frac{e^{-x} - e^{x}}{2}\right)$$

$$= -\frac{1}{i}\sin hx$$

$$= i \sin hx$$

(iii) $\sin(x \pm iy) = \sin x \cos hy \pm i \sin hy \cos x$

Proof:

$$\sin(x + iy) = \sin x \cos(iy) + \cos x \sin(iy)$$

$$= \sin x \cos hy + \cos xi \sin hy$$

$$= \sin x \cos hy + i \cos x \sin hy$$

Similarly,

$$\sin(x - iy) = \sin x \cos hy - i \cos x \sin hy$$

(iv) $\cos(x \pm iy) = \cos x \cos hy \pm i \sin hy \sin x$

Proof:

$$\cos(x + iy) = \cos x \cos(iy) - \sin x \sin(iy)$$

$= \cos x.\cos hy - \sin x.i \sin hy$

$= \cos x \cos hy - i \sin x \sin hy$

Similarly,

$$\cos(x - iy) = \cos x \cos hy + i \sin x \sin hy$$

To find the Logarithm of a Complex Number

Let $z = x + iy$ be a complex number.

Let $z = re^{i\theta}$ where $r = \sqrt{x^2 + y^2}$ and $\theta = \tan^{-1}\left(\frac{y}{x}\right)$

Taking log on both sides,

$$\log_e z = \log(re^{i\theta})$$

$$= \log r + \log(e^{i\theta})$$

$$= \log_e z = \log(r + i\theta)$$

$$\therefore \log_e(x + iy) = \log\sqrt{x^2 + y^2} + i\tan^{-1}\frac{y}{x}$$

Note 1:

The above formula is useful for finding the logarithm of negative numbers also.

Note 2:

$$\log(-1) = \log(1 + i\pi)$$

$$\log(-1) = i\pi$$

EXERCISES

1. Plot the graphs of the functions:
   (i) $r = \sin\varphi$,
   (ii) $r = \cos 5\varphi$,
   (iii) $r = \varphi$,
   (iv) $r = \dfrac{1}{\cos(\varphi - \alpha)}$, α constant.
   (v) $r = \sin 6\varphi$,
2. Find the polar equation of
   (*a*) The circle with radius *a* with centre at the origin,

(*b*) The circle with radius a with centre (a, ϕ_0);

(*c*) The general straight line.

3. Use De Moivre's theorem to express cos 2θ and sin 2θ in terms of sin θ and cos θ; similarly, for cos 3θ, sin 3θ, cos 5θ, sin 5θ.

   Prove that cos $n\theta$ is a polynomial in cos θ, and also that, if n is odd, sin $n\theta$ is a polynomial in sin θ.

4. Work out the following expressions and state the modulus and amplitude of each of the numbers involved and of the answers:

   (*a*) -3, $2i$

   (*b*) $(4+4i)\left(\frac{1}{4}-\frac{1}{4}\sqrt{3i}\right)$

   (*c*) $(1 + i)(1 - i)$

   (*d*) $(\sqrt{3}-i)^2$,

   (*e*) $f^{-1/2}$,

   (*f*) $(1 + i)^{1/2}$,

   (*g*) $1^{1/3}$,

   (*h*) $(16i)^{1/4}$

6. Prove that, if $c = \cos\frac{2\pi}{n} + i\sin\frac{2\pi}{n}$, where n is an integer greater than 1,

$$g^v + e^{2v} + g^{2v} + \ldots + e^{nv} = \begin{cases} 0 \text{ if } n \text{ is not a factor of } v. \\ n \text{ if } n \text{ is not a factor of } v \end{cases}$$

Chapter 6

Topology of Complex Number

INTRODUCTION

In this section we investigate some basic ideas concerning sets of points in the plane. The first concept is that of a curve. Intuitively, we think of a curve as a piece of string placed on a flat surface in some type of meandering pattern. More formally, we define a curve to be the range of a continuous complex-valued function $z(t)$ defined on the interval $[a, b]$. That is, a curve C is the range of a function given by $z(t) - (x(t), y(t)) = x(t) + i\,y(t)$, for $a \le t \le b$, where both $x(t)$ and $y(t)$ are continuous real-valued functions. If both $x(t)$ and $y(t)$ are differentiable, we say that the curve is smooth. A curve for which $x(t)$ and $y(t)$ are differentiable except for a finite number of points is called piecewise smooth.

CURVE

In analytic geometry, a curve is continuous map from a one-dimensional space to an n-dimensional space. Loosely speaking, the word "curve" is often used to mean the function graph of a two- or three-dimensional curve. The simplest curves can be represented parametrically in n-dimensional space as

$$\begin{aligned} x_1 &= f_1(t) \\ x_2 &= f_2(t) \\ \vdots &= \\ x_n &= f_n(t) \end{aligned} \qquad ...(1)$$

Other simple curves can be simply defined only implicitly, i.e., in the form

$$f(x_1, x_2, ...) = 0 \qquad ...(2)$$

Definition (Curve).

We specify a curve C as

$$C : z(t) = x(t) + i\,y(t) \text{ for } a \le t \le b,$$

and say that $z(t)$ is a parameterization for the curve C. Notice that with this parametrization, we are specifying a direction the curve C, and we say that C is a curve that goes from the initial point $z(a) = (x(a), y(a)) = x(a) + i\,y(a)$ to the terminal point $z(b) = (x)(b), y(b) = x(b) + i\,y(b)$. If we had another function whose range was the same set of points as $z(t)$ but whose initial and final points were reversed, we would indicate the curve this function defines by $-C$.

Example 1 Find parameterizations for C and $-C$, where C is the straight line segment beginning at $z_0 = x_0 + i\,y_0$ and ending at $z_1 = x_1 + i\,y_1$.

Solution. Refer to Figure 1. The vector form of a line shows that the direction of C is $z_1 - z_0$. As z_0 is a point on C, its vector equation is

$$C : z(t) = z_0 + (z_1 - z_0)\,t \text{ for } 0 \le t \le 1, \text{ or} \qquad ...(3)$$

$$C : z(t) = (x_0 + x_1 - x_0)\,t) + i\,(y_0 + (y_1 - y_0)\,t) \text{ for } 0 \le t \le 1.$$

Clearly one parameterization for $-C$ is

$$-C : \gamma(t) = z_1 + (z_0 - z_1)\,t \text{ for } 0 \le t \le 1, \text{ or}$$

Remark. Note that $\gamma(t) = z(1-t)$, which illustrates a general principle: If C is a curve parameterized by $z(t)$ for $0 \le t \le 1$, then one parameterization for $-C$ will be $\gamma(t) = z(1-t)$, for $0 \le t \le 1$.

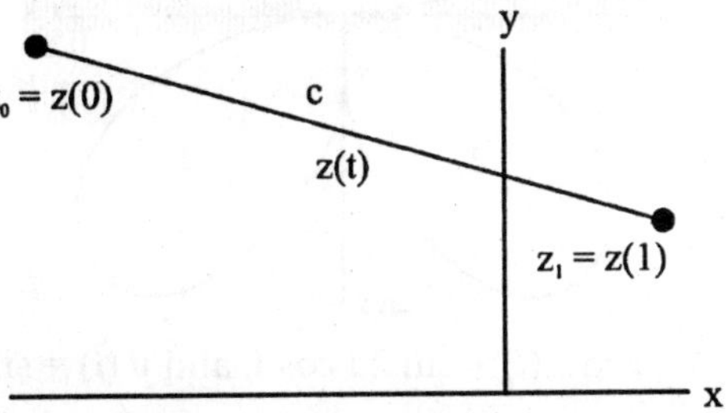

Fig. 1 The Straight-line Segment C Joining z_0 to z_1.

Example 2. Find the equation of the line segment with the initial point $z_0 = -3 + 2\,i$ and the terminal point $z_1 = 1 + i$.

Solution: Enter the initial point $z_0 = -3 + 2\,i$ and the terminal point $z_1 = 1 + i$ and construct the line segment.

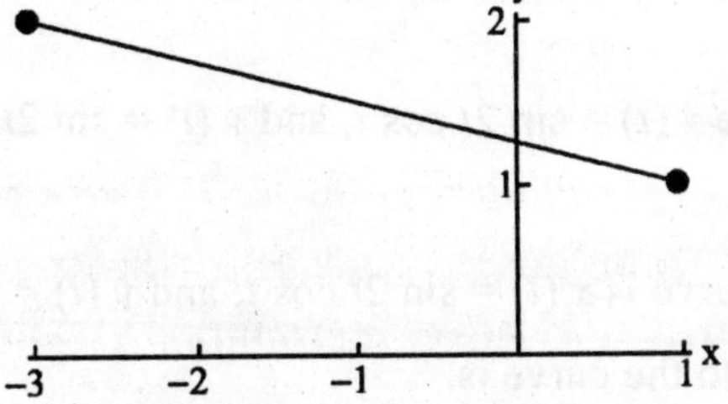

Equation of aline segment :

$$z[t] = -3 + i(2 - t) + 4t \text{ for } 0 \le t \le 1$$

Initial point (green) $z[0] = -3 + 2i$

Terminal point (red) $z[1] = 1 + i$

A curve C having the property that $z(a) = z(b)$ is said to be a closed curve. The line segment (3) is not a closed curve. The range of $z(t) = x(t) + i\,y(t)$, where $x(t) = \sin 2t \cos t$, and $y(t) = \sin 2t \sin t$ for $0 \le t \le 2\pi$ is a closed curve because $z(0) = (0, 0) = z(2\pi)$. The range of is the four-leaved rose shown in Figure 2 Note that, as t goes from 0 to $\frac{\pi}{2}$, the point is on leaf 1; from $\frac{\pi}{2}$ to π to, it is on leaf 2 ; between π and $\frac{3\pi}{2}$, it is on leaf 3; and finally, for t between $\frac{3\pi}{2}$ and 2π, it is on leaf 4.

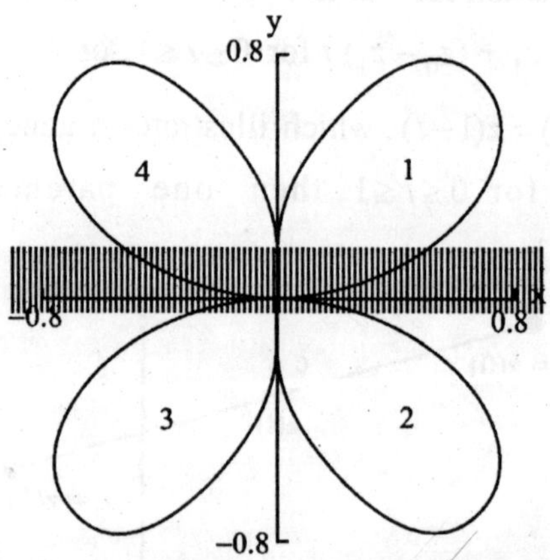

Fig. 2 The Curve $x(t) = \sin 2t \cos t$, and $y(t) = \sin 2t \sin t$ for $0 \le t \le 2\pi$, which forms a Four-leaved Rose.

Note further, in Figure 2, that at the point (0, 0), the curve has crossed over itself (at points other than those corresponding with $t = 0$ and $t = 2\pi$; we want to be able to distinguish when a curve does not cross over itself in this way. The curve C is called simple if it does not cross over itself, except possibly at its initial and terminal points. In other words, the curve C : $z(t)$, for $a \le t \le b$, is simple provided that $z(t_1) \ne z(t_2)$ whenever $t_1 \ne t_2$, except possibly when $t_1 = a$ and $t_2 = b$.

Example 3. The curve $x(t) = \sin 2t \cos t$, and $y(t) = \sin 2t \sin t$ for $0 \le t \le 2\pi$.

Solution:

The equation of the curve is $x(t) = \sin 2t \cos t$, and $y(t) = \sin 2t \sin t$.

Curve corresponding to the curve is.

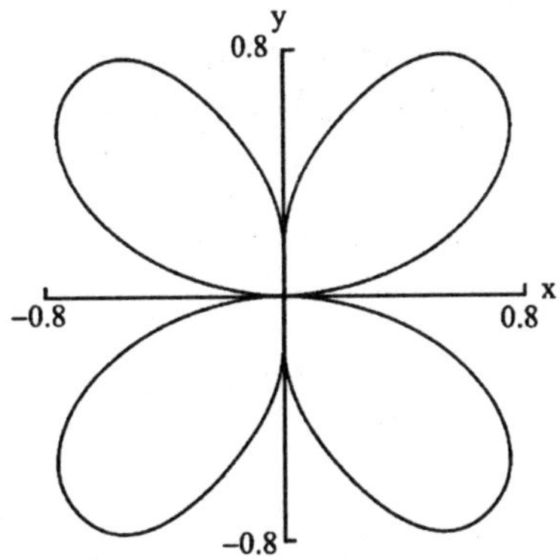

$x = \cos[t] \sin[2t]$

$y = \sin[t] \sin[2t]$ for $0 \le t \le 2\pi$

Remark. The curve looks like a "four leafed rose".

Example 4. Show that the circle C with center and radius can be parameterized to form a simple closed curve.

Solution: Note that the required parametrization is

$$C : z(t) = z_0 + Re^{it}$$

$$C : z(t) = x_0 + i\, y_0 + Re^{it}$$

$$C : z(t) = (x_0 + R\cos t) + i\,(y_0 + R\sin t),$$

for $0 \le t \le 2\pi$

Figure 3 shows that, as t varies from 0 to 2π, the circle is traversed counter clockwise. If you were traveling around the circle in this manner, its interior would be on your left. When a simple closed curve is paiametrized in this fashion, we say that the curve has a positive orientation. We will have more to say about this idea shortly.

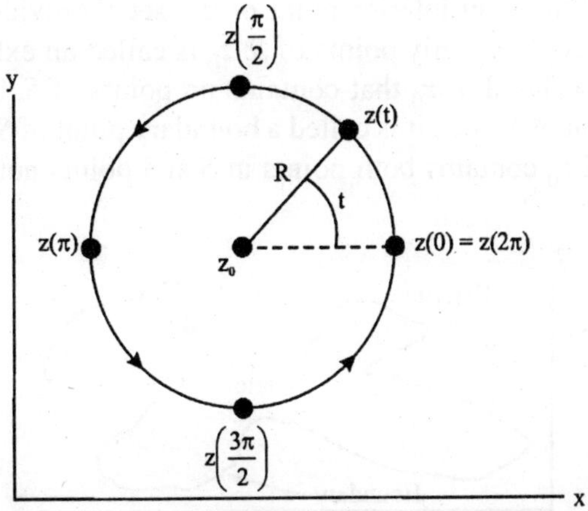

Fig. 3 The Simple Closed Curve $z(t) = z_0 + Re^{it}$, for $0 \le t \le 2\pi$.

We need to develop some vocabulary that will help describe sets of points in the plane. One fundamental idea is that of an $\in$neighborhood of the point z_0. It is the open disk of radius $\in >$ 0 about z_0 shown in Figure 4. Formally, it is the set of all points satisfying the inequality $D_\in(z_0) = \{z : | z - z_0 | < \in\}$ and is denoted $D_\in(z_0)$.

That is, (1-49) $D_\in(z_0) = \{z : | z - z_0 | < \in\}$.

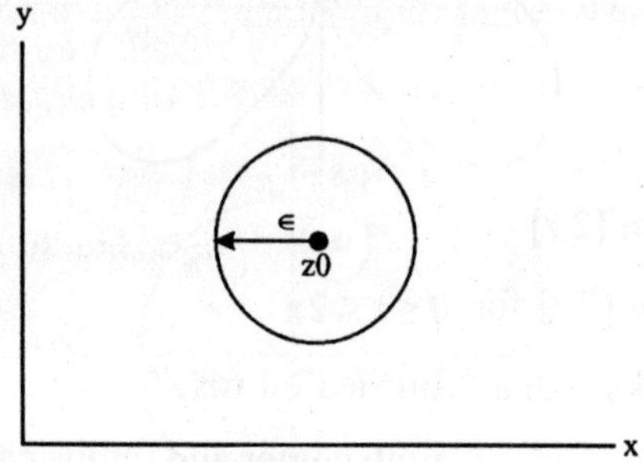

Fig. 4 An $\in$ Neighborhood of the Point z_0.

Example 5. The solution sets of the inequalities $| z | < 1$, $| z - i | < 2$, and $| z + 1 + 2 i | <$ 3 are neighborhoods of the points 0, i, and $- 1 - 2 i$, with radii 1, 2, and 3 respectively. They can also be expressed as D_1 (0), D_2 (i), and D_2 ($- 1 - 2 i$).

We also define $\overline{D_\in}(z_0)$, the closed disk of radius $\in$centered at z_0, and $D^*_\in(z_0)$, the punctured disk of radius $\in$centered at z_0 as

$$\overline{D_\in}(z_0) = \{z : | z - z_0 | \leq \in\}, \text{ and} \quad ...(4)$$

$$D^*_\in(z_0) = \{z : 0 < | z - z_0 | < \in\}. \quad ...(5)$$

Definitions (Interior Point, Exterior Point, Boundary Point).

The point z_0 is said to be an interior point of the set S provided that there exists an $\in$ neighborhood of z_0 that contains only points of S; z_0 is called an exterior point of the set S if there exists an $\in$ neighborhood of z_0 that contains no points of S. If z_0 is neither an interior point nor an exterior point of S, then it is called a boundary point of S and has the property that each $\in$ neighborhood of z_0 contains both points in S and points not in S. Figure 5 illustrates this situation.

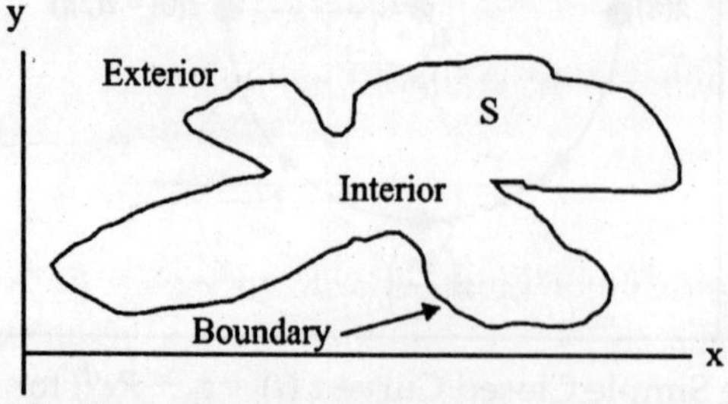

Fig. 5 The Interior, Exterior, and Boundary of a set S.

The boundary of $D_R(z_0)$ is the circle depicted in Figure 6. We denote this circle $C_R(z_0)$ and refer to it as the circle of radius R centered at z_0. Thus

$$C_R(z_0) = \{z : |z - z_0| R\}. \quad ...(6)$$

We use the notation $C_R^+(z_0)$ to indicate that the parametrization we chose for this simple closed curve resulted in a positive orientation; $C_R^-(z_0)$ denotes the same circle, but with a negative orientation. (In both cases, counterclockwise denotes the positive direction.) Using notation that we have already introduced, we get $C_R^-(z_0) = -C_R^+(z_0)$.

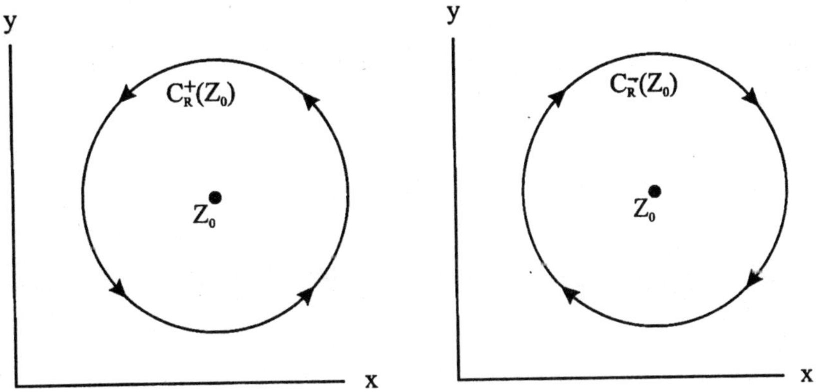

Fig. 6 The Positively Oriented Curve $C_R^+(z_0)$ and its Opposite $C_R^-(z_0) = -C_R^+(z_0)$.

Example 6. Let S = $\{z : |z| < 1\}$.

(a) Find the interior of *S*. **(b)** Find the exterior of *S*. **(c)** Find boundary of S.

Solution: We show that every point of *S* is an interior point of *S*. Let be a point of *S*. Then $|z_0| < 1$, and we can choose $\in = 1 - |z_0| > 0$. We claim that $D_\in(z_0) c\, S$. If $z \in D_\in(z_0)$, then

$$|z| = |z - z_0 + z_0| \le |z - z_0| + |z_0| < \in + |z_0| = 1 - |z_0| + |z_0| = 1$$

Hence the $\in$ neighborhood of z_0 is contained in *S*, which shows that z_0 is an interior point of *S*.

It follows that the interior of *S* is the set *S* itself.

Similarly, it can be shown that the exterior of *S* is the set $\{z : |z| > 1\}$. The boundary of *S* is the unit circle $C_1(0) = \{z : |z| = 1\}$. This condition is true because if $z_0 = e^{i\theta_0}$ is any point on the circle, then any $\in$-neighborhood of z_0 will contain the point $\left(1 - \frac{\in}{2}\right)e^{i\theta_0}$, which belongs to *S*, and $\left(1 + \frac{\in}{2}\right)e^{i\theta_0}$, which does not belong to *S*. We leave the details as an exercise.

Solution :

(a). Find the interior of *S*.

Let z_0 be a point of *S*. Then $|z_0| < 1$ so that we can choose $\in = 1 - |z_0| > 0$. If *z* lies in the disk $|z - z_0| < \in$, then

$$|z| = |z_0 + z - z_0| \le |z_0| + |z - z_0| < |z_0| + \in < 1.$$

Hence the $\in$-neighborhood of z_0 is contained in *S*, and z_0 is an interior point of *S*. It follows that the interior of *S* is the open unit disk.

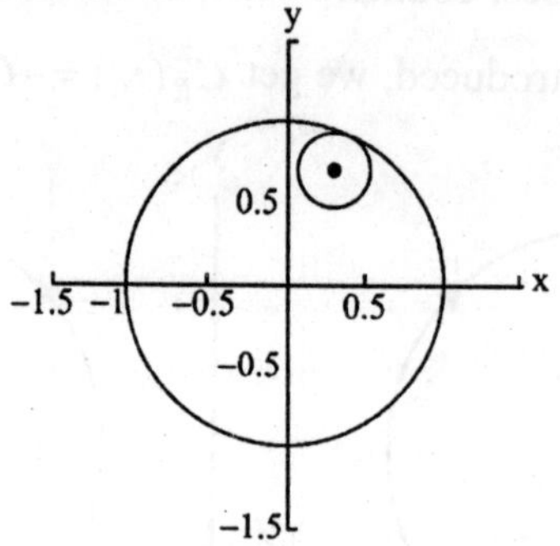

The $\in$-neighbourhood of z_0 is contained in $S = \{z : |z| < 1\}$.

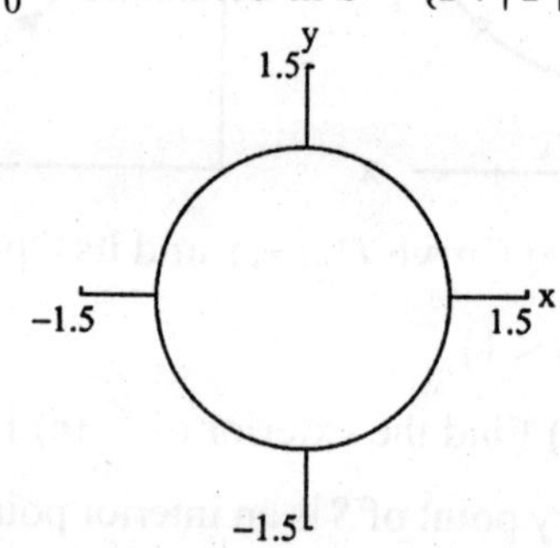

The interior of the set *S* is the unit disk $\{z : |z| < 1\}$.

(b). Find the exterior off *S*.

If $|z_0| > 1$ then a similar argument will produce an $\in$-neighborhood of z_0 that lies entirely in the region $|z| > 1$, hence z_0 is an exterior point of *S*. It follows that the exterior of *S* is the open region $|z| > 1$.

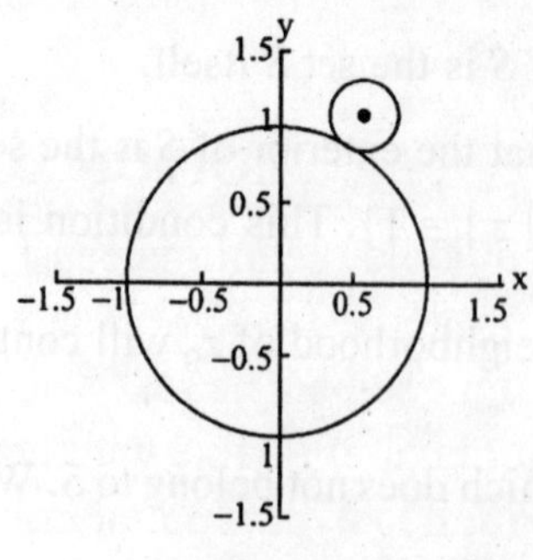

The $\in$-neighborhood of z_0 is contained in the region $|z| > 1$.

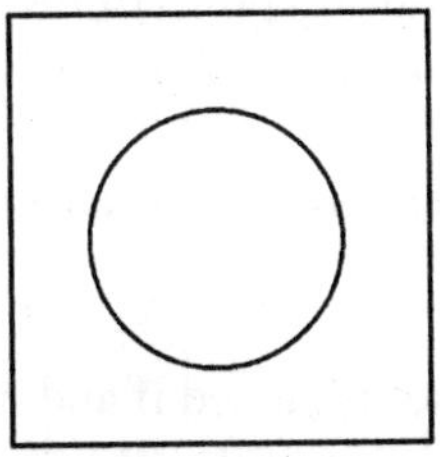

The exterior of the set S is the open region $\{z : |z| > 1\}$.

(c). Find the boundary of S.

If $z_0 = e^{i\theta_0}$ is any point on the unit circle, then any $\in$-neighborhood of z_0 will contain the point $\left(1-\frac{\in}{2}\right)e^{i\theta_0}$, which belongs to S, and $\left(1+\frac{\in}{2}\right)e^{i\theta_0}$ which does not belong to S. It follows that the boundary of S is the unit circle $|z| = 1$.

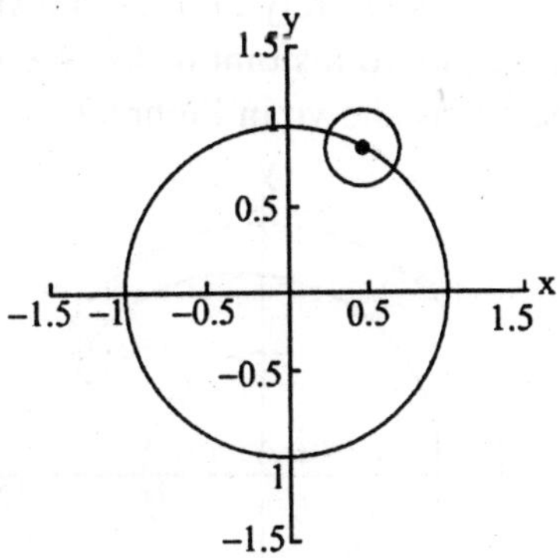

Every $\in$-neighborhood of z_0 contains points of S and points that do not belong to S.

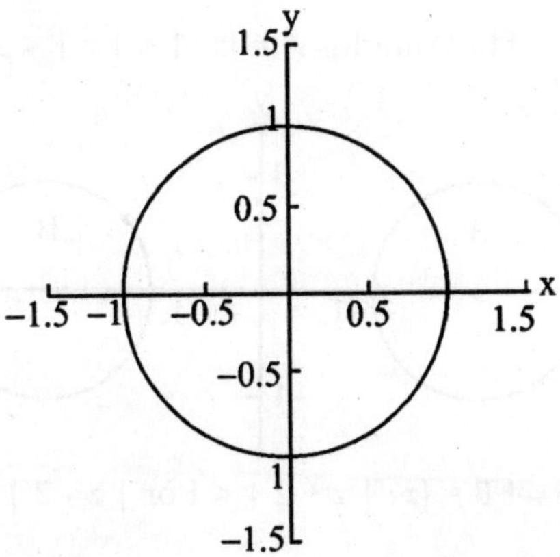

The boundary of the set $S = \{z : |z| < 1\}$ is the unit circle $\{z : |z| = 1\}$.

The point z_0 is called an accumulation point of the set S if, for each $\in$, the punctured disk contains at least one point of S. We ask you to show in the exercises that the set of accumulation points of $D_1(0)$ is $\overline{D_1}(0)$, and that there is only one accumulation point of $S = \left\{\frac{i}{n} : n = 1, 2, ...\right\}$, namely, the point 0.

We also ask you to prove that a set is closed if and only if it contains all of its accumulation points.

A set S is called an open set if every point of S is an interior point of S. Thus, Example 6 shows that $D_1(0)$ is open. A set S s called a closed set if it contains all its boundary points. A set S is said to be a connected set if every pair of points z_1 and z_2 contained in S can be joined by a curve that lies entirely in S. Roughly speaking, a connected set consists of a "single piece." The unit disk $D_1(0) = \{z : |z| < 1\}$ is a connected open set. We ask you to verify in the exercises that, if z_1 and z_2 lie in $D_1(0)$, then the straight-line segment joining them lies entirely in $D_1(0)$. The annulus $A = \{z : 1 < |z| < 2\}$ is a connected open set because any two points in A can be joined by a curve C that lies entirely in A, as shown in Figure 6. The set $B = \{z : |z + 2| < 1 \text{ or } |z - 2| < 1\}$ consists of two disjoint disks. We leave it as an exercise for you to show that the set is not connected, as shown in Figure 7.

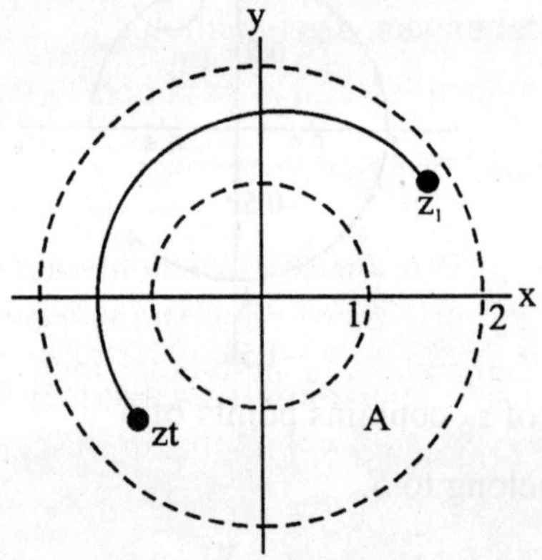

Fig. 7 The Annulus $A = \{z : 1 < |z| < 2\}$ is a Connected set.

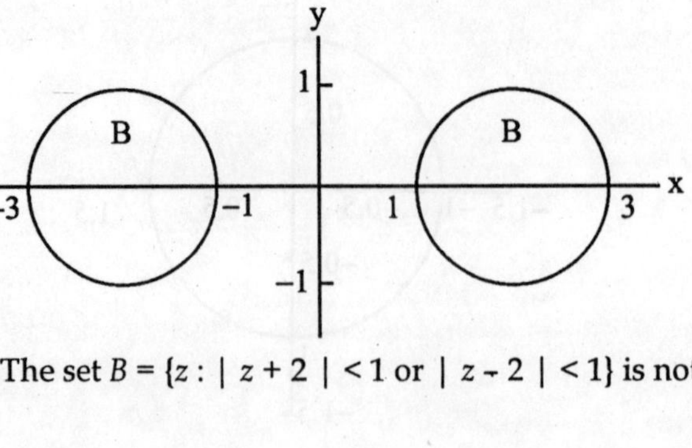

Fig. 8 The set $B = \{z : |z + 2| < 1 \text{ or } |z - 2| < 1\}$ is not a connected set.

We call a connected open set a domain. In the exercises we ask you to show that the open unit disk $D_1(0) = \{z : | z || < 1\}$ is a domain and that the closed unit disk $\overline{D_1}(0) = \{z : | z | \leq 1\}$ is not a domain. The term domain is a noun and is a type of set. In Chapter 2 we note that it also refers to the set of points on which a function is defined. In the latter context, it does not necessarily mean a connected open set.

A domain, together with some, none, or all its boundary points, is called a region. For example, the horizontal strip $\{z : 1 < \text{In}(z)\}$ is a region. A set formed by taking the union of a domain and its boundary is called a closed region; thus $\{z : 1 \leq \text{In}(z) \leq 2\}$ is a closed region. A set S is said to be a bounded set if it can be completely contained in some closed disk, that is, if there exists an $R > 0$ such that for each z in S we have $| z | \leq R$. The rectangle given by $\{z : | x | \leq 4 \text{ and } | y | \leq 3\}$ is bounded because it is contained inside the disk $\overline{D_5}(0) = \{z : | z | \leq 5\}$. A set that cannot be enclosed by any closed disk is called an unbounded set.

We mentioned earlier that a simple closed curve is positively oriented if its interior is on the left when the curve is traversed.

JORDAN CURVE THEOREM

If J is a simple closed curve in $\mathbb{R}^2$, then the Jordan curve theorem, also called the Jordan-Brouwer theorem states that $\mathbb{R}^2 - J$ has two components (an "inside" and "outside"), with J the boundary of each.

The Jordan curve theorem is a standard result in algebraic topology with a rich history.

Theorem (Jordan Curve Theorem).

The compliment of any simple closed curve C can be partitioned into two mutually exclusive domains I and E in such a way that I is bounded, E is unbounded, and C is the boundary for both I and E. In addition $I \cup E \cup C$ is the entire complex plane. The domain I is called the interior of C, and the domain E is called the exterior of C.

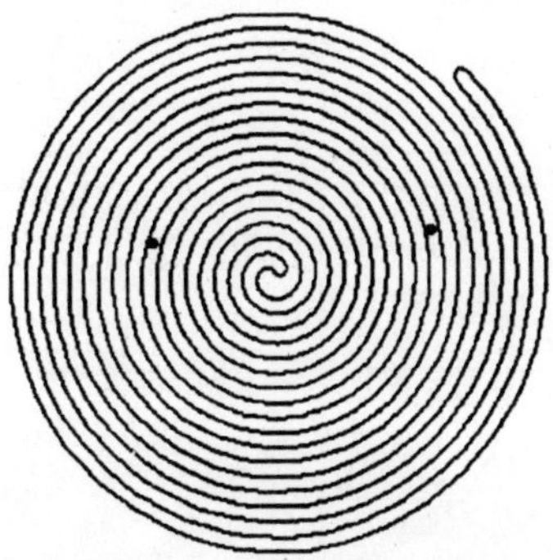

Fig. 9 Are the Points z_1 (blue), and z_2 (red) in the Interior or Exterior of this Simple Closed Curve?

EXERCISE

1. Let S be the open set consisting of all points z such that $|z| < 1$ *or* $|z-2| < 1$. State why S is not connected.
2. Show that a set S is open if and only if each point in S is an interior point.
3. Determine the accumulation points of each of the following sets:

   (a) $z_n = i^n \, (n = 1, 2, ...)$;

   (b) $z_n = i^n / n \, (n = 1, 2, ...)$;

   (c) $0 \le \arg z < \pi/2 \, (z \ne 0)$;

   (d) $z_n = (-1)^n (1+i)(n-1)/2 \, (n = 1, 2, ...)$.

Ans. (a) None; (b) 0; (c) $\pm(1+i)$.

4. Prove that if a set contatins each of its accumulation points, then it must be a closed set.
5. Show that any point z_0 of a domain is an accumulation point of that domain.
6. Prove that a finite set of points z_1, z_2, ..., z_n cannot have any accumulation points.
7. Suppose f is analytic on a connected open set D, and $f(z) = 0$ for all $z\hat{I}D$. Prove that f is constant.
8. Suppose D is a bounded domain. For any e > 0, let D_e denote the larger domain containing D and every point in C whose distance from D is less than e. Give an example of a simply connected domain D such that D_e is not simply connected for any e > 0.

Chapter 7

Functions of Complex Number

INTRODUCTION

We have already the topics such as Set, Relations, Functions on real number. Now we are going to discuss about functions of Complex numbers.

Once we have mastered using i as an exponent, we can advance to functions of complex numbers. Exponential functions, of course, are simply functions which use i as an exponent. In thgis chapter we are going to learn more about other than exponential functions such as polynomial functions, rational functions and so on.

Polynomial Functions

It is defined by $w = a_0 z^n + a_1 z^{n-1} + \ldots + a_{n-2} z^2 + a_{n-1} z + a_n = P(z)$

where the coefficients $a_0 \neq 0, a_1, \ldots, a_{n-2}, a_{n-1}$ and a_n are all constants and n is a positive integer called degree of the polynomial P(z).

The transformation $w = az + b$ is called a linear transformation.

Rational Algebraic Function

It is defined by $w = \dfrac{P(z)}{Q(z)}$

Where P(z) and Q(z) are polynomials. We sometimes call as a rational transformations

The special case $w = \dfrac{az+b}{cz+d}$, where $ad - bc \neq 0$ is often called a bilinear transformation.

Exponential Functions

It is defined by

$$z = x + iy = r(\cos\theta + i\sin\theta) = re^{i\theta} = re^{i(\theta + 2n\pi)}$$

where n is an integer.

Common Operations of Complex Numbers

$$z_1 z_1 = r_1 e^{i\theta_1} r_2 e^{i\theta_2} = r_1 r_2 e^{i(\theta_1+\theta_2)}$$

$$z_1 / z_2 = r_1 e^{i\theta_1} / r_2 e^{i\theta_2} = \frac{r_1}{r_2} e^{i(\theta_1-\theta_2)}$$

$$z^p = \left(re^{i\theta}\right)^p = r^p e^{ip\theta}$$

$$\sqrt[p]{z} = \sqrt[p]{re^{i\theta}} = \left(re^{i\theta}\right)^{\frac{1}{p}} = \left[re^{i(\theta+2n\pi)}\right]^{\frac{1}{p}} = \sqrt[p]{r} e^{\frac{i(\theta+2n\pi)}{p}}$$

where n is an integer.

Logarithm Functions

Since a complex number can be written in exponential form as re^{ip}, a logarithm is almost immediate. We get

$$\ln z = \ln\left(re^{i\theta}\right) = \ln\left(re^{i(\theta+2n\pi)}\right) = \ln r + i\theta + 2n\pi i$$

which gives the principal value of the natural logarithm. But the angle P is not unique, as it can vary by multiples of 2π. Therefore, the formula for the natural logarithm of a complex number is

$$\ln z = \ln\left(re^{i\theta}\right) = \ln\left(re^{i(\theta+2n\pi)}\right) = \ln r + i\theta + 2n\pi i$$

For example, we can find the natural logarithm of i. Its principal value is

$$\ln i = \ln\left(e^{i\pi/2}\right) = \frac{\pi}{2} i$$

Another interesting example is the natural logarithm of negative one. Its principal value is $\ln(-1) = \ln(1e^{\pi}) = \pi i$. (And you thought you couldn't take logarithms of negative numbers! You can, but the answers are not real numbers.)

We can use the change-of-base formula to find other base logarithms (including complex bases). For example,

$$\log_i e = \frac{\ln e}{\ln i} = \frac{1}{\frac{\pi}{2} i} = -\frac{2i}{\pi}.$$

Trigonometric Functions

From formula $e^{ix} = \cos x + i \sin x$, we can immediately obtain some formulas for the sine and cosine functions, in terms of complex exponentials. These formulas are often used as definitions of the trigonometric functions for complex numbers.

$$\cos x = \frac{e^{ix} + e^{-ix}}{2}, \sin x = \frac{e^{ix} - e^{-ix}}{2i}$$

Substituting the quantity ix in place of the variable x, we can produce formulas for pure imaginary numbers. Notice that the results are related to the hyperbolic functions.

$$\cos ix = \frac{e^{x} + e^{-x}}{2} = \cosh x, \sin ix = \frac{e^{-x} - e^{x}}{2i} i \sinh x$$

Then, using the formulas for the sine and cosine of the sum of two angles, we can obtain formulas for the sine and cosine of complex numbers.

$\cos (a + bi) = \cos a \cosh b - i \sin a \sinh b$

$\sin (a + bi) = \sin a \cosh b + i \cos a$ singh b

As an example, consider the cosine of i.

We get

$$\cos i = \cos 0 \cosh 1 - i \sin 0 \sinh 1 = \cosh 1 = \frac{e^{1} + e^{-1}}{2} \approx 1.5431$$

Hyperbolic Functions

Using the connection between hyperbolic functions and trigonometric functions, the results for hyperbolic functions are almost immediate.

$\cosh(a + bi) = \cosh a \cos b + i \sinh a \sin b$

$\sinh(a + bi) = \sinh a \cos b + i \cosh a \sin b$

As an example, consider the hyperbolic cosine of i. We get

$\cosh i = \cosh 0 \cos 1 + i \sinh 0 \sin 1 = \cos 1 \approx 0.5403$

Inverse Trigonometric Functions

Since $\cos x = \frac{e^{ix} + e^{-ix}}{2}$, we can find the inverse by replacing the function with the variable y, swapping x and y, then solving for y. After the variable swap, we have $x = \frac{e^{iy} + e^{-iy}}{2}$. Multiplying both sides by $2e^{iy}$ gives the equation $0 = e^{2iy} - 2xe^{iy} + 1$. This can be solved using the quadratic formula, to get $e^{iy} = \frac{2x \pm \sqrt{4x^2 - 4}}{2}$. After simplifying, we can take the natural logarithm of both sides and divide by i. Then, the result is $\cos^{-1} x = -i \ln \left[x \pm i\sqrt{1 - x^2} \right]$. (A factor of i typically appears as the coefficient of the square root so that the similarities between the inverse cosine and inverse sine formulas are more apparent.)

By similar arguments, the inverse sine formula is found to be $\sin^{-1} x = -i \ln \left[ix \pm \sqrt{1-x^2}\right]$, and the inverse tangent formula is $\tan^{-1} x = \frac{i}{2} \ln \frac{i+x}{i-x}$. As an example, consider the inverse cosine of *i*. We shall find only one of its many values, by using the plus sign in the formula. We get: $\cos^{-1} i = -i \ln \left[i + i\sqrt{1-i^2}\right] = -i \ln\left[\left(1+\sqrt{2}\right)i\right]$ Then we use the logarithm formula to continue:

$\cos^{-1} i = -i \ln \left[\left(1+\sqrt{2}\right)e^{i\pi/2}\right] = -i\left[\ln\left(1+\sqrt{2}\right)\right] + \frac{\pi}{2} i$ This can be simplified, and we get $\cos^{-1} i = \frac{\pi}{2} - i \ln\left(1+\sqrt{2}\right) \approx 1.57 - 0.88$

Inverse Hyperbolic Functions

The formulas are identical to those previously found for inverse hyperbolic functions:

$$\cosh^{-1} x = \ln \left[x \pm \sqrt{x^2 - 1}\right]$$

$$\sinh^{-1} x = \ln \left[x \pm \sqrt{x^2 + 1}\right]$$

For example, one of the values of the inverse hyperbolic cosine of *i* (using the plus sign in the formula) is:

$\cosh^{-1} i = \ln \left[i + \sqrt{i^2 - 1}\right] = \ln\left[\left(1+\sqrt{2}\right)i\right]$. Then using the logarithm formula, we get:

$$\cosh^{-1} i = \ln \left(1+\sqrt{2}\right) + \frac{\pi}{2} i \approx 0.88 + 1.57i$$

The Function

z^{α}, α may be complex, is defined as $e^{\alpha \ln z}$. Similarly if *f*(*z*) and *g*(*z*) are two given functions of *z*, we can define $f(z)^{g(z)} = e^{g(z) \ln f(z)}$. In general such functions are multiple-valued.

Algebraic and Transcendental Functions

If *w* is a solution of the polynomial equation

$$P_0(z)w^n + P_1(z)w^{n-1} + P_2(z)w^{n-2} + P_3(z)w^{n-3} + \ldots + P_{n-1}(z)w^n + P_n(z) = 0 \ldots(2)$$

Where $P_0 \neq 0, P_1(z), P_2(z), \ldots, P_n(z)$ are polynomials in *z* and n is a positive integer then $w = f(z)$ is called an algebraic function of z.

Example. $w = z^{\frac{1}{2}}$ is a solution of the equation $w^2 - z = 0$ and so is an algebraic function of z.

Any function which cannot be expressed as a solution of (6) is called a transcendental function. The logarithmic, trigonometric and hyperbolic function and their corresponding inverses are example of transcendental functions.

BRANCH POINT AND BRANCH CUT

In complex analysis, a branch is a portion of the range of a multiple-valued function over which the function is single-valued. Combining all the branches gives the full structure of the function. It is often convenient to choose a particular branch of a function to work with, and this choice is often designated the principal branch.

In complex analysis, a branch point may be thought of informally as a point z_0 at which a "multiple-valued function" changes values when one winds once around z_0.

A branch point of an analytic function is a point in the complex plane whose complex argument can be mapped from a single point in the domain to multiple points in the range. For example, consider the behavior of the point z = 0 under the power function

$$f(z) = z^a \qquad ...(3)$$

Examples

- 0 is a branch point of the square root function. Suppose $w = \sqrt{2}$ and z starts at 4 and moves along a circle of radius 4 centered at 0. The dependent variable w changes while

  depending on z in a continuous manner. When z has made one full circle, going from 4 back to 4 again, w will have made one half-circle, going from the positive square root of 4, i.e., from 2, to the negative square root of 4, i.e., –2.
- 0 is also a branch point of the natural logarithm. Since e^0 is the same as $e^{\pi i}$, both 0 and $2\pi i$ are among the multiple values of Log (1). As z moves along a circle of radius 1 centered at 0, w = Log (z) goes from 0 to $2\pi i$.
- In trigonometry, since tan (π/4) and tan (5π/4) is both equal to 1, the two numbers π/4 and 5π/4 are among the multiple values of arc-tan (1). The imaginary units i and $-i$ are branch points of the arctangent function. That this is so may be seen by observing that the derivative $\frac{d}{dz}\{\text{arc-tan }(z)\} = 1/(1 + z^2)$ has simple poles at these two points i.e., at z = i, $-i$ $\frac{d}{dz}(\tan^{-1} z) = 0$ since the denominator is zero at those points.
- If the derivative f' of a function f has a simple pole at a point a, then f has a branch point at a.

The branch cut device may appear arbitrary (it is); but it is very useful, for example in the theory of special functions. An invariant explanation of the branch phenomenon is developed in Riemann surface theory (of which it is historically the origin), and more generally in the ramification and monodromy theory of algebraic functions and differential equations.

Branch Cut

A branch cut is a curve (with ends possibly open, closed, or half-open) in the complex plane across which an analytic multiple-valued function is discontinuous. For convenience, branch cuts are often taken as lines or line segments. Branch cuts (even those consisting of curves) are also known as cut lines, slits, or branch lines.

For example, consider the function z^2 which maps each complex number z to a well-defined number z^2. Its inverse function $\sqrt{z}$, on the other hand, maps, for example, the value $z = 1$ to $\sqrt{1} = \pm 1$. While a unique principal value can be chosen for such functions, the choices cannot be made continuous over the whole complex plane. Instead, lines of discontinuity must occur. The most common approach for dealing with these discontinuities is the adoption of so-called branch cuts. In general, branch cuts are not unique, but are instead chosen by convention to give simple analytic properties. Some functions have a relatively simple branch cut structure, while branch cuts for other functions are extremely complicated. An alternative to branch cuts for representing multiple-valued functions is the use of Riemann surfaces.

In addition to branch cuts, singularities known as branch points also exist. It should be noted, however, that the endpoints of branch cuts are not necessarily branch points.

Branch cuts do not arise for the single-valued trigonometric, hyperbolic, integer power, and exponential functions. However, their multiple-valued inverses do require branch cuts.

Function Name	Function	Branch cut(s)
inverse cosecant	$\csc^{-1} z$	$(-1, 1)$
inverse cosine	$\cos^{-1} z$	$(-\infty, -1)$ and $(1, \infty)$
inverse cotangent	$\cot^{-1}$ z	$(-i, i)$
inverse hyperbolic cosecant	csch^{-1}	$(-i, i)$
inverse hyperbolic cosine	$\cosh^{-1}$	$(-\infty, 1)$
inverse hyperbolic cotangent	$\coth^{-1}$	$(-1, 1)$
inverse hyperbolic secant	sech^{-1}	$(-\infty\ 0)$ and $(1, \infty)$
inverse hyperbolic sine	$\sinh^{-1}$	$(-i\ \infty, -i)$ and $(i, i\ \infty)$
inverse hyperbolic tangent	$\tanh^{-1}$	$(-\infty, -1)$ and $(1, \infty)$
inverse secant	$\sec^{-1} z$	$(-1, 1)$
inverse sine	$\sin^{-1} z$	$(-\infty, -1)$ and $(1, \infty)$
inverse tangent	$\tan^{-1} z$	$(-i\ \infty, -i)$ and $(i, i\ \infty)$

natural logarithm	$\ln z$	$(-\infty, 0)$
power	$z^n, n \notin \mathbb{Z}$ $(-\infty, 0)$ for R $[n] > 0$	$(-\infty, 0)$ for R $[n] \leq 0$;
square root	$\sqrt{z}$	$(-\infty, 0)$

Next we discuss about 'Riemann surface'.

RIEMANN SURFACE

A Riemann surface is a surface-like configuration that covers the complex plane with several, and in general infinitely many, "sheets." These sheets can have very complicated structures and interconnections. Riemann surfaces are one way of representing multiple-valued functions; another is branch cuts.

For example, they can look like a sphere or a torus or a couple of sheets glued together.

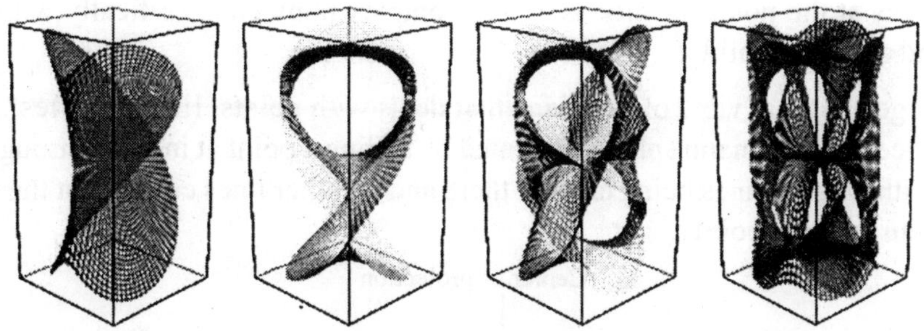

The above plot shows Riemann surfaces for solutions of the equation

$[w(z)]^d + w(z) + z^{d-1} = 0$

with $d = 2, 3, 4$, and 5, where $w(z)$ is the Lambert W-function.

The main point of Riemann surfaces is that holomorphic functions may be defined between them. Riemann surfaces are nowadays considered the natural setting for studying the global behavior of these functions, especially multi-valued functions such as the square root or the logarithm.

Examples

- The complex plane C is perhaps the most trivial Riemann surface. The map $f(z) = z$ (the identity map) defines a chart for C, and $\{f\}$ is an atlas for C. The map $g(z) = \bar{z}$ (the conjugate map) also defines a chart on C and $\{g\}$ is an atlas for C. The charts f and g are not compatible, so this endows C with two distinct Riemann surface structures. In fact, given a Riemann surface X and its atlas A, the conjugate atlas $B = \{f^* : f \to A\}$ is never compatible with A, and endows X with a distinct, incompatible Riemann structure.

- Every open subset of the complex plane can be viewed as a Riemann surface in a natural way. More generally, every open subset of a Riemann surface is a Riemann surface.

Every non-compact Riemann surface admits non-constant holomorphic functions (with values in). In fact, every non-compact Riemann surface is a Stein manifold.

In constrast, on a compact Riemann surface every holomorphic function with value in $\mathbb{C}$ is constant due to the maximum principle. However, there always exists non-constant meromorphic functions .

Riemann Sphere

The Riemann sphere, is the unique way of viewing the extended complex plane (the complex plane plus a point at infinity) so that it looks exactly the same at the point infinity as at any complex number. The main application is to deal with extended complex functions (which may be defined at the point infinity and/or take the value infinity, in addition to complex numbers) in the same way at the point infinity as at any complex number, specifically with respect to continuity and differentiability.

From the geometrical view of the plane that deals with points, lines, circles and angles but not distances, the Riemann sphere is created by adding a point at infinity through which all lines cross, with parallel lines being tangent there and all other lines crossing at the same angle as they do at an existing point.

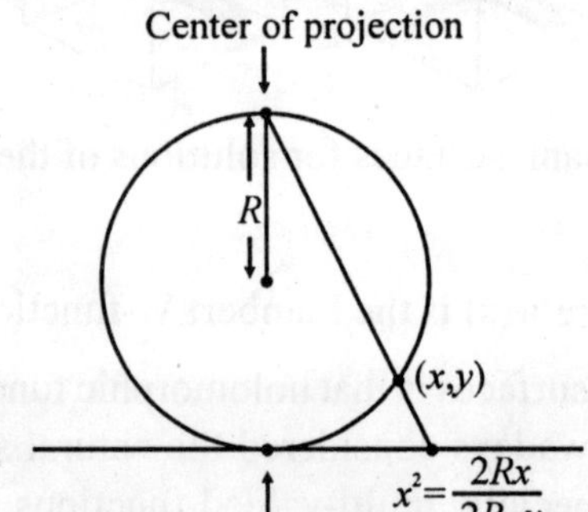

Topologically, the Riemann sphere is the one-point compactification of the complex plane.

In all of these viewpoints, the point at infinity acquires an identical role to any point in the complex plane.

EXERCISE

1. Evaluate each of the following limits:

(a) $\lim\limits_{z \to i} (4z^3 - 6z + 3)$

(b) $\lim_{z \to 1-i} \left(z^2 - 3z\right)$

(c) $\lim_{w \to 3i} \left(\frac{w^2+9}{w-3i}\right)$

(d) $\lim_{z \to i} \left(\frac{z^4-1}{z^2+1}\right)$

2. Suppose $f(z) = 3xy + i\left(x - y^2\right)$. Find $\lim_{z \to 3+2i} f(z)$ or explain carefully why it does not exist.
3. Prove that if f has a derivative at z, then f is continuous at z.

4. Suppose that f(z) = $\frac{a_0 + a_1 z + a_2 z^2}{b_0 + b_1 z + b_2 z^2}$

where $a_0, a_1, a_2, b_0, b_1, b_2 \in C$. Examine the behaviour of f(z) at z = 0 and at z = ∞.

5. Each of the following functions is not defined at z = z0. What value must f(z0) take to ensure continuity at z = z0?

   (a) f(z) = $\frac{z - z_0}{z - z_0}$

   (b) f(z) = $\frac{z^3 - z_0^{\,3}}{z - z_0}$

   (c) f(z) = $\frac{1}{z - z_0}\left(\frac{1}{z} - \frac{1}{z_0}\right)$

   (d) f(z) = $\frac{1}{z - z_0}\left(\frac{1}{z^3} - \frac{1}{z_0^{\,3}}\right)$

6. Prove that polynomial is continuous everywhere in the finite plane.

Chapter 8

Derivative of Complex Number

INTRODUCTION

In the earlier chapter you have studied on complex number and some their properties. Also you have learnt about some operation on complex number. We ahve also studird more about functions of complex number in the previous chapter. In this chapter you will learn about derivative of complex number.

FUNCTIONS OF A COMPLEX VARIABLE

Functions of (x, y) i.e., $f(x,y)$ that depend only on the combination $(x + i\,y)$ are called functions of a complex variable. The combination $(x + i\,y)$ is generally called z, and we can define such functions as z^n, $\exp(z)$, $\sin z$ and all the standard functions of z as well as of x.A symbol, such as z, which can stand for any one of a set of complex numbers is called a *complex variable*.

They are defined in exactly the same way the only difference being that they are actually complex valued functions, that is, they are vectors in this two dimensional complex number space, each with a real and an imaginary part (or component).

Since we can multiply z by itself and by any other complex number, we can form any polynomial in z and any power series as well. We define the exponential and sine functions of z by their power series expansions which converge everywhere in the complex plane.

We can produce all the standard functions of a complex variable by the same steps as go to producing standard functions of real variables.If to each value which a complex variable z can assume there corresponds one or more values of a complex variable w, say that w is a function of z and write $w = f(z) = f(x+iy) = u(x,y)+iv(x,y) = u+iv$.

Single-valued Function

A single-valued complex function of a complex variable is a complex function $f: C \to C$ that has the same value (unique value) at every point z_0 independent of the path along which it is reached by analytic continuation

Example. If $w = f(z) = z$, then to each value of z there is only one value of w. Hence $w = f(z)$ single-valued function.

Multiple-valued Function

A multiple-valued function, also known as a multiple-valued function, is a "function" that assumes two or more distinct values in its range for at least one point in its domain. While these "functions" are, not functions in the normal sense of being one-to-one or many-to-one.

The trigonometric, hyperbolic, exponential, and integer power functions are all single-valued functions, their inverses are multiple-valued.

Example. If $w = z^{\frac{1}{2}}$, then to each value of z there are two values of w. Hence $w = f(z) = z^{\frac{1}{2}}$ is a multiple-valued function.

The discontinuities of multiple-valued functions in the complex plane are commonly handled through the adoption of branch cuts, but use of Riemann surfaces is another possibility.

Inverse Functions

If $w = f(z)$, then there exist the function f^{-1} is often called the inverse function corresponding to f. Thus $w = f(z) =$ and $w = f^{-1}(z)$ are inverse function of each other.

MAPPING

If $w = u + i\,v$ (where u and v are real) is a single-valued function of $z = x + i\,y$, we can write $w = f(z) = u + i\,v = f(x + i\,y)$. By equating real and imaginary parts we get,

$$u = u\,(x, y),\ v = v\,(x, y). \qquad ...(1)$$

Thus, for given a point P(x, y) in the z plane such as P in Figure there corresponds a point P'(u, v) in the z plane. We say that z is mapped or transformed into (u, v) by means of the mapping (transformation) and call P' the image of P.

(a) (b)

Fig. 1 Plane

LIMIT AND CONTINUITY OF COMPLEX NUMBER .

Complex Number System

The complex plane with metric $d(x,y) := | x - y |$ is also a metric space. There are two different types of limits when we consider complex-valued functions.

Limit of a Function at a Point

Suppose f is a complex-valued function, then we write

$$\lim_{x \to p} f(x) = L$$

if and only if

for every $\varepsilon > 0$ there exists a $\delta > 0$ such that for all real numbers x with $0<|x-p|<\delta$, we have $|f(x)-L| < \varepsilon$

It is just a particular case of functions over metric spaces with both M and N are the complex plane.

Limit of a function of more than one Variable

By noting that $|x-p|$ represents a distance, the definition of a limit can be extended to functions of more than one variable.

$$\lim_{(x,y)\to(p,q)} f(x, y) = L$$

if and only if

for every $\varepsilon > 0$ there exists a $\delta > 0$ such that for all real numbers x with $0 < \|(x, y)-(p,q)\| < \delta$, we have $|f(x, y)-L| < \varepsilon$

where $\|(x, y)-(p, q)\|$ represents the Euclidean distance. This can be extended to any number of variables.

Functions of a Complex Variable

The real excitement begins when we consider function $f: D \to C$ in which the domain D is a subset of the complex numbers. In some sense, these too are familiar to us from elementary calculus—they are simply functions from a subset of the plane into the plane:

$$f(z) = f(x,y) = u(x,y) + iv\,(x,y) = (u(x,y), y(x,y))$$

Thus $f(z) = z^2$ looks like $f(z) = z^2 = (x + iy)^2 = x^2 - y^2 + 2xyi$. In other words, $u(x,y) = x^2 - y^2$ and $v(x,y) = 2xy$. The complex perspective, as we shall see, generally provides richer and more profitable insights into these function.

The definition of the limit of a function f at a point $z = z_0$ is essentially the same as that which we learned in elementary calculus:

$$\lim_{z \to z0} f(z) = L$$

means that given and ε > 0, there is a δ so that |f(z) – L| < e whenever $0 < |z - z_0| < \delta$. As you could guess, we say that is continuous at z_0 if it is true $\lim_{z\to z0} f(z) = f(z_0)$. If *f* is continuous at each point of its domain, we say simply that *f* is continuous.

Properties of limits:

Suppose both $\lim_{z\to z0} f(z)$ and $\lim_{z\to z_0} g(z)$ exist. Then the following properties are easy to establish:

$$\lim_{z\to z_0}\left[f(z)\pm g(z)\right] = \lim_{z\to z_0} f(z)\pm \lim_{z\to z_0} g(z)$$

$$\lim_{z\to z_0} f(z)g(z) = \lim_{z\to z_0} f(z)\lim_{z\to z_0} g(z)$$

and,
$$\lim_{z\to z_0}\frac{f(z)}{g(z)} = \frac{\lim_{z\to z_0} f(z)}{\lim_{z\to z_0} g(z)}$$

provided, of course, that $\lim_{z\to z_0} g(z) \neq 0$.

It now follows at once from these properties that the sum, difference, product, and quotient of two functions are continuous at z_0 are z_0. (We must, as usual, except the dereaded 0 in the denominator.)

It should not be too difficult to convince yourself that if $z = (x,y)$, $z_0 = (x_0,y_0)$, and $f(z) = u(x,y) + iv(x,y)$, then

$$\lim_{z\to z_0} f(z) = \lim_{(x,y)\to(x_0,y_0)} u(x,y) + i \lim_{(x,y)\to(x_0,y_0)} v(x,y)$$

Thus *f* is continuous at $z_0 = (x_0,y_0)$ precisely when *u* and *v* are.

Let u = u(*x,y*) be a real-valued function of the two real variables x and y. Recall that u has the limit u_0 as (*x,y*) approaches (x_0, y_0) provided that the value of u (x,y) can be made to get as close as we please to the value u_0 by taking (x, y) to be sufficiently close to (x_0, y_0). When this happens we write

$$\lim_{(x,y)\to(x_0,y_0)} u(x, y) = u_0.$$

In more technical language, u has the limit u_0 as (x,y) approaches (x_0, y_0) iff $|u(x, y) - u_0|$ can be made arbitrarily small by making both $|x-x_0|$ and $|y-y_0|$ small. This condition is like the definition of a limit for functions of one variable. The point (x, y) is in the x-y plane, and the distance between (x, y) and (x_0, y_0) is $\sqrt{(x-x_0)^2+(y-y_0)^2}$. With this perspective we can now give a precise definition of a limit.

***Definition* (Limit of u(x, y)):** The expression $\lim_{(x,y)\to(x_0,y_0)} u(x, y)$ means that for each number $\in > 0$, there corresponds a number $\delta > 0$ such that

$|u(x, y) - u_0| < \in$ whenever

$0 < \sqrt{(x-x_0)^2+(y-y_0)^2} < \delta$.

Example 1. Show, if $u(x, y) = \frac{2x^2}{x^2+y^2}$,

then $\lim_{(x,y)\to(0,0)} u(x, y) = 0$.

Solution. If $x = r \cos \theta$ and $y = r \sin \theta$, and $z = x + iy \neq 0$ then

$$u(x, y) = \frac{2r^3 \cos\theta^3\theta}{r^2\cos^2\theta + r^2\sin^2\theta} = 2r\cos^3\theta.$$

Because $\sqrt{(x-0)^2+(y-0)^2} = r$ and ...(2)

because $|\cos^3\theta| < 1$, we have

$|u(r, \theta)| = 2r|\cos^3\theta < \in$ whenever

$\sqrt{(x-0)^2+(y-0)^2} = r < \frac{\in}{2}$.

Hence, for any $\in > 0$, Inequality is satisfied for $\delta = \frac{\in}{2}$; that is, u (x, y) has the limit $u_0 = 0$ as (x, y) approaches (0, 0).

Example 2. Show that the function $u(x, y)\ \frac{xy}{x^2+y^2}$ does not have a limit as (x, y) approaches (0, 0).

Solution: If we let (x,y) approach (0,0) along the x axis, then

$$\lim_{(x,0)\to(0,0)} u(x, y) = \lim_{(x,0)\to(0,0)} \frac{(x)(o)}{x^2+o^2} = 0$$

But if we let (x,y) approach (0, 0) along the line y = x, then

$$\lim_{(x,y)\to(0,0)} u(x, y) = \lim_{(x,x)\to(0,0)} \frac{(x)(x)}{x^2+x^2} = \frac{1}{2}$$

Because the value of the limit differs depending on how (x,y) approaches (0, 0), we conclude that u (x, y) does not have a limit as (x, y)approaches (0, 0).

Let f(z) be a complex function of the complex variable z that is defined for all values of z in some neighborhood of z_0, except perhaps at the point z_0. We say that f has the limit w_0 as z approaches z_0 provided that the value f(z) can be made as close as we please to the value w_0 by taking z to be sufficiently close to z_0. When this happens we write

$\lim_{x \to z_0} f(z) = w_0$

The distance between the points z and z_0 can be expressed by $|z - z_0|$, so we can give a precise definition similar to the one for a function of two variables.

***Definition* (Limit of *f*(*z*)):** The expression $\lim_{z \to z_0} f(z) = w_0$ means that for each number $\in > 0$, there exists a real number $\delta > 0$ such that

$|f(z) - w_0| < \in$ whenever $O < |z - z_0| < \delta$.

We can also express the above relationship as

$f(z) \in D_\in(w_0)$ whenever $z \in D_\delta^*(z_0)$.

The formulation of limits in terms of open disks provides a good context for looking at this definition. It says that for each disk of radius $\in > 0$ about the point w_0 (represented by $D \in (w_0)$ there is a punctured disk of radius $\delta > 0$ about the point z_0 (represented by $D_\delta^*(z_0)$ such that the image of each point in the punctured δ-disk lies in the $\in$-disk. The image of the δ-disk does not have to fill up the entire $\in$-disk; but if z approaches z_0 along a curve that ends at z_0, then $w = f(z)$ approaches w_0.

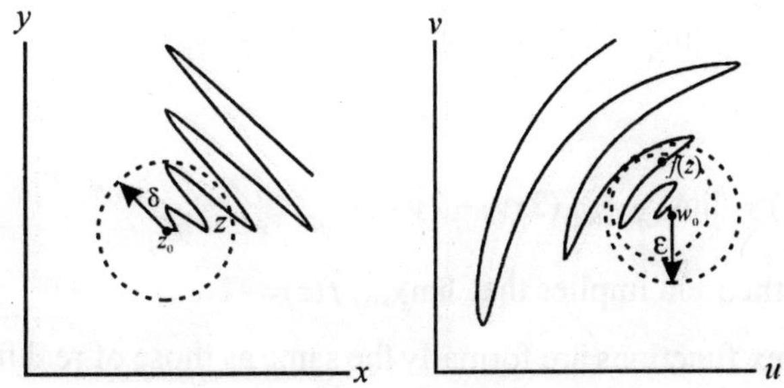

Fig. 2 The Limit $f(z) \to w_0$ as $z \to z_0$

Example 3. Show that if $f(z) = \bar{z}$, then

$\lim_{z \to z_0} f(z) = \bar{z}_0$, where z_0 is any complex number.

Solution: As f merely reflects points about the y axis, we suspect that any $\in$-disk about the point $\bar{z}_0$ would contain the image of the punctured δ-disk about z_0 if $\delta = \in$. To confirm thisconjecture, we let $\in$ be any positive number and set $\delta = \in$. Then we suppose that $z \in D_{\delta}^{*}(z_0) = D_{\in}^{*}(z_0)$, which means that $0 <| \overline{z - z_0} < \in$. The modulus of a conjugate is the same as the modulus of the number itself, so the last inequality implies that $0 <| z - z_0 | < \in$. This is the same as $0 <| \bar{z} - \bar{z}_0 | < \in$. Since $f(z) = \bar{z}$ and $w_0 = \bar{z}_0$, this is the same as $0 <| f(z) - w_0 | < \in$, which in turn is the same as $f(z) \in D_{\in}(w_0)$, which is what we needed to show.

Theorem 1. Let f (z) = u (x, y) + ṅ v(x, y) be a complex function that is defined in some neighborhood of z_0, except perhaps at $z_0 = x_0 + \dot{n} y_0$. Then

$$\lim_{z \to z_0} f(z) = w_0 = u_0 + \dot{n} v_0$$

iff both

$$\lim_{(x,y)\to(x_0,y_0)} u(x,y) = u_0 \text{ and } \lim_{(x,y)\to(x_0,y_0)} v(x,y) = v_0$$

Example 4. Show that $\lim_{z \to 1+i} (z^2 - 2z + 1) = -1$

Solution: We have

$$f(z) = 1 - 2z + z^2$$

$$f(x + \dot{n}y) = (x^2 - y^2 - 2x + 1) + \dot{n}(2xy - 2y)$$

$$u(x, y) = x^2 - y^2 - 2x + 1$$

$$v(x, y) = 2xy - 2y$$

Computing the limits for u and v, we obtain

$$\lim_{(x,y)\to(1,1)} u(x,y) = \lim_{(x,y)\to(1,1)} (x^2 - y^2 - 2x + 1) = -1$$

and

$$\lim_{(x,y)\to(1,1)} v(x,y) = \lim_{(x,y)\to(1,1)} (2xy) - 2y = 0,$$

so our previous theorem implies that $\lim_{z \to 1+i} f(z) = -1$.

Limits of complex functions are formally the same as those of real functions, and the sum, difference, product, and quotient of functions have limits given by the sum, difference, product, and quotient of the respective limits. We state this result as a theorem and leave the proof as an exercise.

Theorem 2. Suppose that $\lim_{z \to z_0} f(z) = A$ and $\lim_{z \to z_0} g(z) = B$ Then

(i) $\lim_{z \to z_0} f(z) + g(z) = A + B$, *and*

(ii) $\lim_{z \to z_0} f(z) - g(z) = A - B$,

(iii) $\lim_{z\to z_0} f(z)\,g(z) = A\,B,$

(iv) $\lim_{z\to z_0} \dfrac{f(z)}{g(z)} = \dfrac{A}{B}$ *where* $B \neq 0.$

CONTINUITY

Refer to Figure. The graph of f is unbroken at a i.e., it's continuous there. It has a hole at b i.e., it's broken, or discontinuous (ie, not contnuous) there; we notice that $f(b)$ doesn't exit. The graph jumps abruptly at c; it's discontinuous there; we notice that $f(c)$ exists but $\lim_{x\to c} f(x)$ doesn't exist (both $\lim_{x\to c+} f(x)$ and $\lim_{x\to c-} f(x)$) exist but aren't equal). The graph has a hole at d; it's discontinuous there; we notice that $f(d)$ exists and $\lim_{x\to d} f(x)$ exists, but they aren't equal.

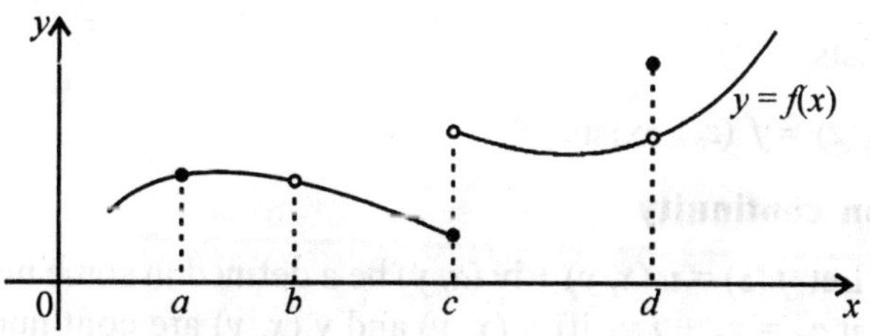

Fig. 3 f **is Continuous at** a, **discontinuous at** b, c, **and** d.

This means that:

$\lim\limits_{x\to a} f(x)$, exists and equals $f(a)$.

All that's just been examined forms the basis for the definition of continuity.

***Definition of Continuity*:** Three conditions for continuity of $f(x)$ at a point $x = c$:

a. the function must be defined at c, or $f(c)$ must be defined.

b. $\lim\limits_{x\to c} f(x)$ must exist.

c. $\lim\limits_{x\to c} f(x) = f(c)$

***Definition* (Continuity of $u(x, y)$):** Let u(x,y) be a real-valued function of the two real variables x and y. We say that u is continuous at the point (x_0, y_0) if the three conditions are satisfied:

(i) $\lim\limits_{(x,y)\to(x_0,y_0)} u(x,y)$ exists.

(ii) $u(x_0, y_0)$ exists,

(iii) $\lim\limits_{(x,y)\to(x_0,y_0)} u(x,y) = u(x_0 y_0)$

Condition (iii) actually implies Conditions (i) and (ii) because the existence of the quantity on each side of Equation (iii) is implicitly understood to exist. For example, if $u(x, y) = \frac{x^2}{x^2 + y^2}$ when $(x, y) \neq (0, 0)$ and if $u(0, 0) = 0$, then $\lim_{(x, y)\to(0,0)} u(x, y) = 0$ so that Conditions (i), (ii), and (iii) are satisfied. Hence $u(x, y)$ is continuous at (0, 0).

There is a similar definition for complex valued functions.

Continuity of Complex Valued Function

Definition. **(Continuity of *f*(z)):** Let f(z) be a complex function of the complex variable z that is defined for all values of z in some neighborhood of z_0. We say that f is continuous at z_0 if three conditions are satisfied:

(i) $\lim_{z\to z_0} f(z)$ exists,

(ii) $f(z_0)$ exists,

(iii) $\lim_{z\to z_0} f(z) = f(z_0)$ exists.

Theorems on continuity

Theorem 3. Let $f(z) = u(x, y) + iv(x, y)$ be a defined in some neighborhood of z_0. Then $f(z)$ is continuous at $z_0 = x_0 + i y_0$ iff $u(x, y)$ and $v(x, y)$ are continuous at (x_0, y_0).

Theorem 4. Suppose that f (z) and g (z) are continuous at the point z_0. Then the following functions are continuous at z_0.

(i) The sum f (z) + g (z),

(ii) The difference f (z) – g (z),

(iv) The product f (z) g (z),

(v) The quotient $\frac{f(z)}{g(z)}$, provided that $g(z_0) \neq 0$.

The composition f (g(z)), provided that f (z) is continuous in a neighborhood of the point g (z_0).

Example 5. Show that the polynomial $p(z) = 1 - z - z^2 + z^3 - z^4 + z^5$ is continuous at the point $z_0 = 1$ in the complex plane.

Solution: We have $p(z) = 1 - z - z^2 + z^3 - z^4 + z^5$ and find the limit.

$$p(z) = 1 - z - z^2 + z^3 - z^4 + z^5$$

$$p(1) = 0$$

$$\lim_{z\to 1} p(z) = \lim_{x\to 1} (1 - z - z^2 + z^3 - z^4 + z^5)$$

$$\lim_{z\to 1} p(z) = 0$$

Therefore, f(z) is continous at z0 = 1.

Example 6. Show that the polynomial $p(z) = 1 - z - z^2 + z^3 - z^4 + z^5$ is continuous at the point $z_0 = \frac{1}{2}$ in the complex plane.

Solution: We have, $p(z) = 1 - z - z^2 + z^3 - z^4 + z^5$ and find the limit.

$$p(z) = 1 - z - z^2 + z^3 - z^4 + z^5$$

$$p\left[\frac{1}{2}\right] = \frac{11}{32}$$

$$\lim_{z \to \frac{1}{2}} p(z) = \lim_{z \to \frac{1}{2}} (1 - z - z^2 + z^3 - z^4 + z^5)$$

$$\lim_{z \to \frac{1}{2}} p(z) = \frac{11}{32}$$

Therefore, f(z) is continous $z_0 = \frac{1}{2}$

If we let P and Q be polynomials and if $Q(z_0) \neq 0$, then

$$\lim_{z \to z_0} \frac{P(z)}{Q(z)} = \frac{p(z_0)}{Q(z_0)}.$$

Another technique, involves factoring polynomials. If both $p(z_0) = 0$ and $Q(z_0) = 0$, then P and Q can be factored as

$$p(z) = (z - z_0)\, p_1(z)$$

and $Q(z) = (z - z_0)\, Q_1(z)$. If $Q(z_0) \neq 0$

$Q(z) = (z - z_0)\, Q_1(z)$. If $Q(z_0) \neq 0$, then the limit is

$$\lim_{z \to z_0} \frac{P(z)}{Q(z)} = \frac{(z - z_0)P_1(z)}{(z - z_0)Q_1(z)} = \frac{P_1(z_0)}{Q_1(z_0)}.$$

Example 7. Show that $\lim\limits_{z \to 1+i} \dfrac{z^2 - 2i}{z^2 - 2z + 2} = 1 - i$

Solution: Here P and Q can be factored in the form

$$P(z) = (z - 1 - i)(z + 1 + i),$$

and

$$Q(z) = (z - 1 - i)(z - 1 + i),$$

so that the limit is obtained by the calculation

$$\lim_{x\to 1+i}\frac{z^2-2i}{z^2-2z+2} = \lim_{z\to 1+i}\frac{(z-1-i)(z+1+i)}{(z-1-i)(z-1+i)}$$

$$= \lim_{z\to 1+i}\frac{z+1+i}{z-1+ia}$$

$$= \frac{(1+i)+1+i}{(1+i)-1+i}$$

$$= \frac{2+2i}{2i} = 1-i$$

DIFFERENTIATION OF COMPLEX NUMBER

Earlier we have learnt about the derivative of real valued functions, here we study the derivative of complex-valued functions.A complex function is a function in which the independent variable and the dependent variable are both complex numbers. More precisely, a complex function is a function whose domain Ù is a subset of the complex plane and whose range is also a subset of the complex plane.

For any complex function, both the independent variable and the dependent variable may be separated into real and imaginary parts:

$$z = x+iy \text{ and } w = f(z) = f(x+iy) = u(x,y)+iv(x,y) = u+iv$$

where $x,\ y \in R$ and $u(z), v(z)$ are real-valued functions.

In other words, the components of the function $f(z)$, $u(x,y)$ and $v(x,y)$ can be interpreted as real valued functions of the two real variables, x and y.

THE DERIVATIVE OF A FUNCTION OF A COMPLEX VARIABLE

Let f be a function whose domain of definition contains a neigbourhood of a point z_0. The derivative of f at z_0 written $f'(z_0)$ is defined by the equation

$$f'(z_0) = \lim_{z\to z_0}\frac{f(z)-f(z_0)}{z-z_0},$$

provided this limits exists. The function f is said to be differentiable at z_0 when its derivable at z_0 exists.

To begin with, the usual calculus definition of a derivative can be applied to complex functions just as well as to real functions. It is just that the norm $|\Delta z|$ of an increment must be sent to zero, rather than the absolute value of a real difference, $|\Delta x|$.

$$\frac{d}{dz}f(z) = c\lim_{|\Delta z|\to 0}\frac{f(z+\Delta z)-f(z)}{\Delta z}, \qquad \text{...(3)}$$

supposing that the limit exists.

From this, the usual rules governing the derivatives of sums, products, differences and quotients can be deduced. Derivatives of powers can be obtained from the binomial theorem

$$(z+\Delta z)^n = \sum_{i=0}^{n} \frac{n!}{i!(n-i)!} z^{n-i} \Delta z^i \qquad ...(4)$$

in the usual way with the result

$$\frac{d}{dz} z^n = n\, z^{n-1} \qquad ...(5)$$

The chain rule for composite functions is applicable, so that the derivatives of inverse functions can be obtained. For the exponenial, the derivative of products and the definition holding that

$$e^z = \lim_{x \to \infty} \left(1 + \frac{z}{n}\right)^n \qquad ...(6)$$

SUFFICIENT CONDITIONS FOR DIFFRENTIABILITY

Theorem: Let the function

$$f(z) = u(x,y) + iv(x,y)$$

be defined throughout some ε neighborhood of a point $z_0 = x_0 + iy_0$. Suppose that the first-order partial derivatives of the functions *u* and *v* with respect to *x* and *y* exist everywhere in that neighborhood and that they are continuous at (x_0, y_0). Then if those partial derivatives satisfy the Cauchy-Riemann equations

$u_x = v_y,\ u_y = -v_x$

at (x_0, y_0), the derivative $f'(z_0)$ exists.

To check that the derivative is well-defined, separate the complex function into the sum of two real functions, just as *z* can be written as the sum of a real and an imaginary part:

$$z = x + i\,y \qquad ...(7)$$

$$f = u(x,y) + i\,v(x,y). \qquad ...(8)$$

By first taking Δz real,

$$\frac{df}{dz} = \lim_{\Delta z \to 0} \frac{f(z+\Delta x)}{\Delta x} \qquad ...(9)$$

$$= \lim_{\Delta x \to 0} \frac{u(x+\Delta x,y) + iv(x+\Delta x,y) - (u(x,y) + iv(x,y))}{\Delta x} \qquad ...(10)$$

$$= \underset{\Delta x \to 0}{lt}\ \underset{\Delta x \to 0}{lt} \frac{u(x+\Delta xy)-u(x,y)}{\Delta x + iu}$$

$$\underset{\Delta x \to 0}{lt} \frac{(x+\Delta x, y)-u(x,y)}{\Delta x}$$

$$f'(t) = \frac{\partial u}{\partial x} + i\frac{\partial v}{dx}, \quad ...(11)$$

and then imaginary,

$$\frac{df}{dz} = \lim_{\Delta z \to 0} \frac{f(z+i\Delta y)}{i\Delta y} \quad ...(12)$$

$$= \lim_{\Delta y \to 0} \frac{u(x,y+\Delta y)+iv(x,y+\Delta y)-(u(x,y)+iv(x,y))}{i\Delta y} \quad ...(13)$$

$$\underset{\Delta y \to 0}{lt} \frac{u(x,y+\Delta y)-u(x,y)}{i\Delta y} + \underset{\Delta y \to 0}{lt} \frac{u(x,y+\Delta y)-u(x,y)}{\Delta y}$$

$$f'(t) = -i\frac{\partial u}{\partial y} + \frac{dv}{dy}. \quad ...(14)$$

CAUCHY RIEMAN EQUATIONS

Supposing that the derivative is independent of phase, at least in the context taken here, comparison shows

$$\frac{\partial u}{\partial x} = \frac{\partial v}{\partial y} \quad ...(15)$$

$$-\frac{\partial v}{\partial x} = \frac{\partial u}{\partial y} \quad ...(16)$$

The members of this pair of equations are known as the Cauchy - Riemann equations, relating the real and imaginary parts of a complex function which is supposed to have a derivative.

Consider the Jacobian matrix for the real functions $u(x,y)$ and $v(x,y)$ of the real variables x and y,

$$\frac{d(u,v)}{d(x,y)} = \begin{bmatrix} \frac{\partial u}{\partial x} & \frac{\partial v}{\partial x} \\ \frac{\partial u}{\partial y} & \frac{\partial v}{\partial y} \end{bmatrix} \quad ...(17)$$

$$= \begin{bmatrix} \frac{\partial u}{\partial x} & \frac{\partial v}{\partial x} \\ -\frac{\partial v}{\partial x} & \frac{\partial v}{\partial y} \end{bmatrix} \quad ...(18)$$

$$= \frac{\partial u}{\partial x}1 + \frac{\partial v}{\partial x}i \quad ...(19)$$

The impact of the Cauchy-Riemann equations is to give the Jacobian matrix the form of a complex number in quaternion.

By knowing one of the pair, one effectively knows the other. That is,

$$v(x,y) - v(a,y) = \int_a^x \frac{\partial v(\sigma,y)}{\partial \sigma} d\sigma \quad ...(20)$$

$$= \int_a^x \frac{\partial u(\sigma,y)}{\partial y} d\sigma \quad ...(21)$$

Knowing *u*, take its derivative and integrate to get *v*. For example, according to Euler's formula,

$$e^z = e^x(\cos y + i \sin y) \quad ...(22)$$

Then

$$u = e^x \cos y \quad ...(23)$$

$$\frac{\partial u}{dy} = -e^x \sin y \quad ...(24)$$

$$-\int_a^b e^\sigma \sin y d\sigma = e^\sigma \sin y\Big|_a^x \quad ...(25)$$

$$v(x,y) - v(a,y) = e^x \sin y - e^a \sin a \quad ...(26)$$

which is consistent with

$$v(x,y) = e^x \sin y \quad ...(27)$$

So trying to think of something like e^x (cos *y* + *i* tan *y*), as an analytic function just wouldn't work.

Harmonic Functions of Real Variables

Statement: If *D* is region in the plane, a real-valued function *u*(*x*,*y*) having continuous second partial derivatives is said to be harmonic on *D* if it satisfies Laplace's equation on *D*:

$$\frac{\partial^2 u}{\partial x^2} + \frac{\partial^2 u}{\partial y^2} = 0. \quad ...(28)$$

There is an intimate relationship between harmonic functions and analytic functions. Suppose f is analytic on D, and let $f(z) = u(x,y) + iv(x,y)$. Since, we know

$$\frac{\partial u}{\partial x} = \frac{\partial v}{\partial y}$$

and, $$\frac{\partial u}{\partial y} = \frac{\partial v}{\partial x}.$$

If we differentiate the first and the second with respect to x and respectively y, and then add the two results, we have

$$\frac{\partial^2 u}{\partial x^2} + \frac{\partial^2 u}{\partial y^2} = \frac{\partial^2 y}{\partial x dy} - \frac{\partial^2 y}{\partial y \partial x} = 0.$$

Thus the real part of any analytic function is harmonic! Next, if we differentiate the first of the Cauchy-Riemann equations with respect to y and the second with respect to x, and then substract the second from the first, we have

$$\frac{\partial^2 v}{\partial x^2} + \frac{\partial^2 v}{\partial y^2} = 0, \qquad ...(29)$$

and we see that the imaginary part of an analytic function is also harmonic.

There is even more excitement. Suppose we are given a function φ harmonic in a simply connected region D. Then there is a function f analytic on D which is such that Ref = φ. First, define g by

$$g(z) = \frac{\partial \varphi}{\partial x} - i\frac{\partial \varphi}{\partial y}.$$

We'll show that g is analytic by verifying that the real and imaginary parts satisfy the Cauchy-Riemann equations:

$$\frac{\partial}{\partial x}\left(\frac{\partial \varphi}{\partial x}\right) = \frac{\partial^2 \varphi}{\partial x^2} = -\frac{\partial^2 \varphi}{\partial y^2} = \frac{\partial}{\partial y}\left(-\frac{\partial \varphi}{\partial y}\right),$$

since φ is harmonic. Next,

$$\frac{\partial}{\partial y}\left(\frac{\partial \varphi}{\partial x}\right) = \frac{\partial^2 \varphi}{\partial y \partial x} = \frac{\partial^2 \varphi}{\partial x \partial y} = -\frac{\partial}{\partial x}\left(-\frac{\partial \varphi}{\partial y}\right).$$

since g is analytic on the simply connected region D, we know that the intgral of g around any closed curve is zero, and so it has an antiderivative $G(z) = u + iv$. This antiderivative is, of course, analytic on D, and we know that

$$G'(z) = \frac{\partial u}{\partial x} - i\frac{\partial u}{\partial y} = \frac{\partial \varphi}{\partial x} - i\frac{\partial \varphi}{\partial y}.$$

Thus, $u(x, y) = \varphi(x, y)\ h(y)$. From this,

$$\frac{\partial u}{\partial y} = \frac{\partial \varphi}{\partial y} + h'(y),$$

and so $h'(y) = 0$, or h = constant, from which it follows that $u(x, y) = \varphi(x, y) + c$. In other words, Re $G = u$, as we promised to show.

An important practical consequence is that such functions never have maxima nor minima. By writing the second derivatives in finite difference form

$$\frac{d^2 f}{dx^2} = \frac{f(x - \Delta x) - 2f(x) + f(x + \Delta x)}{\Delta x^2}$$

so that

$$u(x, y) = \frac{1}{4}\{u(x + \Delta x, y) + u(x - \Delta x, y) + u(x, y + \Delta y) + u(x, y + \Delta y)\} \quad ...(30)$$

or, in other words, each central value is the average of its neighbors. No average can exceed all the values it is averaging, nor fall below all of them, either.

Example 8. The Function $\varphi(x, y) = x^3 - 3xy^2$ is harmonic everywhere.

Solution: We shall find function G that Re $G = \varphi$. We know that $G(z) = x^3 - 3xy^2 + iv$, and so from the Cauchy-Riemann equations:

$$\frac{\partial v}{\partial x} = \frac{\partial u}{\partial y} = 6xy$$

Hence,

$$v(x, y) = 3x^2 y + k(y).$$

To find $k(y)$ differentiate with respect to y :

$$\frac{\partial v}{\partial y} = 3x^2 + k'(y) = \frac{\partial u}{\partial x} = 3x^2 - 3y^2,$$

and so,

$k'(y) = -3y^2$, or

$k(y) = -y^3$ + any constant.

If we choose the constant to be zero, this gives us

$$v = 3x^2 y + k(y) = 3x^2 y - y^3,$$

and finally,

$$G(z) = u + iv = (x^3 - 3xy^2) + i(3x^2 y - y^3).$$

The Schwartz Derivative

Consider the basic linear fractional transformation

$$w = \frac{az+b}{cz+d} \qquad \text{...(31)}$$

which is equivalent to the vanishing of the implicit expression

$$cwz + dw - az - b = 0. \qquad \text{...(32)}$$

To get a relation between w and z which does not depend on a, b, c and d, take three derivatives with respect to a still unspecified parameter:

$$c(wz)' + dw' - az' = 0 \qquad \text{...(33)}$$

$$c(wz)'' + dw'' - az'' = 0 \qquad \text{...(34)}$$

$$c(wz)''' + dw''' - az''' = 0. \qquad \text{...(35)}$$

In equivalent matrix form,

$$\begin{bmatrix} (wz)' & w' & -z' \\ (wz)'' & w'' & -z'' \\ (wz)''' & w''' & -z''' \end{bmatrix} \begin{bmatrix} c \\ d \\ a \end{bmatrix} = \begin{bmatrix} 0 \\ 0 \\ 0 \end{bmatrix}. \qquad \text{...(36)}$$

The determinant of this matrix, which is a Wronskian, must vanish to get a nontrivial solution. But

$$\begin{vmatrix} (wz)' & w' & -z' \\ (wz)'' & w'' & -z' \\ (wz)''' & w''' & -z''' \end{vmatrix}$$

$$= \begin{vmatrix} w'z + wz' & w' & -z' \\ w''z + 2w'z' + wz'' & w'' & -z'' \\ w'''z + 3w''z' + 3w'z'' + wz''' & w''' & -z''' \end{vmatrix}$$

Operate $C_1, (-z)z + w3$, we get

$$= \begin{vmatrix} 0 & w' & -z' \\ 2w'z' & w'' & -z'' \\ 3w''z' + 3w'z'' & w''' & -z''' \end{vmatrix}$$

Expowding along first raw, we get

$$= -w'z' \begin{vmatrix} 0 & w' & -z' \\ 2 & w'' & -z'' \\ 3\frac{w''}{w'} + 3\frac{z''}{z'z} & w''' & -z''' \end{vmatrix}$$

$$= -w'z' \left\{ -w' \begin{vmatrix} 2 & z'' \\ 3\left(\frac{w''}{w'} + \frac{z''}{z'}\right) & z''' \end{vmatrix} - z' \begin{vmatrix} 2 & w'' \\ 3\left(\frac{w''}{w'} + \frac{z''}{z'}\right) & w''' \end{vmatrix} \right\}.$$

$$= -w'z'\left\{-w'\left(2z''' - 3z''\left(\frac{w''}{w'}+\frac{z''}{z'}\right)\right)+z'\left(2w''' - 3w''\left(\frac{w''}{w'}+\frac{z''}{z'}\right)\right)\right\}$$

$$= -w'z'\left\{-2w'z''' + 3w'z''\left(\frac{w''}{w'}+\frac{z''}{z'}\right)+2z'w''' - 3w''\left(\frac{w''}{w'}+\frac{z''}{z'}\right)\right\}$$

$$= -(w'z')^2\left\{-2\frac{z'''}{z'}+3\frac{z''}{z'}\left(\frac{w''}{w'}+\frac{z''}{z'}\right)+2\frac{w'''}{w'} - 3w''\left(\frac{w''}{w'}+\frac{z''}{z'}\right)\right\}$$

Since this expression vanishes, the factor $(w'z')^2$ can be discarded (unless it happens to vanish everywhere) to leave

$$2\frac{w''}{w'} - 3\left(\frac{w''}{w'}\right)^2 = 2\frac{z'''}{z'} - 3\left(\frac{z''}{z'}\right)^2, \qquad \text{...(37)}$$

which is a relationship which is expected to hold after applying a linear fractional transformation to a variable, no matter what the linear fractional transformation. Besides, if z is the variable, the right hand is zero, which leaves the left hand side as a constraint on any function that supports invariance. Some rearrangement can make the invariant read

$$\frac{2w'w''' - 3(w'')^2}{(w')^2}$$

half of which is called a *Schwartz derivative*. The symbol for the Schwartz derivative of w with respect to z is $\{w,z\}$, some variants of whose definition are:

$$\{w,z\} = \frac{w''}{w'} - \frac{3}{2}\left(\frac{w''}{w'}\right)^2, \qquad \text{...(38)}$$

$$= \left\{\frac{w''}{w'}\right\}' - \frac{1}{2}\left(\frac{w''}{w'}\right)^2, \qquad \text{...(39)}$$

$$= \frac{d^2}{dz^2}\left[\ln\frac{dw}{dz}\right] - \frac{1}{2}\left\{\frac{d}{dz}\left[\ln\frac{dw}{dz}\right]\right\}^2. \qquad \text{...(40)}$$

The Schwartz derivative is not really a derivative (it is a differential invariant) because it doesn't follow the product rule when applied to products, nor does it even work linearly on sums, and it is furthermore immune to non-zero multiplicative constants. As the derivative was derived, w and z depended on a parameter with respect to which two differential expressions were identical. When z itself was the parameter, one result was obviously zero, forcing the other to assume the same value; which of the two was chosen creates an asymmetry by which $\{w,z\} \neq \{z,w\}$.

Nevertheless, the invariant can be calculated for various w's all depending on the same z, for which the following rules and particular results can be verified through calculation:

$$\left\{\frac{aw+b}{cw+d}, z\right\} = \{w, z\}, \quad \text{...(41)}$$

$$\left\{w, \frac{az+b}{cz+d}\right\}\frac{(ad-bc)}{(cz+d)^4} = (w, z), \quad \text{...(42)}$$

$$\{w, z\} = \{w, u\}\left(\frac{du}{dz}\right)^2 + \{u, z\}, \quad \text{...(43)}$$

$$\{w, z\} = -\{z, w\}\left(\frac{dw}{dz}\right)^2 \quad \text{...(44)}$$

$$\{w, z\} = 0 \Rightarrow w = \frac{az+b}{cz+d'} \quad \text{...(44)}$$

$$\{z^n, z\} = \frac{1-n^2}{2z^2}, \quad \text{...(45)}$$

$$\{e^{\lambda z}, z\} = -\frac{1}{2}\lambda^2. \quad \text{...(46)}$$

The last three lines suggest the task of inverting the Schwartz derivative; that is, of finding $w(z)$ for which $\{w,z\} = Q(z)$, for some prescribed function Q. The answer is that one should solve the differential equation

$$y''(z) + \frac{1}{2}Q(z)y(z) = 0, \quad \text{...(47)}$$

for which one can expect to obtain two linearly independent solutions $y_1(z)$ and $y_2(z)$. Their quotient

$$w(z) = \frac{y_2(z)}{y_1(z)} \quad \text{...(48)}$$

will then solve the Schwartzian equation.

HIGHER ORDER DERIVATIVES

Note that a partial derivative is itself a function of two variables, and so further partial derivatives can be calculated. We write

$$\frac{\partial}{\partial x}\frac{\partial f}{\partial x} = \frac{\partial^2 f}{\partial x^2}, \quad \frac{\partial}{\partial x}\frac{\partial f}{\partial y} = \frac{\partial^2 f}{\partial x \partial y}, \quad \frac{\partial}{\partial y}\frac{\partial f}{\partial x} = \frac{\partial^2 f}{\partial y \partial x} \text{ and } \frac{\partial}{\partial y}\frac{\partial f}{\partial y} = \frac{\partial^2 f}{\partial y^2} \quad \text{...(49)}$$

This notation generalises to more than two variables, and to more than two derivatives in the way you would expect. There is a complication that does not occur when dealing with functions of a single variable; there are *four* derivatives of second order, as follows:

$$\frac{\partial^2 f}{\partial x^2}, \frac{\partial^2 f}{\partial x \partial y}, \frac{\partial^2 f}{\partial y \partial x} \text{ and } \frac{\partial^2 f}{\partial y^2}.$$

Fortunately, when f has mild restrictions, the order in which the differentiation is done doesn't matter.

Proposition 1. Assume that all second order derivatives of f exist and are continuous. Then the mixed second order partial derivatives of f are equal. i.e.

$$\frac{\partial^2 f}{\partial x \partial y} = \frac{\partial^2 f}{\partial y \partial x}.$$

Example 9. Suppose that $f(x, y)$ is written in terms of u and v where $x = u + \log v$ and $y = u - \log v$. Show that, with the usual convention,

$$\frac{\partial^2 f}{\partial u^2} = \frac{\partial^2 f}{\partial x^2} + 2\frac{\partial^2 f}{\partial x \partial y} + \frac{\partial^2 f}{\partial y^2}$$

and

$$v^2\frac{\partial^2 f}{\partial u^2} = \frac{\partial f}{\partial y} - \frac{\partial f}{\partial x} + \frac{\partial^2 f}{\partial x^2} + 2\frac{\partial^2 f}{\partial x \partial y} + \frac{\partial^2 f}{\partial y^2}$$

You may assume that all second order derivatives of f exist and are continuous.

Solution: Using the chain rule, we have

$$\frac{\partial f}{\partial u} = \frac{\partial f}{\partial x}\frac{\partial x}{\partial u} + \frac{\partial f}{\partial y}\frac{\partial y}{\partial u} = \frac{\partial f}{\partial x} + \frac{\partial f}{\partial y}$$

and $\frac{\partial f}{\partial v} = \frac{\partial f}{\partial x}\frac{\partial x}{\partial v} + \frac{\partial f}{\partial y}\frac{\partial y}{\partial v} = \frac{1}{v}\frac{\partial f}{\partial x} - \frac{1}{v}\frac{\partial f}{\partial y}$

Thus using both these and their operator form, we have

$$\frac{\partial^2 f}{\partial v^2} = \frac{\partial}{\partial u}\left(\frac{\partial f}{\partial u}\right) = \frac{\partial}{\partial u}\left(\frac{\partial f}{\partial x} + \frac{\partial f}{\partial y}\right)$$

$$= \frac{\partial}{\partial x}\left(\frac{2f}{\partial x} + \frac{\partial f}{\partial y}\right)\frac{\partial x}{\partial u} + \frac{\partial}{\partial y}\left(\frac{\partial f}{\partial x} + \frac{\partial f}{\partial y}\right)\frac{\partial y}{\partial u}$$

$$= \left(\frac{\partial^2 f}{\partial x^2} + \frac{\partial^2 f}{\partial y}\right)^1 + \left(\frac{\partial^2 f}{\partial y \partial x} + \frac{\partial^2 f}{\partial y^2}\right)^1$$

$$= \frac{\partial^2 f}{\partial x^2} + \frac{\partial^2 f}{\partial x \partial y} + \frac{\partial^2 f}{\partial y \partial x} + \frac{\partial^2 f}{\partial y^2}$$

While differentiating with respect to u, we have

$$\frac{\partial^2 f}{\partial v^2} = \frac{\partial}{\partial u}\left(\frac{\partial f}{\partial u}\right) = \frac{\partial}{\partial v}\left(\frac{1}{v}\frac{\partial f}{\partial x} - \frac{1}{v}\frac{\partial f}{\partial y}\right)$$

$$= \frac{-1}{v^2}\frac{\partial f}{\partial x} + \frac{1}{v}\frac{\partial}{\partial v}\left(\frac{\partial f}{\partial x}\right) + \frac{1}{v^2}\frac{\partial f}{\partial y} - \frac{1}{v}\frac{\partial}{\partial v}\left(\frac{\partial f}{\partial y}\right)$$

$$= \frac{-1}{v^2}\frac{2f}{2x} + \frac{1}{v}\left(\frac{\partial}{2x}\left(\frac{\partial f}{\partial x}\right)\frac{\partial x}{\partial v} + \frac{\partial}{\partial y}\left(\frac{\partial f}{\partial y}\right)\frac{\partial y}{\partial v}\right)$$

$$+ \frac{1}{v^2}\frac{\partial f}{\partial y}\frac{-1}{v}\left(\frac{\partial}{\partial x}\left(\frac{\partial f}{\partial y}\right)\frac{\partial x}{\partial v}\right)$$

$$= \frac{-1}{v^2}\frac{2f}{2x} + \frac{1}{v}\left(\frac{\partial}{2x}\left(\frac{\partial f}{\partial x}\right)\frac{\partial x}{\partial v} + \frac{\partial}{\partial y}\left(\frac{\partial f}{\partial y}\right)\frac{\partial y}{\partial v}\right)$$

$$= \frac{1}{v^2}\left(\frac{\partial f}{\partial y} - \frac{\partial f}{\partial x} + \frac{\partial^2 f}{\partial x^2} - 2\frac{\partial^2 f}{\partial x \partial y} + \frac{\partial^2 f}{\partial y^2}\right)$$

Theorem 1: If $f(z)$ is analytic in a region $\Re$, so also are $f'(z), f''(z), \ldots$ analytic in $\Re$, i.e., all higher derivatives exists in $\Re$.

L' Hospital's rule

Let $f(z)$ and $g(z)$ be an analytic in a region containing the point z_0 and suppose that $f(z_0) = g(z_0) = 0$, but $g'(z_0) \neq 0$. Then L' Hospital's rule state that

$$\lim_{z \to z_0} \frac{f(z)}{g(z)} = \frac{f'(z_0)}{g'(z_0)}$$

In case $f'(z_0) = g'(z_0) = 0$, the rule may be extended.

Limits represented by so-called indeterminate forms $\frac{\infty}{\infty}, 0 \circ \infty, \infty^0, 0^0, 1^\infty$ and $\infty - \infty$ can be evaluated by appropriate modifications of L' Hospital's rule.

EXERCISE

1. Evaluate each of the following limits:

(a) $\lim\limits_{z \to i}\left(4z^3 - 6z + 3\right)$

(b) $\lim_{w \to 3i} \left(\frac{w^2 + 9}{w - 3i} \right)$

(c) $\lim_{z \to i} \left(\frac{z^4 - 1}{z^2 + 1} \right)$

2. For each of the following, find $\lim_{n \to \infty} z_n$. Also, plot $z_1, z_2, z_3, \ldots, z_{15}$ in the complex plane:

(a) $z_n = \left(\frac{3-n}{n} \right) + i \left(\frac{n+1}{2n+3} \right)$

(b) $z_n = \left(\frac{2-n}{n^2} \right) - i \left(4 + \frac{6}{n} \right)$

(c) $z_n = 3e^{i\frac{\pi}{n}}$

(d) $z_n = e^{i\frac{\pi(n-1)}{n}}$

3. Suppose $f(z) = 3xy + i(x - y^2)$. Find $\lim_{z \to 3+2i} f(z)$ or explain carefully why it does not exist.
4. Prove that if *f* has a derivative at *z*, then *f* is continuous at *z*.
5. Suppose that f(z) = $\frac{a_0 + a_1 z + a_2 z^2}{b_0 + b_1 z + b_2 z^2}$

where $a_0, a_1, a_2, b_0, b_1, b_2 \in C$. Examine the behaviour of f(z) at z = 0 and at z = ∞.

6. Each of the following functions is not defined at z = z0. What value must f(z0) take to ensure continuity at z = z0?

(a) f(z) = $\frac{z - z_0}{z - z_0}$

(b) f(z) = $\frac{z^3 - z_0^3}{z - z_0}$

(c) f(z) = $\frac{1}{z - z_0} \left(\frac{1}{z} - \frac{1}{z_0} \right)$

7. Prove that polynomial is continuous everywhere in the finite plane.
8. Find all points at which the valued function *f* defined by $f(z) = \bar{z}$ has a derivative.

9. Find all points at which the valued function f defined by $f(z) = (2 + i)z^3 - iz^2 + 4z - (1 + 7i)$ has a derivative.

10. At what points is the function f given by $f(z) = x^3 + i(1 - y)^3$ analytic? Explain.

11. Do the real and imaginary parts of the function f satisfy the Cauchy-Riemann equations at $z = 0$? What do you make of your answer?

12. Find all points at which $f(z) = \dfrac{x}{x^2 + y^2} - i\dfrac{y}{x^2 + y^2}$ is differentiable. At what points is f analytic? Explain.

13. (a)Show that if f = u + iv is a complex function, where u and v are the real part and the imaginary part of f respectively, then the Cauchy-Riemann equation $\dfrac{\partial f}{\partial \bar{z}} = 0$ is equivalent to $u_x = v_y$, $u_y =$ " v_x. (1.11)

    (b) Use (1.11) to check that the function $f(x, y) = e^x \cos y + ie^x \sin y$ is analytic.

    (c) Use the previous exercise or other methods to show that in polar coordinates is

    $$u_r = \frac{1}{r} v_\theta v_r = -\frac{1}{r} u_\theta$$

14. For each of the following functions f(z), determine whether the Cauchy-Riemann equations are satisfied:

    (a) $f(z) = x^3 - 3y^2 + 2x + i(3x^2y - y^3 + 2y)$

    (b) $f(z) = \log(x^2 - y^2) + 2i \tan^{-1}(y/x)$

15. Show that a real valued analytic function is constant.

Chapter 9

Singularities of Complex Number

INTRODUCTION

In the previous chapters we have studied basic concept of complex number, laws associated with complex number, operations on complex number and also derivative of complex number. Before going to study integration of complex numbers, we have to know about the concept of singularaties of complex numbers.

SINGULARITIES, ZEROS AND POLES OF COMPLEX NUMBER

Here we are going to discuss about singularties, zeroes and poles of complex number.

Singular point: The point $z = \alpha$ is called a singular point, or singularity of the complex function *f*(*z*) if f is not analytic at $z = \alpha$, but every neighborhood $D_R(\alpha)$ of α contains at least one point at which *f*(*z*) is analytic.

For example, the function $f(z) = \frac{1}{1-z}$ is not analytic at $z = 1$, but is analytic for all other values of z. Thus the point $z = 1$ is a singular point of *f*(*z*). As another example, consider $g(z) = \text{Log}(z)$. We saw in Section 5.2 that $g(z)$ is analytic for all z except at the origin and at all points on the negative real-axis. Thus, the origin and each point on the negative real axis is a singularity of $g(z) = \text{Log}(z)$.

The point α is called a isolated singularity of the complex function *f*(*z*) if f is not analytic at $z = \alpha$, but there exists a real number $R > 0$ such that *f*(*z*) is analytic everywhere in the punctured disk $D_R^*(\alpha)$. The function $f(z) = \frac{1}{1-z}$ has an isolated singularity at $z = 1$.

The function $g(z) = Log(z)$, however, the singularity at $z = 0$ (or at any point of the negative real axis) that is not isolated, because any neighborhood of contains points on the negative real axis, and $g(z) = log(z)$ is not analytic at those points. Functions with isolated singularities have a Laurent series because the punctured disk $D_R^*(\alpha)$ is the same as the annulus $A(\alpha,0,R)$.The logarithm function $g(z)=Log(z)$ does not have a Laurent series at any point $z=-a$ on the negative real-axis. We now look at this special case of Laurent's theorem in order to classify three types of isolated singularities.

Types of Singularaties

Here we are going to discuss about types of singularties.

Let *f*(*z*) have an isolated singularity at α with Laurent series expansion

$$\alpha(z)=\sum_{n=-\infty}^{\infty} C_n(z-\alpha)^n \text{ valid for } z\in A(\alpha,0,R) \qquad ...(1)$$

Then we distinguish the following types of singularities at α.

(i) If $C_n=0\, for\, n=-1,-2,-3,...,$ then we say that *f*(*z*) has a removable singularity at α.

(ii) If *k* is a positive integer such that $C_{-k}\neq 0$ but $C_n=0$, then we say that *f*(*z*) has a pole of order k at α.

(iii)If $C_n\neq 0$ for infinitely many negative integers *n*, then we say that *f*(*z*) has an essential singularity at $z=\alpha$.

Let's investigate some examples of these three cases.

(i) If *f*(*z*)has a removable singularity at $z=\alpha$, then it has a Laurent series

$$f(z)=\sum_{n=0}^{\infty} C_n(z-\alpha)^n \text{ valid for } z\in A(\alpha,0,R).$$

If we use this series to define $f(\alpha)=C_0$, then the function *f*(*z*)becomes analytic at $z=\alpha$, removing the singularity.

For example, consider the function $f(z)=\dfrac{\sin(z)}{z}$. It is undefined at $z=0$ and has an isolated singularity at $z=0$, as the Laurent series for *f*(*z*)is

$$f(z)=\frac{1}{z}\left(z-\frac{z^3}{3!}+\frac{z^5}{5!}-\frac{z^7}{7!}+\frac{z^9}{11!}+...\right)$$

$$=1-\frac{z^2}{3!}+\frac{z^4}{5!}-\frac{z^6}{7!}+\frac{z^8}{9!}-\frac{z^{10}}{11!}+...$$

valid for $|z|>0$.

We can remove this singularity if we define $f(0)=1$, for then $f(z)$ will be analytic at $z=0$

Here we are giving some example of functions having some type of singularity:

(i). Another example is $g(z)=\dfrac{\cos(z)-1}{z^2}$, which has an isolated singularity at the point $z=0$, as the Laurent series for $g(z)$ is

$$g(z)=\frac{1}{z^2}\left(-\frac{z^2}{2!}+\frac{z^4}{4!}-\frac{z^6}{6!}+\frac{z^8}{8!}-\frac{z^{10}}{10!}+\ldots\right)$$

$$=-\frac{1}{2!}+\frac{z^2}{4!}-\frac{z^4}{6!}+\frac{z^6}{8!}-\frac{z^8}{10!}+\ldots$$

valid for $|z|>0$. If we define $f(0)=-\frac{1}{2}$, then $g(z)$ will be analytic for all z.

(ii). If $f(z)$ has a pole of order k at $z=\alpha$, the Laurent series for $f(z)$ is

$$f(z)=\sum_{n=-k}^{\infty}C_n(z-\alpha)^n \text{ valid for } z\in A(\alpha,0,R).$$

where $C_{-k}\neq 0$.

(iii) The following example will help this concept. Consider the function $f(z)=\cot z$. The leading term in the Laurent series expansion $S(z)$ is $\frac{1}{z}$ and $S(z)$ goes to ∞ *as* $z\to 0$ in the same manner as $\cot z$.

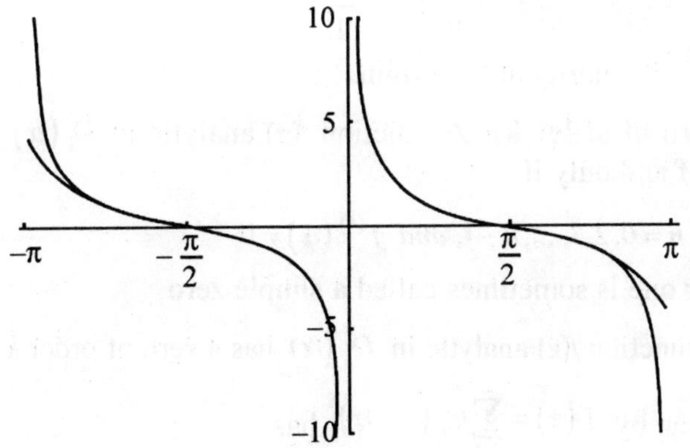

Fig. 1

Another example **is;**

$$f(z)=\frac{\sin(z)}{z^3}$$

$$=\frac{1}{z^3}\left(z-\frac{z^3}{3!}+\frac{z^5}{5!}-\frac{z^9}{9!}-\frac{z^{11}}{11!}+...\right)$$

$$=\frac{1}{z^2}-\frac{1}{3!}+\frac{z^2}{5!}-\frac{z^4}{7!}+\frac{z^6}{9!}-\frac{z^{10}}{11!}+...$$

has a pole of order $k = 2$ at $z = 0$.

If $f(z)$ has a pole of order 1 at $z=\alpha$, we say that $f(z)$ has a simple pole at $z=\alpha$.

For example,

$$g(z)=\frac{1}{2}e^z=\frac{1}{z}\sum_{n=0}^{\infty}\frac{1}{n!}z^n$$

$$=\sum_{n=0}^{\infty}\frac{1}{n!}z^{n-1}$$

$$=\frac{1}{z}+1+\frac{z}{2!}+\frac{z^2}{3!}+\frac{z^3}{4!}+\frac{z^4}{5!}+\frac{z^5}{6!}+...$$

has a simple pole at $z = 0$.

(iv). If infinitely many negative powers of $(z=\alpha)$ occur in the Laurent series, then $f(z)$ has an essential singularity at $z=\alpha$. For example,

$$f(z)=z^2\sin\left(\frac{1}{z}\right)$$

$$=z^2\left(\frac{1}{z}-\frac{\left(\frac{1}{z}\right)^3}{3!}+\frac{\left(\frac{1}{z}\right)^5}{5!}-\frac{\left(\frac{1}{z}\right)^7}{7!}+\frac{\left(\frac{1}{z}\right)^9}{9!}-\frac{\left(\frac{1}{z}\right)^{11}}{11!}+...\right)$$

$$=z-\frac{1}{3!}z^{-1}+\frac{1}{5!}z^3-\frac{1}{7!}z^{-5}+\frac{1}{9!}z^{-7}-\frac{1}{11!}z^{-9}+...$$

has an essential singularity at the origin.

Definition **(Zero of order k).** A function $f(z)$ analytic in $D_r(\alpha)$ has a zero of order k at the point $z=\alpha$ if and only if

$f^{(n)}(\alpha)=0$ *for* $n=0,2,3,...,k-1,$ *and* $f^{(k)}(\alpha)\neq 0.$

A zero of order one is sometimes called a simple zero.

Theorem 1. A function $f(z)$ analytic in $D_R(\alpha)$ has a zero of order k at the point $z=\alpha$ iff its Taylor series given by $f(z)=\sum_{n=0}^{\infty}C_n(z-\alpha)^n$ has

$C_0=C_1=...=C_{k-1}=0$ *and* $C_k\neq 0.$

Example 1. From Theorem 2 we see that the function

$$f(z)=z\sin\left(z^2\right)=z^3-\frac{z^7}{3!}+\frac{z^{11}}{5!}-\frac{z^{15}}{7!}+...$$

has a zero of order $k = 3$ at $z = 0$ Definition 7.6 confirms this fact because

$$= \frac{1}{z^3}\left(z - \frac{z^3}{3!} + \frac{z^5}{5!} - \frac{z^9}{9!} - \frac{z^{11}}{11!} + \ldots\right) \qquad f'(z) = 2z^2 Cos\, z^2 + Sin\, z^2$$

$$f''(z) = 6z\, Cos\, z^2 - 4z^3 \sin z^2$$

$$= \frac{1}{z^2} - \frac{1}{3!} + \frac{z^2}{5!} - \frac{z^4}{7!} + \frac{z^6}{9!} - \frac{z^{10}}{11!} + \ldots \qquad f'''(z) = 6 Cos\, z^2 - 8z^4 Cos\, z^2 - 24 z^2 Sin\, z^2$$

Then, $f(0) = f''(0) = f''(0) = 0,$

but $f'''(0) = 6 \neq 0.$

Theorem 3. Suppose *f*(*z*) is analytic in $D_R(\alpha)$. Then *f*(*z*) has a zero of order *k* at the point $z = \alpha$ if and only if it can be expressed in the form

$$f(z) = (z - \alpha)^k g(z),$$

where *g*(*z*) is analytic at $z = \alpha$ *and* $g(\alpha) \neq 0.$

Corollary1: If *f*(*z*) and *g*(*z*) are analytic at $z = \alpha$, and have zeros of orders *m* and *n*, respectively at $z = \alpha$, then their product $h(z) = f(z)g(z)$ has a zero of order $m + n$ *at* $z = \alpha.$

Example 2. Let $f(z) = z^3 \sin z$. Then *f(z)* can be factored as the product of z^3 and sin *z*, which have zeros of orders $m = 3$ and $n = 1$, respectively, at $z = 0$.

Hence $z = 0$ is a zero of order 4 of *f(z)*.

Theorem 1 gives a useful way to characterize a pole.

Theorem 2. A function *f(z)* analytic in the punctured disk $D_R*(\alpha)$ has a pole of order k at $z = \alpha$ if and only if it can be expressed in the form

$$f(z) = \frac{h(z)}{(z - \alpha)^k}, \tag{2}$$

where the function *h(z)* is analytic at the point $z = \alpha$ *and* $h(\alpha) \neq 0.$

Corollary2. If *f(z)* is analytic and has a zero of order *k* at the point $z = \alpha$, then $g(z) = \frac{1}{f(z)}$ has a pole of order *k* at $z = \alpha$.

Corollary 3. If *f(z)* has a pole of order *k* at the point $z = \alpha$, then $g(z) = \frac{1}{f(z)}$ has a removable singularity at $z = \alpha$. If we define $g(\alpha) = 0$, then *g(z)* has a zero of order *k* at $z = \alpha$.

Corollary 4. If *f(z)* and *g(z)* have poles of orders *m* and *n*, respectively at the point $z=\alpha$, then their product $h(z)=f(z)g(z)$ has a pole of order $m+n\,at\ z=\alpha$.

Corollary 5. Let *f(z)* and *g(z)* be analytic with zeros of orders *m* and *n*, respectively at $z=\alpha$. Then their quotient $h(z)=\dfrac{f(z)}{g(z)}$ has the following behavior:

(i) If m > n, then *h(z)* has a removable singularity at $z=\alpha$. If we define $h(\alpha)=0$, then *h(z)* has a zero of order $m-n\ at\ z=\alpha$.

(ii) If m < n, then *h(z)* has a pole of order $n-m\ at\ z=\alpha$.

(iii) If m = n, then *h(z)* has a removable singularity at $z=\alpha$, and can be defined so that *h(z)* is analytic at $z=\alpha$, by $h(\alpha)=\lim\limits_{z\to\alpha}h(z)$.

Example 3. Locate the zeros and poles of $h(z)=\dfrac{\tan z}{z}$, and determine their order.

Solution: As we know that the zeros of $f(z)=\sin z$ occur at the points $z=n\pi$, where *n* is an integer. Because $f'(n\pi)\ =\cos n\pi\neq 0$, the zeros of *f(z)* are simple. Similarly, the function $g(z)=z\cos z$ has simple zeros at the points $z=0$ and $z=\left(n+\dfrac{1}{2}\right)\pi$, where *n* is an integer. From the information given, we find that $h(z)=\dfrac{f(z)}{g(z)}$ behaves as follows:

(i) *h(z)* has simple zeros at $z=n\pi$, where $n=\pm1,\pm2,...$,

(ii) *h(z)* has simple poles at $z=\left(n+\dfrac{1}{2}\right)\pi$, where *n* is an integer; and

(iii)*h(z)* is analytic at $z=0$ if we define $h(0)=\lim\limits_{z\to0}h(z)=1$.

Example 4. Locate the poles of $g(z)=\dfrac{1}{5z^4+26z^2+5}$, and specify their order.

Solution: The roots of the quadratic equation $5z^2+26z+5=0$ occur at the points $z=-5\,and\ z=-\dfrac{1}{5}$. If we replace *z* with z^2 in this equation, the function $f(z)=5z^4+26z^2+5$ has simple zeros at the points $z=\pm\sqrt{5}\ and\ z=\pm\dfrac{I}{\sqrt{5}}$. Corollary 2 implies that *g(z)* has simple poles at $z=\pm\sqrt{5}\ and\ z=\pm\dfrac{I}{\sqrt{5}}$.

Example 5. Locate the zeros and poles of $g(z) = \dfrac{\pi \cot(\pi z)}{z^2}$, and determine their order.

Solution: The function $f(z) = z^2 \sin \pi z$ has a zero of order k=3 at z=0 and simple zeros at the points $z = \pm 1, \pm 2, \ldots$ Corollary 7.5 implies that $g(z)$ has a pole of order 3 at the point z=0 and simple poles at the points $z = \pm 1, \pm 2, \ldots$.

EXERCISE

1. Find the residue at z = 0 of the function

   (a) $\dfrac{1}{z+z^2}$;

   (b) $z \cos\left(\dfrac{1}{2}\right)$;

   (c) $\dfrac{z - \sin z}{z}$;

   (d) $\dfrac{\cot}{z^4}$;

   (e) $\dfrac{\sinh z}{z^4\left(1-z^2\right)}$.

2. In each case, write the principal part of the function at its isolated singular point and determine whether that point is a pole, a removable singular point, or an essential singular point:

   (a) $z \exp\left(\dfrac{1}{z}\right)$;

   (b) $\dfrac{z^2}{1+z}$;

   (c) $\dfrac{\sin z}{z}$;

   (d) $\dfrac{\cos z}{z}$;

   (e) $\dfrac{1}{(2-z)^3}$.

3. Show that the singular point of each of the following functions is a pole. Determine the order m of that pole and the corresponding residue B.

(a) $\dfrac{1-\cosh z}{z^3}$;

(b) $\dfrac{1-\exp(2z)}{z^4}$;

(c) $\dfrac{e\text{xp}(2z)}{(z-1)^2}$.

4. In each case, show that the singular points of the function are poles. Determine the order m of each pole, and find the corresponding residue B.

(a) $\dfrac{z^2+2}{z-1}$; (b) $\left(\dfrac{z}{2z+1}\right)^3$; (c) $\tanh z$; (d) $\dfrac{\exp z}{z^2+\pi^2}$.

5. Show that

(a) $\underset{z=-1}{res}\dfrac{z^{\frac{1}{4}}}{z+1}=\dfrac{1+i}{\sqrt{2}}\left(|z|>0, 0<\arg z<2\pi\right)$;

(b) $\underset{z=i}{res}\dfrac{Logz}{\left(z^2+1\right)^2}=\dfrac{\pi+2i}{8}$;

(c) $\underset{z=z_n}{res}\left(z\sec z\right)=(-1)^{i+1}\, z_n, where\ z_n=\dfrac{\pi}{2}+n\pi\ (n=0,\pm 1,\pm 2,\ldots)$.

$$\log|f(0)| = \frac{1}{2\pi}\int_0^{2\pi}\log\left|f\left(\mathrm{R}e^{i\theta}\right)\right|d\theta .$$

Chapter 10

Integration of Complex Number

INTRODUCTION

We have already studied derivative of complex number. Now we are going to learn about integration of complex number. The complex integration is elegant, powerful, and a useful tool for physicists and engineers. It also connects widely with other branches of mathematics. For example, even though the ideas presented here belong to the general area of mathematics known as analysis, you will see as an application of them one of the simplest proofs of the fundamental theorem of algebra.

COMPLEX INTEGRALS

In this chapter we gave our attention to the problem of integrating complex functions. We will find that integrals of analytic functions are well behaved and that many properties from calculus carry over to the complex case.

We now introduce the integral of a complex function by defining the integral of a complex-valued function of a real variable.

Integration over Curves in the Complex Plane

The definite integral considered so far represents area in the xy plane. We defined it however by dividing the interval in x into small subintervals (say of width d) and taking the sum of an estimate of the area of each subinterval, namely $f(x)'$d where x' is a point in that subinterval.

Let C be a curve in the complex plane, and let $f(z)$ be a function of the variable z with $z = x + iy$. We can define integration along that curve of an integrand f(z) by dividing the curve into small pieces, and summing $f(z')(z_i - z_{i-1})$ where z' is a point in the i-th interval whose endpoints are z_i and z_{i-1}, over all the pieces.

This integral along a curve will no longer represent area, since neither f nor the difference between z's will be real numbers.

Such an entity is called a contour integral in the complex plane. We denote it as follows

$$\int_c f(z)dz \qquad ...(1)$$

Integration over Curves in Euclidean Space

We now break the curve into small pieces and summing the difference between the endpoints of each piece multiplied by some function, to define integration along a curve C in any Euclidean space.

If f is a scalar field, that is, a function defined on points in our space, we could sum $f(r')$ (r_1-r_{i-1}) where the r_i represent subdivision points on the path, and again r′ is a point in the i-th subinterval. This can be defined, but here the difference in r′s will be a vector, and the sum will be a vector field.

A much more useful concept is obtained when we replace f here by a vector field w(r) and sum $w(r')\bullet(r_i - r_{i-1})$ over the intervals on the curve C that are defined by the endpoints r_i The result is then a number rather than a vector.

It is defined exactly as area is in the definite integral. The standard notation for it is

$$\int w\bullet dl \qquad ...(2)$$

COMPLEX LINE INTEGRAL

Let f be a continuous complex-valued function of a complex variable, and let C be a smooth curve in the complex plane parametrized by

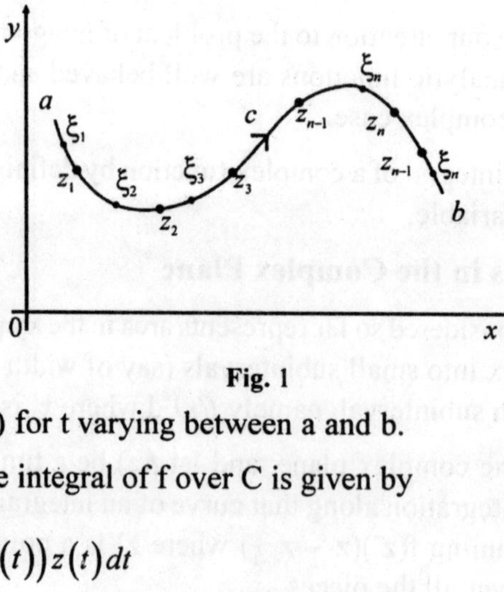

Fig. 1

z(t) = x(t) + i y(t) for t varying between a and b.

Then the complex line integral of f over C is given by

$$\int_C f(x)dx = \int_a^b f(z(t))z(t)dt \qquad ...(3)$$

Note that the "smooth" condition guarantees that z' is continuous and, hence, that the integral exists.

Statement 1. If γ : D→C is simply a function on a real interval D = [a, b], then the intergal $\int_{\alpha}^{\beta} \gamma(t)dt$ is, simply an ordered pair of everyday 3rd grade calculus integrals:

$$\int_{\alpha}^{\beta} \gamma(t)\,dt = \int_{a}^{\beta} x(t)\,dt + i\int_{\alpha}^{\beta} y(t)\,dt, \quad ...(4)$$

where $\gamma(t) = x(t) + iy(t)$. Thus, for example,

$$\int_{0}^{1} \left[\left(t^2+1\right)+it^3\right]dt = \frac{4}{3}+\frac{i}{4}.$$

Rieman sum

Suppose f is a complex-valued function on a subset of the complex plane and suppose a and b are *complex* numbers in the domain of f. Let C be a path from a to b, and we must also require that C be a subset of the domain of f.

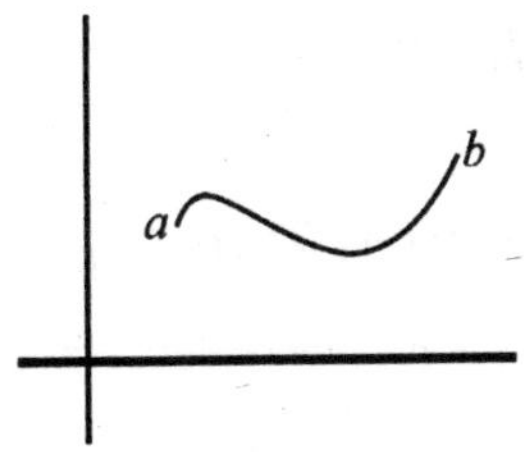

Figure 2.

(Note we do not even require that $a \neq b$; but in case $a = b$, we must specify an *orientation* for the closed path C.) Next, let P be a partition of the curve; that is, $P = \{z_0, z_1, z_2, \ldots, z_n\}$. is a finite subset of C, such that $a = z_0$, $b = z_n$, and such that zj comes immediately after z_{j-1} as we travel along C from a to b.

A Riemann sum associated with the partition P is just what it is in the real case:

$$S(P) = \sum_{j=1}^{n} f\left(z_j^*\right)\Delta z_j, \quad ...(5)$$

where z_j^* is a point on the are between z_{j-1} and z_j, and $\Delta z_j = z_j - z_{j-1}$. given partition P, there are many $S(P)$—depending on how the point z_j^* are chosen.) If there is a number L so that given any $\varepsilon > 0$, there is a partition P_s of C such that

$$|S(p) - L| < \varepsilon \quad ...(6)$$

whenever $P \supset P\varepsilon$, then f is said to be integrable on C and the number L is called integral of f on C. This number L is usually written $\int_C f(z)dz$.

Some properties of integrals are more or less evident from looking at Riemann sums:

$$\int_C cf(z)dz = c\int_C f(z)dz \quad ...(7)$$

for any complex constant c.

$$\int_C (f(z)+g(z))dz = \int_C f(z)dz + \int_C g(z)dz \quad ...(8)$$

EVALUATING INTEGRALS

Let $\gamma : [\alpha, \beta] \to C$ be a complex description of the curve C. Take partition of C by partitioning the interval in the usual way: $\alpha = t_0 < t_1 < t_2 <... < t_n = \beta$.

Then $\{a = \gamma(\alpha), \gamma(t_1), \gamma(t_2), ..., \gamma(\beta) = b\}$ is partition of C. (Recall we assume that $\gamma'(t) \neq 0$ for a complex description of a curve C.) A corresponding Riemann sum looks like

$$S(P) = \sum_{j=1}^{n} f\left(\gamma\left(t_j^*\right)\right)\left(\gamma(tj) - \gamma\left(t_{j-1}\right)\right). \quad ...(9)$$

We have chosen the points $z_j^* = \gamma\left(t_j^*\right)$, where $t_{j-1} \le t_j^* \le t_j$., multiply each term in the sum by 1:

$$S(P) = \sum_{j=1}^{n} f\left(\gamma\left(t_j^*\right)\right)\left(\frac{\gamma\left(t_j\right) - \gamma\left(t_{j-1}\right)}{t_j - t_{j-1}}\right)\left(t_j - t_{j-1}\right). \quad ...(10)$$

After applying the limit, we have

$$\int_C f(z)dz = \int_\alpha^\beta f(\gamma(t))\gamma'(t)dt. \quad ...(11)$$

Relation between Real and Complex line integrals

If $f(z) = u(x, y) = u + iv$ the complex line integral can be expressed in terms of real line integral as

$$\int_C f(z)dz = \int_C (u+iv)(dx+idy) = \int_C udx - vdy + i\int_C vdx - udy \quad ...(12)$$

Definition 1. (Definite Integral of a Complex Integrand)

Let $f(t) = u(t) + i\, v(t)$ where $u(t)$ and $v(t)$ are real-valued functions of the real variable t for $a \le t \le b$. Then

$$\int_a^b f(t)\,dt = \int_a^b \left(u(t) + iv(t)\right) dt$$

$$= \int_a^b u(t)\,dt + i\int_a^b v(t)\,dt \qquad \text{...(13)}$$

We generally evaluate integrals of this type by finding the antiderivatives of $u(t)$ and $v(t)$ and evaluating the definite integrals on the right side of Equation (13). That is, if $U'(t) = u(t)$ and $v'(t)$, we have

$$\int_a^b f(t)\,dt = \int_a^b u(t)\,dt + i\int_a^b v(t)\,dt \qquad \text{...(14)}$$

$$= U(b) - i\,(v(b) - v(a)) \qquad \text{...(15)}$$

$$\int_a^b \left(f(t) + g(t)\right) dt = \int_a^b f(t)\,dt + \int_a^b g(t)\,dt \qquad \text{...(16)}$$

If we divide the interval $a \le t \le b$ into $a \le t \le c$ and $c \le t \le b$ and integrate $f(t)$ over these subintervals by using (16), then we get

$$\int_a^b f(t)\,dt = \int_a^c f(t)\,dt + \int_a^b f(t)\,dt \qquad \text{...(17)}$$

Similarly, if $\alpha + i\beta$ denotes a complex constant, then

$$\int_a^b (c + id)\,f(t)\,dt = (c + id)\int_a^b f(t)\,dt \qquad \text{...(18)}$$

If the limits of integration are reversed, then

$$\int_a^b f(t)\,dt = -\int_b^a f(t)\,dt \qquad \text{...(19)}$$

The integral of the product $f(t)g(t)$ becomes

$$\int_a^b f(t)g(t)\,dt = \int_a^b \left(u(t)p(t) - v(t)q(t)\right) dt \text{ s} \qquad \text{...(20)}$$

$$+\ i\int_a^b \left(u(t)q(t) + v(t)p(t)\right) dlt$$

$$\int_a^b f(t)\,dt = F(t)\Big|_{t=a}^{t=b} = F(b) - f(a) \qquad \text{...(21)}$$

where $F'(t) = f(t)$.

Example 1. Show that $\int_0^1 (t - i)^3\,dt = -\frac{5}{4}$.

Solution: We write the integrand in terms of its real and imaginary parts, i.e., $f(t) = (t-i)^3 = t^3 - 3t + i(-3t^2 + 1)$.

Here,

$u(t) = t^3 - 3t$ and $v(t) = -3t^3 + 1$. The intergrals of $u(t)$ and $v(t)$ are

$$\int_0^1 u(t)\,dt = \int_0^1 (t^3 - 3t)\,dt = \left(\frac{t^4}{4} - \frac{3t^2}{2}\right)\Big|_{t=0}^{t=1} = -\frac{5}{4}, \text{ and}$$

$$\int_0^1 v(t)\,dt = \int_0^1 (-3t^2 + 1)\,dt = (-t^3 + t)\Big|_{t=0}^{t=1} = 0$$

Hence, $\int_0^1 (t-i)^3\,dt = \int_0^1 u(t)\,dt + i\int_2^1 v(t)\,dt$

$$= -\frac{5}{4} + 0i = -\frac{5}{4}$$

Example 2. Show that

$$\int_0^{\frac{\pi}{2}} e^{t+it}\,dt = \frac{1}{2}\left(e^{\frac{\pi}{2}} - 1\right) + \frac{1}{2}\left(e^{\frac{\pi}{2}} - 1\right) + \frac{e}{2}\left(e^{\frac{\pi}{2}} + 1\right)$$

Solution: We use the method suggested by Definitions (1).

$$\int_0^{\pi/2} e^{t+it}\,dt = \int_0^{\pi/2} e^t e^{it}\,dt = \int_0^{\pi/2} e^t(\cos t + i\sin t)\,dt$$

$$= \int_0^{\pi/2} e^t \cos t\,dt + i\int_0^{\pi/2} e^t \sin t\,dt$$

We can evaluate each of the integrals via integration by parts. For example,

$$\int_0^{\pi/2} e^t \cos t\,dt = (e^t \sin t)\Big|_{t=0}^{t=\pi/2} - \int_0^{\pi/2} \sin t e^t\,dt$$

$$= \left(e^{\pi/2}\sin\frac{\pi}{2} - e^0 \sin 0\right) - \int_0^{\pi/2} \sin t e^t\,dt$$

$$= (e^{\pi/2}1 - 1 * 0) - \int_0^{\pi/2} e^t \sin t\,dt$$

$$= e^{\pi/2} - \int_0^{\pi/2} e^t \sin t\,dt$$

$$= e^{\pi/2} - (e^t(-\cos t))\Big|_{t=0}^{t=\pi/2} + \int_0^{\pi/2} (-\cos t e^t)\,dt$$

$$= e^{\pi/2} + \left(e^t \cos t\right) \Big|_{t=0}^{t=\pi/2} - \int_0^{\pi/2} e^t \cos t dt$$

$$= e^{\pi/2} + e^{\pi/2} * 0 - 1 * 1 - \int_0^{\pi/2} e^t \cos t dt$$

$$= e^{\pi/2} - 1 - \int_0^{\pi/2} e^t \cos t dt$$

Adding $\int_0^{\pi/2} e^t \cos t dt$ to both sides of this equation and then dividing by 2 gives

$$\int_0^{\pi/2} e^t \cos t dt = \frac{1}{2}\left(e^{\pi/2} - 1\right).$$

Likewise, $i\int_0^{\pi/2} e^t \sin t dt = i\frac{1}{2}\left(e^{\pi/2} + 1\right)$. Therefore,

$$\int_0^{\pi/2} e^{t+it} dt = \frac{1}{2}\left(e^{\pi/2} - 1\right) + \frac{i}{2}\left(e^{\pi/2} + 1\right).$$

Complex integrals have properties that are similar to those of real integrals. We now trace through several commonalities.

Let $f(t) = u(t) + iv(t)$ and $g(t) = p(t) + iq(t)$ be continuous on $a \le t \le b$.

Here, we can easily show that the integral of their sum is the sum of their integrals, that is.

Example 3. Show that $\int_a^b (v + id) f(t) dt = (c + id)\int_a^b f(t) dt$.

Solution: Since, $(c + id) f(t) = (c + id)\left(u(t) + iv(t)\right)$

$$= cu(t) - dv(t) + i(cv(t) + du(t))$$

Using Definition (1), we write the left side of Equation (22) as

$$\int_a^b (c + id)(u(t) + iv(t)) dt$$

$$= c\int_a^b u(t) dt - d\int_a^b v(t) dt + ic\int_a^b v(t) dt + id\int_a^b u(t) dt$$

which is equivalent to

$$\int_a^b (c + id)\left(u(t) + iv(t)\right) dt$$

$$= c\int_a^b u(t) dt + id\int_a^b u(t) dt + ci\int_a^b v(t) + idi\int_a^b v(t) dt$$

$$= (c + id)\int_a^b u(t) dt + (c + id)i\int_a^b v(t) dt$$

$$= (c+id)\left(\int_a^b u(t)dt + i\int_a^b v(t)dt\right)$$

Therefore, $\int_a^b (v+id)f(t)dt = (c+id)\int_a^b f(t)dt$.

Example 4. Show that $\int_0^{\pi/2} e^{t+it}dt = \frac{1}{2}(e^{\pi/2}-1)+\frac{i}{2}(e^{\pi/2}+1)$..

Solution: We seek a function F with the property that $F'(t) = e^{(1+i)t}$. We note that $F(t) = \frac{1}{1+i}e^{(1+i)t}$ satisfies this requirement, so

$$\int_0^{\pi/2} e^{t+it}dt = \left(\frac{1}{1+i}e^{(1+i)t}\right)\Big|_{t=0}^{t=\pi/2} = \frac{1}{1+i}e^{(1+i)\pi/2} - \frac{1}{1+i}e^0$$

$$= \frac{1}{1+i}e^{\pi/2}e^{i\pi/2} - \frac{1}{1+i} = \frac{1}{1+i}e^{\pi/2}i - \frac{1}{1+i}$$

$$= \frac{1}{1+i}(ie^{\pi/2}-1) = \frac{1}{2}(1-i)(-1+ie^{\pi/2})$$

$$= \frac{1}{2}(e^{\pi/2}-1)+\frac{i}{2}(e^{\pi/2}+1).$$

Example 5. Find the integral of $f(z) = (x^2 + y) + i(xy)$ $a = 0$ to $b = 1 + i$ along three different contours.

Solution: First, let C_1 be the part of the parabola $y = x^2$ connecting the two point. A complex description of C_1 is $\gamma_1(t) = t + it^2$, $0 \le t \le 1$:

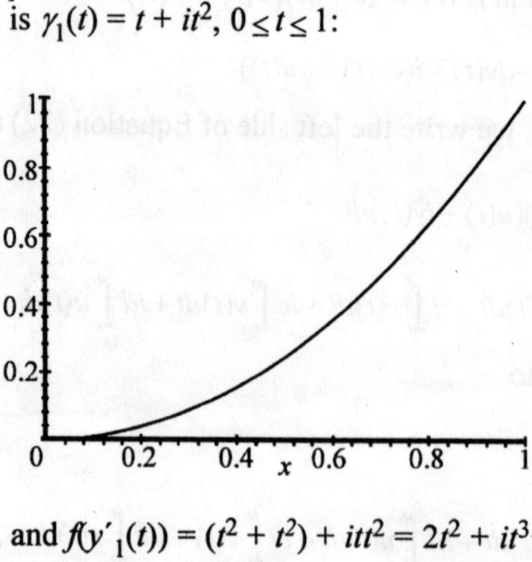

Now, $y'_1(t) = 1 + 2ti$, and $f(y'_1(t)) = (t^2 + t^2) + itt^2 = 2t^2 + it^3$.

Hence, $\int_{C_1} f(z)dz = \int_0^1 f(\gamma_1(t))\gamma'_1(t)dt$

$$\int_{C_1} f(z)dz = \int_0^1 f(\gamma_1(t))\gamma_1'(t)dt$$
$$= \int_0^1 (2t^2 + it^3)(1+2ti)dt$$
$$= \int_0^1 (2t^2 - 2t^4 + 5t^3 i)dt$$
$$= \frac{4}{15} + \frac{5}{4}i$$

Next, let's integrate along the straight line segment *C*2 joining 0 and 1+ *i*.

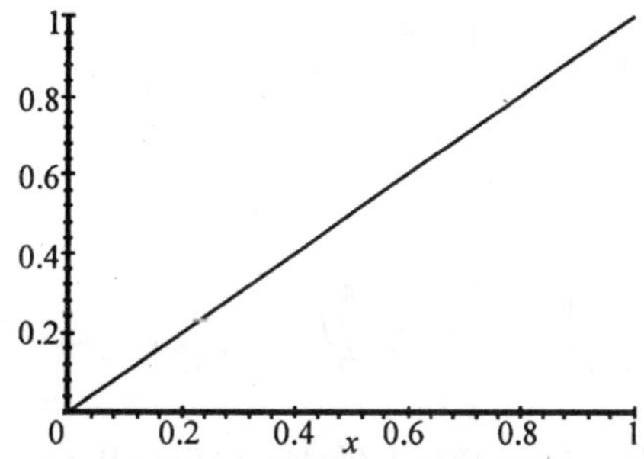

Here we have $\gamma_2(t) = t + it$, $0 \le t \le 1$. Thus, $\gamma_2'(t)$ 1+ *i*, and our integral looks like

$$\int_{C_2} f(z)dz = \int_0^1 f(\gamma_2(t))\gamma_2'(t)dt$$
$$= \int_0^1 \left[(t^2 + t) + it^2\right](1+i)dt$$
$$= \int_0^1 \left[t + i(t + 2t^2)\right]dt$$
$$= \frac{1}{2} + \frac{7}{6}i \qquad ...(22)$$

Finally, let's integrate along C_3, the path consisting of the line segment from 0 to 1 together with the segment from 1 to 1 + *i*.

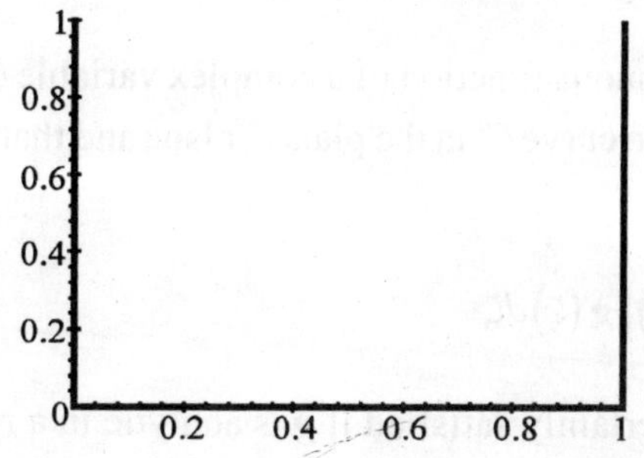

Then we have

$$\int_{C_3} f(z)dz = \int_{C_{32}} f(z)dz + \int_{C_{32}} f(z)dz.$$

For C_{31} we have $\gamma(t)$ t, $0 \le t \le 1$. Hence,

$$\int_{C_{32}} f(z)dz = \int_0^1 dt = \frac{1}{3}.$$

For C_{32} we have $\gamma(t) = 1 + it$, $0 \le t \le 1$. Hence,

$$\int_{C_{32}} f(z)dz = \int_0^1 (1+t+it)itdt = -\frac{1}{3}+\frac{5}{6}i.$$

Thus,

$$\int_{C_3} f(z)dz = \int_{C_{31}} f(z)dz + \int_{C_{32}} f(z)dz = \frac{5}{6}i.$$

Suppose there is a number M so that $|f(z)| \le M$ for all $z \in C$. Then

$$\left|\int_C f(z)dz\right| = \left|\int_\alpha^\beta f(\gamma(t))\gamma'(t)dt\right|$$

$$\le \int_\alpha^\beta |f(\gamma(t))\gamma'(t)|dt$$

$$\le M \int_\alpha^\beta |\gamma'(t)|dt = ML,$$

Where $L = \int_\alpha^\beta |\gamma'(t)|dt$ is the length of C.

CHANGE OF VARIABLES

Let $z = g(\zeta)$ be a continuous function of a complex variable $\zeta = u + iv$. Suppose that curve C in the z plane corresponds to curve C' in the plane ζ plane and that derivative $g'(\zeta)$ is continuous on C'. Then

$$\int_C f(z) = \int_C f\{g(\zeta)\} g'(\zeta)d\zeta \qquad ...(23)$$

These conditions are certainly satisfied if g is analytic in a region containing curve C'.

SIMPLY AND MULTIPLY CONNECTED REGIONS

A region $\mathfrak{R}$ is called a simply-connected if any simple closed curve which lies in $\mathfrak{R}$ can be shrunk to a point without leaving $\mathfrak{R}$. A region $\mathfrak{R}$ which is not simply-connected is called multiply-connected.

For example, suppose $\mathfrak{R}$ is the region defined by $|z| < 2$ shown shaded in the Fig 3(a) given below. If Γ is any simple closed curve lying in $\mathfrak{R}$, we see that it can be shrunk to a point which lies in $\mathfrak{R}$, and thus does not leave $\mathfrak{R}$, so that $\mathfrak{R}$ is simply connected .

On the other hand if $\mathfrak{R}$ is the region defined by $1 < |z| < 2$, shown in shaded in the Fig 3(b), then there is a simple closed curve Γ lying in $\mathfrak{R}$ which cannot possibly be shrunk to a point without leaving $\mathfrak{R}$, so that $\mathfrak{R}$ is multiply-connected.

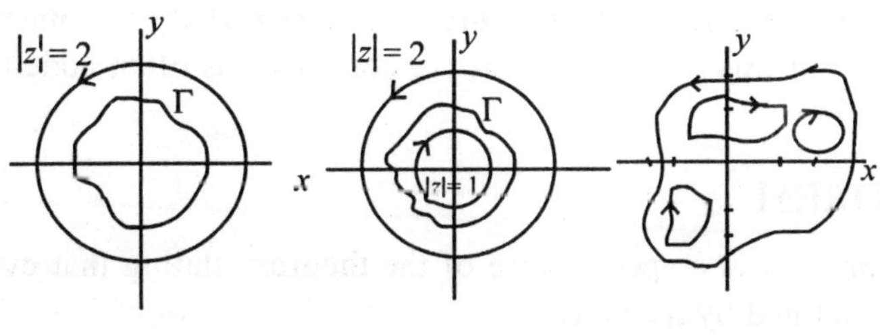

Fig. 3(a) **Fig.** 3(b) **Fig.** 3(c)

Intuitively, a simply-connected region is one which does which one does not have any "holes" in it, while a multiply-connected region is one which does. Thus the multiply-connected regions of Fig 3(b) and Fig 3(c) have respectively one and three holes in them.

JORDAN CURVE THEOREM

***Statement*: 2.** If is C a simple, closed curve in the plane $\mathfrak{R}^2$, that C is, $C \subset \mathfrak{R}^2$ and C is homeomorphic to S_1, then R^2 the complement of $R^2 - C$ has two components, each sharing C as boundary.

A Jordan curve, or simple, closed curve, is a subset C of R^2 that is homeomorphic to a circle. A Jordan arc, or simple arc, is a subset of R^2 homeomorphic to a closed line segment in R. A choice of homeomorphism gives a parameterization of the Jordan curve or arc, $\alpha: [0, 1] \to R^2$, as the composite of the homeomorphism f: $S_1 \to C \subset R^2$ with w: $[0, 1] \to S_1$, given by w(t) = (*cos* 2πt, *sin* 2πt). A Jordan curve will have many choices of parameterization α and so relevant properties of the curve C must be shown to be independent of the choice of α. Notice that the parameterisation, $\alpha: [0, 1] \to R^2$ is one-one on $[0, 1)$ and $\alpha(0) = \alpha(1)$.

Separation Theorem for Jordan arcs: A Jordan arc Λ does not separate the plane, that is, $R^2 - \Lambda$ is connected. Since R^2 is locally path-connected, the complement of Λ is connected if and only if it is path-connected. An intuitive argument to establish the separation theorem begins with a pair of points P and Q in $R^2 - \Lambda$. We can join P and Q by a path in R^2, and then try to show that the path can be modified to a path that avoids Λ. It may be the case that Λ is very complicated, and a general proof requires great care to show that you can always find such a path.

Jordan Curve Theorem

Any simple closed curve C divides the points of the plane not on C into two distinct domains (with no points in common) of which C is the common boundary.

Indefinite integrals

If $f(z)$ and $F(z)$ are analytic in a region $\Re$ and such that $F'(z)= f(z)$, then $F(z)$ is called an indefinite integral of $f(z)$ denoted by

$$F(z) = \int f(z)dz \qquad ...(24)$$

Since the derivative of any constant is zero, it follows that any two indefinite integrals can differ by a constant. For this reason an arbitrary constant c is often added to the right of the above equation.

CAUCHY THEOREM

The Cauchy theorem is a special case of the theorem stating that every closed convex surface is uniquely defined by its metric.

Some Consequences of Cauchy's Theorem

Theorem 1: **(Cauchy's theorem)** If γ is a simple closed anticlockwise curve in the complex plane and $f(z)$ is analytic on some open set that includes all of the curve γ and all points inside γ, then

$$\int_\gamma f(z) = 0 \qquad ...(25)$$

Proof. We write $dz = dx + i\,dy$ and use Green's theorem on

$$\int_\gamma f(z)dz = \int_\gamma f(x)dx + (if(z))dy \qquad ...(26)$$

$$= \iint_R \left(\frac{\partial(if(z))}{\partial x} - \frac{\partial f(z)}{\partial y} \right) dxdy \qquad ...(27)$$

(with R denoting the interior of γ). we egt

$$f'(z) = \frac{\partial f(z)}{\partial x} = \frac{1}{i}\frac{\partial f(z)}{\partial y}$$

It follows that the integrand of the double intergal we got from Green's theorem

$$\frac{\partial(if(z))}{\partial x}-\frac{\partial f(z)}{\partial y}=i\left(\frac{\partial f(z)}{\partial y}-\frac{1}{i}\frac{\partial f(z)}{\partial y}\right)=0 \qquad ...(28)$$

and so we get $\int_{\gamma} f(z)dz = 0$.

Corollary 2. Suppose we have two anticlockwise simple closed curves γ_1 and γ_2 with one entirely contained in the interior of the other. Suppose f(z) is analytic on some open set that includes both γ_1 and γ_2 and the region between the two curves. Then $\int_{\gamma_1} f(z) = \int_{\gamma_2} f(z)$

CAUCHY INTEGRAL THEOREM

Cauchy's integral theorem allows for integration over a complex variable. It deals only with analytic functions.

Cauchy's integral theorem also deals with functions which are not analytic, usually because of a discontinuity caused by attempting to divide by zero somewhere in the function's range. In this case, the result of the integration around the closed curve is 2 times pi times the square root of negative one (also denoted as i), times the residue of the function at the non-analytic point. (The point of non-analyticity is sometimes called a pole.) The residue is calculated by decomposing the function into an analytic and a non-analytic part, and then taking the value of the analytic part at the pole.

If $f(z)$ is analytic in some simply connected region R, then

$$\oint_r f(z)dz = 0 \qquad ...(29)$$

for any closed contour γ completely contained in R. Writing z as

$$z = x + iy \qquad ...(30)$$

and $f(z)$ as

$$f(z) = u + iv \qquad ...(31)$$

then gives

$$\oint_r f(z)dz = \int_r (u+iv)(dx+idy) \qquad ...(32)$$

$$= \int_r udx - vdy + i\int_r vdx + udy. \qquad ...(33)$$

From Green's theorem,

$$\int_r f(x,y)dx - g(x,y)dy = -\iint\left(\frac{\partial g}{\partial x}+\frac{\partial f}{\partial y}\right)dxdy \qquad ...(34)$$

$$\int_r f(x,y)dx + g(x,y)dy = \iint\left(\frac{\partial g}{\partial x}-\frac{\partial f}{\partial y}\right)dxdy, \qquad ...(35)$$

so (32) becomes

$$\oint_r f(z)dz = -\iint\left(\frac{\partial v}{\partial x}+\frac{\partial u}{\partial y}\right)dxdy + i\iint\left(\frac{\partial u}{\partial x}-\frac{\partial v}{\partial y}\right)dxdy. \qquad ...(36)$$

But the Cauchy-Riemann equations require that

$$\frac{\partial u}{\partial x} = \frac{\partial v}{\partial y} \qquad ...(37)$$

$$\frac{\partial u}{\partial y} = -\frac{\partial v}{\partial x} \qquad ...(38)$$

so

$$\oint_r f(z)dz = 0, \qquad ...(39)$$

For a multiply connected region,

$$\oint_{C_1} f(z)dz = \oint_{C_2} f(z)dz, \qquad ...(40)$$

Cauchy Integral Formula

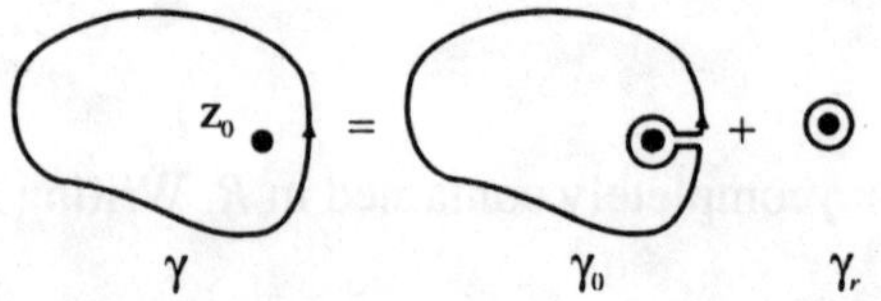

Cauchy's integral formula states that

$$f(z_0) = \frac{1}{2\pi i}\oint_r \frac{f(z)dz}{z-z_0}, \qquad ...(41)$$

where the integral is a contour integral along the contour γ enclosing the point z_0.

It can be derived by considering the contour integral

$$\oint_r \frac{f(z)dz}{z-z_0}, \qquad ...(42)$$

defining a path γ_r as an infinitesimal counterclockwise circle around the point z_0, and defining the path γ_0 as an arbitrary loop with a cut line (on which the forward and reverse contributions cancel each other out) so as to go around z_0. The total path is then

$$\gamma = \gamma_0 + \gamma_r, \quad \text{...(43)}$$

so

$$\oint_\gamma \frac{f(z)dz}{z-z_0} = \oint_{\gamma_0} \frac{f(z)dz}{z-z_0} + \oint_{\gamma_r} \frac{f(z)dz}{z-z_0}, \quad \text{...(44)}$$

From the Cauchy integral theorem, the contour integral along any path not enclosing a pole is 0. Therefore, the first term in the above equation is 0 since γ_0 does not enclose the pole, and we are left with

$$\oint_r \frac{f(z)dz}{z-z_0} = \oint_{\gamma_r} \frac{f(z)dz}{z-z_0}, \quad \text{...(45)}$$

Now, let $z = z_0 + re^{i\theta}$, so $dz = ire^{i\theta}d\theta$. ...(46)

Then

$$\oint_r \frac{f(z)dz}{z-z_0} = \oint_{\gamma_r} \frac{f(z_0 + re^{i\theta})}{re^{i\theta}} ire^{i\theta}d\theta$$

$$= \oint_{\gamma_r} f(z_0 + re^{i\theta})id\theta.$$

But we are free to allow the radius γ to shrink to 0, so

$$\oint_r \frac{f(z)dz}{z-z_0} = \lim_{r\to 0} \oint_{\gamma_r} f(z_0 + re^{i\theta})id\theta$$

$$= \oint_{\gamma_r} f(z_0)id\theta \quad \text{...(47)}$$

$$if(z_0)\oint_{\gamma_r} d\theta = 2\pi i f(z_0), \quad \text{...(48)}$$

If multiple loops are made around the point z_0, then equation (48) becomes

Winding Number

$$n(\gamma, z_0) f(z_0) = \frac{1}{2\pi i} \oint_r \frac{f(z)dz}{z-z_0}, \quad \text{...(49)}$$

where $n(\gamma, z_0)$ is the contour winding number.

A similar formula holds for the derivatives of $f(z)$,

$$f'(z_0) = \lim_{h\to 0}\frac{f(z_0+h)-f(z_0)}{h} \qquad ...(50)$$

$$= \lim_{h\to 0}\frac{1}{2\pi ih}\left[\oint_r \frac{f(z)dz}{z-z_0-h} - \oint_r \frac{f(z)dz}{z-z_0}\right]$$

$$= \lim_{h\to 0}\frac{1}{2\pi ih}\oint_r \frac{f(z)\left[(z-z_0)-(z-z_0-h)\right]dz}{(z-z_o-h)(z-z_0)}$$

$$= \lim_{h\to 0}\frac{1}{2\pi ih}\oint_r \frac{hf(z)dz}{(z-z_0-h)(z-z_0)}$$

$$= \frac{1}{2\pi i}\oint_r \frac{f(z)dz}{(z-z_0)^2}, \qquad ...(51)$$

Iterating again,

$$f''(z_0) = \frac{2}{z\pi i}\oint_r \frac{f(z)dz}{(z-z_0)^3}, \qquad ...(52)$$

Continuing the process and adding the contour winding number n,

$$n(\gamma, z_0)f^{(r)}(z_0) = \frac{r!}{2\pi i}\oint_r \frac{f(z)dz}{(z-z_0)^{r+1}} \qquad ...(53)$$

Now we are going to learn about some theorems:

(i) Morera's theorem

Statement 3. if the integral of a continuous complex-valued function f of a complex variable along every simple closed curve within an open set is zero, that is, if

$$\int_C f(z)dz = 0$$

for C any simple closed curve, then f is differentiable or "holomorphic" at every point in that open set.

Statement 4. If $f(z)$ is continuous in a region D and satisfies

$$\oint_r f dz = 0$$

for all closed contours γ in D, then $f(z)$ is analytic in D.

Morera's theorem does not require simple connectedness. Morera's theorem states that if the integral of a continuous complex-valued function f of a complex variable along every simple closed curve within an open set is zero, that is, if

$$\int_C f(z)dz = 0$$

for C any simple closed curve, then f is differentiable or "holomorphic" at every point in that open set.

Morera's theorem can be used to show the analyticity of functions defined by sums or integrals, such as the Riemann zeta function $\zeta(s) = \sum_{n=1}^{\infty} \frac{1}{n^s}$.

(ii) Liouville's Theorem

It states that every bounded entire function must be constant. That is, every function f for which there exists a real number M such that $|f(z)|$ d $\leq M$ for all z in C and which is holomorphic, or defined on the whole complex plane C, is constant.

This theorem can be generalized as follows: if M is a parabolic Riemann surface (such as the complex plane C) and N is a hyperbolic one (such as an open disk), then every holomorphic function $f: M \to N$ must be constant.

Proof:

Given f, we have

$$f(z) = \sum_{k=0}^{\infty} a_k z^k$$

by Taylor series about 0, which implies

$$|a_k| = \left|\frac{1}{2\pi i}\oint_{C_r} \frac{f(z)}{(z-0)^{k+1}}dz\right| = \frac{1}{2\pi}\left|\oint_{C_r} \frac{f(z)}{z^{k+1}}dz\right|$$

where C_r is the circle about 0 of radius r. By moving the absolute value inside of the integral, we find

$$\left|\oint_{C_r} \frac{f(z)}{z^{k+1}}dz\right| \leq \oint_{C_r} \frac{|f(z)|}{|z|^{k+1}}dz.$$

Now we can use the assumption that $|f(z)|$ d $\leq M$ for all z (since f is given to be bounded), and the fact that $|z|=r$ on the circle C_r. We get

$$\oint_{C_r} \frac{|f(z)|}{|z|^{k+1}}dz \leq \oint_{C_r} \frac{M}{\gamma^{k+1}}dz.$$

Then,

$$|a_k| \leq \frac{1}{2\pi}\frac{M}{\gamma^{k+1}}2\pi\gamma = \frac{\gamma M}{\gamma^{k+1}} = \frac{M}{\gamma^k}. \quad ...(54)$$

Let r now tend to infinity so the circle C_r gets ever larger. If k is greater than 0, M/r^k tends to zero and so a_k must be zero.

However, if $k = 0$, $r^0 = 1$ ($r \neq 0$ as r tends to infinity), so a_0 is the only term left in the Taylor series.

Statement: A bounded analytic function is constant

Proof: Suppose that $|f(z)| < M$ and that Cauchy's integral formula is used to compare the values of $f(z)$ at two points. Then

$$f(z_2) - f(z_1) = \frac{i}{2\pi i}\oint\left\{\frac{f(z)}{z-z_1} - \frac{f(z)}{z-z_2}\right\}dz.$$

A little rearrangement produces

$$\frac{f(z_2)-f(z_1)}{z_2-z_1} = \frac{1}{2\pi i}\oint\frac{f(z)\,dz}{(z-z_0)^2}\left\{\frac{z-z_0}{z-z_1}\right\}\left\{\frac{z-z_0}{z-z_2}\right\},$$

which could very easily be used in an alternative derivation of the Cauchy formula for a derivative.

(iii) Fundamental Theorem of Algebra

Statement: Every polynomial equation having complex coefficients and degree ≥ 1 has at least one complex root.

Let us suppose that the polynomial

$f(z) = a_n z^n + a_{n-1} z^{n-1} + \ldots + a_1 z^1 + a_0$

has no root, so that for every complex number z

$f(z) \neq 0$.

(iv) Gauss's Mean-Value Theorem

Let $f(z)$ be an analytic function in $|z-\alpha| < R$. Then

$$f(z) = \frac{1}{2\pi}\int_0^{2\pi} f\left(z + re^{i\theta}\right)d\theta \qquad \ldots(55)$$

for $0 < \gamma < R$.

(v) Maximum Modulus Principle

Let the contour be a circle of radius ρ centered at the origin. Then, supposing that

$$M(\rho) = \max_{0 \leq \theta \leq 2\pi} f(\rho,\theta) \qquad \ldots(56)$$

we get for the function

$$|f(0)| \leq \frac{1}{2\pi}\int_0^{2\pi}\frac{|f(\rho,\theta)|}{\rho}d\theta \qquad ...(57)$$

$$\leq \frac{M(\rho)}{\rho}, \qquad ...(58)$$

for its derivative,

$$|f'(0)| \leq \frac{n!}{2\pi}\int_0^{2\pi}\frac{|f(\rho,\theta)|}{\rho^n}d\theta \qquad ...(59)$$

$$\leq \frac{n!M(\rho)}{\rho^n}. \qquad ...(60)$$

and in general

$$|f^{(n)}(0)| \leq \frac{n!}{2\pi}\int_0^{2\pi}\frac{|f(\rho,\theta)|}{\rho^n}d\theta$$

$$\leq \frac{n!M(\rho)}{\rho^n}. \qquad ...(61)$$

This series of inequalities relates the value of the function (as well as its derivatives) at the center of a circle of analyticity to the maximum value on the circumference of the circle. Any bound would establish an inequality, but the bound used in this series of equations is optimal, given that it actually occurs somewhere on the perimeter of the bounding circle.

Since the extreme case of equality in equation (59) reduces $|f(\rho,\theta)|$ to the real constant $M(\rho)$, equations (60) through (61) and onwards experience the same reduction, establishing equality along the line.

Supposing a Maclaurin series for $f(z)$

$$f(z) = \sum_{n=0}^{\infty}\frac{|f(n)(0)|}{n!}|z|^n.$$

taking absolute values would lead to an inequality

$$|f(z)| \leq \sum_{n=0}^{\infty}\frac{|f(n)(0)|}{n!}|z|^n. \qquad ...(62)$$

Substituting the extreme case equalities gives an exact replacement, but the series can be summed whether extreme or not:

$$|f(z)| \leq \sum_{n=0}^{\infty} M(\rho)\left(\frac{|z|}{\rho}\right)^n, \qquad ...(63)$$

$$\leq M(\rho)\frac{\rho}{\rho-|z|}. \qquad ...(64)$$

which is consistent with what we have already learned at $z = 0$, but but thoroughly ineffective when $|z| = \rho$.

There are corresponding minimum inequalities, for functions which do not vanish within the circle; it is only necessary to apply the foregoing reasoning to the then finite-valued $g(z) = 1/f(z)$ in place of $f(z)$.

For a single circle, these results are of limited utility, but it is easy enough to extend them to more general domains, as long as they accomodate the intermediate circles needed in the following argument. The main concern is with taking the bounding modulus from anywhere within the domain, rather than just from the boundary.

The maximum absolute value of an analytic function lies on the boundary of any domain of analyticity.

(vi) (Maximum modulus theorem) Let $G \subset C$ be a connected open set and $f : G \to C$ analytic. If there is any $a \in G$ with $|f(a)| \geq |f(z)|$ for all $z \in G$, then f is constant.

Proof: (Another way to state this is that $|f(z)|$ cannot have a maximum in G, unless f is constant.)

Choose $\delta > 0$ so that $D(a; \delta) \subset G$. Fix $0 < r < \delta$ and then we have (by the Cauchy integral

Formula)

$$|f(a)| = \frac{1}{2\pi i}\int_{|x-a|=r}\frac{f(z)}{z-a}dz. \qquad ...(65)$$

Write this out in terms of a parametrisation $z = a + re^{i\theta}$ with $0 \leq \theta \leq 2\pi, dz = ire^{i\theta}d\theta.$

$$f(a) = \frac{1}{2\pi i}\int_0^{2\pi}\frac{f\left(a+re^{i\theta}\right)}{ire^{i\theta}}d\theta\frac{1}{2\pi}\int_0^{2\pi} f\left(a+re^{i\theta}\right)d\theta. \qquad ...(66)$$

Hence

$$|f(a)| \leq \frac{1}{2\pi}\int_0^{2\pi}\left|f\left(a+re^{i\theta}\right)\right|d\theta$$

$$\le \frac{1}{2\pi}\int_0^{2\pi}|f(a)|d\theta = |f(a)|,$$

Using $\left|f\left(a+re^{i\theta}\right)\right| \le |f(a)| \forall \theta.$ ...(67)

We must therefore have equality in the inequalities. Since the integrand $\left|f\left(a+re^{i\theta}\right)\right|$ is a continuous function of θ, this implies $\left|f\left(a+re^{i\theta}\right)\right| = |f(a)|$ for all a.

Put $|f(a)|$ = Arg(*f*(*a*)). Now

$$|f(a)| = e^{-i\alpha}f(a)$$

$$= \frac{e^{-i\alpha}}{2\pi}\int_0^{2\pi} f\left(a+re^{i\theta}\right)d\theta$$

$$= \frac{1}{2\pi}\int_0^{2\pi} e^{-i\alpha} f\left(a+re^{i\theta}\right)d\theta \quad ...(68)$$

$$\Re|f(a)| = |f(a)| = \frac{1}{2\pi}\Re\int_0^{2\pi} e^{-i\alpha} f\left(a+re^{i\theta}\right)d\theta$$

$$= \frac{1}{2\pi}\int_0^{2\pi} \Re\left(e^{-i\alpha} f\left(a+re^{i\theta}\right)\right)d\theta$$

$$\le \frac{1}{2\pi}\int_0^{2\pi} \left|e^{-re} f\left(a+re^{i\theta}\right)\right|d\theta$$

using $\Re w \le |w|$ for $w \in \mathbb{C}$

$$\le \frac{1}{2\pi}\int_0^{2\pi} \left|f\left(a+re^{i\theta}\right)\right|d\theta$$

$$\le \frac{1}{2\pi}\int_0^{2\pi} |f(a)|d\theta = |f(a)| \quad ...(69)$$

Thus we must again have equality in all the inequalities and so

$$\Re\left(e^{-i\alpha} f\left(a+re^{i\theta}\right)\right) = \left|e^{-i\alpha} f\left(a+re^{i\theta}\right)\right| = |f(a)|$$

for all θ. Thus $\Im\left(e^{-i\alpha} f\left(a+re^{i\theta}\right)\right) = 0$ and

$$e^{-i\alpha} f\left(a+re^{i\theta}\right) = |f(a)| \text{ or } f\left(a+re^{i\theta}\right) = e^{i\alpha}|f(a)|. \quad ...(70)$$

Thus f(z) is constant for z in the infinite compact subset $\{z : |z-a| = r\}$ of G.

Note: The maximum modulus theorem is not always true on an unbounded domain

(vii) Argument Principle and Rouche's Theorem

If $f(z)$ is meromorphic in a region R enclosed by a contour γ, let N be the number of complex roots of $f(z)$ in γ, and P be the number of poles in γ, with each zero and pole counted as many times as its multiplicity and order, respectively. Then

$$N - P = \frac{1}{2\pi i}\int_\gamma \frac{f'(z)\,dz}{f(z)}. \qquad ...(71)$$

Defining $w = f(z)$ and $\rho = f(\gamma)$ gives

$$N - P = \frac{1}{2\pi i}\int_\rho \frac{dw}{w}.$$

Corollary 3. Suppose that f(z) is analytic in the simply connected domain D. Let C be a simple closed positively oriented contour in D such that for $z \in C, f(z) \neq 0$. Then

$$\frac{1}{2\pi i}\int_C \frac{f'(z)}{f(z)}dz = Z_f \qquad ...(72)$$

where Z_f is the number of zeros of f(z) that lie inside C.

Recall that a branch of the logarithm function, $\log_\alpha(z)$, is defined by

$$\log_\alpha(z) = \ln|z| + i\arg_\alpha(z) = \ln r + i\phi \qquad ...(73)$$

where $z = re^{i\phi}$ and $0 < \phi \leq 2\pi$. Loosely speaking, suppose that for some branch of the logarithm, the composite function $\log_\alpha(f(z))$ were analytic in a simply connected domain D containing the contour C. This would imply that $\log_\alpha(f(z))$ is an anti-derivative of the function $\frac{f'(z)}{f(z)}$ for all $z \in D$.

In the given figure, we are indicating that $\frac{1}{2\pi i}\int_C \frac{f'(z)}{f(z)}dz$ counts how many times the curve $f(C)$ winds around the origin.

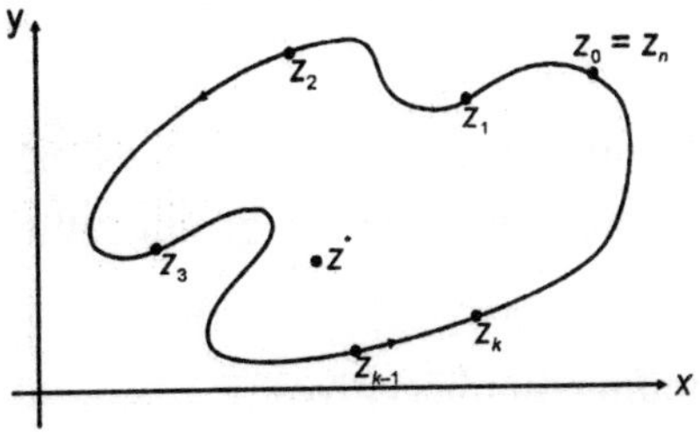

Fig. 4 The points Z_k on the contour C that winds around Z^*.

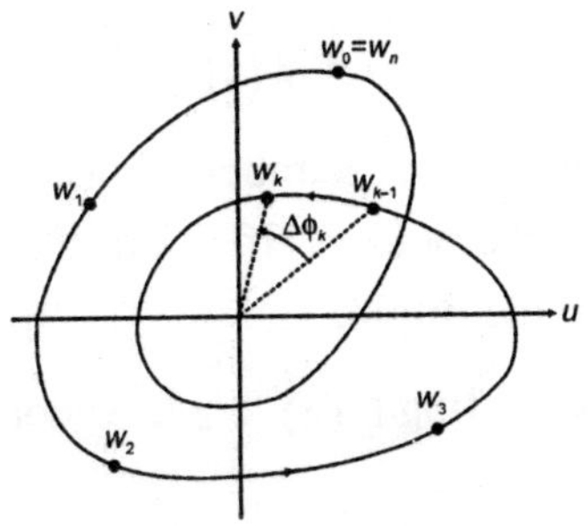

Fig. 5 The points w_k on the contour $f(C)$ that winds around 0.

Example 6. Find the residue of $f(z) = z^2 + z$ define for the curve $\{(x, y) = (4\cos 2t + 2\cos t, 4\sin 2t + 2\sin t) : 0 \le t \le 2\pi\}$

Solution: Since, the image of the curve $f(C_2(0))$ winds twice around the origin. Therefore

$$\frac{1}{2\pi i}\int_{C_2^+(0)} \frac{f'(z)}{f(z)} dz = \frac{1}{2\pi i}\int_{C_2^+(0)} \frac{2z+1}{z^2+z} dz.$$

Also, the residues of the integrand are at 0 and -1. Thus

$$\frac{1}{2\pi i}\int_{C_2^+(0)} \frac{2z+1}{z^2+z} dz = \text{Res}\left[\frac{2z+1}{z^2+z}, 0\right] + \text{Res}\left[\frac{2z+1}{z^2+z}, 1\right]$$

$$= 1 + 1 = 2$$

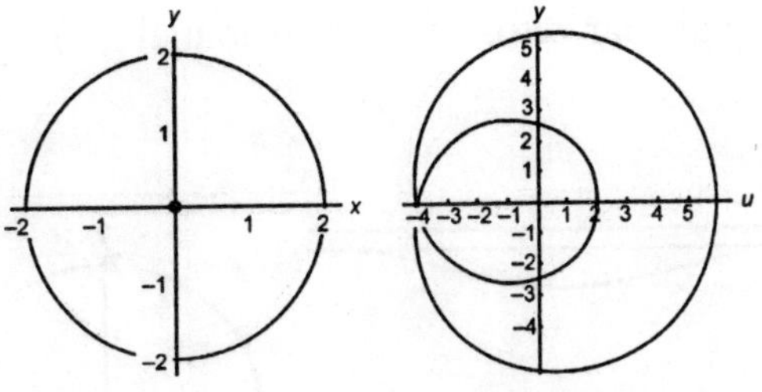

Fig. 6 The image curve $f(C_2(0))$ under $w = f(z) = z^2 + z$.

Finally, we note that if $g(z) = f(z) - a$, then $g'(z)$ f = (z), and thus we can generalize what we've just said to compute how many times the curve $f(c)$ winds around the point a. Theorem 8.9 summarizes our discussion.

***Theorem* 2. (Winding Number).** Suppose $f(z)$ is meromorphic in the simply connected domain D. If C is a simple closed, positively oriented contour in D such that for $z \in C, f(z) \ne 0$ and $f(z) \ne \infty$, then

$$w(f(c),a) = \frac{1}{2\pi i}\int_C \frac{f'(z)}{f(z)-a}dz, \tag{74}$$

known as the winding number of $f(C)$ about a, counts the number of times the curve $f(C)$ winds around the point a. If $a = 0$, the integral counts the number of times the curve $f(C)$ winds around the origin.

Remark. Letting $f(z) = z$ in Theorem (Winding number) gives

$$W(c, a) = \frac{1}{2\pi i}\int_C \frac{1}{z-a}dz = \begin{cases}1, \text{ if a lies inside C, or}\\ 0, \text{ If a lies outside C,}\end{cases} \quad \text{...(75)}$$

which counts the number of times the curve C winds around the point a. If C is not a simple closed curve, but crosses itself perhaps several times, we can show (but omit the proof) that $W(c, a)$ still gives the number of times the curve C winds around the point a. Thus winding number is indeed an appropriate term.

***Theorem* 3. (Rouche's Theorem).** Suppose that $f(z)$ and $g(z)$ are meromorphic functions defined in the simply connected domain D, that C is a simply closed contour in D, and that $f(z)$ and $g(z)$ have no zeros or poles for $z \in C$. If the strict inequality $|f(z)+g(z)| < |f(z)|+|g(z)|$ holds for all $z \in C$, then $z_f - P_f = Z_g - P_g$.

Corollary 4. Suppose that f(z) and g(z) are analytic functions defined in the simply connected domain D, that C is a simple closed contour in D, and that f(z) and g(z) have no zeros for $z \in C$. If the strict inequality $|f(z) + g(z)| < |f(z)| + |g(z)|$–(60)

holds for all $z \in C$, then $z_f = z_g$.

Example 7. Show that al four zeros of the polynomial $g(z) = z^4 - 7z - 1$ lie in the disk $D_2(0) = \{z : |z| < 2\}$.

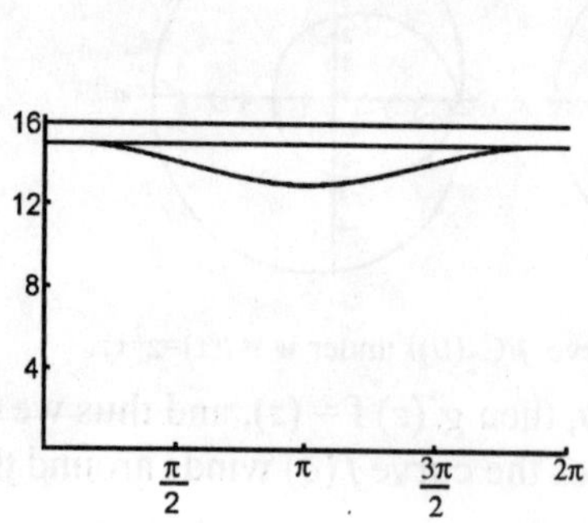

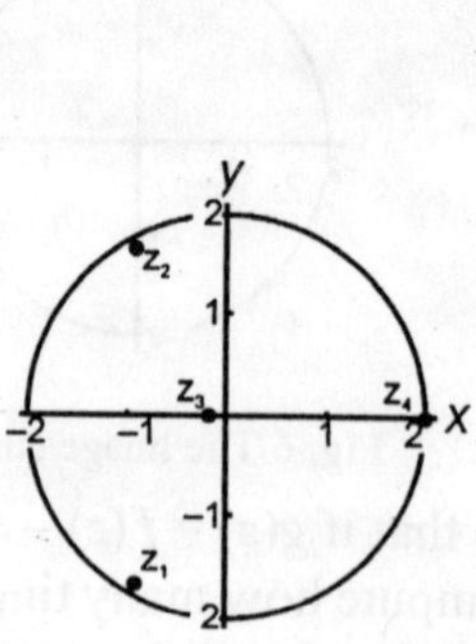

Solution: Let $f(z) = -z^4$. Then $f(z) + g(z) = -7z - 1$, and at points on the circle $C_2(0) = \{z : |z| = 2\}$ we have the relation

$$\begin{aligned}|f(z)+g(z)| &= |-7z-1|\\ &\le |-7z|+|-1|\\ &= 7|z|+1 = 7\times 2+1\\ &= 15 < 16\\ &= |z^4| = |-z^4| = |f(z)|\end{aligned} \qquad ...(76)$$

Of course, if $|f(z)+g(z)| < |f(z)|$, then as we indicated in the above remark, we certainly have $|f(z)+g(z)| < |f(z)|+|g(z)|$, so that the conditions for applying above corollary are satisfied on the circle $C_2(0)$. The function *f*(*z*) has a zero of order 4 at the origin, so *g*(*z*) must have four zeros inside $D_2(0)$.

(viii) Poisson Integral Formulas for Circle

In complex analysis, let $u : U \to \Re$ be a harmonic function on a neighborhood of the closed disk $\bar{D}(0,1)$, then for any point z_0 in the open disk $D(0,1)$,

$$u(z_0) = \frac{1}{2\pi}\int_0^{2\pi} u(e^{i\varphi})\frac{1-|z_0|^2}{|z_0-e^{i\varphi}|^2}\,d\varphi. \qquad ...(77)$$

In polar coordinates on $\bar{D}(0,R)$,

$$u(z_0) = \frac{1}{2\pi}\int_0^{2\pi} K(r,\theta)\phi(z_0+re^{i\theta})\,d\theta, \qquad ...(78)$$

where $R=|z_0|$ and $K(r,\theta)$ is the Poisson kernel. For a circle,

$$u(x,y) = \frac{1}{2\pi}\int_0^{2\pi} u(\alpha\cos\phi,\ \alpha\sin\phi)\,\frac{a^2-R^2}{a^2+R^2-2aR\cos(\theta-\phi)}\,d\phi. \qquad ...(79)$$

For a sphere,

$$u(x,y,z) = \frac{1}{4\pi a}\iint_S u\frac{a^2-R^2}{\left(a^2+R^2-2aR\cos\theta\right)^{\frac{3}{2}}}\,ds, \qquad ...(80)$$

where

$$\cos\theta = \mathrm{x}.\xi \qquad ...(81)$$

Poisson Kernel

The integral kernel in the Poisson integral, given by

$$k(\psi) = \frac{1}{2\pi}\frac{1-|z_0|^2}{|z_0 - e^{i\psi}|^2} \quad ...(82)$$

for the open unit disk $D(0,1)$. Writing $z_0 = re^{i\theta}$ and taking $D(0,R)$ gives

$$K(r,\theta) \equiv \frac{1}{2\pi}R\left[\frac{R+re^{i\theta}}{R-re^{i\theta}}\right]$$

$$= \frac{1}{2\pi}R\left[\frac{(R+re^{i\theta})(R-re^{-i\theta})}{(R-re^{i\theta})(R-re^{-i\theta})}\right]$$

$$= \frac{1}{2\pi}R\left[\frac{R^2 - rR(e^{i\theta}-e^{-i\theta})-r^2}{R^2 - rR(e^{i\theta}+e^{-i\theta})+r^2}\right]$$

$$= \frac{1}{2\pi}R\left[\frac{R^2 + 2irR\sin\theta - r^2}{R^2 - 2Rr\cos\theta + r^2}\right]$$

$$= \frac{1}{2\pi}\frac{R^2 - r^2}{R^2 - 2Rr\cos\theta + r^2} \quad ...(83)$$

Poisson's Equation

In mathematics, Poisson's equation is a partial differential equation with broad utility in electrostatics, mechanical engineering and theoretical physics.

The Poisson equation is

$$\Delta\varphi = f \quad ...(84)$$

where Δ is the Laplace operator, and f and $f(E)$ are real or complex-valued functions on a manifold. When the manifold is Euclidean space, the Laplace operator is often denoted as ∇^2 and so Poisson's equation is frequently written as

$$\nabla^2\varphi = f \quad ...(85)$$

In three-dimensional Cartesian coordinates, it takes the form

$$\left(\frac{\partial^2}{\partial x^2}+\frac{\partial^2}{\partial y^2}+\frac{\partial^2}{\partial z^2}\right)\varphi(x,y,z) = f(x,y,z). \quad ...(86)$$

For vanishing *f*, this equation becomes Laplace's equation

$$\Delta\varphi = 0.$$

The Poisson equation may be solved using a Green's function.

Theorem **(Poisson's Integral Formula).** Let U be a real-valued function that is piecewise continuous and bounded for all real t. The function

$$\phi(x,y) = \frac{Y}{\pi}\int_{-\infty}^{\infty}\frac{U(t)}{(x-t)^2+Y^2}dt \qquad ...(87)$$

is harmonic in the upper half plane Im(z) > 0 and has the boundary values

$$\phi(x, 0) = U(x)$$

wherever U(x) is continuous.

Example 8. Find the function $\phi(x, y)$ that is harmonic in the upper half-plane Im(z) > 0, which takes on the boundary values

$\phi(x, 0) = 1$, for $-1 \le x \le 1$;

$\phi(x, 0) = 0$, for $|x| > 1$.

Solution: Using poisson integral formula, we get

$$\phi(x,y) = \frac{Y}{\pi}\int_{-1}^{1}\frac{1}{(x-t)^2+Y^2}dt$$

Since, $\displaystyle\int\frac{Y}{(x-t)^2+Y^2}dt = \text{Arc}\tan\left(\frac{Y}{x-t}\right)$.

Therefore, $\displaystyle\phi(x,y) = \frac{1}{\pi}\left(\text{Arc}\tan\frac{Y}{x-t}\right)\Bigg|_{t=-1}^{t-1}$

$$= \frac{1}{\pi}\text{Arc}\tan\left(\frac{Y}{x-1}\right) - \frac{1}{\pi}\text{Arc}\tan\left(\frac{Y}{x+1}\right)$$

CONTOUR INTEGRATION

Contour integration is the process of calculating the values of a contour integral around a given contour in the complex plane. As a result of a truly amazing property of holomorphic functions, such integrals can be computed easily simply by summing the values of the complex residues *inside* the contour.

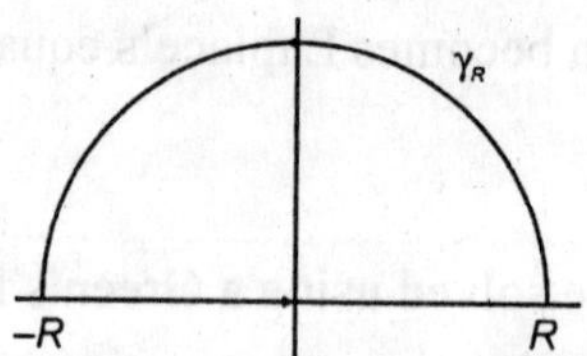

Let $P(x)$ and $Q(x)$ be polynomials of polynomial degree n and respectively m with coefficients $b_n, ..., b_0$ and $C_m, ..., C_0$. Take the contour in the upper half-plane, replace x by z, and write $z =$ Then

$$\int_{-\infty}^{\infty} \frac{P(z)dz}{Q(z)} = \int_{-R}^{R} \frac{P(z)dz}{Q(z)} \quad ...(88)$$

Define a path γR which is straight along the real axis from $-R$ to R and make a circular half-arc to connect the two ends in the upper half of the complex plane. The residue theorem then gives

$$\lim_{R\to\infty} \int_{\gamma R} \frac{P(z)dz}{Q(z)} = \lim_{R\to\infty} \int_{-R}^{R} \frac{P(z)dz}{Q(z)} + \lim_{R\to\infty} \int_0^{\pi} \frac{P(\mathrm{Re}^{i\theta})}{Q(\mathrm{Re}^{i\theta})} i\,\mathrm{Re}^{i\theta}\, d\theta \quad ...(89)$$

$$= 2\pi i \sum_{I[z]>0} \mathrm{Res}\left[\frac{P(z)}{Q(z)}\right] \quad ...(90)$$

where Res[z] denotes the complex residues. Solving,

$$\lim_{R\to\infty} \int_{-R}^{R} \frac{P(z)dz}{Q(z)} = 2\pi i \sum_{I[z]>0} \mathrm{Res}\frac{P(z)}{Q(z)} - \lim_{R\to\infty} \int_0^{\pi} \frac{P(\mathrm{Re}^{i\theta})}{Q(\mathrm{Re}^{i\theta})} i\,\mathrm{Re}^{i\theta}\, d\theta \quad ...(91)$$

Define

$$I_R = \lim_{R\to\infty} \int_0^{\pi} \frac{P(\mathrm{Re}^{i\theta})}{Q(\mathrm{Re}^{i\theta})} i\,\mathrm{Re}^{i\theta}\, d\theta \quad ...(92)$$

$$\lim_{R\to\infty} \int_0^{\pi} \frac{b_n(\mathrm{Re}^{i\theta})^n + b_{n-1}(\mathrm{Re}^{i\theta})^{n-1} + ... + b_0}{c_m(\mathrm{Re}^{i\theta})^m + c_{m-1}(\mathrm{Re}^{i\theta})^{m-1} + ... + c_0} iRd\theta \quad ...(93)$$

$$= \lim_{R\to\infty} \int_0^{\pi} \frac{b_n}{c_m} (\mathrm{Re}^{i\theta})^{n-m} iRd\theta \quad ...(94)$$

$$\lim_{R\to\infty} \int_0^{\pi} \frac{b_n}{c_m} R^{n+1-m} i(e^{i\theta})^{n-m} d\theta \quad ...(95)$$

and set

$$\in = -(n+1-m), \quad ...(96)$$

then equation (95) becomes

$$I_R \equiv \lim_{R\to\infty} \frac{i}{R^{\epsilon}} \frac{b_a}{c_m} \int_0^{\pi} e^{i[n-m]} d\theta. \qquad ...(97)$$

Now,

$$\lim_{R\to\infty} R^{-\epsilon} = 0 \qquad ...(98)$$

for $\epsilon > 0$. That means that for $-n-1+m \geq 1$, or

$m \geq n+2, I_R = 0$, so

$$\int_{-\infty}^{\infty} \frac{P(z)dz}{Q(z)} = 2\pi i \sum_{I[z]>0} \text{Res}\left[\frac{P(z)}{Q(z)}\right] \qquad ...(99)$$

for $m \geq n+2$. Apply Jordan's lemma with $f(x) = P(x)/Q(x)$. We must have

$$\lim_{x\to\infty} f(x) = 0, \qquad ...(100)$$

so we require $m \geq n+1$.

Then

$$\int_{-\infty}^{\infty} \frac{P(z)}{Q(z)} e^{iaz} dz = 2\pi i \sum_{I[z]>0} \text{Res}\left[\frac{P(z)}{Q(z)} e^{iaz}\right] \qquad ...(101)$$

for $m \geq n+1$ and $\alpha > 0$. Since this must hold separately for real and imaginary parts, this result can be extended to

$$\int_{-\infty}^{\infty} \frac{P(x)}{Q(x)} \cos(\alpha x)dx =$$

$$\int_{-\infty}^{\infty} \frac{P(x)}{Q(x)} \cos(ax)dx = 2\pi \mathrm{R}\left\{\sum_{I(z)>0} \text{Res}\left[\frac{P(z)}{Q(z)} e^{iaz}\right]\right\} \qquad ...(102)$$

$$\int_{-\infty}^{\infty} \frac{P(x)}{Q(x)} \sin(ax)dx = 2\pi \mathrm{I}\left\{\sum_{I(z)>0} \text{Res}\left[\frac{P(z)}{Q(z)} e^{iaz}\right]\right\} \qquad ...(103)$$

CONTOUR

An integral obtained by contour integration. The particular path in the complex plane used to compute the integral is called a contour.

The notation $\int^{(a+)} f(z)dz$ is use to denote the contour integral of $f(z)$ with contour encircling the point α once in a counterclockwise direction.

Here, we define and evaluate integrals of the form $\int_{c} f(z)dz$, where f(t) is complex-valued and C is a contour in the plane (so that z is complex, with $z \in C$).

Recall that for evaluating contour integral represent the curve C we used the parametric notation

$$C : z(t) = x(t) + iy(t) \text{ for } a \le t \le b, \ldots \qquad \ldots(104)$$

where x(t) and y(t) are continuous functions. We now place a few more restrictions on the type of curve to be described. The following discussion leads to the concept of a contour, which is a type of curve that is adequate for the study of integration.

Simple Curve: C is simple if it does not cross itself, which means that $z(t_1) \neq z(t_2)$ whenever $t_1 \neq t_2$, except possibly when $t_1 = a$ and $t_2 = b$.

Closed Curved: A curve C with the property that, $z(a) = z(b)$ is a closed curve. If $z(a) = z(b)$, is the only point of intersection, then we say that C is a simple closed curve. As the parameter t increases from the value $t = a$ to the value $t = b$, the point $z(t)$ starts at the initial point $z(a)$, moves along the curve C, and ends up at the terminal point $z(b)$. If C is simple, then $z(t)$ moves continuously from $z(a)$ to $z(b)$ as t increases and the curve is given an orientation, which we indicate by drawing arrows along the curve. Figure 1 illustrates how the terms simple and closed describe a curve.

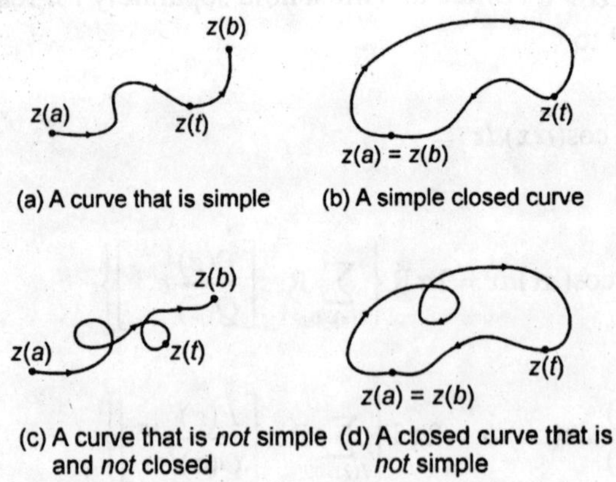

Fig. 7 The terms simple and closed used to describe curves.

The complex-valued function $z(t) = x(t) + iy(t)$ is said to be differentiable on $[a, b]$ if both $x(t)$ and $y(t)$ are differentiable for $a \le t \le b$. Here we require the one-sided derivatives of $x(t)$ and $y(t)$ to exist at the endpoints of the interval. The derivative $z'(t)$ is

$$z'(t) = x'(t) + i\,y'(t) \text{ for } a \le t \le b.$$

The curve C defined by abobve equation is said to be a smooth curve if $z'(t)$ is continuous and nonzero on the interval. If C is a smooth curve, then C has a nonzero tangent vector at each point $z(t)$, which is given by the vector $z'(t)$. If $x'(t_0) = 0$, then the tangent vector $z'(t_0) = iy'(t_0)$ is vertical. If $x'(t_0) \neq 0$, then the slope $\frac{dy}{dx}$ of the tangent line to C at the point $z(t_0)$ is given by $\frac{y'(t_0)}{x'(t_0)}$. Hence for a smooth curve the angle of inclination $\theta(t)$ of its tangent vector $z'(t)$ is defined for all values of $t \in [a, b]$ and is continuous. Thus a smooth curve has no corners or cusps, which is illustrated in the figure given below..

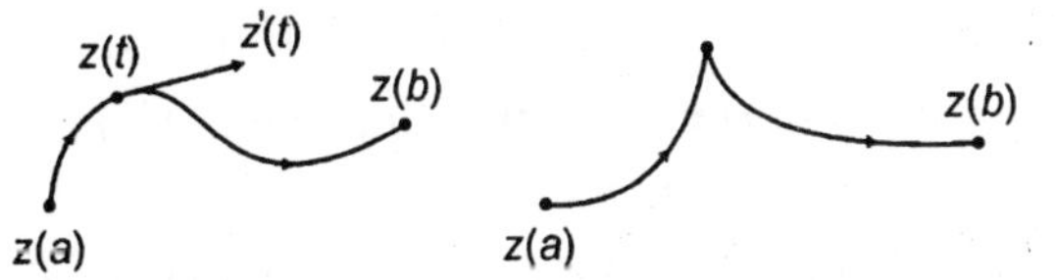

(a) A smooth curve (b) A curve that is *not* smooth

Fig. 8 The term smooth used to describe curves.

If C is a smooth curve, then ds, the differential of arc length, is given by

$$ds = \sqrt{[x'(t)]^2 + [y'(t)^2]}\,dt = |z'(t)|\,dt. \qquad ...(105)$$

The function $s(t) = \sqrt{[x'(t)^2 + [y'(t)^2]]}$ is continuous, as $x'(t)$ and $y'(t)$ are continuous functions, so the length $L(C)$ of the curve C is

$$L(C) = \int_a^b \sqrt{[x'(t)]^2 + [y'(t)]^2}\,dt = \int_a^b |z'(t)|\,dt \qquad ...(106)$$

Now consider C to be a curve with parameterization

$$C : z_1(t) = x(t) + iy(t) \text{ for } a \le t \le b. \qquad ...(107)$$

The opposite curve $-C$ traces out the same set of points in the plane, but in the reverse order, and has the parametrization

$$-C : z_2(t) = x(-t) + iy(-t) \text{ for } a \le t \le b. \qquad ...(108)$$

Since $z_2(t) = z_1(-t)$, $-C$ is merely C traversed in the opposite sense, as illustrated in figure given below..

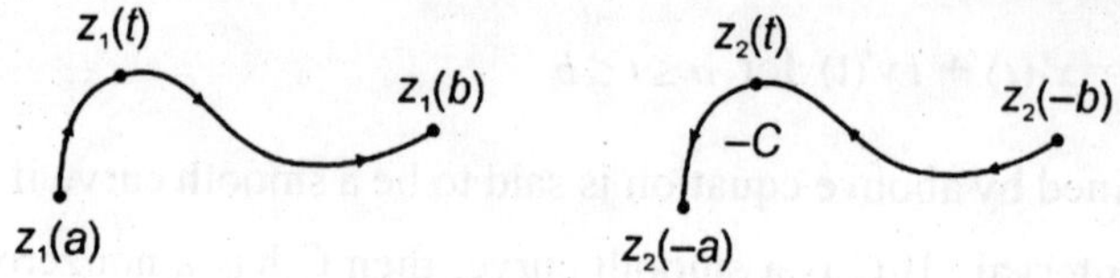

Fig. 9 The curve C and its opposite curve $-C$.

A curve C that is constructed by joining finitely many smooth curves end to end is called a contour. Let $C_1, C_2, ..., C_n$ denote n smooth curves such that the terminal point of the curve C_k coincides with the initial point of C_{k+1} for $k = 1, 2, ..., n-1$. We express the contour C by the equation

$$C = C_1 + C_2 + ... + C_n.$$

A synonym for contour is path.

Example 9. Find a parameterization of the polygonal path $C = C_1 + C_2 + C_3$ from $-1+i$ to $3-i$ shown in Figure 4. Here C_1 is the line from $-1+i$ to -1, C_2 is the line from -1 to $1 + i$, and C_3 is the line from $1 + i$ to $3 - i$.

Solution: We express C as three smooth curves, or $C = C_1 + C_2 + C_3$. If we set $z_0 = -1 + i$ and $z_1 = -1$, we can use equation of cvurve to get a formula for the straight-line segment joining two points:

$$C_1 : z_1(t) = z_0 + t(z_1 - z_0) = (-1+i) + t(-1-(-1+i))$$

for $0 \le t \le 1$. When simplified, this formula becomes

$$C_1 : z_1(t) = -1 + i(1-t), \text{ for } 0 \le t \le 1.$$

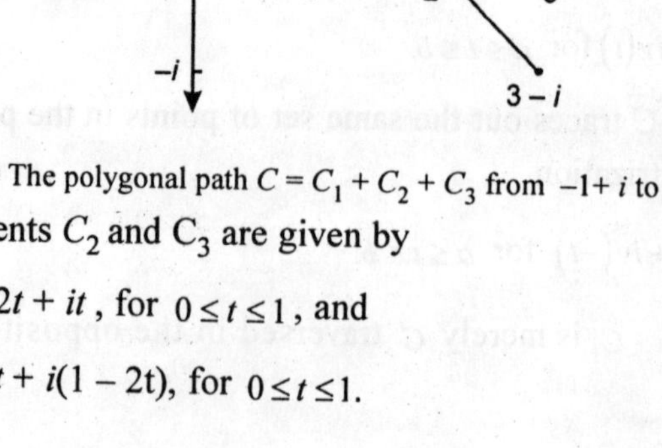

Fig. 10 The polygonal path $C = C_1 + C_2 + C_3$ from $-1+i$ to $3-i$.

Similarly, the segments C_2 and C_3 are given by

$$C_2 : z_2(t) = -1 + 2t + it, \text{ for } 0 \le t \le 1, \text{ and}$$

$$C_3 : z_3(t) = 1 + 2t + i(1 - 2t), \text{ for } 0 \le t \le 1.$$

We are now ready to define the integral of a complex function along a contour C in the plane with initial point A and terminal point B. Our approach is to mimic what is done in calculus. We create a partition $Pn = \{z_0 = A, z_1, z_2, ..., z_n = B\}$ of points that proceed along C from A to B and form the differences $\Delta z_k = z_k - z_{k-1}$ for $k = 1,2,...,n$. Between each pair of partition points z_{k-1} and z_k we select a point C_k on C, as shown in Figure 5, and evaluate the function $f(c_k)$.

These values are used to make a Riemann Sum for the partition:

$$S(P_n) = \sum_{k=1}^{n} f(c_k)(z_k - z_{k-1}) = \sum_{k=1}^{n} f(c_k)\Delta z_k. \qquad ...(109)$$

Assume now that there exists a unique complex number L that is the limit of every sequence $\{S(P_n)\}$ of Riemann sums given in

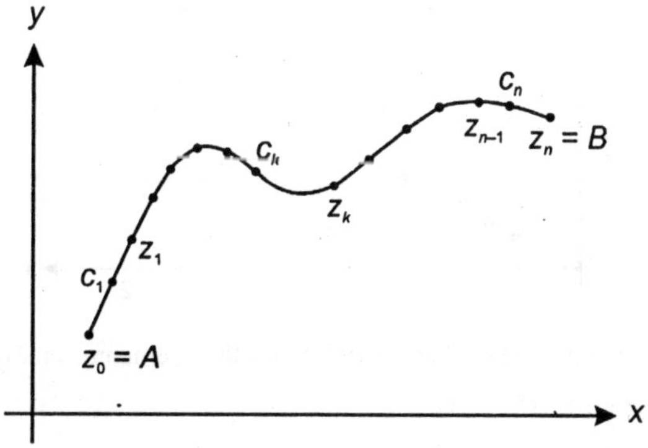

Fig. 11 Partition points $\{z_k\}$ and Function Evaluation Points $\{c_k\}$ for a Riemann Sum Along the Contour C from $z = A$ to $Z = B$.

Equation (109), where the maximum of $|\Delta Z_k|$ tends toward 0 for the sequence of partitions. We define the number L as the value of the integral of the function $f(z)$ taken along the contour C.

***Definition 2* (Complex Integral).** Let C be a contour. Then

$$\int_c f(z)dz = \lim_{n\to\infty} \sum_{k=1} f(c_k)\Delta z_k,$$

provided that the limit exists in the sense previously discussed.

Note that in Definition 1, the value of the integral depends on the contour.

Example 10. Use a Riemann sum to get an approximation for the integral $\int_c e^z dz$ where C is a the line segment joining the point $A = 0$ to $B = 2 + \frac{i\pi}{4}$.

Solution: Set $n = 8$ in Equation (109) and form the partition

$$P_8 = \left\{ z_k = \frac{k}{4} + i\frac{k\pi}{32}, \text{ for } k = 0,1,2,....,8 \right\}.$$

For this situation, we have a uniform increment $\Delta z_k = \frac{1}{4} + i\frac{\pi}{32}$. For convenience we select

$$c_k = \frac{z_{k-1} + z_k}{2} = \frac{2k-1}{8} + i\frac{\pi(2k-1)}{64} \qquad ...(110)$$

for $k = 0, 1, 2, ..., 8$. Figure 12 shows the points $\{z_k\}$ and $\{c_k\}$.

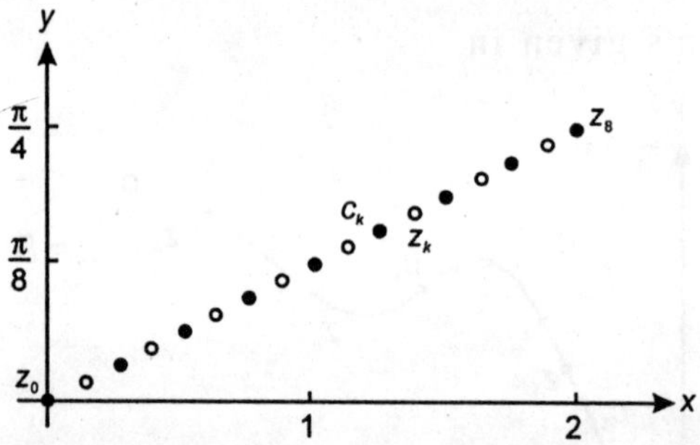

Fig. 12 Partition and evaluation points for the Riemann sum $S(P_8)$.

One possible Riemann sum, then, is

$$S(P_8) = \sum_{k=1}^{\infty} f(c_k)\Delta z_k = \sum_{k=1}^{\infty} \exp\left(\frac{2x-1}{8} + i\frac{\pi(2k-1)}{64}\right)\left(\frac{1}{4} + i\frac{\pi}{32}\right) \qquad ...(111)$$

By rounding the terms in this Riemann sum to five decimal digits, we obtain an approximation for the integral:

$$\begin{aligned} S(P_8) \approx\ & 0.27749 + 0.12501i + 0.33885 + 0.19467i \\ & + 0.40850 + 0.29140i + 0.48532 + 0.42378i \\ & + 0.56683 + 0.60261i + 0.64848 + 0.84138i \\ & + 0.72276 + 1.15676i + 0.77798 + 1.56912i \end{aligned}$$

$$S(P_8) \approx 4.226201 + 5.20473i$$

This result compares favorably with the precise value of the integral, which you will soon see equals

$$\exp\left(2 + \frac{i\pi}{4}\right) - 1 = -1 + e^2\frac{\sqrt{2}}{2} + ie^2\frac{\sqrt{2}}{2} = 4.22485 + 5.22485i.$$

Suppose that we have a parametrization of the contour C given by the function $z(t)$ for $a \le t \le b$. That is, C is the range of the function $z(t)$ over the interval $[a,b]$, as Figure 13 shows.

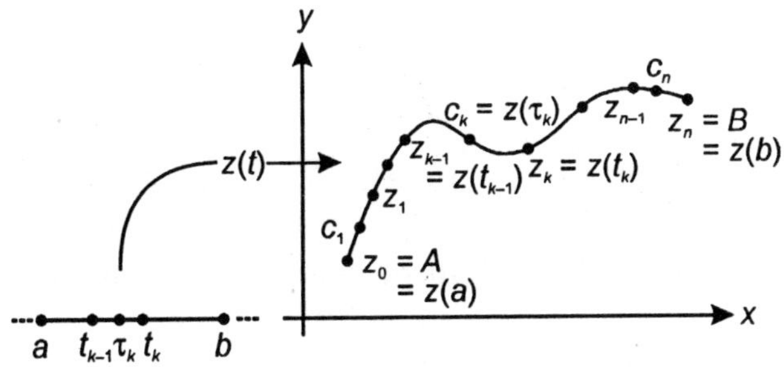

Fig. 13 A parametrisation of the contour C by $z(t)$ for $a \le t \le b$.

It follows that

$$\lim_{n\to\infty}\sum_{k=1}^{n} f(c_k)\Delta z_k = \lim_{n\to\infty}\sum_{k=1}^{n} f(c_k)(z_k - z_{k-1})$$

$$= \lim_{n\to\infty}\sum_{k=1}^{n} f\left(z(\tau_k)z(t_k) - z(t_{k-1})\right)$$

where τ_k and t_k are the points contained in the interval $[a, b]$ with the property that $c_k = z(\tau_k)$ and $z_k = z(t_k)$, as is also shown in Figure 7. If for all k we multiply the kth term in the last sum by $\dfrac{t_k - t_{k-1}}{t_k - t_{k-1}}$, then we get

$$\lim_{n\to\infty}\sum_{k=1}^{n} f(c_k)\Delta z_k = \lim_{n\to\infty}\sum_{k=1}^{n} f\{z(\tau_k)\}\left(\frac{z(t_k) - z(t_{k-1})}{t_k - t_{k-1}}\right)(t_k - t_{k-1})$$

$$= \lim_{n\to\infty}\sum_{k=1}^{n} f(z(\tau_k))\left(\frac{z(t_k) - z(t_{k-1})}{t_k - t_{k-1}}\right)\Delta t_k \tag{112}$$

The quotient inside the last summation looks suspiciously like a derivative, and the entire quantity looks like a Riemann sum.

Theorem 4. Suppose that $f(z)$ is a continuous complex-valued function defined on a set containing the contour C. Let $z(t)$ be any parametrization of C for $a \le t \le b$ Then

$$\int_c f(z)dz = \int_a^b f(z(t))z'(t)dt$$

Example 10. Give an exact calculation of the integral $\int_c e^z dz$ where C is a the line segment joining the point $A = 0$ to $B = 2 + \dfrac{i\pi}{4}$.

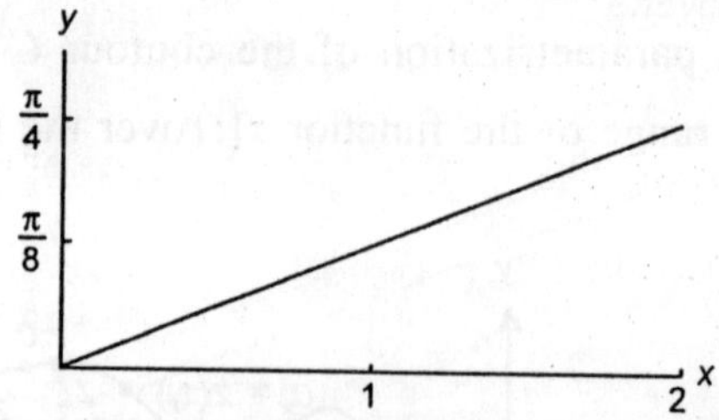

Solution. We must compute $\int_c e^z dz = \int_c \exp(z)dz$, where C is the line segment joining $A = 0$ to $B = 2+\frac{i\pi}{4}$. According to equation of curve, we can parametrize C by $z(t) = \left(2+i\frac{\pi}{4}\right)t$, for $0 \le t \le 1$. As $z'(t) = \left(2+i\frac{\pi}{4}\right)$, Theorem 6.1 guarantees that

$$\int_c e^z dz = \int_c \exp(z)dz = \int_0^1 \exp[z[t]]z'[t]dt$$

$$= \int_0^1 \exp\left[\left(2+i\frac{\pi}{4}\right)t\right]\left(2+i\frac{\pi}{4}\right)dt$$

$$= \left(2+i\frac{\pi}{4}\right)\int_0^1 \exp\left[\left(2+i\frac{\pi}{4}\right)t\right]dt$$

$$= \left(2+i\frac{\pi}{4}\right)\int_0^1 e^{2t}e^{\frac{i\pi}{4}t}dt$$

$$= \left(2+i\frac{\pi}{4}\right)\int_0^1 e^{2t}\left(\cos\left[\frac{\pi t}{4}\right]+i\sin\left[\frac{\pi t}{4}\right]\right)dt$$

$$= \left(2+i\frac{\pi}{4}\right)\left(\int_0^1 e^{2t}\cos\left[\frac{\pi t}{4}\right]dt + i\int_0^1 e^{2t}\sin\left[\frac{\pi t}{4}\right]dt\right)$$

$$= \left(2+i\frac{\pi}{4}\right)\left(\frac{2\sqrt{2}e^2(\pi+8)}{\pi^2+64} - \frac{32}{\pi^2+64} + i\left(\frac{-2\sqrt{2}e^2(\pi-8)}{\pi^2+64} + \frac{4\pi}{\pi^2+64}\right)\right)$$

$$= 4.2248516741 + 5.2248516741i$$

Example 11. Evaluate the contour integral $\int_C \frac{1}{z-2}dz$ where C is a the upper semicircle with radius 1 centered at $x = 2$.

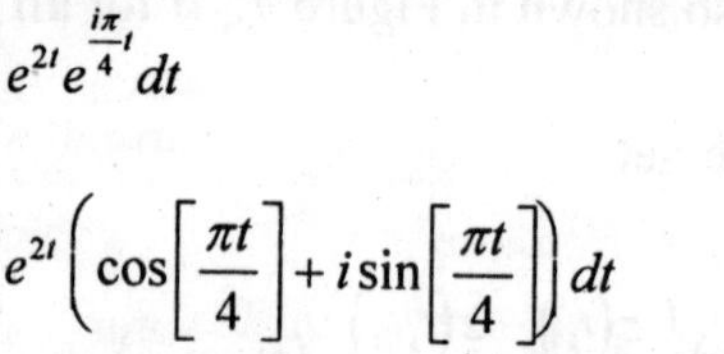

Solution: The function $z(t) = 2 + e^{it}$, for $0 \le t \le \pi$ is a parametrisation for *C*. We apply Theorem 6.1 with $f(z) = \frac{1}{z-2}$. (Note: $f(z(t)) = \frac{1}{z(t)-2}$, and $z'(t) = ie^{it}$.) Hence

$$\int_C \frac{1}{z-2}dz = \int_0^{\pi} \frac{1}{z(t)-2} z'(t)\,dt$$

$$= \int_0^{\pi} \frac{1}{2+e^{it}-2} ie^{it}\,dt = \int_0^{\pi} i\,dt = i\pi$$

Since, $\int_C \frac{1}{z-2}dz = \int_a^b \frac{1}{z(t)-2} z'(t)\,dt$, therefore you can symbolically equate *z* with *z*(*t*) and *dz* with *z*′(*t*)*dt*.

Using these identities *z* is supposed to be a point on the contour *C* parametrized by *z*(*t*), and $\frac{dz}{dt} = z'(t)$, according to the Leibniz notation for the derivative.

If $z(t) = x(t) + iy(t)$, then by the preceding paragraph we have

$$dz = z'(t)\,dt = (x'(t) + iy'(t))\,dt = dx + i\,dy \qquad ...(113)$$

where *dx* and *dy* are the differentials for *x*(*t*) and *y*(*t*), respectively (i.e., *dx* is equated with $x'(t)\,dt$, etc.). The expression *dz* is often called the complex differential of *z*. Just as *dx* and *dy* are intuitively considered to be small segments along the *x* and *y* axes in real variables, we can think of *dz* as representing a tiny piece of the contour *C*. Moreover, if we write

$$|dz| = |(x'(t) + iy'(t))\,dt| = |(x'(t) + iy'(t))|\,dt$$

$$= \sqrt{[x'(t)]^2 + [y'(t)]^2}\,dt$$

we can put Equation (6-11) into the form

$$L(C) = \int_a^b \sqrt{\left[x'(t)^2\right] + \left[y'(t)^2\right]}\,dt = \int_C |dz|. \qquad ...(114)$$

so we can think of $|dz|$ as representing the length of *dz*.

Suppose $f(z) = u(z) + iv(t)$, and $z(t) = x(t) + iy(t)$ is a parametrization for the contour C. Then

$$\int_C f(z)\,dz = \int_a^b f(z(t))\,z'(t)\,dt$$

$$= \int_a^b \left(u(z(t)) + iv(z(t))\right)\left(x'(t) + iy'(t)\right)dt$$

$$= \int_a^b \left[u(z(t))x'(t) - v(z(t))y'(t)\right]dt + i\int_a^b \left[v(z(t))x'(t) + u(z(t))y'(t)\right]$$

$$= \int_a^b (ux' - vy')\,dt + i\int_a^b (vx' + uy')\,dt \qquad ...(115)$$

where we are equating $u(t)$ with $u(z(t))$, x' with $x'(t)$, and so on.

Wwe can write above equation in terms of line integrals of the real-valued functions $u(t)$ and $v(t)$, giving

$$\int_C f(z)\,dz = \int_a^b udx - vdy + i\int_a^b vdx + udy, \qquad ...(116)$$

which is easy to remember if we recall that symbolically

$$f(z)\,dz = (u + iv)(dx + idy)$$
$$= udx - vdy + i(vdx + udy).$$

Example 12. Show that $\int_{C_1} zdz = \int_{C_2} zdz = 4 + 2i$,

where C_1 is the line segment from $-1 - i$ to $3 + i$, and C_2 is the portion of the parabola $x = y^2 + 2y$ joining $-1 - i$ to $3 + i$, as indicated in Figure 8.

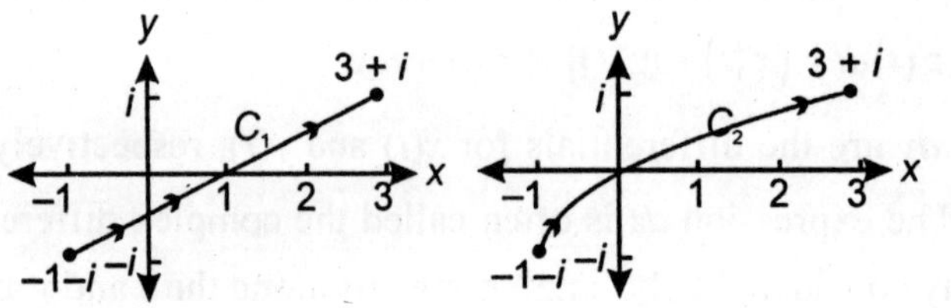

(a) The line segment (b) The portion of the parabola

Fig. 14 The two contours C_1 and C_2 joining $-1 - i$ to $3 + i$.

Solution: The line segment joining (–1, – 1) to (3, 1) is given by the slope intercept formula $y = \frac{1}{2}x - \frac{1}{2}$, which can be written as $x = 2y + 1$.

If we choose the parametrization $y = t$ and $x = 2t + 1$, we can write segment C_1 as

$$C1 : z(t) = 2t + 1 + it \text{ and } dz = (z + i)\,dt$$

for $-1 \le t \le 1$.

Along C_1 we have $f(z(t)) = 2t + 1 + it$. Applying Theorem 1 gives

$$\int_{C_1} zdz = \int_a^b \frac{1}{z(t) - 2} z'(t)\,dt = \int_{-1}^{1} (2t + 1 + it)(2 + i)\,dt .$$

We now multiply out the integrand and put it into its real and imaginary parts:

$$\int_{C_1} zdz = \int_{-1}^{1} (2t + 1 + it)(2 + i)dt$$

$$= \int_{-1}^{1}\left(3t+2+i\left(4t+1\right)\right)dt.$$

$$= \int_{-1}^{1}\left(3t+2\right)dt + i\int_{-1}^{1}\left(4t+1\right)dt$$

$$= 4+2i \qquad ...(117)$$

Similarly, we can parametrize the portion of the parabola $x = y^2 + 2y$ joining (–1, –1) to (3, 1) by $y = t$ and $x = 2t + 1$ and $x = t^2 + 2t$ so that

$C_2 : z(t) = t^2 + 2t + it$ and $dz = (2t + 2 + i)\, dt$ for $-1 \le t \le 1$. Along C_2 we have $f(z(t)) = t^2 + 2t + it$. Theorem 6.1 now gives

$$\int_{C_2} zdz = \int_{-1}^{1}\left(t^2+2t+it\right)\left(2t+2+i\right)dt$$

$$= \int_{-1}^{1}\left(2t^3+6t^2+3t+i\left(3t^2+4t\right)\right)dt$$

$$= \int_{-1}^{1}\left(2t^3+6t^2+3t\right)dt + i\int_{-1}^{1}\left(3t^2+4t\right)dt$$

$$= 4+2i$$

Example 13. Evaluate the contour integrals of $f(z) = z^2$ starting at the points $z_1 = -1 - 1$ and $z_2 = 3 + i$.

(a) Use the line segment joining the points.

(b) Use a portion of a parabola joining the points.

Solution.

(a) Use the straight line segment connecting the points $z_1 = -1 - i$ and $z_2 = 3 + i$, set up the parameterization and compute the contour integral.

Here the function $f[z] = z^2$.

$$f[z] = z^2$$

$$z_1[t] = 1+(2+i)t$$

$$f[z_1[t]] = (1+(2+i)t)^2$$

$$z_1'[t] = 2+i$$

$$f[z_1[t]]z_1'[t] = 2+6t+2t^2+i\left(1+8t+11t^2\right)$$

$$u[t] = 2+6t+2t^2$$

$$v[t] = 1 + 8t + 11t^2$$

The given integral can be evaluated using by the following method:

The integral can be computed using the real and imaginary parts.

$$v[t] = 2 + 6t + 2t^2$$

$$v[t] = 1 + 8t + 11t^2$$

$$\int u[t]dt + i\int v[t]dt = 2t + 3t^2 + \frac{2t^3}{3} + i\left(t + 4t^2 + \frac{11t^3}{3}\right)$$

$$u[t] = 2t + 3t^3 + \frac{2t^3}{3}, \qquad v[t] = t + 4t^2 + \frac{11t^3}{3}$$

$$U[1] = \frac{17}{2}, \; V[1] = \frac{26}{3}$$

$$U[-1] = \frac{1}{3}, \; U[-1] = -\frac{2}{3}$$

$$\int_{-1}^{1} u[t]dt + i\int_{-1}^{1} v[t]dt = U[1] - U[-1] + i(V[1] - V[-1])$$

$$\int_{-1}^{1} u[t]dt + i\int_{-1}^{1} v[t]dt = \frac{16}{3} + i\left(\frac{28}{3}\right)$$

$$\int_{-1}^{1} u[t]dt + i\int_{-1}^{1} v[t]dt = \frac{16}{3} + \frac{28i}{3}$$

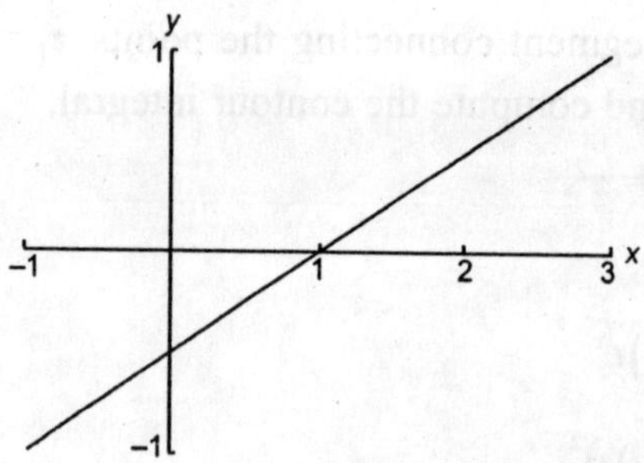

$$\int_C z^2 dz = \frac{16}{3} + \frac{28i}{3}$$

$$C : z_1[t] = 1 + (2 + i)t, \text{ for } -1 \le t \le 1.$$

Solution: (b) Use the portion of a parabola connecting the points $z_1 = -1 - i$ and $z_2 = 3 + i$, set up the parameterization and compute the contour integral.

$$z_2[t] = (2+i)t + t^2$$

$$f[z_2[t]] = ((2+i)t + t^2)^2$$

$$z_2'[t] = (2+i) + 2t$$

$$f[z_2[t]]z_2'[t] = 2t^2 + 12t^3 + 10t^4 + 2t^5 + i(11t^2 + 16t^3 + 5t^4)$$

$$u[t] = 2t^2 + 12t^3 + 10t^4 + 2t^5$$

$$v[t] = 11t^2 + 16t^3 + 5t^4$$

The integral can be computed using the real and imaginary parts.

$$u[t] = 2t^2 + 12t^3 + 10t^4 + 2t^5$$

$$v[t] = 11t^2 + 16t^3 + 5t^4$$

$$\int u[t]dlt + i\int v[t]dt = \frac{2t^3}{3} + 3t^4 + 2t^5 + \frac{t^6}{3} + i\left(\frac{11t^3}{3} + 4t^4 + t^4\right)$$

$$U[t] = \frac{2t^3}{3} + 3t^4 + 2t^5 + \frac{t^6}{3}, \quad V[t] = \frac{11t^3}{3} + 4t^4 + t^5$$

$$U[1] = 6, \qquad V[1] = \frac{26}{3}$$

$$U[-1] = \frac{2}{3}, \qquad V[-1] = -\frac{2}{3}$$

$$\int_{-1}^{1} u[t]dt + i\int_{-1}^{1} v[t]dt = U[1] - U[-1] + i(V[1] - V[-1])$$

$$\int_{-1}^{1} u[t]dt + i\int_{-1}^{1} v[t]dt = \frac{16}{3} + i\left(\frac{28}{3}\right)$$

$$\int_{-1}^{1} u[t]dt + i\int_{-1}^{1} v[t]dt = \frac{16}{3} + \frac{28i}{3}$$

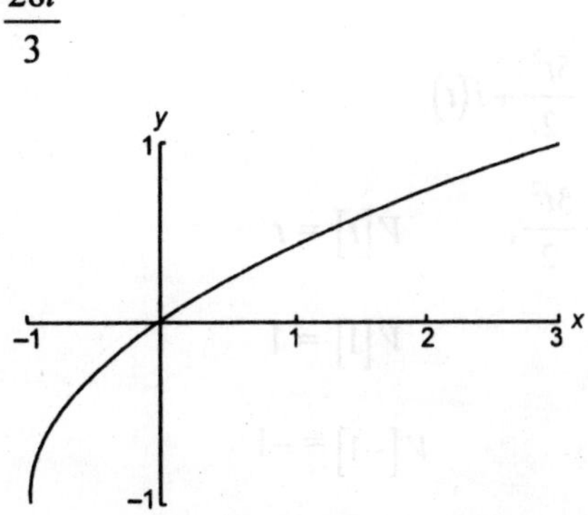

$$\int_C z^2 dz = \frac{16}{3} + \frac{28i}{3}$$

$$C : z_2[t] = (2+i)t + t^2, \text{ for } -1 \le t \le 1.$$

Example 14. Evaluate the contour integrals of $f(z) = \overline{z}$ starting at the points $z_1 = -1 - i$ and $z_2 = 3 + i$.

(a) Use the line segment joining the points.

(b) Use a portion of a parabola joining the points.

Solution:

(a) Use the straight line segment connecting the points $z_1 = -1 - i$ and $z_2 = 3 + i$, set up the parameterization and compute the contour integral. Take the function $f(z) = \overline{z}$.

$$f(z) = \text{conjugate}[z]$$

$$f[x+iy] = x - iy$$

$$z_1[t] = 1 + (2+i)t$$

$$f[z_1[t]] = 1 + (2+i)t$$

$$f[z_1[t]] = 1 + (2-i)t$$

$$z_1'[t] = 2 + i$$

$$f[z_1[t]]z_1'[t] = (2+i) + 5t$$

$$u[t] = 2 + 5t$$

$$v[t] = 1$$

The integral can be computed using the real and imaginary parts.

$$u[t] = 2 + 5t$$

$$v[t] = 1$$

$$\int u[t]dt + i\int v[t]dt = 2t + \frac{5t^2}{2} + i(t)$$

$$U[t] = 2t + \frac{5t^2}{2}, \qquad V[t] = t$$

$$U[1] = \frac{9}{2}, \qquad V[1] = 1$$

$$U[-1] = \frac{1}{2}, \qquad V[-1] = -1$$

$$\int_{-1}^{1} u[t]dt + i\int_{-1}^{1} v[t]dt = U[1]-U[-1]+i(V[1]-V[-1])$$

$$\int_{-1}^{1} u[t]dt + i\int_{-1}^{1} v[t]dt = 4+i(2)$$

$$\int_{-1}^{1} u[t]dt + i\int_{-1}^{1} v[t]dt = 4+2i$$

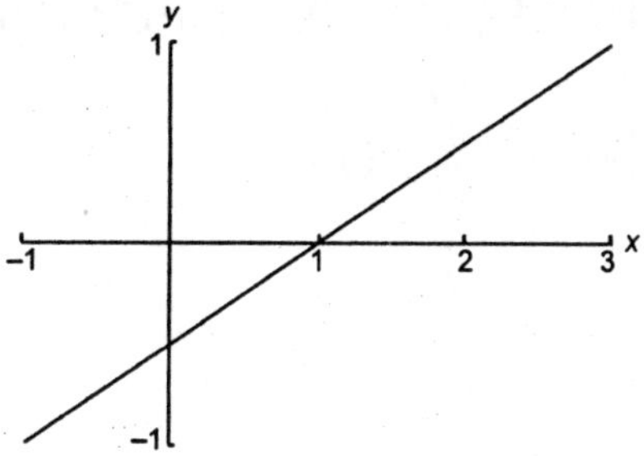

$$\int_C \text{conjugate}[z]dz = 4 + 2i.$$

$$C : z_1[t] = 1+(2+i)t, \text{ for } -1 \le t \le 1.$$

Solution: (b) Use the portion of a parabola connecting the points $z_1 = -1 - i$ and $z_3 = 3 + i$, set up the parameterization and compute the contour integral.

$$z_1[t] = (2+i)t + t^2$$

$$f[z_2[t]] = (2-i)t + t^2$$

$$z_1'[t] = (2+i) + 2t$$

$$f[z_2[t]]z_2'[t] = 5t + (6-i)t^2 + 2t^3$$

$$u[t] = 5t + 6t^2 + 2t^3$$

$$v[t] = -t^2$$

The integral can be computed using the real and imaginary parts

$$u[t] = 5t + 6t^2 + 2t^2$$

$$v[t] = -t^2$$

$$\int u[t]dt + i\int v[t]dt = \frac{5t^2}{2} + 2t^3 + \frac{t^4}{2} + i\left(-\frac{t^3}{3}\right)$$

$$U[t] = \frac{5t^2}{2} + 2t^3 + \frac{t^4}{2}, \quad V[t] = -\frac{t^3}{3}$$

$$U[1] = 5, \qquad V[1] = -\frac{1}{3}$$

$$U[-1] = 1, \qquad V[-1] = \frac{1}{3}$$

$$\int_{-1}^{1} u[t]dt + i\int_{-1}^{1} v[t]dt = U[1] - U[-1] + i(V[1] - V[-1])$$

$$\int_{-1}^{1} u[t]dt + i\int_{-1}^{1} v[t]dt = 4 + i\left(-\frac{2}{3}\right)$$

$$\int_{-1}^{1} u[t]dt + i\int_{-1}^{1} v[t]dt = 4 - \frac{2i}{3}$$

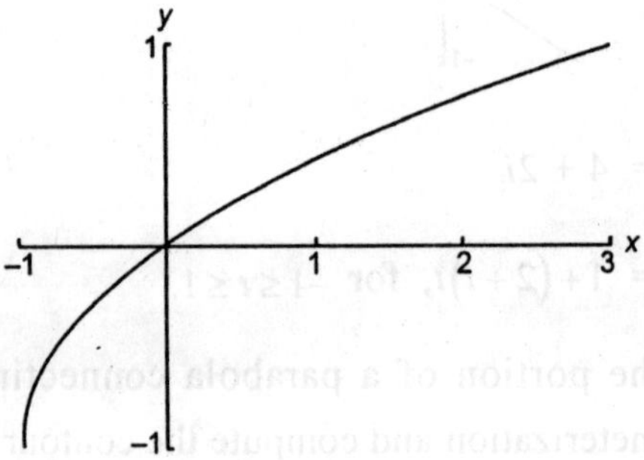

$$\int_c conjugate[z]dz = 4 - \frac{2i}{3}$$

$$C : z_2[t] = (2+i)t + t^2, \text{ for } -1 \le t \le 1.$$

GREEN'S THEOREM

It relates the surface integral over some bounded region to a line integral over the boundary curve of the region.

Statement: Let C be a positively oriented, simply closed, piecewise-smooth curve in R^2; D be the region bounded by C and P and Q be functions having continuous partial derivatives on an open region containing D, then

$$\int_C Pdx + Qdy = \iint_D \left(\frac{\partial Q}{\partial x} - \frac{\partial P}{\partial y}\right) dxdy. \qquad \text{...(118)}$$

As a use of Green's theorem one can show that the following result holds true

$$\iint_D \left(f\nabla^2 g + \nabla f.\nabla g\right) dA = \oint f\nabla g.n\, ds \qquad \text{...(119)}$$

where R is the region enclosed by the simply closed curve C, ds is the arc length line element of C, and n is the outward normal to the curve at $(x(s),y(s))$.

THE CAUCHY-GOURSAT THEOREM

The Cauchy-Goursat theorem states that within certain domains the integral of an analytic function over a simple closed contour is zero. An extension of this theorem allows us to replace integrals over certain complicated contours with integrals over contours that are easy to evaluate.

Recall from previous section that each simple closed contour C divides the plane into two domains. One domain is bounded and is called the interior of C; the other domain is unbounded and is called the exterior of C. Figure given below, illustrates this concept, which is known as the Jordan curve theorem.

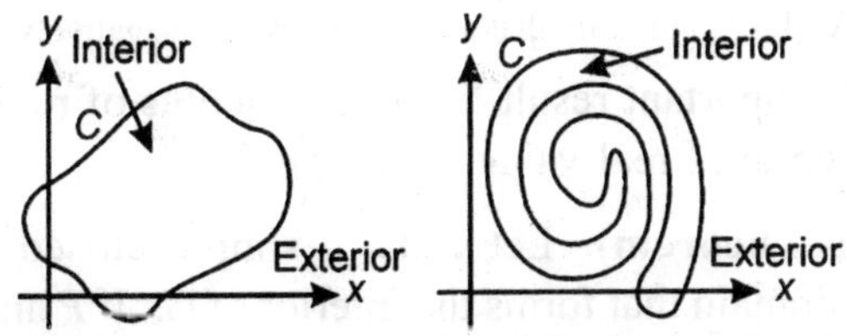

Fig. 15 The interior and exterior of simple closed contours.

Recall also that a domain D is a connected open set. In particular, if z_1 and z_2 are any pair of points in D, then they can be joined by a curve that lies entirely in D. A domain D is said to be a simply connected domain if the interior of any simple closed contour C contained in D is contained in D. In other words, there are no "holes" in a simply connected domain. A domain that is not simply connected is said to be a multiply connected domain.

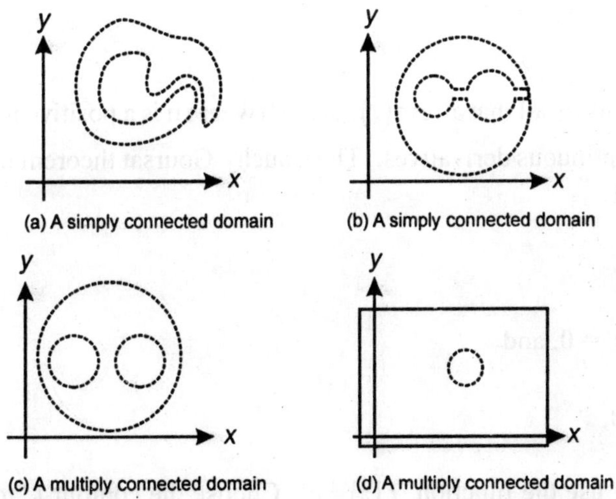

Fig. 16 Simply connected and multiply connected domains.

Let the simple closed contour C have the parametrization $C: z(t) = x(t) + iy(t)$ for $a \le t \le b$. Recall that if C is parametrized so that the interior of C is kept on the left as $z(t)$

moves around C, then we say that C is oriented positively (counterclockwise); otherwise, C is oriented negatively (clockwise). If C is positively oriented, then $-C$ is negatively oriented. Figure 17 illustrates the concept of positive and negative orientation.

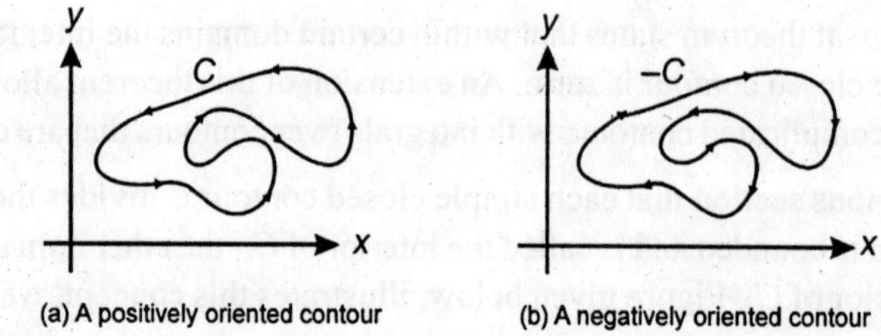

Fig. 17 Simple closed contours that are positively and negatively oriented.

Green's theorem is an important result from the calculus of real variables. It tells you how to evaluate the line integral of real-valued functions.

***Theorem* 5. (Greens Theorem).** Let C be a simple closed contour with positive orientation and let R be the domain that forms the interior of C. If P and Q are continuous and have continuous partial derivatives P_x, P_y, Q_x, and Q_y at all points on C and R, then

$$\int_C P(x,y)dx + \int_C Q(x,y)dy = \iint_R \left(Q_x(x,y)\right) - P_y(x,y)\,dxdy \qquad ...(120)$$

***Theorem* 6. (Cauchy-Goursat Theorem).** Let $f(z)$ be analytic in a simply connected domain D. If C is a simple closed contour that lies in D, then

$$\int_C f(z)dz = 0. \qquad ...(121)$$

Example 16. Let us recall that $e^z, \cos(z),$ *and* z^n (where n is a positive integer) are all entire functions and have continuous derivatives. The Cauchy-Goursat theorem implies that, for any simple closed contour,

(a) $\int_C e^z dz = 0,$

(b) $\int_C \cos(z)dz = 0,$ and

(c) $\int_C z^n dz = 0.$

Solution. (a) We use the function $f(z) = e^z$. Choose the contour C to be the unit circle C: $|z| = 1$ with positive orientation.

$$f[z] = \text{ez}$$

$$z[\theta] = e^{i\theta}$$

$$f[z[\theta]] = e^{e^{i\theta}}$$

$$z'[\theta] = ie^{i\theta}$$

$$f[z[\theta]]z'[\theta] = e^{e^{i\theta}} ie^{i\theta}$$

$$F[\theta] = \int f[z[\theta]]z'[\theta]d\theta = e^{e^{i\theta}}$$

$$F[2\pi] = e, \text{ and } F[0] = e$$

$$\int_0^{2\pi} f[z[\theta]]z'[\theta]d\theta = \int_0^{2\pi} e^{e^{i\theta}} ie^{i\theta} d\theta = e - e = 0$$

$$\int_0^{2\pi} f[z[\theta]]z'[\theta]d\theta = 0$$

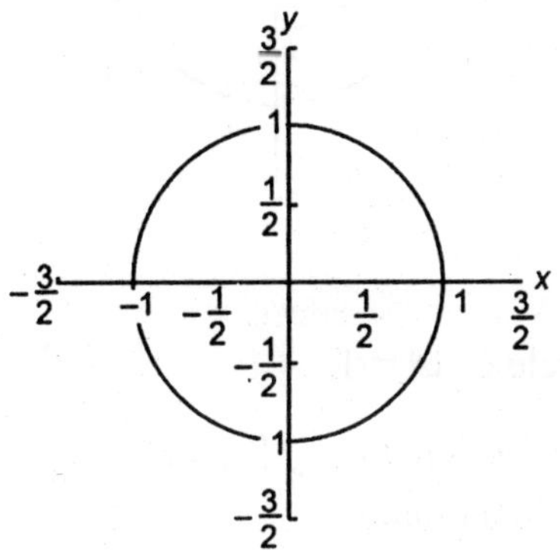

$$\oint_C e^z dz = 0$$

The contour is the unit circle C: $|z| = 1$.

(b) We use the function $f(z) = \cos(z)$. Choose the contour C to be the unit circle C: $|z|$ = 1 with positive orientation.

$$f|z| = \cos[z]$$

$$z[\theta] = e^{i\theta}$$

$$f[z[\theta]] = \cos[e^{i\theta}]$$

$$z'[\theta] = ie^{i\theta}$$

$$f[z[\theta]]z'[\theta] = ie^{i\theta}\cos[e^{i\theta}]$$

$$f[\theta] = \int f[z[\theta]]z'[\theta]d\theta = \sin[e^{i\theta}]$$

$$F[2\pi] = \sin[1], \text{ and } F[0] = \sin[1]$$

$$\int_0^{2\pi} f[z[\theta]]z'[\theta]d\theta = \int_0^{2\pi}\cos[e^{i\theta}]ie^{i\theta}d\theta = \sin[1]-\sin[1] = 0$$

$$\int_0^{2\pi} f[z[\theta]]z'[\theta]d\theta = 0$$

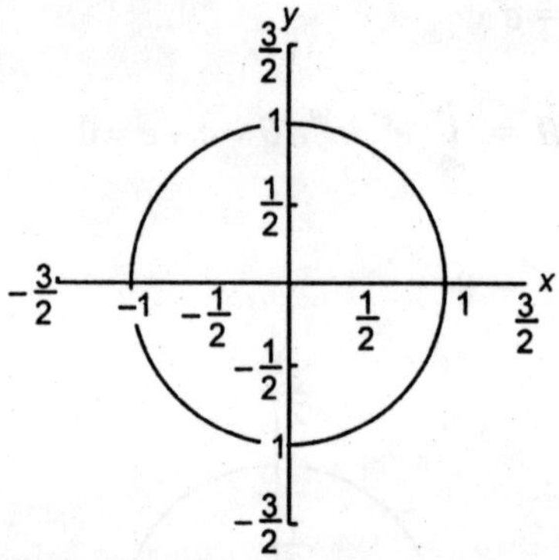

$$\oint_C \cos[z]dz = 0$$

The contour is the unit circle C: $|z| = 1$.

(c) We use $n = 5$ and use the function $f(z) = z^5$. Choose the contour C to be the unit circle C: $|z| = 1$ with positive orientation.

$$f[z] = z^5$$

$$z[\theta] = e^{i\theta}$$

$$f[z[\theta]] = e^{5i\theta}$$

$$z'[\theta] = ie^{i\theta}$$

$$f[z[\theta]]z'[\theta] = ie^{6i\theta}$$

$$F[\theta] = \int f[z[\theta]]z'[\theta]d\theta = \frac{1}{6}e^{6i\theta}$$

$$F[2\pi] = \frac{1}{6}, \text{ and } F[0] = \frac{1}{6}$$

$$\int_0^{2\pi} f[z[\theta]]z'[\theta]d\theta = \int_0^{2\pi} ie^{6i\theta}d\theta = \frac{1}{6} - \frac{1}{6} = C$$

$$\int_0^{2\pi} f[z[\theta]]z'[\theta]d\theta = 0$$

$$\oint_C z^5 dz = 0$$

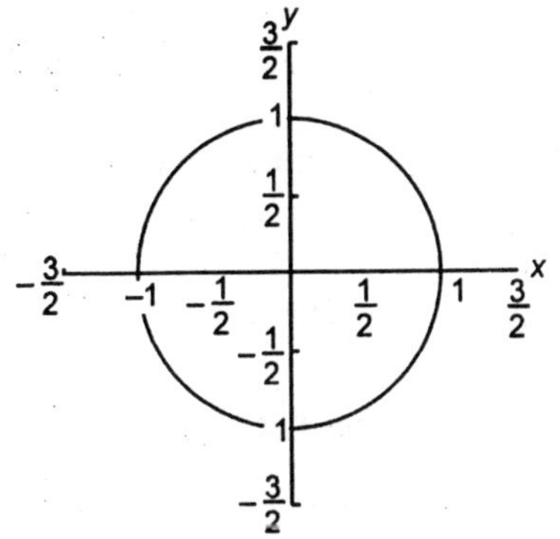

The contour is the unit circle C: $|z| = 1$.

We want to be able to replace integrals over certain complicated contours with integrals that are easy to evaluate. If C_1 is a simple closed contour that can be "continuously deformed" into another simple closed contour C_2 without passing through a point where f is not analytic, then the value of the contour integral of f over C_1 is the same as the value of the integral of f over C_2. To be precise, we state the following result.

***Theorem* 7. (Deformation of Contour).** Let C_1 and C_2 be two simple closed positively oriented contours such that C_1 lies interior to C_2. If $f(z)$ is analytic in a domain D that contains both C_1 and C_2 and the region between them, as shown in Figure 23, then

$$\int_{C_1} f(z)dz = \int_{C_2} f(z)dz$$

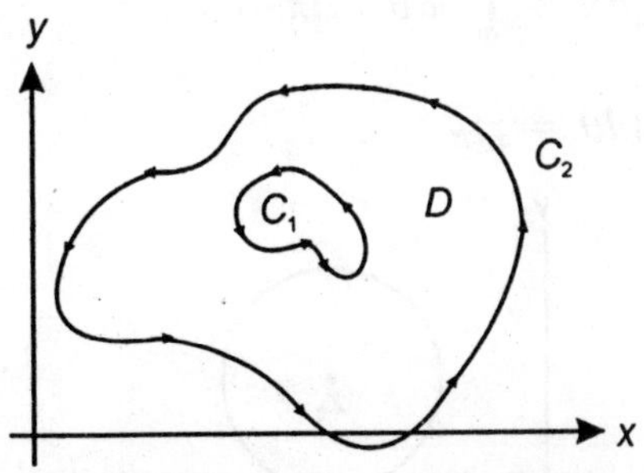

Fig. 18 The domain D that contains the simple closed contours C_1 and C_2 and the region between them.

We now state as a corollary an important result that is implied by the deformation of contour theorem. This result occurs several times in the theory to be developed and is an important tool for computations.

Example 15. Let Z_0 denote a fixed complex value. If C is a simple closed contour with positive orientation such that Z_0 lies interior to C, then show that

(i) $\int_C \frac{1}{z-z_0} dz = 2\pi i$, and

(ii) $\int_C \frac{1}{(z-z_0)^n} dz = 0$, where n is any integer except $n = 1$.

Solution: (i) Here , we show that $\int_c \frac{1}{z-z_0} dz = 2\pi i$.

Let consider a circle of radius 1 centered at Z_0.

$$f[z] = \frac{1}{z-z_0}$$

$$z[\theta] = e^{i\theta} + z_0$$

$$f[z[\theta]] = e^{-i\theta}$$

$$z'[\theta] = ie^{i\theta}$$

$$f[z[\theta]]z'[\theta] = i$$

$$F[\theta] = \int f[z[\theta]]z'[\theta]d\theta = i\theta$$

$$F[2\pi] = 2i\pi, \text{ and } F[0] = 0$$

$$\int_0^{2\pi} f[z[\theta]]z'[\theta]d\theta = \int_0^{2\pi} id\theta = 2i\pi - 0 = 2i\pi$$

$$\int_0^{2\pi} d[z[\theta]]z'[\theta]d\theta = 2i\pi$$

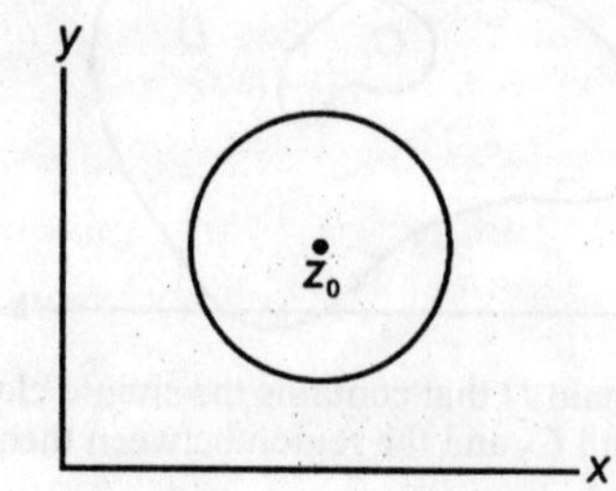

$$\oint_C \frac{1}{z-z_0}dz = 2i\pi \tag{122}$$

The contour is the unit circle $C: |z-z_0| = 1$, with positive orientation.

(ii) Here, we sShow that $\int_C \frac{1}{(z-z_0)^n}dz = 0.$ (123)

Here take n = 5 and a circle of radius 1 centered at z_0.

$$f[z] = \frac{1}{(z-z_0)^5}$$

$$z[\theta] = e^{i\theta} + z_0$$

$$f[z[\theta]] = e^{-5i\theta}$$

$$z'[\theta] = ie^{i\theta}$$

$$f[z[\theta]]z'[\theta] = ie^{-4i\theta}$$

$$F[\theta] = \int f[z[\theta]]z'[\theta]d\theta = -\frac{1}{4}e^{-4i\theta}$$

$$F[2\pi] = -\frac{1}{4}, \text{ and } F[0] = -\frac{1}{4}$$

$$\int_0^{2\pi} f[z[\theta]]z'[\theta]d\theta = \int_0^{2\pi} ie^{-4i\theta}d\theta = -\frac{1}{4} - -\frac{1}{4} = 0$$

$$\int_0^{2\pi} f[z[\theta]]z'[\theta]d\theta = 0$$

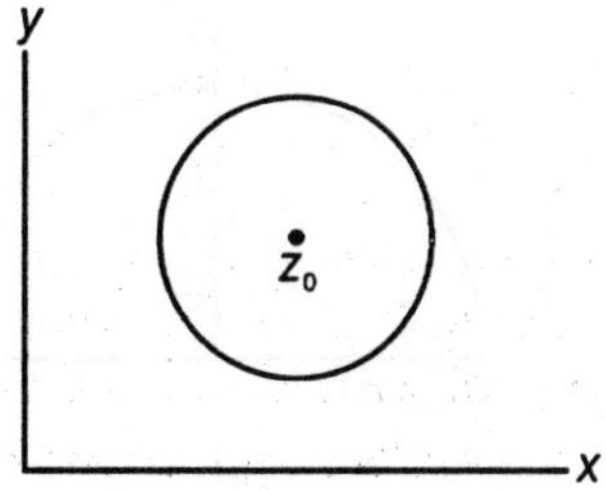

$$\oint_C \frac{1}{(z-z_0)^5}dz = 0$$

The contour is the unit circle C: $|z-z_0| = 1$, with positive orientation.

The deformation of contour theorem is an extension of the Cauchy-Goursat theorem to a doubly connected domain. Let D be a domain that contains C_1 and C_2 and the region between them, as shown in Figure 19. Then the contour $C = C_2 - C_1$ is a parametrization of the boundary of the region R that lies between C_1 and C_2 so that the points of R lie to the left of C as a point $z(t)$ moves around C. Hence C is a positive orientation of the boundary of R, and

$$\int_C f(z)\,dz = 0. \qquad ...(124)$$

***Theorem 8.*. (Extended Cauchy-Goursat Theorem).** Let $C, C_1, C_2, ..., C_n$ be simple closed positively oriented contours with the property that C_k lies interior to C for $k = 1, 2, ..., n$ and the set of interior to C_k has no points in common with the set interior to C_j if $k \neq j$. Let $f(z)$ be analytic on a domain D that contains all the contours and the region between C and $C_1 + C_2 + ... + C_{n,}$ as Then

$$\int_C f(z)\,dz = \sum_{k=1}^{n} \int_{C_k} f(z)\,dz \qquad ...(125)$$

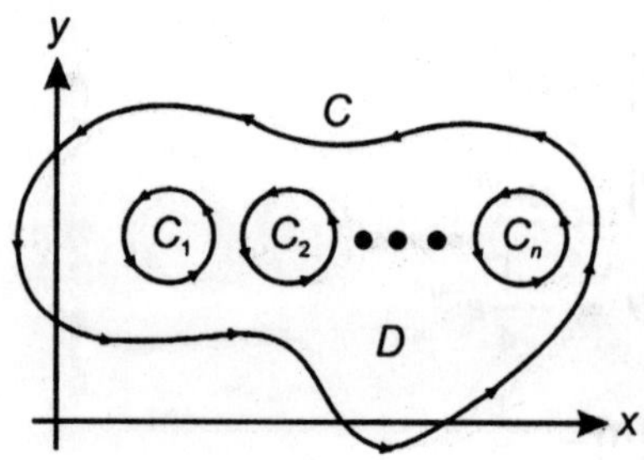

Figure 19. The multiply connected domain D and the contours $C, C_1, C_2, ..., C_n$ in the statement of the extended Cauchy-Goursat theorem.

Example 16. Show that $\int_C \frac{2z}{z^2+2}dz = 4\pi i$, where C is the circle $C: |z| = 2$ taken with positive orientation.

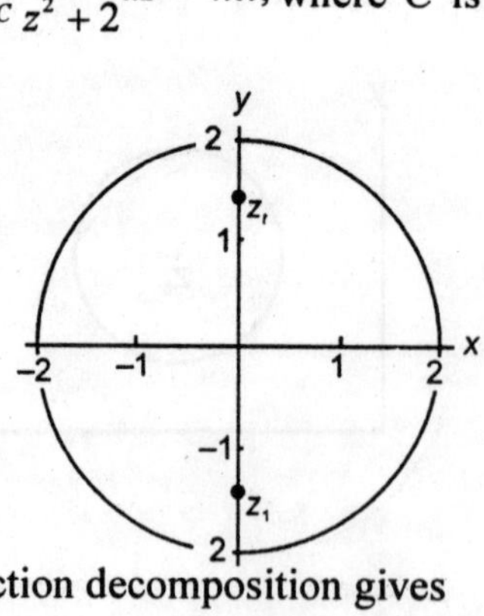

Solution: Using partial fraction decomposition gives

$$\frac{2z}{z^2+2} = \frac{2z}{\left(z+i\sqrt{2}\right)\left(z-i\sqrt{2}\right)} = \frac{1}{z+i\sqrt{2}} + \frac{1}{z-i\sqrt{2}}, \text{ so}$$

$$\int_C \frac{2z}{z^2+2}dz = \int_C \frac{1}{z+i\sqrt{2}}dz + \int_C \frac{1}{z-i\sqrt{2}}dz \qquad ...(126)$$

The points $z = \pm i\sqrt{2}$ lie interior to C, so Corollary 1 implies that $\int_C \frac{1}{z \pm i\sqrt{2}}dz$ = $2\pi i$

Substituting these values into Equation (126) yields

$$\int_C \frac{2z}{z^2+2}dz = \int_C \frac{1}{z+i\sqrt{2}}dz + \int_C \frac{1}{z-i\sqrt{2}}dz$$

$$= 2\pi i + 2\pi i$$

$$= 4\pi i$$

Example 17. Show that $\int_C \frac{2z}{z^2+2}dz = 2\pi i$, where C is the circle $C : |z - i| = 1$ taken with positive orientation.

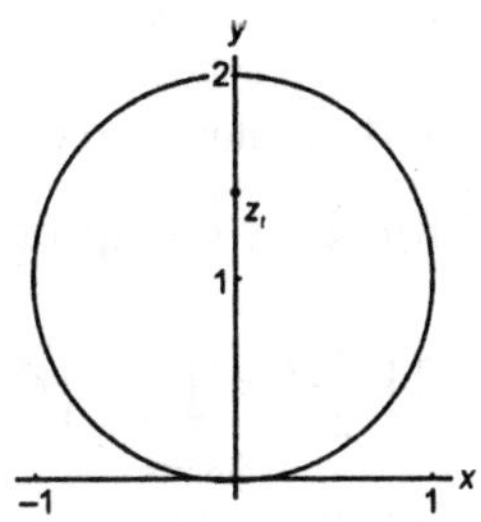

Fig. 20 The circle $C : |z - i| = 1$ and the points $z = \pm i\sqrt{2}$.

Solution. Using partial fractions again, we have

$$\int_C \frac{2z}{z^2+2}dz = \int_C \frac{1}{z+i\sqrt{2}}dz + \int_C \frac{1}{z-i\sqrt{2}}dz$$

In this case, $z = i\sqrt{2}$ lies interior to C but $z = -i\sqrt{2}$ does not, the second integral on the right side of the above equation has the value $2\pi i$. The first integral equals zero by the Cauchy-Goursat theorem because the function $f(z) = \frac{1}{z+i\sqrt{2}}$ is analytic on a simply connected domain that contains C. Thus

$$\int_C \frac{2z}{z^2+2}dz = \int_C \frac{1}{z+i\sqrt{2}}dz + \int_C \frac{1}{z-i\sqrt{2}}dz$$

$$= 0 + 2\pi i$$

$$= 2\pi i$$

Example 18. Show that $\int_C \frac{z-2}{z^2-z}dz = -6\pi i$, where C is the "figure eight" contour shown inin fFigure given below.

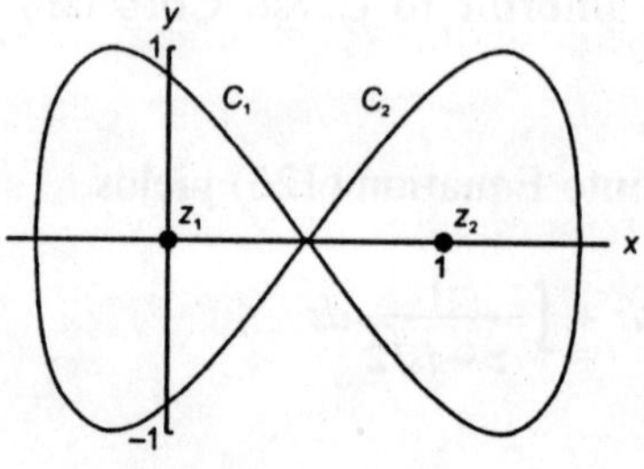

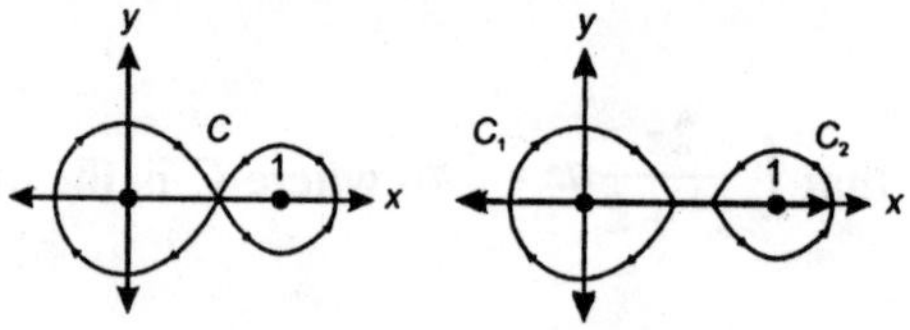

(a) The figure eight contour C (b) The contour C_1 and C_2

Fig. 21 The contour $C = C_1 + C_2$.

Solution: Again, we use partial fractions to express the integral:

$$\int_C \frac{z-2}{z^2-z}dz = \int_C \left(\frac{2}{z} - \frac{1}{z-1}\right)dz = 2\int_C \frac{1}{z}dz - \int_C \frac{1}{z-1}dz \qquad ...(127)$$

Using the Cauchy-Goursat theorem, (with $z_0 = 0$), we compute the value of the first integral on the right side of above equation (127):

$$\begin{aligned} 2\int_C \frac{1}{z}dz &= 2\int_{C_1} \frac{1}{z}dz + 2\int_{C_2} \frac{1}{z}dz \\ &= -2\int_{-C_1} \frac{1}{z}dz + 0 \\ &= -2(2\pi i) \\ &= -4\pi i \end{aligned}$$

Similarly, we find that

$$\begin{aligned} -\int_C \frac{1}{z-1}dz &= -\int_{C_1} \frac{1}{z-1}dz + -\int_{C_2} \frac{1}{z-1}dz \\ &= 0 - 2\pi i \\ &= -2\pi i \end{aligned}$$

If we substitute the results of the last two equations into Equation (6-39) we get

$$\int_C \frac{z-2}{z^2-z}dz = 2\int_C \frac{1}{z}dz - \int_C \frac{1}{z-1}dz$$
$$= -4\pi i - 2\pi i$$
$$= -6\pi i$$

Example 19. (a) Show that $\int_{C_1} \bar{z}dz = -\pi i$ but that $\int_{C2} \bar{z}dz = -4i$,

where C_1 is the semicircular path from –1 to 1, in the upper half plane, and C_2 is the polygonal path from –1 to –1 + *i* to 1 + *i* to 1, respectively, shown in Figure 22.

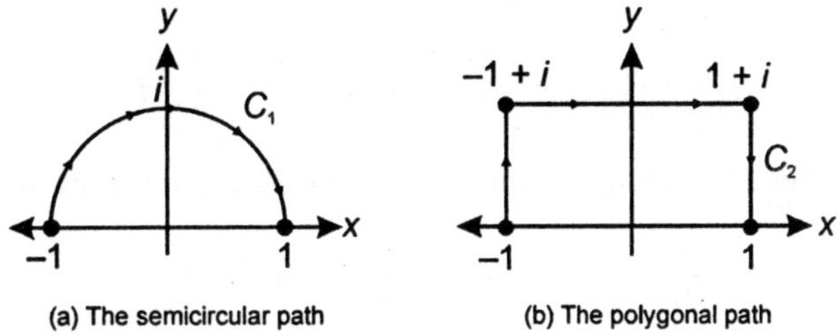

Figure 22. The two contours C_1 and C_2 joining –1 to 1.

Solution. We parametrize the semicircle C_1 as

$C_1 : z(t) = -\cos t + i\sin t$ and $dz = (\sin t + i\cos t)dt$,

for $0 \le t \le \pi$.

We have $f(z) = \bar{z}$, so

$$f(z(t)) = \overline{z(t)} = \overline{-\cos t + i\sin t}$$
$$= -\cos t - i\sin t$$

and

$$\int_{C_1} \bar{z}dz = \int_0^\pi (-\cos t - i\sin t)(\sin t + i\cos t)dt$$
$$= \int_0^\pi \left(-i\left(\cos^2 t + \sin^2 t\right)\right)dt$$
$$= -i\int_0^\pi (1)\,dt$$
$$= -\pi i$$

We parametrise C_2 in three parts, one for each line segment:

$z_1(t) = -1 + it,\ dz = idt,$ and $f(z_1(t)) = -1 - it;$

$z_2(t)=-1+2t+i,\ dz=2dt,$ and $f(z_2(t))=-1+2t-i;$

$z_3(t)=1+i(1-t),\ dz=-idt,$ and $f(z_3(t))=1-i(1-t),$

where $0 \le t \le 1$ in each case. Integrating over these three line segments we obtain

$$\int_0^1 z_1(t)z_1'(t)dt = \int_0^1(-1-it)(i)dt = \int_0^1(t-i)dt = \frac{1}{2}-i,$$

$$\int_0^1 z_2(t)z_2'(t)dt = \int_0^1(-1+2t-i)(2)dt \int_0^1(-2+4t-2i)dt = -2,i$$

$$\int_0^1 z_3(t)z_3'(t)dt = \int_0^1(1-i(1-t))(-i)dt = \int_0^1(-1+t-i)dt = -\frac{1}{2}-i$$

We get our answer by adding the three integrals along the three segments:

$$\int_{C_2} \bar{z}dz = \int_0^1(t-i)dt + \int_0^1(-2+4t-2i)dt + \int_0^1(-1+t-i)dt$$

$$= \frac{1}{2}-i-2i-\frac{1}{2}-i = -4i$$

Note that the value of the contour integral along C_1 isn't the same as the value of the contour integral along C_2, although both integrals have the same initial and terminal points.

Contour integrals have properties that are similar to those of integrals of a complex function of a real variable, which you studied in previous section. If C is given by above equation , then the integral for the opposite contour $-C$ is

$$\int_{-C} f(z)dz = \int_{-b}^{-a} f(z(-\tau))(-z'(-\tau))d\tau$$

Using the change of variable $t=-\tau$ in this last equation and the property that $\int_a^b f(t)dt = -\int_b^a f(t)dt$, we obtain

$$\int_{-C} f(z)dz = -\int_C f(z)dz. \qquad \text{...(128)}$$

If two functions f and g can be integrated over the same path of integration C, then their sum can be integrated over C, and we have the familiar result

$$\int_C (f(z)+g(z))dz = \int_C f(z)dz + \int_C g(z)dz$$

Constant multiples also behave as we would expect:

$$\int_C (a+ib)f(z)dz = (a+ib)\int_C f(z)dz.$$

If two contours C_1 and C_2 are placed end to end so that the terminal point of C_1 coincides with the initial point of C_2, then the contour C = $C_1 + C_2$ is a continuation of C_1, and

$$\int_{C_1+C_2} f(z)dz = \int_{C_1} f(z)dz + \int_{C_2} f(z)dz. \qquad ...(129)$$

If the contour C has two parametrizations

$$C : z_1(t) = x_1(t) + iy_1(t) \text{ for } a \le t \le b, \text{ and}$$

$$C : z_2(\tau) = x_2(\tau) + iy_2(\tau) \text{ for } c \le \tau \le d,$$

and there exists a differentiable function $\tau = \phi(t)$ such that

$$c = \phi(a),\ d = \phi(b),\ \int_a^b f(z_2(\tau))z_2'(\tau)d\tau. \text{ and } \phi'(t) > 0 \text{ for } a \le t \le b, \qquad ...(130)$$

then we say that $z_2(\tau)$ is a reparametrization of the contour *C*. If f is continuous on *C*, then we have

$$\int_a^b f(z_1(t))z_1'(t)dt = \int_a^b f(z_2(\tau))z_2'(\tau)d\tau. \qquad ...(131)$$

$$\int_a^b f(z_1(t))z_1'(t)dt = \int_c^d f(z_2(\tau))z_2'(\tau)d\tau.$$

Equation (45) shows that the value of a contour integral is invariant under a change in the parametric representation of its contour if the reparametrization satisfies Equations (44).

We are going to give two important inequalities relating to complex integrals.

***Theorem* 9. (Absolute Value Inequality).** If $f(t) = u(t) + iv(t)$ is a continuous function of the real parameter *t*, then

$$\left|\int_b^b f(t)dt\right| \le \int_a^b |f(t)dt.| \qquad ...(132)$$

***Theorem* 10. (ML - Inequality).** If $f(z) = u(x, y) + iv(x, y)$ is continuous on the contour *C*, then

$$\int_C |f(z)|dz \le \text{ML}. \qquad ...(133)$$

Example 20. Show that $\left|\int_C \frac{1}{z^2+1}dz\right| \le \frac{1}{2\sqrt{5}}$,

where ***C*** is the straight-line segment from 2 to 2 + *i*.

Solution: Here $|z^2 + 1| = |z - i|\,|z + i|$, and the terms $|z - i|$ and $|z + i|$ represent the distance from the point *z* to the points *i* and –*i*, respectively. Referring to Figure 23 and using a geometric argument, we get

$|z - i| \geq 2$ and $|z + i| \geq \sqrt{5}$, for z on C.

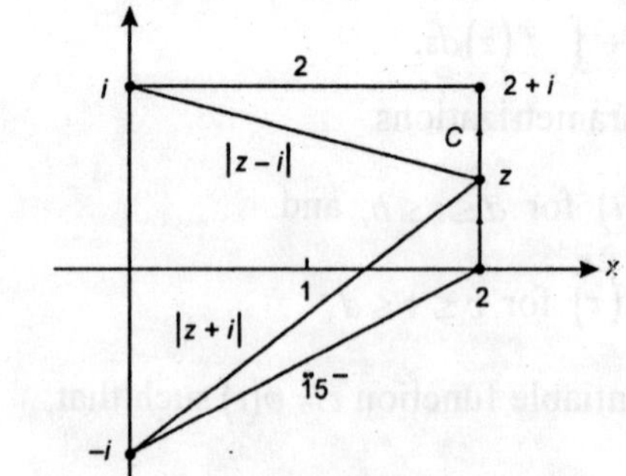

Fig. 23 The distances $|z - i|$ and $|z + i|$ for z on C.

Thus we have

$$|f(z)| = \left|\frac{1}{z^2+1}\right| = \frac{1}{|z-i||z+i|} \leq \frac{1}{2\sqrt{5}} = \text{M}$$

Because L, the length of C, equals 1, Inequality (6-23) implies that

$$\left|\int_C \frac{1}{z^2+1} dz\right| \leq \text{ML} \leq \frac{1}{2\sqrt{5}} \times 1 = \frac{1}{2\sqrt{5}}$$

EXERCISE

1. Find the value of the integral

$$\int_c \frac{3z^3+2}{(z-1)(z^2+9)} dz,$$

taken counterclockwise around circle (a) $|z - 2| = 2$; (b) $|z| = 4$.

2. Find the value of the integral

$$\int_C \frac{dz}{z^3(z+4)},$$

taken counterclockwise around the circle

(a) $|z| = 2$; (b) $|z + 2| = 3$.

3. Use an antiderivative to show that, for every contour C extending from a point z_1 to a point z_2.

$$\int_C z^n = \frac{1}{n+1}\left(z_2^{n+1} - z_1^{n+1}\right) \qquad (n = 0, 1, 2, ...).$$

4. By finding an antiderivative, evaluate each of these integrals, where the path is an arbitrary contour between the indicatd limits of integration:

(a) $\int_{i}^{i/2} e^{\pi z}\, dz$ (b) $\int_{0}^{\pi+2i} \cos\left(\frac{z}{2}\right) dz$

(c) $\int_{1}^{3} (z-1)^3\, dz$

5. Apply the Cauchy-Goursat theorem to show that

$$\int_C f(z)\, dz = 0$$

6. By using the branch

$\log z = \ln r + iq$

(r > 0, 0 < q < 2p)

of the logarithmic function as an antiderivative of 1/z, show that

$$\int_{-2i}^{2i} \frac{dz}{z} = -\pi i$$

7. Show that

$$\int_{-1}^{1} z^i\, dz = \frac{1+e^{-\pi}}{2}(1-i),$$

where z^i denotes the principal branch

$z^i = \exp(i \operatorname{Log} z)$

$(|z| > 0, -\pi < \operatorname{Arg} z < \pi)$

and where the path of integraion is any contour from $z = -1$ to $z = 1$ that, except for its end points, lies above the real axis.

8. Let C denote the positively oriented boundary of the square whose sides lie along the lines x = ± 2 and y = ± 2. Evaluate each of these integrals:

(a) $\int_C \frac{e^{-z}\, dz}{z-(\pi i/2)}$; (b) $\int_C \frac{\cos z}{z\left(z^2+8\right)} dz$; $\int_C \frac{z\,dz}{2z+1}$.

Ans. (a) a, 2π; (b) $\pi i/4$; (c) $-\pi i/2$

9. Find the value of the integral of $g(z)$ around the circle $|z - i| = 2$ in the positive sense when

(a) $g(z) = \frac{1}{z^2+4}$; (b) $\frac{1}{(z^2+4)^2}$.

Ans. (a) $\pi/2$; (b) $\pi/16$

10. Let C be the circle $|z| = 3$, described in the positive sense. Show that if

$$g(w) = \int_C \frac{2z^2 - z - 2}{z - w} dz \quad (|w| \neq 3),$$

then $g(2) = 8\pi i$. What is the value $g(w)$ when $|w| > 3$?.

11. Let C be any simple closed contour, described in the positive sense in the z plane, and write

$$g(w) = \int_C \frac{z^3 + 2z}{(z-w)^3} dz.$$

Show that $g(w) = 6\pi i w$ when w is inside C and that $g(w) = 0$ when w is outside C.

12. Show that if f is analytic within and on a simple closed contour C and z_0 is not on C, then

$$\int_C \frac{f'(z)dz}{z - z_0} = \int_C \frac{f(z)dz}{(z - z_0)^2}.$$

13. Evaluate the integral, $\int_C \bar{z} dz$ where C is the parabola $y = x^2$ from 0 to 1 +? i.

14. Evaluate $\int_C f(z)dz$ where C is the curve $y = x^3$ from $-1 - i$ to $1 + i$, and

$$f(z) = \begin{cases} 1 & \text{for } y < 0 \\ 4y & \text{for } y \geq 0 \end{cases}$$

15. Let C be the part of the circle $g(t) = e^{it}$ in the first quadrant from $a = 1$ to $b = i$.

Find as small an upper bound as you can for $\left|\int_C \left(z^2 - \bar{z}^4 + 5\right) dz\right|$

16. Evaluate $\int_C f(z)dz$ where $f(z) = z + 2\bar{z}$ and C is the path from $z = 0$ to $z = 1 + 2i$ consisting of the line segment from 0 to 1 together with the segment from 1 to 1 + $2i$.

17. Suppose that $D \in \mathbb{C}$ is a domain. For every contour C lying in D and for every $a \in D$, write $n(C, a) = \dfrac{1}{2\pi i}\int_C \dfrac{dz}{z - a}$.

(a) Suppose that D is simply connected. Explain why $n(C, a) = 0$.

(b) Suppose that D is not simply connected. Show that there exists a contour C lying in D and a point $a \in D$ such that $n(C, a) \neq 0$.

18. Suppose that $f(z)$ is analytic and non-zero in the disc $\{z : |z| -R\}$, where $R > 0$ is fixed. Prove the following special case of Jensen's formula, that

$$\log |f(0)| = \frac{1}{2\pi}\int_0^{2\pi} \log\left|f\left(\mathrm{Re}^{i\theta}\right)\right| d\theta .$$

19. Show that $\int_0^{\infty} \frac{\cos mx}{x^2+1}dx = \frac{\pi}{2}e^{-m},\ m > 0$

20. Show that $\int_0^{2\pi} \frac{\cos 3\theta}{5-4\cos\theta}d\theta = \frac{\pi}{12}$

21. (a) Integrate $\frac{50z}{(z+4)(z-1)^2}$ clockwise along the unit circle with centre at $z=1/2$.

(b) Show that $\int_0^{\infty} \frac{x^2}{x^6+64}dx = \frac{\pi}{48}$

22. Evaluate the integrals

(a) $\int_0^{2\pi} \frac{\cos\theta}{3+\sin\theta}d\theta$ (b) $\int_{-\infty}^{\infty} \frac{dx}{(x^2+1)(x^2+9)}$

Chapter 11

Sequence and Series of Complex Number

INTRODUCTION

In the previous chapters we have already studied derivative and integration of complex number. In this chapter we will be discuss about sequences and (infinite) series of complex number. In this chapter we also study abot laurent and Taylor series, which help in the expansion of series of complex numbers. Series do play an important role in the field of ordinary differential equations and without series large portions of the field of partial differential equations would not be possible.

SEQUENCE AND SERIES IN COMPLEX NUMBER

Sequence of Complex Numbers

A sequence is nothing more than a list of numbers written in a specific order. The list may or may not have an infinite number of terms in them General sequence terms are denoted as follows,

$$
\begin{aligned}
&a_1 - \text{first term}\\
&a_2 - \text{second term}\\
&\quad\vdots\\
&a_n - n^{\text{th}} \text{ term}\\
&a_n + 1 - (n+1)^{\text{st}} \text{ term}\\
&\quad\vdots
\end{aligned}
$$

There is a variety of ways of denoting a sequence. Each of the following are equivalent ways of denoting a sequence.

$$\{a_1, a_2,\ldots,a_n, a_n+1,\ldots\} \ \{a_n\} \ \{a_n\}_{n-1}^{\infty}$$

In the second and third notations above a_n is usually given by a formula.

Let's take a look at a couple of sequences.

(a) $\left\{\frac{n+1}{n^2}\right\}_{n-1}^{\infty}$ (b) $\left\{\frac{(-1)^{n+1}}{2^n}\right\}_{n-0}^{\infty}$

$\{b_n\}_{n-1}^{\infty}$ where $b_n = n^{\text{th}}$ digit of π

Definition:A sequence of complex number is a function g: $Z_+ \to C$ from the positive ingegers into the complex numbers. It is traditional to use subscripts to indicato the values of the function. Thus we write $g(n) \equiv z_n$ and an explicit name for the sequence is seldom used; we write simply (z_n) to stand for the sequence g which is such that $g(n) = z_n$. For example, $\left(\frac{i}{n}\right)$ is the sequence g for which $g(n) = \frac{i}{n}$

The number L is a limit of the sequence (z_n) if given an $\varepsilon > 0$, there is an integer N_ε such that $|z_n - L| < \varepsilon$ for all $n \geq N_\varepsilon$. If L is a limit of (z_n), we sometimes say that (z_n) converges to L. We frequently write $\lim(z_n) = L$. It is relatively easy to see that is the complex sequence $(z_n) = (u_n + iv_n)$ converges to L, then the two real sequences (u_n) and (v_n) each have a limit: (u_n) converges to ReL and (v_n) converges to Im L. Conversely, if the two real sequences (u_n) and (v_n) each have a limit, then so also does the complex sequence $(u_n + iv_n)$. All the usual nice properties of limits of sequences are thus true:

Now if $z_n = x_n + y_n i$ and $L = a + bi$, then $\lim_{n\to\infty} z_n = L$ if and only if

$$\lim_{n\to\infty} |z_n - L| = \lim_{n\to\infty} \sqrt{(x_n - a)^2 + (y_n - b)^2} = 0,$$

the latter of which occurs if and only if $\lim_{n\to\infty} x_n = a$ and $\lim_{n\to\infty} y_n = b$. Hence we have the following useful result.

Proposition Let $z_n = x_n + y_{ni}$ and $L = a + bi$. Then

$$\lim_{n\to\infty} z_n = L$$

if and only if

$$\lim_{n\to\infty} x_n = a \text{ and } \lim_{n\to\infty} y_n = b.$$

if and only if

$$\lim_{n\to\infty} x_n = a \text{ and } \lim_{n\to\infty} y_n = b.$$

Thus to determine the limiting behaviour of a sequence $\{z_n\}$ of complex numbers, we need only condider the behaviour of the two sequences of real numbers, $\{\Re(z_n)\}$ and $\{\Im(z_n)\}$.

A complex sequence is a function whose domain is the positive integers and whose range is a subset of the complex numbers. The following are examples of sequences:

$$f(n) = \left(2 - \frac{1}{n}\right) + \left(5 + \frac{1}{n}\right)i \ (n = 1, 2, 3, ...): \quad ...(1)$$

$$g(n) = e^{i}\frac{\pi n}{4} \ (n = 1, 2, 3, ...); \quad ...(2)$$

$$h(k) = 5 + 3\,i + \left(\frac{1}{1+i}\right)^k \ (k = 1, 2, 3, ...); \quad ...(3)$$

$$r(n) = \left(\frac{1}{4} + \frac{i}{2}\right)^n \ (n = 1, 2, 3, ...). \quad ...(4)$$

For convenience, we use the term sequence rather than complex sequence. The sequence r given by Equation (4) could be written in a variety of ways:

$$\left\{\left(\frac{1}{4} + \frac{i}{2}\right)^n\right\}_{n=1}^{\infty}, \left\{\left(\frac{1}{4} + \frac{i}{2}\right)^n\right\}_1^{\infty}, \left\{\left(\frac{1}{4} + \frac{i}{2}\right)^n\right\}_{n=1},$$

$$\left\{\left(\frac{1}{4} + \frac{i}{2}\right)^{n+2}\right\}_{n=-2}^{\infty}, \left\{\left(\frac{1}{4} + \frac{i}{2}\right)^k\right\}_{k=1}^{\infty}$$

The sequences f and g given by Equations (1) and (2) behave differently as n gets larger. The terms in Equation (1) approach $2 + 5\,i = (2, 5)$, but those in Equation (2) do not approach any particular number, as they oscillate around the eight eighth roots of unity on the unit circle. Informally, the sequence $\{z_n\}_1^{\infty}$ has ζ as its limit as n approaches infinity, provided the terms z_n can be made as close as we want to ζ by making n large enough. When this happens, we write

$$\lim_{n\to\infty} z_n = \zeta \text{ or } z_n \to \zeta \text{ as } n \to \infty \quad ...(5)$$

Definition1 **(Limit of a Sequence)** $\lim_{n\to\infty} z_n = \zeta$ means that for any real number $\in > 0$ there corresponds a positive integer $N_{\in}$ (which depends on $\in$) such that $z_n \in D \in (\zeta)$ whenever n > $N_{\in}$. That is $|\zeta - z_n| > \in$ whenever. Figure 1 illustrates a convergent sequence.

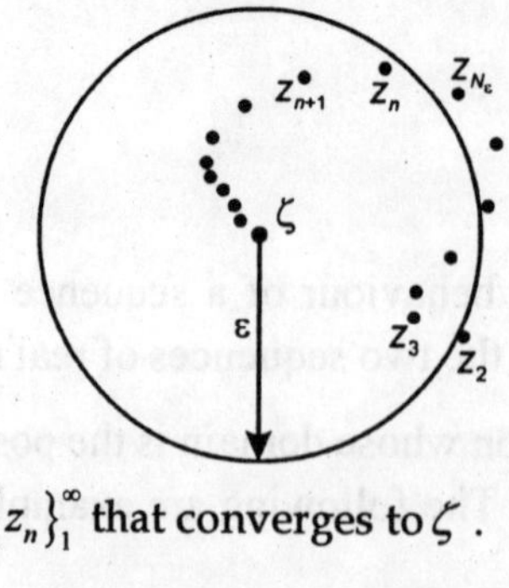

Fig. 1 A sequence $\{z_n\}_1^{\infty}$ that converges to ζ. (If n > $N_{\in}$ then $z_n \in D_{\in}$ (ζ))

Remark 1. The reason we use the notation N_ϵ is to emphasize the fact that this number depends on our choice of ϵ. Sometimes it will be convenient to drop the subscript.

Theorem 1. Let $z_n = x_n + i\,y_n$ and $\zeta = u + i\,v$. Then

$$\lim_{n\to\infty} z_n = \zeta \text{, iff} \qquad \ldots(6)$$

$$\lim_{n\to\infty} x_n = u \text{ and } \lim_{n\to\infty} y_n = v. \qquad \ldots(7)$$

If $\lim_{n\to\infty} z_n = \zeta$, we say that the sequence $\{z_n\}_1^\infty$ converges to ζ.

***Definition 2.* (Bounded Sequence).** A complex sequence $\{z_n\}$ is bounded provided that there exists a positive real number R and an integer N such that $|z_n| < R$ for all $n > N$. In other words, for $n > N$, the sequence $\{z_n\}$ is contained in the disk $D_R(0)$.

Convergent sequences are always bounded.

Theorem 2. If $\{z_n\}$ is a convergent sequence, then $\{z_n\}$ is bounded.

***Definition 3.* (Cauchy Sequence).** The sequence $\{z_n\}$ is said to be a Cauchy sequence if for every $\epsilon > 0$ there exists a positive integer N_ϵ, such that if n, $m > N_\epsilon$ then $|z_n - z_m| < \epsilon$, or, equivalently, $z_n - z_m \in D_\epsilon(0)$.

***Theorem 3.* (Cauchy Sequences Converge).** If $\{z_n\}$ is a Cauchy sequence, then $\{z_n\}$ converges.

Example 1. Evaluate the limit of $z_n = \dfrac{3n-1}{2n+2} + \dfrac{n+1}{n-1}i$, for $n = 1, 2, 3, \ldots$

Solution: $\lim_{n\to\infty} \Re(z_n) = \lim_{n\to\infty} \dfrac{3n-1}{2n+2} = \lim_{n\to\infty} \dfrac{3-\frac{1}{n}}{2+\frac{2}{n}} = \dfrac{3}{2}$

and $\lim_{n\to\infty} \Im(z_n) = \lim_{n\to\infty} \dfrac{n+1}{n-1} = \lim_{n\to\infty} \dfrac{1+\frac{1}{n}}{1-\frac{1}{n}} = 1,$

so $\lim_{n\to\infty} z_n = \dfrac{3}{2} + i.$

Example 2. Evaluate the limit of $z_n = \dfrac{1}{n}\left(\cos\left(\dfrac{n\pi}{3}\right) + \sin\left(\dfrac{n\pi}{3}\right)i\right)$ for $n = 1, 2, 3, \ldots$

Solution: Now, $\lim_{n\to\infty} \Re(z_n) = \lim_{n\to\infty} \dfrac{\cos\left(\frac{n\pi}{3}\right)}{n} = 0$

and $\lim_{n\to\infty} \Im(z_n) = \lim_{n\to\infty} \dfrac{\sin\left(\frac{n\pi}{3}\right)}{n} = 0,$

so $\lim_{n\to\infty} z_n = 0.$

Geometrically, since $|z_n| = \frac{1}{n}$ and $\arg(z_n) = \frac{n\pi}{3}$, the points in this sequence are converging to 0 along a spiral path, as shown in the figure

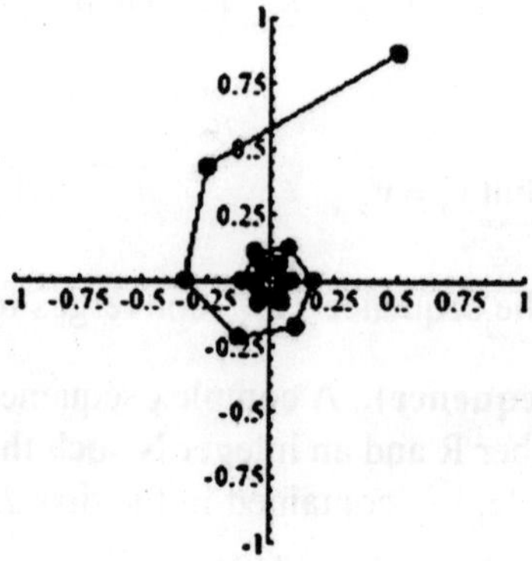

Figure. 1 Showing the plots of the points $z_n = \frac{1}{n}\left(\cos\left(\frac{n\pi}{3}\right) + \sin\left(\frac{n\pi}{3}\right)i\right)$, $n = 1, 2, 3, \ldots 20$

THEOREM ON LIMITS:

(i) $\lim(z_n \pm w_n) = \lim(z_n) \pm \lim(w_n)$;

(ii) $\lim(z_n w_n) = \lim(z_n)\lim(w_n)$; and

(iii) $\lim\left(\frac{z_n}{w_n}\right) = \frac{\lim(z_n)}{\lim(w_n)}$.

Provided that $\lim(z_n)$ and $\lim(w_n)$ exist. (And in the last equation, we must, of course, insist that $\lim(w_n) \neq 0$.)

A necessary and sufficient condition for the convergence of a sequence (a_n) is the celebrated Cauchy criterion: given $\varepsilon > 0$, there is an integer N_ϵ so that $|a_n - a_m| < \epsilon$ whenever $n, m > N_\epsilon$.

Convergence of sequence

A sequence (f_n) of functions on a domain D is the obvious thing: a function from the positive integers into the set of complex functions on D. Thus, for each $z \in D$, we have an ordinary sequence $(f_n(z))$. If each of the sequences $(f_n(z))$ converges, then we say the sequence of functions (f_n) converges to the function f defined by $f(z) = \lim(f_n(z))$.

Definition4: *The sequence* (f_n) *is said to converge to* f *uniformly on a set* S *if given an* $e > 0$, *there is an integer* N_ϵ *so that* $|f_n(z) - f(z)| < \epsilon$ *for all* $n \geq N_\epsilon$ *and all* $z \in S$.

Note that this is possible for a sequence of continuous functions to have a limit function that is *not* continuous. This cannot happen if the convergence us uniform. To see this, suppose the sequence (f_n) of continuous functions converges uniformly to f on a domain D, let $Z_0 \in D$, and let $\varepsilon > 0$. We need to show there is a δ so that $|f(z_0) - f(z)| < \varepsilon$ whenever.

$|z_0 - z| < \delta$. Let's do it. First, choose N so that $|f_N(z) - f(z)| < \frac{\varepsilon}{3}$. We can do this because of the uniform convergence of the sequence (f_n). Next, choose δ so that $|f_N(z_0) - f_N(z)| < \frac{\varepsilon}{3}$ whenever $|z_0 - z| < \delta$, This is possible because f_N is continuous. Now then, when $|z_0 - z| < \delta$, we have

$$
\begin{aligned}
|f(z_0) - f(z)| &= |f(z_0) - f_N(z_0) + f_N(z_0) - f_N(z) + f_N(z) - f(z)| \\
&\le |f(z_0) - f_N(z_0)| + |f_N(z_0) - f_N(z)| + |f_N(z) - f(z)| \\
&< \frac{\varepsilon}{3} + \frac{\varepsilon}{3} + \frac{\varepsilon}{3} = \varepsilon,
\end{aligned}
$$

Now suppose we have a sequence (f_n) of continuous functions which converges uniformly on a contour C to the function f. Then the sequence $\left(\int_C f_n(z)dz\right)$ converges to $\int_C f(z)dz$. This is easy to see. Let $\varepsilon > 0$. Now let N be so that $|f_n(z) - f(z)| < \frac{\varepsilon}{A}$ for $n > N$, where A is the length of C. Then,

$$
\begin{aligned}
\left|\int_C f_n(z)dz - \int_C f(z)dz\right| &= \left|\int_C (f_n(z) - f(z))dz\right| \\
&< \frac{\varepsilon}{A} A = \varepsilon
\end{aligned}
$$

whenever $n > N$.

Now suppose (f_n) is a sequence of functions each *analytic* on some region D, and suppose the sequence converges uniformly on D to the function f. Then f is analytic.

Example 3. Find the limit of the function $\{z_n\} = \left\{\frac{\sqrt{n} + i(n+1)}{n}\right\}$.

Solution: We write $z_n = x_n + i\,y_n = \frac{1}{\sqrt{n}} + i\frac{n+1}{n}$

Using results concerning sequences of real numbers, we find that

$$\lim_{n\to\infty} x_n = \lim_{n\to\infty} \frac{1}{\sqrt{n}} = 0 \text{ and } \lim_{n\to\infty} y_n = \lim_{n\to\infty} \frac{n+1}{n} = 1.$$

Therefore $\lim_{n\to\infty} z_n = \lim_{n\to\infty} \frac{\sqrt{n}+i(n+1)}{n} = 0 + i = i.$

Example 4. Show that the sequence $\{z_n\} = \{(1 + i)^n\}$ diverges.

Solution: We have

$$z_n = (1 + i)^n = x_n + i\, y_n$$

$$z_n = \left(\sqrt{2}\right)^n \cos\frac{n\pi}{4} + i\left(\sqrt{2}\right)^n \sin\frac{n\pi}{4}$$

The real sequences $x_n = \left(\sqrt{2}\right)^n \cos\frac{n\pi}{4}$ and $y_n = \left(\sqrt{2}\right)^n \sin\frac{n\pi}{4}$ both exhibit divergent oscillations, so we conclude that $z_n = (1 + i)n$ diverges.

SERIES OF COMPLEX NUMBER

Consider a sequence $\{z_n\}$, and form a new sequence $\{S_n\}$, called the sequence of partial sums, as follows.

$$S_1 = z_1,$$
$$S_1 = z_1 + z_2,$$
$$S_2 = z_1 + z_2 + z_3,$$
$$\vdots$$
$$S_n = z_1 + z_2 + \ldots + z_n = \sum_{k=1}^{n} z_k,$$
$$\vdots$$

***Definition 5.* (Infinite Series).** The formal expression $\sum_{k=1}^{\infty} z_k = z_1 + z_2 + \ldots + z_n + \ldots$ is called an infinite series, and $z_1, z_2, \ldots, z_n, \ldots$, are called the terms of the series.

If there is a complex number S for which

$$S = \lim_{n\to\infty} S_n = \lim_{n\to\infty} \sum_{k=1}^{n} z_k$$

we will say that the infinite series $\sum_{k=1}^{\infty} z_k$ converges to S, and that S is the sum of the infinite series. When this occurs, we write

$$S = \sum_{k=1}^{\infty} z_k$$

The series $\sum_{k=1}^{\infty} z_k$ is said to be absolutely convergent provided that the (real) series of magnitudes $\sum_{k=1}^{\infty} | z_k |$ converges.

If a series does not converge, we say that it diverges.

Remark 2. The first finitely many terms of a series do not affect its convergence or divergence and, in this respect, the beginning index of a series is irrelevant. Thus, we will without comment conclude that if a series $\sum_{k=n+1}^{\infty} z_k$ converges, then so does $\sum_{k=1}^{\infty} z_k$, where z_1, z_2, ..., z_n is any finite collection of terms. A similar remark holds for determining divergence of a series.

We now give some more standard theorems for complex series, along with examples of how they are used.

Theorem 4. Let $z_n = x_n + i\,y_n$ and $S = U + i\,V$.

$$S = \sum_{n=1}^{\infty} z_n = \sum_{n=1}^{\infty} (x_n + i\,y_n)\,(\text{converges}) \qquad ...(8)$$

if and only if both

$$U = \sum_{n=1}^{\infty} x_n \text{ and } V = \sum_{n=1}^{\infty} y_n \text{ (converge).}$$

Theorem 5. If $\sum_{n=1}^{\infty} z_n$ is a convergent complex series, then $\lim_{n\to\infty} z_n = 0$.

Example 5. Show that the series

$$\sum_{n=1}^{\infty} \frac{1 + in(-1)^n}{n^2} = \sum_{n=1}^{\infty} \left[\frac{1}{n^t} + i\frac{(-1)^n}{n}\right] \text{ is convergent.}$$

Solution: Recall that the real series

$\sum_{n=1}^{\infty} \frac{1}{n^t}$ and $\sum_{n=1}^{\infty} \frac{(1)^n}{n}$ are convergent. Hence, Theorem 4.implies that the given series is convergent.

Example 6. Show that the series

$$\sum_{n=1}^{\infty} \frac{(-1)^n + i}{n} = \sum_{n=1}^{\infty} \left[\frac{(-1)^n}{n} + i\frac{1}{n}\right] \text{ is divergent.}$$

Solution: We know that the real series $\sum_{n=1}^{\infty} \frac{1}{n}$ is divergent. Hence, theorem 4 implies that the given series is divergent.

Example 7. Show that the series $\sum_{n=1}^{\infty} (1+i)^n$ is divergent.

Solution: Here we set $z_n = (1 + i)^n$ and observe that

$$\lim_{n\to\infty} |z_n| = \lim_{n\to\infty} |(1+i)^n| = \lim_{n\to\infty}\left(\sqrt{2}\right)^n = \infty$$

Thus $\lim_{n\to\infty} z_n \neq 0$, and theorem 5 implies that the series is not convergent; hence it is divergent.

Theorem 6. Let $\sum_{n=1}^{\infty} z_n$ and $\sum_{n=1}^{\infty} w_n$ be convergent series, and let c be a complex number. Then

$$\sum_{n=1}^{\infty} cZ_n = \sum_{n=1}^{\infty} c z_n = c\sum_{n=1}^{\infty} z_n \quad ..(9)$$

and

$$\sum_{n=1}^{\infty} (z_n + w_n) = \sum_{n=1}^{\infty} z_n + \sum_{n=1}^{\infty} w_n \quad ...(10)$$

Definition 6. **(Cauchy Product of Series).**

Let $\sum_{n=1}^{\infty} a_n$ and $\sum_{n=1}^{\infty} b_n$ be convergent series, where a_n and b_n are complex numbers. The Cauchy product of the two series is defined to be the series $\sum_{n=1}^{\infty} c_n$, where $c_n \sum_{k=1}^{n} a_k b_{n-k}$.

Theorem 7. If the Cauchy product converges, then

$$\sum_{n=1}^{\infty} c_n = \left(\sum_{n=1}^{\infty} a_n\right)\left(\sum_{n=1}^{\infty} b_n\right) \quad ...(11)$$

where, $c_n = \sum_{k=1}^{n} a_k\, b_{n-k}$ ...(12)

Theorem 8. **(Comparison Test).** Let $\sum_{n=1}^{\infty} M_n$ be a convergent series of real nonnegative terms. If $\{z_n\}$ is a sequence of complex numbers and $|z_n| \leq M_n$ holds for all n, then $\sum_{n=1}^{\infty} z_n$ converges.

Corollary 1. If $\sum_{n=1}^{\infty} | z_n |$, then $\sum_{n=1}^{\infty} z_n$ converges. In other words, absolute convergence implies convergence for complex series.

Example 8. Show that the series $\sum_{n=1}^{\infty} \frac{(3+4i)^n}{5^n n^t}$ is convergent.

Solution: We calculate $|z_n|$ $\left|\frac{(3+4i)^n}{5^n n^t}\right| = \frac{1}{n^t} = M_n$. Using the comparison test and the fact that $\sum_{n=1}^{\infty} \frac{1}{n^t}$ converges, we determine that $\sum_{n=1}^{\infty} \left|\frac{(3+4i)^n}{5^n n^t}\right|$ converges and hence, by Corollary .1, so does $\sum_{n=1}^{\infty} \frac{(3+4i)^n}{5^n n^t}$

Geometric Series and Convergence Theorems

We begin this section by presenting a series of the form $\sum_{n=0}^{\infty} z^n$, which is called a geometric series.

Theorem 9. **(Geometric Series).** If $|z| < 1$, the series $\sum_{n=0}^{\infty} z^n$ converges to $f(z) = \frac{1}{1-z}$. That is, if $|z| < 1$ then

$$\sum_{n=0}^{\infty} z^n = 1 + z + z^2 + \ldots + z^k + \ldots = \frac{1}{1-z}. \qquad \ldots(13)$$

If $|z| \geq 1$, the series diverges.

Corollary 2. If $|z| > 1$, the series $\sum_{n=1}^{\infty} z^{-n}$ converges to $f(z) = \frac{1}{1-z}$. That is, if $|z| > 1$ then

$$\sum_{n=1}^{\infty} z^{-n} = z^{-1} + z^{-2} + \ldots + z^{-4} + \ldots = \frac{1}{z-1},$$

or equivalently,

$$-\sum_{n=1}^{\infty} z^{-n} = -z^{-1} - z^{-2} - \ldots - z^{-4} - \ldots = \frac{1}{1-z}.$$

If $|z| \leq 1$, the series diverges.

Corollary 3. If $z \neq 1$, then for all n,

$$\frac{1}{1-z} = 1 + z + z^2 + \ldots + z^{n-1} + \frac{z^n}{1-z}.$$

Example 9. Show that $\sum_{n=0}^{\infty} \frac{(1-i)^n}{2^n} = 1 - i.$.

Solution: If we set $z = \frac{1-i}{2}$, then $|z| = \left|\frac{1-i}{2}\right| = \frac{\sqrt{2}}{2} < 1$. By using theorem 9, you get the sum is $1/\left(1 - \frac{1-i}{2}\right) = \frac{2}{2-1+i} = \frac{2}{1+i} = 1 - i$

Example 10. Evaluate $\sum_{n=3}^{\infty} \frac{i^n}{2^n}$.

Solution: We can put this expression in the form of a geometric series:

$$\begin{aligned}
\sum_{n=3}^{\infty} \frac{i^n}{2n} &= \sum_{n=3}^{\infty} \left(\frac{i}{2}\right)^n = \sum_{n=3}^{\infty} \left(\frac{i}{2}\right)^3 \left(\frac{i}{2}\right)^{n-3} \\
&= \left(\frac{i}{2}\right)^3 \sum_{n=3}^{\infty} \left(\frac{i}{2}\right)^{n-3} = \left(\frac{i}{2}\right)^3 \frac{1}{1-\frac{i}{i}} \\
&= -\frac{i}{8} \frac{1}{1-\frac{i}{i}} = \frac{-i}{8-4i} \\
&= \frac{-i}{8-4i} \frac{8+4i}{8+4i} = \frac{4-8i}{80} \\
&= \frac{1}{20} - \frac{i}{10}
\end{aligned}$$

Remark 3. If we think of z as the "ratio" by which a given term of the series is multiplied to generate successive terms, we see that the sum of a geometric series equals $\frac{\text{first term}}{1-\text{ratio}}$, provided | ratio | < 1.

The geometric series is used for determining the convergence or divergence of series. The proof is similar to the one used for real series, and we leave it for you to do.

Theorem 10. **(D'Alembert's Ratio Test).** If $\sum_{n=0}^{\infty} \zeta_n$ is a complex series with the property that

$$\lim_{n \to \infty} \frac{|\zeta_{n+1}|}{|\zeta_n|} = L, \qquad \ldots(14)$$

then the series converges absolutely if L < 1 and diverges if L > 1.

Example 11. Show that $\sum_{n=0}^{\infty}\frac{(1-i)^n}{n!}$ converges.

Solution: Using the ratio test, we find that

$$\lim_{n\to\infty}\left|\frac{(1-i)^{n+1}}{(n+1)!}\Big/\frac{(1-i)^n}{n!}\right| = \lim_{n\to\infty}\left|\frac{n!(1-i)}{(n+1)!}\right|$$

$$= \lim_{n\to\infty}\left|\frac{n!(1-i)}{(n+1)!}\right|$$

$$= \lim_{n\to\infty}\frac{n!|1-i|}{(n+1)!}$$

$$= \lim_{n\to\infty}\frac{|1-i|}{n+1}$$

$$= \lim_{n\to\infty}\frac{\sqrt{2}}{n+1}$$

$$= 0 = L$$

Because $L < 1$, the series converges.

Example 12. Show that $\sum_{n=0}^{\infty}\frac{(z-i)^n}{2^n}$ converges for all z in the disk $|z-i|<2$.

Solution: Using the ratio test, we find that

$$\lim_{n\to\infty}\left|\frac{(z-i)^{n+1}}{2^{n+1}}\Big/\frac{(z-i)^n}{2^n}\right| = \lim_{n\to\infty}\left|\frac{z-i}{2}\right| = \frac{|z-i|}{2} = L$$

If $|z-i|<2$, then $\frac{|z-i|}{2} = L < 1$, and the series converges.

If $|z-i|>2$, then $L > 1$, and the series diverges.

Definition 6. **(Limit Supremum).** Let $\{t_n\}$ be a sequence of positive real numbers. The limit supremum of the sequence (denoted by $\limsup_{x\to\infty} t_n$) is the smallest real number L with the property that for any $\in > 0$ there are at most finitely many terms in the sequence that are larger than L + $\in$. If there is no such number L, then we set $\limsup_{x\to\infty} t_n = \infty$.

Theorem 11. **(Root Test).** Suppose that the series $\sum_{n=0}^{\infty}\zeta_n$, has $\limsup_{n\to\infty}$ has $|\zeta_n|^{\frac{1}{n}} = L$

$$\left(\text{i.e}\limsup_{n\to\infty}\sqrt[n]{|\zeta_n|} = L\right) \quad ...(15)$$

Then the series is absolutely convergent if $L < 1$ and divergent if $L < 1$.

Example 13. Find the radius of convergence of the infinite series $S(z) = 1 + \frac{z}{2} + \frac{z^2}{3^2} + \frac{z^3}{2^3} + \frac{z^4}{3^4} + \frac{z^5}{3^5} + \frac{z^6}{3^6} + \frac{z^7}{3^7} + \frac{z^8}{3^8} + \frac{z^9}{3^9} + \frac{z^{10}}{3^{10}} + \frac{z^{11}}{3^{11}} + ...$

Solution: The sequence of coefficients is $\left\{1, \frac{1}{2}, \frac{1}{3^2}, \frac{1}{2^3}, \frac{1}{3^4}, \frac{1}{2^5}, \frac{1}{2^7}, \frac{1}{3^8}, \frac{1}{2^9}, \frac{1}{3^{10}}, \frac{1}{2^{11}}, ...\right\}$.

The coefficients with even subscripts are $a_2k\frac{1}{3^2k}$, and the coefficients with odd subscripts are $a_{2k+1} = \frac{1}{2^{2k+1}}$.

So that we have

$$|a_{2k}|^{\frac{1}{2k}} = \left(\frac{1}{3^{2k}}\right)^{\frac{1}{2k}} = \frac{1}{3} \text{ and } |a_{2h+1}|^{\frac{1}{2k+1}} = \left(\frac{1}{2^{2k+1}}\right)^{\frac{1}{2k+1}} = \frac{1}{2}.$$

It follows that $\limsup_{n\to\infty} \sqrt[n]{|a_n z^n|} = \limsup_{n\to\infty}\left\{1, \frac{1}{2}, \frac{1}{3}, \frac{1}{2}, \frac{1}{3}, \frac{1}{2}, \frac{1}{3}, ...\right\} = \frac{1}{2}$.

Hence $\rho = \frac{1}{\limsup_{n\to\infty} |c_n|^{\frac{1}{n}}} = \frac{1}{\frac{1}{2}} = 2$

Hence, 2 is the required radius of convergence.

POWER SERIES OF COMPLEX NUMBER

In mathematics, a power series (in one variable) is an infinite series of the form

$$f(z) = \sum_{n=0}^{\infty} a_n (z-c)^n = a_0 + a_1(z-c) + a_2(z-c)^2 + a_3(z-c)^3 + ...$$

where a_n represents the coefficient of the nth term, c is a constant, and x varies around c (for this reason one sometimes speaks of the series as being *centered* at c). This series usually arises as the Taylor series of some known function.

In many situations c is equal to zero, for instance when considering a Maclaurin series. In such cases, the power series takes the simpler form

$$f(z) = \sum_{n=0}^{\infty} a_n z^n = a_0 + a_1 z + a_2 z^2 + a_3 z^3 + ... \qquad ...(16)$$

Remember: Any polynomial can be easily expressed as a power series around any center c, albeit one with most coefficients equal to zero.

Suppose that we have a series $\sum_{n=0}^{\infty} \xi_n$ where $\xi_n = c_n(z - \alpha)^n$. If α and the collection of c_n are fixed complex numbers, we will get different series by selecting different values for z. For example, if $\alpha = 2$ and $c_n = \frac{1}{n!}$ for all n, we get the series $\sum_{n=0}^{\infty} \frac{1}{n!}\left(\frac{i}{2} - 2\right)^n$ if $z = \frac{i}{2}$, and $\sum_{n=0}^{\infty} \frac{1}{n!}(2+i)^n$ if $z = 4 + i$. Note that when $\alpha = 0$ and $c_n = 1$ for all n, we get a geometric series. The collection of points for which the series $\sum_{n=0}^{\infty} c_n(z-\alpha)^n$ converges is the domain of a function $f(z) = \sum_{n=0}^{\infty} c_n(z-\alpha)^n$, which we call a power series function. The series $\sum_{n=0}^{\infty} c_n(z-\alpha)^n$ is really compact notation for $c_0 + \sum_{n=1}^{\infty} c_n(z-\alpha)^n$.

***Definition 7.* (Power Series).** The function $f(z) = \sum_{n=0}^{\infty} c_n(z-\alpha)^n$ is called a power series, with center $z = \alpha$.

Theorem 12. Suppose $f(z) = \sum_{n=0}^{\infty} c_n(z-\alpha)^n$. Then the set of points z for which the series converges is one of the following:

(i) The single point $z = \alpha$.

(ii) The disk $D_\rho(\alpha) = \{z: | - \alpha | < r\}$, along with part (either none, or some or all) of the circle $c_r(\alpha) = \{z: | z - \alpha| = \rho\}$.

(iii) The entire complex plane.

The power series $f(z) = \sum_{n=0}^{\infty} c_n(z-\alpha)^n$ converges if $| z - \alpha | < \rho$ and diverges if $| z - \alpha | > \rho$. We call the number ρ the radius of convergence of the power series (see Figure 3). For case (i) of Theorem 12, we say that the radius of convergence is zero and that the radius of convergence is infinity for case (iii).

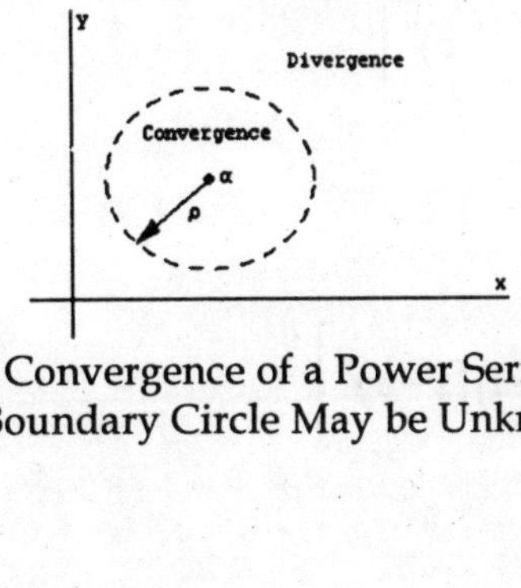

Fig. 3 The Radius of Convergence of a Power Series. What happens on the Boundary Circle May be Unknown.

***Theorem 13*. (Radius of Convergence).** For the power series function $f(z) = \sum_{n=0}^{\infty} c_n(z-\alpha)^n$, we can find ρ, its radius of convergence, by any of the following methods:

(i) Cauchy's Root Test: $\rho = \dfrac{1}{\lim_{n\to\infty} |c_n|^{\frac{1}{n}}}$ ...(17)

(provided that the limit exists.)

(ii) Cauchy-Hadamard Formula: $\rho = \dfrac{1}{\limsup\limits_{n\to\infty} |c_n|^{\frac{1}{n}}}$ ...(18)

(the limit superior always exists.)

(iii) d'Alembert's Ratio Test: $\rho = \dfrac{1}{\lim_{n\to\infty} |\frac{c_n+1}{c_n}|}$ ...(19)

(provided that the limit exists.)

We set $\rho = \infty$ if the limit equals 0, and $\rho = 0$ if the limit equals ∞.

OPERATIONS ON POWER SERIES

Addition and Subtraction

When two functions *f* and *g* are decomposed into power series around the same center *c*, the power series of the sum or difference of the functions can be obtained by termwise addition and subtraction. That is, if:

$$f(z) = \sum_{n=0}^{\infty} a_n z^n \text{ and } g(z) = \sum_{n=0}^{\infty} b_n z^n$$

$$\text{then } (f \pm g)(z) = \sum_{n=0}^{\infty} (a_n \pm b_n) z^n \quad ...(20)$$

Multiplication and Division

With the same definitions above, for the power series of the product and quotient of the functions can be obtained as follows:

$$f(z)(z) = \left[\sum_{n=0}^{\infty} a_n (z-c)^n\right]\left[\sum_{n=0}^{\infty} b_n (z-c)^n\right] \quad ...(21)$$

$$= \sum_{i=0}^{\infty}\sum_{j=0}^{\infty} a_n b_n (z-c)^{i+j} \quad ...(22)$$

$$= \sum_{i=0}^{\infty}\left[\sum_{j=0}^{n} (a_n b_n - i)(z-c)^n\right] \quad ...(23)$$

The sequence $m_n = \sum_{i=0}^{n} a_i b_n - i$ is known as the convolution of the sequence a_n and b_n.

For division, observe:

$$\frac{f(z)}{g(z)} = \frac{\sum_{n=0}^{\infty} a_n (z-c)^n}{\sum_{n=0}^{\infty} b_n (z-c)^n} = \sum_{j=0}^{n} d_n (z-c)^n \quad ...(24)$$

$$f(z) = \left[\sum_{n=0}^{\infty} a_n (z-c)^n\right]\left[\sum_{n=0}^{\infty} d_n (z-c)^n\right] \quad ...(25)$$

and then use the above, comparing coefficients.

Differentiation and Integration

Once a function is given as a power series, it is continuous wherever it converges and is differentiable on the interior of this set. It can be differentiated and integrated quite easily, by treating every term separately:

$$f'(z) = \sum_{n=1}^{\infty} a_n (z-c)^{n-1} \quad ...(26)$$

$$\int f(z) = \sum_{n=0}^{\infty} \frac{a_n (z-c)^{n+1}}{n+1} + C \quad ...(27)$$

Both of these series have the same radius of convergence as the original one.

ANALYTIC FUNCTIONS

A function f defined on some open subset U of R or C is called analytic if it is locally given by power series. This means that every a–U has an open neighborhood V–U, such that there exists a power series with center a which converges to $f(x)$ for every x –V.

Every power series with a positive radius of convergence is analytic on the interior of its region of convergence. All holomorphic functions are complex-analytic. Sums and products of analytic functions are analytic, as are quotients as long as the denominator is non-zero.

If a function is analytic, then it is infinitely often differentiable, but in the real case the converse is not generally true. For an analytic function, the coefficients a_n can be computed as

$$a_n = \frac{f^{(n)}(c)}{n!} \quad ...(28)$$

where $f^{(n)}(c)$ denotes the nth derivative of f at c, and $f^{(0)}(c) = f(c)$. This means that every analytic function is locally represented by its Taylor series.

The power series expansion of the inverse function of an analytic function can be determined using the Lagrange inversion theorem.

Example 14. The infinite series $f(z) = \sum_{n=0}^{\infty}\left(\frac{n+2}{3n+1}\right)^n (z-4)^n$ has radius of convergence ρ = 3 Cauchy's root test because

$$\lim_{n\to\infty} |c_n|^{\frac{1}{n}} = \lim_{n\to\infty}\left(\left(\frac{n+2}{3n+1}\right)^n\right)^{\frac{1}{n}} = \lim_{n\to\infty}\frac{n+2}{3n+1} = \frac{1}{3},$$

$$\text{Hence} \quad \rho = \frac{1}{\lim_{n\to\infty}|c_n|^{\frac{1}{n}}} = \frac{1}{\frac{1}{3}} = 3.$$

Example 15. By using the Cauchy-Hadamard formula we get known that the infinite series

$f(z) = \sum_{n=1}^{\infty} c_n z^n = 4z + 5^2 z^2 + 4^3 z^3 + 5^4 z^4 + 4^5 z^5 + 5^6 z^6 + \ldots$ has radius of convergence $\rho = \frac{1}{5}$. Also using property of supremum, you get

$$\left\{|c_n|^{\frac{1}{n}}\right\} = \{4, 5, 4, 5, 4, 5, \ldots\}, \text{ so}$$

$$\limsup_{n\to\infty}|c_n|^{\frac{1}{n}} = \limsup_{n\to\infty}\{4, 5, 4, 5, 5, \ldots\} = 5,$$

$$\text{Hence } \rho = \frac{1}{\limsup_{n\to\infty}|c_n|^{\frac{1}{n}}} = \frac{1}{5}.$$

Example 16. The infinite series $f(z) = \sum_{n=0}^{\infty}\frac{z^n}{n!}$ has radius of convergence ρ = 8 by the ratio test because

$$\lim_{n\to\infty}\left|\frac{c_n+1}{c_n}\right| = \lim_{n\to\infty}\left|\frac{n!}{(n+1)!}\right| = \lim_{n\to\infty}\frac{1}{n+1} = 0$$

Since the limit equals 0, we set ρ = 8.

Example 17. Find the radius of convergence of the infinite series.

$$S(z) = 1 + \frac{z}{2} + \frac{z^2}{3^2} + \frac{z^3}{2^3} + \frac{z^4}{3^4} + \frac{z^5}{2^5} + \frac{z^6}{3^6} + \frac{z^7}{2^7} + \frac{z^8}{3^8} + \frac{z^9}{2^9} + \frac{z^{10}}{3^{10}} + \frac{z^{11}}{2^{11}} + \ldots$$

Solution: The sequence of coefficients is

$$\left\{1, \frac{1}{2}, \frac{1}{3^2}, \frac{1}{2^3}, \frac{1}{3^4}, \frac{1}{2^5}, \frac{1}{3^6}, \frac{1}{2^7}, \frac{1}{3^8}, \frac{1}{2^9}, \frac{1}{3^{10}}, \frac{1}{2^{11}}, \ldots\right\}$$

The coefficients with even subscripts are $a_{2k} = \frac{1}{3^{2k}}$, and the coefficients with odd subscripts are $a_{2k+1} = \frac{1}{2^{2k+1}}$.

So that we have

$$|a_{2k}|^{\frac{1}{2k}} = \left(\frac{1}{3^{2k}}\right)^{\frac{1}{2k}} = \frac{1}{3} \text{ and } |a_{2k+1}|^{\frac{1}{2k+1}} = \left(\frac{1}{2^{2k+1}}\right)^{\frac{1}{2k+1}} = \frac{1}{2}$$

It follows that

$$\limsup_{n\to\infty} \sqrt[n]{|a_n z^n|} = \limsup_{n\to\infty} \left\{1, \frac{1}{2}, \frac{1}{3}, \frac{1}{2}, \frac{1}{3}, \frac{1}{2}, \frac{1}{3}, \ldots\right\} = \frac{1}{2}$$

$$\text{Hence } \rho = \frac{1}{\limsup\limits_{n\to\infty} |c_n|^{\frac{1}{n}}} = \frac{1}{\frac{1}{2}} = 2.$$

Hence, the required radius of convergence is 2.

Theorem 14. Suppose the function $f(z) = \sum_{n=0}^{\infty} c_n (z-\alpha)^n$ has radius of convergence $\rho > 0$. Then

(i) $f(z)$ is infinitely differentiable for all $z \in D\rho\ (\alpha)$. In fact ...(29)

(ii) for all $k, f(h)\ (z) = \sum_{n=0}^{\infty} n(n-1)\ldots(n-k+1)c_n(z-\alpha)^{n-k}$; and

(iii) $c_k = \dfrac{f^{(k)}(\alpha)}{k!}$ ...(30)

where $f(k)\ (\alpha)$ denotes the Kth derivative of f. (When $k = 0$, $f^{(k)}$ denotes the function f itself so that $f^{(0)}(z) = f(z)$ for all z.)

UNIFORM CONVERGENCE OF COMPLEX NUMBER

Here, we begin by discussing an important property associated with power series-uniform convergence.

Recall that, for a function f(z) defined on a set T, the sequence of functions $\{S_n(z)\}$ converges to the function $f(z)$ at the point $z_0 \in T$ provided that $\lim\limits_{n\to\infty} S(z_0) = f(z_0)$. Thus, for the particular point z_0, we know that for each $\in > 0$, there exists a positive integer $N_{\in, z_0}$ (which depends on both $\in$ and z_0) such that

if $n \geq N_{\in, z_0}$ then $| S(z_0) - f(z_0) | < \in$ ...(31)

If $S_n(z)$ is the nth partial sum of the series $\sum_{k=0}^{\infty} c_k (z - \alpha)^k$, Statement (40) becomes

If $n \geq N_{\in, z_0}$, then $\left| \sum_{k=0}^{n-1} c_k (z_0 - \alpha)^k - f(z_0) \right| < \in$ ...(32)

For a given value of $\in$, the integer $N_{\in, z_0}$ needed to satisfy Statement (40) often depends on our choice of z_0. This is not the case if the sequence $\{S_n(z)\}$ converges uniformly. For a uniformly convergent sequence, it is possible to find an integer $N_\in$ (depending only on $\in$) that guarantees Statement (40) no matter what value for $z_0 \in T$ we pick. In other words, if n is large enough, the function $S_n(z)$ is uniformly close to the function $f(z)$ for all $z \in T$.

Definition 8. **(Uniform Convergence)**,. The sequence $\{S_n(z)\}$ converges uniformly to $f(z)$ on the set T if for every $\in > 0$, there exists a positive integer $N_\in$ (Depending only on $\in$) such that

if $n \geq N$, then $| S_n(z) - f(z) < \in$ for all $z \in T$. ...(33)

If $S_n(z)$ is the nth partial sum of the series $\sum_{k=0}^{\infty} c_k (z - \alpha)^k$, we say that the series $\sum_{k=0}^{\infty} c_k (z - \alpha)^k$ converges uniformly to $f(z)$ on the set T.

Example 17. The sequence $\{S_n(z)\} = \left\{ e^z + \frac{1}{n} \right\}$ converges uniformly to the function $f(z) = e^z$ on the entire complex plane because for any $\in > 0$, statement (42) is satisfied for all z for $n \geq N_\in$, where $N_\in$ is any integer greater than $\frac{1}{\in}$.

The given sequence $S_n(z)$ diverges, if there exists $\in > 0$, such that for all positive integers N, there is some $n \geq N$ and some $z_0 \in T$

such that $| S_n(s_0) - f(z_0) | \geq \in$.

The geometric series do not converge uniformly to $f(z) = \frac{1}{1-z}$ for $z \in D_1(0)$, which is shown in the figure below:

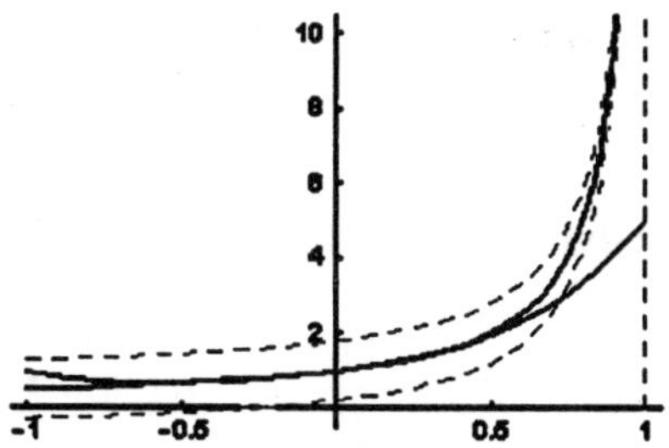

Fig. 4 The Geometric Series does not Converge Uniformly on (-1,1).

The Weierstrass M-test can be used to determine whether an infinite series is uniformly convergent or not .

***Theorem 15*. (Weierstrass M-Test)** Suppose the infinite series $\sum_{k=0}^{\infty} u_k(z)$ has the property that for each k, we have $|u_k(z)| \le M_k$ for all $z \in T$. If $\sum_{k=0}^{\infty} M_k$ converges, then $\sum_{k=0}^{\infty} u_k(z)$ converges uniformly on T.

***Theorem 16*.** Suppose the power series $\sum_{k=0}^{\infty} c_k (z-\alpha)^k$ has radius of convergence $\rho > 0$. Then for each *r*, (where $0 < r < \rho$), the series converges uniformly on the closed disk $\overline{D_r}(\alpha) = \{z : |z-\alpha| \le r\}$.

***Corollary 4*.** For each r, (where 0 < r < 1), the geometric series $\sum_{n=0}^{\infty} z^n$ converges uniformly on the closed disk $\overline{D_r}(0) = \{z : |z| \le r\}$.

***Theorem 17*.** Suppose $\{z_k(z)\}$ is a sequence of continuous functions defined on a set T containing the contour C.

If $\{S_k(z)\}$ converges uniformly to $f(z)$ on the set T, then

(i) $f(z)$ is continuous on T, and (ii) $\lim_{k\to\infty} \int_C S_k(z)dz = \int_C \lim_{k\to\infty} S_k(z)dz = \int_C f(z)dz.$ ...(34)

***Corollary 5*.** If the series $\sum_{n=0}^{\infty} c_n (z-\alpha)^n$ converges uniformly to $f(z)$ on the set T, and C is a contour contained in T, then $\sum_{n=0}^{\infty} \int_C c_n (z-\alpha)^n dz = \int_C \sum_{n=0}^{\infty} c_n (z-\alpha)^n dz = \int_C f(z)dz$.

TAYLOR AND LAURENT SERIES

Taylor series

Definition 8. If $f(z)$ is analytic at $z = \alpha$, then the series

$$f(\alpha)+f'(\alpha)(z-\alpha)+\frac{f^{(2)}(\alpha)}{2!}(z-\alpha)^2+\frac{f^{(2)}(\alpha)}{3!}(z-\alpha)^3+...=\sum_{k=0}^{\infty}\frac{f^{(k)}(\alpha)}{k!}(z-\alpha)^k \quad ...(35)$$

is called the Taylor series for $f(z)$ centered at $z = \alpha$. When the center is $\alpha = 0$, the series is called the Maclaurin series for $f(z)$.

Definition 9. Suppose f is analytic on the open disk $:|z - z_0| < r$. Let z be any point in this disk and choose C to be the positively oriented circle of radius r, where $|z - z_0| < \rho < r$. Then for $s \in C$ we have

$$\frac{1}{s-z}=\frac{1}{(s-z_0)(z-z_0)}=\frac{1}{s-z_0}\left[\frac{1}{1-\frac{z-z_0}{s-z_0}}\right]=\sum_{j=0}^{\infty}\frac{(z-z_0)^j}{(s-z_0)^{j+1}} \quad ...(36)$$

since $\left|\frac{z-z_0}{s-z_0}\right|<1$ The convergence is uniform, so we may integrate

$$\int_C\frac{f(s)}{s-z}ds=\sum_{j-0}^{\infty}\left(\int_C\frac{f(s)}{(s-z_0)^{j+1}}ds\right)(z-z_0)^j, \text{ or} \quad ...(37)$$

$$f(z)=\frac{1}{2\pi i}\int_C\frac{f(s)}{s-z}ds=\sum_{j-0}^{\infty}\left(\frac{1}{2\pi i}\int_C\frac{f(s)}{(s-z_0)^{j+i}}ds\right)(z-z_0)^j. \quad ...(38)$$

We have thus produced a power series having the given analytic function as a limit:

$$f(z)=\sum_{j-0}^{\infty}c_j(z-z_0)^j, |z-z_0|<r, \quad ...(39)$$

where

$$C_j=\frac{1}{2\pi i}\int_C\frac{f(s)}{(s-z_0)^{j+1}}ds. \quad ...(40)$$

We know we may differentiate the series to get

$$f(z)=\sum_{j-1}^{\infty}jc_j(z-z_0)^{j-1} \quad ...(41)$$

and this one converges uniformly where the series for f does. We can thus differentiate again and again to obtain

$$f^{(n)}(z) = \sum_{j-n}^{\infty} j(j-1)(j-2)....(j-n+1)_{cj}(z-z_0)^{j-n}. \qquad ...(42)$$

Hence,

$$f^{(n)}(z_0) = n!c_n, \text{ or} \qquad ...(43)$$

$$c_n = \frac{f^{(n)}(z_0)}{n!}. \qquad ...(44)$$

But we also know that

$$c_n = \frac{1}{2\pi i}\int_C \frac{f(s)}{(s-z_0)^{n+1}}ds. \qquad ...(45)$$

This gives us

$$f^{(n)}(z_0) = \frac{n!}{2\pi i}\int_C \frac{f(s)}{(s-z_0)^{n+1}}ds, \text{ for } n = 0, 1, 2, ... \qquad ...(46)$$

which is the Cauchy Integral Formula. Suppose we have

$$f(z) = \sum_{j-0}^{\infty} c_j (z-z_0)^j, \qquad ...(47)$$

and the radius of convergence is R. Then we know, of course, that the limit function f is analytic for $|z - z_0| < R$.

Lemma 1. If z, z_0 and α are complex numbers with $z \neq z_0$, and $z \neq \alpha$, then

$$\frac{1}{z-z_0} = \frac{1}{z-\alpha} + \frac{z_0-\alpha}{(z-\alpha)^2} + \frac{(z_0-\alpha)^2}{(z-\alpha)^3} + ... + \frac{(z_0-\alpha)^n}{(z-\alpha)^{n+1}} + \frac{1}{z-z_0}\frac{(z_0-\alpha)^{n+1}}{(z-\alpha)^{n+1}} \qquad ...(48)$$

where n is a positive integer.

***Theorem 18.* (Taylor's Theorem).** Suppose $f(z)$ is analytic in a domain G, and that $D_R(\alpha) = \{z : |z - \alpha| < R\}$ is any disk contained in G. Then the Taylor series for $f(z)$ converges to $f(z)$ for all z in $D_R(\alpha)$; that is,

$$f(z) = \sum_{k=0}^{\infty} \frac{f^{(k)}(\alpha)}{k!}(z-\alpha)^k \text{ for all } Z \in DR(\alpha).$$

Furthermore, for any r, $0 < r < R$, the convergence is uniform on the closed subdisk $\overline{Dr}(\alpha) = \{z : |z - \alpha| \leq r\}$ for $0 < r < R$.

A singular point of a function is a point at which the function fails to be analytic. In other words a non-removable singular point of a function has the property that it is impossible to redefine the value of the function at that point so as to make it analytic there. For example, the function $f(z) = \frac{1}{1-z}$ has a nonremovable singularity at $z = 1$.

Corollary 6. Suppose that $f(z)$ is analytic in the domain G that contains the point $z = \alpha$. Let z_0 be a nonremovable singular point of minimum distance to the point $z = \alpha$. If $|z_0 - \alpha| = R$, then

(i) the Taylor series $\sum_{n=0}^{\infty} \frac{f^{(n)}(\alpha)}{n!}(z-\alpha)^n$ converges to $f(z)$ on all of $D_R(\alpha)$, and

(ii) if $|z_1 - \alpha| = S > R$, the series $\sum_{n=0}^{\infty} \frac{f^{(n)}(\alpha)}{n!}(z_1-\alpha)^n$ does not converges to $f(z_1)$.

***Remark*.** In the following figure you observe that their interval of convergence is the open interval (–1, 1).

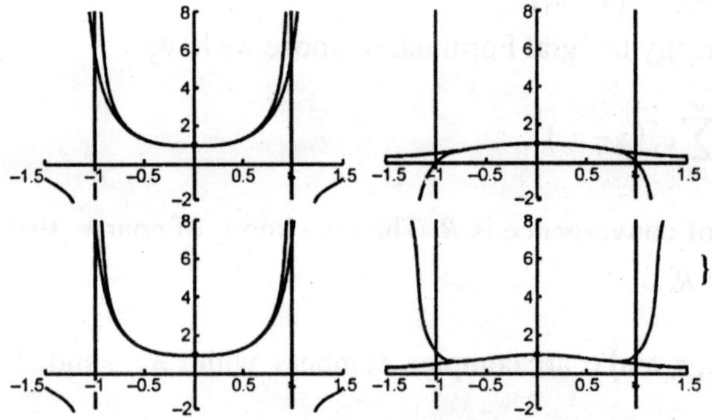

$$\frac{1}{1-x^2} \approx \sum_{n=0}^{5} x^{2n}, \sum_{n=0}^{6} x^2 n \quad \frac{1}{1+x^2} \approx \sum_{n=0}^{5} (-1)^n x^{2n}, \sum_{n=0}^{6} (-1)^n x^{2n}$$

Fig. 5 The Maclaurin series $\sum_{n=0}^{\infty} (-1)^n x^{2n}$ does not converge uniformly on (-1, 1).

***Theorem 19.* (Uniqueness of Power Series).** Suppose that in some disk $D_r(a)$ we have

$$f(z) = \sum_{n=0}^{\infty} a_n (z-\alpha)^n = \sum_{n=0}^{\infty} b_n (z-\alpha)^n \quad \text{...(49)}$$

Then $a_n = b_n$ for $n = 0, 1, 2, \ldots$

Example 18. Find the Maclaurin series for $f(z) = \sin^3 z$.

Solution: Computing derivatives for $f(z)$ would be an onerous task. Fortunately, we can make use of the trigonometric identity

$$\sin^3 z = \frac{3}{4}\sin z - \frac{1}{4}\sin 3z.$$

Recall that the series for sin *z* (valid for all *z*) is $\sin z = \sum_{n=0}^{\infty} \frac{(-1)^n}{(2n+1)!} z^{2n+1}$. Using the identity for $\sin^3 z$, we obtain

$$\sin^3 z = \frac{3}{4}\sum_{n=0}^{\infty} \frac{(-1)^n}{(2n+1)!} z^{2n+1} - \frac{1}{4}\sum_{N=0}^{\infty} \frac{(-1)^n}{(2n+1)!}(3z)^{2n+1}$$

$$= \frac{3}{4}\sum_{n=0}^{\infty} \frac{(-1)^n}{(2n+1)!} z^{2n+1} - \frac{3}{4}\sum_{n=0}^{\infty} \frac{(-1)^n g^n}{(2n+1)!} z^{2n+1}$$

$$= \sum_{n=0}^{\infty} (-1)^n \frac{3(1-g^n)}{4(2n+1)!} z^{2n+1}$$

By the uniqueness of power series, this last expression is the Maclaurin series for $\sin^3 z$.

Theorem 20. Let f(z) and g(z) have the power series representations

$$f(z) = \sum_{n=0}^{\infty} (z-\alpha)^n \text{ for } z \in D_{r1}(\alpha), \qquad \text{...(50)}$$

and

$$g(z) = \sum_{n=1}^{\infty} b_n (z-\alpha)^n \text{ for } z \in D_{r2}(\alpha) \qquad \text{...(51)}$$

If $r = \min\{r_1, r_2\}$ and β is any complex constant, then

$$\beta f(z) = \sum_{n=0}^{\infty} \beta a_n (z-\alpha)^n \text{ for } z \in D_{r1}(\alpha), \qquad \text{...(52)}$$

$$f(z) + g(z) = \sum_{n=0}^{\infty} (a_n + b_n)(z-\alpha)^n \text{ for } z \in D_r(\alpha), \text{and} \qquad \text{...(53)}$$

$$f(z) + g(z) = \sum_{n=0}^{\infty} c_n (z-\alpha)^n \text{ for } z \in D_r(\alpha), \text{ where} \qquad \text{...(54)}$$

$$c_n = \sum_{k=0}^{n} a_n b_{n-k}. \qquad \text{...(55)}$$

Identity (52) is known as the Cauchy product of the series for *f*(*z*) and *g*(*z*).

Example 19. Use the Cauchy product of series to show that

$$\frac{1}{(1-z)^2} = \sum_{n=0}^{\infty} (n+1) z^n \text{ for } z \in D_1(0).$$

Solution: We let $f(z) = g(z) = \frac{1}{1-z}$, for $z \in D_1(0)$. In terms of Theorem 20, we have $a_n = b_n = 1$, for all n, and thus Equation (52) gives

$$\frac{1}{(1-z)^2} = h(z) = f(z)\,g(z) = \sum_{n=0}^{\infty}\left(\sum_{k=0}^{n} a_k b_{n-k}\right) z^n$$

$$= \sum_{n=0}^{\infty}\left(\sum_{k=0}^{n} 1\right) z^n$$

$$= \sum_{n=0}^{\infty}(n+1) z^n$$

Example 20. Use the Cauchy product of series to show that

$$\frac{z}{(z-1)^4} = \sum_{n=0}^{\infty}\frac{n(1+n)(2+n)}{6} z^n \text{ for } z \in D_1(0).$$

Solution: Use the following facts $\frac{z}{(z-1)^2} = \sum_{n=0}^{\infty} nz^n$ for $z \in D_1(0)$ and $\frac{z}{(z-1)^2} = \sum_{n=0}^{\infty}(n+1)z^n$

for $z \in D_1(0)$.

Take the coefficients $a_n = n$ and b_n and $b_n = n+1$ and let carry out the calculations.

$$a_n = n$$
$$bn = 1 + n$$

$$f[z] = \sum_{n=0}^{\infty} a_n z^n = \sum_{n=0}^{\infty}(k) z^n = \frac{z}{(-1+z)^2}$$

$$g[z] = \sum_{n=0}^{\infty} a_n z^n = \sum_{n=0}^{\infty}(1-k+n) z^n = \frac{1}{(-1+z)^2}$$

$$c_n = \sum_{k=0}^{n} a_k b_{n-k}$$

$$C_n = \sum_{k=0}^{n}(k)(1-k+n) = \sum_{k=0}^{n}(k(1-k+n))$$

$$\sum_{k=0}^{n}(k(1-k+n)) = \frac{1}{6}n(1+n)(2+n)$$

$$c_n = \frac{1}{6}n(1+n)(2+n)$$

$$h[z] = \sum_{n=0}^{\infty}\left(\frac{1}{6}n(1+n)(2+n)z^n\right) = \frac{z}{(-1+z)^4}$$

Example 21. Show that $\dfrac{z}{(1-z)^2} = \sum_{n=1}^{\infty} nz^n$ for $z \in D_1(0)$

Solution: Use the result of Example 20 and equation (48) , we obtain

$$\frac{z}{(1-z)^2} = \frac{z-1+1}{(1-z)^2}$$

$$= \frac{1}{(1-z)^2} + \frac{-1}{(1-z)}$$

$$= \sum_{n=0}^{\infty}(n+1)z^n - \sum_{n=0}^{\infty} z^n = \sum_{n=0}^{\infty} nz^n$$

We could calculate the sum of the series, $\sum_{n=1}^{\infty} nz^n = \dfrac{z}{(-1+z)^2}$.

LAURENT SERIES REPRESENTATIONS

Suppose $f(z)$ is not analytic in $D_R(\alpha)$, but is analytic in $D_R^*(\alpha)' = \{z: 0 < |z - \alpha| < R\}$.For example, the function $f(z) = \dfrac{1}{z^3}e^z$ is not analytic when $z = 0$ but is analytic for $|z| > 0$. Clearly, this function does not have a Maclaurin series representation. If we use the Maclaurin series for $g(z) = e^z$, however, and formally divide each term in that series by z^3, we obtain the representation

$$f(z) = \frac{1}{z^3}e^3 = \frac{1}{z^3} + \frac{1}{z^2} + \frac{1}{2!z} + \frac{1}{3!} + \frac{z}{4!} + \frac{z^2}{5!} + \frac{z^3}{5!} + ... \qquad ...(56)$$

that is valid for all z such that $|z| > 0$.

Definition 10. **(Laurent Series).** Let C_n be a complex number for $n = 0, \pm 1, \pm 2, \pm 3,$ ..... The doubly infinite series $\sum_{n=-\infty}^{\infty} c_n (z-\alpha)^n$, called a Laurent series, is defined by

$$\sum_{n=-\infty}^{\infty} c_n (z-\alpha)^n = \sum_{n=1}^{\infty} c_{-n}(z-\alpha)^{-n} + \sum_{n=0}^{\infty} c_n (z-\alpha)^n, \qquad ...(57)$$

provided the series on the right-hand side of this equation converge.

Example 22. Find the series for $f(z) = \dfrac{e^z}{z^3}$

Solution: Enter the function $f(z) = \dfrac{e^z}{z^3}$ and find the first few terms of the series in the punctured plane $0 < z$.

$$f[z] = \frac{e^z}{z^3}$$

$$S[z] = \left(1+z+\frac{z^2}{2}+\frac{z^3}{6}+\frac{z^4}{24}+\frac{z^5}{120}+\frac{z^6}{720}+\frac{z^7}{5040}+\frac{z^8}{40320}+\frac{z^9}{362880}+\ldots\right)/(z^3)$$

$$L[z] = \frac{1}{6}+\frac{1}{z^3}+\frac{1}{z^2}+\frac{1}{2z}+\frac{z}{24}+\frac{z^2}{120}+\frac{z^3}{720}+\frac{z^4}{5040}+\frac{z^5}{40320}+\frac{z^6}{362880}+\ldots$$

The latter is a series is valid for $0 < |z|$. Sum up infinitely many terms to verify that we have things right.

$$S_g[z] = \frac{1}{6}+\frac{1}{z^3}+\frac{1}{2^2}+\frac{1}{2z}+\frac{z}{24}+\frac{z^2}{120}+\frac{z^3}{720}+\frac{z^4}{5040}+\frac{z^5}{40320}+\frac{z^6}{362880}$$

$$S_\infty[z] = \sum_{n=0}^{\infty}\frac{z^{n-3}}{n!} = \frac{e^z}{z^3}$$

Remark 3. Recall that $\sum_{n=0}^{\infty} c_n(z-\alpha)^n$ is a simplified expression for the sum $c_{n0} + \sum_{n=1}^{\infty} c_n(z-\alpha)^n$. At times it will be convenient to write $\sum_{n=-\infty}^{\infty} c_n(z-\alpha)^n$ as

$$\sum_{n=-\infty}^{\infty} c_n(z-\alpha)^n = \sum_{n=-\infty}^{-1} c_n(z-\alpha)^n + \sum_{n=0}^{\infty} c_n(z-\alpha)^n, \qquad \ldots(58)$$

***Definition 11.*(Annalus):** Given $0 \le r \le R$, we define the annulus centered at α with radii r and R by.

$$A(a, r, R) = \{z : r < |z - a| < R\}. \qquad \ldots(59)$$

The closed annulus centered at with radii r and R is denoted by.

$$\bar{A}(\alpha, r, R) = \{z : r \le |z-\alpha| \le R\}. \qquad \ldots(60)$$

Figure 2 illustrates these terms.

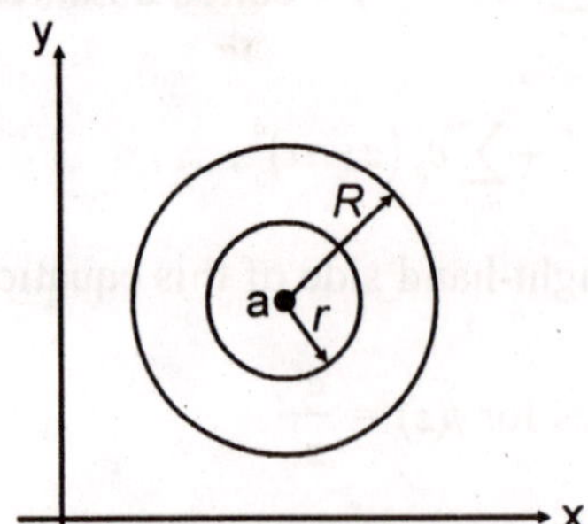

Fig. 6 The Closed Annulus $\bar{A}$ (*a*, *r*, R).The Shaded Portion is the Open Annulus *A*(*a*, *r*, R).

Theorem 21. Suppose that the Laurent series $\sum_{n=-\infty}^{\infty} c_n (z-\alpha)^n$ converges on an annulus $A(\alpha, r, R) = \{z : r < |z-\alpha| < R\}$. Then the series converges uniformly on any closed subannulus $\overline{A}(\alpha, s, t) = \{z : s \leq |z-\alpha| \leq t\}$ where $r < s < t < R$.

In Laurent series, we will use symbols of the form $C_\rho^+(\alpha)$, which we remind you to designate the positively oriented circle with radius ρ and center α. That is, $C_\rho^+(\alpha) = \{z : |z - \alpha| < \rho\}$, oriented counterclockwise.

***Theorem 22.* (Laurent's Theorem).** Suppose $0 \leq r \leq R$, and that $f(z)$ is analytic in the annulus $A = A(\alpha, r, R) = \{z : r < |z-\alpha| < R\}$. If ρ is any number such that $r < \rho < R$, then for all $z_0 \in A(\alpha, r, R)$ the function value $f(z_0)$ has the Laurent series representation

$$f(z_0) = \sum_{n=-\infty}^{\infty} c_n (z_0-\alpha)^n = \sum_{n=1}^{\infty} c_{-n} (z_0-\alpha)^{-n} + \sum_{n=0}^{\infty} c_n (z_0-\alpha)^n, \quad \text{...(61)}$$

where for $n = 0, 1, 2, \ldots$, the coefficients c_{-n} and c_n are given by

$$c_{-n} = \frac{1}{2\pi i} \int_{C_\rho^+(\alpha)} \frac{f(z)}{(z-\alpha)^{-n+1}} dz \text{ and} \quad \text{...(62)}$$

$$c_n = \frac{1}{2\pi i} \int_{C_\rho^+(\alpha)} \frac{f(z)}{(z-\alpha)^{n+1}} dz. \quad \text{...(63)}$$

Remark 4. If $f(z)$ is analytic in the disk $D_R(\alpha)$ then the coefficient C_n for the positive power $(z_0 - \alpha)^n$ equals $\frac{f(n)(z_0)}{n!}$ by using Cauchy's integral formula for derivatives. Hence, the series in equation (30) involving the positive powers of $(z_0 - \alpha)$ is actually the Taylor series for $f(z)$.

***Theorem 23.* (Uniqueness and differentiation of Laurent Series).** Suppose that $f(z)$ is analytic in the annulus $A(\alpha, r, R)$, and has the Laurent series representation

$$f(z) = \sum_{n=-\infty}^{\infty} c_n (z-\alpha)^n \text{ for all } z \in A(\alpha, r, R).$$

(i) If $f(z) = \sum_{n=-\infty}^{\infty} b_n (z-\alpha)^n$ for all $z \in A(\alpha, r, R)$, then $b_n = c_n$ for all n. (In other words, the Laurent series for $f(z)$ in a given annulus is unique.)

(ii) For all $z \in A(\alpha, r, R)$, the derivatives for $f(z)$ may be obtained by termwise differentiation of its Laurent series.

Example 23. Find three different Laurent series representations for the function $f(z)=\dfrac{3}{2+z-z^2}$ involving powers of z.

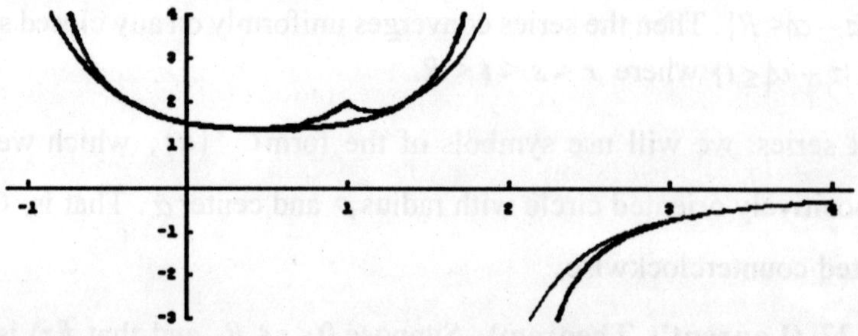

Solution: The function $f(z)$ has singularities at $z=-1, 2$ and is analytic in the disk $D:|z|<1$, in the annulus $A:1<|z|<2$, and in the region $R:1\ |z|>2$. We want to find a different Laurent series for $f(z)$ in each of the three domains D, A, and R. We start by writing $f(z)$ in its partial fraction form:

$$f(z)=\frac{3}{(1+z)(2-z)}=\frac{1}{1+z}+\frac{1}{2-z}=\frac{1}{1+z}+\frac{1}{2}\frac{1}{1-\frac{z}{2}} \qquad ...(64)$$

We use previous Theorem and Corollary to obtain the following representations for the terms on the right side of Equation (33):

$$\frac{1}{1+z}=\sum_{n=0}^{\infty}(-1)^n z^n \qquad ...(65)$$

valid for $|z|<1$,

$$\frac{1}{1+z}=\sum_{n=1}^{\infty}\frac{(-1)^{n+1}}{z^n} \qquad ...(66)$$

valid for $|z|>1$,

$$\frac{1}{2}\frac{1}{1-\frac{z}{2}}=\sum_{n=0}^{\infty}\frac{z^n}{2^{n+1}} \qquad ...(67)$$

valid for $|z|<2$,

$$\frac{1}{2}\frac{1}{1-\frac{z}{2}}=\sum_{n=1}^{\infty}\frac{-2^{n-1}}{z^n} \qquad ...(68)$$

valid for $|z|>2$.

Representations (64) and (66) are both valid in the disk $D:|z|<1$, and thus we have

$$f(z)=\sum_{n=0}^{\infty}(-1)^n z^n+\sum_{n=0}^{\infty}\frac{z^n}{z^{n+1}}=\sum_{n=0}^{\infty}\left((-1)^n+\frac{1}{2^{n+1}}\right)z^n$$

valid for $|z|<1$, which is a Laurent series that reduces to a Maclaurin series.

In the annulus $A:1<|z|<2$, representations (65) and (66) are valid; hence we get

$$f(z) = \sum_{n=1}^{\infty} \frac{(-1)^{n+1}}{z^n} + \sum_{n=0}^{\infty} \frac{z^n}{2^{n+1}} \qquad ...(69)$$

valid for $1 < |z| < 2$.

Finally, in the region $R : |z| > 2$ we use Representations (65) and (67) to obtain6

$$f(z) = \sum_{n=1}^{\infty} \frac{(-1)^{n+1}}{z^n} + \sum_{n=1}^{\infty} \frac{-2^{n-1}}{z^n} = \sum_{n=1}^{\infty} \frac{(-1)^{n+1} - 2^{n-1}}{z^n}$$

valid for $|z| > 2$.

EXERCISE

1. Give a sequence (f_n) of functions continuous on a set D with a limit that is not continuous.
2. Prove that a sequence cannot have more than one limit.
3. Give an example of a sequence that does not have a limit, or explain carefully why there is no such sequence.
4. Give an example of a bounded sequence that does not have a limit, or explain carefully why there is no such sequence.
5. Give a sequence of real functions differentiable on an interval which converges uniformly to a non-differentiable function.
6. Prove that a sequence cannot have more than one limit.
7. Give an example of a sequence that does not have a limit, or explain carefully why there is no such sequence.
8. Give an example of a bounded sequence that does not have a limit, or explain carefully why there is no such sequence.
9. Give a sequence (f_n) of functions continuous on a set D with a limit that is not continuous.
10. Give a sequence of real functions differentiable on an interval which converges uniformly to a non-differentiable function.
11. Find the set D of all z for which the sequence $\dfrac{z_n}{z_n - 3^n}$ has a limit. Find the limit.
12. Prove that the series $\sum_{j=1}^{n} a_j$ converges if and only if both the series $\sum_{j=1}^{n} \text{Im}\, a_j$ and $\sum_{j=1}^{n} \text{Im}\, a_j$ converge.

13. Explain how you know that the series $\sum_{j=1}^{n}\left(\frac{1}{z}\right)^{j}$ converges uniformly on the set |z| ³ 5.

14. Suppose the sequence of real numbers (a_j) has a limit. Prove that lim $\sup(a_j)$ = $\lim(a_j)$.

For each of the following, find the set D of points at which the series converges:

15. $\sum_{j=0}^{n} j!z^{j}$ 11. $\sum_{j=0}^{n} jz^{j}$

16. $\sum_{j=0}^{n} \frac{j^2}{3^j} z^{j}$

13. $\sum_{j=0}^{n} \frac{(-1)^j j^2}{2^{2j}(j!)^2} z^{2j}$

17. Find the limit of $\sum_{j=0}^{n}(j+1)z^{j}$ For what values of z does the series converge?

18. Find the limit of $\sum_{j=1}^{n}\frac{z^j}{j}$. For what values of z does the series converge?

19. Find a power series $\sum_{j=0}^{n} c_j(z-1)^{j}$ such that $\frac{1}{z}=\sum_{j=0}^{\infty} c_j(z-1)^{j}$ for $|z - 1| < 1$.

20. Find a power series such that for $|z - 1| < 1$, Log z = for $|z - 1| < 1$.

21. Show that

$$\frac{1}{1-z}=\sum_{j=0}^{\infty}\frac{(z-i)^j}{(1-i)^{j+1}} \text{ for } |z-i|<\sqrt{2}.$$

22. If $f(z)=\frac{1}{1-z}$ what is $f^{(10)}(i)$?

23. Suppose f is analytic at $z = 0$ and $f(0)=f'(0)=f''(0)$. Prove there is a function g analytic at 0 such that $f(z)=z^3g(z)$ in a neighborhood of 0.

24. Find the Taylor series for $f(z)=\sin z$ at $z_0 = 0$.

25. Show that the function f defined by

$$f(z)=\begin{cases}\dfrac{\sin z}{z} & \text{for } z\neq 0\\ 1 & \text{for } z=0\end{cases}$$

is analytic at $z = 0$, and find f ¢(0).

26. Find two Laurent series in powers of z for the function f defined by

$$f(z)=\frac{1}{z^2(1-z)}$$

and specify the regions in which the series converge to $f(z)$.

27. Find two Laurent series in powers of z for the function f defined by

$$f(z)=\frac{1}{z(1+z^2)}$$

and specify the regions in which the series converge to $f(z)$.
which is valid for $|z| > 0$.

28. Show in two ways that the sequence

$$z_n=-2+i\frac{(-1)^n}{n^2}$$

$(n = 1, 2, ...)$

29. Show that

if $\lim_{n\to\infty} z_n = z$, then $\lim_{n\to\infty}|z_n|=|z|$.

30. By considering the remainders $\rho N(z)$, verify that

$$\sum_{n=1}^{\infty} z_n=\frac{z}{1-z} \quad \text{when } |z| < 1.$$

31. Show that a limit of a convergent sequence of complex numbers is unique by appealing to the corresponding result for a sequence of real numbers.

32. Let c denote any complex number and show that

if $\sum_{n=1}^{\infty} z_n = S$, then $\sum_{n=1}^{\infty} cz_n = cS$.

33. Obtain the Maclaurin series representation

$$z\cosh(z^2)=\sum_{n=0}^{\infty}\frac{z^{4n+1}}{(2n)!} \qquad (|z|<\infty).$$

34. Show that

$$e^z e\sum_{n=0}^{\infty}\frac{(z-1)^n}{n!} \qquad (|z|<\infty).$$

35. Find the Maclaurin series expansion of the function

$$f(z)=\frac{z}{z^4+9}=\frac{z}{9}\cdot\frac{1}{1+\left(z^{4/9}\right)}.$$

36. Derive the Taylor series representation

$$\frac{1}{1-z}=\sum_{n=0}^{\infty}\frac{(z-i)^n}{(1-i)^{z+1}} \qquad \left(|z-1|<\sqrt{2}\right).$$

37. Find the Laurent series that represents the function

$$f(z)=z^2\sin\left(\frac{1}{z^2}\right)$$

in the domain $0<|z|<\text{x}\ \infty$.

38. Derive the Laurent series representation

$$\frac{e^2}{(z+1)^2}=\frac{1}{e}\left[\sum_{n=0}^{\infty}\frac{(z+1)^n}{(n+2)!}+\frac{1}{z+1}+\frac{1}{(z+1)^2}\right] \qquad 3\left(0<|z+1|<\infty\right).$$

39. Find a representation for 1(1 + z) in negative powers of *z* that is valid in the domin $1<|z|<\infty$.

40. Give two Laurent series expansions *n* powers of *z* for the function

$$f(z)=\frac{1}{z^2(1-z)},$$

and specify the regions in which those expansions are valid.

Chapter 12

Curvilinear Coordinates

INTRODUCTION

Any linear coordinates which are not Cartesian coordinates; frequently used curvilinear coordinates are polar coordinates and cylindrical coordinates.

Curvilinear Coordinates

Curvilinear coordinates are a coordinate system based on some transformation of the standard coordinate system. We need the same number of coordinates. If we consider the 2D case, then instead of Cartesian coordinates x and y we use e.g. p and q; the level curves of p and q in the xy-plane, as well as those of x and y in the pq-plane are in general curved. Required is that the transformation is locally invertible at each point. This means that we can convert a point given in one coordinate system to its curvilinear coordinates and back. Depending on the application, a curvilinear coordinate system may be simpler than the Cartesian coordinate system. This also has consequences that we can express many of the concepts in vector calculus which are given in Cartesian or spherical coordinates or any other arbitrary coordinate system, also in curvilinear coordinates.

WHY CURVILINEAR COORDINATES?

Because the world is not rectangular!

How can you calculate:

(i) Diffusion from a cylinder,

(ii) Flow around a sphere,

(iii) Field of a charged ellipsoid,

(iv) Many other exciting things?

Definition:

Let x, y and z be functions of some variables u_1, u_2 and u_3:

$$x = x(u_1, u_2, u_3)$$

$$y = y(u_1, u_2, u_3)$$

$$z = z(u_1, u_2, u_3)$$

or, in vector notation

$$\mathrm{r} = \mathrm{r}\,(u_1, u_2, u_3)$$

u_1, u_2 and u_3 are called *curvilinear coordinates.*

In R^3, for example, we have some transformation:

$x_i = x_i(x'_1, x'_2, x'_3); i = 1,2,3$ giving *curvilinear coordinates* x_1', x_2', x_3', for x_1, x_2, x_3.

If this transformation is locally invertible everywhere, the Jacobian determinant

$$\frac{\partial(x_1, x_2, x_3)}{\partial(x'_1, x'_2\, x'_3)}$$

is nonzero, and for this to happen, the vectors

$$\frac{\partial x}{\partial x'_i}$$

must form a basis for R^3.

From these basis vectors, we define *scale factors*

$$hx'_1 = h = \left|\frac{\partial \mathrm{x}}{\partial x'_1}\right| = \sqrt{\sum_{k=0}^{3}\left(\frac{\partial x_k}{\partial x'_i}\right)^2}$$

and thus arrive at the unit basis vectors for the curvilinear coordinates to be

$$ex'_i = 1/h_i \frac{\partial \mathrm{x}}{\partial x'_1}$$

Note that the coordinate system we choose need not be orthogonal, but for the purposes of this article, they are treated as being so. The system is orthogonal iff

$$e_{x'_i} . e_{x'_j} = \delta_{ij}$$

where δ_{ij} is the Kronecker delta.

Coordinate curves:

Let us held u_2 and u_3 constant, and change u_1. The vector **r** will change along a curve. This curve is called *coordinate curve* for u_1. Analogously we can define coordinate curves for u_2 and u_3. Examples:

1. Cylindrical coordinates:
2. Only ρ changes--lines parallel to *xy* plane and containing the point (0, 0, *z*)

3. Only ϕ changes--circles parallel to *xy* plane with the center in the point (0, 0, *z*)
4. Only *z* changes–lines parallel to *z* axis
5. Spherical coordinates:
6. Only *r* changes–lines containing origin
7. Only θ changes–circles with the center in origin an normal to *xy* plane
8. Only ϕ changes–circles parallel to *xy* plane with the center in the point (0, 0, z)

COORDINATE SURFACES

If we take u_1 constant and allow u_2 and u_3 change, we will have *coordinate surface*. Analogously we can define other two surfaces. Examples:

1. Cylindrical coordinates:
2. $\rho = const$: cylinders with the axis *z* (hence the name!)
3. $\phi = const$: planes containing the axis *z*
4. $z = const$: planes normal to the axis *z*
5. Spherical coordinates:
6. $r = const$: spheres with the center at the origin (hence the name!)
7. $\theta = const$: cones with the center at origin and the axis *z*
8. $\phi = const$: planes containing the axis *z*

Cartesian coordinates x_1, x_2, x_3 which have the scalar product, are called Euclidean coordinates. It is often convenient to associate the points of Euclidean space with vectors, for example, with each point P we associate the vector (or arrow) with its tail at the origin of coordinates and its tip at P. This vector is called the radius vector with components (x_1, x_2, x_3). At any point P of Euclidean space we can construct the small line element

$$dx = (dx_1, dx_2, dx_3)$$

which is vector too. Two vectors $h = (x_1, x_2, x_3)$ and $f = (y_1, y_2, y_3)$ from the same origin can be added and result is the vector with coordinates $(x_1 + y_1, x_2 + y_2, x_3 + y_3)$. A vector can also be multiplied by any real number. The Euclidean scalar product of two (real) vectors is the number

$$\langle f, h \rangle = \sum_{i=1}^{n} x_i y_i$$

The scalar product of the vector with itself give the square of the vector length. The square of the length of a line element in space with scalar product is called the metric of the space. The metric of Euclidean space is

$$< dx, dx > = dx_1^2 + dx_2^2 + dx_3^2.$$

The same Euclidean metric in curvilinear coordinates is

$$< dx, dx >= \sum_{k=1}^{3} \frac{\partial x_k \partial x_k}{\partial x'_i \partial x'_j} dx'_i \, dx'_j$$

The symmetric tensor

$$g_{i,j}(x'_i, x'_j) = \sum_{k=1}^{3} \frac{\partial x_k \partial x_k}{\partial x'_i \partial x'_j}$$

are called the fundamental (or metric) tensor of the Euclidean space in curvilinear coordinates. Connection between fundamental tensor and Lamé coefficients is

$$g_{i,i}(x'_i, x'_j) = h_i^2$$

Example

If we consider polar coordinates for R^2, note that

$$(x, y) = (r\cos\theta, r\sin\theta)$$

(r, θ) are the curvilinear coordinates, and the Jacobian determinant of the transformation $(r, \theta) \to (r\cos\theta, r\sin\theta)$ is r.

The basis vectors are $b_r = (\cos\theta, \sin\theta)$, $b_\theta = (-r\sin\theta, r\cos\theta)$, with unit basis vectors $e_r = (\cos\theta, \sin\theta)$, $e_\theta = (-\sin\theta, \cos\theta)$ with scale factors $h_r = 1$ and $h_\theta = r$. The fundamental tensor is $g_{1,1} = 1$, $g_{2,2} = r^2$, $g_{1,2} = g_{2,1} = 0$.

EXAMPLES OF CURVILINEAR COORDINATES

Scaling Coefficients and Local Basis

Let e_i be the tangents to coordinate curves. Vectors e_i are proportional to $\partial r / \partial u_i$:

$$e_i = \frac{1}{h_i}\frac{\partial r}{\partial u_i}, \quad h_i = \left|\frac{\partial r}{\partial u_i}\right|$$

Vectors e_i are local basis, numbers h_i are *scaling coefficients*

Taylor expansion:

$$dr = \frac{\partial r}{\partial u_1}du_1 + \frac{\partial r}{\partial u_2}du_2 + \frac{\partial r}{\partial u_3}du_3 = e_1h_1du_1 + e_2h_2du_2 + e_3h_1du_3 \qquad ...(1)$$

Cylindrical coordinates:

$$r = \begin{pmatrix} \rho\cos\phi \\ \rho\sin\phi \\ z \end{pmatrix}$$

Differentiating:

$$\text{Along } \rho : \frac{\partial r}{\partial \rho} = \begin{pmatrix} \cos\phi \\ \sin\phi \\ 0 \end{pmatrix}$$

$$h_\rho = 1, e_\rho = \begin{pmatrix} \cos\phi \\ \sin\phi \\ 0 \end{pmatrix}$$

Along ϕ:

$$\frac{\partial r}{\partial \phi} = \begin{pmatrix} -\rho\sin\phi \\ \rho\cos\phi \\ 0 \end{pmatrix}$$

$$h_\phi = \rho,\ e_\phi = \begin{pmatrix} -\sin\phi \\ \cos\phi \\ 0 \end{pmatrix},$$

Along z:

$$\frac{\partial r}{\partial z} = \begin{pmatrix} 0 \\ 0 \\ 1 \end{pmatrix}$$

$$h_z = 1,\ e_z = \begin{pmatrix} 0 \\ 0 \\ 1 \end{pmatrix}$$

Spherical coordinates:

$$r = \begin{pmatrix} r\sin\theta\cos\phi \\ r\sin\theta\sin\phi \\ r\cos\theta \end{pmatrix}$$

Differentiating:

Along r:

$$\frac{\partial r}{\partial r} = \begin{pmatrix} \sin\theta\cos\phi \\ \sin\theta\sin\phi \\ \cos\theta \end{pmatrix}$$

$$h_r = 1,\ e_r = \begin{pmatrix} \sin\theta\cos\phi \\ \sin\theta\sin\phi \\ \cos\theta \end{pmatrix}$$

Along θ:

$$\frac{\partial r}{\partial \theta} = \begin{pmatrix} r\cos\theta\cos\phi \\ r\cos\theta\sin\phi \\ -r\sin\theta \end{pmatrix}$$

$$h_\theta = e_\theta = \begin{pmatrix} \cos\theta\cos\phi \\ \cos\theta\sin\phi \\ -\sin\theta \end{pmatrix}$$

Along ϕ :

$$\frac{\partial r}{\partial \phi} = \begin{pmatrix} -r\sin\theta\sin\phi \\ r\sin\theta\cos\phi \\ 0 \end{pmatrix}$$

$$h_\phi = r\sin\theta,\ e_\phi = \begin{pmatrix} -\sin\phi \\ \cos\phi \\ 0 \end{pmatrix}$$

Circular Cylindrical Coordinates:

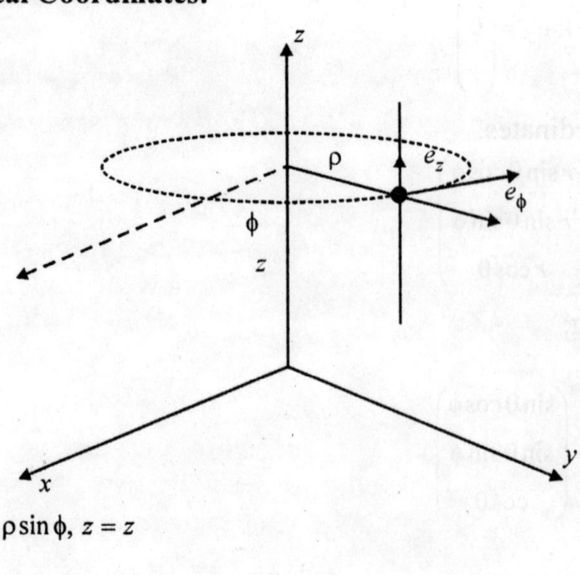

$x = \rho\cos\phi,\ y = \rho\sin\phi,\ z = z$

Spherical Coordinates:

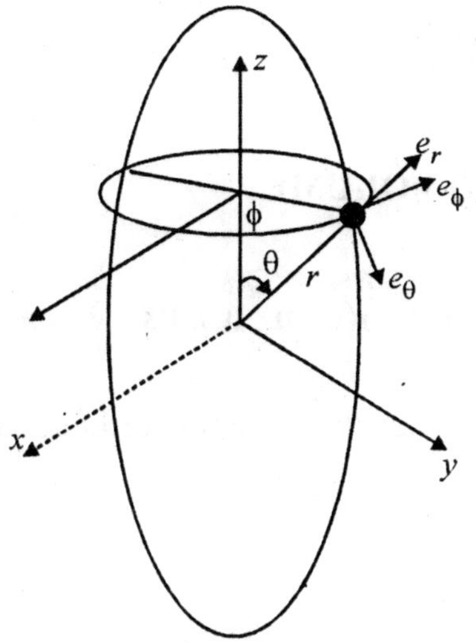

$$x = r\sin\theta\cos\phi,\ \ y = r\sin\theta\sin\phi,\ z = r\cos\theta$$

PHYSICAL COMPONENTS OF VECTORS AND TENSORS

Vectors:

Suppose we have a vector a. At any point we can expand it in the local basis:

$$a = a_1e_1 + a_2e_2 + a_3e_3$$

The numbers a_1, a_2 and a_3 are called *physical components* of the vector a.

Example. flow and sphere. Suppose we have a flow with the constant velocity $v = ve_z$. What is the flow on the surface of the sphere r = const??

Solution: in Cartesian coordinates

$$v = \begin{pmatrix} 0 \\ 0 \\ v \end{pmatrix}$$

We can write down:

$$v = v_r e_r + v_\theta e_\theta + v_\phi e_\phi$$

Multiplying this equation by e_r, e_θ and e_ϕ, we obtain:

$$v_r = e_r.v = v\cos\theta,\ v_\theta = e_\theta.v = -v\sin\theta,$$

$$v_\phi = e_\phi.v = 0$$

Physical Components of Vectors:

Once again, use local basis:

$$\hat{U} = u_{ij} e_i \otimes e_j$$

LENGTH, SURFACE AND VOLUME

Element of length:

Suppose we have a curve r = r(u_1, u_2, u_3). Its element of length:

$$dl^2 = dr^2$$

Use Taylor expansion . Since the system is orthogonal,

$$dl^2 = h_1^2 du_1^2 + h_2^2 du_2^2 + h_3^2 du_3^2$$

Examples.

1. Cylindrical coordinates:

$$dl^2 = d\rho^2 + r^2 d\phi^2 + dz^2$$

2. Spherical coordinates:

$$dl^2 = dr^2 + r^2 d\theta^2 + r^2 \sin^2 \theta d\phi^2$$

Meaning of scaling coefficients: they measure the real length of du_i.

Element of surface:

The surface S is a vector with the components:

$$dS = e_1 h_2 h_3 du_2 du_3 + e_2 h_1 h_3 du_1 du_3 + e_3\ h_2 h_1 du_2 du_1$$

Element of volume:

Scalar

$$dV = h_1 h_2 h_3 du_1 du_2 du_3$$

LINE AND SURFACE INTEGRALS

Since we use curvilinear coordinates to aid in the calculation in vector calculus, there are adjustments we need to make in the calculation of line, surface and volume integrals.

Line integrals

Normally in the calculation of line integrals we are interested in calculating

$$\int_C f ds = \int_a^b f(x(t)) \left| \frac{\partial x}{\partial t} \right| dt$$

where x(*t*) parametrizes C in Cartesian coordinates. In curvilinear coordinates, the term

$$\left|\frac{\partial x}{\partial t}\right| = \left|\sum \frac{\partial x}{\partial x'_i}\frac{\partial x'_i}{\partial t}\right|$$

by the chain rule. But from the definition of the curvilinear coordinates,

$$\frac{\partial x}{\partial x'_i} = h_i e x'_i$$

and thus

$$\left|\frac{\partial x}{\partial t}\right| = \sqrt{\sum h_i e x'_i \frac{\partial x'_i}{\partial t}}$$

and we can proceed normally.

Surface integrals

Likewise, if we are interested in a surface integral, the relevant calculation, with the parameterization of the surface in Cartesian coordinates is:

$$\int_S f ds = \iint_T f(x(s,t)) \left|\frac{\partial x}{\partial s} \times \frac{\partial x}{\partial t}\right| dx dt$$

Again, in curvilinear coordinates, the term

$$\left|\frac{\partial x}{\partial s} \times \frac{\partial x}{\partial t}\right| = \left|\frac{\partial x}{\partial x'_i}\frac{\partial x'_i}{\partial s} \times \frac{\partial x}{\partial x'_i}\frac{\partial x'_i}{\partial t}\right|$$

and we make use of the definition of curvilinear coordinates again to yield

$$\frac{\partial x}{\partial x'_i}\frac{\partial x'_i}{\partial s} = \sum \frac{\partial x'_i}{\partial s} hx'_i e_{x'_i}$$

and

$$\frac{\partial x}{\partial x'_i}\frac{\partial x'_i}{\partial t} = \sum \frac{\partial x'_i}{\partial t} hx'_i e_{x'_i}$$

where the cross product, in terms of curvilinear coordinates, will be:

$$\begin{vmatrix} e_{x'_1} & e_{x'_2} & e_{x'_3} \\ h_1 \frac{\partial x'_1}{\partial s} & h_2 \frac{\partial x'_2}{\partial s} & h_3 \frac{\partial x'_3}{\partial s} \\ h_1 \frac{\partial x'_1}{\partial t} & h_2 \frac{\partial x'_2}{\partial t} & h_3 \frac{\partial x'_3}{\partial t} \end{vmatrix}$$

GRAD, CURL, DIV

Gradient

Gradient of a function f is such a vector that grad $f = g = g_1 e_1 + g_2 e_2 + g_3 e_3$ that $g.dr = df$

But, using expansion (1) and chain rule

$$g.dr = g_1h_1du_1 + g_2h_2du_2 + g_3h_3du_3$$

$$df = \frac{\partial f}{\partial u_1}du_1 + \frac{\partial f}{\partial u_2}du_2 + \frac{\partial f}{\partial u_3}du_3$$

We have:

$$gi = \frac{1}{h_i}\frac{\partial f}{\partial u_i}$$

or $$\nabla = e_1\frac{1}{h_1}\frac{\partial}{\partial u_1} + e_2\frac{1}{h_2}\frac{\partial}{\partial u_2} + e_3\frac{1}{h_3}\frac{\partial}{\partial u_3} \quad ...(2)$$

Examples:

1. Cylindrical coordinates:

$$\text{grad } f = e_\rho\frac{\partial f}{\partial \rho} + e_\phi\frac{1}{\rho}\frac{\partial f}{\partial \phi} + e_z\frac{\partial f}{\partial z}$$

2. Spherical coordinates:

$$\text{grad } f = e_r\frac{\partial f}{\partial r} + e_\theta\frac{1}{r}\frac{\partial f}{\partial \theta} + e_\phi\frac{1}{\sin\theta}\frac{\partial f}{\partial \phi}$$

Divergence

Suppose we have a vector a. In the local basis

$$a = a_1e_1 + a_2e_2 + a_3e_3$$

What is a?

Simple idea:

We can multiply ∇ from equation (2) by a:

$$\text{div a} = e_1\frac{1}{h_1}\frac{\partial}{\partial u_1}(a_1e_1) + ...$$

PROBLEM

Both a_i and e_i depend on u!. So

$$e_1\frac{1}{h_1}\frac{\partial}{\partial u_1}(a_1e_1) = \frac{1}{h_1}\frac{\partial a_1}{\partial u_1} + e_1a_1\frac{1}{h_1}\frac{\partial e_1}{\partial u_1}$$

We do not know what to do with the second term!

A useful theorem:

There are two simple relations:

$$\text{curl}\frac{e_1}{h_1}=0,\ \text{div}\frac{e_1}{h_2h_3}=0$$

Proof:

1. First, let us take grad u_1. From the general formula

$$\nabla u_1 = e_1\frac{1}{h_1}\frac{\partial u_1}{\partial u_1}=\frac{e_1}{h_1}$$

Since e_1/h_1 is a gradient, its curlis zero.

2. We have

$$\frac{e_1}{h_2h_3}=\frac{e_2\times e_3}{h_2h_3}=\frac{e_2}{h_2}=\frac{e_3}{h_3}$$

This is a vector product of two gradients. Each of them has zero curl. From the formula

$$\nabla.(a\times b)=b = \text{b. curla a} - \text{a . curl b}$$

we conclude that the divergence of this product is zero.

Calculation of div a:

We can write down:

$$a_1e_1 = a_1h_2h_3\frac{e_1}{h_2h_3}$$

but

$$e_1\frac{1}{h_1}\frac{\partial}{\partial u_1}\left(\frac{e_1}{h_2h_3}\right)=\text{div}\frac{e_1}{h_2h_3}=0$$

and therefore

$$e_1\frac{1}{h_1}\frac{\partial}{\partial u_1}\left(a_1h_2h_3\frac{e_1}{h_2h_3}\right)=\frac{1}{h_1h_2h_3}\frac{\partial}{\partial u_1}(h_2h_3a_1)$$

We obtained

$$\text{div a} = \frac{1}{h_1h_2h_3}\left[\frac{\partial}{\partial u_1}(h_2h_3a_1)+\frac{\partial}{\partial u_3}(h_1h_3a_2)+\frac{\partial}{\partial u_3}(h_1h_2a_3)\right]$$

Examples.

1. Cylindrical Coordinates:

$$\text{div a} = \frac{1}{\rho}\left[\frac{\partial}{\partial\rho}(\rho a_\rho)+\frac{\partial}{\partial\phi}(a\phi)+\frac{\partial}{\partial_z}(\rho a_z)\right]=\frac{1}{\rho}\frac{\partial}{\partial\rho}(\rho a_\rho)+\frac{1}{\rho}\frac{\partial a_\phi}{\partial\phi}+\frac{\partial a_z}{\partial_z}$$

2. Spherical coordinates:

$$\text{div a} = \frac{1}{r^2 \sin\theta}\left[\frac{\partial}{\partial r}(r^2 \sin\theta\, a_r) + \frac{\partial}{\partial \theta}(r \sin\theta a_\theta) + \frac{\partial}{\partial \phi}(r a_\phi)\right] = \frac{1}{r^2}\frac{\partial}{\partial r}(r^2 a_r) + \frac{1}{r}\frac{\partial}{\partial \theta}(\sin\theta a_\theta)$$

Laplacian

Question:

What happens if the vector a is ∇?

Answer:

Laplacian:

$$\nabla = \frac{1}{h_1 h_2 h_3}\left[\frac{\partial}{\partial u_1}\left(\frac{h_2 h_3}{h_1}\frac{\partial}{\partial u_1}\right) + \frac{\partial}{\partial u_2}\left(\frac{h_1 h_2}{h_2}\frac{\partial}{\partial u_2}\right) + \frac{\partial}{\partial u_3}\left(\frac{h_1 h_2}{h_3}\frac{\partial}{\partial u_3}\right)\right]$$

Examples:

1. Cylindrical coordinates:

$$\Delta = \frac{1}{\rho}\frac{\partial}{\partial_\rho}\left(\rho\frac{\partial}{\partial_\rho}\right) + \frac{1}{\rho^2}\frac{\partial^2}{\partial \phi^2} + \frac{\partial^2}{\partial z^2}$$

2. Spherical coordinates:

$$\Delta = \frac{1}{r^2}\frac{\partial}{\partial r}\left(r^2\frac{\partial}{\partial r}\right) + \frac{1}{r^2 \sin\theta}\frac{\partial^2}{\partial \theta}\sin\theta\frac{\partial}{\partial \theta} + \frac{1}{r^2 \sin^2\theta}\frac{\partial^2}{\partial \phi^2}$$

Curl *a*

Let us calculate *a*. The same idea as for gradient: we will factor out the term with zero curl!

Result:

$$\text{curl } a = \frac{1}{h_1 h_2 h_3}\begin{vmatrix} h_1 e_1 & h_2 e_2 & h_2 e_2 \\ \frac{\partial}{\partial u_1} & \frac{\partial}{\partial u_2} & \frac{\partial}{\partial u_3} \\ h_1 a_1 & h_2 a_2 & h_3 a_3 \end{vmatrix}$$

Examples:

1. Cylindrical coordinates:

$$\text{curl } a = \frac{1}{\rho}\begin{vmatrix} e_\rho & \rho e_\phi & e_z \\ \frac{\partial}{\partial \rho} & \frac{\partial}{\partial \phi} & \frac{\partial}{\partial z} \\ a_\rho & r a_\phi & a_z \end{vmatrix}$$

2. Spherical coordinates:

$$\text{curl } a = \frac{1}{r^2 \sin\theta} \begin{vmatrix} e_r & re_\theta & r\sin\theta e_\phi \\ \frac{\partial}{\partial r} & \frac{\partial}{\partial \theta} & \frac{\partial}{\partial \phi} \\ a_r & r\,a_\theta & r\sin\theta a_\phi \end{vmatrix}$$

In orthogonal curvilinear coordinates, one can express the gradient, curl and divergence of a function or vector field as follows:

$$\nabla f = \sum \frac{1}{h} \frac{\partial f}{\partial x_i} \hat{e}_{x_i}$$

$$\nabla \times \vec{v} = \frac{1}{\Omega h_i} \begin{pmatrix} h_1 \partial_1 \\ \vdots \\ h_n \partial_n \end{pmatrix} \times \vec{v}$$

$$\nabla . \vec{v} = \sum \frac{1}{\Omega} \frac{\partial}{\partial_{x_i}} \left(\frac{\Omega v_i}{h_i} \right),$$

where Ω is the product of all h_i.

VECTOR CALCULUS

Vector calculus (also called *vector analysis*) is a field of mathematics concerned with multivariate real analysis of vectors in two or more dimensions. It consists of a suite of formulas and problem solving techniques very useful for engineering and physics. Vector analysis has its origin in quaternion analysis, and was formulated by the American scientist, J. Willard Gibbs [1].

Vector calculus is concerned with scalar fields, which associate a scalar to every point in space, and vector fields, which associate a vector to every point in space. For example, the temperature of a swimming pool is a scalar field: to each point we associate a scalar value of temperature. The water flow in the same pool is a vector field: to each point we associate a velocity vector.

Three operations are important in vector calculus:

Gradient: measures the rate and direction of change in a scalar field; the gradient of a scalar field is a vector field.

Curl: measures a vector field's tendency to rotate about a point; the curl of a vector field is another vector field.

Divergence: measures a vector field's tendency to originate from or converge upon a given point.

A fourth operation, the Laplacian, is combination of the divergence and gradient operations.

Likewise, there are three important theorems related to these operators:

Gradient theorem

Stokes' theorem

Divergence theorem

Most of the analytic results are easily understood, in a more general form, using the machinery of differential geometry, of which vector calculus forms a subset.

DIFFUSION AROUND A SPHERE

Let us discuss a sphere with radius R with the concentration of some reagent c_0 on the surface. The diffusion coefficient is D. What is the steady flow of reagent?

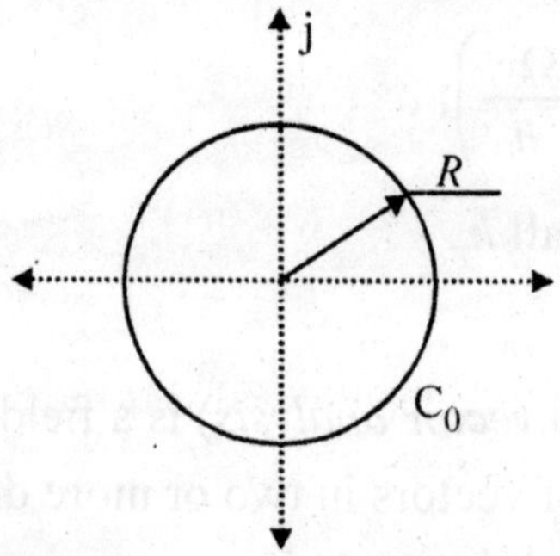

Solution: the diffusion equation

$$\frac{\partial c}{\partial t} = D\Delta c$$

Steady state: no time dependence,

$$\Delta c = 0$$

Let us use spherical coordinates. Obviously c depends only on r, and diffusion equation is

$$\frac{1}{r^2}\frac{\partial}{\partial r}\left(r^2\frac{\partial c}{\partial r}\right) = 0 \quad ...(3)$$

Its solution should satisfy:

$$c(R) = c_0,\ c(\infty) = 0$$

Let us seek the solution in the form

$$c = \frac{A}{r}$$

1. Equation 3 is satisfied
2. Obviously $c(\infty) = 0$
3. $c(R) = c_0$. if $A = c_0 R$

So the distribution of reagent is:

$$c(r) = c_0 \frac{R}{r}$$

The flow:

$$j = -D \text{ grad } c$$

We have:

$$j_r = D\frac{R}{r^2}e_r,\ j_\phi = j_\phi = 0$$

The total flow over a sphere of radius *r* is

$$J = 4\pi r^2 j_r = 4\pi D R$$

PLUMBING SYSTEM

Let us discuss a cylindrical pipe of radius *r*. Incompressible liquid with viscosity coefficient η is pumped with the pressure gradient $\nabla p = -G = const$. How much liquid flows through the pipe in the unit time?

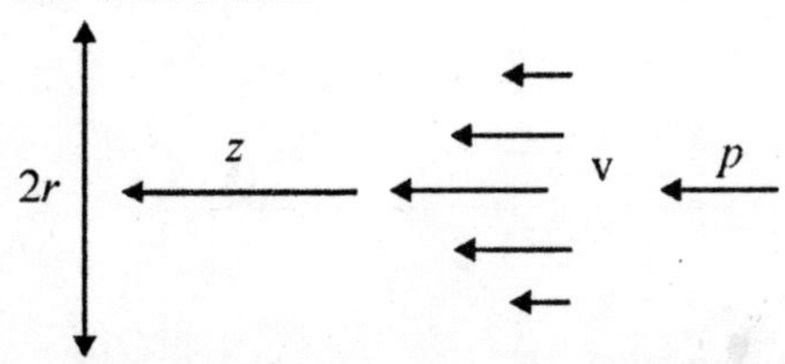

Let us introduce cylindrical coordinates with the axis *z* along the pipe axis. The velocity *v* is along the *z* axis and does not depend on ϕ or *z*:

$$v = v(\rho)e_z$$

The amount of liquid flowing through the tube in the init time is the integral through the cross-section of the tube:

$$Q = \iint_S v\, dS$$

Since

$$dS = \rho \, d\rho \, d\phi$$

we have

$$Q = \int_0^r d\rho \int_0^{2\pi} d\phi \, p v(\rho) = 2\pi \int_0^r p v(\rho) \, d\rho$$

Now let us calculate $v(\rho)$. Use the Navier-Stokes equation (see lectures about hydrodynamics) and the following considerations:

(i) Steady flow, so $\partial v / \partial t = 0$

(ii) In this particular geometry $(v.\nabla)v = 0$

Therefore

$$-\text{grad}\, p + \eta \Delta v = 0$$

This equation has only z component:

$$G + \frac{\eta}{\rho} \frac{\partial}{\partial \rho}\left(\rho \frac{\partial v}{\partial \rho} \right) = 0$$

The velocity v must be zero near the wall:

$$v(r) = 0$$

Let us seek the solution in the form:

$$v(\rho) = a - b\rho^2$$

Then

$$\frac{\eta}{\rho} \frac{\partial}{\partial \rho}\left(\rho \frac{\partial v}{\partial \rho} \right) = -4b\eta$$

and therefore

$$b = \frac{G}{4\eta}$$

From the boundary condition

$$a = br^2 = \frac{G}{4\eta} r^2$$

We obtained

$$Q = 2\pi \int_0^r \rho v(\rho) \, d\rho = \frac{\pi G}{8\eta} r^4$$

The total amount of liquid

$$Q = 2\pi \int_0^r \rho v(\rho) \, d\rho = \frac{\pi G}{8\eta} r^4$$

The answer is proportional to r^4. It means that enlarging pipe diameter two times means increasing the capacity $2^4 = 16$ times.

EXRECISE

1. Show that $xi + yj + zk$ is a irrotational vector.
2. Show that $(z^2+2x+3y)i+(3x+2y+z)j+(y+2zx)k$ is a irrotational vector.
3. Consider the transformation $x = u^2 - v^2,\ y = 2uv,\ z = w$
   (i) Show that (u, v, w) from orthogonal curvilinear coordinates.
   (ii) Express $\nabla^2 g(u, v, w)$.
4. Find the divergence and curl of the vector field
   $F = w\hat{e}_1 + u\hat{e}_2 + v\hat{e}_3$
5. Express $\nabla^2\phi$ in orthogonal curvilinear coordinates.
6. Compute gradf in orthogonal curvilinear coordinates system
   $x = u - v,\ y = u + v,\ z = w^2$ for $\phi(u, v, w) = uv + w^2$
7. Show that $r^a R$ is an irrotational vector for any value of a, but is solenoidal only if a + 3 = 0, where R = xI + yJ + zK and r is the magnitude of R.
8. If f is a solution of the Laplace equation, prove that $\nabla\phi$ is both solenoidal and irrotational.
9. If A and B are irrotational, prove that A × B is solenoidal.
10. Express the following vectors in cylindrical coordinates
    (i) $2yI - zJ + 3xK$. (ii) $2xI - 3y^2J + zxK$.
11. Express the following vectors in spherical polar coordinates
    (i) $xi + 2yJ + yzK$ (ii) $xyI + yzJ + zxK$
12. Evaluate $\nabla\phi$ = xyz in cylindrical coordinates.
13. Show that $\nabla(r/\sin\theta)\nabla\theta = \nabla\phi$.
14. Prove that $V = \dfrac{\cos\theta}{r^3}\left(T_r/\sin\theta - T_\theta/\cos\theta + r^4 T_\phi\right)$ is solenoidal.
15. If u, v, w are orthogonal curvilinear coordinates with h_1, h_2, h_3 as scale factors, prove that

$$\left[\frac{\partial R}{\partial u}, \frac{\partial R}{\partial v}, \frac{\partial R}{\partial w}\right] = \frac{1}{[\nabla v, \nabla \mathbf{v}, \nabla w]} = h_1 h_2 h_3.$$

Encyclopaedia of

COMPLEX ANALYSIS

Encyclopaedia of
COMPLEX ANALYSIS

Volume 2

R.K. Pandey

ANMOL PUBLICATIONS PVT. LTD.
NEW DELHI - 110 002 (INDIA)

ANMOL PUBLICATIONS PVT. LTD.
H.O.: 4374/4B, Ansari Road, Daryaganj,
New Delhi-110 002 (India)
Ph.: 23278000, 23261597
B.O.: No. 1015, Ist Main Road, BSK IIIrd Stage
IIIrd Phase, IIIrd Block,
Bangalore - 560 085 (India)
Visit us at: www.anmolpublications.com

Encyclopaedia of Complex Analysis

First Edition, 2009
ISBN 978-81-261-3925-5 (Set)

PRINTED IN INDIA

Printed at Mehra Offset Press, Delhi.

Contents

Preface

The encyclopedia of complex analysis is designed for students in engineering, physics, and mathematics. Without sacrificing ease and clarity of proofs, mathematical preciseness and rigour are stressed. Cross references are used to justify almost every step in each proof.

The authors have made an effort to present some of the deeper and more interesting results, for example, Taylor's theorems, Riemann mapping theorem, Schwarz lemma in few chapters. However, the very basic theory is nevertheless given a thorough treatment so that readers should never feel lost. Each chapter finishes with exercises.

During the preparation of this encyclopedia there is stress on the new topics so as to cover as references for all the universities. This encyclopedia is a condensed correlation of mathematical and engineering concepts. Comments from reader are most invited for better revision of this encyclopedia.

Author

List of Symbols

$\pm$	$\not\subset$	$\subset$	$\subseteq$	$\in$	$\notin$	$\Re$	∂	$\neq$	$\approx$	$\angle$	∇
Σ	Π	Ω	ϕ	α	β	χ	ε	γ	η	λ	μ
ν	$\perp$	π	θ	ρ	σ	τ	$\Re$	ω	ξ	ψ	ζ
Υ		$\exists$	$\int$	$\iint$	$\iiint$	$\times$	$\sqrt{}$	$\because$	$\geq$	$\oint$	$\Rightarrow$

Chapter 1

Derivative of Vector Point Function

INTRODUCTION

Derivative of vector point function is most important tool of Mathematics it play a vital role in Mechanics, Statics, Comploex number, Calculus of variation. We have already studied about vectors and basic operation on vectors such as addition, subtraction and multiplication. In this chapter we learn about differentiation of vectors and their application.

SCALAR AND VECTOR VALUED FUNCTIONS

Scalar Function

A function $f(x_1, x_2, \ldots, x_n)$ of one or more variables, whose range is one-dimensional, as compared to a vector function, whose range is three-dimensional (or in general n-dimensional).

Vector Functions

A vector function is a function that produces a vector. The typical example is the trajectory of a particle in space. At each time "t" the particle has position vector r(t). In a given coordinate system we can write the components of r(t) in terms of the basis vectors i,j and k.

In rectangular coordinates system, the vector function f can be expressed in component form as:

$$f = (f_1, f_2, f_3) = f_1 i + f_2 j + f_3 k$$

f_1, f_2 and f_3 are scalar functions of t are called the components of f.

Note: Sometimes we denote the vector function f of the scalar variable t by r, and write $r = f(t)$, then r may or may not be the position vector of a point.

LIMIT AND CONTINUITY

Let a vector function f be defined for all values of t in some neighborhood about a point t_0 possibly at t0. Then, a vector a is called the limit of f as $t \to t0$ if given any $\varepsilon > 0$, however small, there exists a real number δ > 0 such that

$|f(t)-a|<\varepsilon$, whenever $0<|t-t_0|<\delta$

and we write $\lim_{t\to t_0} f(t)=a$ or $\underset{t\to t_0}{Lt} f(t)=a$.

The same definition holds for the limit of a scalar function if *f* is replaced by a scalar function *f*, and vector *a* by a scalar *a*.

We state below some important theorems on limits of a vector function.

Theorem 1. If $f= f_1i+f_2j+f_3k$ and $a=a_1i+a_2j+a_3k$, then

$\lim_{t\to t_0} f(t)=a$ if and only if $\lim_{t\to t_0} f_i(t)=a_i, i=1,\ 2,\ 3$. ...(1)

Theorem 2. Let *f* and *g* be vector functions of a scalar variable t. Let f be a scalar function of t.

Then,

(i) $\underset{t\to t_0}{Lt}\left[f(t)+g(t)\right]=\underset{t\to t_0}{Lt} f(t)+\underset{t\to t_0}{Lt} g(t)$ ...(2)

(ii) $\underset{t\to t_0}{Lt}\left[f(t)-g(t)\right]=\underset{t\to t_0}{Lt} f(t)-\underset{t\to t_0}{Lt} g(t)$ ...(3)

(iii) $\underset{t\to t_0}{Lt}\left[f(t)\cdot g(t)\right]=\left\{\underset{t\to t_0}{Lt} f(t)\right\}\cdot\left\{\underset{t\to t_0}{Lt} g(t)\right\}$ ...(4)

(iv) $\underset{t\to t_0}{Lt}\left[f(t)\times g(t)\right]=\left\{\underset{t\to t_0}{Lt} f(t)\right\}\times\left\{\underset{t\to t_0}{Lt} g(t)\right\}$ ...(5)

(v) $\underset{t\to t_0}{Lt}\left[\phi(t)\times g(t)\right]=\left\{\underset{t\to t_0}{Lt} \phi(t)\right\}\times\left\{\underset{t\to t_0}{Lt} g(t)\right\}$ ...(6)

A vector function *f* is said to be continuous at $t = t_0$ if it is defined in some neighborhood of t_0 and $\underset{t\to t_0}{Lt} f(t)=\underset{t\to t_0}{Lt} f(t)$

Theorem 3. If $f= f_1i+f_2j+f_3k$ is a function of t, then *f* is continuous at t_0 if and only if f_1, f_2 and f_3 are all continuous at t_0.

DIFFRENTIATION OF VECTOR-VALUED FUNCTIONS

A curve C is deûned by r = r(t), a vector-valued function of one (scalar) variable. Let us imagine that C is the path taken by a particle and t is time. The vector r(t) is the position vector of the particle at time t and r(t + h) is the position vector at a later time t + h. The average velocity of the particle in the time interval [t, t + h] is then

$$\frac{\text{displacement vector}}{\text{lengh of time interval}} = \frac{r(t+h)-r(t)}{h} \qquad ...(7)$$

In terms of the components of r this is

$$\left(\frac{x(t+h)-x(t)}{h}, \frac{y(t+h)-y(t)}{h}, \frac{z(t+h)-z(t)}{h}\right) \qquad ...(8)$$

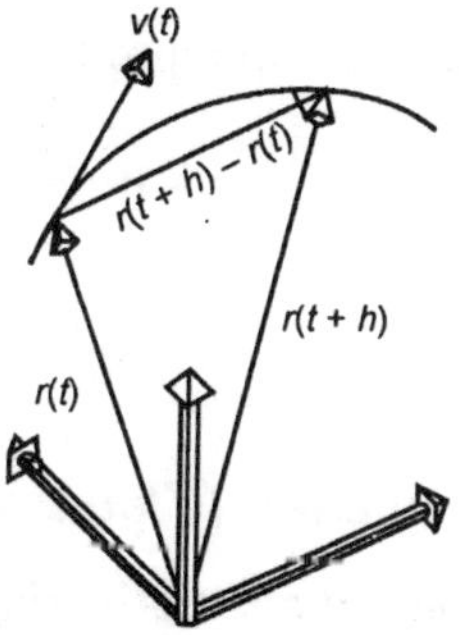

If each of the scalar function x, y and z are diûerentiable, then this vector has a limit

$$\left(\frac{dx}{dt}, \frac{dy}{dt}, \frac{dz}{dt}\right) = (\dot{x}, \dot{y}, \dot{z}) \qquad ...(9)$$

which is the instantaneous velocity of the particle v(t). This means that (if the motion is smooth) then

$$v(t) = \lim_{h \to} \frac{r(t+h)-r(t)}{h} = \frac{d}{dt} r(t) \qquad ...(10)$$

Rules for Differentiation

To differentiate the vector product of vector function, the order of factors must be preserved because the vector product is not commutative.. If f, g and h are differentiable vector functions and ϕ is a scalar function of a scalar variable t, then

1. $\frac{d}{dt}(f \pm g) = \frac{df}{dt} \pm \frac{dg}{dt}$ ...(11)

2. $\frac{d}{dt}(f \cdot g) = f \cdot \frac{dg}{dt} + g \cdot \frac{df}{dt}$ ...(12)

   [Since f.g = g.f, \ in evaluating $\frac{d}{dt}(f \cdot g)$, we do not care about the order of f and g]

3. $\frac{d}{dt}(f \times g) = f \times \frac{dg}{dt} + \frac{df}{dt} \times g$ ...(13)

[Since f × g = –g ′ f, \ in evaluating $\frac{d}{dt}(f \times g)$, we do not care about the order of f and g]

4. $\frac{d}{dt}(\phi f) = \phi\frac{df}{dt} + f\frac{d\phi}{dt}$ ...(14)

5. $\frac{d}{dt}[fgh] = \left[\frac{df}{dt}gh\right] + \left[f\frac{dg}{dt}h\right] + \left[fg\frac{dh}{dt}\right]$ ...(15)

Where [fgh] denotes the scalar triple product of f, g and h.

[In evaluating $\frac{d}{dt}[fgh]$, we must keep the cyclic order of f, g and h in tact]

6. $\frac{d}{dt}(f \times (g \times h)) = \frac{df}{dt} \times (g \times h) + f \times \left[\frac{dg}{dt} \times h\right] + f \times \left[g \times \frac{dh}{dt}\right]$ ...(16)

Derivative of a Function of a Function (Chain rule)

Let r be a differentiable function of a scalar variables s and s be a differentiable function of another scalar variable t, then r is a differentiable function of t.

Let δt be a small change in t and δs, δr denote the corresponding changes in s and r respectively.

Now, $\frac{\delta r}{\delta t} = \frac{\delta r}{\delta s} \cdot \frac{\delta s}{\delta t}$ ...(17)

Taking limits as $\delta t \to 0$ (and hence $\delta s \to 0$), we have

$$\underset{\delta t \to 0}{Lt} \frac{\delta r}{\delta t} = \left(\underset{\delta t \to 0}{Lt} \frac{\delta r}{\delta s}\right) \cdot \left(\underset{\delta t \to 0}{Lt} \frac{\delta s}{\delta t}\right) \text{ or } \frac{dr}{dt} = \frac{dr}{ds} \cdot \frac{ds}{dt} \quad ...(18)$$

Constant Vectors

A vector is said to be constant if both its magnitude and direction do not change.

Partial Derivatives of Vector Functions of more than one Variable

So far, we have considered vector functions of a single scalar variable t. But, we come across with many vector functions of more than one independent scalar variable.

If f is a vector function of independent variables x, y and z, we write f = f (x, y, z).

Let δx be a small change in x, while other variables remain unchanged. Let f be the corresponding change in f,

Then, $\delta f = f(x + \delta x, y, z) - f(x, y, z)$

Dividing by δx both sides we get

$$\frac{\delta f}{\delta x}=\frac{f(x+\delta x,y,z)-f(x,y,z)}{\delta x} \quad ...(19)$$

Taking limits $\delta x \to 0$, we have

$$\underset{\delta x\to 0}{Lt}\frac{\delta f}{\delta x}=\underset{\delta x\to 0}{Lt}\frac{f(x+\delta x,y,z)-f(x,y,z)}{\delta x}$$

If this limit exists, we denote it by $\frac{\partial f}{\partial x}$ or f_x and is called the partial derivative of f w.r.t x i.e.,

$$\frac{\partial f}{\partial x}=\underset{\delta x\to 0}{Lt}\frac{\delta f}{\delta x}$$

$$=\underset{\delta x\to 0}{Lt}\frac{f(x+\delta x,y,z)-f(x,y,z)}{\delta x} \quad ...(20)$$

Similarly we define the partial derivatives

$$\frac{\partial f}{\partial y}=\underset{\delta y\to 0}{Lt}\frac{\delta f}{\delta y}=\underset{\delta y\to 0}{Lt}\frac{f(x,y+\delta y,z)-f(x,y,z)}{\delta y} \quad ...(21)$$

and $$\frac{\partial f}{\partial z}=\underset{\delta z\to 0}{Lt}\frac{\delta f}{\delta z}=\underset{\delta z\to 0}{Lt}\frac{f(x,y,z+\delta z)-f(x,y,z)}{\delta z}$$

Higher –order partial derivatives may also be defined in the same way. For example,

$$f_{xx}=\frac{\partial^2 f}{\partial x^2}=\frac{\partial}{\partial x}\left(\frac{\partial f}{\partial x}\right) \qquad f_{yy}=\frac{\partial^2 f}{\partial y^2}=\frac{\partial}{\partial y}\left(\frac{\partial f}{\partial y}\right) \quad ...(22)$$

$$f_{xy}=\frac{\partial}{\partial y}\left(\frac{\partial f}{\partial x}\right)=\frac{\partial^2 f}{\partial y\partial x}, \qquad f_{yx}=\frac{\partial}{\partial x}\left(\frac{\partial f}{\partial y}\right)=\frac{\partial^2 f}{\partial x\partial y}$$

If f has continuous partial derivatives of second order, then

$$\frac{\partial^2 f}{\partial x\partial y}=\frac{\partial^2 f}{\partial y\partial x}. \quad ...(23)$$

Rules for Partial Differentiation

Rules for partial differentiation of vectors are similar to those used in elementary calculus for scalar functions. Thus, if f and g are differentiable vector functions of x, y, z and ϕ is a differentiable scalar function of *x*, *y*, z then (using x as an example)

1. $\frac{\partial}{\partial x}(f\pm g)=\frac{\partial f}{\partial x}\pm\frac{\partial g}{\partial x}$ ...(24)

2. $\dfrac{\partial}{\partial x}(f \cdot g) = f \cdot \dfrac{\partial g}{\partial x} + g \cdot \dfrac{\partial f}{\partial x}$ ...(25)

3. $\dfrac{\partial}{\partial x}(f \times g) = f \times \dfrac{\partial g}{\partial x} + \dfrac{\partial f}{\partial x} \times g$ ...(26)

4. $\dfrac{\partial}{\partial x}(\phi f) = \phi \dfrac{\partial f}{\partial x} + f \dfrac{\partial \phi}{\partial x}$ ...(27)

5. $\dfrac{\partial^2}{\partial y \partial x}(f \cdot g) = \dfrac{\partial}{\partial y}\left\{\dfrac{\partial}{\partial x}(f \cdot g)\right\} = \dfrac{\partial}{\partial y}\left\{f \cdot \dfrac{\partial g}{\partial x} + g \cdot \dfrac{\partial f}{\partial x}\right\}$

$= \dfrac{\partial}{\partial y}\left(f \cdot \dfrac{\partial g}{\partial x}\right) + \dfrac{\partial}{\partial y}\left(g \cdot \dfrac{\partial f}{\partial x}\right)$ ...(28)

$= f \cdot \dfrac{\partial^2 g}{\partial y \partial x} + \dfrac{\partial f}{\partial x} \cdot \dfrac{\partial g}{\partial x} + \dfrac{\partial f}{\partial x} \cdot \dfrac{\partial g}{\partial x} + \dfrac{\partial^2 f}{\partial y \partial x} \cdot g$

6. $\dfrac{\partial^2}{\partial x \partial y}(f \times g) = \dfrac{\partial}{\partial x}\left\{\dfrac{\partial}{\partial y}(f \times g)\right\} = \dfrac{\partial}{\partial x}\left\{f \times \dfrac{\partial g}{\partial y} + \dfrac{\partial f}{\partial y} \times g\right\}$

$= \dfrac{\partial}{\partial x}\left(f \times \dfrac{\partial g}{\partial y}\right) + \dfrac{\partial}{\partial x}\left(\dfrac{\partial f}{\partial y} \times g\right)$ ...(29)

$= f \times \dfrac{\partial^2 g}{\partial x \partial y} + \dfrac{\partial f}{\partial x} \times \dfrac{\partial g}{\partial y} + \dfrac{\partial f}{\partial y} \times \dfrac{\partial g}{\partial x} + \dfrac{\partial^2 f}{\partial x \partial y} \times g$

If $f(x,y,z) = f_1(x,y,z)i + f_2(x,y,z)j + f_3(x,y,z)k$, then

$\dfrac{\partial f}{\partial x} = \dfrac{\partial f_1}{\partial x} i + \dfrac{\partial f_2}{\partial x} j + \dfrac{\partial f_3}{\partial x} k$ etc ...(30)

If a vector function f is a function of x, y, z and the variables x, y, z are expressed in terms of new variables s, t, then the function will be a function of s and t. Then, we can write

$\dfrac{\partial f}{\partial t} = \dfrac{\partial f}{\partial x} \cdot \dfrac{\partial x}{\partial t} + \dfrac{\partial f}{\partial y} \cdot \dfrac{\partial y}{\partial t} + \dfrac{\partial f}{\partial z} \cdot \dfrac{\partial z}{\partial t}$ etc. ...(31)

Total Differentials

Differentials of vectors follow rules similar to those of elementary calculus. For example,

1. If $f(x,y,z) = f_1(x,y,z)i + f_2(x,y,z)j + f_3(x,y,z)k$, then

$df(x,y,z) = df_1(x,y,z)i + df_2(x,y,z)j + df_3(x,y,z)k$ ...(32)

2. $d(f \cdot g) = f \cdot dg + g \cdot df$ ...(33)

3. $d(f \times g) = f \times dg + df \times g$ ...(34)

4. If f = f(x, y, z), then df = $\frac{\partial f}{\partial x}dx + \frac{\partial f}{\partial y}dy + \frac{\partial f}{\partial z}dz$, etc ...(35)

Space Curve

If r = f(t) denotes a vector function of a scalar variable t and r is the position vector of a point P(x, y, z), in space, then as t varies, so does r and therefore the end point P or r describes a continuous curve. Therefore, a curve C can always be represented by expressing the position vector r of any point on it as a function of some scalar variable t.

If a $\leq$ t $\leq$ b, the end point of the vector r(a) is called initial point of the curve C and the end point of the vector r(b) is called the end point of C. if r (a) = r (b), then curve C is a closed curve. That is, a curve is closed if its initial and end point coincide. If there exist a < t_1 and t_2 < b such that r (t_1) = r (t_2) = r_0 (say), then r is called a multiple point. Geometrically, a multiple point occurs where curve C crosses itself or is tangent to itself. A curve C is called a simple curve if it has no multiple points. A curve which is both simple and closed, is called a simple closed curve.

Let P be a fixed point on the curve having position vector r corresponding to some t. Let Q be the point corresponding to t+δ t and have position vector r + δr. Then,

$$\underset{\delta t \to 0}{Lt} \frac{PQ}{\delta t} = \underset{\delta t \to 0}{Lt} \frac{\delta r}{\delta s} = \frac{dr}{dt}$$

is a vector tangent to the space curve C at P. This vector is called the tangent vector to C at P. A point t on a space curve C is called singular or non-singular according as the value of $\frac{dr}{dt}$ is zero or non-zero at t. The direction of the space curve C at a non-singular point P is the same as that of the tangent vector to C at P.

A vector function which has a continuous derivative and no singular points is called a smooth vector function.

If the parameter t is the time, then represents $\frac{dr}{dt}$ the velocity v, with which the end point of r describes the curve and $\frac{dv}{dt} = \frac{d^2r}{dt^2}$ represents its acceleration a along the curve.

ILUSTRATIVE EXAMPLES

1. $If r = t^2 i + \sin tj + (1-t)k \ find \frac{dr}{dt} and \frac{d^2r}{dt^2}$

Sol. $\frac{dr}{dt} = 2ti + \cos ij - k$ and $\frac{d^2r}{dt^2} = 2i - \sin tj$

2. Show that if *a, b, c are constant vectors when* $r = at^2 + bt + c$ *is the path of a point moving with constant acceleration.*

Sol. The velocity of the point $= \frac{dr}{dt} = 2at = b$ and the acceleration $= \frac{d^2r}{dt^2} = 2a =$ constant vector. Thus the point whose path is r=at²=dt+c is moving with constant acceleration.

3. *If* $r = a\cos ti + a\sin tj + tk$, *find* $\frac{dr}{dt}, \frac{d^2r}{dt^2}$ *and* $\left|\frac{d^2r}{dt^2}\right|$

$$\frac{dr}{dt} = \frac{d(a\cos t)}{dt} i + \frac{d}{dt}(a\sin t) j + \frac{d}{dt}(t) k = -a\sin ti + a\cos tj + k$$

$$\frac{d^2r}{dt^2} = \frac{d}{dt}(a - \sin t) i + \frac{d(a\cos t) j}{dt} + \frac{d}{dt}(l) k = -a\cos ti - a\sin tj$$

$$\therefore \left|\frac{d^2r}{dt^2}\right| = \sqrt{[(-a\cos t)^2 + (-a\sin t)^2]} = a\sqrt{[\cos t^2 + \sin t^2]} = a$$

4. *If* $r = a\cos ti + a\sin tj + at\tan ak$ *find* $\left|\frac{dr}{dt} \times \frac{d^2r}{dr^2}\right|$ *and* $\left|\frac{dr}{dt}\frac{d^2r}{dt^2}\frac{d^3r}{dt^3}\right|$.

Sol. $\frac{dr}{dt} = -a\sin tj + a\tan ak$ $\frac{d^2r}{dt^2} = -a$ cos it-a sin tj+0 as a tan α =constant.

$$\frac{d^3r}{dt^3} = a\sin ti - a\cos tj.$$

$$\therefore \frac{dr}{dt}\frac{d^2r}{dt^2} = \begin{vmatrix} i & j & k \\ -a\sin t & a\cos t & a\tan\alpha \\ -a\cos t & -a\sin t & 0 \end{vmatrix}$$

$$= i(-a^2\sin t\tan\alpha) + j(-a^2\cos t\tan\alpha) + k(a^2\sin^2 t + a^2\cos^2 t)$$

$$\therefore \left|\frac{dr}{dt} \times \frac{d^2r}{dt^2}\right| = \sqrt{[(a^2\sin t\tan a^2 + (a^2\cos t\tan\alpha)^2 + a^4]}$$

$$= \sqrt{[a^4\tan^2\alpha(\sin^2 t + \cos^2 t) + a^4]} = a^2\sec\alpha.$$

Again, we know that $[abc]=\begin{vmatrix} a_1 & a_2 & a_3 \\ b_1 & b_2 & b_3 \\ c_1 & c_2 & c_3 \end{vmatrix}$.

Therefore

$$\left[\frac{dr}{dt}\frac{d^2r}{dt^2}\frac{d^3r}{dt^3}\right]=\begin{vmatrix} -a\sin t & a\cos t & a\tan\alpha \\ -a\cos t & -a\sin t & 0 \\ a\sin t & -a\cos t & 0 \end{vmatrix}=a\tan\alpha\,(a^2\cos^2 t+a^2\sin^2 t)=a^3\tan\alpha.$$

5. *Find the unit tangent vector at any point on the curve $x=t^2+2$, $y=4t-5$, $z=2t^2-6t$ where t is any variable. Also determine the unit tangent vector at the point $t=2$.*

Sol. Let r be the position vector of any point (x, y, z) on the given curve, then we have

$$r=xi+yj+zk=(t^2+2)i+(4t-5)j+(2t^2-6t)k$$

$$\Rightarrow \frac{dr}{dr}=2ti+4j+(4t-6)k \text{ and } a=\frac{d^2r}{dt^2}=\frac{d^2x}{dt^2}i+\frac{d^2y}{dt^2}j$$

Then multiply (5) scalarly by $\hat{t}=\dfrac{\frac{dr}{dt}}{\left|\frac{dr}{dt}\right|}=\dfrac{2ti+4j+(4t-6)k}{2\sqrt{5t^2-12t+13}}=\dfrac{ti+2j+(2t-3)k}{\sqrt{(5t^2-12t+13)}}$

Then the unit tangent vector at the point $t=2$ is given by

$$\frac{2i+2j+(2\times 2-3)k}{\sqrt{(5\times 4-12\times 2+13)}}=\frac{1}{3}(2i+2j+k).$$

6. *Find the angle between the tangents to the curve $r=t^2i-2tj+t^3k$ at the points $t=1$ and $t=2$.*

Sol. $r=t^2i=2tj+t^3k \;\Rightarrow\; \frac{dr}{dt}=2ti-2j+3t^2k$

Let T_1 and T_2 be the vectors along the tangents at $t=1$ and $t=2$ respectively, then we have

$$T_1=\left(\frac{dr}{dt}\right)_{t=1}=2i-2j+3k \text{ ; and } T_1=\left(\frac{dr}{dt}\right)_{t=2}=4i-2j+12k.$$

Let q be the required angle between the tangents, then we have $T_1 \cdot T_2=|T_1||T_2|\cos q$

$$\Rightarrow (2)(4)+(-2)(-2)+(3)(12)=\sqrt{[2^2+(-2)^2+3^2]}=\sqrt{[4^2+(-2)^2+12^2]}\cos\theta$$

$$\Rightarrow 48 = \sqrt{(17)}\ \sqrt{(164)}\cos q \Rightarrow \cos q = \frac{48}{\sqrt{(17)}\sqrt{(164)}} \Rightarrow \theta = \cos^{-1}\left\{\frac{48}{\sqrt{(17)}\sqrt{(164)}}\right\}$$

7. *Find the curvature and torsion of the curve x= a cos t, y=a sin t, z=bt.*

(***Note.*** Any curve drawn on a circular cylinder and cutting its generators at a constant angle is known as a circular helix.)

Sol. The vector equation of the curve is given by $r = a\cos ti + a\sin tj + btk \therefore (dr/dt)$

$$= -a\sin ti + a\cos tj + bk$$

Its arc length from $P_0(t=0)$ to any point $P(t)$ is s

$$= \int_0^t |dr/dt|\,dt = \sqrt{(a^2+b^2)t} \Rightarrow \frac{ds}{dt} = \sqrt{(a^2+b^2)}$$

Also $\hat{t} = \frac{dr}{dt} = \frac{dr}{dt}/\frac{ds}{dt} = \frac{-a\sin ti + a\cos tj + bk}{\sqrt{(a^2b^2)}}$ and $= \frac{dt}{ds} = \frac{d\hat{t}}{dt}/\frac{ds}{dt} = \frac{-a(\cos ti + \sin tj)}{a^2+b^2}$

$$\Rightarrow k = \left|\frac{d\hat{t}}{ds}\right| = \frac{a}{a^2+b^2}$$

Also $\hat{n} = -(\cos ti + \sin tj)$ and $\hat{b} = \hat{t}\times\hat{n} = (b\sin ti - b\cos tj + ak)/\sqrt{(a^2+b^2)}$

$$\Rightarrow \frac{d\hat{b}}{ds} = \frac{db}{dt}\Big/\frac{ds}{dt} = b\,(\cos ti + \sin tj)/(a^2+b^2) = -\tau\hat{n} = \tau\,(\cos ti = \sin tj)$$

$$\Rightarrow \qquad \tau = \frac{b}{a^2+b^2}$$

EXERCISE

1. If r = a cos wt + b sin wt

   Show that $r \times \frac{dr}{dt} = \omega a \times b.$

   and $\frac{d^2r}{dt^2} = -\omega^2 r$

   a, b, w being constants.

2. If r = a cos ti + a sin t j + at tan ak.

   find $\left|\frac{dr}{dt} \times \frac{d^2r}{dt^2}\right|$ and $\left|\frac{dr}{dt}\frac{d^2r}{dt^2}\frac{d^2r}{dt^2}\right|$.

3. Show that, if R = A sin wt + B cos wt, where A, B, w are constants, then $\frac{d^2R}{dt^2} = -\omega^2 R$ and $R \times \frac{dR}{dt} = -$ wA × B.
4. Given R = t^m A + t^n B, where A, B are constant vectors, show that, if R and d^2R/dt^2 are parallel vectors, then m + n = 1, unless m = n.
5. If P = $5t^2$I + t^3J – tK and Q = 2I sin t – J cos t + 5tK, find (i) $\frac{d}{dt}$(P.Q); (ii) $\frac{d}{dt}$(P × Q).
6. If $\frac{dU}{dt}$ = W × U and $\frac{dV}{dt}$ = W × V, prove that $\frac{d}{dt}$(U × V) = W × (U × V).
7. If A = x^2yz I – $2xz^3$ J + xz^2 K and B = 2z I + yJ – x^2K, find $\frac{\partial^2}{\partial x \partial y}$(A × B) at (1, 0, –2).
8. If R = (a cos t) I + (a sin t) J + (at tan a) K, find the value of

   (i) $\left|\frac{dR}{dt} \times \frac{d^2R}{dt^2}\right|$

   (ii) $\left|\frac{dR}{dt}, \frac{d^2R}{dt^2}, \frac{d^3R}{dt^3}\right|$

   also find the unit tantent vector at any point t of the vurve.
9. Find the unit tangent vector at any point on the vurve x = t^2 + 2, y = 4t – 5, z = $2t^2$ – 6t, where t is any variable. Also determine the unit tangent vector at the point t = 2.
10. A particle moves along a curve x = e^{-t}, y = 2 cos 3t, z = 2 sin 3t, where t is the time variable. Determine its velocity and acceleration vectors and also the magnitudes of velocity and acceleration at t = 0.
11. The position vector of a particle at time t is R = cos (t – 1) I + sinh (t – 1) J + at^3K. Find the condition imposed on a by requiring that at time t = 1, the acceleration is normal to the position vector.

Chapter 2

Vector Integration and Application

INTRODUCTION

Vector integration play a vital role in Fluid mechanics, Aerodynamics and some other subject also.

We have alraedy learnt about vector point functions differentiation and their properties. In this chapter we study the integration of vector valued function and its application..

Integration of Vector valued Function

If f and F are two vector functions of the scalar variable t such that

$$\frac{dF}{dt} = f, \quad \text{...(1)}$$

Then F is called the indefinite integral of w. r .t. and we write

The function f, which is to be integrated is called the integrand.

If c is an arbitrary constant vector independent of t, then

$$\frac{d}{dt}(F+c) = \frac{dF}{dt} = f \quad \text{...(2)}$$

$$\therefore \int f dt = F + c \quad \text{...(3)}$$

Thus, the integral F is indefinite to the extent of an additive arbitrary constant vector c.

If $\frac{dF}{dt} = f$ for all values of t in the interval [a, b], the definite integral between the limits $t = a$ and $t = b$ is written as

$$\int_a^b f dt = [F]_a^b = F(b) - F(a) \quad ...(4)$$

Theorem. If $f = f_1\, i + f_2 j + f_3\, k$, where, f_1, f_2, f_3 are scalars functions of the scalar variable t, then

$$\int f dt = i\int f_1 dt + j\int f_2 dt + k\int f_3 dt \quad ...(5)$$

Thus, to integrate a vector function, we should integrate its components.

Some Standard Results

We are going to dicuss some standard results for integration:

Let f and g be two vector function of scalar variable t.

1. Since $\frac{d}{dt}(f \cdot g) = \frac{df}{dt} \cdot g + f \cdot \frac{df}{dt}$ ...(6)

   Integrating both side, we get

   $$\int\left(\frac{df}{dt}.g + f.\frac{df}{dt}\right)dt = f.g + c \quad ...(7)$$

   Where c is the constant of integration and is a scalar because the integrand is a scalar.

2. Since $\frac{d}{dt}(f \cdot f) = \frac{d}{dt}(f \cdot f) = \frac{df}{dt} \cdot f + f \cdot \frac{d}{dt} = 2f \cdot \frac{df}{dt}$

   $$\therefore \int 2f \cdot \frac{df}{dt} = f^2 + c \quad ...(8)$$

   Where c is the constant of integration and is a scalar because the integrand is a scalar.

3. Since $\frac{d}{dt}\left(\frac{df}{dt}\right)^2 = 2\frac{df}{dt} \cdot \frac{d^2 f}{dt^2}$

   $$\therefore \int\left(2\frac{df}{dt} \cdot \frac{d^2 f}{dt^2} g\right)dt = \left(\frac{df}{dt}\right) + c \quad ...(9)$$

   Where c is the constant of integration and is a scalar because the integrand is a scalar.

4. Since $\frac{d}{dt}(f \times g) == \frac{df}{dt} \times g + f \times \frac{dg}{dt}$ ...(10)

$$\therefore \quad \int\left(\frac{df}{dt}\times g+f\times\frac{dg}{dt}\right)dt = f\times g+c \qquad ...(11)$$

Where c is the constant of integration and is a scalar because the integrand is a scalar

5. Since $\frac{d}{dt}(a\times g) = a\times\frac{df}{dt}$, where a is constant vector.

$$\therefore \quad \int\left(a\times\frac{df}{dt}\right)dt = a\times f+c \qquad ...(12)$$

Where c is the constant of integration.

6. If c is a constant scalar, $\int cfdt = c\int fdt$ ...(13)

7. $\int(f\pm g)dt = \int fdt \pm \int gdt$ ...(14)

Remember: The constant of integration is of the same nature as the integrand

Example: If $r = (7t^2)\, i + 2\, t\, j - 3\, t^2\, k$, find $\int_2^3 r\times\frac{d^2r}{dt^2}dt$.

Solution. Since, $\frac{d}{dt}\left[r\times\frac{dr}{dt}\right] = r\times\frac{d^2r}{dt^2}+\frac{dr}{dt}\times\frac{dr}{dt}$

$$\Rightarrow \quad \frac{d}{dt}\left[r\times\frac{dr}{dt}\right] = r\times\frac{d^2r}{dt^2}$$

$\therefore$ Integrating both sides w. r. t., we have

$$\int\left(r\times\frac{d^2r}{dt^2}\right)dt = r\times\frac{dr}{dt}+c$$

where c is the constant of integration.

$$\therefore \quad \int_2^3\left(r\times\frac{d^2r}{dt^2}\right)dt = \left[r\times\frac{dr}{dt}\right]_2^3 \qquad ...(15)$$

Let us find $r\times\frac{dr}{dt}$

Since, $r = (7t^2)\, i + 2\, t\, j - 3\, t^2\, k$

$$\therefore \frac{dr}{dt} = \frac{d}{dt}\{(7t^2)\, i + 2\, t\, j + (-3\, t^2)\, k\}$$

$$\therefore \frac{dr}{dt} = 14\,t\,I + 2\,j - 6\,t\,k$$

$$\therefore \qquad \left[r \times \frac{dr}{dt}\right] = \begin{vmatrix} i & j & k \\ 7t^2 & 2t & -3t^2 \\ 14t & 2 & -6t \end{vmatrix} = i\left(-12t^2 + 6t^2\right)$$

$$+ j\left(-42t^3 + 42t^3\right) + k\left(14t^2 - 28t^2\right)$$

$$\Rightarrow \qquad \left[r \times \frac{dr}{dt}\right] = i\left(-6t^2\right) + j(0) + k\left(-14t^2\right) \qquad ...(16)$$

Using (2) in (1), we get

$$\therefore \int_2^3 \left(r \times \frac{d^2r}{dt^2}\right) dt = \left[i\left(-6t^2\right) + j(0) + k\left(-14t^2\right)\right]_2^3$$

$$= \left[i\left\{\left(-6.3^2\right) - \left(-6.2^2\right)\right\} + k\left\{\left(-14.3^2\right) - \left(-14.2^2\right)\right\}\right]$$

Hence, $\displaystyle\int_2^3 r \times \frac{d^2r}{dt^2}\,dt = -30\,\text{i} - 70\,\text{k}$

EXERCISE

1. Evaluate the line integral $\displaystyle\int_C \left[\left(x^2 + xy\right)dx + \left(x^2 + y^2\right)dy\right]$ where C is the square formed by the lines y = ± 1 and x = ± 1.
2. If F = $(5xy - 6x^2)$I + (2y – 4x)J, evaluate $\displaystyle\int_C F.dR$ along the curve C in the xy-plane, y= x^3 from the point (1, 1) to (2, 8).
3. A vector field is given by F = (sin y)I + x(1 + cos y)J. Evaluate the line integral over a circular path given by $x^2 + y^2 = a^2$, z = 0.
4. Compute the line integral $\displaystyle\int_C \left(y^2dx - x^2dy\right)$ about the triangle whose vertices are (1, 0), (0, 1) and (–1, 0).
5. If velocity vector is F = yI + 2J + xzK m/sec., show that the flux of water through the parabolic cylinder $y = x^2$, $0 \le x \le 3$, $0 \le z \le 2$ is 69 m^3/sec.

6. Evaluate $\int_S F.dS$ were F = x I + (z² – zx) J – xy K and S is the triangular surface with vertices (2, 0, 0), (0, 2, 0) and (0, 0, 4).

APPLICATION OF VECTOR INTEGRATION

CURVE IN A VECTOR FIELD

Let *I* be an interval of real numbers (i.e. a non-empty connected subset of R). Then a curve γ is a continuous mapping $\gamma : I \to X$, where *X* is a real space.

The curve γ is said to be simple if for all *x*, *y* in *I*, we have $\gamma(x) = \gamma(y) \to x = y$. If *I* is a closed bounded interval [a, b] and $\gamma(a) = \gamma(b)$ then the curve is closed simple curve). If $\gamma(x) = \gamma(y)$ for some x $x \neq y$ (other than the extremities of *I*), then $\gamma(x)$ is called a double (or multiple) point of the curve.

A curve γ is said to be closed or a loop if I = (a, b) and if $\gamma(a) = \gamma(b)$. A closed curve is thus a continuous mapping of the circle S_1; a simple closed curve is also called a Jordan curve.

LINE INTEGRALS

Suppose that a wire has as density $f(x, y, z)$ at the point (x, y, z) on the wire. Then the line integral will equal the total mass of the wire. Below is the definition in symbols.

Definition of the Line Integral

Let f be a function defined on a curve *C* of finite length. Then the *line integral* of *f* along *C* is

$$\int_C f(x,y)ds = \lim_{n\to\infty}\sum_{i=1}^{n} f(x_i, y_i)\Delta s_i \text{ (for two dimension)} \quad ...(1)$$

$$\int_C f(x,y,z)ds = \lim_{n\to\infty}\sum_{i=1}^{n} f(x_i, y_i, z_i)\Delta s_i \text{ (for three dimension)} \quad ...(2)$$

Evaluating Line Integrals

Let, $r(t) = x(t)i + y(t)j$

be a differentiable vector valued function. Then

$$ds = \|r'(t)\|dt = \sqrt{x'(t)^2 + y'(t)^2}\,dt$$

We are now ready to state the theorem that shows us how to compute a line integral.

Theorem: Line Integrals Over Vector Valued Functions

Let $r(t) = x(t)i + y(t)j \quad a \le t \le b$

be a differentiable vector valued function that defines a smooth curve C. Then

$$\int_C f(x,y)ds = \int_a^b f(x(t),y(t))\sqrt{x'(t)^2 + y'(t)^2}\,dt \qquad ...(3)$$

and for three dimensions, if

$$r(t) = x(t)i + y(t)j + z(t)k \qquad a \le t \le b$$

then $$\int_C f(x,y,z)ds = \int_a^b f(x(t),y(t),z(t))\sqrt{x'(t)^2 + y'(t)^2 + z'(t)^2}\,dt \qquad ...(4)$$

Line Integrals in Differential Form

We can rewrite r′(t) dt as

$$\frac{dr}{dt}dt = \left(\frac{dx}{dt}i + \frac{dy}{dt}j + \frac{dz}{dt}k\right)dt \qquad ...(5)$$

$\Rightarrow$ $$dr = dxi + dyj + dzk \qquad ...(6)$$

So that if

$$F = Mi + Nj + Pk \qquad ...(7)$$

then

$$F \cdot r'(t)dt = Mdx + Ndy + Pdz \qquad ...(8)$$

This is called the differential form.

Let $r = xi + yj + zk$ be the position vector of appoint $P(x, y, z)$ on a curve C, where x, y, z are functions of scalar parameter t, $a \le t \le b$.

Let $t = a$ at A and $t = b$ at B.

Let $f = f_1 i + f_2 j + f_3 k$ be a vector point function defined and continuous along curve C.

If s denotes the arc length of the curve C, then $\frac{dr}{dt} = t$ is a unit vector along the tangent to the curve C at P. The component of the vector f along the tangent at P is

$$f_t = f \cdot t = f \cdot \frac{dr}{ds} \qquad ...(9)$$

For points on C, f_t is a unction of s and $\int_A^B f_t ds$ is known as the tangent line integral or just line integral of f along C.

But $f_t = f \cdot t = f \cdot \frac{dr}{ds}$ ds = f .dr ...(10)

$$\therefore \int_A^B f_t ds = \int_A^B f \cdot tds = \int_A^B f \cdot dr \text{ and is also written as } \int_C f \cdot dr$$

Note that $\int_A^B f \cdot dr = -\int_B^A f \cdot dr$

The differential displacement vector dr along the curve C is defined by

$$dr = dxi + dyj + dzk$$

$$f \cdot dr = (f_1 i + f_2 j + f_3 k) \cdot (dxi + dyj + dzk) \quad ...(11)$$

$$= f_1 dx + f_2 dy + f_3 dz \quad ...(12)$$

Therefore , in component form the above line can be written as

$$\int_C f \cdot dr = \int_C (f_1 i + f_2 j + f_3 k) \cdot (dxi + dyj + dzk)$$

$$= \int_C (f_1 dx + f_2 dy + f_3 dz) \quad ...(13)$$

If the given curve is given by the parametric equations:

$x = x(t), y = y(t), z = z(t)$, then can be write

$$\int_C f \cdot dr = \int_a^b \left(f_1 \frac{dx}{dt} + f_2 \frac{dy}{dt} + f_3 \frac{dz}{dt} \right) dt, \quad ...(14)$$

Where a, b are the values of the parameter t at A and B.

Circulation

If C is a simple closed curve, then the tangent line integral of f around C is called the circulation of about C and is often denoted by

$$\oint_C f \cdot dr = \oint_C (f_1 dx + f_2 dy + f_3 dz) \quad ...(15)$$

Applications of line integral

The main application of line integrals is finding the work done on an object in a force field. The work done W along each piece will be approximately equal to

$$dW = F \cdot T\, ds \quad ...(16)$$

Now recall that

$$T = \frac{r'(t)}{\|r'(t)\|} \quad ...(17)$$

and that

$$ds = \|r'(t)\| dt \quad ...(18)$$

Hence,

$$dW = F \cdot r'(t) dt \quad ...(19)$$

As usual, we add up all the small pieces of work and take the limit as the pieces get small to end up with an integral.

Definition

Let F *be a vector field and* C *be a curve defined by the vector valued function* r. *Then the work done by* F *on an object moving along* C *is given by*

$$\text{Work done} = \int_C F \cdot dr = \int_a^b F(x(t), y(t), z(t)) \cdot r'(t)\, dt \qquad ...(20)$$

SURFACE INTEGRAL

Let S be a surface in three-dimensional space R^3 and let f be a function defined on this surface. Then the *integral of f with respect to the surface area* is usually notated as

$$\int_S f(u,v) d^2 A \qquad ...(21)$$

To compute above integral, one may use the following formula to convert it to a double integral:

$$\int_S f(u,v) d^2 A = \int f(u,v) \sqrt{\left(\frac{\partial(x,y)}{\partial(u,v)}\right)^2 + \left(\frac{\partial(y,z)}{\partial(u,v)}\right)^2 + \left(\frac{\partial(z,x)}{\partial(u,v)}\right)^2}\, dudv \qquad ...(21)$$

If the surface is described as the graph of a function g, then we may also use the following formula:

$$\int_S f(x,y) d^2 A = \int f(x,y) \sqrt{1 + \left(\frac{\partial g}{\partial x}\right)^2 + \left(\frac{\partial g}{\partial y}\right)^2}\, dxdy \qquad ...(22)$$

Evaluating Surface Integral

For a scalar function f over a surface parameterized by u and v, the surface integral is given by

$$\phi = \int_S f d\alpha = \int_S f(u,v) |T_u \times T_v| dudv \qquad ...(23)$$

where T_u and T_v are tangent vectors and $a \times b$ is the cross product.

Let P be any point on the surface S and let n be the unit vector at P in the direction of the outward drawn normal to the surface S at P.

Then, $f \cdot n$ is the normal component of f at P. The integral of $f \cdot n$ over S is called the (normal) surface integral of f over S and is denoted by

$$\iint_S f \cdot n dS \qquad ...(24)$$

It is also called the flux of *f* over *S*.

If we associate with the differential of surface area *dS* a vector dS whose magnitude is *dS* and whose direction is that of n, then

$$dS = ndS \quad ...(25)$$

where $f_n = f \cdot n$ denotes the component of *f* in the direction of the outward normal at *P*.

If the outward drawn normal to the surface *S* at *P* makes angles α, β, γ with the positive direction of *x*, *y*, *z* axes respectively, and *l*, *m*, *n* are the direction cosines of this outward normal, then

$$l = \cos\alpha, m = \cos\beta, n = \cos\gamma \quad ...(26)$$

$$\text{Also, } \hat{n} = \cos\alpha\, i + \cos\beta\, j + \cos\gamma\, k = li + mj + nk \quad (27)$$

If

$$f = f_1 i + f_2 j + f_3 k, \text{ then } f \cdot \hat{n}$$

$$= (f_1 i + f_2 j + f_3 k) \cdot (li + mj + nk)$$

$$= f_1 l + f_2 m + f_3 n$$

$$= f_1 \cos\alpha + f_2 \cos\beta + f_3 \cos\gamma \quad (28)$$

$$\iint_S f \cdot \hat{n}\, dS = \iint_S (f_1 \cos\alpha + f_2 \cos\beta + f_3 \cos\gamma) dS \quad (29)$$

Now, *dS* cos α = orthogonal projection of the area of the dS on the *yz* – plane = *dy dz*

dS cos β = orthogonal projection of the area of the *dS* on the *yz* – plane = *dz dx*: and

dS cos γ = orthogonal projection of the area of the *dS* on the *yz* –plane = *dx dy*

$$\therefore \quad \iint_S f \cdot \hat{n}\, dS = \iint_S (f_1 dydz + f_2 dzdx + f_3 dxdy) \quad ...(30)$$

Also, we can evaluate the surface integral using the method given below.

$$\phi = \int_S F \cdot da = \int_S (F \cdot \hat{n}) \cdot d\alpha \quad ...(31)$$

$$= \int_S (f_x dy\, dz + f_y dz\, dx + f_z dx\, dy) \quad ...(32)$$

where $a \cdot b$ is a dot product and $\hat{n}$ is a unit normal vector.

If $z = f(x, y)$, then *d a* is given explicitly by

$$da = \pm\left(-\frac{\partial z}{\partial x}\hat{x} - \frac{\partial z}{\partial y}\hat{y} + \hat{z}\right) dxdy \quad ...(33)$$

If the surface is surface parameterized using u and v, then

$$\phi = \int_S F\cdot(T_u \times T_v)\, dudv \qquad ...(34)$$

Let S be a surface such that any line perpendicular to xy – plane meets S in not more than one point. Then, the equation of the surface S can be written as $z = h(x, y)$.

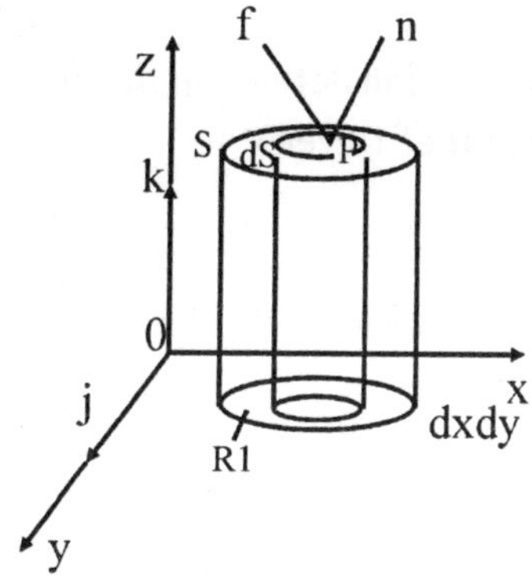

Fig. 1

Let the projection of S on xy– plane = $dS\cos\gamma$,

where γ is the acute angle which the normal n at P to the surface S makes with z- axis.

$$\therefore \qquad dx = \cos\gamma = dxdy$$

Also, $\cos\gamma = |n\cdot k|$, where k is unit vector along z-axis

$$\therefore \qquad dS = \frac{dxdy}{|n\cdot k|} \qquad ...(35)$$

Hence,
$$\iint_S f\cdot n dS = \iint_{R_1} f\cdot n\frac{dxdy}{|n\cdot k|}. \qquad ...(36)$$

Similarly,

$$\iint_S f\cdot n dS = \iint_{R_2} f\cdot n\frac{dydz}{|n\cdot i|} = \iint_{R_2} f\cdot n\frac{dzdx}{|n\cdot j|} \qquad ...(37)$$

Volume Integral

Let be a scalar point function defined throughout a given region of volume V. Divide the given region into n sub-regions of volumes $\delta V_1, \delta V_2, \ldots, \delta V_n$. Let $P(x_i, y_i, z_i)$ be any point inside or on the boundary of the sub-regions of volume δV_i. From the sum $\sum_{i=1}^{n}\phi(P_i)\delta V_i$ where $\phi(P_i) = \phi(x_i, y_i, z_i)$. If as $n\to\infty$ in such a way that each δV_i tends to zero, the sum tends to a finite limit and is independent of the mode of sub-division, then this limit is called the volume integral of ϕ over the volume V and is denoted by

$$\iiint_V \phi \, dV \quad ...(38)$$

If we sub-divide the region of volume V into small a cuboids by drawing lines parallel to the coordinate axes, then $dV = dx\, dy\, dz$ and

$$\iiint_V \phi \, dxdydz \quad ...(39)$$

Likewise if f is a vector point function defined in the given region of volume V, we define and denote the vector volume integral of f over V by

$$\iiint_V f \, dV . \quad ...(40)$$

A triple integral over three coordinates giving the volume within some region G,

$$V = \iiint_G \phi \, dxdydz \quad ...(41)$$

Volume Element

A volume element is the differential element dV whose volume integral over some range in a given coordinate system gives the volume of a solid,

$$V = \iiint_G \phi \, dxdydz \quad ...(42)$$

In $\Re^n$, the volume of the infinitesimal n-hypercube bounded by $dx_1, dx_2, dx_3, \ldots, dx_n$ has volume given by the wedge product

$$dV = dx_1 \wedge dx_2 \wedge dx_3 \wedge \ldots \wedge dx_n$$

The use of the antisymmetric wedge product instead of the symmetric product $dx_1, dx_2, dx_3, \ldots, dx_n$ is a technical refinement often omitted in informal usage. Dropping the wedges, the volume element for curvilinear coordinates in $\Re^3$ is given by

$$dV = \left|(h_1 u_1 du_1)\cdot(h_2 u_2 du_2)\times(h_3 u_3 du_3)\right| \quad ...(43)$$

$$= h_1 h_2 h_3 \, du_1 du_2 du_3 \quad ...(44)$$

$$= \left|\frac{\partial r}{\partial u_1}\cdot\frac{\partial r}{\partial u_2}\times\frac{\partial r}{\partial u_3}\right| du_1 du_2 du_3 \quad ...(45)$$

$$= \begin{vmatrix} \frac{\partial x}{\partial u_1} & \frac{\partial x}{\partial u_2} & \frac{\partial x}{\partial u_3} \\ \frac{\partial y}{\partial u_1} & \frac{\partial y}{\partial u_2} & \frac{\partial y}{\partial u_3} \\ \frac{\partial z}{\partial u_1} & \frac{\partial z}{\partial u_2} & \frac{\partial z}{\partial u_3} \end{vmatrix} du_1 du_2 du_3 \quad ...(46)$$

$$= \left| \frac{\partial(x, y, z)}{\partial(u_1, u_2, u_3)} \right| du_1 du_2 du_3 \quad ...(47)$$

where the latter is the Jacobian and the h_i are scale factors.

Reduction of Volume to Surface Integrals

One of the most important theorems in vector analysis is known as the Divergençe Theorem, which is also sometimes called Gauss' Theorem. This is essentially just an application of the fundamental theorem of calculus

$$\int_a^b \frac{df(x)}{dx} dx = f(b) - f(a) \quad ...(48)$$

This enables us to express the integral of the quantity $\frac{df}{dx}$ along an interval in terms of the values of f itself at the endpoints of that interval. Now suppose we are given a scalar function f (x, y, z) throughout a region V enclosed by a surface S, and we want to evaluate the integral of the quantity $\frac{\partial f}{\partial x}$ over this entire region. This can be written as

$$\int_V \frac{\partial f}{\partial x} dV = \iiint_{xyz} \frac{\partial f}{\partial x} \, dxdydz \quad ...(49)$$

where the three integrals are evaluated over suitable ranges to cover the entire region V. For any fixed values of $y = y_0$ and $z = z_0$ the function $f(x, y_0, z_0)$ is totally dependent on x, so we can evaluate the integral along any line parallel to the x axis through the region V for any particular y_0, z_0 using the fundamental theorem of calculus

$$\int_a^b \frac{df(x, y_0, z_0)}{dx} dx = f(b, y_0, z_0) - f(a, y_0, z_0) \quad ...(50)$$

where a,b are the x values at which the line intersects the surface S. Hence, we can cover all the contributions by integrating the value of the function $f(x, y, z)$ over the entire surface S.

Consequently, we can write our original integral as

$$\int_V \frac{\partial f}{\partial x} dV = \int_S f \cos(n \cdot i) \, dS \quad (51)$$

where cos (n, i) denotes the cosine of the angle between a unit vector normal to the surface and a basis vector in the positive x direction. Naturally, since the coordinate axes are symmetrical, we could arrive at analogous results with x replaced by either y or z, and i replaced by j or k respectively.

Now suppose we are given a vector field $F(x, y, z)$ with the components $F_x(x, y, z)$, $F_y(x, y, z)$, and $F_z(x, y, z)$. Using the above results and simple additivity we have

$$\int_V \left(\frac{\partial F_x}{\partial x} + \frac{\partial F_y}{\partial y} + \frac{\partial F_z}{\partial z} \right) dV = \int_S [F_x \cos(n \cdot i) + F_y \cos(n \cdot j) + F_z \cos(n.k)] dS \quad ...(52)$$

The integrand on the left side is $\nabla \times F$, i.e. the divergence of F. Also, notice that $\cos(n \cdot i)$, $\cos(n \cdot j)$, and $\cos(n \cdot k)$ are the components of the normal unit vector n, so the integrand on the right side is simply $F \times n$, i.e., the dot product of F and the unit normal to the surface. Hence we can express the Divergence Theorem in its familiar form

$$\int_V \nabla \cdot F dV = \int_S F \cdot n \, dS \quad ...(53)$$

Gauss's Divergence Theorem

Statement: The surface integral of the normal component of a vector f taken over a closed surface is equal to the integral of the divergence of f taken over the volume enclosed by the surface.

Symbolically: If f is a continuously differentiable vector point function and S is closed surface enclosing the volume V with an outward pointing unit normal, n, the Divergence Theorem says:

$$\iint_S F \cdot n \, dS = \iiint_V div \text{ F dV} \quad ...(54)$$

Or

$$\iint_S F \cdot n \, d\sigma = \iiint_V div \text{ F dV} \quad ...(55)$$

We take $F = Mi + Nj + Pk$

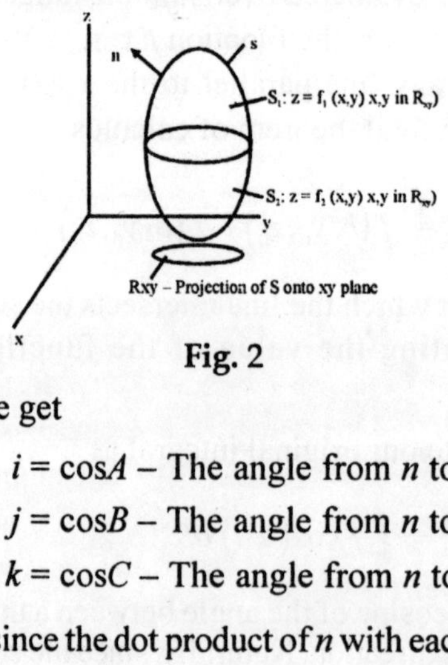

Fig. 2

Using the above figure, we get

$n \cdot i = \cos A$ – The angle from n to the x-axis

$n \cdot j = \cos B$ – The angle from n to the y-axis

$n \cdot k = \cos C$ – The angle from n to the z-axis

$n = \cos A i + \cos B j + \cos C k$ since the dot product of n with each basis vector is that component of n.

Now we restate the Divergence Theorem using the components of the vector field F.

$$\iint_S [M \cos A + N \cos B + P \cos C] \, d\sigma = \iiint_V \left(\frac{\partial M}{\partial x} + \frac{\partial N}{\partial y} + \frac{\partial P}{\partial z} \right) dV \quad ...(56)$$

Taking just the z terms

$$\iint_S P\cos C \, d\sigma = \iiint_V \frac{\partial P}{\partial z} dV \quad ...(57)$$

Splitting into integrals over the two surfaces S_1 and S_2

$$\iint_{S_1} P(x, y, f_1(x, y)) \cos C(x, y, f_1(x, y))\, dxdy$$
$$+ \iint_{S_2} P(x, y, f_2(x, y)) \cos C(x, y, f_2(x, y))\, dxdy \quad ...(58)$$

Project onto R_{xy} $dS = \dfrac{dA}{\|n \cdot k\|} = \dfrac{dA}{|\cos C|}$ ...(59)

On top(S_1) $\cos C \geq 0 \Rightarrow |\cos C| = \cos C$ ...(60)

On bottom (S_2) $\cos C \leq 0 \Rightarrow |\cos C| = -\cos C$ ...(61)

$$\iint_{R_{xy}} P(x, y, f_1(x, y)) \cos C \frac{dA}{\cos C} - \iint_{S_2} P(x, y, f_2(x, y)) \cos C \frac{dA}{\cos C}$$

$$= \iint_{R_{xy}} \left\{ \int_{f_2(x,y)}^{f_1(x,y)} \frac{\partial P(x, y, z)}{\partial z} dz \right\} dA \quad ...(62)$$

The last step involves taking the integral of the derivative of P. This is like multiplying by one. It doesn't change anything, you can do it anytime.

Looking at the last term, what we really have is a volume integral. Evaluating the inner integral, we have $[f_1(x, y) - f_2(x, y)] \times dA$. This is the volume of a differential element of the region. Adding all of them up gives us the volume of the region:

$$\iiint_V \frac{\partial P}{\partial V} dV$$

The other terms work out in a similar manner. If the region is not a convex region, we can construct it by placing convex regions next to each other and summing up the results. Thus, we have proved the Divergence Theorem for a large number of regions

Deduction from Gauss's Theorem

1. $\iiint_V f \cdot \nabla\phi \, dV = \iint_S \phi f \cdot n dS - \iiint_V \phi \operatorname{div} f \, dV$ ...(63)

2. $\iint_S f \times n \, dS = -\iiint_V curl f \, dV$ ...(64)

3. $\iint_S \phi \, n \, dS = \iiint_V grad \phi \, dV$ ...(65)

GREEN'S THEOREM

Let S be a sufficiently nice region in the plane, and let δS be its boundary; then we have

$$\iint_S \left(\frac{\partial v_2}{\partial x} - \frac{\partial v_1}{\partial y}\right) dxdy = \oint_{\delta S} \left(v_1 dx + v_2 dy\right) \quad ...(66)$$

where the boundary, δS is traversed counterclockwise on its outside cycle, (and clockwise on any internal cycles).

Statement: If ϕ and ψ are two scalar point functions having continuous second order derivatives in a region V bounded by a closed surface S, then

$$= \iiint_V \left(\phi\nabla^2\psi - \psi\nabla^2\phi\right) dv = \iint_S \left(\phi\nabla\psi - \psi\nabla\phi\right)\cdot n\, dS \quad (67)$$

Proof: By Gauss divergence theorem, we have

$$\iiint_V \nabla\cdot f dv = \iint_S f\cdot n\, dS \quad (68)$$

Putting $f = \phi\nabla\psi$, we get

$$\begin{aligned} \nabla\cdot f &= \nabla\cdot\left(\phi\nabla\psi\right) = \phi\left(\nabla\cdot\nabla\psi\right) + \left(\nabla\phi\right)\cdot\left(\nabla\psi\right) \\ &= \phi\left(\nabla^2\psi\right) + \left(\nabla\phi\right)\cdot\left(\nabla\psi\right) \end{aligned}$$

Also, $\quad f\cdot n = \left(\phi\nabla\psi\right)\cdot n$

$\therefore$ (68), becomes

$$\iiint_V \left[\phi\nabla^2\psi + \left(\nabla\phi\right)\cdot\left(\nabla\psi\right)\right] dv = \iint_S \left(\phi\nabla\psi\right)\cdot n\, dS \quad ...(69)$$

This is called the Green's first theorem (or identity)

Interchanging ϕ and ψ in (69), we get

$$\iiint_V \left[\psi\nabla^2\phi + \left(\nabla\psi\right)\cdot\left(\nabla\phi\right)\right] dv = \iint_S \left(\psi\nabla\phi\right)\cdot n\, dS \quad ...(70)$$

Operate (70) – (71) we get,

$$\iiint_V \left(\phi\nabla^2\psi - \psi\nabla^2\phi\right) dv = \iint_S \left(\phi\nabla\psi - \psi\nabla\phi\right)\cdot n\, dS \quad ...(71)$$

This is called the Green's second theorem (or identity)

Hence, the Green's theorem can also be stated as

$$\iiint_V \left(\phi\nabla^2\psi - \psi\nabla^2\phi\right) dv = \iint_S \left(\phi\frac{\partial\psi}{\partial n} - \psi\frac{\partial\phi}{\partial n}\right) dS \quad ...(72)$$

Where $\frac{\partial \psi}{\partial n}$ and $\frac{\partial \phi}{\partial n}$ denote the directional derivative of ϕ and ψ respectively along the outward drawn normal at any point of *S*.

Other forms of Green's Theorem

Green's theorem is a vector identity which is equivalent to the curl theorem in the plane. Over a region *D* in the plane with boundary ∂D, Green's theorem states

$$\int_{\partial D}\left[f(x,y)dx + g(x,y)dy\right] = \iint_D \left(\frac{\partial g}{\partial x} - \frac{\partial f}{\partial y}\right) dxdy \qquad ...(73)$$

where the left side is a line integral and the right side is a surface integral. This can also be written compactly in vector form as

$$\int_{\partial D} F \cdot dS = \iint_D (\nabla \times F) \cdot \mathrm{da} \qquad ...(74)$$

If the region *D* is on the left when traveling around ∂D, then area of *D* can be computed using the elegant formula

$$\mathrm{A} = \frac{1}{2}\int_{\partial D} xdy - ydx \qquad ...(75)$$

giving a surprising connection between the area of a region and the line integral around its boundary. For a plane curve specified parametrically as $(x(t), y(t))$, equation (75) becomes

$$A = \frac{1}{2}\int\left(x\frac{dy}{dt} - y\frac{dx}{dt}\right)dt$$

$$A = \frac{1}{2}\int(xy' - yx')\,dt \qquad ...(76)$$

which gives the *signed* area enclosed by the curve.

GREEN'S THEOREM FOR AREA IN PLANE RELATING TO ITS BOUNDING CURVE

The analog of the "*Fundamental Theorem of Differential and Integral Calculus*" for a region $\Re$ bounded in a plane with normal *k* that is bounded by a curve $\partial \Re$ is:

$$\iint_R \left(\nabla \times \vec{F}\right) \cdot \hat{k}\; dxdy = \oint_{\partial R} \vec{F} \cdot d\vec{r} \qquad ...(77)$$

Within the domain, the rotational flow (Ñ′ *F*) from one cell moves into its neighbors; however, at the edges the local rotation is a net loss or gain. The local net loss or gain is

$$\vec{F} \cdot (dx, dy).$$

The generalization of this idea to a surface $\partial\Re$ bounding a domain $\Re$ results in Stokes' theorem.

Reduction of Surface Integral to Line Integral

Let the region R on the surface be break up into tiny pieces each one looking planar, apply Green's theorem on each and add up the mess to get the same theorem on all of R. The result is called *Stokes' Theorem.*

Theorem. With R an arbitrary reasonably smooth and nice region on a surface that is locally planar, and with δ R its boundary, we have

$$\iint_R \overline{\nabla \times v dS} = \oint_{\delta R} \nabla \cdot d\bar{l} \qquad ...(78)$$

We can use it as a tool for rewriting integrals of one kind as integrals of another and for proving peculiar facts about irrotational fields (those for which curl = 0).

It can be stated in the following form:The circulation integral of v over a bounding curve is the flux integral of the curl of v over any surface it bounds.

Notice that this theorem immediately implies that if R is simply connected and curl $v = 0$ throughout it, then the potential f (with $v = \nabla f$) defined as path integral is path independent.

Stokes' Theorem

The divergence theorem is used to find a surface integral over a closed surface and Green's theorem is use to find a line integral that encloses a surface (region) in the xy-plane. This theorem relates the surface integral to a line integral.

Let S be a oriented surface with unit normal vector N and let C be the boundary of S. Then C is *positively oriented* if its orientation follows the right hand rule, that is if you right hand curls around N in the direction of C's orientation, then your thumb will be pointing in the direction of N.

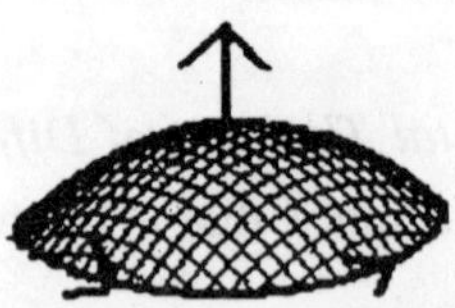

Fig. 5

Statement: Let S be an oriented surface with unit normal vector n and C be the positively oriented boundary of S. If F is a vector field with continuous first order partial derivatives then

$$\int_C F \cdot dr = \iint_S (Curl F) \cdot n dS \qquad ...(79)$$

In other words: The line integral of the tangential component of a vector f taken around a simple closed curve *C* is equal to the surface integral of the normal component of the curl of f taken over any surface *S* having *C* as its boundary.

Cartesian form of Stokes' Theorem for plane

Let us take a system of Cartesian rectangular coordinate axes such that the plane of the given surface *S* is the *xy*-plane and *z*-axis lies along the direction of the normal vector *n*.

Here, the normal vector is constant

$$\text{Let } F = F_i i + F_2 j + F_3 k$$

$$\therefore F \cdot dr = (F_i i + F_2 j + F_3 k) \cdot (dxi + dyj + dzk)$$

$$\Rightarrow F \cdot dr = F_i dx + F_2 dy + F_3 dz$$

Now,

$$\iint_S (Curl F) \cdot n dS = \iint_S (Curl F) \cdot k dS$$

$$= \iint_S \left(\frac{\partial F_2}{\partial x} - \frac{\partial F_1}{\partial y} \right) dxdy \qquad ...(80)$$

$\therefore$ Stokes' Theorem for plane becomes

$$\oint_C (F_i dx + F_2 dy) = \iint_S \left(\frac{\partial F_2}{\partial x} - \frac{\partial F_1}{\partial y} \right) dxdy \qquad ...(81)$$

Let us, now prove it .

Let the region *S* be such that any line parallel to either axis meet the boundary *C* in at the most two points.

Let the region be $x = a, y = \phi_1(x), y = \phi_2(x)$, where $\phi_2(x) \geq \phi_1(x)$.

The boundary *C* is split into two arc C_1 and C_2 as shown.

$$\therefore \iint_S \frac{\partial F_1}{\partial y} dxdy = \int_a^b \left[\int_{\phi_1}^{\phi_2} \frac{\partial F_1}{\partial y} dy \right] dx$$

$$= \int_a^b \left[F_1(x, \phi_2(x)) - F_1(x, \phi_1(x)) \right] dx$$

$$= -\iint_{C_2} F_1(x, y) dx - \iint_{C_1} F_1(x, y) dx$$

$$= -\oint_C F_1(x, y) dx \qquad ...(82)$$

$$\text{Similarly, } \therefore \iint_S \frac{\partial F_2}{\partial y} dxdy = -\oint_C F_2(x, y) dx \qquad ...(83)$$

∴ From (82) and (83), we get

$$\oint_C \left(F_i dx + F_2 dy\right) = \iint_S \left(\frac{\partial F_2}{\partial x} - \frac{\partial F_1}{\partial y}\right) dxdy$$

Hence the result.

If the region S is such that a line parallel to either axes meets the boundary C in more than tow points, we divide S into a finite number of sub-regions such that the boundary of each is met in at the most two points by any line parallel to either axis. By applying the above result to each sub-region and then adding the results, we find that the theorem is true for S. This because the line integrals along the boundary curves will cancel in pairs.

Cartesian form of Stokes' Theorem in space

Let $f = f_1 i + f_2 j + f_3 k$. Let the outward drawn normal vector n makes angles α, β, γ with the positive direction of x, y, z axes respectively so that

$$n = \cos\alpha i + \cos\beta j + \cos\gamma k$$

Then,
$$Ñ' f = \begin{vmatrix} i & j & k \\ \frac{\partial}{\partial x} & \frac{\partial}{\partial y} & \frac{\partial}{\partial z} \\ f_1 & f_2 & f_3 \end{vmatrix} \quad ...(84)$$

$$= \left[i\left(\frac{\partial f_3}{\partial x} - \frac{\partial f_2}{\partial z}\right) + j\left(\frac{\partial f_1}{\partial z} - \frac{\partial f_3}{\partial x}\right) + k\left(\frac{\partial f_2}{\partial x} - \frac{\partial f_1}{\partial y}\right)\right]$$

$$(\nabla \times f)\cdot n = \left[i\left(\frac{\partial f_3}{\partial x} - \frac{\partial f_2}{\partial z}\right) + j\left(\frac{\partial f_1}{\partial z} - \frac{\partial f_3}{\partial x}\right) + k\left(\frac{\partial f_2}{\partial x} - \frac{\partial f_1}{\partial y}\right)\right] \cdot (i\cos\alpha + j\cos\beta + k\cos\gamma)$$

$$= \left[\left(\frac{\partial f_3}{\partial x} - \frac{\partial f_2}{\partial z}\right)\cos\alpha + \left(\frac{\partial f_1}{\partial z} - \frac{\partial f_3}{\partial x}\right)\cos\beta + \left(\frac{\partial f_2}{\partial x} - \frac{\partial f_1}{\partial y}\right)\cos\gamma\right] \quad ...(85)$$

Also, $f \cdot dr = \left(f_1 i + f_2 j + f_3 k\right)\cdot\left(dxi + dyj + dzk\right)$

$$= f_1 dx + f_2 dy + f_3 dz$$

∴ Stokes' Theorem becomes

$$\int_C \left(f_1 dx + f_2 dy + f_3 dz\right)$$

$$= \iint \left[\left(\frac{\partial f_3}{\partial x} - \frac{\partial f_2}{\partial z}\right)\cos\alpha + \left(\frac{\partial f_1}{\partial z} - \frac{\partial f_3}{\partial x}\right)\cos\beta + \left(\frac{\partial f_2}{\partial x} - \frac{\partial f_1}{\partial y}\right)\cos\gamma\right] dS \quad ...(86)$$

Curl and Circulation

Just as the divergence theorem assisted us in understanding the divergence of a function at a point, Stokes' theorem helps us understand what the Curl of a vector field is. Let P be a point on the surface and C_e be a tiny circle around P on the surface.

$$\int_C F \cdot dr$$

The measures the amount of circulation around p. You can see this by noticing that if F flows in the direction of the tangent vector, then $F \cdot dr$ will be positive. If it flows in the opposite direction, then it will be negative. The stronger the force field in the direction of the tangent vector, the greater the circulation.

Since the region enclosed by C_e is tiny, the surface integral can be approximated by

$$\iint_S (CurlF \cdot n)dS \approx CurlF(P) \cdot n(\text{Area of C}) \qquad ...(87)$$

or

Curl $F.n$ = Circulation per unit area

ILLUSTRATIVE EXAMPLES

(i)

1. *Evaluate* $\int\int_S F \bullet \hat{n}\ ds$ *where S is the surface of the sphere* $x^2+y^2+z^2=16$ *and* $F=3xi+4yj+5zk$

Sol. By Gauss's Divergence theorem, we have $\int\int_S F \bullet \hat{n}\ ds = \int\int\int_V \nabla \bullet F\ dv$

$$= \int\int\int_V \left[\frac{\partial}{\partial x}(3x) + \frac{\partial}{\partial z}(4y) + \frac{\partial}{\partial z}(5z)\right] dV = \int\int\int_V (3+4+5)dV = 12\int\int\int_V dv$$

= 12 V, where V is the voloume of the given sphere.$=12.\frac{4}{3}\pi$. Ex.2. *Prove that*

2. *Prove that* $\iiint_V \frac{1}{r^2} dV = \iint_S \frac{r \bullet \hat{n}}{r^2} ds$

where the symbols have their usual meanings.

Sol. By Gauss's Divergence theorem $\iiint_V \nabla \bullet F dV = \iint_S F \bullet \hat{n} ds$

Here $F = \frac{r}{r^2}$ But $\nabla \bullet F = \nabla \bullet \frac{r}{r^2} = div\left(\frac{r}{r^2}\right) = \frac{1}{r^2} div$

$$r + grad\left(\frac{r}{r^2}\right) \bullet r = \frac{3}{r^2} + \left(\frac{-2}{r^4}r\right) \bullet r$$

$$\left[\because grad\, r^n = nr^{n-2}r \because gradr^{-2} = \frac{-2}{r^4}r\right]$$

$$= \frac{3}{r^2} - \frac{2}{r^4}r^2 = \frac{1}{r^2}$$

$$= \frac{3}{r^2} - \frac{2}{r^4}(r \bullet r)$$

Then (1) implies $\iiint_V \frac{1}{2} dv = \iint_S \frac{r}{r^2} \bullet \hat{n}\, ds.$

3. *Prove that* $\iiint_V (u\nabla^2 w - w\nabla^2 u) dv = 3 \iiint_S (u\nabla w - w\nabla u) \bullet ds.$

Sol. By divergence theorem, we know that $\iiint_V \nabla \bullet F\, dv = \iint_S F \bullet \hat{n}\, ds$

Now let $F = u\nabla w. \therefore \iiint_V \nabla \bullet \{u\nabla w\} dv \iint_S (u\nabla w) \bullet \hat{n}\, ds.$

But we already know that

$$\nabla \bullet (u\nabla w) = u(\nabla \bullet \nabla w) + (\nabla u) \bullet (\nabla w) = u\nabla^2 w + (\nabla u) \bullet (\nabla w).$$

Putting this value of $\nabla \bullet (u\nabla w)$ in (1), we

get $*\iiint_V \{u\nabla^2 w + (\nabla u) \bullet (\nabla w)\} dv = \iint_S (u73\tilde{o}w) \bullet ds.$

Result (2) is known as Green's first identity or theorem.

Now interchanging *u* and *w* in (2,) we get

$$\iiint_V \{w\nabla^2 u + (\nabla w) \bullet (\nabla u)\} dv = \iint_S (w\nabla u) \bullet ds.$$

Subtracting (3) from (2), we get $\iiint_V \{u\nabla^2 w - w\nabla^2 u\} dv = \iint_S \{u\nabla w - w\nabla u\} \bullet ds.$

This is called Green's second identity or symmetrical theorem.

4. *Verify Gauss's Theorem and show that* $\int_S \left[(x^3 - yz)i - 2x^2 yj + 2k\right] \bullet \hat{n}\, ds = \frac{1}{3}a^5,$

where S denotes the surface of the cube bounded by the planes x=0, x=a, y=0 y=a, z=0, z=a.

Sol. Let $F=(x^3-yz)i-2x^2yj+2k$. $\therefore$ div

$$A=\nabla\bullet F=\left(i\frac{\partial}{\partial x}+j\frac{\partial}{\partial y}+k\frac{\partial}{\partial z}\right)\bullet\cdot\ \{(x^3-yz)i-2x^{3}{}'yj+2k\}$$

$$=\frac{\partial}{\partial x}(x^3-yz)+\frac{\partial}{\partial y}(-2x^2y)+\frac{\partial}{\partial z}(2)=3x^2-2x^2=x^2.$$

Now applying Gauss's divergence Theorem, we have

$\int F\bullet\hat{n}\,ds=\int \textbf{div}\,F dv=\int\int\int x^2 dx\,dy\,dz,$

{using (A)}

$=\left[\int_a^0 x^2dx\right]\left[\int_a^0 dy\right]\left[\int_a^0 dz\right]=\frac{1}{3}a^2.a.a=\frac{1}{3}a^5$ Verification.On the faces *PQRS*

$\int F\bullet\hat{n}\,ds=\int\int\{(x^3-yz)i-2x^2yj+2k\}\bullet i dy\,dz$ $(\because\hat{n}=i)$

$=\int_0^a\int_0^a(x^3-yz\,dy\,dz)=\int_{z=0}^a\int_{y=0}^a(a^3-yz)dy\,dz,$ $(\because i\cdot k=0\,\textbf{etc.})(\because x=0)$

$$=a^3\int_0^a dy\int_0^a dz-\int_0^a ydy\int_0^a zdz=a^3\cdot a\cdot a-\frac{1}{2}a^2\frac{1}{2}a^2=a^2-\frac{1}{4}a^4$$

On the face *OQ'R'R* $\int F\bullet\hat{n}ds=\int\int(x^3-yz)i-2x^2j+2k)\bullet(-idy\,dz)73$ $(\because\hat{n}=-i)$

$=\int_{y=0}^a\int_{z=0}^a(x^3-yz)dy\,dz=-\int_0^a\int_0^a yz\,dy\,dz,$ $(\because x=o)$

$$=\frac{1}{2}a^2.a^2=\frac{1}{4}a^4.$$

On the face *QQ'R'R'* $\int F\bullet\hat{n}\,ds=\int\int\int\{x^3-yz)i-2x^2yj=2k\}\bullet jdxdz$ $(\because\hat{n}=j)$

$=-\int_{y=0}^a\int_{z=0}^a 2x^2ydx\,dz=-2a\int_0^a x^2dx\int_0^a dz$ $(\because j\cdot k=0=i\cdot j;\because y=a)$

$$=-2a\left(\frac{1}{2}a^3\right)(a)=-\frac{2}{3}a^5.$$

$\int F\bullet\hat{n}ds=\int\int\int\{x^3-yz)i-2x^2yj+2k\}\bullet(-jdx\,dz)$OS'SP $(\because\hat{n}=-j)$

$=\int_0^a\int_0^a 2x^2ydx\,dz=0,(\because y=0)$ $(\because\hat{n}=k)$

$$\int_{RR'S'S} F \bullet \hat{n}\, ds = \iint \{(x^3 - yz)2 - 2x^2 yj - 2k\{\bullet k\, dx\, dy = \int_{x=0}^{a}\int_{y=0}^{a} 2dx\, dy = 2a(a) = 2a^2$$

$$\int_{OPQQ'} F \bullet \hat{n}\, ds = \iint \{(x^3 - yz)i - 2x^2 yj + 2k\} \bullet (-kdx\, dy) \qquad (\because \hat{n} = -k)$$

$$= -\int_o^a \int_0^a 2dx\, dy = -2(a)(a) = -2a^2$$

get $\int_s F \bullet \hat{n}\, ds = a^5 - \frac{1}{4}a^4 + \frac{1}{4}a^4 - \frac{2}{3}a^5 + 0 + 2a^2 = \frac{1}{3}a^5$.

So the theorem is verified by the ddirect integration method.

EXERCISE

1. Evaluate $\int_S F.Nds$ where F = 6z I – 4 J + y K and S is the portion of the plane 2x + 3y + 6z = 12 in the first octant.

2. Verify Green's theorem for $\iint_C \left[\left(3x - 8y^2\right)dx + \left(4y - 6xy\right)dy\right]$ where C is the boundary of the region bounded by x = 0, y = 0 and x + y = 1.

3. Applying Green's theorem, evaluate $\iint_C \left[(y - \sin x)dx + \cos x dy\right]$ where C is the plane triangle enclosed by the lines y = 0, x = p/2 and y = $\frac{2}{\pi}x$.

4. Apply Green's theorem to prove that the area enclosed by a plane curve is $\frac{1}{2}\int_C (xdy - ydx)$. Hence find the area of an ellipse whese semi-major and semi-minor axes are of lengths a and b.

5. Verify Stoke's theorem for the vector field F = $(s^2 - y^2)$I + $2xy$J
   (i) Integrated round the rectangle in the plane $z = 0$ and bounded by the lines $x = 0$, $y = 0$, $x = $ a and y = b.
   (ii) Over the box bounded by the planes x = 0, $x =$ a; $y = 0$, $y =$ b; $z = 0$, $z =$ c; if the face $z =$ o is cut.

6. Verify Stoke's theorem for a vector field defined by F = $-y^3$I + x^3J, in the region $x^2 + y^2 \le 1$, z = 0.

7. Verify Stoke's theorem for the vector field $F = (2x - y)I - yz^2J - y^2zK$ over the upper half surface of $x^2 + y^2 + z^2 = 1$, bounded by its projection on the xy-plane.
8. Verify divergence theorem for $F = 4xzI - y^2J + yzK$, taken over the cube bounded by $x = 0, x = 1; y = 0, y = 1; z = 0, z = 1$.
9. Verify Gauss divergence theorem for the function $F = yI + xJ + z^2K$ over the cylindrical region bounded by $x^2 + y^2 = 0$, $z = 0$ and $z = 2$.
10. Using divergence theorem, prove that

    (i) $\int_S R.dS = 3V.$

    (ii) $\int_S \nabla r^2.dS = 6V.$

    where S is any closed surface enclosing a volume V and $r^2 = x^2 + y^2 + z^2$.
11. Using divergence theorem, evaluate $\int_S R.dS$ where S is the surface of the sphere $x^2 + y^2 + z^2 = 0$.
12. Use divergence theorem to evaluate $\int_S F.dS$, where $F = x^3I + + y^3J + y^3K$, and s is the surface of the sphere $x^2 + y^2 + z^2 = a^2$.

Chapter 3

Application of Vector Calculus

INTRODUCTION

The divergence and Stokes' theorems (and their related results) supply fundamental tools which can be used to derive equations which can be used to model a number of physical situations. Essentially, these theorems provide a mathematical language with which to express physical laws such as conservation of mass, momentum and energy. The resulting equations are some of the most fundamental and useful in engineering and applied science.

In the following sections the derivation of some of these equations will be outlined. The goal is to show how vector calculus is used in applications. Generally speaking, the equations are derived by first using a conservation law in integral form, and then converting the integral form to a diûerential equation form using the divergence theorem, Stokes' theorem, and vector identities. The diûerential equation forms tend to be easier to work with, particularly if one is interested in solving such equations, either analytically or numerically.

THE HEAT EQUATION

Consider a solid material occupying a region of space V. The region has a boundary surface, which we shall designate as S. Suppose the solid has a density ñ and a heat capacity c. If the temperature of the solid at any point in V is T (r, t), where

$$\vec{r} = x\hat{i} + y\hat{j} + z\hat{k} \qquad ...(1)$$

is the position vector (so that T depends upon *x*, *y*, *z* and *t*), then the total heat energy contained in the solid is

$$\iiint_V \rho CTdV \qquad ...(2)$$

Heat energy can only get in or out of the region V by flowing across the boundary S. Let's suppose that the heat flux vector is called $\vec{q}$. This vector measures rate of energy flow past a point per unit area. Thus, if we have a small element of surface area d S with outward unit normal vector n, the rate of energy flow outward through this element of surface area is

$$\vec{q} \cdot \hat{n} dS \qquad ...(3)$$

Integrating over the entire surface, the total rate of energy flow (i.e., the flux) out of the region is therefore

$$\oiint_S \vec{q} \cdot \hat{n} dS \qquad ...(4)$$

Conservation of energy then requires that the rate at which the energy within the region V and the rate at which energy crosses the boundary of region must balance, i.e.,

$$\frac{d}{dt}\left(\iiint_V \rho CT dV \right) + \oiint_S \vec{q} \cdot \hat{n} dS = 0 \qquad ...(5)$$

The signs of the terms must appear in this manner, because if the flux of energy out of the region is positive, then the total energy inside the region must be decreasing, hence its derivative will be negative.

To simplify this, we take the time derivative inside the volume integral.

$$\frac{d}{dt} \iiint_V \rho CTV dV = \iiint_V \frac{\partial}{\partial t}(\rho CT) dV \qquad ...(6)$$

We can do this because the region V over which the integral is being taken is independent of time; V is the same region for all time. Thus, at any fixed time, integration over V just gives a single number, i.e., the energy inside V at that time. The energy will change with time because T (and possibly ñ and c) changes with time. Note that the time derivative must be turned into a partial derivative when it is moved inside the volume integral, however, because the functions being integrated depend upon both position and time. Similarly, we apply the divergence theorem to the flux integral

$$\oiint_S \vec{q} \cdot \vec{n} dS = \iiint_V \vec{\nabla} \cdot \vec{q} dV \qquad ...(7)$$

The result is

$$\iiint_V \frac{\partial}{\partial t}(\rho CT) dV + \iiint_V \vec{\nabla} \cdot \vec{q} dV \qquad ...(8)$$

Or equivalently,

$$\iiint_V \left[\frac{\partial}{\partial t}(\rho CT) + \vec{\nabla} \cdot \vec{q} \right] dV = 0 \qquad ...(9)$$

It would be nice to say at this point that the quantity in brackets is zero; this would allow us to get rid of the integral. This would be true, but the right argument is needed, because an integral can be zero either because the integrand is zero or because it is equally positive and negative, and when integrated there is perfect cancellation.

In the present case, however, there is one more fact: the integral must be zero for arbitrary volumes V. This forces the integrand to be zero everywhere. The proof is by contradiction. First, assume that the integrand is not zero everywhere. Pick a point at which it is positive, and then choose a region V that is entirely within the region where it's positive. Since we are integrating a quantity that is entirely positive, the resulting integral must be zero. This can't be true, however, since we know the integral must be zero. Therefore, we must have

$$\frac{\partial}{\partial t}(\rho CT) + \vec{\nabla} \cdot \vec{q} = 0 \qquad ...(10)$$

This is the partial diûerential equation form of the fundamental principle of energy conservation for heat transfer.

More can be done if one has an explicit result for the heat flux _q in terms of the temperature. In many cases, observations show that the heat flux is given by Fick's law,

$$\vec{q} = -k\vec{\nabla}T \qquad ...(11)$$

i.e., the heat transfer is proportional to the negative of the temperature gradient. Thus, heat tends to flow from hotter regions to colder regions. While this is true in most cases, it is not always true; in particular, if phase transitions are possible it can be violated.

If we use Fick's law in the equation for conservation of heat energy, it becomes

$$\frac{\partial}{\partial t}(\rho CT) = \vec{\nabla} \cdot \left(k\vec{\nabla}T\right) \qquad ...(12)$$

Finally if r, *C* and *k* are all constants, this equation simplifies to the heat equation.

$$\frac{\partial T}{\partial t} = k\nabla^2 T \qquad ...(13)$$

where

k = *k*/r*C* ...(14)

is the thermal diûusivity.

A special case of this equation results if the object is in thermal equilibrium, i.e., the temperature doesn't change with time:

Ñ^2 T = 0 ...(15)

This is known as Laplace's equation.

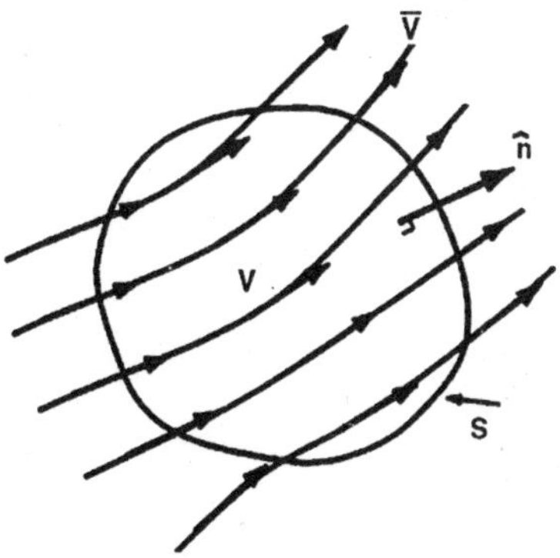

Fig. 1 Schematic Diagram Indicating the Region V, the Boundary Surface S, the Normal to the Surface ˆn, the Fluid Velocity Vector Field *v*, and the Particle Paths (Dashed Lines).

FLUID MECHANICS

Conservation of mass

Conservation laws are also the basis for the equations of fluid mechanics. Let ñ be the fluid density and $\vec{v}$ be the fluid velocity. Then the mass in a region V is given by

$$\iiint_V \rho dV \qquad ...(16)$$

and the rate at which mass leaves the region through its boundary S is

$$\oiint_S \rho\vec{v}\cdot\vec{n}dS \qquad ...(17)$$

see Fig. 2 for a diagram of the situation. Here ñ *v* is the mass flux vector; this has units of kg/m^3 × m/sec = kg/m^{2}sec, i.e., mass per unit area per second.

The rate of change of the mass in V must be balanced by the rate at which mass leaves through S,

$$\frac{d}{dt}\iiint_V \rho dV + \oiint_S \rho\vec{v}\cdot\vec{n}dS = 0 \qquad ...(18)$$

Pulling the time derivative inside the V integral, and converting the surface integral into a volume integral using the divergence theorem, we have

$$\iiint_V \left[\frac{\partial\rho}{\partial t} + \vec{\nabla}\cdot(\rho\vec{v})\right] dV = 0 \qquad ...(19)$$

As before, because the region V is arbitrary, we must have the terms between the brackets be identically zero, which gives

$$\frac{\partial\rho}{\partial t} + \vec{\nabla}\cdot(\rho\vec{v}) = 0 \qquad ...(20)$$

This is the partial diûerential equation form of conservation of mass.

Equation (20) can be simplified somewhat by using the vector identity

$$\vec{\nabla}\cdot(\rho\vec{v})=\vec{\nabla}\rho\cdot\vec{v}+\rho\vec{\nabla}\cdot\vec{v}=\vec{v}\cdot\vec{\nabla}\rho+\rho\vec{\nabla}\cdot\vec{v}$$

Which gives.

$$\frac{\partial\rho}{\partial t}+\vec{v}\cdot\vec{\nabla}\rho+\rho\vec{\nabla}\cdot\vec{v}=0 \qquad ...(21)$$

The combination $\frac{\partial}{\partial t}+\vec{v}\cdot\vec{\nabla}$ occurs so much in fluid mechanics that it is given a special notation,

$$\frac{\partial}{\partial t}+\vec{v}\cdot\vec{\nabla}\equiv\frac{D}{Dt}$$

This is called the material or substantial derivative, and physically it represents taking the time derivative 'following the fluid', i.e., taking the time derivative of something transported along by the fluid. One can see this by considering ñ(x, t) not at a fixed position *x*, but instead ñ(x(t), t), where d*x*/d*t* = *v*; because the position _x moves at the fluid velocity *v*, this is the density of a specific fluid element. Taking the time derivative using the chain rule,

$$\rho(\vec{x},t)=\vec{\nabla}\rho\cdot\frac{d\vec{x}}{dt}+\frac{\partial\rho}{\partial t}=\frac{\partial\rho}{\partial t}+\frac{d\vec{x}}{dt}\cdot\vec{\nabla}\rho=\frac{\partial\rho}{\partial t}+\vec{v}\cdot\vec{\nabla}\rho=\frac{D\rho}{Dt}$$

In any event, with this notation we have

$$\frac{D\rho}{Dt}+\rho\vec{\nabla}\cdot\vec{v}=0$$

This has a nice physical interpretation: if $\vec{\nabla}\cdot\vec{v}>0$, so that the fluid velocity is locally expanding, then Dñ/Dt < 0, i.e., the local density is decreasing. Similarly, if $\vec{\nabla}\cdot\vec{v}<0$, so the fluid velocity is locally compressing, then Dñ/Dt > 0 and the density is increasing.

We can also consider the case of an incompressible fluid : if ñ is constant, then Dñ/Dt = 0 and therefore $\vec{\nabla}\cdot\vec{v}=0$

Water is an example of a fluid that is almost incompressible.

CONSERVATION OF MOMENTUM

The next main conservation law is momentum. Actually, of course, the law is that the time rate of change of momentum is equal to force, d(momentum)/dt = force. To apply it to a fluid, we need to identify all of the sources of momentum and all of the forces involved.

First, the momentum density (per unit volume) of the fluid is ñv. Therefore, the momentum in a volume V of fluid is

$$\iiint_V \rho\vec{v}dV$$

Of course, what we really need is the time rate of change of this momentum,

$$\frac{d}{dt}\iiint_V \rho\vec{v}dV$$

Another way that the momentum in V can change is if the fluid transports momentum out of the region across the boundary S .

$$\oiint_S \rho\vec{v}(\vec{v}\cdot\vec{n})dS$$

This is similar to the previous case of conservation of mass, where we had mass density times the normal component of the velocity, giving mass flux. In this case, we have momentum density times the normal component of the velocity, giving momentum flux.

We also need the forces acting on V. First of all, we have the net pressure force, which is

$$-\oiint_S p\vec{n}dS$$

since the pressure acts normally to each element of the surface (with an inward force when the pressure is positive, hence the minus sign). Similarly, if we have any body forces, such as gravity, we have to include them. It's traditional to use $\vec{f}$ as the body force per unit mass, so that

$$\iiint_V \vec{f}\,\vec{v}dV$$

is the total body force.

Putting everything together, we have conservation of momentum:

$$\frac{d}{dt}\iiint_V \rho\vec{v}dV+\oiint_S \rho\vec{v}(\vec{v}\cdot\vec{n})dS=-\oiint_S p\vec{n}dS+\iiint_V \vec{f}\,\vec{v}dV$$

Note that other forces have been neglected, such as forces due to viscosity, where a bit of fluid exerts a drag force on nearby fluid elements.

The next step should be to convert the surface integrals to volume integrals using the divergence theorem. Unfortunately, the form of the surface integrals is not right for the divergence theorem, because the integrands of the surface integrals are vectors, not scalars. Thus, we need to somehow modify the divergence theorem. Let's look first at the force due to pressure. Here, we are missing a vector. We can fix this by taking the dot product of the integral (neglecting the " sign for the moment) with a constant vector a:

$$\vec{a}\cdot\oint\!\!\oint_S p\vec{n}dS = \oint\!\!\oint_S p\vec{a}\cdot\vec{n}dS = \oint\!\!\oint_S (p\vec{a})\cdot\vec{n}dS$$

Now we can apply the divergence theorem

$$\oint\!\!\oint_S (p\vec{a})\cdot\vec{n}dS = \iiint_V \vec{\nabla}\cdot(p\vec{a})dV$$

which give the desired result.

DYNAMICS OF IDEAL FLUIDS

Most of our knowledge regarding the motion of fluids comes from laboratory and terrestrial flows. We will use this opportunity to review elementary fluid dynamics with an emphasis on application to everyday flows.

THE EQUATIONS OF MOTION

First, we will discuss abstract fluid motion and then consider some special cases. To describe the motion of a fluid, we must give its properties at every point. For example, at different places, the fluid is moving with different velocities. To specify the character of the flow, therefore, we must give the three components of velocity at every point and for any time. If we can find the equations that determine the velocity, then we would know how the liquid moves at all times. The velocity, however, is not the only property that the fluid has which varies from point to point. We have just discussed the variation of the pressure from point to point. And there are still other variables. There may also be a variation of density from point to point. In addition, the fluid may carry an electric current whose density varies from point to point in magnitude and direction. There may be a temperature which varies from point to point, or a magnetic field, and so on. So the number of fields needed to describe the complete situation will depend on how complicated the problem is. There are interesting phenomena when currents and magnetism play a dominant part in determining the behavior of the fluid; this subject is called magnetohydrodynamics.

We will take the situation where there is no magnetic field and no conductivity, and we will not worry about the temperature because we will suppose that the density and pressure determine in a unique manner the temperature at any point. We will reduce the complexity by making the assumption that the density is a constant--we imagine that the fluid is essentially incompressible. Putting it another way, we are supposing that the variations of pressure are so small that the changes in density produced thereby are negligible. If that is not the case, we would encounter phenomena additional to the ones we will be discussing here, for example, the propagation of sound or of shock waves. It is easy to determine when the approximation of constant ρ is a good one if we can say that if the velocities of flow are much less than the speed of sound.

In the general theory of fluids one must begin with an equation of state for the fluid which connects the pressure to the density. In our approximation this equation of state is simply

ρ = const. ...(22)

This is the first relation for our variables. The escape that fluids makes in our attempts at understanding it is not related to the approximation of constant density.

THE EQUATION OF CONTINUITY

The equation of continuity expresses the conservation of matter--if matter flows away from a point, there must be a decrease in the quantity remaining. If the fluid velocity is u, then the mass which flows in a unit time across a unit area of surface is the component of ρu normal to the surface. There is a similar relation for electric charge and current. We know from the divergence of the current gives the rate of decrease of the density per unit time. If instead of charge we consider the change of mass in a volume V, this is balanced by the flow out of that volume

$$\frac{d}{dt}\int_v \rho dV = -\int_s \rho u.n dA, \quad ...(23)$$

where the volume V is fixed in space, and the minus sign is necessary because when u and the unit vector n points in the same direction, i.e., if u is outward and $u \cdot n > 0$, the mass within V decreases.

Since the volume is fixed in space, the limits of integration time independent so the time differentiation may be included in the integrand on the LHS, and by Gauss's divergence theorem the second term can be writtten as a volume integral:

$$\int_v \frac{\partial \rho}{\partial t} dV + \int_s \nabla.(\rho u) dV = 0 \quad ...(24)$$

Since the volume is arbitrary, the continuity equation,

$$\frac{\partial \rho}{\partial t} + \nabla.(\rho u) = 0 \quad ...(25)$$

must hold everywhere. If ρ = const. then the equation of continuity simplifies to

$$\nabla.u = 0 \quad ...(26)$$

The velocity field of an incompressible fluid, like the magnetic field B, has zero divergence. The hydrodynamic equations are often closely analogous to the electrodynamics equations, but electrodynamics is much easier than hydrodynamics

THE EQUATION OF MOTION

We will get our next equation from Newton's law which tells us how the velocity changes because of the forces. The mass of an element of volume of the fluid times its acceleration, a, must be equal to the force on the element. Taking an element of unit volume, and writing the force per unit volume as f, we have

$$\rho a = f \qquad ...(27)$$

We will write the force density as the sum of three terms. We have already considered the pressure force per unit volume, $-\nabla p$. Then there are the body forces which act at a distance, e.g., gravity or electricity. When they are conservative forces with a potential per unit mass, ϕ, they give a force density

$$-\rho\nabla\phi. \qquad ...(28)$$

If the external forces are not conservative, we would have to write f_{ext} for the external force per unit volume. Then there are the surface forces per unit volume, which is due to the fact that in a flowing fluid there can also be a shearing stress. This is called the viscous force, which we will write f_{visc}. Our equation of motion is

$$\rho a = -\nabla p - \rho\nabla\phi + f_{visc}. \qquad ...(29)$$

For now we set $f_{visc} = 0$ and by ignoring viscosity we will be making an approximation which describes an ideal fluid. The mathematician and computer scientist John von Neumann was well aware of the profound impact of this approximation, and he was also aware that, during most of the development of hydrodynamics until about 1900, almost the main interest was in solving beautiful mathematical problems with this approximation which had almost nothing to do with real fluids. He characterized the theorist who worked on this problem as scientists who studied "dry water." Such analyses leave out an essential property of the fluid.

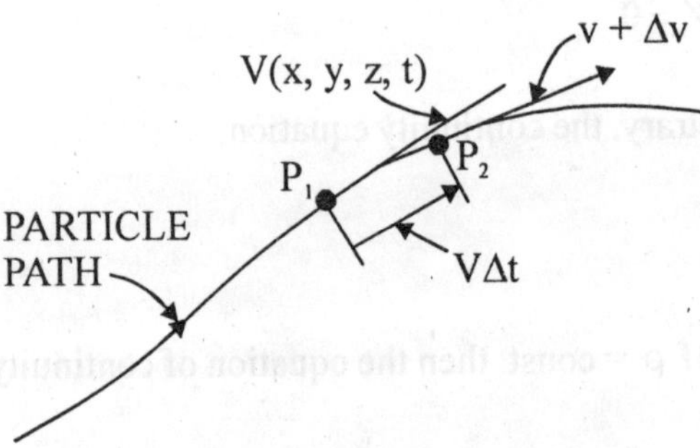

Fig. 2 The Acceleration of a Fluid Element.

If we leave out f_{visc}, we have everything we need except an expression for the acceleration. Naïvely, you might think that the formula for the acceleration of a fluid particle would be very simple, for it seems obvious that if u is the velocity of a fluid particle at some place in the fluid, the acceleration would just be $\partial u/\partial t$; it is not.

The derivative $\partial u/\partial t$, is the rate at which the velocity $u(x, t)$ changes at a fixed point in space. What we need is how fast the velocity changes for a given fluid element. Imagine that we mark one of the drops of fluid with ink so we can follow it. In a small interval of time Δt, this drop will move from P_1 to a different location P_2.

$$u_j(x_i + \Delta x_i, t + \Delta t) = u_j(x_i, t) + \frac{\partial u_j}{\partial t}\Delta t + \frac{\partial u_j}{\partial x_i}\Delta x_i, \qquad \text{...(30)}$$

where repeated indices implies summation. Using $\Delta x_i = u_i \Delta t$

$$u_j(x_i + \Delta x_i, t + \Delta t) = u_j(x_i, t) + \left(\frac{\partial u_j}{\partial t} + u_j \frac{\partial u_j}{\partial x_i}\right)\Delta t \qquad \text{...(31)}$$

From the definition of the derivative

$$a_j \lim_{\Delta t \to 0}\left(\frac{\Delta u_j}{\Delta t}\right) = \frac{\partial u_j}{\partial t} + u_i \frac{\partial u_j}{\partial x_i} \qquad \text{...(32)}$$

and so the acceleration of a fluid element is

$$\boxed{a = \frac{\partial u}{\partial t} + (u.\nabla)u.} \qquad \text{...(33)}$$

We define the differential operator

$$\frac{D}{Dt} \equiv \frac{\partial}{\partial t} + (u.\nabla),$$

which is known as the material derivative. The notation *D*/*Dt* is used to emphasize that the material time derivative includes both space and time partial derivatives. The material derivative is a scalar operator so that the material derivative of a scalar variable, such as density, is a scalar quantity. The material derivative of a vector is a vector.

Thus, there can be acceleration of fluid even when the velocity at a given point is constant, i.e., even if $\partial u / \partial t = 0$. It is possible that $Du / Dt \neq 0$ because $(u.\nabla)u \neq 0$. For steady flow, ($\partial u / \partial t = 0$), *u* is a static vector field and we can speak of streamlines, which are the trajectories followed by fluid elements. To understand describe a streamline mathematically consider two points on a streamline *r* and $r + \delta r$ (see Fig 1). The velocity of a fluid element on such a streamline is perpendicular to *r* and parallel to δr so that

$$u \times \delta r = 0\cdot$$

The cross product can be written as

$$\begin{vmatrix} i & j & k \\ u_x & u_y & u_z \\ dx & dy & dz \end{vmatrix} = 0$$

Thus, the differential equation describing a streamline is

$$\frac{dx}{u_x}=\frac{dy}{u_y}=\frac{dz}{u_z}.$$

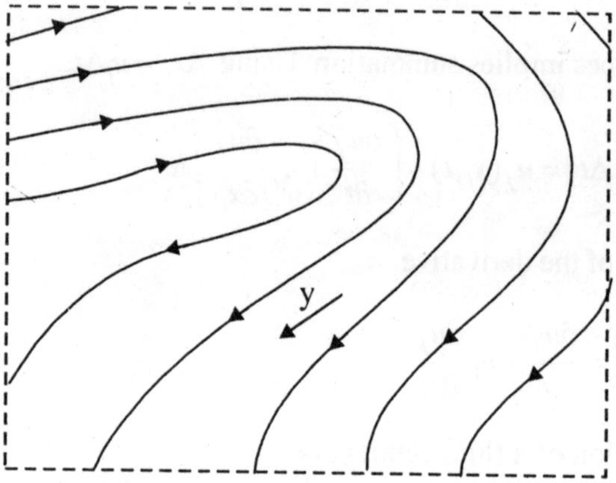

Fig, 3 Streamlines in a Steady Fluid flow Track the Path of Fluid Elements.

The term $(u.\nabla)u$ represents the change of u in the direction of u, and vanishes only when there is no variation of u in the direction of u. If all the fluid elements follow straight streamlines then this acceleration term is zero. Acceleration occurs where streamlines change direction. As an example, water flowing through a converging nozzle can be in steady state so that at any location the velocity remains constant, however the water accelerates as the cross-section of the nozzle decreases and the streamlines converge. Of course, a fluid element can also accelerate by changing its speed along a straight streamline. The physical origin of this term becomes clearer in terms of curvilinear coordinates attached to a streamline. Let s be the coordinate measured along the streamline, $V(s)$ the speed $|u|$, and $ds = V\,dt$.

$$(u.\nabla)u = V\frac{dV}{ds}\hat{e}_t - \frac{V^2}{R}\hat{e}_n, \quad \text{...(34)}$$

where R is the radius of curvature of the streamline, and $\hat{e}_t$ and $\hat{e}_n$ are unit vectors tangential and normal to the streamline. The first part of this expression is the rate of change of speed $dV/dt = V\,dV/ds$, while the second is the centripetal acceleration, which is directed towards the center of curvature of the streamline and is associated with the change in direction of the velocity of a particle.

Now we have all we need for the equation of motion, and the rest is just math,

$$\rho\left[\frac{\partial u}{\partial t}+(u.\nabla)u\right] = -\nabla p - \rho\nabla\phi \quad \text{...(35)}$$

or

$$\frac{\partial u}{\partial t} + (u.\nabla)u = -\frac{1}{\rho} - \nabla p - \nabla\phi \qquad ...(36)$$

where viscosity has been omitted. It is often useful to use the vector identity

$$(u.\nabla)u \equiv (\nabla \times u) \times u + \frac{1}{2}\nabla |u|^2 \qquad ...(37)$$

and identify a new vector field $w = \nabla \times u$ so that

$$(u.\nabla)u \equiv w \times u + \frac{1}{2}\nabla |u|^2 \qquad ...(38)$$

Giving the final form for the equation of motion

$$\frac{\partial u}{\partial t} + w \times u + \frac{1}{2}\nabla |u|^2 = -\frac{1}{\rho}\nabla p - \nabla\phi, \qquad ...(39)$$

which is Euler equation.

The vector field *w* is called the vorticity. If the vorticity is zero everywhere, we say that the flow is irrotational. The circulation, Γ around any closed loop, C, in a fluid is the line integral of the fluid velocity, at a given instant of time, around that loop,

$$\Gamma = \oint_c u.dl. \qquad ...(40)$$

The circulation per unit area is, by Stokes' theorem, equal to $\nabla \times u$ in a simply connected region. So the vorticity is the circulation round a unit area in the direction perpendicular to the surface. The vorticity is a measure of local rotation. It has nothing to do with global rotation. It follows that if you put a speck of dirt, i.e., not an infinitesimal point, at any place in the liquid it will spin with the angular velocity *w*/2.

A simple example of vorticity and circulation is a shear flow described by

$$v_x = \beta_y \qquad ...(41)$$

In this case, from the definition of the curl,

$$w_x = -\frac{\partial v_x}{\partial y} = -\beta \qquad ...(42)$$

We can also use the vorticity to calculate that angular velocity of a fluid element. Since *w* is the circulation per unit area

$$w = \frac{\oint u.dl}{\pi r^2} = \frac{2\pi r v}{\pi r^2} \quad ...(43)$$

which means that

$$w = \frac{2v}{r} \quad ...(44)$$

If the angular velocity, $\Omega = v/r$, then the angular velocity of a fluid element is,

$$\Omega = \frac{w}{2} \quad ...(45)$$

POTENTIAL FLOW AND ELECTROSTATICS

If we are interested only in the velocity field, we can eliminate the pressure from our equations. Taking the curl of both sides of the Euler equation, remembering that the curl of any gradient is zero and that ρ is a constant so that $\nabla.u = 0$, we get

$$\frac{\partial w}{\partial t} + \nabla \times (w \times u) = 0 \quad ...(46)$$

This equation, together with

$$w = \nabla \times u \quad ...(47)$$

and

$$\nabla.\, u = 0 \quad ...(48)$$

completely describe the velocity field. Mathematically speaking, if we know *w* at some time, then we know the curl of the velocity vector, and we also know that its divergence is zero, so given the physical situation we have all we need to determine *u* everywhere. It is just like the situation in magnetism where we had

$$\nabla.\, B = 0, \;\; \nabla \times B = \frac{4\pi}{c} j \quad ...(49)$$

Thus, a given *w* determines *u* just as a given *j* determines *B*. We can therefore hijack the Biot-Savart law from electromagnetic theory to invert the relationship $w = \nabla.\, u$. That is to say, in infinite domains,

$$u = \frac{1}{4\pi} \int \frac{w(x') \times r}{r^3} d^3x', \; r = x - x'. \quad ...(50)$$

Note that, like *B*, the vorticity field is solenoidal, $\nabla . w = 0$, since it is the curl of another vector. Consequently, we may invoke the idea of vortex tubes, which are analogous to magnetic flux tubes or streamtubes. This procedure gives the velocity field only; we have lost all information about the pressure. The absence of information regarding the pressure is not surprising since pressure acts normally and cannot change the angular momentum of a fluid element.

We point out one special consequence of our equation. If $w = 0$ everywhere at any time *t*, $\frac{\partial w}{\partial t}$ also vanishes, so that *w* is still zero everywhere at $t + \Delta t$. We have a solution to the equation; the flow is permanently irrotational. If a flow was started with zero rotation, it would always have zero rotation. The equations to be solved then are

$$\Delta . u = 0, \quad \nabla \times u = 0. \qquad ...(51)$$

The problem is mathematically equivalent to magnetostatics or electrostatics, and can be reduced to finding the solution of Laplace's equation,

$$\nabla^2 \phi = 0 \qquad ...(52)$$

where the velocity is related to a scalar potential

$$u = \nabla \phi \qquad ...(53)$$

Hence, incompressible, irrotational flow is called potential flow.

Steady flow

Consider Euler's equation of motion in the limit of steady flow in which the velocity picture remains unchanged and *u* is a static vector field, i.e.,

$$\frac{\partial u}{\partial t} = 0 \qquad ...(54)$$

The fluid velocity is constant at all locations; the fluid is not static, but is always replaced by new fluid moving in exactly the same way. In analogy field lines in magnetostatics, we can draw lines, called streamlines, which are always tangent to the fluid velocity as shown in Fig. 2.2. For steady flow, they represent the paths of fluid elements. In unsteady flow the streamline pattern changes in time, and the streamlines do not represent the path of a fluid element at any instant.

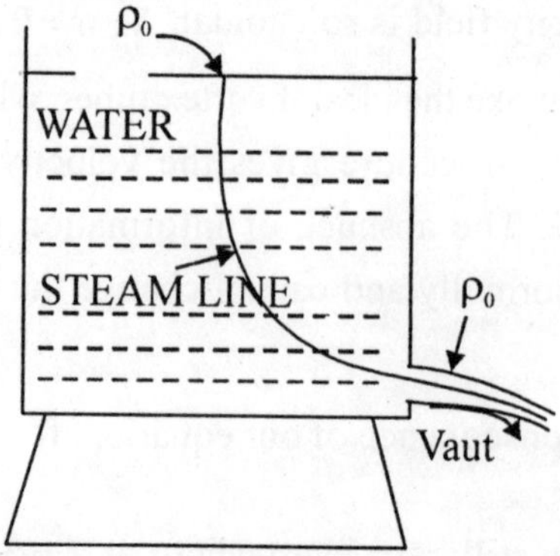

Fig. 4 Flow from a Tank. The Stream Contracts to 62% of the Area of the Aperture.

Starting with the Euler equation,

$$\frac{\partial u}{\partial t} + w \times u + \frac{1}{2}\Delta |u|^2 = -\frac{1}{\rho}\nabla_p - \nabla\phi. \qquad ...(55)$$

and setting time derivative to zero for steady flow,

$$w \times u + \frac{1}{2}\Delta |u|^2 = -\frac{1}{\rho}\nabla_p - \nabla\phi. \qquad ...(56)$$

Now take the dot product with u, and since

$$u \,.\, (w \times u) = 0 \qquad ...(57)$$

we are left with

$$u.\nabla\left(\frac{|u|^2}{2} + \frac{p}{\rho} + \phi\right) = 0 \qquad ...(58)$$

for incompressible flow. For steady flow we have that $\frac{\partial u}{\partial t}$, thus the material derivative D/Dt becomes $(u.\nabla)$, meaning that the quantity in parenthesis does not change for a material fluid element as is passes through the flow. Another way of thinking about this is to note that the quantity in parenthesis does not change for a small displacement in the direction of the fluid velocity. In other words, the quantity

$$\boxed{\frac{|u|^2}{2} + \frac{p}{\rho} + \phi = \text{const.}} \qquad ...(59)$$

along a stream line. This is Bernoulli's theorem (pronounced Ber-noo-yee). Furthermore, for an irrotational flow, $w = 0$, Euler's equation of motion gives us for incompressible flow,

$$\frac{1}{2}\nabla|u|^2 = -\frac{1}{\rho}\nabla_p - \nabla\phi. \qquad ...(60)$$

or

$$\nabla\left(\frac{|u|^2}{2} + \frac{p}{\rho} + \phi\right) = 0, \qquad ...(61)$$

that is

$$\frac{|u|^2}{2} + \frac{p}{\rho} + \phi = \text{const.} \qquad ...(62)$$

everywhere. Bernoulli's theorem is simply a statement of conservation of energy. Note that the fluid's internal energy $\epsilon = c_v T$ does not appear. This is because we have derived Bernoulli's theorem for the restrictive case where ρ = const. Under these conditions no pdV work can be done on the fluid and, since the flow is also isentropic (no heat flow and no energy dissipation by friction) the internal energy remains constant.

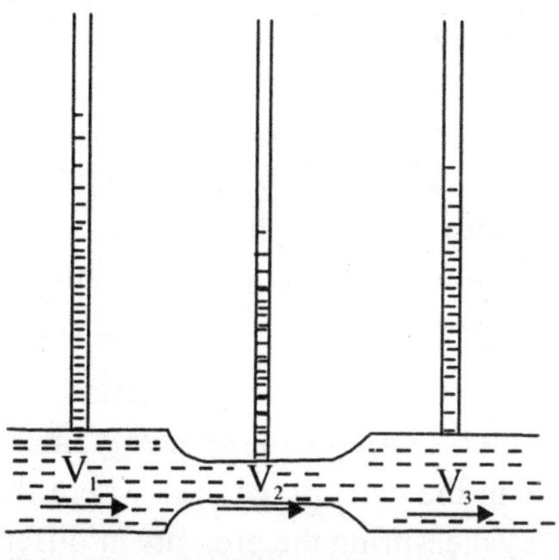

Fig. 5 According to Bernoulli's theorem the Pressure is Lowest where the Velocity is Highest.

Bernoulli's theorem can be demonstrated by holding two pieces of paper close together and trying to blow them apart? Try it! They come together. The reason, of course, is that the air has a higher speed going through the constricted space between the sheets than it does when it gets outside. The pressure between the sheets is lower than atmospheric pressure, so they come together rather than separating.

There is an important class of problems in astrophysics associated with accretion and mass loss known as the Bondi problem that uses Bernoulli's theorem. See Shu Chapter 6 and problem set 2. But first consider some examples where the Bernoulli integral gives us a description of the flow. Suppose we have water flowing out of a hole near the bottom of a tank, as drawn in Fig. 3. We take a situation in which the flow speed u_{out} at the hole is much larger than the flow speed near the top of the tank; in other words, we imagine that the diameter of the tank is so large that we can neglect the drop in the liquid level. At the top of the tank the pressure is p_0, the atmospheric pressure, and the pressure at the sides of the jet is also p_0. Now we write our Bernoulli equation for a streamline, such as the one shown in the figure. At the top of the tank, we take $u = 0$ equal to zero and we also take the gravity potential $\phi = 0$. At the speed u_{out}, and $\phi = -gh$, so that

$$p_0 = p_0 + \frac{1}{2}\rho u^2{}_{out} - pgh. \quad \text{...(63)}$$

Hence,

$$u_{out} = \sqrt{2gh}. \quad \text{...(64)}$$

This velocity is just what we would get for something which falls the distance *h*. It is not too surprising, since the water at the exit gains kinetic energy at the expense of the potential energy of the water at the top.

You cannot figure out the rate that the fluid flows out of the tank by multiplying this velocity by the area of the hole. The fluid velocities as the jet leaves the hole are not all parallel to each other but have components inward toward the center of the stream; the jet is converging. After the jet has gone a little way, the contraction stops and the velocities do become parallel. So the total flow is the velocity times the area at that point. In fact, if we have a discharge opening which is just a round hole with a sharp edge, the jet contracts to 62% of the area of the hole. The reduced effective area of the discharge varies for different shapes of discharge tubes, and experimental contractions are available as tables of efflux coefficients. If the discharge tube is re-entrant, it is possible to prove that the efflux coefficient is exactly 50%.

Consider a horizontal pipe with changing cross section, as shown in Fig. 4, with water flowing in one end and out the other. Conservation of energy, in the form of Bernoulli's formula, says that the pressure is lower in the constricted area where the velocity is higher. We can easily demonstrate this effect by measuring the pressure at different cross sections with small vertical columns of water attached to the flow tube through holes small enough so that they do not disturb the flow. The pressure is then measured by the height of water in these vertical columns. The pressure is found to be less at the constriction than it is on either side. If the area beyond the constriction comes back to the same value it had before the constriction, the pressure rises again.

Bernoulli's formula predicts that the pressure downstream of the constriction should be the same as it was upstream, but actually it is noticeably less. The reason that our prediction is wrong is that we have neglected the frictional, viscous forces which cause a pressure drop along the tube. Despite this pressure drop the pressure is definitely lower at the constriction (because of the increased speed) than it is on either side of it-as predicted by Bernoulli. The speed u_2 must certainly exceed u_1 to get the same amount of water through the narrower tube.

So the water accelerates in going from the wide to the narrow part. The force that gives this acceleration comes from the drop in pressure. We can check our results with another simple demonstration. Suppose we have on a tank a discharge tube which throws a jet of water upward as shown in Fig. 6

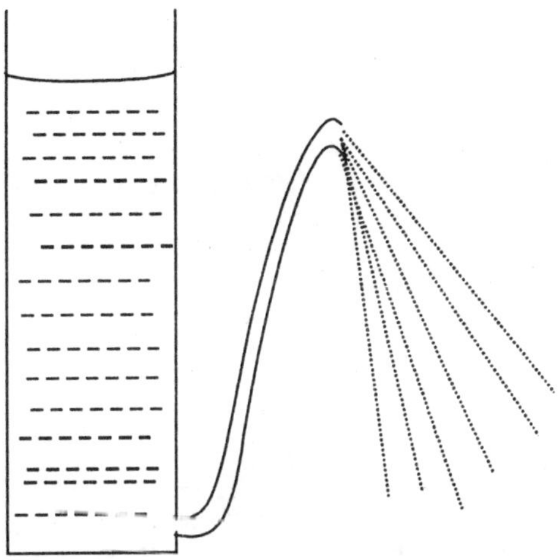

Fig. 6 The jet Does not Reach the Level of the Free Liquid Surface Hence the Jet Speed is Less than $\sqrt{2gh}$ because of Friction Due to Viscosity.

If the efflux velocity were exactly $\sqrt{2gh}$, the discharge water should rise to a level even with the surface of the water in the tank. Experimentally, it falls somewhat short. Our prediction is roughly right, but again viscous friction which has not been included in our energy conservation formula has resulted in a loss of energy.

CIRCULATION

The flow of an incompressible fluid with no vorticity satisfies:

$$\nabla . u = 0 \quad \nabla \times u = 0 \qquad ...(65)$$

which are identical to the equations of electrostatics or magnetostatics in a vacuum. The divergence of the E is zero when there are no charges, and the curl of the E always zero. The curl of B is zero if there are no currents, and the divergence of B is always zero. Therefore, incompressible, irrotational flow has the same solutions as the equations for B in electrostatics or for B in magnetostatics. For example, the electrostatic analog of flow past a sphere is a uniform electric field plus a dipole. The dipole field is so adjusted that the flow velocity normal to the surface of the sphere is zero. The same problem for the flow past a cylinder can be worked out in a similar way by using a suitable line dipole plus a uniform flow field. This solution holds for a situation in which the fluid velocity at large distances is constant, both in magnitude and direction. The solution is sketched in Fig. 7.

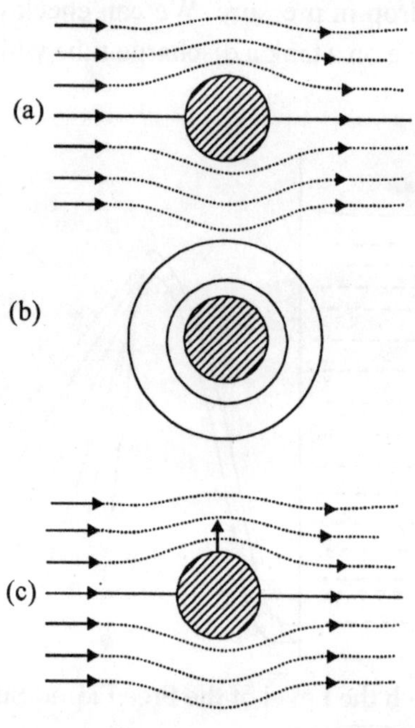

Fig. 7 Ideal flow Past a Cylinder. (b) Circulation Around a Cylinder ($w = 0$). (c) The Superposition of (a) and (b).

There is a solution for the flow around a cylinder when the fluid at large distances moves in circles around the cylinder. The flow is, then, circular everywhere, as in Fig. 7. Such a flow has a circulation around the cylinder, although $\nabla \times u = 0$ in the fluid. How can there be circulation without curl? We have a circulation around the cylinder because the line integral of u around any loop enclosing the cylinder is not zero. At the same time, the line integral of u around any closed path which does not include the cylinder is zero. The circumstances are the same for the magnetic field around a wire. The curl of B is zero outside of the wire, although a line integral of B around a path which encloses the wire did not vanish. The velocity field in an irrotational circulation around a cylinder is identical to B around a wire carrying a current. For a circular path with its center at the center of the cylinder, the line integral of the velocity is

$$\oint_c u.dl = 2\pi r u_\theta \quad ...(66)$$

For irrotational flow the integral must be independent of r. Call this constant value Γ, then we have that

$$u_\theta = \frac{\Gamma}{2\pi r} \quad ...(67)$$

where u_θ is the tangential velocity, and r is the distance from the axis.

There is a nice demonstration of a fluid circulating around a hole. Take a large clear plastic Coke bottle and cut off the base. Fill this funnel with water, stir up some circulation with a stick, and remove the cap. You get the pretty effect shown in Fig. 8. Although you put in some w at beginning, it soon dies down because of viscosity and the flow becomes irrotational-although still with some circulation around the hole.

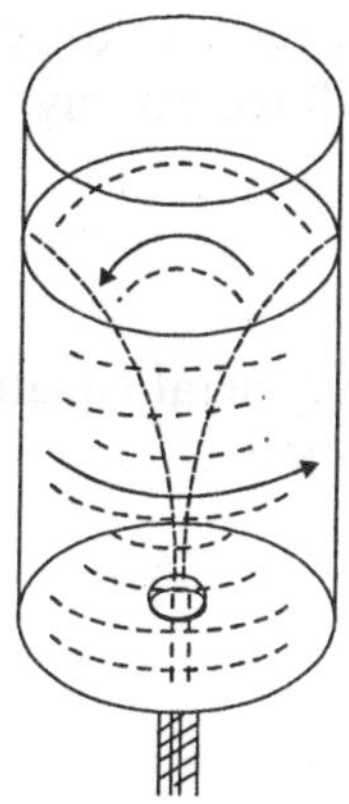

Fig. 8 Water with circulation draining from a cylindrical funnel.

From the theory, we can calculate the shape of the surface of the water. As a particle of the water moves inward it picks up speed. The tangential velocity goes as $1/r$ is just a manifestation of the conservation of angular momentum. The radial velocity also goes as $1/r$. In cylindrical coordinates, $\nabla.u$ is written as

$$\frac{1}{r}\frac{\partial}{\partial r}(ru_r)+\frac{1}{r}\frac{\partial}{\partial\theta}u_\theta+\frac{\partial}{\partial z}u_z=0 \quad ...(68)$$

If we assume azimuthal symmetry, $(\partial u_\theta/\partial\theta)=0$, and that the hole in the funnel is small compared to the diamter of the vessel so that $u_z \simeq 0$, we have water going radially inward toward a hole it follows that the radial velocity is proportional to $1/r$. So the total velocity also increases as $1/r$, and the water goes in along Archimedean spirals. The air-water surface is at atmospheric pressure, so from Bernoulli's theorem it must satisfy

$$gz+\frac{1}{2}u^2=\text{const.} \quad ...(69)$$

But u is proportional to $1/r$, so the shape of the surface is

$$(z-z_0)=\frac{k}{r^2}. \quad ...(70)$$

An interesting point-which is for incompressible, irrotational flow-is that if we have one solution and a second solution, then the sum is also a solution. This is true because the incompressible, irrotational fluid equations, like those of electrostatics or magnetostatics are linear. The complete equations of hydrodynamics are highly non-linear. For the irrotational flow about the cylinder, however, we can superpose the flows and get the new flow pattern shown. This flow is of special interest. The flow velocity is higher on the upper side of the cylinder than on the lower side. The pressures are therefore lower on the upper side than on the lower side. So when we have a combination of a circulation around a cylinder and a net horizontal flow, there is a net vertical force on the cylinder called a lift force. Of course, if there is no circulation, there is no net force on any body according to our theory of "dry" water.

VORTEX LINES

We have already written down the general equations for the flow of an incompressible fluid when there may be vorticity. They are

I. $\nabla . u = 0,$

II. $w = \nabla \times u,$

III. $\dfrac{\partial w}{\partial t} + \nabla \times (w \times u) = 0.$

The physical content of these equations has been described in words by Helmholtz in terms of three theorems. First, imagine that in the fluid we were to draw vortex lines rather than streamlines. By vortex lines we mean field lines that have the direction of w and have a density in any region proportional to the magnitude of w. From II the divergence of w is always zero (the divergence of a curl is always zero). Vortex lines are like lines of B, they never start or stop, and tend to go in closed loops. Helmholtz described III in words by the following statement: vortex lines move with the fluid. This means that if you mark the fluid particles along some vortex lines then as the fluid moves and carries those particles along, they will always mark the new positions of the vortex lines. In whatever way the atoms of the liquid move, the vortex lines move with them.

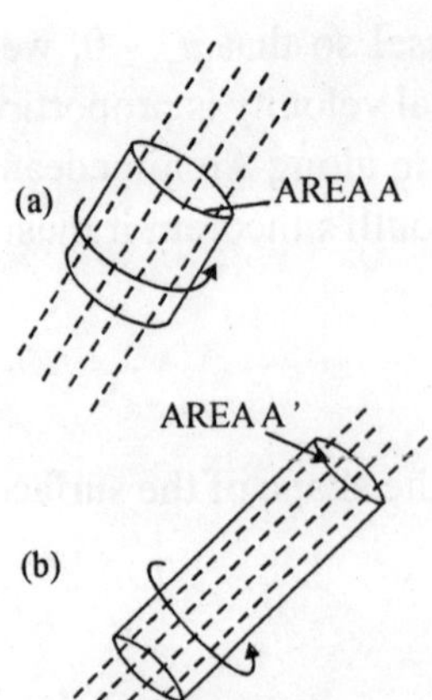

Fig. 9 (a) A Group of Vortex Lines at t; (b) the same Group of Lines Later at t'.

That is one description of the laws that also suggests a method for solving any problem. Given the initial flow pattern, say u everywhere, then you can calculate w. From the u you can also tell where the vortex lines are going to be a little later, they move with the speed u. With the new w you can use I and II to find the new u. That's just like the problem of finding B, given the currents. If we are given the flow pattern at one instant we can in principle calculate it for all subsequent times. We have the general solution for nonviscous flow.

We would like to show how Helmholtz's statement III can be partly understood. It is just the law of conservation of angular momentum applied to the fluid. Imagine a small cylindrical fluid element whose axis is parallel to the vortex lines, as in Fig 9 (a). At some time later, this same piece of fluid will be somewhere else. Generally it will occupy a cylinder with a different diameter and be in a different place. It may also have a different orientation, say as in Fig. 9 (b). If the diameter has changed, however, the length will have increased to keep the volume constant (since we are assuming an incompressible fluid). Also, since the vortex lines are stuck with the material, their density will go up as the cross-sectional area goes down. The product of the vorticity w and area A of the cylinder will remain constant, so according to Helmholtz, we should have

$$w_1A_2 = w_1A_1. \qquad ...(71)$$

Notice that with zero viscosity all the forces on the surface of the cylinder (or any volume) are normal to the surface. The pressure forces can cause the volume to be moved from place to place, or can cause it to change shape; but with no tangential forces the magnitude of the angular momentum of the material inside cannot change. The angular momentum of the liquid in the little cylinder is its moment of inertia I times the angular velocity of the liquid, which is proportional to the vorticity w. For a cylinder, the moment of inertia is proportional to mr^2. From the conservation of angular momentum, we would conclude that

$$(M_1R_1^2)w_1 = (M_2R_2^2)w_2 \qquad ...(72)$$

But the mass is the same, $M_1 = M_2$, and the areas are proportional to R^2, so this reduces to $w_2A_2 = w_1A_1$. Helmholtz's statement which is equivalent to III is just a consequence of the fact that in the absence of viscosity the angular momentum of an element of the fluid cannot change.

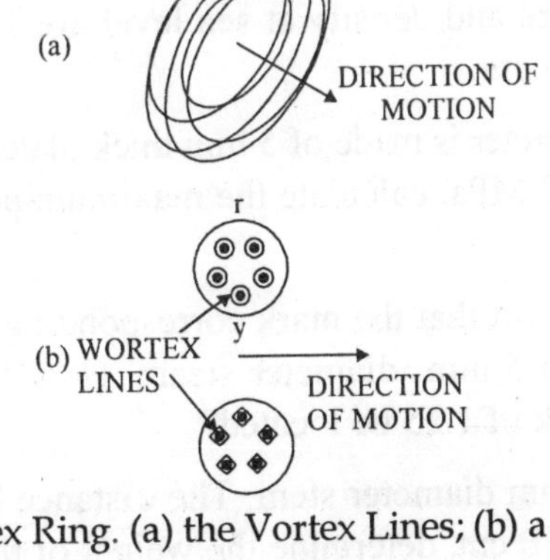

Fig. 10 A Moving Vortex Ring. (a) the Vortex Lines; (b) a Cross-section of the Ring.

A smoke ring is a torus-shaped bundle of vortex lines, as shown in Fig. 10 (a). Since $w = \nabla \times u$, these vortex lines represent also a circulation of u as shown in part (b) of the figure. We can understand the forward motion of a vortex ring in the following way: The circulating velocity around the bottom of the ring extends up to the top of the ring, having there a forward motion. Since the lines of w move with the fluid, they also move ahead with the velocity u. Of course, the circulation of u around the top part of the ring is responsible for the forward motion of the vortex lines at the bottom.

We must now mention a serious difficulty. We have already noted that III says that, if w is initially zero, it will always be zero. This result is a great failure of the theory of ``dry" water, because it means that once w is zero it is always zero -- it is impossible to produce any vorticity under any circumstance. Yet, in our simple demonstration with the drum, we can generate a vortex ring starting with air which was initially at rest. Also, we all know that we can start some vorticity in a lake with a paddle. Clearly, we must go to a theory of "wet" water to get a complete understanding of the behavior of a fluid.

Another feature of the "dry" water theory which is incorrect is the supposition we make regarding the flow at the boundary between it and the surface of a solid. When we discussed the flow past a cylinder we permitted the fluid to slide along the surface of the solid. The velocity at a solid surface could have any value depending on how it got started, and we did not consider any friction between the fluid and the solid. It is an experimental fact, however, that the velocity of a real fluid always goes to zero at the surface of a solid object. Therefore, our solution for the cylinder, with or without circulation, is wrong, as is our result regarding the generation of vorticity.

EXERCISE

1. A bourdon gauge connected to a closed tank indicates 3f kPa at a place where the atmospheric pressure is 95.480 kPa. If the tank is now carried to ahigher elevation where the atmospheric pressure is 70.11 kPa, what reading will be indicated by the gauge?

2. Calculate the pressure and density of air at an elevation of 4500 m above sea level, if the atmospheric pressure and density at sea level are 101.35 kPa and 1.2255 kg/m^3. Assume isothermal process.

3. A cylinder of 0.8 m diameter is made of 5 mm thick plates. If the maximum permissible stress in the plate is 100 MPa, calculate the maximum permissible pressure inside the cylinder.

4. A hydrometer is to be built that the mark corresponding to relative density of 1.0 and 1.5 are 8 cm apart on a 5 mm diameter steam. How far from the mark of 1.0 will the relative density mark of 1.25 be located?

5. A hydrometer has a 6 mm diameter stem. The distance between markings of relative density 1.0 and 0.90 is 10 cm. determine the weight of the hydrometer.

6. A solid cube of side 0.5 m is made of a material of relative density 0.5. The cube floats in aliquid of relative density 0.95 with two of its faces horizontal. Examine its stability.

7. The velocity in a flow is found to vary as $\upsilon_x = V_o/\left(1-\frac{x}{L}\right)^2$. Determine the acceleration at a point $x = 0.5$ m when $L = 1.5$ m and V_o=3.0 m/s.

8. A 40 cm diameter pipe is reduced uniformly to 20 cm diameter in a length of 0.5 m. If the steady discharge of water through this pipe is 100 L/s, determine the acceleration at 10 cm from the 20 cm diameter and (i) when the flow is steady and (ii) when the flow is increasing at a rate of 600 L/s per minute.

9. A pump has a 30 cm diameter suction pipe and 25 cm diameter delivery pipe. When 220 L/s of water was being pumped, the pressure on the suction side of the pump was 4 m of vacuum and on the delivery side the pressure was 100 kPa. Assuming an efficiency of 50% for the pump-motor set, estimate the electrical power consumed.

10. A mercury-water defferential manometer connected to the 15 cm diameter suction pipe and 12 cm delivery pipe of a pump shows a deflection of 40 cm. The centrelines of the suction and delivery pipes are at the same level. If the pump is discharging 70 L/s of water, estimate the head developed by the pump.

11. A 15 cm diameter jet of water with a velocity of 15 m/s strikes a plane normally. If the plate is moving with a velocity of 6 m/s in the direction of the jet calculate the work done per second on the plate and plate and the efficiency (η) of energy transfer.

Chapter 4

Differential Equations

INTRODUCTION

Ordinary differential equations are equations to be solved in which the unknown element is a function, rather than a number, and in which the known information relates that function to its derivatives. Few such equations admit an explicit answer, but there is a wealth of qualitative information describing the solutions and their dependence on the defining equation. There are many important classes of differential equations for which detailed information is available. Applications to engineering and the sciences abound. Numerical solutions are actively studied.

Many economic models with a temporal dimension involve relationships between the values of variables at a given point in time and the changes in these values over time. A model of economic growth, for example, typically contains a relationship between the change in the capital stock and the value of output. When time is modeled as a discrete variable (taking the values 1, 2, 3, ..., for example), the relationships may be modeled as difference equations. When time is modeled as a continuous variable, they may be modeled as differential equations. This type of equations comes up often in many different branches of mathematics. They are also especially important in many problems in physics and engineering.

Some Basics Terms

Equation: Equations describe the relations between the dependent and independent variables. An equal sign "=" is required in every equation.

Differential Equation: A *differential equation* is an equation involving an unknown function of one or more variables, its derivatives and the independent variables.

There are many types of differential equations.

Ordinary Differential Equation: Differential equations that involve only *ONE independent variable* are called *ordinary* differential equations.

Partial Differential Equation: Differential equations that involve *two or more independent variables* are called *partial* differential equations.

Order and Degree

Order: The order of a differential equation is the *highest derivative* that appears in the differential equation.

Degree: The degree of a differential equation is the power of the *highest derivative* term.

Linear, Non-linear, and Quasi-linear

Linear: A differential equation is called linear if there are no multiplications among dependent variables and their derivatives. In other words, all coefficients are functions of independent variables.

Non-linear: Differential equations that do not satisfy the definition of linear are non-linear.

Quasi-linear: For a non-linear differential equation, if there are no multiplications among all dependent variables and their derivatives in the highest derivative term, the differential equation is considered to be quasi-linear.

Homogeneous

Homogeneous: A differential equation is homogeneous if every single term contains the dependent variables or their derivatives.

Non-homogeneous: Differential equations which do not satisfy the definition of homogeneous are considered to be non-homogeneous.

Types of Solutions

General Solution: Solutions obtained from integrating the differential equations are called general solutions. The general solution of a nth order ordinary differential equation contains n arbitrary constants resulting from integrating n times.

Particular Solution: Particular solutions are the solutions obtained by assigning specific values to the arbitrary constants in the general solutions.

Singular Solutions: Solutions that can not be expressed by the general solutions are called singular solutions.

Conditions

Initial Condition: Constrains that are specified at the initial point, generally time point, are called initial conditions. Problems with specified initial conditions are called initial value problems.

Boundary Condition: Constrains that are specified at the boundary points, generally space points, are called boundary conditions. Problems with specified boundary conditions are called boundary value problems.

ORDINARY DIFFERENTIAL EQUATIONS

An ordinary differential equation takes the form

$$G(t, x(t), x'(t), x''(t), \ldots) = 0 \text{ for all } t,$$

where t is a scalar variable (typically but not necessarily interpreted as "time"), G is a known function, x is an unknown function, $x'(t)$ is the derivative of x with respect to t, $x''(t)$ is the second derivative of x with respect to t, and so on. (In expositions of the theory of differential equations, derivatives are usually denoted by dots over the variable rather than primes beside the variable. The word "ordinary" indicates that the unknown function x has only one argument; an equation involving the partial derivatives of a function of more than one variable is known as a "partial differential equation".

Ordinary differential equations may be classified according to the highest degree of derivative involved. If the highest degree is n, the equation is called an nth order ordinary differential equation. A first-order ordinary differential equation, for example, takes the form $G(t, x(t), x'(t)) = 0$, and may alternatively be written as

$$x'(t) = F(t, x(t)) \text{ for all } t.$$

Similarly a second-order ordinary differential equation takes the form $G(t, x(t), x'(t), x''(t)) = 0$. Note that an nth order equation may or may not involve derivatives of degree less than n.

For example, following areordinary differential equations:

$x'(t) - 1 = 0$ for all t (first-order)

$x''(t) - 1 = 0$ for all t (second-order)

$x''(t) - 2tx(t) = 0$ for all t (second-order)

$x''(t)(x'(t))^{1/2} - t/x(t) = 0$ for all t (second-order).

Definition: An *ordinary differential equation* (ODE) is a differential equation where the unknown function depends on a single variable. A general ODE has the form

$$F(x, f(x), f'(x), \ldots, f^{(n)}(x)) = 0 \tag{1}$$

where the unknown f is usually understood to be a real or complex valued function of x, and x is usually understood to be either a real or complex variable. The *order* of a differential equation is the order of the highest derivative appearing in Eq. (1). In this case, assuming that F depends nontrivially on $F^{(n)}(x)$, the equation is of nth order.

If a differential equation is satisfied by a function which identically vanishes (i.e. $f(x) = 0$ for each x in the domain of interest), then the equation is said to be *homogeneous*. Otherwise it is said to be *non-homogeneous*. Many differential equations can be expressed in the form

$$L[f] = g(x),$$

where L is a differential operator (with $g(x) = 0$ for the homogeneous case). If the operator L is linear in f, then the equation is said to be a *linear* ODE and otherwise *nonlinear*.

Other types of differential equations involve more complicated relations involving the unknown function. A *partial differential equation* (PDE) is a differential equation where the unknown function depends on more than one variable. In a *delay differential equation* (DDE), the unknown function depends on the state of the system at some instant in the past.

Solving differential equations is a difficult task. Three major types of approaches are possible:

- *Exact* methods are generally restricted to equations of low order and/or to linear systems.
- *Qualitative* methods do not give explicit formula for the solutions, but provide information pertaining to the asymptotic behavior of the system.
- Finally, *numerical* methods allow to construct approximated solutions.

Examples

A common example of an ODE is the equation for simple harmonic motion

$$\frac{d^2u}{dx^2} + ku = 0$$

This equation is of second order. It can be transformed into a system of two first order differential equations by introducing a variable $v = du/dx$. Indeed, we then have

$$\frac{dv}{dx} = -ku$$

$$\frac{du}{dx} = v\,.$$

A common example of a PDE is the wave equation in three dimensions

$$\frac{\partial^2 u}{\partial x^2} + \frac{\partial^2 u}{\partial y^2} + \frac{\partial^2 u}{\partial z^2} = c^2 \frac{\partial^2 u}{\partial t^2}$$

Solutions of Differential Equation

A solution of an ordinary differential equation is a *function* x that satisfies the equation for all values of t. A solution of the equation $x'(t) - 1 = 0$, for example, is the function x defined by $x(t) = t$ for all t. You will immediately notice that this equation has many other solutions: for any value of C, the function x defined by $x(t) = t + C$ for all t is a solution. This multiplicity of solutions is normal: most differential equations with at least one solution have many solutions. A function that includes parameters (like C in the previous example) is a "general solution" if every solution is equal to the function for some value of the parameters.

Definition: The function f of the variable t and a vector C of parameters is a general solution of an ordinary differential equation if for every solution x of the equation there is a value of C such that $x(t) = f(t, C)$ for all t.

For example, The general solution of the differential equation $x'(t) - 1 = 0$ is the function f defined by $f(t, C) = t + C$ for all t, where C is a scalar.

A common though slightly informal way to express the conclusion of this example is that "the general solution of the equation is $x(t) = t + C$." That is, the symbol C in a solution is taken to be a constant that may take any value. The same is true for the symbols C_1, C_2, and so on.

For example, Consider the differential equation $x''(t) - 1 = 0$. From the previous example we know that $x'(t) = t + C_1$, and hence the general solution of the equation is $x(t) = t^2/2 + C_1 t + C_2$.

Initial Value Problems

Many models specify both that a function satisfy a differential equation and that the value of the function, or the values of the derivatives of the function, take certain values for some values of the variable. A model may specify, for example, that $x'(t) - 1 = 0$ and $x(0) = 1$, or that $x''(t) - 1 = 0$, $x'(1) = 4$, and $x(1) = 1$. The additional conditions on the values of the function are referred to as *initial conditions* (though these conditions may specify the value of x or the value of its derivatives at any value of t, not necessarily the "first" value). A differential equation together with a set of initial conditions is called an initial value problem.

If the initial conditions are sufficiently restrictive, an initial value problem may have a unique solution, as the following examples illustrate.

For example, Consider the initial value problem $x'(t) - 1 = 0$ and $x(0) = 1$. The general solution of the differential equation is $x(t) = t + C$. Given $x(0) = 1$ we deduce that $C = 1$, so that the unique solution of the problem is $x(t) = t + 1$

For Example, Consider the initial value problem $x''(t) - 1 = 0$, $x'(1) = 4$, and $x(1) = 1$. The general solution of the differential equation is $x(t) = t^2/2 + C_1 t + C_2$. We have $x'(t) = t + C_1$, so that $x'(1) = 1 + C_1$ and hence in any solution of the initial value problem we have $C_1 = 3$. We have also $x(1) = 1/2 + 3 + C_2$, so the condition $x(1) = 1$ implies that $C_2 = -5/2$. Thus the unique solution of the problem is $x(t) = t^2/2 + 3t - 5/2$.

These examples illustrate a general principle: an initial value problem in which the differential equation involves only the first derivative has a unique solution if it has one initial condition, and an initial value problem in which the differential equation involves only the first and second derivatives has a unique solution if it has two initial conditions.

Equilibrium and Stability

If for some initial conditions a differential equation has a solution that is a constant function (independent of t), then the value of the constant is called an equilibrium state or stationary state of the differential equation. If, for all initial conditions, the solution of the differential equation converges to the equilibrium as t increases without bound, then the equilibrium is globally stable. If, for all initial conditions sufficiently close to the equilibrium the solution converges to the equilibrium as t increases without bound, then the equilibrium is locally stable. Otherwise the equilibrium is unstable.

Example1. Consider the differential equation

$$x'(t) + x(t) = 2.$$

Solution:

The general solution of this equation, as we shall see later, is

$$x(t) = Ce^{-t} + 2.$$

Thus for the initial condition $x(0) = 2$, the solution of the problem is $x(t) = 2$ for all t. Thus the equilibrium state of the system is 2.

As t increases without bound, e^{-t} converges to zero, so the equilibrium is globally stable.

First-order Differential Equations

A first-order ordinary differential equation may be written as

$$x'(t) = F(t, x(t)) \text{ for all } t.$$

Before trying to find a solution of such an equation, it is useful to know whether a solution exists.

Existence of a Solution

We may think about the question of the existence of a solution with the help of a diagram known as a direction field or integral field. We plot t on the horizontal axis and x on the vertical axis, and for each of a set of pairs (t, x) draw a short line segment through (t, x) with slope equal to $F(t, x(t))$, which is equal to $x'(t)$.

For example, The direction field of the equation

$$x'(t) = x(t)t$$

is shown in the figure below. For example, at every point $(t, 0)$ and every point $(0, x)$ we have $x'(t) = 0$, so the slope of the line segment through every such point is 0. Similarly, at every point $(1, x)$ we have $x'(t) = x$, so the slope of the line segment through $(1, x)$ is x for each value of x. (The grid size in the figure is 1/2.)

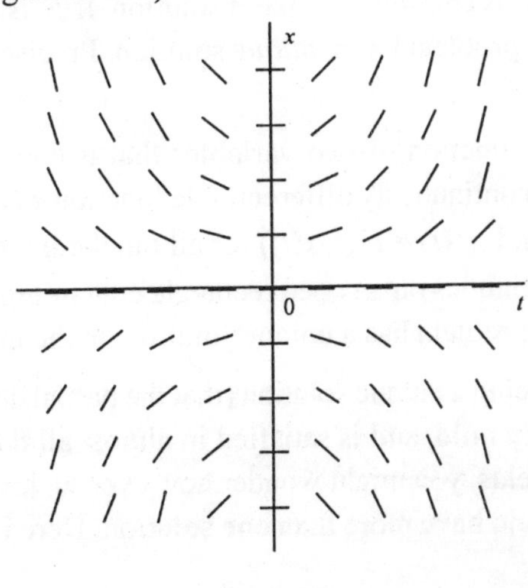

If the differential equation has a solution that passes through (t, x), then the slope $x'(t)$ of the solution at this point is equal to the slope of the line segment through (t, x). Thus we can construct a rough solution starting from any initial condition by following the slopes of the line segments. The blue line in the following figure is a sketch of such a solution for the equation in the previous example.

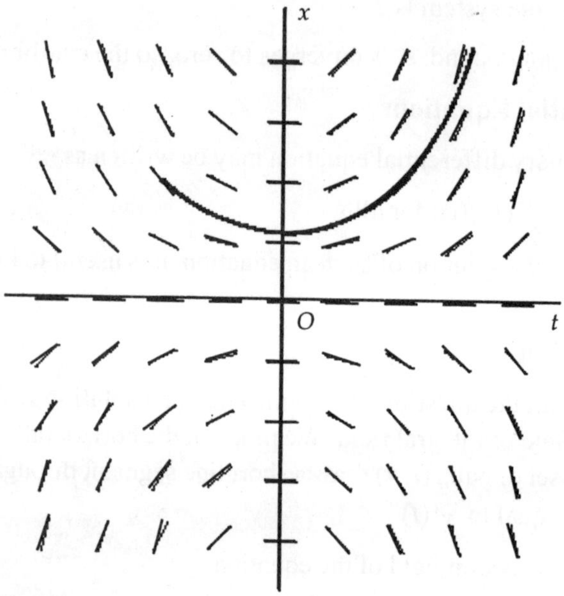

This construction suggests that any initial value problem

$$x'(t) = F(t, x(t))$$
$$x(t_0) = x_0$$

in which the slopes of the line segments in the direction field change continuously as (t, x) changes---that is, in which F is continuous---has a solution. If F is continuously differentiable, then in fact the initial value problem has a *unique* solution. Precisely, the following result may be shown.

Proposition: If F is a function of two variables that is continuous at (t_0, x_0) then there exists a number $a > 0$ and a continuously differentiable function x defined on the interval $(t_0 - a, t_0 + a)$ such that $x(t_0) = x_0$ and $x'(t) = F(t, x(t))$ for all t in the interval. If the partial derivative of F with respect to x is continuous on an open rectangle containing (t_0, x_0) then there exists $a > 0$ such that the initial value problem has a unique solution on the interval $(t_0 - a, t_0 + a)$.

The condition guaranteeing a unique solution (that the partial derivative of F with respect to x be continuous) is relatively mild, and is satisfied in almost all the examples we study. After looking at some direction fields, you might wonder how even an initial value problem that does not satisfy the condition could have more than one solution. Here is an example.

For example, Consider the initial value problem

$$x'(t) = (x(t))^{1/2}$$

$$x(0) = 0.$$

This problem does not satisfy the condition in the proposition for a unique solution, because the square root function is not differentiable at 0.

The problem has two solutions: $x(t) = 0$ for all t, and $x(t) = (t/2)^2$ for all t.

Example 2. Draw direction fields for the following differential equations.

(i) $x'(t) = x/t$. (ii) $x'(t) = -t/x$.

Solution: The direction diagram for the first equation is given in the following figure.

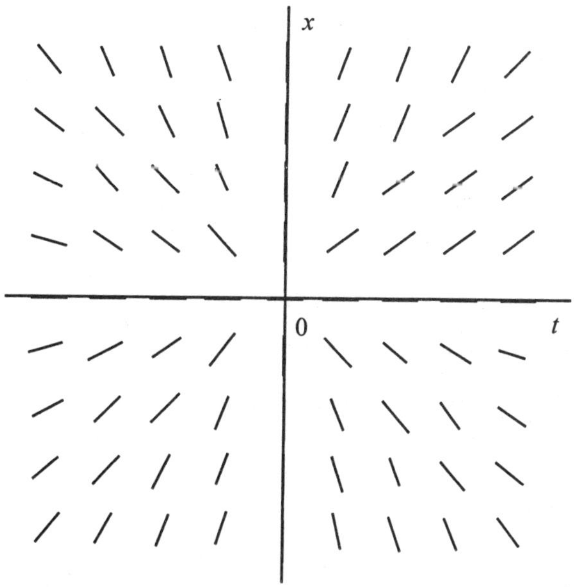

a. In the diagram for the second equation, the line segments point around concentric circles.

Separable equations

A first-order ordinary differential equation that may be written in the form

$$x'(t) = f(t)g(x) \text{ for all } t$$

is called separable.

For example, The equation

$$x'(t) = [e^{x(t)+t}/x(t)]\sqrt{(1+t^2)}$$

is separable because we can write it as

$$x'(t) = [e^{x(t)}/x(t)] \cdot \left[e^t \sqrt{(1+t^2)}\right].$$

For example, The equation

$$x'(t) = F(t) + G(x(t))$$

is not separable unless either F or G is identically 0: it cannot be written in the form $x'(t) = f(t)g(x)$.

If the function g is a constant, independent of x, then the general solution of the equation is simply the indefinite integral of f.

Even if g is not constant, the equation may be easily solved. Assuming that $g(x) \neq 0$ for all values that x assumes in a solution, we may write it as

$$dx/g(x) = f(t)dt.$$

Then we may integrate both sides, to get

$$\int^x (1/g(x))dx = \int^t f(t)dt,$$

where $\int^z h(y)dy$ is the indefinite integral of h evaluated at z. Assuming we can perform the integrations, the resulting expression defines a solution.

If $g(x^*) = 0$ for some x^* then $x(t) = x^*$ for all t is also a solution.

For example, Consider the differential equation

$$x'(t) = x(t) \cdot t,$$

for which we drew a direction field previously. First write the equation as

$$dx/x = t\, dt.$$

Then integrate both sides, to obtain

$$\ln x = t^2/2 + C.$$

(I have consolidated the constants of integration in C on the right-hand side.)

Finally, isolate x to obtain

$$x(t) = Ce^{t^2/2} \text{ for all } t.$$

The C in this equation is equal to e^C from the previous equation. I follow standard practice in using the same letter C to denote the new constant.

This argument is valid only if $x(t) \neq 0$ for all t; thus in all the solutions it generates we need $C \neq 0$. However, looking at the original differential equation we see that the function x defined by $x(t) = 0$ for all t is also a solution.

If we have an initial condition $x(t_0) = x_0$ then the value of C is determined by the equation

$$x_0 = Ce^{(t_0)^2/2}.$$

Example 3. Consider the differential equation

$$x'(t) = -2(x(t))^2 t.$$

Solution:

We may write this equation as

$$-dx/x^2 = 2t\,dt,$$

which may be integrated to yield

$$1/x = t^2 + C,$$

yielding the set of solutions

$$x(t) = 1/(t^2 + C) \text{ for all } t.$$

In addition, $x(t) = 0$ for all t is a solution. As before, an initial condition determines the exact solution. If $x(0) = 1$, for example, we need

$$1 = 1/C$$

so $C = 1$. Thus for this initial condition the solution is

$$x(t) = 1/(t^2 + 1).$$

To take another example, if $x(0) = 0$, then the solution is $x(t) = 0$ for all t.

Example 4. Consider the differential equation

$$x'(t) = t/((x(t))^4 + 1).$$

Solution: Separating the variables yields

$$(x^4 + 1)dx = t\,dt.$$

The value of $x^4 + 1$ is not 0 for any value of x, so all the solutions of the equation are obtained by integrating both sides of this equation, which leads to

$$x^5/5 + x = t^2/2 + C.$$

Thus a solution x satisfies

$$(x(t))^5/5 + x(t) = t^2/2 + C \text{ for all } t.$$

In this case we cannot isolate $x(t)$---we cannot find the solutions explicitly, but know only that they satisfy this equation.

Example 5. (Solow's model of economic growth)

Output is produced from capital, denoted K, and labor, denoted L, according to the production function $AK^{1-a}L^a$, where A is a positive constant and $0 < a < 1$. A constant fraction s of output is "saved" (with $0 < s < 1$), and used to augment the capital stock. Thus the capital stock changes according to the differential equation

$$K'(t) = sA(K(t))^{1-a}(L(t))^a$$

and takes the value K_0 at $t = 0$. The labor force is $L_0 > 0$ at $t = 0$ and grows at a constant rate λ, so that

$$L'(t)/L(t) = \lambda.$$

One way of finding a solution of this model is to first solve for L, then substitute the resulting expression into the equation for $K'(t)$ and solve for K.

The equation for L is separable, and may be written as

$$dL/L = \lambda\, dt.$$

The equation yields $\ln L = \lambda t + C$, or $L(t) = Ce^{\lambda t}$. Given the initial condition, we have $C = L_0$.

Substituting this result into the equation for $K'(t)$ yields

$$K'(t) = sA(K(t))^{1-a}(L_0e^{\lambda t})^a = sA(L_0)^a e^{a\lambda t}(K(t))^{1-a}.$$

This equation is separable, and may be written as

$$K^{a-1}dK = sA(L_0)^a e^{a\lambda t}dt.$$

Integrating both sides, we obtain

$$K^a/a = sA(L_0)^a e^{a\lambda t}/a\lambda + C,$$

so that

$$K(t) = (sA(L_0)^a e^{a\lambda t}/\lambda + C)^{1/a}.$$

Given $K(0) = K_0$, we conclude that $C = (K_0)^a - sA(L_0)^a/\lambda$. Thus

$$K(t) = [sA(L_0)^a(e^{a\lambda t} - 1)/\lambda + (K_0)^a]^{1/a} \text{ for all } t.$$

An economically interesting feature of the model is the evolution of the capital-labor ratio $K(t)/L(t)$. We have

$$K(t)/L(t) = [sA(L_0)^a(e^{a\lambda t} - 1)/\lambda + (K_0)^a]^{1/a}(L_0e^{\lambda t}) \text{ for all } t.$$

As t grows without bound, this fraction converges to

$$(sA/\lambda)^{1/a}.$$

A more general version of this model is studied in a later example.

LINEAR FIRST-ORDER DIFFERENTIAL EQUATIONS

General form

A linear first-order differential equation takes the form

$$x'(t) + a(t)x(t) = b(t) \text{ for all } t$$

for some functions a and b.

Coefficient of $x(t)$ Constant

Consider the case in which $a(t) = a \neq 0$ for all t, so that

$$x'(t) + ax(t) = b(t) \text{ for all } t.$$

If the left-hand side were the derivative of some function and we could find the integral of b then we could solve the equation by integrating each side. Now, the left-hand side looks something like the derivative of a product. But for it to be *exactly* the derivative of a product of the form $f(t)x(t)$ we would need $f(t) = 1$ and $f'(t) = a$ for all t, which is clearly not possible.

Now suppose that we multiply both sides by $g(t)$ for each t. Then we have

$$g(t)\,x'(t) + ag(t)x(t) = g(t)b(t) \text{ for all } t.$$

For the left-hand side of *this* equation to be the derivative of a product of the form $f(t)x(t)$ we need $f(t) = g(t)$ and $f'(t) = ag(t)$. Is there any function f with this property? Yes! If $f(t) = e^{at}$ then $f'(t) = ae^{at} = af(t)$.

Thus if we set $g(t) = e^{at}$, so that we have

$$e^{at}\,x'(t) + ae^{at}x(t) = e^{at}b(t),$$

then the integral of the left-hand side is

$$e^{at}x(t),$$

and hence the solution of the equation is given by

$$e^{at}x(t) = C + \int^t e^{as}b(s)ds,$$

(where $\int^t f(s)ds$ is, as before, the indefinite integral of $f(s)$ evaluated at t), or

$$x(t) = e^{-at}[C + \int^t e^{as}b(s)ds].$$

This conclusion is summarized in the following result.

Proposition: The general solution of the differential equation

$$x'(t) + ax(t) = b(t) \text{ for all } t,$$

where a is a constant and b is a continuous function, is given by

$$x(t) = e^{-at}[C + \int^t e^{as}b(s)ds] \text{ for all } t.$$

Because multiplying the original equation by e^{at} allows us to integrate the left-hand side, we call e^{at} an integrating factor.

If $b(t) = b$ for all t then the solution simplifies to

$$x(t) = Ce^{-at} + b/a.$$

Looking at the original equation we see that $x'(t) = 0$ if and only if $x(t) = b/a$. Thus $x = b/a$ is an equilibrium state.

For the initial condition $x(t_0) = x_0$ we have $x_0 = Ce^{-at_0} + b/a$, so that $C = (x_0 - b/a)e^{at_0}$, and hence the solution is

$$x(t) = (x_0 - b/a)e^{a(t_0 - t)} + b/a.$$

Thus as t increases without bound, $x(t)$ converges to b/a if $a > 0$, and grows without bound if $a < 0$ and $x_0 \neq b/a$. That is, the equilibrium is globally stable if $a > 0$ and unstable if $a < 0$.

Example 6. Consider the differential equation

$$x'(t) + 2x(t) = 6.$$

Solution: The general solution of this equation is

$$x(t) = Ce^{-2t} + 3.$$

For the initial condition $x(0) = 10$, we obtain $C = 7$, so that the solution is

$$x(t) = 7e^{-2t} + 3.$$

This solution is stable, because $2 > 0$.

Example 7. When the price of a good is p, the total demand is $D(p) = a - bp$ and the total supply is $S(p) = \alpha + \beta p$, where a, b, α, and β are positive constants.

Solution: When demand exceeds supply, price rises, and when supply exceeds demand it falls. The speed at which the price changes is proportional to the difference between supply and demand. Specifically,

$$p'(t) = \lambda[D(p) - S(p)]$$

with $\lambda > 0$. Given the forms of the supply and demand functions, we thus have

$$p'(t) + \lambda(b + \beta)p(t) = \lambda(a - \alpha).$$

The general solution of this differential equation is

$$p(t) = Ce^{-\lambda(b+\beta)t} + (a - \alpha)/(b + \beta).$$

The equilibrium price is $(a - \alpha)/(b + \beta)$, and, given $\lambda(b + \beta) > 0$, this equilibrium is globally stable.

Coefficient of $x(t)$ Dependent on t

We may generalize these arguments to the differential equation

$$x'(t) + a(t)x(t) = b(t),$$

in which a is a function. For $g(t)$ times the left-hand side of this equation to be the derivative of $f(t)x(t)$ we need $f(t) = g(t)$ and $f'(t) = a(t)g(t)$ for all t. The function $g(t) = e \int ta(s)ds$ has this property, because the derivative of $\int^t a(s)ds$ is $a(t)$ (by the fundamental theorem of calculus). Multiplying the equation by this factor, we have

$$e^{\int^t a(s)ds}x'(t) + a(t)e^{\int^t a(s)ds}x(t) = e^{\int^t a(s)ds}b(t),$$

or

$$(d/dt)[x(t)e^{\int^t a(s)ds}] = e^{\int^t a(s)ds}b(t).$$

Thus

$$x(t)e^{\int^t a(s)ds} = C + \int^t e^{\int^u a(s)ds}b(u)du,$$

or

$$x(t) = e^{-\int^t a(s)ds}[C + \int^t e^{\int^u a(s)ds}b(u)du].$$

In summary, we have the following result.

Proposition The general solution of the differential equation

$$x'(t) + a(t)x(t) = b(t) \text{ for all } t,$$

where a and b are continuous functions, is given by

$$x(t) = e^{-\int^t a(s)ds}[C + \int^t e^{\int^u a(s)ds}b(u)du] \text{ for all } t.$$

Example 8. Consider the differential equation

$$x'(t) + (1/t)x(t) = e^t.$$

Solution: We have, $\int^t (1/s)ds = \ln t$

so $\quad e^{\int^t (1/t)dt} = t.$

Thus the solution of the equation is

$$x(t) = (1/t)(C + \int^t ue^u du)$$

$$= (1/t)(C + te^t - \int^t e^u du)$$

$$= (1/t)(C + te^t - e^t)$$

$$= C/t + e^t - e^t/t.$$

(I used integration by parts to obtain the second line.) We can check that this solution is correct by differentiating:

$$x'(t) + x(t)/t = -C/t^2 + e^t - e^t/t + e^t/t^2 + C/t^2 + e^t/t - e^t/t^2 = e^t.$$

As before, an initial condition determines the value of C.

Differential Equations: Phase Diagrams for Autonomous Equations

We are often interested not in the exact form of the solution of a differential equation, but only in the qualitative properties of this solution. In economics, in fact, the differential equations that arise in our models usually contain functions whose forms we do not specify explicitly, so there is no question of finding explicit solutions. One way of studying the qualitative properties of the solutions of a differential equation is to construct a phase diagram. I discuss this technique for the class of "autonomous" first-order differential equations.

Autonomous Equations

A first-order differential equation is autonomous if it takes the form $x'(t) = F(x(t))$ (i.e. the value of $x'(t)$ does not depend independently on the variable t).

An equilibrium state of such an equation is a value of x for which $F(x) = 0$. (If $F(x) = 0$ then $x'(t) = 0$, so that the value of x does not change.)

A phase diagram indicates the sign of $x'(t)$ for a representative collection of values of x. To construct such a diagram, plot the function F, which gives the value of x'. For values of x at which the graph of F is above the x-axis we have $x'(t) > 0$, so that x is increasing; for values of x at which the graph is below the x-axis we have $x'(t) < 0$, so that x is decreasing. A value of x for which $F(x) = 0$ is an equilibrium state.

If x^* is an equilibrium and $F'(x^*) < 0$, then if x is slightly less than x^* it increases, whereas if x is slightly greater than x^* it decreases. If, on the other hand, $F'(x^*) > 0$, then if x is slightly less than x^* it decreases, moving further from the equilibrium, and if x is slightly greater than x^* it increases, again moving away from the equilibrium. That is, if x^* is an equilibrium then

- if $F'(x^*) < 0$ then x^* is locally stable
- if $F'(x^*) > 0$ then x^* is unstable.

The example in the following figure has three equilibrium states, a, b, and c. The arrows on the x-axis indicate the direction in which x is changing (given by the sign of $x'(t)$) for each possible value of x.

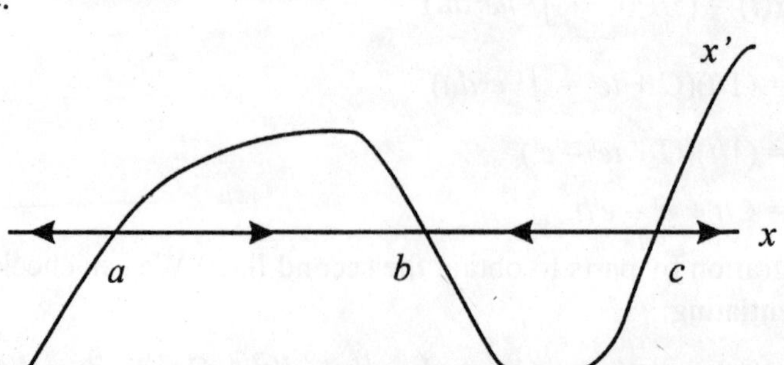

We see that the equilibrium b is locally stable, whereas the equilibria a and c are unstable.

Example 9. (Solow's model of economic growth)

The following model generalizes the one in an earlier example. Suppose that the production function is a strictly increasing and strictly concave function F that is homogeneous of degree 1 (i.e. has "constant returns to scale"), rather than taking the specific form assumed in the earlier example.

We now have, $K'(t) = sF(K(t), L(t))$.

As before, the labor force grows at the constant rate λ, so that

$$L'(t)/L(t) = 1.$$

We may study the behavior of the capital-labor ratio $K(t)/L(t)$ as follows. Let $k = K/L$ and define the function f of a single variable by

$$f(k) = F(k, 1) \text{ for all } k.$$

Since F is increasing and concave we have $f' > 0$ and $f'' < 0$, and since it is homogeneous of degree 1 we have

$$F(K, L) = LF(K/L, 1) = Lf(k).$$

Thus, $K'(t) = sL(t)f(k(t))$.

Now, $k'(t) = [K'(t)L(t) - K(t)L'(t)]/(L(t))^2 = [K'(t) - k(t)L'(t)]/L(t)$

so $k'(t) = sf(k(t)) - \lambda k(t)$.

The phase diagram of this equation is shown in the following figure.

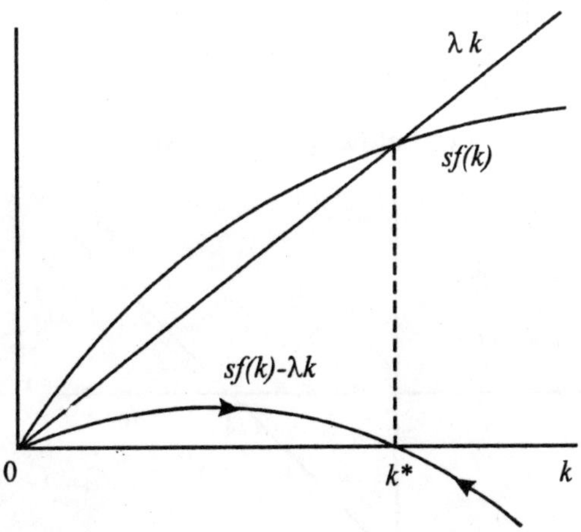

We see that the equation has two equilibria, one in which $k^* = 0$, which is unstable, and one in which $sf(k^*) = \lambda k^*$, which is locally stable.

The stable equilibrium value of k depends on s: the equation

$$sf(k^*) = \lambda k^*$$

implicitly defines k^* as a function of s. Which value of s maximizes the equilibrium value of per capita consumption? Per capita consumption is equal to $(1 - s)F(K, L)/L$, or, given our previous calculations, $(1 - s) f(k)$. Thus in an equilibrium per capita consumption is equal to $(1 - s) f(k^*(s))$. A stationary point of this function satisfies

$$(1 - s)\ f'(k^*(s))\, k^{*\prime}(s) = f(k^*(s)).$$

Now, differentiating the equation defining k^* and rearranging the terms we obtain

$$k^{*\prime}(s) = f(k^*(s))/[\lambda - s\ f'(k^*(s))].$$

Combining the last two equations we deduce that if s maximizes per capita consumption then

$$f'(k^*(s)) = \lambda.$$

That is, the marginal product of capital is equal to the rate of population growth.

Example 10. Draw the phase diagrams associated with the following differential equations and determine the nature of the possible equilibrium states.

a. $x'(t) = x(t) - 1$.

b. $x'(t) + 2x(t) = 24$.

c. $x'(t) = x(t)^2 - 9$.

d. $x'(t) = x(t)^3 + x(t)^2 - x(t) - 1$.

e. $x'(t) = 3x(t)^2 + 1$.

f. $x'(t) = x(t)e^{x(t)}$.

g. $x'(t) = (x(t) + 2)(x(t))^2$.

Solution:

a. The phase diagram is shown below. $x = 1$ is unstable.

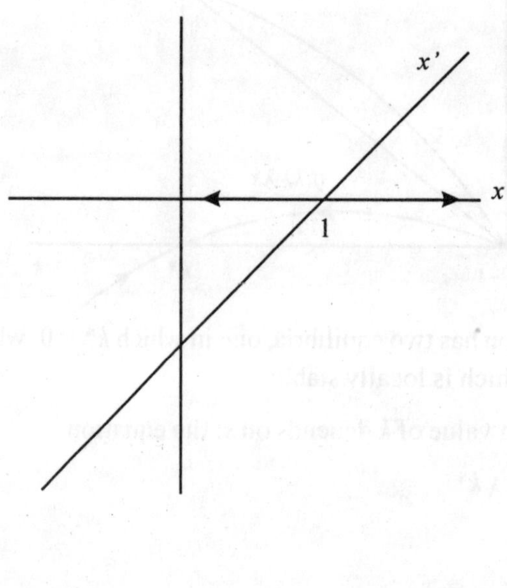

b. The phase diagram is shown below. $x = 12$ is stable.

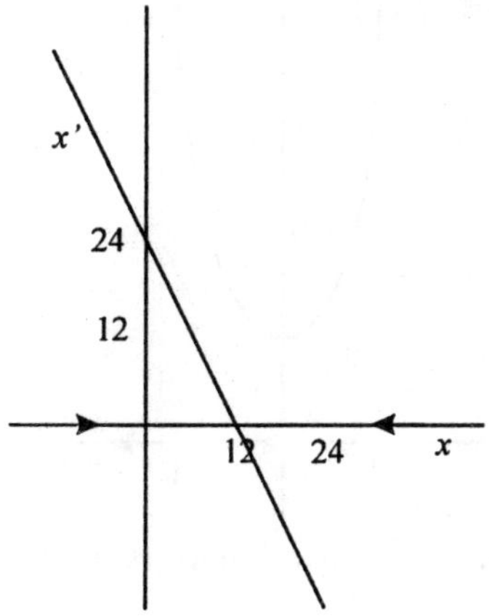

c. The phase diagram is shown below. $x = -3$ is stable; $x = 3$ is unstable.

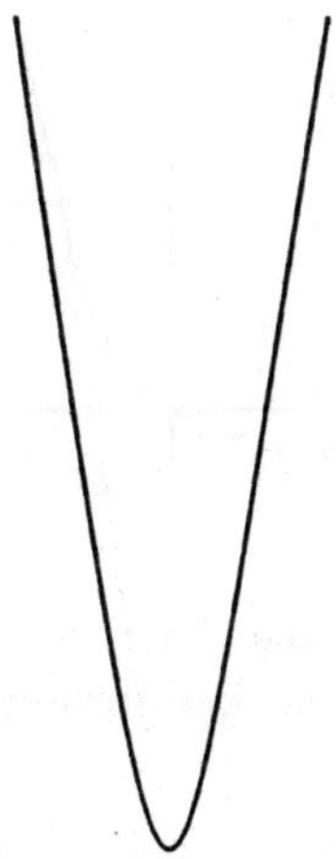

d. Note that $x'(t) = (x(t) - 1)(x(t) + 1)^2$. $x = 1$ is unstable; $x = -1$ is unstable. (It is stable from the right and unstable from the left.) The phase diagram is shown below.

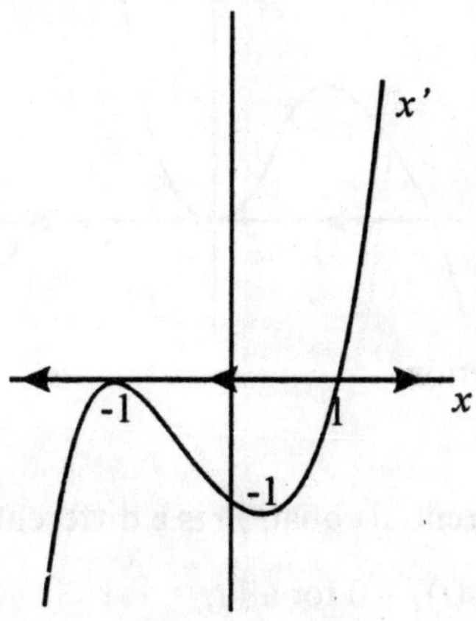

e. The phase diagram is shown below. There are no equilibrium states.

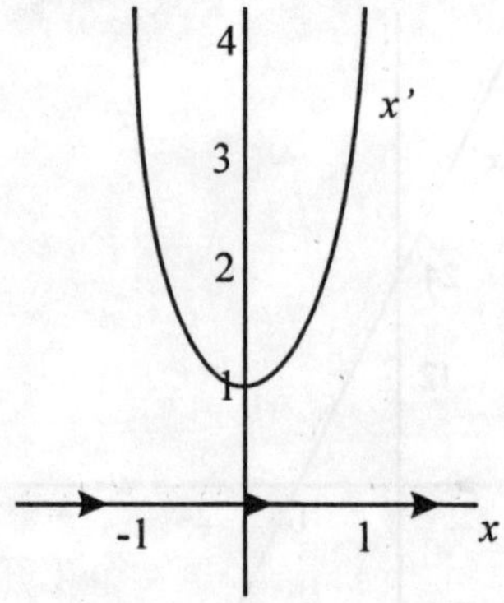

f. The phase diagram is shown below. There is a single equilibrium, $x = 0$, which is unstable.

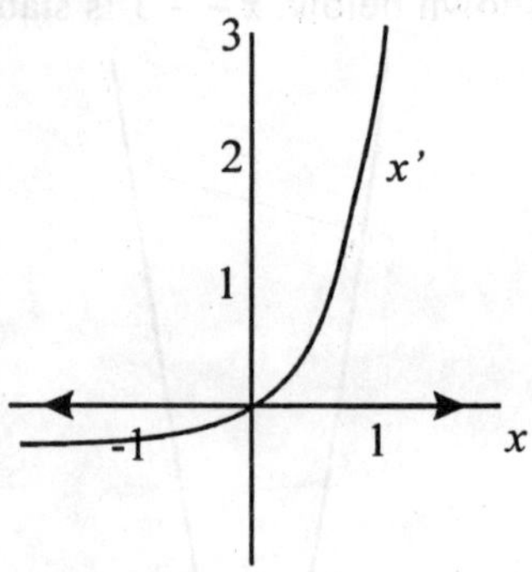

g. The phase diagram is shown below. There are two equilibria, $x = -2$ and $x = 0$. The equilibrium $x = -2$ is unstable; the equilibrium $x = 0$ is unstable. (It is stable from the left, but not from the right.)

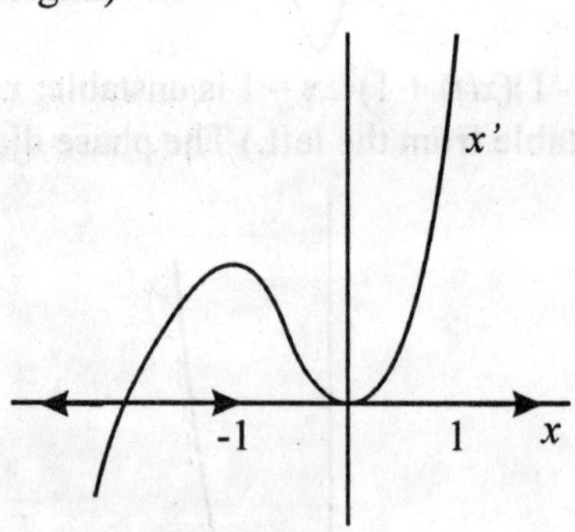

Second-order differential equations

General form

A second-order ordinary differential equation is a differential equation of the form

$$G(t, x(t),\ x'(t),\ x''(t)) = 0 \text{ for all } t,$$

involving only t, $x(t)$, and the first and second derivatives of x. We can write such an equation in the form

$$x''(t) = F(t, x(t), x'(t)).$$

Equations of the form $x''(t) = F(t, x'(t))$

An equation of the form

$$x''(t) = F(t, x'(t)),$$

in which $x(t)$ does not appear, can be reduced to a first-order equation by making the substitution $z(t) = x'(t)$.

Example 11. Let $u(w)$ be a utility function for wealth w. The function

$$\rho(w) = -wu''(w)/u'(w)$$

is known as the *Arrow-Pratt measure of relative risk aversion.* (If $\rho_u(w) > \rho_v(w)$ for two utility functions u and v then u reflects a greater degree of risk-aversion than does v.)

What utility functions have a degree of risk-aversion that is independent of the level of wealth? That is, for what utility functions u do we have

$$a = -wu''(w)/u'(w) \text{ for all } w?$$

Solution: This is a second-order differential equation in which the term $u(w)$ does not appear. (The variable is w, rather than t.) Define $z(w) = u'(w)$. Then we have

$$a = -wz'(w)/z(w)$$

$$\text{or } az(w) = -wz'(w),$$

a separable equation that we can write as

$$a \cdot dw/w = -dz/z.$$

The solution is given by

$$a \cdot \ln w = -\ln z(w) + C,$$

$$\text{or } z(w) = Cw^{-a}.$$

Now, $z(w) = u'(w)$, so to get u we need to integrate:

$$u(w) = \{C \ln w + B, \text{ if } a = 1$$

$$Cw^{1-a}/(1-a) + B, \text{ if } a \neq 1$$

We conclude that a utility function with a constant degree of risk-aversion equal to a takes this form.

Linear Second-order Equations with Constant Coefficients

A linear second-order differential equation with constant coefficients takes the form

$$x''(t) + ax'(t) + bx(t) = f(t)$$

for constants a and b and a function f. Such an equation is homogeneous if $f(t) = 0$ for all t.

Let x_1 be a solution of the equation

$$x''(t) + ax'(t) + bx(t) = f(t),$$

to which I refer subsequently as the "original equation". For any other solution of this equation x, define $z = x - x_1$. Then z is a solution of the homogeneous equation

$$x''(t) + ax'(t) + bx(t) = 0$$

(because $z''(t) + az'(t) + bz(t) = [x''(t) + ax'(t) + bx(t)] - [x_1''(t) + ax_1'(t) + bx_1(t)] = f(t) - f(t) = 0$). Further, for every solution z of the homogeneous equation, $x_1 + z$ is clearly a solution of original equation.

We conclude that the set of all solutions of the original equation may be found by finding *one* solution of this equation and adding to it the general solution of the homogeneous equation. That is, we may use the following procedure to solve the original equation.

Procedure for finding general solution of linear second-order differential equation with constant coefficients

The general solution of the differential equation

$$x''(t) + ax'(t) + bx(t) = f(t)$$

may be found as follows:

1. Find the general solution of the associated homogeneous equation $x''(t) + ax'(t) + bx(t) = 0$.
2. Find a single solution of the original equation $x''(t) + ax'(t) + bx(t) = f(t)$.
3. Add together the solutions found in steps 1 and 2.

I now explain how to perform to the first two steps and illustrate step 3.

1. Finding the general solution of a homogeneous equation

You might guess, based on the solutions we found for first-order equations, that the homogeneous equation has a solution of the form $x(t) = Ae^{rt}$. Let's see if it does. If $x(t) = Ae^{rt}$ then $x'(t) = rAe^{rt}$ and $x''(t) = r^2Ae^{rt}$, so that

$$x''(t) + ax'(t) + bx(t) = r^2Ae^{rt} + arAe^{rt} + bAe^{rt}$$

$$= Ae^{rt}(r^2 + ar + b).$$

Thus for $x(t)$ to be a solution of the equation we need

$$r^2 + ar + b = 0.$$

This equation is knows as the characteristic equation of the differential equation. It has either two distinct real roots, a single real root, or two complex roots. If $a^2 > 4b$ then it has two distinct real roots, say r and s, and we have shown that both $x(t) = Ae^{rt}$ and $x(t) = Be^{st}$, for any values of A and B, are solutions of the equation. Thus also $x(t) = Ae^{rt} + Be^{st}$ is a solution. It can be shown, in fact, that *every* solution of the equation takes this form.

The cases in which the characteristic equation has a single real root ($a^2 = 4b$) or complex roots ($a^2 < 4b$) require a slightly different analysis, with the following conclusions.

Proposition: Consider the homogeneous linear second-order differential equation with constant coefficients

$$x''(t) + ax'(t) + bx(t) = 0 \text{ for all } t,$$

where a and b are numbers. The general solution of this equation depends on the character of the roots of the characteristic equation $r^2 + ar + b = 0$ as follows.

Distinct real roots

If $a^2 > 4b$, in which case the characteristic equation has distinct real roots, the general solution of the equation is

$$Ae^{rt} + Be^{st}.$$

Repeated real root

If $a^2 = 4b$, in which case the characteristic equation has a single root, the general solution of the equation is

$$(A + Bt)e^{rt},$$

where $r = -(1/2)a$ is the root.

Complex roots

If $a^2 < 4b$, in which case the characteristic equation has complex roots, the general solution of the equation is

$$(A\cos(\beta t) + B\sin(\beta t)e^{\alpha_2},$$

where $\alpha = -a/2$ and $\beta = \sqrt{(b - a^2/4)}$. This solution may alternatively be expressed as

$$Ce^{\alpha_2}\cos(\beta t + \omega),$$

where the relationships between the constants C, ω, A, and B are $A = C\cos\omega$ and $B = -C\sin\omega$.

Example 13. Consider the differential equation

$$x''(t) + x'(t) - 2x(t) = 0.$$

Solution: The characteristic equation is

$$r^2 + r - 2 = 0$$

so the roots are 1 and – 2. That is, the roots are real and distinct. Thus the general solution of the differential equation is

$$x(t) = Ae^t + Be^{-2t}.$$

Example 14. Consider the differential equation

$$x''(t) + 6x'(t) + 9x(t) = 0.$$

Solution: The characteristic equation has a repeated real root, equal to – 3. Thus the general solution of the differential equation is

$$x(t) = (A + Bt)e^{-3t}.$$

Example 15. Consider the equation

$$x''(t) + 2x'(t) + 17x(t) = 0.$$

Solution: The characteristic roots are complex. We have $a = 2$ and $b = 17$, so $\alpha = -1$ and $\beta = 4$, so the general solution of the differential equation is

$$[A\cos(4t) + B\sin(4t)]e^{-t}.$$

2. Finding a solution of a non-homogeneous equation

A good technique to use to find a solution of a non-homogeneous equation is to try a linear combination of $f(t)$ and its first and second derivatives. If, for example, $f(t) = 3t - 6t^2$, then try to find values of A, B, and C such that $A + Bt + Ct^2$ is a solution. Or if $f(t) = 2\sin t + \cos t$, then try to find values of A and B such that $f(t) = A\sin t + B\cos t$ is a solution. Or if $f(t) = 2e^{Bt}$ for some value of B, then try to find a value of A such that Ae^{Bt} is a solution.

Example 16. Consider the differential equation

$$x''(t) + x'(t) - 2x(t) = t^2.$$

The function on the right-hand side is a second-degree polynomial, so to find a solution of the equation, try a general second-degree polynomial---that is, a function of the form $x(t) = C + Dt + Et^2$. (The important point to note is that you should not restrict to a function of the form $x(t) = Et^2$.) For this function to be a solution,

$$2E + D + 2Et - 2C - 2Dt - 2Et^2 = t^2 \text{ for all } t,$$

so we need (equating the coefficients of t^2, t, and the constant on both sides),

$$2E + D - 2C = 0$$

$$2E - 2D = 0$$

$$-2E = 1.$$

We deduce that $E = -1/2$, $D = -1/2$, and $C = -3/4$. That is,

$$x(t) = -3/4 - t/2 - t^2/2$$

is a solution of the differential equation.

3. Add together the solutions found in steps 1 and 2

This step is quite trivial!

Example 17. Consider the equation from the previous example, namely

$$x''(t) + x'(t) - 2x(t) = t^2.$$

Solution. We saw above that the general solution of the associated homogeneous equation is $x(t) = Ae^t + Be^{-2t}$.

and that $x(t) = -3/4 - t/2 - t^2/2$

is a solution of the original equation.

Thus the general solution of the original equation is

$$x(t) = Ae^t + Be^{-2t} - 3/4 - t/2 - t^2/2.$$

Stability of Solutions of Homogeneous Equation

Consider the homogeneous equation

$$x''(t) + ax'(t) + bx(t) = 0.$$

If $b \neq 0$, this equation has a single equilibrium, namely 0. (That is, the only constant function that is a solution is equal to 0 for all t.) To consider the stability of this equilibrium, consider separately the three possible forms of the general solution of the equation.

Characteristic Equation has Two Real Roots

If the characteristic equation has two real roots, r and s, so that the general solution of the equation is $Ae^{rt} + Be^{st}$, then the equilibrium is stable if and only if $r < 0$ and $s < 0$.

Characteristic Equation has a Single Real Root

If the characteristic equation has a single real root, then the equilibrium is stable if and only if this root is negative. (Note that if $r < 0$ then tor any value of k, $t^k e^{rt}$ converges to 0 as t increases without bound.)

Characteristic Equation has Complex Roots

If the characteristic equation has complex roots, the form of the solution of the equation is $Ae^{\alpha 2} \cos(\beta t + \omega)$, where $\alpha = -a/2$, the *real* part of each root. Thus the equilibrium is stable if and only if the real part of each root is negative.

The real part of a real root is simply the root, so we can combine the three cases: the equilibrium is stable if and only if the real parts of both roots of the characteristic equation are negative. A bit of algebra shows that this condition is equivalent to $a > 0$ and $b > 0$.

If $b = 0$, then every number is an equilibrium, and none of these equilibria is stable.

In summary, we have the following result.

Proposition: An equilibrium of the homogeneous linear second-order differential equation $x''(t) + ax'(t) + bx(t) = 0$ is stable if and only if the real parts of both roots of the characteristic equation $r^2 + ar + b = 0$ are negative, or, equivalently, if and only if $a > 0$ and $b > 0$.

Example 18. Consider the following macroeconomic model. Denote by Q aggregate supply, p the price level, and p the expected rate of inflation. Assume that aggregate demand is a linear function of p and π, equal to $a - bp + c\pi$ where $a > 0$, $b > 0$, and $c > 0$. An equilibrium condition is

$$Q(t) = a - bp(t) + c\pi(t).$$

Denote by Q^* the long-run sustainable level of output, and assume that prices adjust according to the equation

$$p'(t) = h(Q(t) - Q^*) + \pi(t),$$

where $h > 0$. Finally, suppose that expectations are adaptive:

$$\pi'(t) = k(p'(t) - \pi(t))$$

for some $k > 0$. Is this system stable?

One way to answer this question is to reduce the system to a single second-order differential equation by differentiating the equation for $p'(t)$ to obtain $p''(t)$ and then substituting in for $\pi'(t)$ and $\pi(t)$. We obtain

$$p''(t) - h(kc - b)\,p'(t) + khbp(t) = kh(a - Q^*).$$

We conclude the system is stable if and only if $kc < b$. (Since $k > 0$, $h > 0$, and $b > 0$, we have $khb > 0$.)

In particular, if $c = 0$ (i.e. expectations are ignored) then the system is stable. If expectations are taken into account, however, and respond rapidly to changes in the rate of inflation (k is large), then the system may be unstable.

Systems of First-order Linear Differential Equations

Two Equations in Two Variables

Consider the system of linear differential equations (with constant coefficients)

$$x'(t) = ax(t) + by(t)$$

$$y'(t) = cx(t) + dy(t).$$

We can solve this system using the techniques from the previous section, as follows. First isolate $y(t)$ in the first equation, to give

$$y(t) = x'(t)/b - ax(t)/b.$$

Now differentiate this equation, to give

$$y'(t) = x''(t)/b - ax'(t)/b.$$

Why is this step helpful? Because we can now substitute for $y(t)$ and $y'(t)$ in the second of the two equations in our system to yield

$$x''(t)/b - ax'(t)/b = cx(t) + d[x'(t)/b - ax(t)/b],$$

which we can write as

$$x''(t) - (a + d)x'(t) + (ad - bc)x(t) = 0,$$

an equation we know how to solve!

Having solved this linear second-order differential equation in $x(t)$, we can go back to the expression for $y(t)$ in terms of $x'(t)$ and $x(t)$ to obtain a solution for $y(t)$.

(We could alternatively have started by isolating $x(t)$ in the second equation and creating a second-order equation in $y(t)$.)

Example 19. Consider the system of equations

$$x'(t) = 2x(t) + y(t)$$

$$y'(t) = -4x(t) - 3y(t).$$

Isolating $y(t)$ in the first equation we have $y(t) = x'(t) - 2x(t)$, so that $y'(t) = x''(t) - 2x'(t)$. Substituting these expressions into the second equation we get

$$x''(t) - 2x'(t) = -4x(t) - 3x'(t) + 6x(t),$$

or $x''(t) + x'(t) - 2x(t) = 0.$

We have seen that the general solution of this equation is

$$x(t) = Ae^{t} + Be^{-2t}.$$

Using the expression $y(t) = x'(t) - 2x(t)$ we get

$$y(t) = Ae^{t} - 2Be^{-2t} - 2Ae^{t} - 2Be^{-2t},$$

or $y(t) = -Ae^{t} - 4Be^{-2t}.$

General Linear Systems

We may write the system

$$x'(t) = ax(t) + by(t)$$

$$y'(t) = cx(t) + dy(t),$$

studied above, as

$$\begin{pmatrix} x'(t) \\ y'(t) \end{pmatrix} = \begin{pmatrix} a & b \\ c & d \end{pmatrix} \begin{pmatrix} x(t) \\ y(t) \end{pmatrix}.$$

More generally, we may write a system of n simultaneous homogeneous linear equations in the n variables $x_i(t)$ for $i = 1, ..., n$ as

$$x'(t) = Ax(t),$$

where A is an $n \times n$ matrix.

Now, if $n = 1$, in which case A is simply a number, we know that for the initial condition $x(0) = C$ the equation has a unique solution

$$x(t) = Ce^{At}.$$

Here's a stunning result:

for *every* value of n, the unique solution of the homogeneous system of linear differential equations $x'(t) = Ax(t)$ subject to the initial condition $x(0) = C$ is

$$x(t) = Ce^{At}.$$

You will immediately ask: what is e^{At} when A is a matrix? Recall that if a is a number we have

$$e^a = 1 + a/1! + a^2/2! + ...,$$

or, more precisely,

$$e^a = \sum_{k=0}^{\infty}(A^k / k!).$$

Now, when A is a matrix we may *define*

$$e^A = \sum_{k=0}^{\infty}(A^k / k!),$$

where A^0 is the identity matrix (and $0! = 1$). (You know how to multiply matrices together, so you know how to compute the right hand side of this equation.) That's it! You can now find the solution of any homogeneous system of linear differential equations ... assuming that you can compute the infinite sum in the definition of e^{At}. Therein lies the difficulty. There are techniques for finding e^{At}, but they involve methods more advanced than the ones in this tutorial.

I give only one example, which shows how the trigonometric functions may emerge in the solution of a system of two simultaneous linear equations, which, as we saw above, is equivalent to a second-order equation.

Example

Consider the system

$$\begin{pmatrix} x'_1(t) \\ x'_2(t) \end{pmatrix} = \begin{pmatrix} 0 & -b \\ b & 0 \end{pmatrix} \begin{pmatrix} x_1(t) \\ x_2(t) \end{pmatrix}.$$

We need to find A^k for each value of k, where

$$A = \begin{pmatrix} 0 & -b \\ b & 0 \end{pmatrix}.$$

You should be able to convince yourself (by computing A^2, A^3, and A^4) that if k is odd we have

$$A^k = \begin{pmatrix} 0 & (-1)^{(k+1)/2} b^k \\ (-1)^{(k-1)/2} b^k & 0 \end{pmatrix},$$

whereas if k is even we have

$$A^k = \begin{pmatrix} (-1)^{k/2} b^k & 0 \\ 0 & (-1)^{k/2} b^k \end{pmatrix}.$$

Given these results, we have

$$e^A = \begin{pmatrix} 1 - b^2/2 + b^4/4! - b^6/6! + \ldots & -b + b^3/3! - b^5/5! + \ldots \\ b - b^3/3 + b^5/5! - \ldots & 1 - b^2/2! + b^4/4! + b^6/6! + \ldots \end{pmatrix}.$$

Now, you should recall that

$$\sin b = b - b^3/3! + b^5/5! - \ldots$$

and

$$\cos b = 1 - b^2/2! + b^4/4! - b^6/6! + \ldots$$

Thus

$$e^A = \begin{pmatrix} \cos b & -\sin b \\ \sin b & \cos b \end{pmatrix}.$$

We conclude that the solution of the system of equations given the initial conditions $x_1(0) = C_1$ and $x_2(0) = C_2$ is

$$x_1(t) = C_1 \cos bt - C_2 \sin bt$$
$$x_2(t) = C_1 \sin bt + C_2 \cos bt.$$

You are asked, in an exercise, to verify this solution by using the technique discussed at the start of this section to convert the two-equation system to a single second-order linear differential equation.

You may wonder whether the emergence of the trigonometric functions in this example is accidental. It is not! Exponentials, the trigonometric functions, and complex numbers (which arise as roots of the characteristic equation in the technique of the previous section), are closely related. Specifically, for any value of x we have

$$e^{ix} = \cos x + i \sin x,$$

(where i is the square root of -1). But that is another story.

(An excellent, but advanced, exposition of the material in this section is contained in *Differential equations, dynamical systems, and linear algebra* by Morris W. Hirsch and Stephen Smale.)

Example 20.

1. Find the general solution of the system of equations

$$x'(t) = 4y(t)$$

$$y'(t) = -x(t) + 4y(t).$$

2. Find the general solution of the following system of equations and the particular solution with $x(0) = 1$ and $y(0) = 0$.

$$x'(t) = x(t) - 5y(t)$$

$$y'(t) = 2x(t) - 5y(t).$$

3. Find the solution of the system

$$x'(t) = -by(t)$$

$$y'(t) = bx(t)$$

4. given the initial conditions $x(0) = C_1$ and $y(0) = C_2$.

Solution.

1. Isolating $y(t)$ in the first equation we have $y(t) = x'(t)/4$, so that $y'(t) = x''(t)/4$. Substituting these expressions into the second equation we get

$$x''(t)/4 = -x(t) + x'(t),$$

or

$$x''(t) - 4x'(t) + 4x(t) = 0.$$

The characteristic equation is $(r-2)^2$, which has the repeated root $r = 2$. Thus the general solution of this equation is

$$x(t) = (A + Bt)e^{2t}.$$

Given $y(t) = x'(t)/4$, we thus have

$$y(t) = (1/2)(A + Bt)e^{2t} + (1/4)Be^{2t} = [(1/4)B + (1/2)A + (1/2)Bt]e^{2t}.$$

2. The two equations generate the second-order equation

$$x''(t) + 4x'(t) + 5x(t) = 0.$$

The characteristic equation has complex roots. The solution of the second-order equation is

$$x(t) = e^{-2t}(A \cos t + B \sin t).$$

Calculating x' and substituting into the first equation we obtain

$$y(t) = (1/5)e^{-2t}((3A - B)\cos t + (A + 3B)\sin t).$$

For the given initial conditions we have $A = 1$ and $B = 3$.

3. We have $x''(t) = -b\,y'(t) = -b^2x(t)$, or

$$x''(t) + b^2x(t) = 0.$$

The characteristic equation is $r^2 + b^2 = 0$, which has complex roots. Thus, given the general form of the solution to such an equation, the solution of our equation is

$$x(t) = C_1\cos bt - C_2\sin bt.$$

Using $y(t) = -x'(t)/b$, we have $y(t) = C_1\sin bt + C_2\cos bt$. (Remember that the derivative of sin bt is $b$cos bt, and the derivative of cos bt is $-b$sin bt.)

EXERCISE

1. (i) Solve the following differential equations.

   a. $(x(t))^2 x'(t) = t + 1$.

   b. $x'(t) = t^3 - t$.

   c. $x'(t) = te^t - t$.

   d. $e^{x(t)} x'(t) = t + 1$.

   e. $x'(t) = e^{-x(t)}/tx(t)$.

   f. $x'(t) = 4tx(t) + t$.

   (ii) Solve the following initial value problems.

   a.. $t\,x'(t) = x(t)(1 - t)$, $(t, x) = (1, 1/e)$.

   b. $(1 + t^3)\,x'(t) = t^2x(t)$, $(t, x) = (0, 2)$.

   c. $x(t)\,x'(t) = t$, $(t, x) = (\sqrt{2}, 1)$.

   d. $e^{2t} x'(t) - (x(t))^2 - 2x(t) - 1 = 0$, $(t, x) = (0, 0)$.

   e. $x'(t) = 4tx(t) + t$, $(t, x) = (0, 0)$.

2. Form a differential equation:

   a. $x(t) = Cte^{-t}$; $C = 1$.

   b. $x(t) = C(1 + t^3)^{1/3}$; $C = 2$.

   c. $x(t) = \sqrt{\{t^2 + C\}}$; $C = -1$.

d. $x(t) = (2 - C - e^{-2t})/(C + e^{-2t})$; $C = 1$.

e. $x(t) = Ce^{2t2} - 1/4$; $C = 1/4$.

3. **Solve the following differential equations in general, and subject to the given initial conditions. In each case, determine if the solution is stable or unstable.**

   a. $x''(t) + 3x'(t) - 4x(t) = 12$. **Initial conditions:** $x(0) = 4$, $x'(0) = 2$.

   b. $x''(t) - 2x'(t) + x(t) = 3$. **Initial conditions:** $x(0) = 4$, $x'(0) = 2$.

   c. $x''(t) + 4x'(t) + 8x(t) = 2$. **Initial conditions:** $x(0) = 9/4$, $x'(0) = 4$.

4. **Find a particular solution of the differential equation**

   $x''(t) + 2x'(t) + x(t) = t$.

5. **Solve the differential equation** $x''(t) - 4x'(t) + 4x(t) = 5$ **in general, and subject to the initial conditions** $x(0) = 4$ **and** $x'(0) = 6$.

6. **Find the general solution of** $x'(t) + (1/2)x(t) = 1/4$. **Determine the equilibrium state and examine its stability.**

7. **Find the general solutions of the following differential equations, and in each case solve the associated initial value problem for** $x(0) = 1$.

   (i) $x'(t) - 3x(t) = 5$.

   (ii) $3x'(t) + 2x(t) + 16 = 0$.

   (iii) $x'(t) + 2x(t) = t^2$.

8. **Solve the following differential equations.**

   (i) $tx'(t) + 2x(t) + t = 0$ for $t \neq 0$.

   (ii) $x'(t) - x(t)/t = t$ for $t > 0$.

   (iii) $x'(t) - tx(t)/(t^2 - 1) = t$ for $t > 1$.

   (iv) $x'(t) - 2x(t)/t + 2a^2/t^2 = 0$ for $t > 0$.

9. **In the theory of auctions, an initial value problem of the form**

   $x'(t)G(t) + x(t)G'(t) = tG'(t)$ with $x(t_0) = t_0$,

   where G is a known function, arises. (The variable t is interpreted as a player's valuation of the object for sale in the auction.) Solve this equation, expressing the solution in terms of the function G (but not its derivative). (If you start by converting the equation into the standard form and finding the integrating factor, reflect afterwards on the fact that you did not need to do so.)

Chapter 5

Conformal Mappings

COMPLEX FUNCTIONS

This chapter focuses primarily on very basic functions, their representations, and properties associated with functions such as limits and continuity. You will learn some interesting applications as well as some exciting new ideas.

FUNCTIONS AND LINEAR MAPPINGS

A complex-valued function f of the complex variable z is a rule that assigns to each complex number z in a set D one and only one complex number w. We write $w = f(z)$ and call w the image of z under f. A simple example of a complex-valued function is given by the formula $w = f(z) = z^4$. The set D is called the domain of f, and the set of all images $\{w = f(z) : z \in D\}$ is called the range of f. When the context is obvious, we omit the phrase complex-valued, and simply refer to a function f, or to a complex function f.

We can define the domain to be any set that makes sense for a given rule, so for $w = f(z) = z^4$, we could have the entire complex plane for the domain D, or we might artificially restrict the domain to some set such as $D = D_1(0) = \{z : |z| < 1\}$. Determining the range for a function defined by a formula is not always easy, but we will see plenty of examples later on. In some contexts functions are referred to as mappings or transformations.

In Section 1.6, we used the term domain to indicate a connected open set. When speaking about the domain of a function, however, we mean only the set of points on which the function is defined. This distinction is worth noting, and context will make clear the use intended.

Just as z can be expressed by its real and imaginary parts, $z = x + i\,y$, we write $f(z) = w = v + i\,v$, where u and v are the real and imaginary parts of w, respectively. Doing so gives us the representation.

$$w = f(z) = f(x, y) = f(x + i\,y) = u + i\,v.$$

Because u and v depend on x and y, they can be considered to be real-valued functions of the real variables x and y; that is,

$$u = u(x, y) \text{ and } v = v(x, y).$$

Combining these ideas, we often write a complex function f in the form

$$f(z) = f(x + i\,y) = u\,(x, y) + i\,v\,(x, y). \quad ...(1)$$

Figure 1 illustrates the notion of a function (mapping) using these symbols.

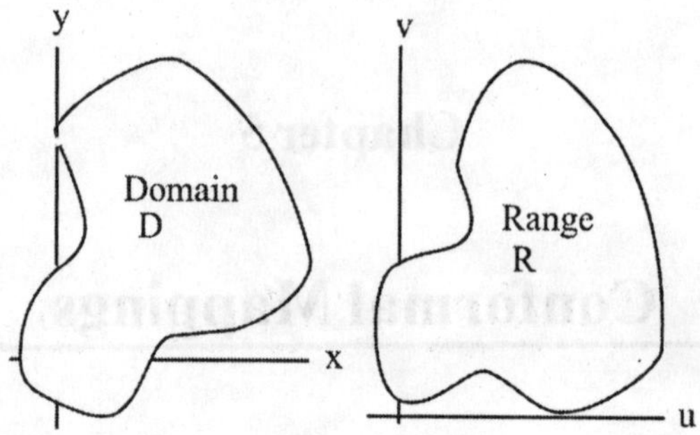

Fig. 1 The mapping $w = f(z) = u\,(x, y) + i\,v\,(x, y)$.

Example 1. Write $f(z) = z^4$ in the for $f(z) = u\,(x, y) + i\,v\,(x, y)$.

Solution. Using the binomial formula, we obtain

$$\begin{aligned} f(z) &= f(x + i\,y) = (x + i\,y)4 \\ &= x^4 + 4x^3 (i\,y) + 6x^2 (i\,y)^2 + 4x\,(i\,y)^3 + (i\,y)^4 \\ &= x^4 + 4\,i\,x^3\,y - 6\,x^2 y^2 - 4\,i\,x\,y^3 + y^4 \\ &= x^4 - 6\,x^2\,y^2 + y^4 + i\,(4\,x^3\,y - 4\,x\,y^3) \\ &= u\,(x, y) + i\,v\,(x, y) \end{aligned}$$

so that $u\,(x, y) = x^4 - 6\,x^2\,y^2 + y^4$ and $v\,(x + y) = 4\,x^3\,y - 4\,x\,y^3$.

Example 2. Express the function $f(z) = z\,\text{Re}\,[z] + z^2\,\text{Im}\,[z]$ in the form $f(z) = u\,(x, y) + v\,(x, y)$.

Solution. Using the elementary properties of complex numbers, it follows that

$$f[z] = (x - i\,y)x + (x + i\,y)^2 + y = (2x^2 - y^2 + y) + i\,(x\,y)$$

so that $u\,(x, y) = 2x^2 - y^2 + y$ and $(x, y) = x\,y$.

Examples 1 and 2 show how to find $u(x, y)$ and $v(x, y)$ when a rule for computing f is given. Conversely, if $u(x, y)$ and $v(x, y)$ are two real-valued functions of the real variables x and y, they determine a complex-valued function

$f(z) = u\,(x, y) + i\,v\,(x, y)$, and we can use the formulas

$$x = \frac{z + \bar{z}}{2} \text{ and } y = \frac{z - \bar{z}}{2i} \quad ...(2)$$

to find a formula for f involving the variables z and $\bar{z}$.

Example 3. Express $f(z) = 4x^2 + i4y^2$ by a formula involving the variables z and $\bar{z}$.

Solution. Calculation reveals that

$$f(z) = 4\left(\frac{z+\bar{z}}{2}\right)^2 + i4\left(\frac{z-\bar{z}}{2i}\right)^2$$

$$= z^2 + 2z\bar{z} + \bar{z}^2 - i(z^2 - 2z\bar{z} + \bar{z}^2)$$

$$= (1-i)z^2 + (2+2i)z\bar{z} + (1-i)\bar{z}^2$$

Using $z = re^{i\theta}$ in the expression of a complex function f may be convenient. It gives us the polar representation

$$f(z) = f(re^{i\theta}) = u(r,\theta) + iv(r,\theta), \qquad \text{...(3)}$$

where U and V are real functions of the real variables r and θ.

Remark. For a given function f, the functions u and v defined above are different from those used previously in $f(z) = f(x + iy) = u(x, y) + iv(x, y)$ which used Cartesian coordinates instead of polar coordinates.

Example 4. Express $f(z) = z^2$ in both Cartesian and polar form.

Solution. For the Cartesian form, a simple calculation gives

$$f(z) = f(x + iy) = (x + iy)^2$$

$$x^2 - y^2 + 2ixy$$

$$u(x, y) + iv(x, y)$$

so that $u(x, y) = x^2 - y^2$ and $v(x, y) = xy$.

For the polar form, we get v

$$f(z) = f(re^{i\theta}) = (re^{i\theta})^2 = r^2e^{i2\theta}$$

$$= r^2(\cos 2\theta + i\sin 2\theta)$$

$$= r^2\cos 2\theta + ir^2\sin 2\theta$$

$$= U(r,\theta) + iV(r,\theta)$$

so that $= u(r,\theta) + r^2\cos 2\theta$ and $v(r,\theta) = r^2\sin 2\theta$.

Remark. Once we have defined u and v for a function f in Cartesian form, we must use different symbols if we want to express f in polar form. As is clear here, the functions u and U are quite different, as are v and V. Of course, if we are working only in one context, we can use any symbols we choose.

For a given function f, the functions u and v defined here are different from those defined by equation (1), because equation (2) involves Cartesian coordinates and equation (3) involves polar coordinates.

Example 5. Express $f(z) = z^5 + 4z^2 - 6$ in polar form.

Solution. We obtain

$$f(z) = f(re^{i\theta}) = (re^{i\theta})^5 + 4(re^{i\theta})^2 - 6$$

$$= r^5 e^{i5\theta} + 4r^2 e^{i2\theta} - 6$$

$$= r^5 \cos 5\theta + 4r^2 \cos 2\theta - 6 + i(r^5 \sin 5\theta + 4r^2 \sin 2\theta)$$

$$= U(r, \theta) + i\,V(r, \theta)$$

so that $U(r, \theta) = r^2 \cos 5\theta + 4r^2 \cos 2\theta - 6$ and $V(r, \theta) = r^2 \sin 5\theta + 4r^2 \sin 2\theta$.

We now look at the geometric interpretation of a complex function. If D is the domain of real-valued functions $u(x, y)$ and $v(x, y)$, the equations

$$u = u(x, y) \text{ and } v = v(x, y)$$

describe a transformation (or mapping) from D in the xy plane into the uv plane, also called the w plane. Therefore, we can also consider the function

$$w = f(z)\, u(x, y) + i\, v(x, y)$$

to be a transformation (or mapping) from the set D in the z plane onto the range R in the w plane. This idea was illustrated in Figure 1. In the following paragraphs we present some additional key ideas. They are staples for any kind of function, and you should memorize all the terms in bold.

If A is a subset of the domain D of f, the set $B = \{w = f(z) : z \in A\}$ is called the image of the set A, and f is said to map A onto B. The image of a single point is a single point, and the image of the entire domain, D, is the range, R. The mapping $w = f(z)$ is said to be from A into S if the image of A is contained in S. Mathematicians use the notation $f : A \to S$ to indicate that a function maps A into S. Figure 2 illustrates a function f whose domain is D and whose range is R. The shaded areas depict that the function maps A onto B. The function also maps A into R, and, of course, it maps D onto R.

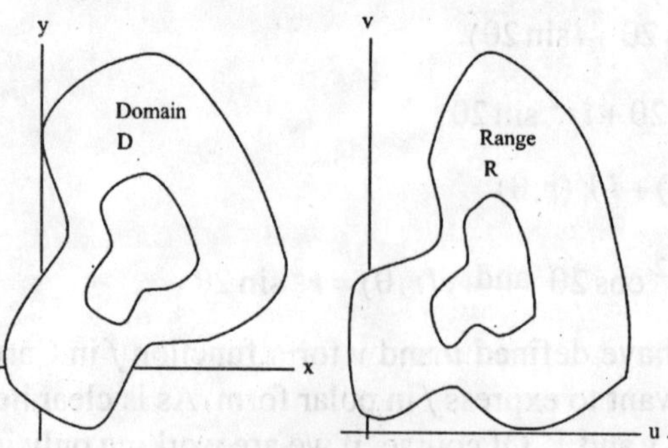

Fig. 2 $w = f(z)$ maps A onto B; $w = f(z)$ maps A into R.

The inverse image of a point w is the set of all points z in D such that $w = f(z)$. The inverse image of a point may be one point, several points, or nothing at all. If the last case occurs then the point w is not in the range of f. For example, if $w = f(z) = i\,z$, the inverse image of the point -1 is the single point i, because $w = f(i) = i\,(i) = -1$, and i is the only point that maps to -1. In the case of $w = f(z) = z^2$, the inverse image of the point -1 is the set $\{i, -i\}$. You will learn that, if $w = f(z) = e^z$, the inverse image of the point 0 is the empty set---there is no complex number z such that $e^z = 0$.

The inverse image of a set of points, S, is the collection of all points in the domain that map into S. If f maps D onto R it is possible for the inverse image of R to be function as well, but the original function must have a special property: a function f is said to be one-to-one if it maps distinct points $z_1 \neq z_2$, onto distinct points $f(z_1) \neq f(z_2)$. Many times an easy way to prove that a function f is one-to-one is to suppose $f(z_1) = f(z_2)$, and from this assumption deduce that z_1 must equal z_2. Thus, $f(z) = i\,z$ is one-to-one because if $f(z_1) = f(z_2)$, then $i\,z_1 = i\,z_2$. Dividing both sides of the last equation by i gives $z_1 = z_2$. Figure 3 illustrates the idea of a one-to-one function: distinct points get mapped to distinct points.

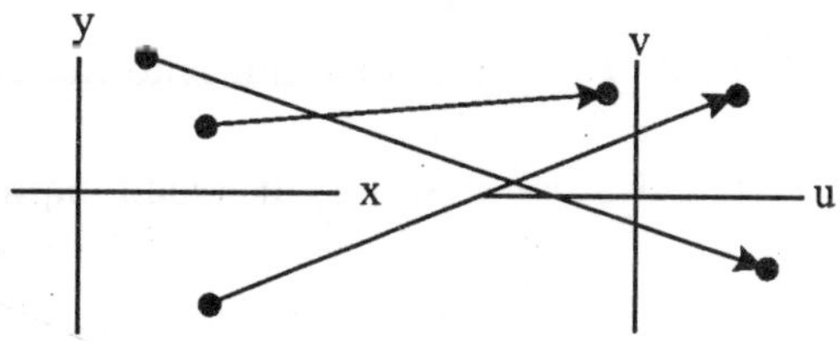

Fig. 3 A Function $w = f(z)$ that is one-to-one.

The function $f(z) = z^2$ is not one-to-one because $-i \neq i$, but $f(i) = f(-i) = -1$. Figure 4 depicts this situation: at least two different points get mapped to the same point.

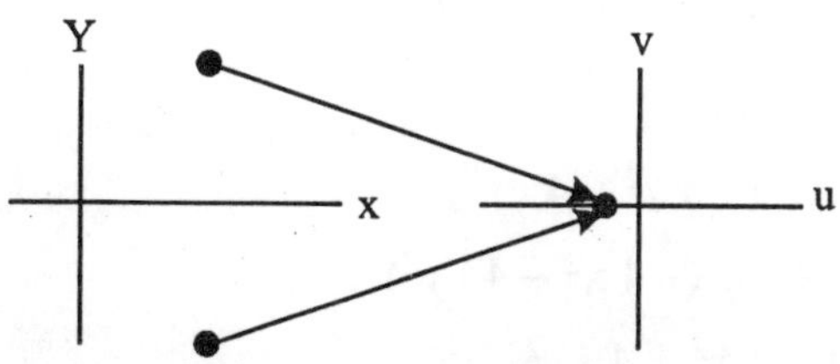

Fig. 4 A Function that is not one-to-one.

In the exercises we ask you to demonstrate that one-to-one functions give rise to inverses that are functions. Loosely speaking, if $w = f(z)$ maps the set A one-to-one and onto the set B, then for each w in B there exists exactly one point z in A A such that $w = f(z)$. For any such value of z we can take the equation $w = f(z)$ and "solve" for z as a function of w. Doing so produces an inverse function $z = g(w)$ where the following equations hold:

$g(f(z)) = z$ for all $z \in A$, and

$f(g(w)) = w$ for all $z \in B$

Conversely, if $w = f(z)$ and $z = g(w)$ are functions that map A into B and B into A, respectively, and the above hold, then f maps the set A one-to-one and onto the set B.

Further, if f is a one-to-one mapping from D onto T and if A is a subset of D, then f is a one-to-one mapping from A onto its image B. We can also show that, if $\zeta = f(z)$ is a one-to-one mapping from A onto B and $w = g(\zeta)$ is a one-to-one mapping from B onto S, then the composite mapping $w = g(f(z))$ is a one-to-one mapping from A onto S.

We usually indicate the inverse of f by the symbol f^{-1}. If the domains of f and f^{-1} are A and B respectively, then we write

$f^{-1}(f(z)) = z$ for all $z \in A$, and $f(f^{-1}(w)) = w$ for all $w \in B$.

Also, for $z_0 \in A$ and $w_0 \in B$.

$$w_0 = f(z_0) \text{ iff } f^{-1}(w_0) = z_0, \text{ and}$$

$$z_0 = f^{-1}(w_0) \text{ iff } f(w_0) = z_0.$$

Example 6. If $w = f(z) = i\,z$ for any complex number z, find $f^{-1}(w)$. Also find $f(f^{-1}(w))$

Solution. We can easily show f is one-to-one and onto the entire complex plane. We solve for z, given $w = f(z) = i$, to get $z = \dfrac{w}{i} = -i\,w$. This result implies that $f^{-1}(w) = -i\,w$ for all complex numbers w.

Since,

$$w = f[z] = i\,z$$

$$\{\{z \to i\,(-4\,x^2 - 4\,i\,y^2)\}\}$$

$$z = f^{-1}[w] = i\,(-4\,x^2 - 4\,i\,y^2)$$

$$f^{-1}[f[z]] = -i\,(i\,z)$$

$$f^{-1}[f[z]] = i\,(-4\,x^2 - 4\,i\,y^2)$$

$$f[f^{-1}[w]] = i\,(i\,(-4\,x^2 - 4\,i\,y^2))$$

$$f[f^{-1}[w]] = 4\,x^2 + 4\,i\,y^2$$

Remark. Once we have specified $f^{-1}(w) = -i\,w$ for all complex numbers w, we note that there is nothing magical about the symbol w. We could just as easily write $f^{-1}(z) = -i\,z$ for all complex numbers z.

We now show how to find the image B of a specified set A under a given mapping $u + iv = w = f(z)$.

The set A is usually described with an equation or inequality involving x and y. Using inverse functions, we can construct a chain of equivalent statements leading to a description of the set B in terms of an equation or an inequality involving u and v.

Example 7. Show that the function $f[z] = i\,z$ maps the line y = x + 1 in the *xy* plane onto the line $v = -u - 1$ in the *w* plane.

Solution.

Method 1: With $A = \{(x, y)\} : y = x + 1$, we want to describe $B = f(A)$. We let $z = x + i\,y \in A$ and get

$$u + i\,v = w = f(z) \in B$$

$$\Leftrightarrow f^{-1}(w) = z = x + i\,y \in A$$

$$\Leftrightarrow i\,w \in A$$

$$\Leftrightarrow v - i\,u \in A$$

$$\Leftrightarrow (v, -u) \in A$$

$$\Leftrightarrow u = v + 1$$

$$\Leftrightarrow v = -u - 1$$

where $\Leftrightarrow$ is the notation for "if and only if." Note what this result says: $u + i\,v = w + B \Leftrightarrow v = -u - 1$. The image of A under f, therefore, is the set $B = \{(u, v) : v = -u - 1\}$.

Method 2: We write $u + i\,v = w = f(z) = i\,(x + i\,y) = -y + i\,x$ and note that the transformation can be given by the equations $u = -y$ and $v = x$. Because A is described by $A = \{x + i\,y : y = x + 1\}$, we can substitute $u = -y$ and $v = x$ into the equation $y = x + 1$ to obtain $-u = v + 1$, which we can rewrite as $v = -u - 1$. If you use this method, be sure to pay careful attention to domains and ranges.

We now look at some elementary mappings. If we let $B = a + ib$ denote a fixed complex constant, the transformation $w = T(z) = z + B = x + a + i\,(y + b)$ is a one-to-one mapping of the *z*-plane onto the *w*-plane and is called a translation.

TRANSLATION. TRANSFORMATION

A transformation consisting of a constant offset with no rotation or distortion. In *n*-dimensional Euclidean space, a translation may be specified simply as a vector giving the offset in each of the *n*coordinates. This transformation can be visualized as a rigid translation whereby the point z is displaced through the vector $B = a + ib$ to its new position $w = T(z)$. The inverse mapping is given by

$$z = T^{-1}(w) = w - B = u - a + i\,(v - b)$$

and shows that T is a one-to-one mapping from the *z*-plane onto the *w*-plane. The effect of a translation is depicted in Figure 5.

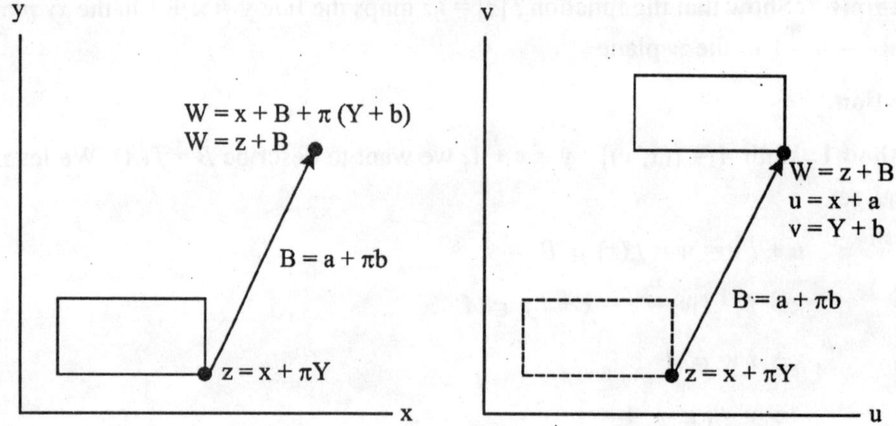

Fig. 5 The Translation $w = T(z) = z + B = x + a + i(y + b)$.

If we let α be a fixed real number, then for $z = re^{i\theta}$, the transformation

$$w = R(z) = ze^{i\alpha} = re^{i\theta}e^{i\alpha} = re^{i(\theta+\alpha)}$$

is a one-to-one mapping of the z-plane onto the w-plane and is called a rotation.

ROTATION TRANSFORMATION

The symmetry operation corresponding to a rotation followed by an inversion operation, also called a roto-inversion. This operation is denoted $\bar{n}$ for an improper rotation by 360°/n, so the crystallography restriction gives only $\bar{1}, \bar{2}, \bar{3}, \bar{4}, \bar{6}$ for crystals. The mirror plane symmetry operation is $(x, y, z) \to (x, y, -z)$, etc., which is equivalent to $\bar{2}$.

The turning of an object or coordinate system by an angle about a fixed point. A rotation is an orientation-preserving orthogonal transformation. Euler's rotation theorem states that an arbitrary rotation can be parameterized using three parameters. These parameters are commonly taken as the Euler angles. Rotations can be implemented using rotation matrices.

Rotation in the plane can be concisely described in the complex plane using multiplication of complex numbers with unit modulus such that the resulting angle is given by arg (z). For example, multiplication by i represents a rotation to the right by 90° and by $-i$ represents rotation to the left by 90°. So starting with $z = 1$ and rotating left twice gives $z = 1(-i)(-i) = i^2 = -1$, which is the same as rotating right twice, $z = 1(i)(i) = i^2 = -1$, and arg $(-1) = \pi = 180°$. For multiplication by multiples of 90°, the possible positions are then concisely represented by $z = 1, i, -1$, and $-i$.

The rotation symmetry operation for rotation by 360°/n is denoted "n." For periodic arrangements of points ("crystals"), the crystallography restriction gives the only allowable rotations as 1, 2, 3, 4, and 6.

It can be visualized as a rigid rotation whereby the point z is rotated about the origin through an angle α to its new position $w = R(z)$. If we use polar coordinates and designate $w = \rho e^{i\phi}$ in the w-plane, then the inverse mapping is

$$z = R^{-1}(w) = we^{-i\alpha} = \rho e^{i\phi} e^{-i\alpha} \rho e^{i(\phi - \alpha)}.$$

This analysis shows that R is a one-to-one mapping from the z-plane onto the w-plane. The effect of rotation is depicted in Figure 6.

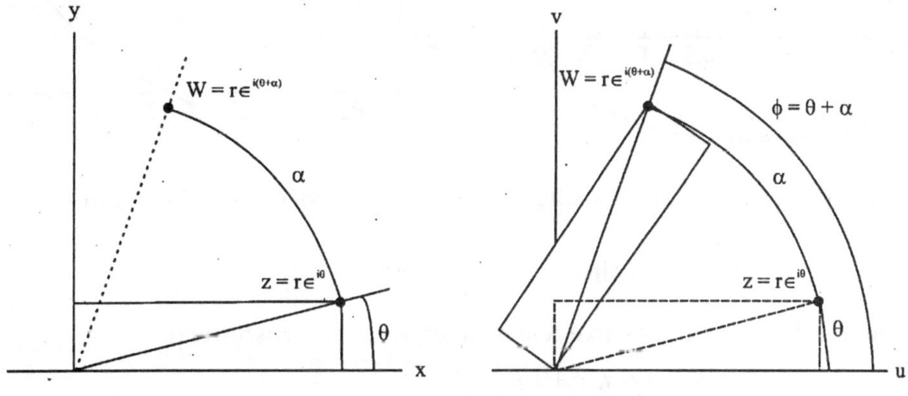

Fig. 6 The rotation w = R (z) = $re^{i(\theta+\alpha)}$.

Example 8. The ellipse centered at the origin with a horizontal major axis of 4 units and vertical minor axis of 2 units can be represented by the parametric equation

$$s(t) = 2\cos t + i\sin t = (2\cos t, \sin t), \text{ for } 0 \le t \le 2\pi.$$

Solution:

Suppose we wanted to rotate the ellipse by an angle of $\frac{\pi}{6}$ radians and shift the center of the ellipse 2 units to the right and 1 unit up. Using complex arithmetic, we can easily generate a parametric equation $r(t)$ that does so:

$$r[t] = s[t]e^{-i\pi/6} + (2+i)$$

$$= (2\cos t + i\sin t)\left(\cos\frac{\pi}{6} + i\sin\frac{\pi}{6}\right) + (2+i)$$

$$= \left(2\cos t\cos\frac{\pi}{6} + -\sin t\sin\frac{\pi}{6}\right) + i\left(2\cos t\sin\frac{\pi}{6} + \sin t\cos\frac{\pi}{6}\right) + (2+i).$$

$$= \left(\sqrt{3}\cos t - \frac{1}{2}\sin t + 2\right) + i\left(\cos t + \frac{\sqrt{3}}{2}\sin t + 1\right)$$

$$\sqrt{3}\cos t - \frac{1}{2}\sin t + 2, \cos t + \frac{\sqrt{3}}{2}\sin t + 1$$

for $0 \leq t \leq 2\pi$. Figure 7 shows parametric plots of these ellipses.

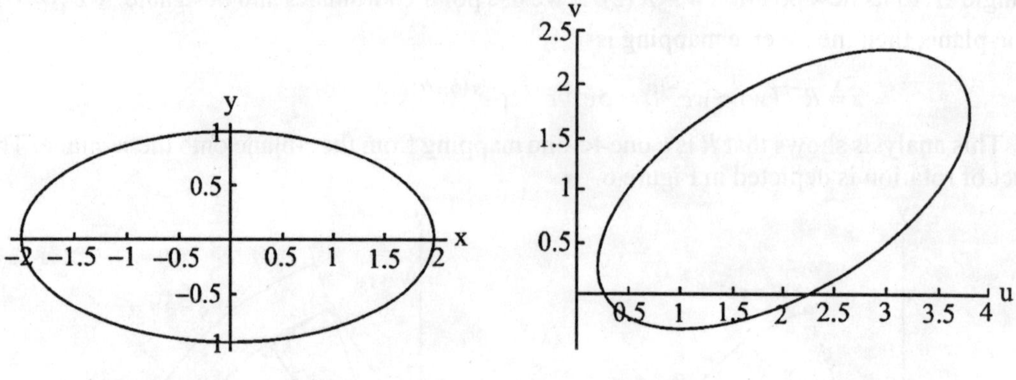

Fig. 7 (*a*) Plot of the Original Ellipse (*b*) Plot of the Rotated Ellipse

$$s(t) = 2\cos t + i \sin t \; r(t) = s(t)\, e^{i\pi/6} + (2+i)$$

If we let $K > 0$ be a fixed positive real number, then the transformation

$$w = S(z) = Kz = Kx + iKy$$

is a one-to-one mapping of the *z*-plane onto the *w*-plane and is called a magnification. If $K > 1$, it has the effect of stretching the distance between points by the factor *K*. If $K > 1$, then it reduces the distance between points by the factor *K*. The inverse transformation is given by

$$z = S^{-1}(w) = \frac{1}{K} w = \frac{1}{K} u + i \frac{1}{k} v$$

and shows that *S* is a one-to-one mapping from the *z*-plane onto the *w*-plane. The effect of magnification is shown in Figure 8.

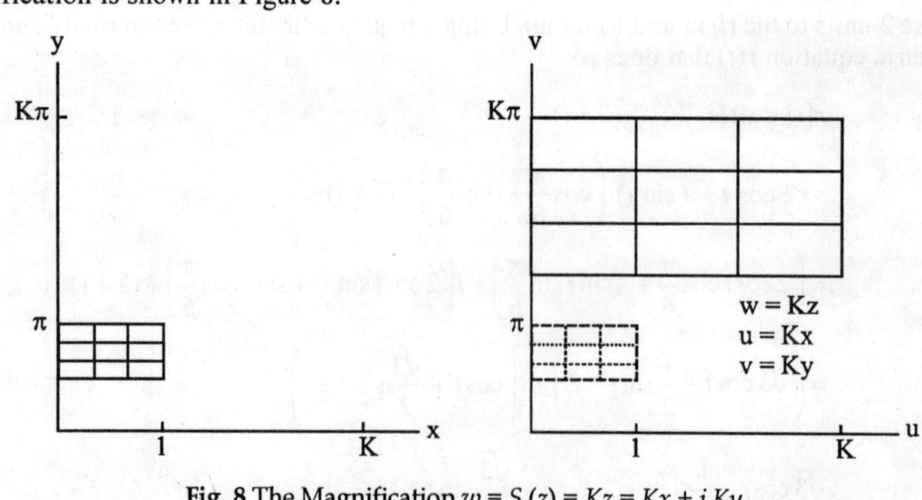

Fig. 8 The Magnification $w = S(z) = Kz = Kx + iKy$.

Finally, if we let $A = Ke^{i\alpha}$ and $B = a + i\,b$, where $K > 0$ is a positive real number, then the transformation

$$w = L\,(z) = A\,z + B$$

is a one-to-one mapping of the z-plane onto the w-plane and is called a linear transformation. It can be considered as the composition of a rotation, a magnification, and a translation. It has the effect of rotating the plane though an angle given by $\alpha = \text{Agr}\,A$, followed by a magnification by the factor $K = |\,A\,|$, followed by a translation by the vector $B = a + ib$. The inverse mapping is given by $z = L^{-1}(w) = \dfrac{1}{A}w - \dfrac{B}{A}$ and shows that L is a one-to-one mapping from the z-plane onto the w-plane.

Example 9. Show that the linear transformation $w = f(z) = i\,z + i$ maps the right half plane Re $(z) > 1$ onto the upper half plane Im $(w) > 2$.

Solution.

Method 1: Let A = {(x, y) : x ≥ 1}. To describe $B = F\,(A)$, we solve $w = i\,z + i$ for z to get $z = \dfrac{w-i}{i} = -iw - 1 = f^{-1}(w)$. We have the following

$$\text{u} + \text{i v} = \text{w} = \text{f (z)} \in \text{B}$$

$$\Leftrightarrow f^{-1}\,(w) = z \in A$$

$$\Leftrightarrow -\,i\,w - 1 \in A$$

$$\Leftrightarrow v - 1 - i\,u \in A$$

$$\Leftrightarrow (v - 1, -\,u) \in A$$

$$\Leftrightarrow v - 1 \geq 1$$

$$\Leftrightarrow v \geq 2$$

Thus $B = \{w = u + iv : v \geq 2\}$, which is the same as saying Im $(w) > 2$.

Method 2: When we write $w = f\,(z)$ in Cartesian form as

$$w = u + iv = i\,(x + i\,y) + i = -\,y + i\,(x + 1),$$

we see that the transformation can be given by the equations $u = -\,y$ and $v = x + 1$. Substituting $x = v - 1$ in the inequality Re $(z) = x \geq 1$ gives $v - 1 \geq 1$, or v ≥ 2, which is the upper half-plane Im $(w) > 2$.

Method 3: The effect of the transformation $w = f(z)$ is a rotation of the plane through the angle $\alpha = \dfrac{\pi}{2}$ (when z is multiplied by i) followed by a translation by the vector $B = i$. The first operation yields the set Im $(w) \geq 1$. The second shifts this set up 1 unit, resulting in the set Im $(w) > 2$. We illustrate this result in Figure 9.

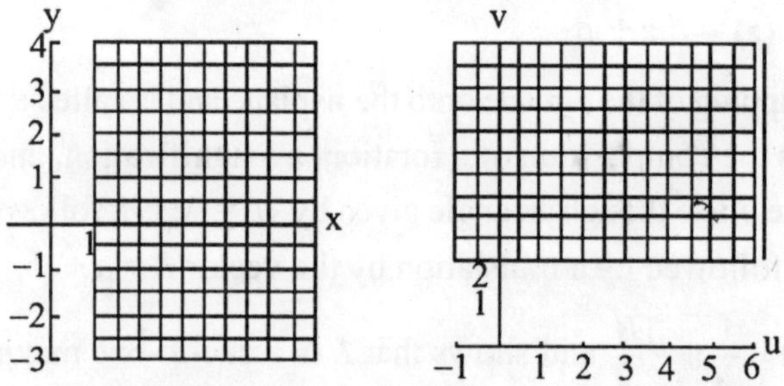

Fig. 9 The linear transformation $w = f(z) = i\,z + i$.

Translations and rotations preserve angles. First, magnifications rescale distance by a factor K, so it follows that triangles are mapped onto similar triangles, preserving angles. Then, because a linear transformation can be considered to be a composition of a rotation, a magnification, and a translation, it follows that linear transformations preserve angles. Consequently, any geometric object is mapped onto an object that is similar to the original object; hence linear transformations can be called similarity mappings.

Example 10. Show that the image of the open disk $D_1(-1-i) = \{z : |z+1+i| < 1\}$ under the linear transformation $w = f(z) = (3-4i)z + 6 + 2i$ is the open disk $D_5(-1+3i) = \{w : |w+1-3i| < 5\}$.

Solution. The inverse transformation is $z = \dfrac{w-6-2i}{3-4i}$, so if we designate the range of f as B, then

$$w = f(z) \in B \Leftrightarrow f^{-1}(w) = z \in D_1(-1-i)$$

$$\Leftrightarrow \frac{w-6-2i}{3-4i} \in D_1(-1-i)$$

$$\Leftrightarrow \left|\frac{w-6-2i}{3-4i} + 1 + i\right| < 1$$

$$\Leftrightarrow \left|\frac{w-6-2i}{3-4i} + 1 + i\right| |3-4i| < 1|3-4i|$$

$$\Leftrightarrow |w-6-2i+(1+i)(3-4i)| < 5$$

$$\Leftrightarrow |w+1-3i| < 5$$

Hence the disk with center $-1-i$ and radius 1 is mapped one-to-one and onto the disk with center $-1+3i$ and radius 5 as shown in Figure 10.

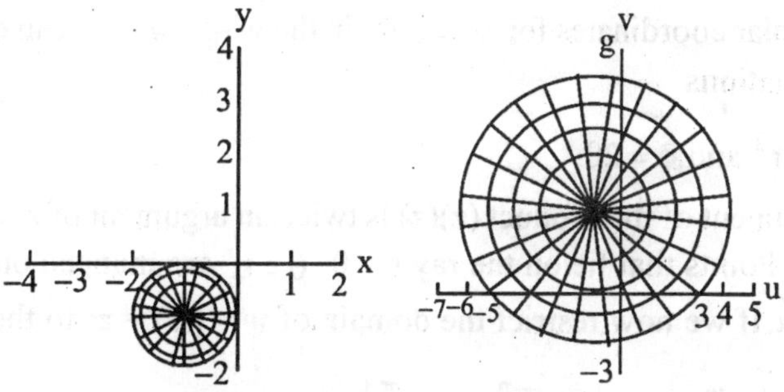

Fig. 10 The mapping $w = f(z) = (3 - 4i)z + 6 + 2i$.

Example 11. Show that the image of the right half plane Re $(z) = x > 1$ under the linear transformation $w = (-1 + i)z - 2 + 3i$ is the half plane $v > u + 7$.

Solution. The inverse transformation is given by

$$z = \frac{w+2-3i}{-1+i} = \frac{u+2+i(v-3)}{-1+i},$$

which we write as

$$x + iy = \frac{-u+v-5}{2} + i\frac{-u-v+1}{2}$$

Substituting $x = \dfrac{-u+v-5}{2}$ into e $\text{Re}(z) = x > 1$ gives $\dfrac{-u+v-5}{2} > 1$, which simplifies $v > u + 7$. Figure 11 illustrates the mapping.

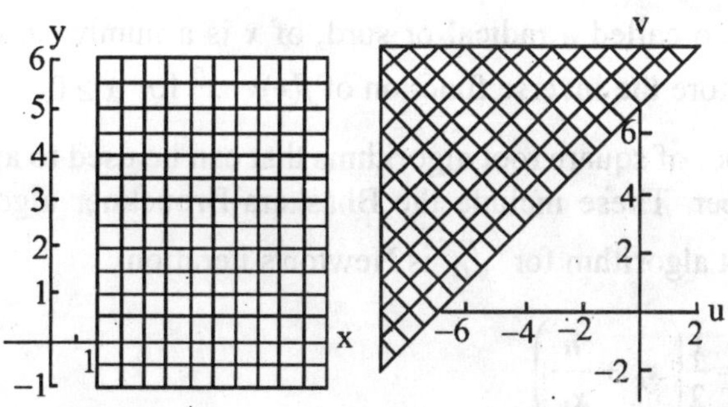

Fig. 11 The the linear transformation

THE MAPPINGS $w = z^n$ AND $w = z^{1/n}$

In this section we turn our attention to power functions. For $z = re^{i\theta}$, we can express the function $w = f(z) = z^2$ in polar coordinates by

7 $$w = f(z) = z^2 r^2 e^{i2\theta}$$

If we also use polar coordinates for $w = \rho e^{i\theta}$ in the w–plane, we can express this mapping by the system of equations

$$\rho = r^2 \text{ and } \phi = 2\theta.$$

Because an argument of the product (z)(z) is twice an argument of z, we say that f doubles angles at the origin. Points that lie on the ray $r > 0,\ \theta = \alpha$ are mapped onto points that lie on the ray $\rho = 0, \phi = 2\alpha$. If we now restrict the domain of $w = f(z) = z^2$ to the region

$$A = \left\{ re^{i\theta} : r > 0 \text{ and } \frac{-\pi}{2} < \theta \le \frac{\pi}{2} \right\},$$

then the image of A under the mapping $w = z^2$ can be described by the set

$$B = \left\{ \rho e^{i\theta} : \rho > 0 \text{ and } -\pi < \phi \le \pi \right\},$$

which consists of all points in the w-plane except the point w = 0.

The inverse mapping of f, which we shall denote by g, is then

$$z = g(w) = w^{1/2} \rho^{1/2} e^{i\phi/2}$$

where $w \in B$. That is

$$z = g(w) = w^{1/2} = |w|^{1/2} e^{iArg(w)/2},$$

where $w \neq 0$. The function g is so important that we call special attention to it with a formal definition.

Square Root Function

A square root, also called a radical or surd, of x is a number r such that $r^2 = x$. The function $\sqrt{x}$ is therefore the inverse function of $f(x) = x^2$ for $x \ge 0$.

There are a number of square root algorithms that can be used to approximate the square root of a given number. These include the Bhaskara-Brouckner algorithm and Wolfram's iteration. The simplest algorithm for $\sqrt{n}$ is Newton's iteration

$$x_{k+1} = \frac{1}{2}\left(x_k + \frac{n}{x_k} \right) \qquad ...(4)$$

with $x_0 = 1$.

Note that any positive real number has two square roots, one positive and one negative. For example, the square roots of 9 are – 3 and + 3, since $(-3)^2 = (+3)^2 = 9$. Any nonnegative real number x has a unique nonnegative square root r; this is called the principal square root and is written $r = x^{1/2}$ or $r = \sqrt{x}$. For example, the principal square root of 9 is $\sqrt{9} = +3$, while the other square root of 9 is $-\sqrt{9} = -3$.

The square roots of a complex number $z = x + i\,y$ are given by

$$\sqrt{x+i\,y} = \pm(x^2+y^2)^{1/4}\left\{\cos\left[\frac{1}{2}\tan^{-1}(x,y)\right] + i\sin\left[\frac{1}{2}\tan^{-1}(x,y)\right]\right\} \quad ...(5)$$

In addition,

$$\sqrt{x+i\,y} = \frac{1}{2}\sqrt{2}\left[\sqrt{\sqrt{x^2+y^2}+x} + x + i\sin(y)\sqrt{\sqrt{x^2+y^2}-x}\right] \quad ...(6)$$

As can be seen in the above figure, the imaginary part of the complex square root function has a branch cut along the negative real axis.

A nested radical of the form $\sqrt{a \pm b\sqrt{c}}$ can sometimes be simplified into a simple square root by equating

$$\sqrt{a \pm b\sqrt{c}} = \sqrt{d} \pm \sqrt{e}$$

Squaring gives

$$a \pm b\sqrt{c} = d + e \pm 2\sqrt{de}$$

so

$$a = d + e$$
$$b^2\,c = 4\,d\,e.$$

Solving for d and e gives

$$d,\, e = \frac{a \pm \sqrt{a^2 - b^2 c}}{2}.$$

For example,

$$\sqrt{5+2\sqrt{6}} = \sqrt{2} + \sqrt{3}$$
$$\sqrt{3-2\sqrt{2}} = \sqrt{2} - 1$$

A counterintuitive property of inverse functions is that

$$\sqrt{z}\sqrt{\frac{1}{z}} = \begin{cases} -1 & \text{for } J[z]=0 \text{ and } R[z]<0 \\ \text{undefined} & \text{for } z = 0 \\ 1 & \text{otherwise,} \end{cases} \quad ...(7)$$

so the expected identity (i.e., canceling of the $\sqrt{z}$ s) does not hold along the negative real axis.

DEFINITION 2.1 (PRINCIPAL SQUARE ROOT FUNCTION).

The function

$$z = g(w) = w^{1/2} = |\,w\,|^{1/2}\, e^{iArg\,(w)/2}, \text{ for } w \neq 0.$$

is called the principal square root function.

It is left as an exercise to show that f and g satisfy equations $g(f(z)) = z$ and $f(g(w)) = w$, and thus are inverses of each other that map the set A one-to-one and onto the set B and the set B one-to-one and onto the set A, respectively. Figure 12. illustrates this relationship.

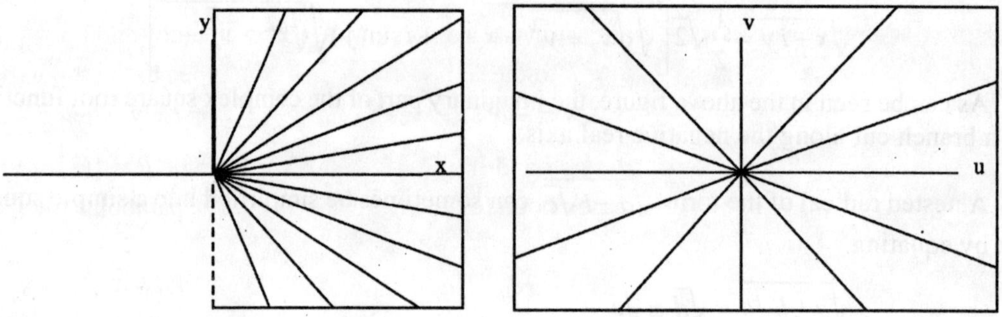

Fig. 12 The Mapping $w = f(z) = z^2$ (and $z = g(w) = w^{1/2}$).

What are the images of rectangles under the mapping $w = z^2$? To find out, we make use of the Cartesian form

$$w = u + i\,v = f(z) = z^2 = x^2 - y^2 + i\,2\,x\,y = (x^2 - y^2, 2\,x\,y) = (u, v) \quad \text{...(8)}$$

and the resulting system of equations

$$u = x^2 - y^2, \text{ and} \quad \text{...(9)}$$

$$v = 2\,x\,y. \quad \text{...(10)}$$

Example 12. Show that the transformation $w = f(z) = z^2$, usually maps vertical and horizontal lines onto parabolas and use this fact to find the image of the rectangle $R = \{(x, y)\} : 0 < x < a$, $0 < x < a$, $0 < y < b\}$. (a) Find the image of the vertical line $x = a$. (b) Find the image of the horizontal line y = b.

Solution. Using the above equations, we determine that the vertical line $x = a$ is mapped onto the set of points given by the equations $u = a^2 - y^2$ and $v = 2\,a\,y$. If a $\neq$ 0, then $y = \dfrac{v}{2a}$ and

$$u = a^2 - \frac{v^2}{2a^2},$$

this equation represents a parabola with vertex at $(a^2, 0)$, oriented horizontally, and opening to the left. If $a > 0$, the set $\{(u, v) : u = a^2 - y^2, v = 2\,a\,y\}$ has $v > 0$ precisely when $y > 0$, so the part of the line $x = a$ lying above the x axis is mapped to the top half of the parabola.

The horizontal line $y = b$ is mapped onto the parabola given by the equations $u = x^2 - b^2$ and

$v = 2\,x\,b$. If $b \neq 0$, then as before we get

$$v = -b^2 + \frac{v^2}{4b^2},$$

this equation represents a parabola with vertex at $-b^2$, oriented horizontally and opening to the right. If $b > 0$, the part of the line $y = b$ to the right of the y axis is mapped to the top half of the parabola because the set $\{(u, v) : u = x^2 - b^2, v = 2\,b\,x\}$ has $v > 0$ precisely when $x > 0$.

Quadrant I is mapped onto quadrants I and II by $w = z^2$, so the rectangle $R = \{(x, y) : 0 < x < a, 0 < y < b\}$ is mapped onto the region bounded by the top halves of the parabolas given by

$u = a^2 - \dfrac{v^2}{4a^2}$ and $v = -b^2 + \dfrac{v^2}{4b^2}$ and the u axis. The vertices 0, a^2, $a^2 + 2\,i\,a\,b - b^2$, and $-b^2$ of the rectangle are mapped onto the four points 0, a^2, $a^2 - b^2 + 2\,i\,a\,b$, and $-b^2$ respectively, as indicated in Figure 13.

Finally, we can easily verify that the vertical line $\{(0, y) : -\infty < y < \infty$ is mapped to the set $\{(-y^2, 0) : -\infty < y < \infty\}$. This is simply the set of non-positive real numbers. Similarly, the horizontal line $\{(x, 0) : -\infty < x < \infty\}$ is mapped to the set $\{(x^2, 0) : -\infty < x < \infty\}$, which is the set of non-negative real numbers.

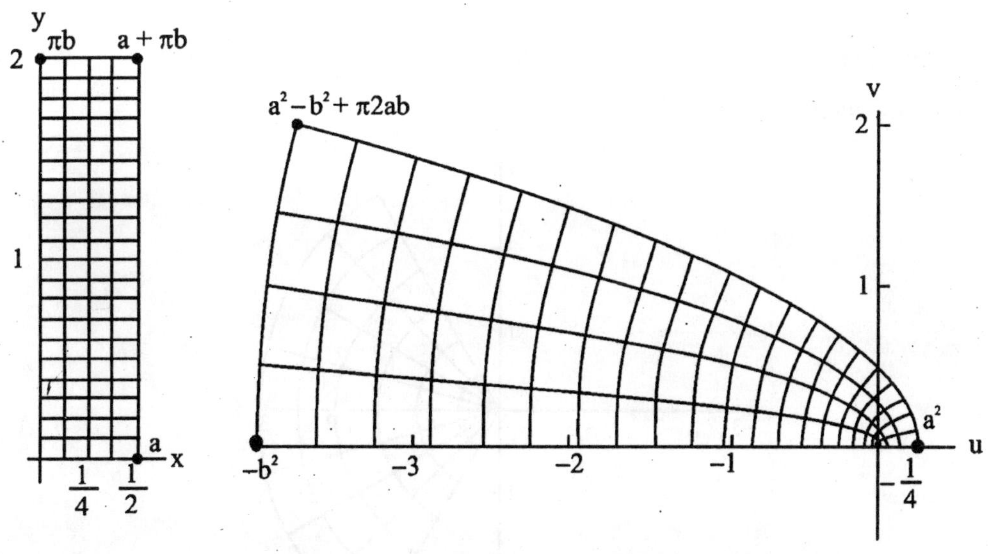

Fig. 13 The Transformation $w = f(z) = z^2$.

Example 13. Consider the mapping $w = z^2$. Find the image of the ray $r > 0$, $\theta = \alpha$ and circle $r = c$.

Solution

Here we find the image of the ray $r > 0$, $\theta = \alpha$ and circle $r = c$.

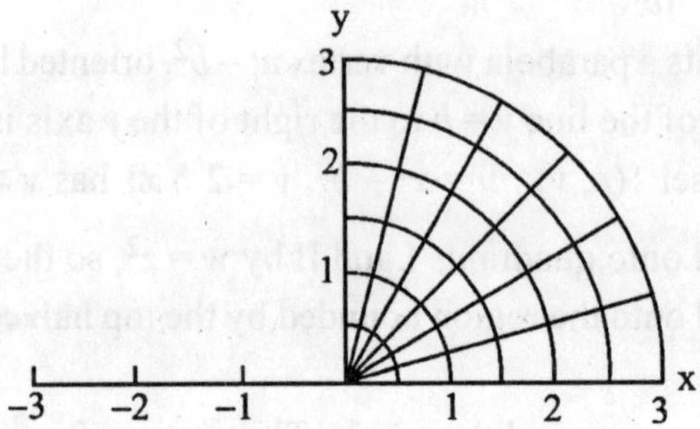

The rays $r > 0$, $\theta = \alpha$ and circles $r = c$.

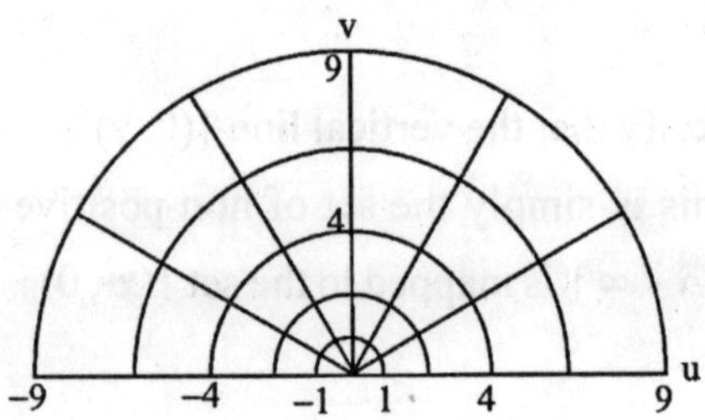

The image rays $r > 0$, $\theta = 2\alpha$, and circles $r = c^2$, under the mapping $w = z^2$

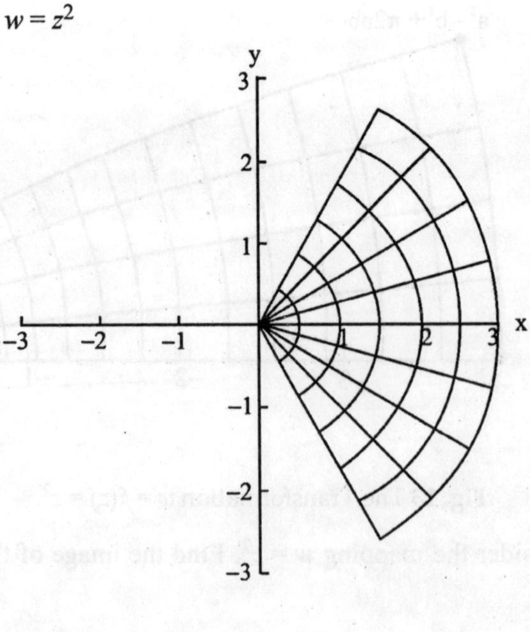

The rays $r > 0,\ \theta = \alpha$ and circles $r = c$.

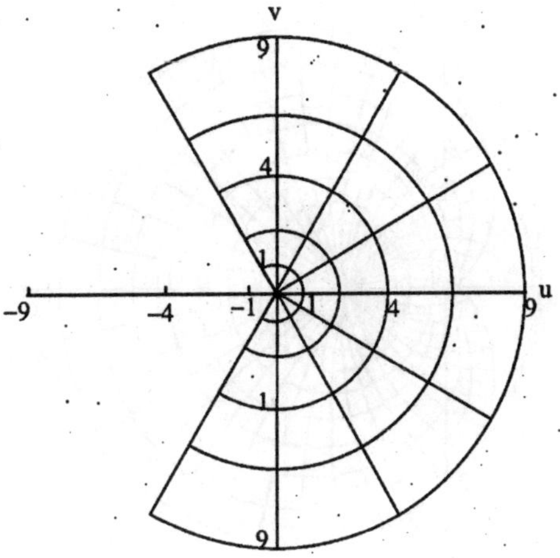

The image $r > 0,\ \theta = 2\alpha$, and circles $r = c^2$,

under the mapping $w = z^2$

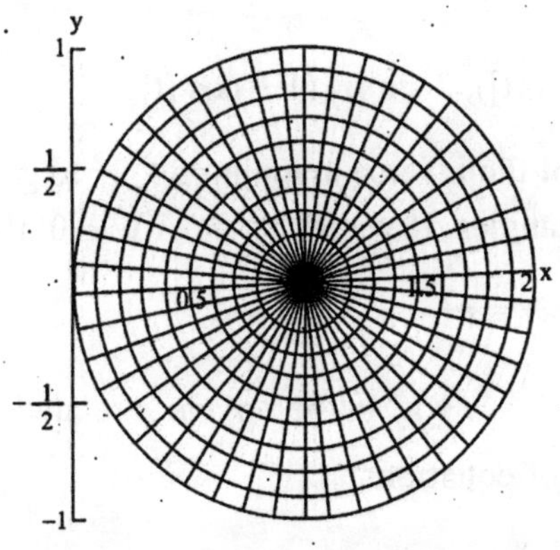

The radii, and circles.

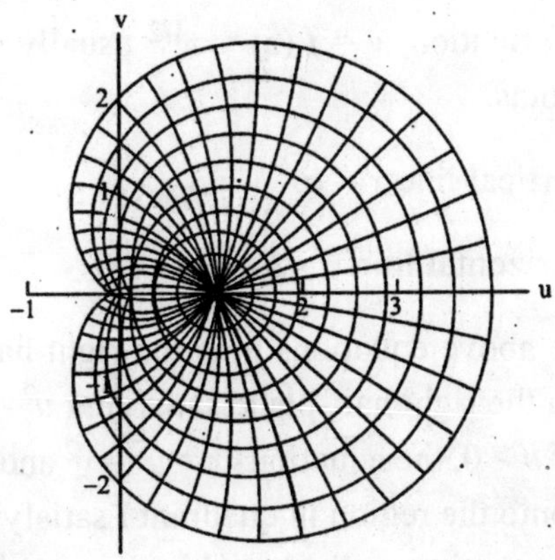

The image curves under the mapping $w = z^2$

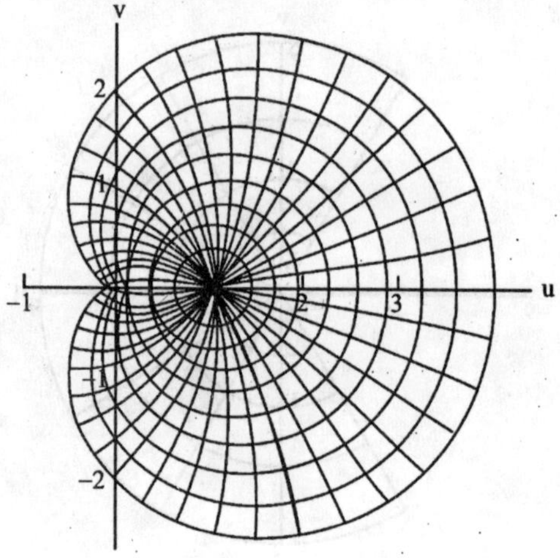

The outer boundary curve is a cardioid.

$(u^2 + v^2 - 2\,u)^2 = 4\,(u^2 + v^2)$

(u, v) = (2 cos [t] (1 + cos [t]), 2 sin [t] (1 + cos [t]))

We can use knowledge of the inverse mapping $z = w^2$ to get further insight into how the mapping $w = z^{1/2}$ acts on rectangles. If we let $z = x + i\,y \neq 0$, then

$$z = x + i\,y = w^2 = u^2 - v^2 + i\,2\,u\,v,$$

and we note that the point $z = x + i\,y$ in the z plane is related to the point $w = u + i\,v = z^{1/2}$ in the w-plane by the system of equations

$$x = u^2 - v^2, \text{ and } y = 2\,u\,v.$$

Example 14. The transformation $w = f(z) = z^{1/2}$ usually maps vertical and horizontal lines onto portions of hyperbolas.

(a) Find the image of the vertical line $x = a$.

(b) Find the image of the horizontal line $y = b$.

Solution. Let $a > 0$. The above equations map the right half-plane given by Re $(z) > a$ (i.e., $x > a$) onto the region in the right half-plane satisfying $u^2 - v^2 > a$ and lying to the right of the hyperbola $u^2 - v^2 = a$. If $b > 0$, the equations $x = u^2 - v^2$ and $y = 2\,u\,v$ map the upper half-plane Im $(z) > b$ (i.e., $y > b$) onto the region in quadrant I satisfying $2\,u\,v > b$ and lying above the hyperbola $2\,u\,v = b$. This situation is illustrated in Figure 14.

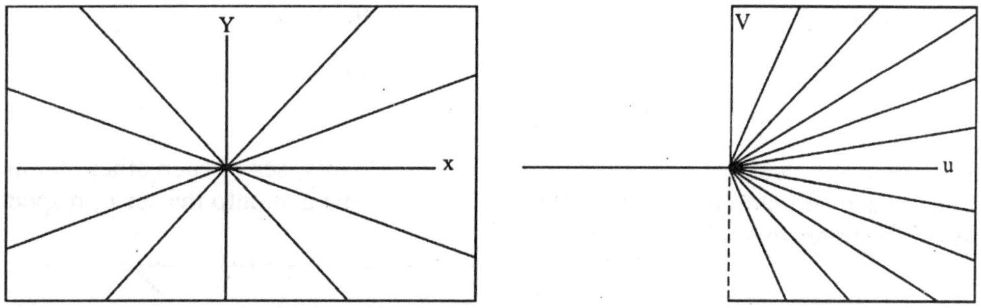

Fig. 14 The Mapping $w = f(z) = z^{1/2}$ and $(z = f^{-1}(w) = w^2)$.

The images of regions under the mapping

$$w = f(z) = z^{1/2} = |z|^{1/2} e^{i\,Arg(z)/2} = r^{1/2} e^{i\theta/2} \text{ for } re^{i\theta} \neq 0,$$

where $-\pi < \theta \leq \pi$. By the use of polar coordinates for $w = \rho e^{i\phi}$ in the w plane, we can represent this mapping by the system

$$\rho = r^{1/2} \text{ and } \phi = \frac{\theta}{2}.$$

This Equation indicate that the argument of $f(z)$ is half the argument of z and that the modulus of $f(z)$ is the square root of the modulus of z. Points that lie on the ray $r > 0$, $\theta = \alpha$ are mapped onto the ray $\rho > 0, \phi = \frac{\alpha}{2}$. The image of the z plane (with the point $z = 0$ deleted) consists of the right half-plane Re $(w) > 0$ together with the positive v axis, which is shown in Figure 15.

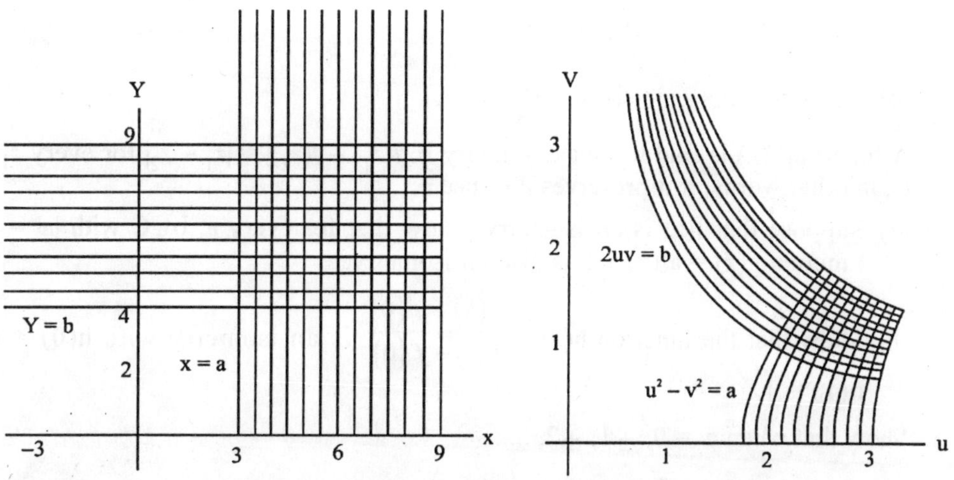

Fig. 15 The Mapping $w = f(z) = z^{1/2}$ (and $z = f^{-1}(w) = w^2$).

Definition 2 (Principal Root Function). The function

$$z = g(w) = w^{1/n} = |w|^{1/n} e^{i\,Arg(\omega)/n}, \text{ for } w \neq 0$$

is called the principal n^{th} root function.

We leave as an exercise to show that f and g are inverses of each other that map the set E one-to-one and onto the set F and the set F one-to-one and onto the set E, respectively. Figure 16 illustrates this relationship.

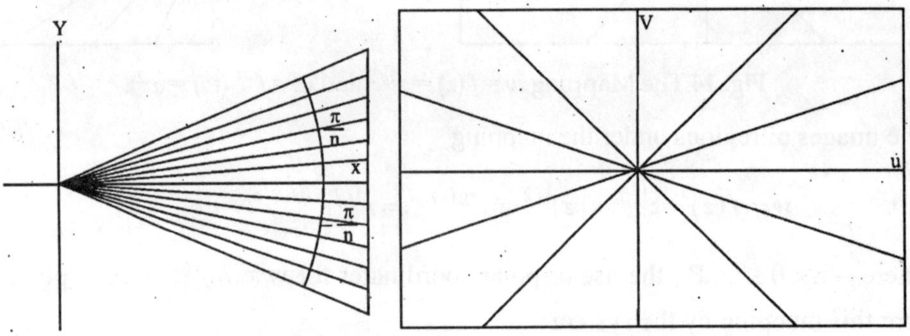

Fig. 16 The mapping $w = f(z) = z^n$ (and $z = g(w) = w^{1/n}$).

EXERCISE

1. Suppose z and w are complex numbers. What is the curve described by $\gamma(t) = (1-t)w + t\,z \;\; 0 \leq t \leq 1$?
2. Find a function γ that describes that part of the curve $y = 4x^3 + 1$ between $x = 0$ and $x = 10$.
3. Find the image of the strip {z: |Rez| < 1} and of the disc {z : |z| < 1} under each of the following mappings:

   (a) $w = (1+i)z + 1$ (b) $w = 2z^2$

   (c) $w = z^{-1}$ (d) $w = \left|\dfrac{z+1}{z-1}\right|$
4. A function f(z) is said to be an isometry if $|f(z_1) - f(z_2)| = |z_1 - z_2|$ for every $z_1, z_2 \in C$; in other words, if it preserves distance.

   (a) Suppose that f(z) is an isometry. Show that for every a, $b \in C$ with $|a| = 1$, the function $g(z) = af(z) + b$ is also an isometry.

   (b) Show that the function $h(z) = \dfrac{f(z) - f(0)}{f(1) - f(0)}$ is an isometry with $h(0) = 0$ and $h(1) = 1$.
5. Show that when n = 0, ±1, ±2, ...

   (a) $(1 + i)i = \exp\left(-\dfrac{\pi}{4} + 2n\pi\right)\exp\left(\dfrac{i}{2} ln 2\right)$;

(b) $(-1)^{1/\pi} = e^{(2n+1)i}$

6. Find the principal value of

   (a) 1^i

   (b) $\left[\frac{e}{2}\left(-1-\sqrt{3i}\right)\right]^{3\pi i}$

   (c) $(1-i)^{4i}$

7. Show that if $z \neq 0$ and a is a rea. number, then $\left|z^a\right| = \exp\left(a\ln|z|\right) = |z|^a$, where the principal value of $|z|^a$ is to be taken.

8. Let $c = a + bi$ be a fixed complex number, where $c \neq 0, \neq 1, \neq 2, \ldots$, and note that i^c is multiple-valued. What restriction must be placed on the constant c so that the values of $\left|i^c\right|$ are all same?

9. Find all the values of

   (a) $\tan^{-1}(2i)$; (b) $\tan^{-1}(1+i)$; (c) $\cosh^{-1}(-1)$; (d) $\tanh^{-1}0$.

10. Solve the equation $\sin z = 2$ for z by equating real parts and imaginary parts in that equation;

11. Solve the equation $\cos z = \sqrt{2}$ for z.

12. $f(x) = (z+2)/z$ and C is

    (a) the semicircle $z = 2e^{i\theta}$ $(0 \leq \theta \leq \pi)$;

    (b) the semicircle $z = 2e^{i\theta}$ $(\pi \leq \theta \leq 2\pi)$;

    (c) the semicircle $z = 2e^{i\theta}$ $(0 \leq \theta \leq 2\pi)$;

13. $f(z) = z - 1$ and C is the arc from $z = 0$ to $z = 2$ consisting of

    (a) the semicircle $z = 1 + e^{i\theta}$ $(\pi \leq \theta \leq 2\pi)$;

    (b) the segment $0 \leq x \leq 2$ of the real axis.

14. $f(z)$ is defined by the equations

$$f(x) = \begin{cases} 1 & \text{wheny } < 0, \\ 4 & \text{ywheny} > 0, \end{cases}$$

Chapter 6

Introduction to Fourier Series

Fourier series are named in honor of Joseph Fourier (1768-1830), who made important contributions to the study of trigonometric series, after preliminary investigations by Madhava, Nilakantha Somayaji, Jyesthadeva, Leonhard Euler, Jean le Rond d'Alembert, and Daniel Bernoulli. He applied this technique to find the solution of the heat equation.

Fourier series serve many useful purposes, as manipulation and conceptualization of the modal coefficients are often easier than with the original function. Areas of application include electrical engineering, vibration analysis, acoustics, optics, signal and image processing, and data compression. Fourier series decomposes a periodic function into a sum of simple oscillating functions, namely sines and cosines. The study of Fourier series is a branch of Fourier analysis. Fourier series were introduced by Joseph Fourier (1768–1830) for the purpose of solving the heat equation in a metal plate, it led to a revolution in mathematics, forcing mathematicians to reexamine the foundations of mathematics and leading to many modern theories such as Lebesgue integration.

Using the tools and techniques of spectroscopy, for example, astronomers can deduce the chemical composition of a star by analyzing the frequency components, or spectrum, of the star's emitted light. Similarly, engineers can optimize the design of a telecommunications system using information about the spectral components of the data signal that the system will carry. The Fourier series is named after the French scientist and mathematician Joseph Fourier, who used them in his influential work on heat conduction. The Fourier Series is a mathematical tool that allows the representation of any periodic signal as the sum of harmonically related sinusoids. This operation seems incredible. How, for instance, could a function that is not smooth be represented as the sum of sinusoids that are smooth? The answer lies in the number of sinusoids that are added to approximate the desired result. As more sinusoids are added, the output more closely resembles the desired output:

The first graphic shows a single wave. The y-axis represents magnitude and the x-axis represents 'n'. 'n' in discrete systems represents the sample number of the waveform. The above three graphics are actually constructed by taking 500 samples of the waveform and plotting them by a "connect the dot" method. If the sample rate is high enough, the resulting waveform will very closely resemble the actual waveform. The waveform in the middle graph is the sum

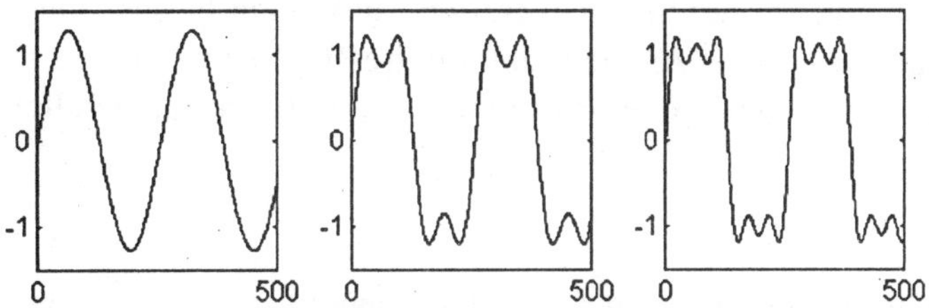

of the waveform in the first graph plus a sine wave that has a frequency that is three times that of the the original sine wave. The waveform in the last graphic contains is the sum of the middle waveform plus a sine wave that has a frequency that is six times that of the original sine wave. This waveform faintly resembles a square wave. The addition of the first ten waveforms is shown in the following animation:

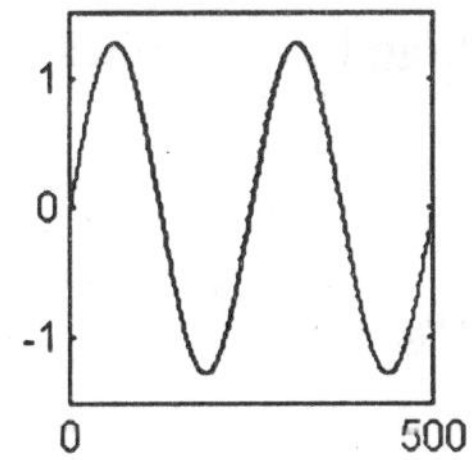

After the addition of ten and twenty sinusoids, the resultant waveform is clearly square in shape:

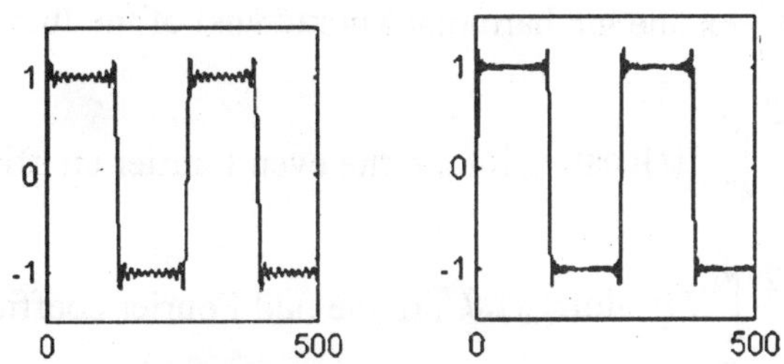

If all the intermediate graphs between the basic sine wave and the 20th iteration were placed evenly spaced on an axis, the result would be a 3-D shape:

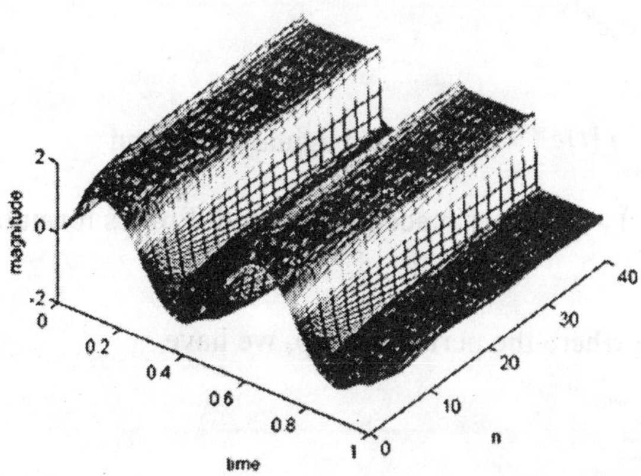

From this graph, the Fourier Series operation can be clearly seen. On the bottom left, the first waveform is the basic sine. As the graph progresses backwards and to the right into the monitor, the waveform becomes more and more square. Each line parallel to the x axis is at distinct intervals.

Definition

Given a complex-valued function f of real argument t, $f : R \rightarrow C$, where $f(t)$ is piecewise smooth and continuous, periodic with period T, and square-integrable over the interval from t_1 to t_2 of length T, that is,

$$\int_{t_1}^{t_2} |f(t)|^2 \, dt > +\infty$$

Where; $T = t_1 - t_2$, is the period, t_1 and t_2 are times to integrate between.

The Fourier series expansion of f is:

$$f(t) = \frac{a_0}{2} + \sum_{n=1}^{\infty} \left[a_n \cos(\omega_n t) + b_n \sin(\omega_n t) \right] \qquad \text{...(1)}$$

where, for any non-negative integer n:

- $\omega_n = n\frac{2\pi}{T}$ is the nth harmonic (in radians) of the function f, ...(2)
- $a_n = \frac{2}{T}\int_{t_1}^{t_2} f(t)\cos(\omega_{nt})\,dt$ are the even Fourier coefficients of f, and ...(3)
- $b_n = \frac{2}{T}\int_{t_1}^{t_2} f(t)\sin(\omega_{nt})\,dt$ are the odd Fourier coefficients of f. 1...(4)

Equivalently, in exponential form,

$$f(t) = \sum_{n=-\infty}^{+\infty} c_n e^{i\omega_n t} \qquad \text{...(5)}$$

$$c_n = \frac{1}{T}\int_{t_1}^{t_2} f(t) e^{-i\omega_n t}\,dt \quad i \text{ is the imaginary unit, and} \qquad \text{...(6)}$$

where: $e^{-i\omega_n t} = \cos(\omega_n t) + i\sin(\omega_n t)$ in accordance with Euler's formula.

Canonical Form

In the special case where the period $T = 2p$, we have

$\omega_n = n$

In this case, the Fourier series expansion reduces to a particularly simple form:

$$f(t) = \frac{a_0}{2} + \sum_{n=1}^{\infty}\left[a_n \cos(nt) + b_n \sin(nt)\right] \quad ...(7)$$

where
$$a_n = \frac{1}{\pi}\int_{-\pi}^{\pi} f(t)\cos(nt)\,dt \quad ...(8)$$

$$b_n = \frac{1}{\pi}\int_{-\pi}^{\pi} f(t)\sin(nt)\,dt \quad ...(9)$$

for any integer n.

or, equivalently:

$$f(t) = \sum_{n=-\infty}^{+\infty} c_n e^{\text{int}} \quad ...(10)$$

where

$$c_n = \frac{1}{2\pi}\int_{-\pi}^{\pi} f(t)e^{-\text{int}}\,dt = \frac{1}{2}(a_n - ib_n) \quad ...(11)$$

Choice of the Form

The form for period T can be easily derived from the canonical one with the change of variable defined by $x = \frac{2\pi}{T}t$. Therefore, both formulations are equivalent. However, the form for period T is used in most practical cases because it is directly applicable. For the theory, the canonical form is preferred because it is more elegant and easier to interpret mathematically, as will later be seen.

PERIODICITY OF FUNCTIONS

Let X designates the domain of a function $y = f(x)$. A function $f(x)$ is said to be *periodic* if its image values are repeated in its domain i.e. if there exists a nonzero real T such that $f(x + T) = f(x)$ for each $x \in X$. The least positive value of those T is said to be the *period* of this function. For example for the sine function we can write sin(x + 4p) = sin x, sin(x + 24 p) = sin x, sin(x - 8p) = sin x etc., so the sine function is a periodic function. But the least positive value for which such an equation holds is 2π and this is the period of the sine function. It is also well known that the cosine function has the period 2π and the tangent function has the period π.

An example of the graph of a periodic function is illustrated in Figure 1

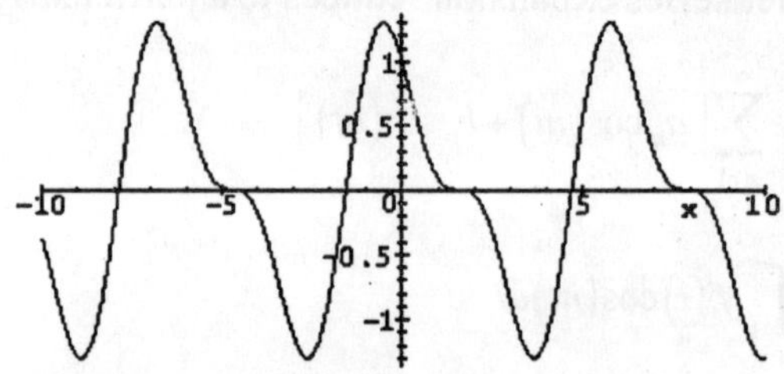

Fig. 1

Further we need two following lemmas

Lemma 1. If the period of the function f(x) is T, then the period of the function f(ax) is $\frac{T}{a}$.

Proof. $f\left[a\left(x+\frac{T}{a}\right)\right] = f(ax+T) = f(ax)$.

According to this lemma the period of the function sin3*x* is $\frac{2\pi}{3}$ and the period of the function $\cos\frac{x}{2}$ is $\frac{2\pi}{\frac{1}{2}} = 4\pi$.

Lemma 2. If the period of the function *f*(*x*) is *T*, then $\int_c^{c+T} f(x)\,dx = \int_0^T f(x)\,dx$

This result says that if we integrate the period of the function over the interval whose length equals to the period of the function then the value does not depend on the choice of the boundaries.

Proof. First we use the property of a definite integral and write $\int_c^{c+T} f(x)\,dx = \int_c^{O} f(x)\,dx + \int_0^T f(x)\,dx + \int_T^{c+T} f(x)\,dx$.

In the last integral we take $x = z + T$. Then $dx = dz$, $f(x) = f(z + T) = f(z)$, if $x = T$ then $z = 0$ and if $x = c + T$ then $z = c$. So we have

$$\int_c^{c+T} f(x)\,dx = \int_c^{O} f(x)\,dx + \int_0^T f(x)\,dx + \int_0^c f(z)\,dx$$

$$= \int_0^c f(x)\,dx + \int_0^T f(x)\,dx + \int_0^c f(z)\,dx$$

$$= \int_0^T f(x)\,dx$$

as the first and third integrals are equal.

FOURIER SERIES OF 2π-PERIODIC FUNCTIONS

In many problems in mechanics and electronics there arises a question, how to represent the behavior of a system by a combination of some simple behaviors. Mathematically – how to represent a function in the form of the functional series $f(x) = \sum_{k=0}^{\infty} c_k \varphi_k(x)$. Here $\varphi_0, \varphi_1, \varphi_2, ..., \varphi_k, ...$ form a base set of functions and $c_0, c_1, c_2, ..., ...c_k, ...$, are said to be the *coefficients* of the expansion. The most familiar expansions are power series of the form $f(x) = \sum_{k=0}^{\infty} c_k x^k$, in which the base set comprises the power functions $1, x, x^2, ... x^k, ...$. Another widely used expansion is the expansion of a periodic function $f(x)$ of a period T in which the base set comprises the cosine functions, giving an expanded representation of the form

$$f(x) = A_0 + \sum_{k=1}^{\infty} A_k \cos\left(\frac{2k\pi x}{T} + \phi_k\right) \quad ...(12)$$

which is called the Fourier series of the function $f(x)$. In engineering the term *circular frequency* is used and it is defined by $\omega = \frac{2\pi}{T}$. The circular frequency is measured in radians per second. It is common to drop the term 'circular' and to refer to this simply as the frequency. Using frequency, we can rewrite (12) in the form

$$f(x) = A_0 + \sum_{k=1}^{\infty} A_k \cos(k\omega x + \phi_k) \quad ...(13)$$

Every term $A_k \cos(k\omega x + \phi_k)$ represents a simple harmonic vibration that is called kth *harmonic*. The first harmonic $A_1 \cos(\omega x + \phi_1)$ has the same frequency ω as the parent function $f(x)$. The kth harmonic has the frequency $k\omega$, which is k times frequency of the first harmonic. A_k denotes the *amplitude* of the kth harmonic and ϕ_k is its *phase angle*. If $f(x)$ is a 2π-periodic function then its frequency $\omega = 1$, and the Fourier series (2) has a more simple form

$$f(x) = A_0 + \sum_{k=1}^{\infty} A_k \cos(kx + \phi_k) \quad ...(14)$$

Here we shall consider the Fourier series of 2π-periodic functions. The kth harmonic can be represented in the form $A_k \cos(kx + \phi_k) = a_k \cos k_x + b_k \sin k_x$, where $a_k = A_k \cos\phi_k$ and $b_k = -A_k \sin\phi_k$. Now the expansion (14) may be written as

$$f(x) = \frac{a_0}{2} + \sum_{k=1}^{\infty} a_k \cos kx + b_k \sin kx \quad ...(15)$$

Later we shall see that taking the constant term as $\frac{a_0}{2}$ enables us to make a_0 fit a general result. The expression (15) is called *Fourier series expansion of a function f(x)*, and a_k and b_k are called the *Fourier coefficients.*

The set of functions $\{1, \cos x, \sin x, \cos 2x, ..., \cos kx, \sin kx, ...\}$ is said to be the trigonometric system of functions. It is possible to show that the functions of the trigonometric system satisfy the orthogonality relations on an interval $[-\pi;\pi]$:

$$\int_{-\pi}^{\pi} \cos nx \cos kx dx = \begin{cases} 0 \text{ for } n \neq k, \\ \pi \text{ for } n = k: \end{cases} \quad ...(16)$$

$$\int_{-\pi}^{\pi} \sin nx \sin kx dx = \begin{cases} 0 \text{ for } n \neq k, \\ \pi \text{ for } n = k: \end{cases} \quad ...(17)$$

$$\int_{-\pi}^{\pi} \sin nx \cos kx dx = 0 \text{ for all n and } k; \quad ...(18)$$

$$\int_{-\pi}^{\pi} \cos kx dx = \begin{cases} 2\pi \text{ for } k = 0, \\ 0 \text{ for } k = 0; \end{cases} \quad ...(19)$$

$$\int_{-\pi}^{\pi} \sin kx dx = 0 \text{ for all } k. \quad ...(20)$$

Assume that term-by-term integration of the series (15) is permissible. Then integrating the series with respect to x over the interval $[-\pi;\pi]$ and using (19) and (20) we find that

$$\int_{-\pi}^{\pi} f(x) dx = \frac{a_0}{2}\int_{-\pi}^{\pi} dx + \sum_{k=1}^{\infty}\left(a_k \int_{-\pi}^{\pi} \cos kx dx + b_k \int_{-\pi}^{\pi} \sin kx dx\right)$$

$$= \frac{a_0}{2}.2\pi + \sum_{k=1}^{\infty}(a_k.0 + b_k.0) = \pi a_0 \quad ...(21)$$

To find the coefficients an a_n (n = 1, 2, ...) we multiply both sides of equality (21) by $\cos nx$. Integrating the result of the multiplication with respect to x over the interval $[-\pi;\pi]$ we have

$$\int_{-\pi}^{\pi} f(x)\cos nx dx = \frac{a_0}{2}\int_{-\pi}^{\pi} \cos nx dx +$$

$$\sum_{k=1}^{\infty}\left(a_k \int_{-\pi}^{\pi} \cos nx \cos kx dx + b_k \int_{-\pi}^{\pi} \cos nx \sin kx dx\right) \quad ....(22)$$

Because of (19), the first term on the right side of the equality equals to zero. By (18) all coefficients of b_k equal zero and by (16) we obtain the result that there is only one nonzero coefficient of a_k in the case $k = n$. We obtain

$$\int_{-\pi}^{\pi} f(x)\cos nx dx = \pi a_n \qquad ...(23)$$

Analogously, multiplying both sides of the equality (15) by $\sin nx$ integrating the result with respect to x over the interval $[-\pi;\pi]$ and using (17), (18) and (20) we find that

$$\int_{-\pi}^{\pi} f(x)\sin nx dx = \pi b_n \qquad ...(24)$$

Now, dividing both sides of (21) – (24) by , and replacing n by k we find respectively.

$$a_0 = \frac{1}{\pi}\int_{-\pi}^{\pi} \int f(x)dx, \qquad ...(25)$$

$$a_k = \frac{1}{\pi}\int_{-\pi}^{\pi} \int f(x)\cos kx\, dx \;\; for\; k=1, 2, ... \qquad ...(26)$$

$$bk = \frac{1}{\pi}\int_{-\pi}^{\pi} \int f(x)\sin kx\, dx \;\; for\; k=1, 2, ... \qquad ...(27)$$

Evidently the formula (25) for finding a_0 fits the formula (26) with k = 0 and instead of two formulae (25) and (26) we can write one formula.

Actually, if we compute the Fourier coefficients according to the formulae (13) - (15) and compose the Fourier series, we don't know whether the obtained series is convergent at all, and if it is convergent then does it converge to the parent function *f*(*x*) or to some different function. In this context there arises a question as to which conditions a function should satisfy so that its Fourier series expansion is convergent to the function *f*(*x*). We shall consider the convergence problems in the next section . Note that the class of functions which are representable by their Fourier series expansion is quite wide and in most of the practical cases we can assume that the convergence conditions are satisfied.

Example 1. Find the Fourier series expansion of the function *f*(*x*) with period 2π defined on the half-interval $[-\pi;\pi]$ as *f*(*x*) = *x*.

Solution:

The graph of this function $f(x) = x$ for $x \in$ (–p, p) is illustrated in Figure 2.

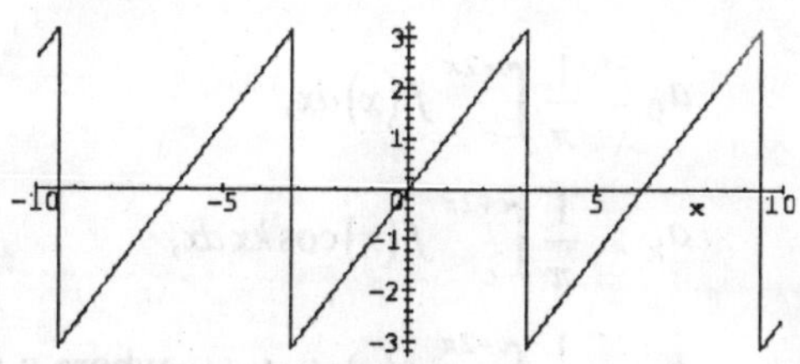

Fig. 2

First we find a_0 using (25):

$$a_0 = \frac{1}{\pi}\int_{-\pi}^{\pi} x dx = \frac{1}{\pi}\frac{x^2}{2}\Big|_{-\pi}^{\pi} = 0$$

Using (26) and integrating by parts we find that

$$a_k = \frac{1}{\pi}\int_{-\pi}^{\pi} x \cos kx\, dx$$

$$= \frac{1}{\pi}\left[\frac{x\sin kx}{k}\Big|_{-\pi}^{\pi} - \int_{-\pi}^{\pi}\sin kx\right] = 0$$

and using (27) and integrating by parts we find that

$$b_k = \frac{1}{\pi}\int_{-\pi}^{\pi} x \sin kx\, dx$$

$$= \frac{1}{\pi}\left[-\frac{x\cos kx}{k}\Big|_{-\pi}^{\pi} + \frac{1}{k}\int_{-\pi}^{\pi}\cos kx dx\right] = (-1)^{k+1}\frac{2}{k}$$

So we have the Fourier series expansion of the function

$$f(x) \sim \left[\frac{\sin 2x}{1} - \frac{\sin 2x}{2} + \frac{\sin 3x}{3} - .... + (-1)^{k+1}\frac{\sin kx}{k} + ...\right]$$

The sign ``~" indicates that we have found the Fourier series expansion according to the formulae (25) – (27), but we don't know whether this expansion converges to this function, so we can't still use the sign ``=". The nth partial sum of this series expansion is

$$Sn(x) = 2\sum_{k=1}^{n(n)}(-1)^{k+1}\frac{\sin kx}{k}$$

The graph of the nth partial sum $S_n(x)$ for $n = 4, 8, 16$ is shown in Figure 3

Remark. According to Lemma 2 we can replace the interval of integration $[-\pi;\pi]$ with an arbitrary chosen interval $[c; c+2\pi]$ i.e. we can use the formulas

$$a_0 = \frac{1}{\pi}\int_{-c}^{c+2\pi} f(x)dx, \qquad ...(28)$$

$$a_k = \frac{1}{\pi}\int_{-c}^{c+2\pi} f(x)\cos kx\, dx, \qquad ...(29)$$

$$b_k = \frac{1}{\pi}\int_{-c}^{c+2\pi} f(x)\sin kx\, dx, \text{ where c is any real number.} \qquad ...(30)$$

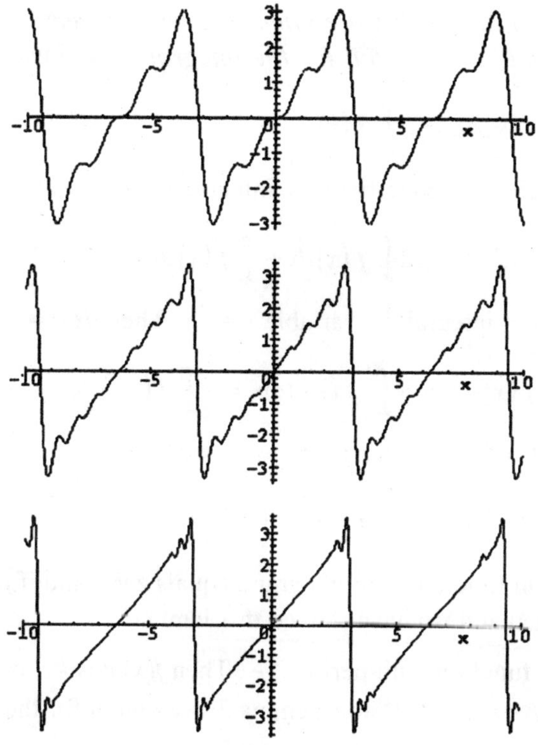

Figure 3

FOURIER SERIES OF EVEN AND ODD FUNCTIONS

If a particular function possesses certain symmetrical properties then some terms are absent from its Fourier series expansion and the expressions determining the remaining coefficients can be simplified.

A function $f(x)$ is said to be an even *function* if for each x from its domain

$$f(-x) = f(x)$$

A function $y = f(x)$ is said to be odd *function* if for each x from its domain

$$f(-x) = -f(x)$$

The graph of an even function is symmetrical about the vertical axis and the graph of an odd function is symmetrical about the origin.

The even and the odd functions satisfy the following properties:

- The product of two even functions is an even function,
- The product of two odd functions is an even function,
- The product of an even and an odd function is an odd funnction.

Lemma 3. *If limits of integration are symmetric with respect to zero then the integral of any odd function equals to zero, but for the integral of any even function f (x)*

$$\int_{-c}^{c} f(x)dx = 2\int_{0}^{c} f(x)dx \qquad ...(31)$$

Proof. Using the additivity property of the integral we have that

$$\int_{-c}^{c} f(x)dx = 2\int_{0}^{c} f(x)dx + \int_{0}^{c} f(x)dx$$

Substituting in the first integral the variable $x = -y$. Then $dx = -dy$ and

$$\int_{-c}^{o} f(x)dx = -\int_{-c}^{o} f(-y)dy = -\int_{o}^{c} f(-y)dy$$

Denoting the variable of integration again by x, we have

$$\int_{-c}^{c} f(x)dy = \int_{O}^{c}\left[f(x)+f(-x)\right]dx$$

Now, if $f(x)$ is an odd function then integrand equals zero, and if $f(x)$ is an even function then integrand equals $2f(x)$ and we have proved this lemma.

Let $f(x)$ be an even function with period 2π. Then $f(x)$ cos kx is even and $f(x)$ sin kx is an odd function for any k =1, 2, ...Using Lemma 3, we obtain for the calculation of Fourier coefficients the formulas

$$a_0 = \frac{2}{\pi}\int_{O}^{\pi} f(x)dx, \qquad ...(32)$$

$$a_k = \frac{2}{\pi}\int_{O}^{\pi} f(x)\cos kx\,dx \text{ for } k = 1, 2, ..., \qquad ...(33)$$

$$b_k = 0 \text{ for } k = 1, 2, ... \qquad ...(34)$$

Consequently the Fourier series expansion of an even periodic function with period 2π does not involve the terms with sines and (26) has the form

$$f(x) \sim \frac{a_o}{2} + \sum_{k=1}^{\infty} a_k \cos kx, \qquad ...(35)$$

where the a-s are calculated using the formulae (32) and (33).

If $f(x)$ is an odd function then $f(x)$cos kx is odd and is even for any k = 1,2,.... By Lemma 3, we have for calculation of Fourier coefficients the formulae

$$a_k = k = 0, 1, 2,..., \qquad ...(36)$$

$$b_k = \frac{2}{\pi}\int_{O}^{\pi} f(x)\sin kx\,dx \textit{ for } k=1,2,.... \qquad ...(37)$$

Thus the Fourier series expansion of the given periodic function with period consists of sine terms only and (26) has the form

$$f(x) \sim \sum_{k=1}^{\infty} b_k \sin kx, \qquad \text{...(38)}$$

where the *b-s* are calculated by (37).

Example 2. Find the Fourier series expansion of the function $f(x)$ with period 2π, which on half-interval $[-\pi;\pi]$ is defined by $f(x) = |x|$.

Solution:

A graph of this function is shown in Figure 4. As this is an even function we have to calculate only the coefficients a_k, $c_k = 1, 2, \ldots$. First using (32) we find

$$a_0 = \frac{2}{\pi}\int_0^{\pi} |x|\,dx = \frac{2}{\pi}\int_0^{\pi} x\,dx = \pi$$

and using (33) and integration by parts we obtain

$$a_k = \frac{2}{\pi}\int_0^{\pi} |x|\cos kx\,dx = \frac{2}{\pi}\int_0^{\pi} x\cos kx\,dx$$

$$= \frac{2}{\pi}\left[\frac{1}{k}x\sin kx\Big|_0^{\pi} - \frac{1}{k}\int_0^{\pi}\sin kx\,dx\right]$$

$$= \frac{2}{\pi k^2}\cos kx\Big|_0^{\pi} = \frac{2}{\pi k^2}\left[(-1)^k - 1\right]$$

$$= \begin{cases} -\frac{4}{\pi k^2} & \text{if } k \text{ is odd} \\ 0 & \text{if } k \text{ is even.} \end{cases}$$

Thus by (35) the Fourier series expansion of this function is

$$f(x) \sim \frac{\pi}{2} - \frac{4}{\pi}\left(\cos x + \frac{1}{9}\cos 3x + \frac{1}{25}\cos 5x + \ldots\right)$$

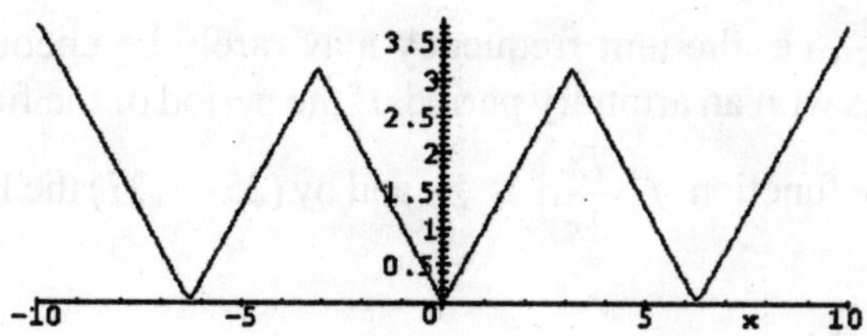

Fig. 4

Example 3. Find the Fourier series expansion of the *square wave function* i.e. the periodic function $f(x)$ with period 2π defined within the period $-\pi < x < \pi$ by

$$f(x) = \begin{cases} -1 & if -\pi < x < 0 \\ 1 & if \ 0 < x < p \end{cases}$$

Solution:

A graph of this square wave function is shown in Figure 5.

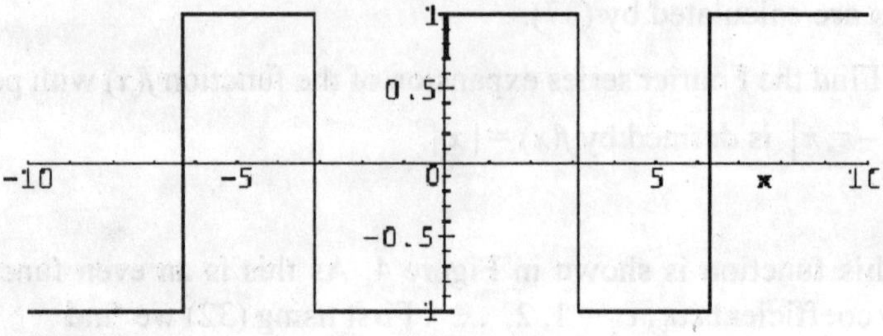

Fig. 5

Clearly $f(x)$ is an odd function and thus its Fourier series expansion consists of sine terms only. By (21) we calculate the coefficients

$$b_k = \frac{2}{\pi}\int_O^{\pi} \sin kx \, dx = -\frac{2}{k\pi}\cos kx \Big|_O^{\pi}$$

$$= -\frac{2}{k\pi}\left[(-1)^k - 1\right] = \begin{cases} \frac{4}{k\pi} & if\ k\ is\ old \\ 0 & if\ k\ is\ even. \end{cases}$$

Thus the Fourier series expansion of $f(x)$ is

$$f(x) \sim \frac{4}{\pi}\left(\sin x + \frac{1}{3}\sin 3x + \frac{1}{5}\sin 5x + ...\right)$$

$$= \frac{4}{\pi}\sum_{k=1}^{\infty}\frac{\sin(2k -}{2k -}$$

FOURIER SERIES OF FUNCTIONS WITH ARBITRARY PERIOD

Since the period 2π, i.e. the unit frequency may rarely be encountered in practice, we have to consider functions with an arbitrary period. If the period of the function $f(x)$ is T, then by Lemma 1 the period of the function $f\left(\frac{Tx}{2\pi}\right)$ is 2π and by (25) – (27) the Fourier series expansion of this function is

$$f\left(\frac{Tx}{2\pi}\right) \sim \frac{a_O}{2} + \sum_{k=1}^{\infty} a_k \cos kx + b_k \sin kx,$$

where $\quad a_0 = \frac{1}{\pi}\int_{-\pi}^{\pi} f\left(\frac{Tx}{2\pi}\right) dx,$

$$a_k = \frac{1}{\pi}\int_{-\pi}^{\pi} f\left(\frac{Tx}{2\pi}\right)\cos kx\,dx, k = 1, 2, ...$$

$$b_k = \frac{1}{\pi}\int_{-\pi}^{\pi} f\left(\frac{Tx}{2\pi}\right)\sin kx\,dx, k = 1, 2, ...$$

Changing the variable $t = \frac{Tx}{2\pi}$, i.e. $x = \frac{2\pi t}{T}$ and $dx = \frac{2\pi}{T}dt$, we obtain that the Fourier series expansion of the function $f(t)$ with period T is

$$f(t) \sim \frac{a_0}{2} + \sum_{k=1}^{\infty} a_k \cos\frac{2k\pi t}{T} + b_k \sin\frac{2k\pi t}{T},$$

where

$$a_0 = \frac{2}{T}\int_{-\frac{T}{2}}^{\frac{T}{2}} f(t)\,dt,$$

$$a_k = \frac{2}{T}\int_{-\frac{T}{2}}^{\frac{T}{2}} f(t)\cos\frac{2k\pi t}{T}\,dt \text{ for } k = 1, 2, ...$$

$$b_k = \frac{2}{T}\int_{-\frac{T}{2}}^{\frac{T}{2}} f(t)\sin\frac{2k\pi t}{T}\,dt \text{ for } k = 1, 2, ...$$

Replacing the variable t by x again, we express the Fourier series expansion in terms of

$$f(x) \sim \frac{a_0}{2} + \sum_{k=1}^{\infty} a_k \cos kwx + b_k \sin kwx,$$

where

$$a_0 = \frac{2}{T}\int_{-\frac{T}{2}}^{\frac{T}{2}} f(x)\,dx, \quad ...(39)$$

$$a_k = \frac{2}{T}\int_{-\frac{T}{2}}^{\frac{T}{2}} f(x)\cos kwx\,dx \text{ for } k = 1, 2, ... \quad ...(40)$$

$$b_k = \frac{2}{T}\int_{-\frac{T}{2}}^{\frac{T}{2}} f(x)\sin kwx dx \text{ for } k = 1, 2, ... \quad ...(41)$$

Remark: In formulae (39) – (41) the interval of integration can be replaced with an arbitrary interval $c; c + T$ with length T.

Now, if the periodic function $f(x)$ with period T is even then from (39) – (41) we obtain

$$a_0 = \frac{4}{T}\int_0^{\frac{T}{2}} f(x)\,dx, \quad ...(42)$$

$$a_k = \frac{4}{T}\int_0^{\frac{T}{2}} f(x)\cos kwx\,dx, \text{ for } k = 1, 2, ..., \quad ...(43)$$

$$b_k = 0 \quad \text{for } k = 1, 2, \ldots \qquad \ldots(44)$$

and the Fourier series expansion of even function $f(x)$ is

$$f(x) \sim \frac{A_0}{2} + \sum_{K=1}^{\infty} a_k \cos kwx.$$

If $f(x)$ is a odd periodic function with period T then from (39) - (41) we have

$$a_k = 0 \text{ for k} = 0, 1, 2, \ldots \qquad \ldots(45)$$

$$b_k = \frac{4}{T}\int_0^{\frac{T}{2}} f(x)\sin kwx\,dx \qquad \ldots(46)$$

and the Fourier series expansion of odd function $f(x)$ is

$$f(x) \sim \sum_{K=1}^{\infty} b_k \sin kwx\,dx.$$

Example 1.9. Find the Fourier series expansion of the function $f(x) = |\sin x|$.

The graph of this function is shown in Figure 6..

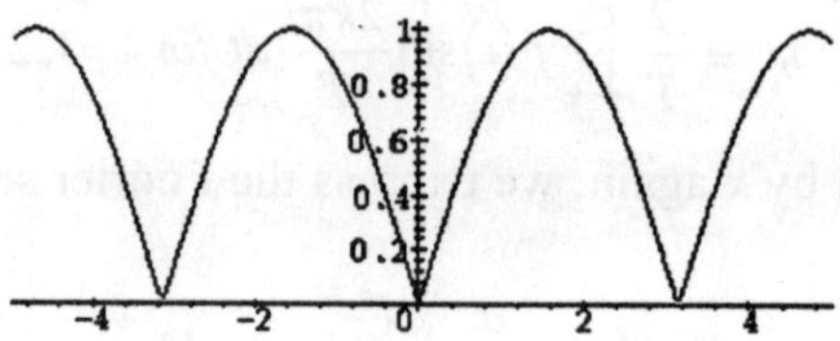

Fig. 6

Solution: This function is clearly even and its period is π (i.e. frequency $\omega = 2$). Therefore its Fourier series expansion consists of cosine terms only and we calculate the coefficients a_k for k=1, 2, ... using the formulae (42) and (43). By (42)

$$a_0 = \frac{4}{T}\int_0^{\frac{T}{2}} |\sin x|\,dx = \frac{4}{\pi}\cos x\Big|_0^{\frac{\pi}{2}} = \frac{4}{\pi}$$

and by (43).

$$a_k = \frac{4}{T}\int_0^{\frac{T}{2}} |\sin x|\cos 2kx\,dx$$

$$= \frac{4}{\pi}\frac{1}{2}\left[\int_0^{\frac{\pi}{2}} \sin(x+2lx)\,dx + \int_0^{\frac{\pi}{2}} \sin(x-2kx)\,dx\right]$$

$$= \frac{2}{\pi}\left[-\frac{1}{2k+1}\cos(2k+1)x\Big|_0^{\frac{\pi}{2}} + \frac{1}{2k-1}\cos(2k-1)x\Big|_0^{\frac{\pi}{2}}\right]$$

$$= \frac{2}{\pi}\left[\frac{1}{2k+1} - \frac{1}{2k-1}\right]$$

$$= \frac{2}{\pi}\frac{2k-12k-1}{4k^2-1} - \frac{4}{\pi}\frac{1}{4k^2-1}.$$

The Fourier series expansion of the given function is

$$|\sin x| \sim \frac{2}{\pi} - \frac{4}{\pi}\sum_{k=1}^{\infty}\frac{\cos 2kx}{4k^2-1}$$

FOURIER SERIES OF FUNCTIONS DEFINED OVER A FINITE INTERVAL

Suppose the given function $f(x)$ is defined only over the finite half-interval a, $a + T$ [(or over a, $a + T$)].

Then we can define the periodic extension $\phi(x)$ of $f(x)$ by

$$\phi(x) = f(x) \text{ if } a \le x < a+T,$$

As the result we obtain the periodic function $\phi(x)$ which on half-interval $[a, a+T]$ is identical with $f(x)$. The periodic function $\phi(x)$ has the Fourier series expansion

$$f(x) \sim \frac{a_0}{2} + \sum_{K=1}^{\infty} a_k \cos kwx + b_k \sin kwx$$

where

$$a_0 = \frac{2}{T}\int_a^{a+T} f(x)dx,$$

$$a_k = \frac{2}{T}\int_a^{a+T} f(x)\cos kwx\, dx,$$

$$b_k = \frac{2}{T}\int_a^{a+T} f(x)\sin kwx\, dx,$$

If this Fourier series expansion represents the function $\phi(x)$ then on the half-interval $[a, a+T]$ it represents also the function $f(x)$.

Now we consider the case if the given function $f(x)$ is defined only over the finite interval $\left[0; \frac{T}{2}\right]$. Its even periodic extension $\phi_1(x)$ is the even periodic function with period T defined by

$$\phi_1(x) = \begin{cases} f(x) \text{ if } 0 < x < \frac{T}{2}, \\ f(-x) \text{ if } -\frac{T}{2} < x < 0, \end{cases}$$

$$\phi_1(x+T) = \phi_1(x)$$

The Fourier series expansion of the function $\phi_1(x)$ consists of cosine terms only and is given by

$$\phi_1(x) \sim \frac{a_0}{2} + \sum_{k=1}^{\infty} a_k \cos kwx$$

called the *cosine series expansion*, where

$$a_k = \frac{4}{T}\int_0^{\frac{T}{2}} f(x)\cos kwx dx \text{ for } k = 0,1,2,... \quad ...(47)$$

For a given function $f(x)$ defined only over the finite interval $\left[0;\frac{T}{2}\right]$, its odd periodic extension $\phi_2(x)$ is theodd periodic function with period T defined by

$$\phi_2(x) = \begin{cases} f(x) \text{ if } 0 < x < \frac{T}{2}, \\ -f(-x) \text{ if } -\frac{T}{2} < x < 0, \end{cases}$$

$$\phi_2(x+T) = \phi_2(x)$$

The Fourier series expansion of the function $\phi_2(x)$ consists of sine terms only and is given by

$$\phi_2(x) \sim \sum_{k=1}^{\infty} b_k \sin kwx$$

called the *sine series expansion*, where

$$b_k = \frac{4}{T}\int_0^{\frac{T}{2}} f(x)\sin kwx dx \text{ for } k = 1,2,... \quad ...(48)$$

Example 5. The function $f(x) = x$ is defined only in the interval [0;4]. Obtain (a) the cosine series expansion and (b) sine series expansion.

Solution:

(a) To obtain the cosine series expansion we define the 3 periodic extension with period.

$$\phi_1(x) = \begin{cases} f(x) \text{ if } 0 < x < 4, \\ -x \text{ if } -4 < x < 0, \end{cases}$$

$$\phi_1(x+8) = \phi_1(x)$$

Using (47) we calculate the coefficients

$$a_0 = \frac{4}{8}\int_0^4 x\,dx = 4$$

and (as $\omega = \frac{\pi}{4}$) $\quad a_k = \frac{4}{8}\int_0^4 x\,cox\frac{k\pi x}{4}dx$

$$= \frac{1}{2}\left[\frac{4\pi}{k\pi}\sin\frac{k\pi x}{4}\bigg|_0^4 - \frac{4}{k\pi}\int_0^4 \sin\frac{k\pi x}{4}dx\right]$$

$$= \frac{1}{2}\frac{16}{k^2\pi^2}\cos\frac{k\pi x}{4}\bigg|_0^4$$

$$= \frac{8}{k^2\pi^2}\left[(-1)^k - 1\right]$$

$$= \begin{cases} -\frac{16}{k^2\pi^2} \text{ if } k \text{ is odd}, \\ 0 \text{ if } k \text{ is even} \end{cases}.$$

The cosine series expansion of the function $\phi_1(x)$ is

$$\phi_1(x) \sim 2 - \frac{16}{\pi^2}\sum_{k=1}^{\infty}\frac{1}{(2k-1)^2}\cos\frac{(2K-1)\pi x}{4}$$

In the interval [0; 4] the function $\phi_1(x)$ is identical with $f(x) = x$. Hence on the interval [0; 4] we have obtained also the cosine series expansion

$$\phi_1(x) \sim 2 - \frac{16}{\pi^2}\sum_{k=1}^{\infty}\frac{1}{(2k-1)^2}\cos\frac{(2K-1)\pi x}{4}.$$

(b) To obtain the sine series expansion we define the odd periodic extension $\phi_2(x)$ of $f(x)$:

$$\phi_2(x) = x \text{ if } -4 < x < 4.$$

$$\phi_2(x+8) = \phi_2(x)$$

Using (32) gives

$$b_k = \frac{4}{8}\int_0^4 x\sin\frac{k\pi x}{4}dx$$

$$= \frac{1}{2}\left[-\frac{4x}{k\pi}\cos\frac{k\pi x}{4}\bigg|_0^4 + \frac{4}{k\pi}\int_0^4 \cos\frac{k\pi x}{4}dx\right]$$

$$= \frac{1}{2}\left[-\frac{16x}{k\pi}(-1)+\frac{16}{k^2\pi^2}\sin\frac{k\pi x}{4}\bigg|_0^4\right]$$

$$= \frac{8}{k\pi}(-1)^{\kappa+1}$$

and the sine series expansion of the function $\phi_2(x)$ is

$$\phi_2(x) \sim \frac{8}{\pi}\sum_{k=1}^{\infty}\frac{(-1)^{k+1}}{k}\sin\frac{k\pi x}{4}.$$

On the interval [0; 4] also the sine series expansion of the given function is

$$x \sim \frac{8}{\pi}\sum_{k=1}^{\infty}\frac{(-1)^{k+1}}{k}\sin\frac{k\pi x}{4}.$$

THE COMPLEX FORM OF FOURIER SERIES

Let us assume that the periodic function $f(x)$ has the period 2π. To obtain the complex form of the Fourier series

$$f(x) \sim \frac{\alpha_0}{2}+\sum_{k=1}^{\infty} a_k \cos kx + \sin kx, \quad \text{...(49)}$$

we use the Euler formulae

$$\cos\varphi = \frac{e^{i\varphi}+e^{-i\varphi}}{2}$$

and

$$\sin\varphi = \frac{e^{i\varphi}+e^{-i\varphi}}{2i}.$$

By these formulae we have for kth harmonic

$$a_k \cos kx + b_k \sin kx = a_k\frac{e^{jkx}+e^{-ikx}}{2}+b_k\frac{e^{jkx}-e^{-ikx}}{2i}$$

$$= a_k \frac{e^{ikx}+e^{-ikx}}{2} - ib_k \frac{e^{ikx}-e^{-ikx}}{2}$$

$$= \frac{a_k - ib_k}{2} e^{ikx} + \frac{a_k + ib_k}{2} e^{-ikx}.$$

Denoting by $c_o = \frac{\alpha_O}{2}, c_k = \frac{\alpha_k - ib_k}{2}$ *and* $c_{-k} = \frac{a_k + ib_k}{2},$

the Fourier series expansion takes the form

$$f(x) \sim c_O + \sum_{k=1}^{\infty} \left(c_k e^{ikx} + c_{-k} e^{-ikx}\right) = \sum_{k=-\infty}^{\infty} c_k e^{ikx}.$$

To calculate the coefficients c_k for k > 0 we use the formulae (14) and (15):

$$c_k = \frac{a_{k-ib_k}}{2}$$

$$= \frac{1}{2\pi} \int_{-\pi}^{\pi} f(x) \cos kx dx - \frac{i}{2\pi} \int_{-\pi}^{\pi} f(x) \sin kx dx$$

$$= \frac{1}{2\pi} \int_{-\pi}^{\pi} f(x) (\cos kx - i \sin kx) dx$$

$$= \frac{1}{2\pi} \int_{-\pi}^{\pi} f(x) e^{-ikx} dx.$$

An immediate extension shows that this formula is also valid for $k = 0$. If k < 0 then $-k > 0$ and

$$c_k = \overline{c - k} = \overline{\frac{1}{2\pi} \int_{-\pi}^{\pi} f(x) e^{ikx} dx.} = \frac{1}{2\pi} \int_{-\pi}^{\pi} f(x) \overline{e^{ikx}} dx.$$

$$= \frac{1}{2\pi} \int_{-\pi}^{\pi} f(x) e^{-ikx} dx.$$

Hence for any $k = 0, \pm 1, \pm 2, \ldots$ we have

$$c_k = \frac{1}{2\pi} \int_{-\pi}^{\pi} f(x) e^{ikx} dx. \qquad \ldots(50)$$

Consequently, the complex form of the Fouries series of 2π-periodic function $f(x)$ is

$$f(x) \sim \sum_{k=-\infty}^{\infty} c_k e^{ikx} \qquad \ldots(51)$$

where the coefficients c_k ($k = 0, \pm 1, \pm 2, \ldots$) are calculated by (34)

Now, if the period of the function $f(x)$ is T then by Lemma 1.1 the period of the function $f\left(\frac{Tx}{2\pi}\right)$ is 2π and by (51) it has the Fourier series expansion

$$f\left(\frac{Tx}{2\pi}\right) \sim \sum_{k=-\infty}^{\infty} c_k e^{ikx} \qquad ...(52)$$

where by (51) the coefficients are

$$c_k = \frac{1}{2\pi}\int_{-\pi}^{\pi} f\left(\frac{Tx}{2\pi}\right) e^{-ikx}dx \text{ for } k = 0, \pm 1, \pm 2,... \qquad ...(53)$$

Changing in (52) and (53) the variable $t = \frac{T_2}{2\pi}$, we have that $x = \frac{2\pi t}{T}$, $dx = \frac{2\pi}{T}dt$, and

$$\text{f(t)} \sim \sum_{k=-\infty}^{\infty} c_k e^{\frac{2k\pi it}{T}}$$

where $$c_k = \frac{1}{T}\int_{-\frac{T}{2}}^{\frac{T}{2}} f(t) e^{\frac{-2k\pi it}{T}} dt$$

Denoting the variable by x again, we have that the complex form of the Fourier series expansion of the function $f(x)$ with period T is

$$f(x) \sim \sum_{k=-\infty}^{\infty} c_k e^{ikwx} \qquad ...(54)$$

where $$c_k = \frac{1}{T}\int_{-\frac{T}{2}}^{\frac{T}{2}} f(x) e^{-ikwx}dx \;\; for\, k = 0, \pm 1, \pm 2,... \qquad ...(55)$$

The complex numbers c_k form an infinite sequence $(c_k)_{k=-\infty}^{\propto}$ which is called the *spectrum* of the periodic function $f(x)$. Let $|c_k|$ designates the magnitude of the complex number c_k and φk its argument, i.e. $c_k = |c_k| e^{i\varphi k}$. The sequence of real numbers $(|c_k|)_k^{\infty} = -\infty$ is called the *amplitude spectrum* and the sequence of real numbers $(\varphi_k)_k^{\infty} = -\infty$ is called the *phase spectrum* of the function $f(x)$.

Example 6. Find the amplitude spectrum and the phase spectrum of the 2π-periodic function $f(x)$ defined over interval $\pi; -\pi$ as $f(x) = e^x$.

Solution:

First we find the spectrum, i.e. the complex Fourier coefficients, of the function $f(x)$ using (50).

$$c_k = \frac{1}{2\pi}\int_{-\pi}^{\pi} e^x e^{-ikx} dx = \frac{1}{2\pi}\int_{-\pi}^{\pi} e(1-ik)^x dx$$

$$= \frac{1}{2\pi(1-ik)} e^{(1-ik)x}\Big|_{-\pi}^{\pi} = \frac{e^{(1-ik)\pi} - e^{(1-ik)\pi}}{2\pi(1-ik)}.$$

Since $e^{\pm ik\pi = \cos k\pi \pm i \sin k\pi = \cos k\pi = (-1)^k}$ we obtain that

$$c_k = \frac{(-1)^k\left(e^{\pi} - e^{-\pi}\right)}{2\pi(1-ik)} = \frac{(-1)^k\left(e^{\pi} - e^{-\pi}\right)}{2\pi\left(1+k^2\right)}(1+ik).$$

Since $|1+ik| = \sqrt{1+k^2}$ we find the amplitude spectrum $\left|c_k = \frac{e^{\pi} - e^{-\pi}}{2\pi\sqrt{1+k^2}}\right|$ and phase spectrum $\varphi_k = \arg c_k = \arctan k$.

EXERCISES

In problem 1- 10 find the Fourier series expansions of the given periodic functions.

1. $f(x) = \begin{cases} 1 & if \quad -\pi \le x < 0 \\ 3 & if \quad 0 < x < \pi \end{cases}$

   $f(x+2\pi) = f(x).$

2. $f(x) = \begin{cases} 2 & if \quad -0 \le x < T \\ 0 & if \quad T < x < 2T \end{cases}$

   $f(x + 2T) = f(x).$

3. $f(x) = x$ if $0 \le x < 2T$

   $f(x+2\pi) = f(x).$

4. $f(x) = \frac{\pi = x}{2}$ $if\ 0 < x \le 2\pi,$

   $f(x+2\pi) = f(x).$

5. $f(x) = |x|$ $if\ -T < x < T,$

   $f(x + 2T) = f(x).$

6. $f(x) = x^2$ if $-\pi \le x \le \pi$,

   $f(x+2\pi) = f(x)$.

7. $f(x) = \pi^2 - x^2$ *if* $-\pi < x < \pi$,

   $f(x+2\pi) = f(x)$.

8. $f(x) = x\sin x$ *if* $-\pi < x < \pi$,

   $f(x+2\pi) = f(x)$.

9. $f(x) = |\cos x|$

10. $f(x) = \sin \pi x$ *if* $-\pi < x < \pi$,

    In problems 11-12 find the cosine series expansion on an interval $0; \pi$.

11. $f(x) = x(\pi - x)$.

12. $f(x) = \sin x$

    In problems 1.13–1.14 find the sine series expansion on an interval $0; \pi$.

13. $f(x) = \cos x$

14. $f(x) = \dfrac{\pi - 2x}{4}$.

15. Find the complex form of the Fourier series of the periodic function

$$f(x) = \begin{cases} x & \textit{if} \quad 0 \le x \le 0 \\ 0 & \textit{if} \quad -\pi < x < 0 \end{cases}$$

$$f(x+2\pi) = f(x).$$

Chapter 7

Application of Fourier Series

INTRODUCTION

In the previous chapter we have learnt about Fourier series. In this chapter we will study the application of Fourier series.

A Fourier series is an expansion of a periodic function $f(x)$ in terms of an infinite sum of sines and cosines. Fourier series make use of the orthogonality relationships of the sine and cosine functions. The computation and study of Fourier series is known as harmonic analysis and is extremely useful as a way to break up an *arbitrary* periodic function into a set of simple terms that can be plugged in, solved individually, and then recombined to obtain the solution to the original problem or an approximation to it to whatever accuracy is desired or practical. Examples of successive approximations to common functions using Fourier series are illustrated below.

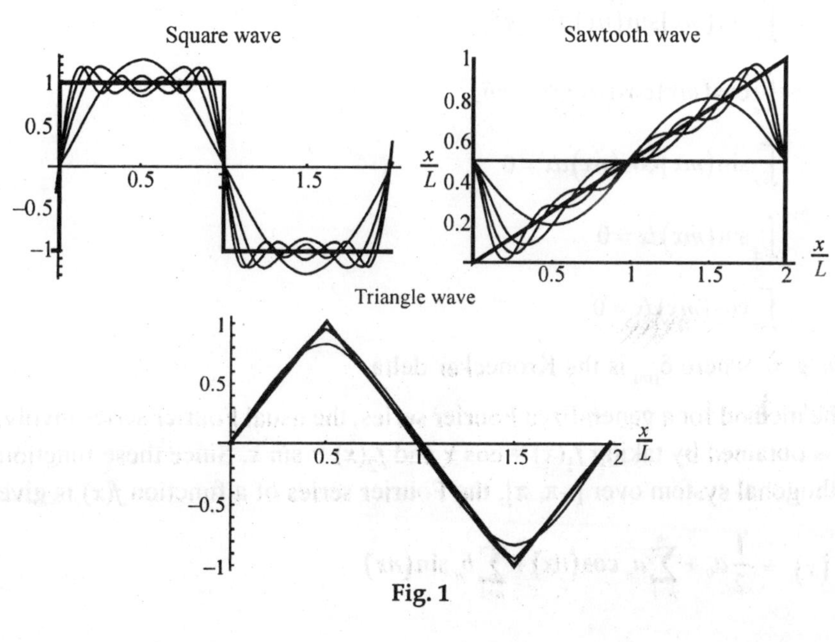

Fig. 1

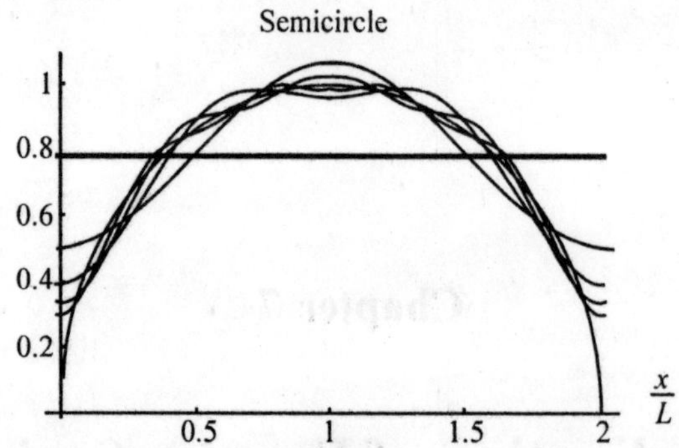

Fig. 2

In particular, since the superposition principle holds for solutions of a linear homogeneous ordinary differential equation, if such an equation can be solved in the case of a single sinusoidal, the solution for an arbitrary function is immediately available by expressing the original function as a Fourier series and then plugging in the solution for each sinusoidal component. In some special cases where the Fourier series can be summed in closed form, this technique can even yield analytic solutions.

Any set of functions that form a complete orthogonal system have a corresponding generalized Fourier series analogous to the Fourier series. For example, using orthogonality of the roots of a Bessel function of the first kind gives a so-called Fourier-Bessel series.

COMPUTATION OF FOURIER SERIES

The computation of the (usual) Fourier series is based on the integral identities

$$\int_{-\pi}^{\pi} \sin(mx)\sin(nx)\,dx = \pi\delta_{mn} \quad ...(1)$$

$$\int_{-\pi}^{\pi} \cos(mx)\cos(nx)\,dx = \pi\delta_{mn} \quad ...(2)$$

$$\int_{-\pi}^{\pi} \sin(mx)\cos(nx)\,dx = 0 \quad ...(3)$$

$$\int_{-\pi}^{\pi} \sin(mx)\,dx = 0 \quad ...(4)$$

$$\int_{-\pi}^{\pi} \cos(mx)\,dx = 0 \quad ...(5)$$

for $m, n \neq 0$, where δ_{mn} is the Kronecker delta.

Using the method for a generalized Fourier series, the usual Fourier series involving sines and cosines is obtained by taking $f_1(x) = \cos x$ and $f_2(x) = \sin x$. Since these functions form a complete orthogonal system over $[-\pi, \pi]$, the Fourier series of a function $f(x)$ is given by

$$f(x) = \frac{1}{2}a_0 + \sum_{n=1}^{\infty} a_n \cos(nx) + \sum_{n=1}^{\infty} b_n \sin(nx) \quad ...(6)$$

where

$$a_0 = \frac{1}{\pi}\int_{-\pi}^{\pi} f(x)dx \qquad ...(7)$$

$$a_n = \frac{1}{\pi}\int_{-\pi}^{\pi} f(x)\cos(nx)dx \qquad ...(8)$$

$$b_n = \frac{1}{\pi}\int_{-\pi}^{\pi} f(x)\sin(nx)dx \qquad ...(9)$$

and $n = 1, 2, 3, ...$ Note that the coefficient of the constant term a_0 has been written in a special form compared to the general form for a generalized Fourier series in order to preserve symmetry with the definitions of a_n and b_n.

A Fourier series converges to the function $\bar{f}$ (equal to the original function at points of continuity or to the average of the two limits at points of discontinuity)

$$\bar{f} = \begin{cases} \frac{1}{2}\left[\lim\limits_{x\to x_0^-} f(x) + \lim\limits_{x\to x_0^+} f(x)\right] & \text{for } -\pi < x_0 < \pi \\ \frac{1}{2}\left[\lim\limits_{x\to -\pi^+} f(x) + \lim\limits_{x\to \pi_-} f(x)\right] & \text{for } x_0 = -\pi, \pi \end{cases} \qquad ...(10)$$

if the function satisfies so-called Dirichlet conditions.

As a result, near points of discontinuity, a "ringing" known as the Gibbs phenomenon, illustrated below, can occur.

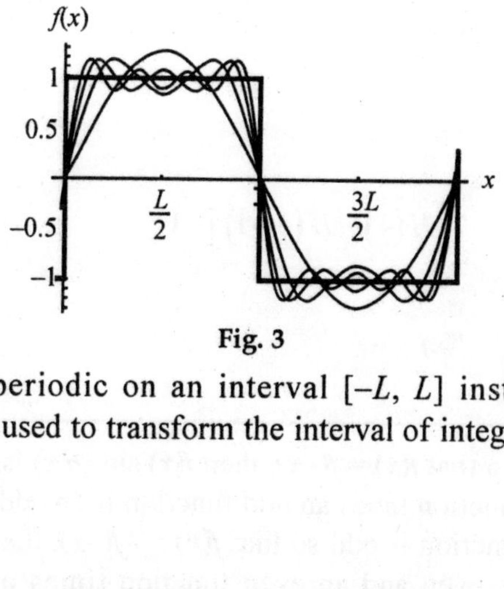

Fig. 3

For a function $f(x)$ periodic on an interval $[-L, L]$ instead of $[-\pi, \pi]$, a simple change of variables can be used to transform the interval of integration from $[-\pi, \pi]$ to $[-L, L]$. Let

$$x = \frac{\pi x'}{L} \qquad ...(11)$$

$$dx = \frac{\pi dx'}{L} \qquad ...(12)$$

Solving for x' gives $x' = Lx/\pi$, and plugging this in gives

$$f(x') = \frac{1}{2}a_0 + \sum_{x=1}^{\infty} a_n \cos\left(\frac{n\pi x'}{L}\right) + \sum_{x=1}^{\infty} b_n \sin\left(\frac{n\pi x'}{L}\right) \quad ...(13)$$

Therefore,

$$a_0 = \frac{1}{L}\int_{-L}^{L} f(x')\,dx' \quad ...(14)$$

$$a_n = \frac{1}{L}\int_{-L}^{L} f(x')\cos\left(\frac{n\pi x'}{L}\right)dx' \quad ...(15)$$

$$b_n = \frac{1}{L}\int_{-L}^{L} f(x')\sin\left(\frac{n\pi x'}{L}\right)dx' \quad ...(16)$$

Similarly, the function is instead defined on the interval [0, 2, L], the above equations simply become

$$a_0 = \frac{1}{L}\int_{0}^{2L} f(x')\,dx' \quad ...(17)$$

$$a_n = \frac{1}{L}\int_{0}^{2L} f(x')\cos\left(\frac{n\pi x'}{L}\right)dx' \quad ...(18)$$

$$b_n = \frac{1}{L}\int_{0}^{2L} f(x')\sin\left(\frac{n\pi x'}{L}\right)dx' \quad ...(19)$$

In fact, for $f(x)$ periodic with period $2L$, *any* interval $(x_0, x_0 + 2L)$ can be used, with the choice being one of convenience or personal preference. One of the most common functions usually analyzed by this technique is the square wave. The Fourier series for a few common functions are summarized in the table below.

function	$f(x)$	Fourier series
Fourier series —saw-tooth wave	$\frac{x}{2L}$	$\frac{1}{2}-\frac{1}{\pi}\sum_{n=1}^{\infty}\frac{1}{2}\sin\left(\frac{n\pi x}{L}\right)$
Fourier series —square wave	$2\left[H\left(\frac{x}{L}\right)-H\left(\frac{x}{L}-1\right)\right]-1$	$\frac{4}{\pi}\sum_{n=1,3,5,...}^{\infty}\frac{1}{n}\sin\left(\frac{n\pi x}{L}\right)$
Fourier series —triangle wave	$T(x)$	$\frac{8}{\pi^2}\sum_{\pi=1,3,5...}^{\infty}\frac{(-1)^{(n-1)/2}}{n^2}\sin\left(\frac{n\pi x}{L}\right)$

If a function is even so that $f(x) = f(-x)$, then $f(x)\sin(n\,x)$ is odd. (This follows since $\sin(n\,x)$ is odd and an even function times an odd function is an odd function.) Therefore, $b_n = 0$ for all n. Similarly, if a function is odd so that $f(x) = -f(-x)$, then $f(x)\cos(n\,x)$ is odd. (This follows since $\cos(n\,x)$ is even and an even function times an odd function is an odd function.) Therefore, $a_n = 0$ for all n.

$a_n = 0$ for all n.

The notion of a Fourier series can also be extended to complex coefficients. Consider a real-valued function $f(x)$. Write

$$f(x) = \sum_{n=-\infty}^{\infty} A_n e^{inx} \qquad \text{...(20)}$$

Now examine

$$\int_{-\pi}^{\pi} f(x) e^{-imx} dx = \int_{-\pi}^{\pi} \left(\sum_{n=-\infty}^{\infty} A_n e^{inx} \right) e^{-imx} dx \qquad \text{...(21)}$$

$$= \sum_{n=-\infty}^{\infty} A_n \int_{-\pi}^{\pi} e^{i(n-m)x} dx \qquad \text{...(22)}$$

$$= \sum_{n=-\infty}^{\infty} A_n \int_{-\pi}^{\pi} \{\cos[(n-m)x] + i\sin[(n-m)x]\} dx \qquad \text{...(23)}$$

$$= \sum_{n=-\infty}^{\infty} A_n 2\pi\delta mn \qquad \text{...(24)}$$

$$= 2\pi A_m, \qquad \text{...(25)}$$

so $$A_n = \frac{1}{2\pi} \int_{-\pi}^{\pi} f(x) e^{-inx} dx \qquad \text{...(26)}$$

The coefficients can be expressed in terms of those in the Fourier series

$$A_n = \frac{1}{2\pi} \int_{-\pi}^{\pi} f(x)[\cos(nx) - i\sin(nx)] dx \qquad \text{...(27)}$$

$$= \begin{cases} \frac{1}{2\pi} \int_{-\pi}^{\pi} f(x)[\cos(nx) + i\sin(nx)] dx & n < 0 \\ \frac{1}{2\pi} \int_{-\pi}^{\pi} f(x) dx & n = 0 \\ \frac{1}{2\pi} \int_{-\pi}^{\pi} f(x)[\cos(nx) - i\sin(nx)] dx & n > 0 \end{cases} \qquad \text{...(28)}$$

$$= \begin{cases} \frac{1}{2}(a_n + ib_n) & \text{for } n < 0 \\ \frac{1}{2} a_0 & \text{for } n = 0 \\ \frac{1}{2}(a_n - ib_n) & \text{for } n > 0 \end{cases} \qquad \text{...(29)}$$

For a function periodic in $[-L/2, L/2]$, these become

$$f(x) = \sum_{n=-\infty}^{\infty} A_n e^{i(2\pi nx/L)} \qquad \text{...(30)}$$

$$A_n = \frac{1}{L}\int_{-L/2}^{L/2} f(x)e^{-i(2\pi nx/L)}dx \qquad ...(31)$$

These equations are the basis for the extremely important Fourier transform, which is obtained by transforming A_n from a discrete variable to a continuous one as the length $L \to \infty$

APPLICATION OF FOURIER SERIES

We have learnt earlier the way how to use Fourier series. Here, we discuss some application of Fourier series. These series play a vital role in Mathematics as well as in other field.

Simple Fourier Series

Let *f* be periodic of period 2π, with *f*(*x*) = *x* for *x* from –π to π. Note that this function is a periodic version of the identity function.

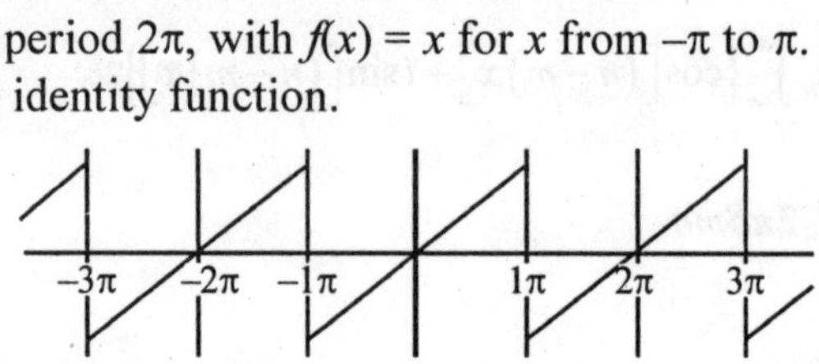

Fig. 4 Plot of a periodic identity function - a sawtooth wave.

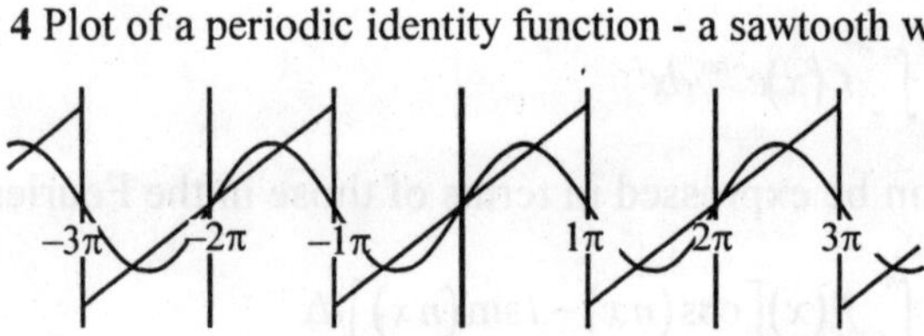

Fig. 5 Plot of the First Five Successive Partial Fourier Series

We will compute the Fourier coefficients for this function.

$$a_n = \frac{1}{\pi}\int_{-\pi}^{\pi} f(x)\cos(nx)dx$$

$$= \frac{1}{\pi}\int_{-\pi}^{\pi} x\cos(nx)dx$$

$$= a_n = 0b_n = \frac{1}{\pi}\int_{-\pi}^{\pi} f(x)\sin(nx)dx$$

$$= \frac{1}{\pi}\int_{-\pi}^{\pi} x\sin(nx)dx$$

$$= \frac{2}{\pi}\int_{0}^{\pi} x\sin(nx)dx$$

$$= \frac{2}{\pi}\left(\left[-\frac{x\cos(nx)}{n}\right]_0^{\pi} + \left[\frac{\sin(nx)}{n^2}\right]_0^{\pi}\right)$$

$$b_n = 2\frac{(-1)^{n+1}}{n}$$

Notice that a_n are 0 because the $x \mapsto x\cos(nx)$ are odd functions. Hence the Fourier series for this function is:

$$f(x) = \frac{a_0}{2} + \sum_{n=1}^{\infty}\left[a_n \cos(nx) + b_n \sin(nx)\right]$$

$$= 2\sum_{n=1}^{\infty}\frac{(-1)^{n+1}}{n}\sin(nx), \forall x \in [-\pi, \pi]$$

One application of this Fourier series is to compute the value of the Riemann zeta function at $s = 2$; by Parseval's theorem, we have:

$$\frac{1}{2\pi}\int_{-\pi}^{\pi} x^2 dx = \frac{1}{2}\sum_{n>0}\left[2\frac{(-1)^n}{n}\right]^2$$

which yields: $\sum_{n>0}\frac{1}{n^2} = \frac{\pi^2}{6}$

Wave Equation

The wave equation governs the motion of a vibrating string, which may be fastened down at its endpoints. The solution of this problem requires the trigonometric expansion of a general function f that vanishes at the endpoints of an interval $x = 0$ and $x = L$. The Fourier series for such a function takes the form

$$f(x) = \sum_{n=1}^{\infty} b_n \sin\left(\frac{n\pi}{L}x\right)$$

where

$$b_n = \frac{2}{L}\int_0^L f(x)\sin\left(\frac{n\pi}{L}x\right)dx$$

Vibrations of air in a pipe that is open at one end and closed at the other are also described by the wave equation. Its solution requires expansion of a function that vanishes at $x = 0$ and whose derivative vanishes at $x = L$. The Fourier series for such a function takes the form

$$f(x) = \sum_{n=1}^{\infty} b_n \sin\left(\frac{(2n+1)\pi}{2L}x\right)$$

where $$b_n = \frac{2}{L}\int_0^L f(x)\sin\left(\frac{(2n+1)\pi}{2L}x\right)dx$$

An Application of Fourier Analysis to the Description and Statistical Analysis of Refractive Error

The description of sphero-cylinder lenses is approached from the viewpoint of Fourier analysis of the power profile. It is shown that the familiar sine-squared law leads naturally to a Fourier series representation with exactly 3 Fourier coefficients, representing the natural parameters of a thin lens. The constant term corresponds to the mean spherical equivalent power while the amplitude and phase of the harmonic correspond to the power and axis of a Jackson cross-cylinder lens, respectively. Expressing the Fourier series in rectangular form leads to the representation of an arbitrary sphero-cylinder lens as the sum of a spherical lens and two cross-cylinders, one at axis 0° and the other at axis 45°. The power of these three component lenses may be interpreted as (x,y,z) coordinates of a vector representation of the power profile. Advantages of this *power vector* representation of a sphero-cylinder lens for numerical and graphical analysis of optometric data are described for problems involving lens combinations, comparison of different lenses, and the statistical distribution of refractive errors.

Various types of finite methods are often used to solve combined conduction and convection problems. These methods are both flexible and accurate but they are also complex to program and relatively slow. At times it can there-fore be advantageous to use an analytical approach. The list of pros and cons for this category of methods is very much a mirror of those for the finite methods. They are easy to program, usually fast but not very flexible.

This article is about a Fourier series solution to the problem of calculating the temperature distribution on a convection cooled plate with discrete heat sources. It is by no means the only way to tackle with this problem. At times it is nevertheless the most convenient one.

The solution has been known in the heat transfer community for several years. As far as the author knows its origin is to find at the University of Poitiers, France. The purpose of this article is therefore not to explain the how it is derived but to discuss some of its application aspects.

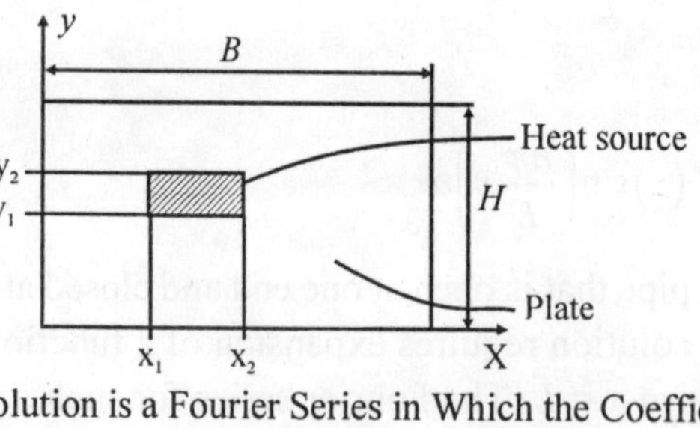

Fig. 6 The 2-D Solution is a Fourier Series in Which the Coefficients are Given by Another Fourier Series.

$$\begin{cases} \vartheta(x,y) = \dfrac{y}{k^2} \cdot \sum\limits_{n=0}^{N} f(n,y) \cdot A_n \cdot \cos(n \cdot \varpi_x \cdot x) \\ f(n,y) = \sum\limits_{m=0}^{M} B_{mn} \cdot \cos(m \cdot \varpi \cdot y) \end{cases}$$

Where, $y = \frac{\dot{q}}{k \cdot \delta}$ $\chi = \sqrt{\frac{1 \cdot h}{k \cdot \delta}}$ $\varpi = \frac{\pi}{B}$ $\varpi = \frac{\pi}{H}$

$$A_0 = \frac{\varpi_x \cdot (x_2 - x_1)}{\pi}$$

$$A_n = \frac{2 \cdot [\sin(n \cdot \varpi_x \cdot x_2) - \sin(n \cdot \varpi_x \cdot x_1)]}{n \cdot \pi}$$

$$B_{0n} = \frac{\varpi_y \cdot (y_2 - y_1)}{\pi \cdot \left[\left(\frac{n \cdot \varpi_x}{\kappa} \right)^2 + 1 \right]}$$

$$B_{mn} = \frac{2 \cdot [\sin(n \cdot \varpi_x \cdot x_2) - \sin(n \cdot \varpi_x \cdot x_1)]}{m \cdot \pi \left[\left(\frac{n \cdot \varpi_x}{\kappa} \right)^2 + \left(\frac{m \cdot \varpi_y}{\kappa} \right)^2 + 1 \right]}$$

$$N \geq \frac{3 \cdot E}{x_2 - x_1}$$

$$M \geq \frac{3 \cdot H}{y_2 - y_1}$$

$\dot{q}$ = heat flux ror heat source (W/m^2)

k = thermal condictivity (W/mK)

δ = plate thickness (m)

h = heat transfer coefficient (W/m^{2}K)

The Solution

The 2-D solution is shown in figure 7. It is basically a Fourier series in which the coefficients are given by another Fourier series. Double sided cooling has been assumed in this particular case. This can however easily be changed by dividing the heat transfer coefficient with a factor 2, (or by suppressing number 2 in the fourth equation). It should also be noted that the solution is the simplest in a family of solutions that originates from a 3-D solution for a plate with axis dependent thermal conductivities. Since their application aspects are similar the discussion that follows is relevant for all members of this family.

The main advantage with the solution is its simplicity. It can programmatically be fitted into a few procedures, which greatly facilitates code writing. The main disadvantage is that its application window is limited to simple cases. It can sometimes also be slow. As this article will reveal there are nonetheless several tricks that can counterbalance these drawbacks.

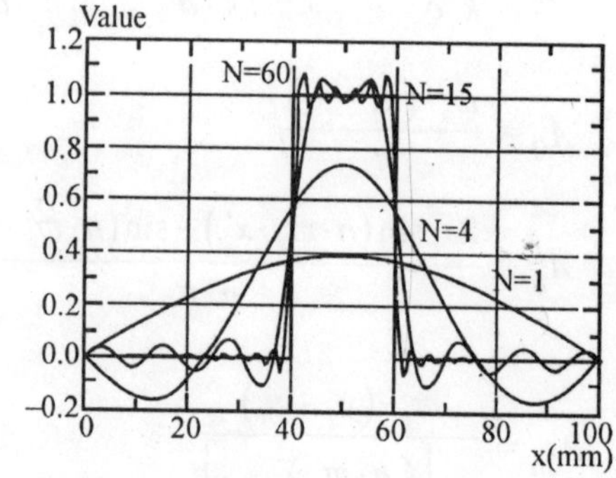

Fig. 7 Fourier Series can be Used to Model Square Wave Functions. The Fit is Strongly Dependent on the Number of Terms Included.

To show the total derivation of the solution would require too much space but the basic principle is to apply two perpendicular square functions to model the heat source flux. These functions take the value 1 above the heat source and the value 0 elsewhere. A standard approach for describing such functions is to use Fourier series. These series can be made to model just about any function but their drawback is that the fit only is perfect if an infinite number of terms are included. One must therefore always settle for an approximation.

To determine the term count needed in a Fourier series is always problematic. The terms do have a tendency to decrease as the series is expanded but not in an orderly and regular manner. It is therefore difficult to use the value of last term as criteria for halting the summation. For the square wave function there is fortunately another possibility. Each term in the modelling series can be looked at as cycloid function with a wavelength that successively decreases. It is evident that a good fit only can be achieved unless the last wavelength is smaller than the width of the square wave. The rule of thumb is to break the summation when the width is covered by 1.5 wavelengths. The equations for N and M in figure 1 reflect this rule and for the particular case in figure 8 it yields N=15.

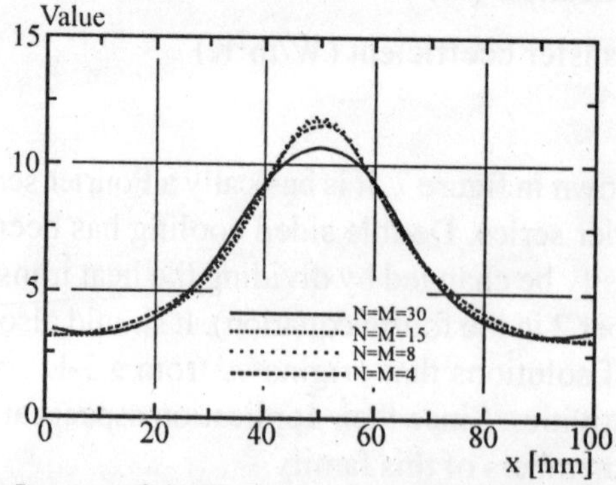

Fig. 8 The Accuracy Impact of the Term Count is Most Important in the Centre of the Heat Source. At Some Distance it is Therefore Possible to Use Half the Value Recommended in figure.

The term count has a large impact on the calculation time. It is therefore highly desirable to reduce it as much as possible. Figure 9 shows the impact on the temperature level for the same example as in figure 8. The accuracy seems to reach saturation for N > 15, which confirms that the equations given for M and N in figure 1 are reasonable. When the maximum summation index is reduced to N = 8 there is some impact inside the heat source but hardly any on its outside. It therefore seems possible to reduce the recommended values for M and N with a factor 2 for points that are at some distance from the heat source. This is important because it reduces the calculation time with a factor 4.

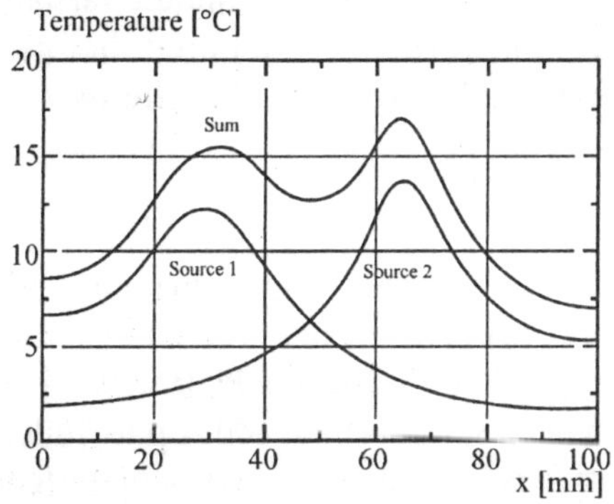

Fig. 9 Superposition can be Used to Simulate the Impact of Several Heat Sources.

Superposition

The super position principle can always be used if the sum of the solutions to an equation also is a solution to that equation. All conduction problems in which the material conductivity is an invariant belong to this category. The superposition principle can therefore be used for almost all conduction problems in thermal design but with a possible exception for half conductors such as silicone. The solution discussed is actually a large scale application of this principle.

Superposition can naturally also be used to simulate the combined impact of several heat sources on a plate, figure 10. This opens up the possibility for fast studies on how a change of a heat source property in a multi-heat source environment impacts the total temperature field. A procedure that manages this has five steps:

1. Calculate the temperature field for the heat source.
2. Subtract it from the total field.
3. Change the properties of the heat source.
4. Calculate the new temperature field for the heat source.
5. Add it to the total field.

The advantage with this procedure is that the temperature field only has to be recalculated for one heat source at a time. If there are 50 of them this can save a lot of calculation time. This technique can for example be used for rapid layout studies on PCBs. The solution discussed is however still is a bit too slow to make this fully enjoyable when the PCBs are large.

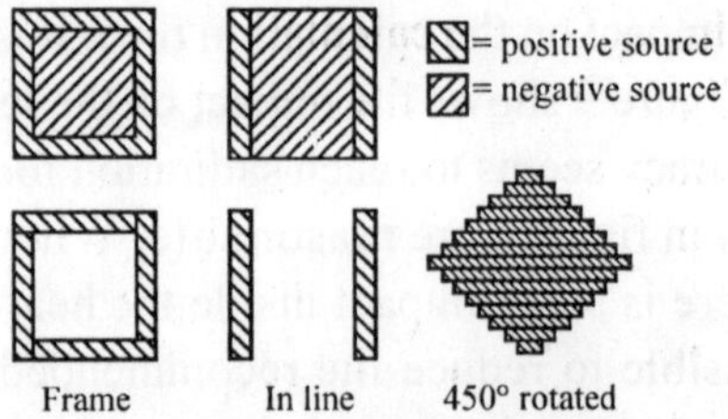

Fig. 10 Various Heat Source Shapes can be Created by Combining Heat Sources.

The superposition principle can also be used to simulate various heat source shapes. Figure shows a few examples. It should nevertheless be noted that using overlapping, although mathematically elegant, always implies the risk that the two calculation errors combine unfavourably.

Execution Speed

The execution speed is highly dependent on the term count. When it is low, such as for the example in figure 9, this is not a concern. For small heat sources on large plates the situation is however radically different. A typical 200 × 200 mm PCB with a 10 × 10 mm heat source would for example require 60 terms in each series which adds up to 3600 terms when the two nested series are combined. If the ambition is to create an overview of the temperature field, calculations must further be made on a grid of the order 20 × 20. All in all this results in 1.4 million terms. If there in addition are some 50 components, it is easy to realise that the task is heavy.

There are nevertheless a couple of measures that can be taken to speed up the process. To reduce the term count with a factor 2 for points outside the heat source has already been mentioned. Another possibility is to save the terms in the second summation, f(n,y), in a list and reuse them for all points that have the same y-value. These two measures boost the calculation speed considerably but for large densely packed PCBs it is still difficult to get down to enjoyable execution times.

The situation is much better for heat sinks. They are usually smaller than PCBs and the heat source count is in most cases limited to less than 10. Overviews can therefore be created in a matter of a few seconds, which is a key issue for successful front-end thermal design. Those who not have tired this method can get an idea of how it works on the web site given at the be-ginning of this article. The full potential of the method can however not be exploited unless the fluid temperature increase problem is reasonably well managed.

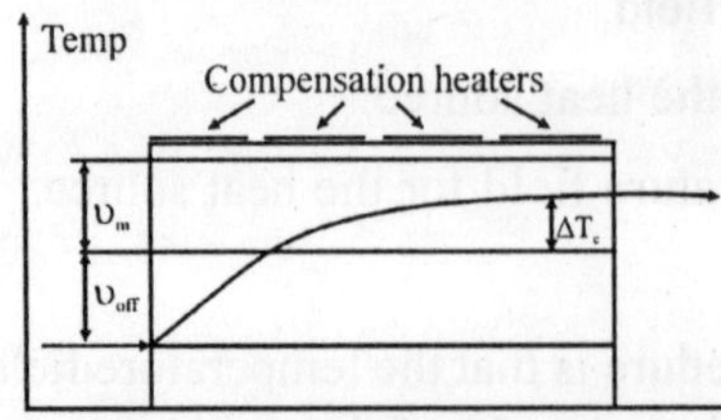

Fig. 11 The Basic Procedure Calculates the Temperature Distribution as if the Fluid Temperature was a Constant. An Array of Large Heaters that Cover the Entire Plate can be Used to Compensate for this Error.

$$\dot{Q}_C = h_m \cdot A_c \cdot \Delta T_c$$

where:

$$\begin{cases} \dot{Q} = \text{compensation heat dissipation (VV)} \\ h_m = \text{average heat transfer coefficient (W/m}^2\text{K)} \\ A_c = \text{compensation element surface (m}^2\text{)} \\ \Delta T_c = \text{temperature compensation (K)} \end{cases}$$

Fluid Temperature Increase Compensation

One problem with the Fourier series solution is that it only is valid for uniform fluid temperatures and heat transfer coefficients. This limitation always causes an error. It is often negligible for small heat sinks but for large ones cooled by small flow rates, it can be considerable.

The above figure shows a case with a high fluid temperature increase. For this case it is obvious the cooling is much more effective at the inlet than at the outlet. This effect is partly caused by a variation of the local temperature difference and partly by a variation of the local heat transfer coefficient.

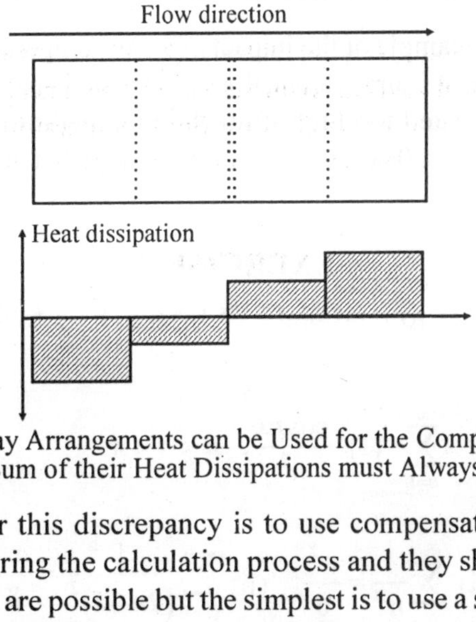

Fig. 12 Different Array Arrangements can be Used for the Compensation Heat Sources But the Sum of their Heat Dissipations must Always be Zero.

One way to rectify for this discrepancy is to use compensation heat sources. They are typically only kept alive during the calculation process and they should cover the entire plate. Various array arrangements are possible but the simplest is to use a single inline layout. Another restriction is that the sum of their heat dissipations always must equal zero.

A first level approach, which for many cases is sufficient, is to initially assume that the plate is isothermal, figure 12. It is a simple task to calculate the corresponding fluid temperature profile. The deviation between the average fluid temperature and the actual fluid temperature determines the strengths of the compensation sources and this is done with a very simple equation. The calculation is performed with the average heat transfer coefficient as parameter and the average fluid temperature as an offset value.

The first level approach basically simulates the fluid temperature profile as if the plate had been isothermal but it is possible to go one step further. The second level approach consists in calculating the average temperature for the compensation sources and use these as a bases for a better approximation of the temperature profile. The basic principle is the same as above but with the complication that the plate now has to be looked at as composed of several isothermal segments. Another difference is that it for this case can make sense to use a two dimensional layout array.

It should also be noted that the compensation sources are large sources that calculate rapidly. Compensating for the fluid temperature increase is therefore not a process that considerably charges the all over execution time.

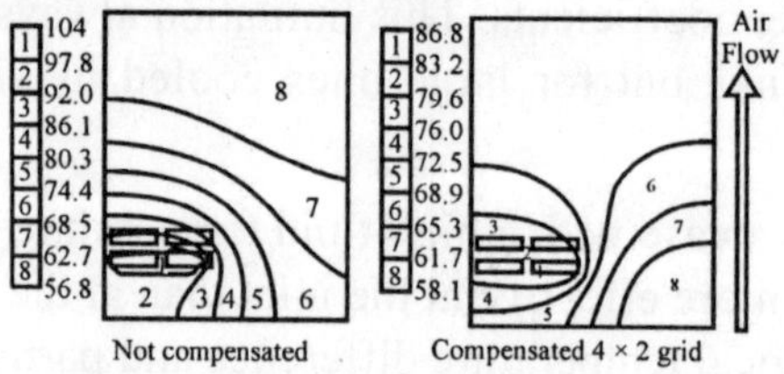

Fig. 13 Comparison of the Temperature Distribution in a 210x260 mm Heat Sink Bottom Plate, with and Without Fluid Temperature Compensation.

Above Figure shows an example of the impact of an air temperature increase compensation for rather extreme case. The heat sources are in this case located near the inlet. Their temperature would consequently be calculated too high if no fluid temperature compensation was made. The result difference is on the 20% level. It is not radical but it considerably expands the usefulness of the solution.

EXERCISE

1. Obtain a Fourier series to represent e^{-ax} from x = –p to x = p. Hence derive series for p/sinh p.

2. Prove that $x^2 = \frac{\pi^2}{3} + 4\sum_{n=1}^{\infty} -(1)^n \frac{\cos nx}{n^2}, -\pi < x < \pi.$

   Hence show that (i) $\sum \frac{1}{n^2} = \frac{\pi^2}{6}$ (ii) $\sum \frac{1}{(2n-1)^2} = \frac{\pi^2}{8}$

3. If $f(x) = \left(\frac{\pi - x}{2}\right)^2$ in the range 0 to 2*p*, show that $f(x) = \frac{\pi^2}{12} + \sum_{n=1}^{\infty} \frac{\cos nx}{n^2}$.

4. Prove that in the range –*p* < x < *p*, cosh ap $\left[\frac{1}{2a^3} + \sum_{n=1}^{\infty} \frac{(-1)^n}{n^2 + a^2} \cos nx\right]$.

5. Expand $f(x) = x \sin x$, $0 < x < 2p$, in a Frourier series.
6. $f(x) = x + x^2$ for $-p$ and $f(x) = p^2$ for $x = \pm p$. Expand $f(x)$ in Fourier series and show that

$$x + x^2 = \frac{\pi^2}{3} + \sum_{n=1}^{\infty} (-1)^n \left\{ \frac{4}{n^2} \cos nx - \frac{2}{n} \sin nx \right\}.$$

7. Show that the even periodic extension of the function

$f(t) = t^4 \qquad 0 < t < T/2$

is given by

$$\frac{T^4}{16\pi^4} \left[\frac{\pi^4}{5} + 8 \sum_{n=1,2,\ldots}^{\infty} (-1)^n \frac{n^2\pi^2 - 6}{n^4} \cos n\omega t \right]$$

Hence deduce that

$$1 + \frac{1}{2^4} + \frac{1}{3^4} + \frac{1}{4^4} + \ldots = \frac{\pi^4}{90}$$

8. Show that the even periodic extension of the function

$f(x) = \cos \gamma x \qquad 0 < x < a$

(with $\gamma \neq 0, \pm \pi/a, \pm 2\pi/a, \ldots$) is given by

$$\frac{2}{\pi} \Gamma \sin(\pi\Gamma) \left[\frac{1}{2\Gamma^2} + \sum_{n=1}^{\infty} \frac{(-1)^{n-1}}{n^2 - \Gamma^2} \cos \frac{n\pi x}{a} \right]$$

where $\Gamma = \gamma a / \pi$. Hence deduce that

$$\sum_{n=1}^{\infty} \frac{(-1)^{n-1}}{n^2 - \Gamma^2} = \frac{1}{2\Gamma^2} \left[\frac{\pi\Gamma}{\sin \pi\Gamma} - 1 \right]$$

and

$$\sum_{n=1}^{\infty} \frac{1}{n^2 - \Gamma^2} = \frac{1}{2\Gamma^2} \left[1 - \frac{\pi\Gamma}{\tan \pi\Gamma} \right]$$

with $\Gamma \neq 0, \pm 1, \pm 2, \ldots$ Show that in the limit of $\Gamma \to 0$ the above summations go over to Eqs. (83) and (84).

9. Show that the odd periodic extension of the function

$f(x) = x^4 - 2lx^3 + l^3 x \qquad 0 < x < 1$

is given by

$$\frac{192\,l^4}{\pi^5}\sum_{n=0}^{\infty}\frac{1}{(2n+1)^5}\sin\left\{(2n+1)\frac{\pi x}{l}\right\}$$

hence deduce that

$$\sum_{n=0}^{\infty}\frac{(-1)^n}{(2n+1)^5}=\frac{5\pi^5}{3072}$$

10. Expand the function

$$f(x)=\begin{cases}-(1-\xi)x \text{ for } 0\le x\le\xi\\ -(1-x)\xi \text{ for } \xi\le x\le 1\end{cases}$$

into a Fourier sine series and show that

$$f(x)=-\frac{2}{\pi^2}\sum_{n=1}^{\infty}\frac{1}{n^2}\sin n\pi x\,\sin n\pi\xi$$

Hint: Extend the definition for $-1 \le x \le 0$ as an odd function, i.e. $f(x) = -f(-x)$.

11. Expand the function

$$f(x)=\begin{cases}\dfrac{-\sin kx\,\sin k(L-\xi)}{k\sin kL} & \text{for } 0\le x\le\xi\\[2ex] \dfrac{-\sin k(L-x)\sin k\xi}{k\sin kL} & \text{for } \xi\le x\le L\end{cases}$$

into a Fourier sine series and show that

$$f(x)=\frac{2}{L}\sum_{n=1}^{\infty}\frac{1}{\left(k^2-\dfrac{n^2\pi^2}{L^2}\right)}\sin\frac{n\pi x}{L}\sin\frac{n\pi\xi}{L}$$

provided

$$k^2\neq\frac{n^2\pi^2}{L^2}$$

Hint: Extend the definition for $-L \le x \le 0$ as an odd function, i.e, $f(x) = -f(-x)$.

Chapter 8

Properties of Fourier Series

In this chapter we discuss the properties of Fourier Series and also, discuss the the application of orthogonal series in some inequality. Also, in this chapter we learnt about some properties of Fourier series.

SQUARE INTEGRABLE FUNCTIONS

A function $f(x)$ is said to be absolutely integrable on an interval [a ; b] if

$$\int_a^b |f(x)|\,dx < \infty.$$

A function f(x) is said to be square integrable on an interval [a ; b] if

$$\int_a^b f^2(x)\,dx < \infty.$$

Taking in Cauchy-Bunyakovsky's inequality

$$\int_a^b |f|(x)g(x) < \sqrt{\int_a^b f^2(x)dx}.\sqrt{\int_a^b g^2(x)dx} \qquad ...(1)$$

$g(x) \equiv 1$, we have that

$$\int_a^b |f(x)|d(x) < \sqrt{\int_a^b f^2(x)dx}.\sqrt{b-a},$$

i.e. each square integrable function on an interval $[a, b]$ is also absolutely integrable. If an integrable function is bounded on the interval $[a, b]$ then it is also square integrable. For an unbounded function that is not true, for example the function $f(x) = \dfrac{1}{\sqrt{x}}$ is integrable on the interval [0, 1] because

$$\int_0^1 \frac{dx}{\sqrt{x}} = 2\sqrt{x}\Big|_0^1 = 2$$

but not square integrable because

$$\int_0^1 \frac{dx}{x} = \lim_{\varepsilon \to 0} \ \text{In} \ |x||_\varepsilon^1 = \infty .$$

From.(1) it also follows that if $f(x)$ and $g(x)$ are two square integrable functions then the product $f(x)g(x)$ is absolutely integrable and therefore integrable. Hence for two given square integrable functions $f(x)$ and $g(x)$ we can define the *inner product*

$$(f; g) = \int_a^b f(x)g(x)dx .$$

The inner product defined this way has all algebraic properties known in case of vectors i.e. $(f; g) = (g; f)$, $(\lambda f; g) = \lambda (f; g)$ for any real λ and $(f; g + h) = (f; g) + (f; h)$.

Two functions $f(x)$ and $g(x)$ are said to be orthogonal in an interval $[a; b]$ if $(f; g) = 0$. So the above conditions mean that the functions of the trigonometric system $\{1, \cos x, \sin x, \cos 2x, \sin 2x, ..., \cos kx, \sin kx, ...\}$ are pairwise orthogonal on the interval $[-\pi, \pi]$. The trigonometric system is said to be an orthogonal system in the interval $[-\pi, \pi]$. In general, the system of square integrable functions

$$\{\varphi_0(x), \varphi_1(x), ..., \varphi_k(x), ...\} \qquad ...(2)$$

is said to be orthogonal in the interval [a ; b] if $\varphi_k(x)$ and $\varphi_n(x)$ are orthogonal for each pair of distinct indices k and n ; that is, if

$$(\varphi_k; \varphi_n) = \int_a^b \varphi_k(x)\varphi_n(x)dx = 0 \ \text{ for } \ k \neq n \qquad ...(3)$$

The real number

$$\|f\| = \left[\int_a^b f^2(x)dx\right]^{\frac{1}{2}}$$

is called the norm of the square integrable function $f(x)$, i.e. $\|f\| = \sqrt{(f; f)}$. The orthogonal system (2) is said to be orthonormal in the interval $[a; b]$ if $\|\varphi_k\| = 1$ for $k = 0, 1, 2, ...$ The trigonometric system is not orthonormal in the interval $[-\pi, \pi]$ because $\|1\| = \sqrt{2\pi}$ and $\|\sin kx\| = \|\cos kx\| = \sqrt{\pi}$ but it is easy to verify that the system of functions

$$\left\{\frac{1}{\sqrt{2\pi}}, \frac{\cos x}{\sqrt{\pi}}, \frac{\sin x}{\sqrt{\pi}}, ..., \frac{\cos kx}{\sqrt{\pi}}, \frac{\sin kx}{\sqrt{\pi}}, ...\right\}$$

is orthonormal in the interval $[-\pi, \pi]$.

ORTHOGONAL SERIES

The trigonometric system is an example of orthogonal systems we have already considered. Examples of other orthogonal systems that are widely used in practice are Legendre polynomials, Bessel functions, Hermite polynomials, Laugrre polynomials, Jacobi polynomials, Chebyshev polynomials.Let the system (2) be orthogonal on the interval $[a, \ b]$. Suppose that it is possible to express $f(x)$ as a series expansion of the form

$$f(x)=\sum_{k=0}^{\infty} c_k\varphi_k(x) \qquad \text{...(4)}$$

Our purpose is to determine the coefficients c_k. Assume term by term integration of this series and the series multiplied by $\varphi_n(x)$ on the interval $[a\ ;\ b]$ to be possible. Multiplying (4) throughout by $\varphi_n(x)$ and integrating term by term, we obtain

$$=\int_a^b f(x)\varphi_n(x)d(x)=\sum_{k=0}^{\infty} c_k\int_a^b \varphi_k(x)\varphi_n(x)dx$$

which, on using (2.3), reduces to

$$\int_a^b f(x)\varphi_n(x)d(x)=c_n\int_a^b \varphi_n^2(x)dx$$

or $$\int_a^b f(x)\varphi_n(x)d(x)=c_n\,\|\varphi_n\|^2 \qquad \text{...(5)}$$

which, after replacing *n* by *k* again, gives

$$c_k=\frac{1}{\|\varphi_k\|^2}\int_a^b f(x)\varphi_k(x)d(x) \text{ for k = 0, 1, 2, ...} \qquad \text{...(6)}$$

The series (4) is called the Fourier series of $f(x)$ with respect to the system (2), and the coefficients c_k, given by (6), are called the Fourier coefficients of $f(x)$ with respect to the same system.

MEAN SQUARE DEVIATION, ITS MINIMUM

Let $f(x)$ be an arbitrary square integrable on the interval $[a\ ;\ b]$ function. Consider a polynomial with respect to the system (2)

$$\sigma_n(x)=\gamma_0\varphi_0(x)+\gamma_1\varphi_1(x)+...+\gamma_n\varphi_n(x) \qquad \text{...(7)}$$

where $\gamma_0,\gamma_1,...,\gamma_n$ are some real constants. Since the functions $\varphi_k(x)$ are square integrable then the difference $f(x)-\sigma_n(x)$ is square integrable also and the integral of square of this difference

$$\delta_n=\int_a^b [f(x)-\sigma_n(x)]^2\,dx$$

is called the mean square deviation between the function $f(x)$ and the polynomial $\sigma_n(x)$. Our purpose is to minimize the mean square deviation, i.e. to determine the coefficients of the polynomial $\sigma_n(x)$ so that mean square deviation δ_n should be minimized. We have

$$\delta_n = \int_a^b f^2(x)\,dx - 2\int_a^b f(x)\sigma_n(x)\,dx + \int_a^b \sigma_n^2(x)\,dx$$

Using (5) and (75), we get

$$\int_a^b f(x)\sigma_n(x)\,dx = \sum_{k=0}^{n} \gamma_k \int_a^b f(x)\varphi_k(x)\,dx = \sum_{k=0}^{n} \gamma_k c_k \|\varphi_k\|^2.$$

Further

$$\int_a^b \sigma_n^2(x)\,dx = \int_a^b \left[\sum_{k=0}^{n} \gamma_k\varphi_k(x)\right]^2 dx = \int_a^b \left[\sum_{k=0}^{n} \gamma_k^2\varphi_k^2(x) + \sum_{\substack{i\neq i \\ 0\le i,j\le n}} \gamma_i\gamma_j\varphi_i(x)\varphi_j(x)\right] dx$$

$$= \sum_{k=0}^{n} \gamma_k^2 \int_a^b \varphi_k^2(x)\,dx + \sum_{\substack{i\neq i \\ 0\le i,j\le n}} \gamma_i\gamma_j \int_a^b \varphi_i(x)\varphi_j(x)\,dx$$

$$= \sum_{k=0}^{n} \gamma_k^2 \|\varphi_k\|^2$$

because in the second sum all the integrals equal to zero. Now

$$\delta_n = \int_a^b f^2(x)\,dx - 2\sum_{k=0}^{n} \gamma_k c_k \|\varphi_k\|^2 + \sum_{k=0}^{n} \gamma_k^2 \|\varphi_k\|^2$$

$$= \int_a^b f^2(x)\,dx + \sum_{k=0}^{n} (c_k - \gamma_k)^2 \|\varphi_k\|^2 - \sum_{k=0}^{n} c_k^2 \|\varphi_k\|^2 \quad \text{...(8)}$$

We have to choose the coefficients γ_k in order to minimize mean square deviation. The first and third term in (8) does not depend on these coefficients. The second term in (8) is least if

$$\sum_{k=0}^{n} (c_k - \gamma_k)^2 \|\varphi_k\|^2 = 0, \quad \text{...(9)}$$

that is for all $k = 0,1,\ldots,n$.

Summarizing, the mean square deviation is minimized if the coefficients of the polynomial (7) are Fourier coefficients of $f(x)$ with respect to the system (2). In other words-the mean square deviation is minimized by the n-th partial sum of Fourier series. If we denote the minimal value of the mean square deviation by Δ_n, we obtain from (8) in virtue of (9) that

$$\Delta_n = \int_a^b f^2(x)dx - \sum_{k=0}^{n} c_k^2 \| \varphi_k \|^2 \quad ...(10)$$

From the definition of δ_n it follows immediately that Δ_n is non-negative and if n increases then Δ_n can only decrease.

Bessel's Inequality

Since in (2.10) $\Delta_n \geq 0$, it follows that

$$\sum_{k=0}^{n} c_k^2 \| \varphi_k \|^2 \leq \int_a^b f^2(x)dx.$$

Hence the sequence $$s_n = \sum_{k=0}^{n} c_k^2 \| \varphi_k \|^2$$

is bounded and non-decreasing and, consequently, has a finite limit, i.e. the series

$$\sum_{k=0}^{\infty} c_k^2 \| \varphi_k \|^2$$

is convergent, and $$\sum_{k=0}^{\infty} c_k^2 \| \varphi_k \|^2 \leq \int_a^b f^2(x)dx \quad ...(11)$$

This result is known as Bessel's inequality. For an orthonormal system (2) $\| \varphi_k \| = 1$ and Bessel's inequality takes the form

$$\sum_{k=0}^{\infty} c_k^2 \leq \int_a^b f^2(x)dx.$$

The norms of the functions of the trigonometric system {1, cos x, sin x, ... cos $k\,x$, sin $k\,x$, ...} are $\| 1 \| = \sqrt{2\pi}$ and $\| \cos k\,x \| = \| \sin k\,x \| = \sqrt{\pi}$ for $k = 1, 2,...$. If $f(x)$ is a square integrable function on the interval $[-\pi\ ;\ \pi]$ then for trigonometric system Bessel's inequality (11) takes the form

$$\left(\frac{a_0}{2}\right)^2 \| 1 \|^2 + \sum_{k=1}^{\infty} (a_k^2 \| \cos k\,x \|^2 + b_k^2 \| \sin k\,x \|^2) \leq \int_{-\pi}^{\pi} f^2(x)dx$$

i.e. $$\frac{a_0^2}{4} 2\pi + \sum_{k=1}^{\infty} (a_k^2 \pi + b_k^2 \pi) \leq \int_{-\pi}^{\pi} f^2(x)dx$$

or $$\frac{a_0^2}{2} + \sum_{k=1}^{\infty} (a_k^2 + b_k^2) \leq \frac{1}{\pi} \int_{-\pi}^{\pi} f^2(x)dx \quad ...(12)$$

This is the form of Bessel's inequality used for trigonometric system . The series on the left side of this inequality is convergent.

Theorem 1. *The series whose terms are the squares of the Fourier coefficients of any square integrable function is always convergent.*

PROPERTIES OF FOURIER SERIES

As mentioned above it should not be assumed that the existence of the coefficients a_k and b_k in itself implies that the associated series converges to the function $f(x)$. In order that the function have a convergent Fourier series expansion it must satisfy a certain set of conditions. To prove a theorem giving the sufficient conditions for convergence of Fourier series we shall start with some auxiliary results.

Riemann Lemma

If $f(x)$ is a square integrable function on the interval $[-\pi, \pi]$ then from Bessel's inequality (12) it follows that $a_k = \lim b_k = 0$. By the formulae (14) and (15) for finding the Fourier coefficients a_k and b_k we have that

$$\lim_{k\to\infty}\int_{-\pi}^{\pi} f(x)\cos kx\,dx = \lim_{k\to\infty}\int_{-\pi}^{\pi} f(x)\sin kx\,dx = 0$$

It is possible to prove that in the last condition the interval $[-\pi, \pi]$ can be replaced with any interval $[a ; b]$ i.e. for each square integrable function $f(x)$ on an interval $[a ; b]$

$$\lim_{k\to\infty}\int_{b}^{a} f(x)\cos kx\,dx = \lim_{k\to\infty}\int_{a}^{b} f(x)\sin kx\,dx = 0$$

A more general result called Riemann lemma holds.

Lemma 1. If the function f(x) is continuous in the interval [a ; b] with the exception of a finite number of points and absolutely integrable in this interval then

$$\lim_{\nu\to\infty}\int_{b}^{a} f(x)\sin \nu x\,dx = \lim_{\nu\to\infty}\int_{a}^{b} f(x)\cos \nu x\,dx = 0$$

Here ν represents not necessarily an integer but a real variable.

Sum of cosines

We need a formula for calculating the sum

$$D_N(u) = \frac{1}{2} + \sum_{k=1}^{N} \cos ku$$

Obviously, multiplying both sides of this equality by $2\sin\frac{u}{2}$, we have

$$2D_N(u)\sin\frac{u}{2} = \sin\frac{u}{2} + 2\cos u \sin\frac{u}{2} + 2\cos 2u \sin\frac{u}{2} + \ldots + 2\cos Nu \sin\frac{u}{2}$$

and after using the formula

$$2\cos\alpha\sin\beta = \sin(\alpha+\beta) - \sin(\alpha-\beta)$$

we find
$$D_n(u)\sin\frac{u}{2} = \sin\frac{u}{2} + \left(\sin\frac{3u}{2} - \sin\frac{u}{2}\right) + \left(\sin\frac{5u}{2} - \sin\frac{3u}{2}\right) + \ldots$$
$$+ \left(\sin\frac{(2N+1)u}{2} - \sin\frac{(2N-1)u}{2}\right)$$

or
$$2D_n(u)\sin\frac{u}{2} = \sin\frac{(2N+1)u}{2}$$

that is

$$D_n(u) = \frac{\sin\frac{(2N+1)u}{2}}{2\sin\frac{u}{2}}$$

So we have the formula

$$\frac{1}{2} + \sum_{k=1}^{N} \cos ku = \frac{\sin\frac{(2N+1)u}{2}}{2\sin\frac{u}{2}} \qquad \ldots(13)$$

Integrating the last equality over the interval [– π ; π] we obtain

$$1 = \frac{1}{\pi}\int_{-\pi}^{\pi} \frac{\sin\frac{(2N+1)u}{2}}{2\sin\frac{u}{2}}\,du$$

Integral Formula of Partial Sums

Next we need an integral formula for the partial sums of a Fourier series. Let $f(x)$ be a 2π-periodic function with its Fourier series expansion

$$f(x) \approx \frac{a_0}{2} + \sum_{k=1}^{\infty} a_k \cos kx + b_k \sin kx$$

Let $S_N(x)$ designates the N th partial sum of this expansion, i.e.

$$S_N(x) = \frac{a_0}{2} + \sum_{k=1}^{N} a_k \cos kx + b_k \sin kx$$

Replacing the coefficients a_0, a_k and b_k by the formulas (13) – (15), we obtain that

$$S_N(x) = \frac{1}{2\pi}\int_{-\pi}^{\pi} f(t)\,dt + \frac{1}{\pi}\sum_{k=1}^{N}\left[\int_{-\pi}^{\pi} f(t)\cos kt\,dt.\cos kx + \int_{-\pi}^{\pi} f(t)\sin kt\,dt.\sin kx\right]$$

$$= \frac{1}{\pi}\int_{-\pi}^{\pi} f(t)\left[\frac{1}{2} + \sum_{k=1}^{N}(\cos kt\,dt.\cos kx)\right]dt$$

$$= \frac{1}{\pi}\int_{-\pi}^{\pi} f(t)\left[\frac{1}{2} + \sum_{k=1}^{N}(\cos k(t-x)\right]dt$$

By the formula for sum of cosines (1)

$$S_N(x) = \frac{1}{\pi}\int_{-\pi}^{\pi} f(t)\,\frac{\sin\frac{(2N+1)(f-x)}{2}}{2\sin\frac{f-x}{2}}\,dt$$

which, after changing the variable $u = t - x$, gives

$$S_N(x) = \frac{1}{\pi}\int_{-\pi-x}^{\pi-x} f(x+u)\,\frac{\sin\frac{(2N+1)u}{2}}{2\sin\frac{u}{2}}\,du$$

or, since the integrand is a periodic function with period 2π,

$$S_N(x) = \frac{1}{\pi}\int_{-\pi}^{\pi} f(x+u)\,\frac{\sin\frac{(2N+1)u}{2}}{2\sin\frac{u}{2}}\,du \qquad ...(14)$$

Convergence of Fourier Series

Now we are ready to prove the theorems giving the sufficient conditions for convergence of Fourier series . Note that there exist various sets of conditions that the given function must satisfy in order that it have a convergent Fourier series expansion. We shall prove one, maybe the simplest, theorem and state some others.

Theorem 2. (Dirichlet's Theorem) If $f(x)$ is an absolutely integrable periodic function with period 2π in the interval $[-\pi, \pi]$ that has a finite number of points of discontinuity, then at any point where f(x) is differentiable its Fourier series expansion converges to the function $f(x)$.

Proof. We have to prove that under these conditions

$$\lim_{N\to\infty} S_N(x) = f(x)$$

Let $f(x)$ be differentiable at the point x. Multiplying (3.2) by $f(x)$ we have

$$f(x)=\frac{1}{\pi}\int_{-\pi}^{\pi}f(x)\frac{\sin\frac{(2N+1)u}{2}}{2\sin\frac{u}{2}}du$$

and subtracting the last equality from (3)

$$S_N(x)-f(x)=\frac{1}{\pi}\int_{-\pi}^{\pi}[f(x+u)-f(x)]\frac{\sin\frac{(2N+1)u}{2}}{2\sin\frac{u}{2}}du$$

$$S_N(x)-f(x)=\frac{1}{\pi}\int_{-\pi}^{\pi}\varphi(u)\sin\left(N+\frac{1}{2}\right)u\,du$$

where $$\varphi(u)=\frac{f(x+u)-f(x)}{2\sin\frac{u}{2}}.$$

It remains to show that the function $\varphi(u)$ satisfies the conditions of the Riemann lemma. We have

$$\varphi(u)=\frac{f(x+u)-f(x)}{u}\frac{u}{2\sin\frac{u}{2}}.$$

Since the function $f(x)$ is differentiable at the point x, there exists

$$f'(x)=\lim_{u\to 0}\frac{f(x+u)-f(x)}{u}.$$

Thus, for $u\to 0$

$$\varphi(u)=\frac{f(x+u)-f(x)}{u}\frac{\frac{u}{2}}{\sin\frac{u}{2}}\to f'(x).1=f'(x)$$

i.e. the function $\varphi(u)$ has at the point $u = 0$ a removable discontinuity. All the other points of discontinuity of $\varphi(u)$ are the same as the points of discontinuity of $f(x+u)$. Therefore, if the conditions of the theorem are satisfied, the function $\varphi(u)$ satisfies the conditions of Riemann lemma and

$$\lim_{N\to\infty}\frac{1}{\pi}\int_{-\pi}^{\pi}\varphi(u)\sin\left(N+\frac{1}{2}\right)u\,du=0,$$

that is $$\lim_{N\to\infty}S_N(x)=f(x).$$

The theorem 2 does not state anything about the convergence at the points of discontinuity. This leads us to state the more general theorem.

The left- and right-hand derivatives of the function $f(x)$ at a point x are defined as the one-sided limits

$$f'(x-)\ \lim_{h\to 0-}\frac{f(x+h)-f(x)}{h}$$

and

$$f'(x+)\ \lim_{h\to 0+}\frac{f(x+h)-f(x)}{h}$$

respectively.

Theorem 3. (Also referred to as Dirichlet's theorem.) If f(x) is an absolutely integrable periodic function with period in the interval $[-\pi,\ \pi]$, that has a finite number of points of discontinuity, then at any point where f(x) has the left- and right-hand derivatives $f'(x-)$ and $f'(x+)$ its Fourier series expansion converges to the function f(x) at all points where f(x) is continuous and to the average of the left- and right-hand limits of f(x), i.e. to $\frac{f(x-)+f(x+)}{2}$, at points where $f(x)$ is discontinuous.

Such a theorem also holds for some different sets of conditions.

Theorem 4. (Also referred to as Dirichlet's theorem.) If f(x) is a bounded periodic function that in the interval $[-\pi,\ \pi]$ has a finite number maxima and minima and a finite number of points of finite discontinuity then the Fourier series expansion of f(x) converges to f(x) at all points where f(x) is continuous and to the average of the left- and right-hand limits of f(x), i.e. to $\frac{f(x-)+f(x+)}{2}$, at points where f(x) is discontinuous.

Every function is differentiable at any interior point of domain. This fact is also easily seen taking in the Fourier series expansion

$$f(x)\approx 2\sum_{k=1}^{\infty}(-1)^{k+1}\frac{\sin kx}{k}$$

$x=-\pi$ or $x=\pi$. In both cases the Fourier series consists of zeros and, of course, equals.

By the Theorem 2.4 the Fourier series expansion of the function converges to the some given function everywhere and thus we can replace the sign '~' with the sign of equality

$$f(x)=\frac{\pi}{2}-\frac{4}{\pi}\sum_{k=1}^{\infty}\frac{\cos(2k-1)x}{(2k-1)^2}. \qquad ...(15)$$

If $x\in[-\pi;\pi]$ then $f(x)=|x|$ and taking in (4) $x=0$ we have

$$0=\frac{\pi}{2}-\frac{4}{\pi}\sum_{k=1}^{\infty}\frac{1}{(2k-1)^2}$$

or $$1+\frac{1}{9}+\frac{1}{25}+\ldots\frac{1}{(2k-1)^2}+\ldots=\frac{\pi^2}{8} \qquad \ldots(16)$$

We can also calculate the sum

$$S=\sum_{k=1}^{\infty}\frac{1}{k^2}$$

because $$S=1+\frac{1}{2^2}+\frac{1}{3^2}+\frac{1}{4^2}+\frac{1}{5^2}+\frac{1}{6^2}+\ldots=$$

$$=1+\frac{1}{3^2}+\frac{1}{5^2}+\ldots\frac{1}{4}\left(1+\frac{1}{2^2}+\frac{1}{3^2}+\ldots\right).$$

Therefore, $$S=\frac{\pi^2}{8}+\frac{1}{4}S$$

and it follows that

$$S=\frac{\pi^2}{6}.$$

Integration of a Fourier Series

Let $f(x)$ be absolutely integrable 2π-periodic function with its Fourier series expansion

$$f(x)\approx\frac{a_0}{2}+\sum_{k=1}^{\infty}a_k\cos kx+b_k\sin kx.$$

Consider the function

$$F(x)=\int_0^x\left[f(x)-\frac{a_0}{2}\right]dx$$

in the interval $[-\pi;\pi]$. The function $F(x)$ is continuous and has an absolutely integrable derivative, moreover

$$F(x+2\pi)=\int_0^{x+2\pi}\left[f(x)-\frac{a_0}{2}\right]dx=\int_0^{x}\left[f(x)-\frac{a_0}{2}\right]dx+\int_0^{x+2\pi}\left[f(x)-\frac{a_0}{2}\right]dx$$

$$=F(x)=\int_{-\pi}^{\pi}f(x)\,dx-\frac{a_0}{2}\int_0^x dx=F(x).$$

Thus, $F(x)$ is a 2π-periodic, absolutely integrable and differentiable function, i.e. it satisfies the conditions of the the Theorem 3.2 and, therefore, is representable by its Fourier series

$$F(x)=\frac{A_0}{2}+\sum_{k=1}^{\infty}A_k\cos kx+B_k\sin kx \qquad \ldots(17)$$

Integration by parts gives the coefficients

$$A_k = \frac{1}{\pi}\int_{-\pi}^{\pi} F(x)\cos k x\, d x = \frac{1}{\pi}\frac{F(x)\sin k x}{k}\Bigg|_{-\pi}^{\pi} - \frac{1}{\pi k}\int_{-\pi}^{\pi}\left[f(x)-\frac{a_0}{2}\right]\sin k x\, d x$$

$$= -\frac{1}{\pi k}\int_{-\pi}^{\pi} f(x)\sin k x + \frac{a_0}{2\pi k}\int_{-\pi}^{\pi}\sin k x\, d x = -\frac{b_k}{k}$$

Analogously $B_k = \frac{u_k}{k}$ and

$$F(x) = \frac{A_0}{2} + \sum_{k=1}^{\infty}\frac{a_k \sin k x - b_k \cos k x}{k}$$

Taking $x = 0$ in the last equality we obtain that

$$0 = \frac{A_0}{2} - \sum_{k=1}^{\infty}\frac{b_k}{k} \text{ or } \frac{A_0}{2} = \sum_{k=1}^{\infty}\frac{b_k}{k}.$$

Thus, $$F(x) = \sum_{k=1}^{\infty}\frac{a_k \sin k x + b_k(1-\cos k x)}{k}$$

or by the definition of $F(x)$

$$\int_0^x f(x)dx - \int_0^x \frac{a_0}{2}dx = \sum_{k=1}^{\infty}\frac{a_k \sin k x + b_k(1-\cos k x)}{k} \qquad ...(18)$$

But $$\frac{a_k \sin k x}{k} = \int_0^x a_k \cos k x\, d x$$

and $$\frac{b_k(1-\cos k x)}{k} = \int_0^x b_k \sin k x\, d x$$

Substituting the last equalities intc (3.7) we have

$$\int_0^x f(x)dx = \int_0^x \frac{a_0}{2}dx + \sum_{k=1}^{\infty}(a_k \cos k x + b_k \sin k x)\, d x$$

In the last equality the lower boundary can be replaced with any real x_0, i.e. for $-\pi \le x_0 < x \le \pi$ we have

$$\int_{x_0}^x f(x)dx = \frac{a_0}{2}(x - x_0) + \sum_{k=1}^{\infty}\int_{x_0}^x (a_k \cos k x + b_x \sin k x)\, dx. \qquad ...(19)$$

Let us summarize the results.

Conclusion 1. The Fourier series expansion of $\int_{x_0}^{x} f(x)dx$ can be obtained on term by term integration of the Fourier series expansion of f(x) and it is independent of the convergence of that Fourier series expansion.

Conclusion 2. It is possible to calculate $\int_{x_0}^{x} f(x)dx$ knowing only the Fourier series expansion of the function f(x) , the function itself can be unknown.

Because of the presence of the term $\frac{a_0}{2}x$ on the right-hand side of (19), this is not clearly a Fourier series expansion of the integral on the left-hand side, the result can be rearranged to be a Fourier series expansion of the function

$$\int_{x_0}^{x} f(x)dx - \frac{a_0}{2}x$$

Example 1. *Integrate term by term the Fourier series expansion for the square wave*

$$f(x) = \begin{cases} -1 \text{ if } -\pi < x \leq \pi \\ 1 \text{ if } 0 < x < x \end{cases}$$

$$f(x + 2\pi) = f(x)$$

Solution:

The Fourier series expansion for $f(x)$ is

$$f(x) \approx \frac{4}{\pi}\sum_{k=1}^{\infty} \frac{\sin(2k-1)}{2k-1} \qquad ...(20)$$

If the function is piecewise-continuous over the interval $[-\pi, \pi]$, then care must be taken to ensure that the integration process is carried out properly over the various subintervals. Now we need to consider separately values of x in the intervals $[-\pi, 0]$ and $[0, \pi]$.

If $-\pi < x < 0$ then, integrating (20) term by term, we have

$$\int_{-\pi}^{x} (-1)dx = \frac{4}{\pi}\sum_{k=1}^{\infty} \int_{-\pi}^{x} \frac{\sin(2k-1)}{2k-1}dx$$

that is

$$-(x + \pi) = -\frac{4}{\pi}\sum_{k=1}^{\infty}\left[\frac{\cos(2k-1)x}{(2k-1)^2}\right]_{-\pi}^{x}$$

$$= -\frac{4}{\pi}\left[\sum_{k=1}^{\infty}\frac{\cos(2k-1)x}{(2k-1)^2} + \sum_{k=1}^{\infty}\frac{1}{(2k-1)^1}\right]$$

Replacing by (16) the last series in square brackets with $\frac{\pi^2}{8}$, the above simplifies to

$$-x = \frac{\pi}{2} - \frac{4}{\pi} \sum_{k=1}^{\infty} \frac{\cos(2k-1)x}{(2k-1)^2} \qquad \text{...(21)}$$

If $0 < x < x$ then, integrating (20) term by term, we have

$$\int_{-\pi}^{x} (-1)\,dx + \int_{0}^{x} 1\,dx = \frac{4}{\pi} \sum_{k=1}^{\infty} \frac{\sin(2k-1)x}{2k-1}\,dx$$

giving

$$x = \frac{\pi}{2} - \frac{4}{\pi} \sum_{k=1}^{\infty} \frac{\cos(2k-1)x}{(2k-1)^2} \text{ if } 0 < x < x. \qquad \text{...(22)}$$

Taking (21) and (22) together, we find that the function

$$g(x) = |x| = \begin{cases} -x & \text{if } -\pi < x < 0 \\ x & \text{if } 0 < x < x \end{cases}$$

$$g(x + 2\pi) = g(x)$$

has the Fourier series expansion

$$g(x) = |x| = \frac{\pi}{2} - \frac{4}{\pi} \sum_{k=1}^{\infty} \frac{\cos(2k-1)x}{(2k-1)^2},$$

The last series expansion converges to $g(x)$ at any point despite the fact that the series expansion (20) does not converge to $f(x)$ at the points of discontinuity.

Differentiation of Fourier series

Let $f(x)$ be a continuous 2π-periodic function having an absolutely integrable derivative (with exeption at the finite number of points) and satisfying the condition $f(-\pi) = f(\pi)$. Then

$$f(x) = \int_{0}^{x} f'(x)\,dx + f(0)$$

Thus, by the results of the previous section the Fourier series expansion of $f(x)$ can be obtained from the Fourier series expansion of

$$f'(x) \sim \sum_{k=1}^{\infty} a'_k \cos kx + b'_k \sin kx \qquad \text{...(23)}$$

on term by term integration. The last expansion does not contain the constant term a_0' because

$$a_0' = \frac{1}{\pi}\int_{-\pi}^{\pi} f'(x)\,dx = \frac{1}{\pi} f(x)|_{-\pi}^{\pi} = 0.$$

If the Fourier series expansion of $f(x)$ can be obtained from the Fourier series expansion of $f'(x)$ on term by term integration then the Fourier series expansion of $f'(x)$ can be also obtained from the Fourier series expansion of $f(x)$ on term by term differentiation, i.e. if the function $f(x)$ satisfies the conditions given above and its Fourier series expansion is

$$f'(x) = \frac{a_0}{2} + \sum_{k=1}^{\infty} a_k \cos kx + b_k \sin kx \qquad ...(24)$$

then the Fourier series expansion of $f'(x)$ is

$$f'(x) \sim \sum_{k=1}^{\infty} kb_k \cos kx - ka_k \sin kx \qquad ...(25)$$

In (24) there is an " = " sign because under the given conditions the Fourier series expansion of $f(x)$ converges to $f(x)$ but in (25) there is an "~" sign because at the points of discontinuity of $f'(x)$ its Fourier series expansion may not converge to $f'(x)$.

Here it is important that $f(x)$ is continuous everywhere, including the end points $-\pi$ and π. To illustrate this point, consider the Fourier series expansion of the function

$$f(x) = x \text{ it} - \pi < x < x,$$
$$f(x + 2\pi) = f(x)$$

which, from Example 1.3, is given by

$$f(x) = 2\sum_{k=1}^{\infty} (-1)^{k+1} \frac{\sin kx}{k}$$

Differentiating term by term, we have

$$f'(x) \sim 2\sum_{k=1}^{\infty} (-1)^{k+1} \cos kx\,.$$

If the differentiation process is valid then $f'(x)$ must be equal to 1 for $-\pi < x < \pi$. Clearly this is not the case, since the series on the right-hand side does not converge for any value of x.

Notice that the Fourier coefficients of the derived expansion are kb_k and $-ka_k$, so, in contrast to the integrated series, the convergence properties of derived series are worse these of the original series expansion.

Example 2. Consider the process of differentiating term by term the Fourier series expansion of the function

$$f(x) = x^2 \text{ if } -\pi < x \leq \pi$$
$$f(x + 2\pi) = f(x).$$

Solution: The Fourier series expansion of *f*(*x*) is

if $-\pi < x \leq \pi.$

Since *f*(*x*) satisfies the conditions of piecewise, we may differentiate above term by term to obtain

$$2x \sim 4\sum_{k=1}^{\infty} \frac{k(-1)^{k+1}}{k^2} \sin kx$$

that is

$$x \sim 2\sum_{k=1}^{\infty} \frac{(-1)^{k+1} \sin kx}{k}$$

which conforms with the Fourier series expansion obtained for the function

$$f(x) = x \text{ if } -\pi < x \leq \pi$$
$$f(x + 2\pi) = f(x)$$

Improvement of Convergence of Fourier Series

If we differentiate term by term the Fourier series expansion then the convergence properties are worse. At the same time, for example solving partial differential equations, we have to differentiate the Fourier series expansion more than once. To provide the convergence of the derived series expansions we separate a part of the original series expansion that converges slowly. If we know the sum of the separated slowly convergent part then we can differentiate it immediately. But the coefficients of the remaining part must be decreasing so fast that we could differentiate term by term the remaining part as many times as needed. In other words we have the following problem. For a given trigonometric series, with the sum *f*(*x*) which may also be unknown,

$$f(x) = \frac{a_0}{2} + \sum_{k=1}^{\infty} a_k \cos kx + b_k \sin kx$$

we have to separate another series with known sum $\varphi(x)$ so that the coefficients of the remainder series determined by the equality

$$f(x) = \varphi(x) + \sum_{k=1}^{\infty} a_k \cos kx + \beta_k \sin kx$$

should be decreasing sufficiently fast. If this problem is solved then we can replace the function *f*(*x*) with the sum of known function $\varphi(x)$ and series with coefficients decreasing sufficiently fast.

The following considerations show that this problem can be solved. For the function $f(x)$ differentiable some times on the interval $[-\pi, \pi]$ its periodic extension is, in general, a discontinuous function or a function with discontinuous derivatives. Thus, it cannot be expected that its Fourier coefficients are decreasing fast. It is easy to see that we can subtract from $f(x)$ a linear function $\varphi(x)$ on the interval $[-\pi, \pi]$ so that the periodic extension of the difference is a continuous function, i.e. the values of this difference are equal at the end points of the interval $[-\pi, \pi]$. For example we can determine the linear function $\varphi(x)$ in the interval $[-\pi, \pi]$ as

$$\varphi(x) = \frac{f(\pi-) - f(-\pi+)}{\pi} x + b$$

where b is any real number. In this case for the difference $g(x) = f(x) - \varphi(x)$

$$g(\pi-) = f(\pi-) - \frac{f(\pi-) - f(-\pi+)}{2\pi}\pi - b = \frac{f(\pi-) + f(-\pi+)}{2} - b$$

and

$$g(-\pi+) = f(-\pi+) - \frac{f(\pi-) - f(-\pi+)}{2\pi} - \pi - b = \frac{f(\pi-) + f(-\pi+)}{2} - b$$

i.e. the periodic extension of the difference $g(x)$ is indeed continuous everywhere. Subtracting from $f(x)$ a certain polynomial we can also guarantee that some derivatives are also continuous at the end points of the interval $[-\pi, \pi]$. Thus, the function itself and its derivatives can be continuously and periodically extended to the whole set of real numbers. As a result, the Fourier coefficients are decreasing quite fast.

The problem is that a series is given, not the function $\varphi(x)$. We have to find the function $\varphi(x)$ starting from the given series. This difficulty can often be overcome and in this case it is said that the convergence of the series is improved.

Let us consider two methods of improvement of the convergence. The first of these two is based on the well-known fact that the difference of two equivalent infinitely decreasing quantities is the infinitely decreasing quantity of higher order i.e. if $\lim y = 0$ and $\lim z = 0$ then $\lim \frac{y-z}{y} = \lim \frac{y-z}{z} = 0$.

Example 3. Improve the convergence of the series

$$f(x) = \sum_{k=2}^{\infty} (-1)^k \frac{k^3}{k^4 - 1} \sin kx.$$

Solution: We use the equivalency $\frac{k^3}{k^4-1} \sim \frac{1}{k}$. The difference of these

$$\frac{k^3}{k^4-1}-\frac{1}{k}=\frac{1}{k^5-k}$$

is the infinitely decreasing quantity of higher order. We have

$$f(x)=\sum_{k=2}^{\infty}(-1)^k\left(\frac{1}{k}+\frac{1}{k^5-k}\right)\sin kx$$

$$=\sum_{k=2}^{\infty}(-1)^k\frac{\sin kx}{k}+\sum_{k=2}^{\infty}(-1)^k\frac{\sin kx}{k^5-k}.$$

If $-\pi<x<\pi$ then, we have the Fourier series expansion

$$\frac{x}{2}=\sum_{k=1}^{\infty}(-1)^{k+1}\frac{\sin kx}{k}=\sin x+\sum_{k=2}^{\infty}(-1)^{k+1}\frac{\sin kx}{k}$$

$$=\sin x-\sum_{k=2}^{\infty}(-1)^k\frac{\sin kx}{k}$$

that is

$$f(x)=\sin x-\frac{x}{2}+\sum_{k=2}^{\infty}(-1)^k\frac{\sin kx}{k^5-k}.$$

The order of Fourier coefficients in the last series is $\frac{1}{k^E}$.

A second method of improvement of convergence is based on the representation of Fourier coefficiénts in the form of the sum

Example 4. Improve the convergence of the series

$$f(x)=\sum_{k=2}^{\infty}\frac{\sin kx}{k+a}\ a\ \text{–consant, a > 0.}$$

Solution:

Obviously

$$\frac{1}{k+a}=\frac{1}{k}\frac{1}{1+\frac{a}{k}}=\frac{1}{k}\left(1-\frac{a}{k}+\frac{a^2}{k^2}-\frac{a^3}{k^3}+.....\right).$$

Summarizing the geomeric sequence after the term $\frac{a^2}{k^2}$ we have

$$\frac{1}{k+a}=\frac{1}{k}\left(1-\frac{a}{k}+\frac{a^2}{k^2}-\frac{\frac{a^3}{k^3}}{1+\frac{a}{k}}\right)=\frac{1}{k}\left(1-\frac{a}{k}+\frac{a^2}{k^2}-\frac{a^3}{k^2(k+a)}\right)$$

$$=\frac{1}{k}-\frac{a}{k^2}+\frac{a^2}{k^3}-\frac{a^3}{k^3(k+a)}.$$

The given Fourier series expansion can be re-written as

$$f(x)=\sum_{k=1}^{\infty}\frac{\sin k}{k}-a\sum_{k=1}^{\infty}\frac{\sin kx}{k^2}+a^2\sum_{k=1}^{\infty}\frac{\sin kx}{k^3}\sum_{k=1}^{\infty}\frac{\sin kx}{k^3(k+a)}.$$

Using the Fourier series expansions

$$\sum_{k=1}^{\infty}\frac{\sin kx}{k}=\frac{\pi-x}{2},\ \sum_{k=1}^{\infty}\frac{\sin kx}{k^3}=\frac{x^3-3\pi x^2+2\pi^2 x}{12}$$

and

$$\sum_{k=1}^{\infty}\frac{\sin kx}{k^2}=-\int_0^x \ln\left(2\sin\frac{x}{2}\right)dx$$

we have

$$f(x)=\frac{\pi-x}{2}+a\int_0^x \ln\left(2\sin\frac{x}{2}\right)dx+\frac{a^2(x^3-3\pi x^2+2\pi^2 x)}{12}-a^3\sum_{k=1}^{\infty}\frac{\sin kx}{k^3(k+a)}.$$

In the last series expansion the order of coefficiets is $\frac{1}{k^4}$.

Complete Systems

Let $\{\varphi_0(x),\varphi_1(x),\ldots,\varphi_k(x),\ldots\}$ (26)

be an orthogonal system of square integrable functions on the interval [*a*, *b*]. It has been proved that the mean square deviation

$$\delta_n=\int_a^b[f(x)-\sigma_n(x)]^2dx$$

is minimized if the coefficients of the polynomial are Fourier coefficients of *f*(*x*) with respect to the system (4.1). This minimal value was denoted by Δ i.e.

$$\Delta_n=\int_a^b\left[f(x)-\sum_{k=0}^{n}c_k\varphi_k(x)\right]^2dx$$

and using (26) gives

$$\int_a^b\left[f(x)-\sum_{k=0}^{n}c_k\varphi_k(x)\right]^2dx=\int_a^b f^2(x)dx-\sum_{k=0}^{n}c_k^2\|\varphi_k\|^2. \qquad ...(27)$$

Definition 1. The system (2.1) is said to be *complete* if the sign "≤" in Bessel's inequality (27) can be replaced with the sign of equality i.e. if for any square integrable function $f(x)$ there holds

$$\int_a^b f^2(x)dx=\sum_{k=0}^{\infty}c_k^2\|\varphi_k\|^2 \qquad ...(28)$$

where c_k (k = 0, 1, 2, ...) are the Fourier coefficients of $f(x)$ with respect to the system (1). The equality (28) is called *Parseval's equality.*

Theorem 5. *The system is complete if for any square integrable function f(x)*

$$\lim_{x\to\infty}\int_a^b\left[f(x)-\sum_{k=0}^{n}c_k\varphi_k(x)\right]^2dx=0 \qquad ...(29)$$

where c_k (k = 0, 1, 2 . . .) are the Fourier coefficients of f(x).

Proof. Parseval's equality can be re-written in the form

$$\lim_{x\to\infty}\int_a^b\left[f^2(x)dx-\sum_{k=0}^{n}c_k^2\|\varphi_k\|^2\right]=0$$

and using (29) completes the proof.

If (29) holds it is said that the Fourier series expansion

$$\sum_{k=0}^{\infty}c_k\varphi_k(x)$$

converges in the mean to the parent function $f(x)$. Now the Theorem 2 states that the system (1) is complete if the Fourier series expansion of any square integrable function $f(x)$ converges in the mean to this function. This does not necessarily guarantee convergence at any particular point, the convergence in the mean represents a generalization of the convergence at a point. It will be clear by showing that the Fourier series expansion can converge in the mean only to one function. Let us assume that in addition.

$$\lim_{n\to\infty}\int_a^b\left[g(x)-\sum_{k=0}^{n}c_k\varphi_k(x)\right]^2dx=0 \qquad ...(30)$$

Using the obvious inequality $(a+b)^2\le 2(a^2+b^2)$ we obtain that

$$0\le\int_a^b[f(x)-g(x)]^2dx=0$$

$$= \int_a^b \left[f(x) - \sum_{k=0}^{n} c_k \varphi_k(x) - g(x) \right]^2 dx \le$$

$$\le \int_a^b [f(x) - g(x)]^2 dx = 0$$

The equalities (4) and (5) give that

$$\int_a^b [f(x) - g(x)]^2 dx = 0$$

i.e. $f(x) = g(x)$ with the exception of a finite number of points of discontinuity. From the point of view of Fourier series we need not consider them as different functions because the Fourier coefficients are calculated by integrals whose values do not depend on the values of the function at the finite number of points.

Conclusion 3. If the system (4.1) is complete then each square integrable function is determined with its Fourier series expansion with the exception of the finite number of points.

In other words - if two functions are different at more than finite number of points then the Fourier series expansions of these functions cannot coincide.

Properties of Complete Systems

Proposition 1. Each continuous function that is orhtogonal to all functions of a complete system (4.1) is identical to zero

Proof. Since f(x) is orthogonal to any function of the system (1) then

$$c_k = \int_a^b f(x)\varphi_k(x)dx = 0,\ k = 0, 1, 2,...$$

From Parseval's equation (4.3) we have that

$$\int_a^b f^2(x)dx \sum_{k=0}^{\infty} c_k^2 \| \varphi_k \|^2 = 0$$

and, since $f(x)$ is assumed to be continuous, $f(x) \equiv 0$

Proposition 2. The Fourier series expansion of each square integrable function f(x) with respect to the complete system can be integrated term by term irrespective of the convergence of this series.

Proof. Let

$$f(x) \sim \sum_{k=0}^{\infty} c_k \varphi_k(x).$$

We have to show that if $a \le x_1 \le x_2 \le b$ then

$$\int_{x_1}^{x_2} f(x)dx = \sum_{k=0}^{\infty} c_k \int_{x_1}^{x_2} \varphi_k(x)dx. \qquad ...(31)$$

Using Cauchy-Bunyakovsky's inequality we obtain that

$$\left|\int_{x_1}^{x_2} f(x)dx - \sum_{k=0}^{n} c_k \int_{x_1}^{x_2} \varphi_k(x)dx\right| \le \int_{x_1}^{x_2} \left|f(x) - \sum_{k=0}^{n} c_k\varphi_k(x)\right| dx \le$$

$$\le \int_a^b \left|f(x) - \sum_{k=0}^{n} c_k\varphi_k(x)\right| dx \le$$

$$\le \left\{\int_a^b \left[f(x) - \sum_{k=0}^{n} c_k\varphi_k(x)\right]^2 . \int_a^b 1dx\right\}^{\frac{1}{2}}$$

$$= \sqrt{b-a}\sqrt{\int_a^b \left[f(x) - \sum_{k=0}^{n} c_k\varphi_k(x)\right]^2 dx}.$$

Since the system is complete then by Theorem 4.2

$$\lim_{n\to\infty} \sum_{k=0}^{n} c_k \int_{x_1}^{x_2} \varphi_k(x)dx = \int_{x_1}^{x_2} f(x)dx$$

Proposition 3. Let $\varepsilon > 0$ be any positive number. If for each continuous on the interval [a; b] function F(x) there exists a polynomial

$$\sigma_n(x) = \gamma_0\varphi_0(x) + \gamma_1\varphi_1(x) + ... + \gamma_n\varphi_n(x)$$

such that

$$\int_a^b [F(x) - \sigma_n(x)]^2 dx < \varepsilon$$

then the system (4.1) is complete.

Completeness of Trigonometric System

We shall use the following theorem called the Weierstrass First Theorem.

Theorem 6. Every continuous on the interval $\lfloor -\pi;\pi \rfloor$ periodic function with period 2π satisfying the equality $f(-\pi) = f(\pi)$ can be uniformly approximated by trigonometric polynomials with any accuracy.

Using the Weierstrass First Theorem it is easy to prove.

Theorem 7. The trigonometric system is complete.

Proof. If f(x) is a continuous periodic function with period 2π on the interval $\lfloor -\pi;\pi \rfloor$ and $f(-\pi) = f(\pi)$ then, by the Weierstrass First Theorem, for any $\varepsilon > 0$ there exists a trigonometric polynomial

$$\sigma_n(x) = \frac{a_0}{2} + \sum_{k=1}^{n} a_k \cos kx + b_k \sin kx$$

such that $$|f(x) - \sigma_n(x)| < \sqrt{\frac{\varepsilon}{2\pi}}$$

uniformly with respect to $x \in \lfloor -\pi;\pi \rfloor$ It follows that

$$\int_{-\pi}^{\pi} (f(x) - \sigma_n(x))^2 < \int_{-\pi}^{\pi} \frac{\varepsilon}{2\pi} dx = \varepsilon$$

and the completeness of the trigonometric system follows from Proposition 4.6.

Thus, all the properties of complete systems hold also for the trigonometric system:

1. The trigonometric Fourier series expansion of any square integrable function $f(x)$ is convergent in the mean to f(x).
2. If the continuous function $f(x)$ is orthogonal to each function of the trigonomeric system then $f(x)$ is identical to zero.
3. The Fourier series expansion of any square integrable function can be integrated term by term.

EXERCISES

1. Verify the convergence of the Fourier series expansion.
2. Integrating term by term the Fourier series expansion of the 2π-periodic function defined on $[-\pi, \pi]$ as $f(x) = x$ obtained in Example 1.3, find the Fourier series expansion of 2π-periodic functions defined on $[-\pi, \pi]$ as $f(x) = x^2$ and $f(x) = x^3$.
3. Consider the function

$$f(x) = \begin{Bmatrix} -A & \text{for} & 2l < x < -l \\ +A & \text{for} & -l < x < +l \\ -A & \text{for} & l < x < 2l \end{Bmatrix}$$

with $f(x + 4l) = f(x)$.

(a) Make of fourier expansion of the above function ad show that

$$f(x) = \frac{4A}{\pi} \sum_{n=0}^{\infty} \frac{(-1)^n}{(2n+1)} \cos\left[\left(n + \frac{1}{2}\right)\pi \frac{x}{l}\right]$$

(b) Using the above series prove that

$$1-\frac{1}{3}+\frac{1}{5}-\frac{1}{7}+...=\frac{\pi}{4}$$

(c) Evaluate the series a $x = \pm l$ and show taht the result is consistent with Eq. (4)

(d) Show that the above series is consistent with Eq. (25) except for a shift of origin.

4. Consider the following output from a half wave rectifier

$$V(t) = 0 \qquad -\frac{T}{2}<t<0$$

$$= V_0 \sin \omega t \qquad 0 < t < T/2$$

with $V(t + T) = V(t)$. Show that we may expand $V(t)$ in the following Fourier series

$$V(t)=\frac{V_0}{\pi}\left[1+\frac{\pi}{2}\sin\omega t-2\left(\frac{1}{1.3}\cos 2\omega t+\frac{1}{3.5}\cos 4\omega t+...\right)\right]$$

From the above series deduce that

$$\frac{1}{1.3}+\frac{1}{3.5}+\frac{1}{5.7}+...=\frac{1}{2}$$

and $\frac{1}{1.3}-\frac{1}{3.5}+\frac{1}{5.7}-...=\frac{\pi}{4}-\frac{1}{2}$

5. Consider the following output from a full wave rectificer

$V(t) = V_0 \mid \sin \omega t \mid$

Show that

$$V(t)=\frac{2}{\pi}V_0\left[1-2\sum_{n=1,2...}\frac{\cos 2n\omega t}{4n^2-1}\right].$$

6. Use the complex form of Fourier series to show that

$$e^{\alpha t}=\frac{1}{\pi}\sin \pi\Omega\sum_{n=-\infty}^{+\infty}(-1)^n\frac{\Omega+in}{\Omega^2+n^2}e^{in\omega t}$$

in the domain $-\pi/\omega<t<\pi/\omega$ and $\Omega=\alpha/\omega$.

7. Show that the odd periodic extension of the function

$$f(t) = t^2 \qquad 0 < t < T/2$$

is given by

$$\frac{T^2}{\pi}\sum_{n=1,2,...}\left[\frac{(-1)^n-1}{\pi^2 n^3}-\frac{(-1)^n}{2n}\right]\sin n\omega t$$

Chapter 9

Introduction to Fourier Transform

INTRODUCTION

In the previous chapter we have learn about Fourier series and its application In this chapter we study the basic tool of Fourier transform the Fourier integral and its properties. Also, we study the Fourier transform and their properties.

FOURIER INTEGRAL

Not all functions are periodic and so we need to extend the ideas of Fourier analysis to deal with non-periodic functions $f(x)$ defined on $-\infty < x < \infty$. One way of achieving this is to consider a portion of a non-periodic function $f(x)$ over a finite interval T and extend this portion periodically over infinite interval $-\infty < x < \infty$ i.e. define, as in previous section, a periodic function with period T as

$$\phi(x) = \begin{cases} f(x) \text{ if } -\frac{T}{2} < x < \frac{T}{2} \\ f(x - nT) \text{ if } nT - \frac{T}{2} < x < nT + \frac{T}{2} \end{cases}$$

Let us assume that the function f(x) has on each finite interval not more than a finite number of points of discontinuity and

$$\int_{-\infty}^{\infty} | f(x) | \, dx < \infty \qquad \text{...(1)}$$

i.e. $f(x)$ is absolutely integrable on the interval $-\infty < x < \infty$. At any point where $\phi(x)$ is differentiable it is representable with its Fourier series expansion

$$\phi(x) = \frac{a_0}{2} + \sum_{k=1}^{\infty} a_k \cos k\omega x$$

and if $-\frac{T}{2} < x < \frac{T}{2}$ then also

$$f(x) = \frac{a_0}{2} + \sum_{k=1}^{\infty} a_k \cos k\omega x + b_k \sin k\omega x.$$

Replacing the Fourier coefficients into the last equality we obtain that

$$f(x) = \frac{1}{T}\int_{-\frac{T}{2}}^{\frac{T}{2}} f(x)dx + \sum_{k=1}^{\infty}\frac{2}{T}\int_{-\frac{T}{2}}^{\frac{T}{2}} f(t)\cos k\omega dt.\cos k\omega x +$$

$$+\sum_{k=1}^{\infty}\frac{2}{T}\int_{-\frac{T}{2}}^{\frac{T}{2}} f(t)k\omega dt.\sin k\omega x$$

$$= \frac{1}{T}\int_{-\frac{T}{2}}^{\frac{T}{2}} f(x)dx + \sum_{k=1}^{\infty}\frac{2}{T}\int_{-\frac{T}{2}}^{\frac{T}{2}} f(t)\cos k\omega(t-x)dt.$$

Since $\omega = \frac{2\pi}{T}$ then

$$f(x) = \frac{1}{T}\int_{-\frac{T}{2}}^{\frac{T}{2}} f(x)dx + \frac{1}{\pi}\sum_{k=1}^{\infty}\omega\int_{-\frac{T}{2}}^{\frac{T}{2}} f(t)\cos k\omega(t-x)dt.$$

If $T \to \infty$ then by (1), we get

$$\frac{1}{T}\int_{-\frac{T}{2}}^{\frac{T}{2}} f(x)dx \to 0$$

Since $\omega \to 0$ and $k\omega \in](k-1)\omega; k\omega]$ then the sum

$$\sum_{k=1}^{\infty}\omega\int_{-\frac{T}{2}}^{\frac{T}{2}} f(t)\cos k\omega(t-x)dt$$

represents an integral sum of the function of the variable ω and it is possible to prove that

$$\sum_{k=1}^{\infty}\omega\int_{-\frac{T}{2}}^{\frac{T}{2}} f(t)\cos k\omega(t-x)dt \to \int_0^{\infty} d\omega\int_{-\infty}^{\infty} f(t)\cos\omega(t-x)dt.$$

We obtain the result

$$f(x) = \frac{1}{x}\int_0^{\infty} d\omega\int_{-\infty}^{\infty} f(t)\cos\omega(t-x)dt \qquad ...(2)$$

which is known as the Fourier integral representation of f(x) and we have the following theorem.

Theorem 1. If the function f(x) is absolutely integrable on the interval $-\infty < x < \infty$ and has at most a finite number of discontinuities in any finite interval then the Fourier integral representation (2) converges to f(x) at all points where f(x) is differentiable.

Theorem 2. If the function f(x) is absolutely integrable on the interval $-\infty < x < \infty$ and it has at most a finite number of maxima and minima and a finite number of discontinuities in any finite interval then the Fourier integral representation (2) converges to f(x) at all points where f(x) is continuous and to the average of the right- and left-hand limits of f(x), i.e. to $\frac{f(x-)+f(x+)}{2}$, at the points of discontinuity of f(x).

The condition (1) means that the area between the *x* axes and the graph of $y = f(x)$ is finite. It is so if $f(x)$ decreases sufficiently fast with x increasing. In general the condition seems to imply a very tight constraint on the nature of $f(x)$. Clearly the functions $f(x)$ = constant, $f(x) = \sin x$, $f(x) = e^x$ etc., defined over $-\infty < x < \infty$, do not meet the requirement. In practice, however, the signals are usually causal and do not last for ever. Also, in practice no signal amplitude goes to infinity. Thus, no practical signal $f(x)$ can have an infinite area under its graph $y = f(x)$ and, therefore, we can assume that (1) holds.

Using the formula

$$\cos\omega(t-x) = \cos\omega t\cos\omega x + \sin\omega t\sin\omega x$$

we obtain that

$$\frac{1}{\pi}\int_{-\infty}^{\infty} f(t)\cos\omega(t-x)dt = \frac{1}{\pi}\int_{-\infty}^{\infty} f(t)\cos\omega t dt.\cos\omega x +$$

$$+\frac{1}{\pi}\int_{-\infty}^{\infty} f(t)\cos\omega t dt.\sin\omega x = a(\omega)\cos\omega w + b(\omega)\sin\omega x$$

where

$$a(\omega) = \frac{1}{\pi}\int_{-\infty}^{\infty} f(t)\cos\omega t dt \qquad ...(3)$$

and

$$b(\omega) = \frac{1}{\pi}\int_{-\infty}^{\infty} f(t)\sin\omega t dt \qquad ...(4)$$

Thus, the Fourier integral representation (4.2) takes the form

$$f(x) = \int_0^{\infty}[a(\omega)\cos\omega x + b(\omega)\sin\omega x]d\omega \qquad ...(5)$$

where $a(\omega)$ and $b(\omega)$ are given by (3) and (4) respectively. Notice that the expression of the Fourier integral is very similar to the Fourier series of a periodic function.

COMPLEX FORM OF FOURIER INTEGRAL: FOURIER TRANSFORM

In mathematics, the Fourier transform is a certain linear operator that maps functions to other functions. Loosely speaking, the Fourier transform decomposes a function into a continuous spectrum of its *frequency components*, and the inverse transform synthesizes a function from its spectrum of frequency components. A useful analogy is the relationship between a series of pure notes (the frequency components) and a musical chord (the function itself). In mathematical physics, the Fourier transform of a signal $x(t)$ can be thought of as that signal in the "frequency domain." This is similar to the basic idea of the various other Fourier transforms including the Fourier series of a periodic function.

Using the Euler formulae we can express the integrand in (5) as

$$a(\omega)\cos\omega x + b(\omega)\sin\omega x = a(\omega)\frac{e^{i\omega s}+e^{-e\omega s}}{2} + b(\omega)\frac{e^{i\omega s}-e^{-i\omega s}}{2i}$$

$$= a(\omega)\frac{e^{i\omega s}+e^{-e\omega s}}{2} - ib(\omega)\frac{e^{i\omega s}-e^{-i\omega s}}{2} =$$

$$= \frac{a(\omega)-ib(\omega)}{2}i^{i\omega s} + \frac{a(\omega)+ib(\omega)}{2}e^{-i\omega s}.$$

Denoting by

$$c(\omega) = \frac{a(\omega)-ib(\omega)}{2} \qquad ...(6)$$

we have by (4.3) and (4.4) that

$$c(-\omega) = \frac{a(\omega)+ib(\omega)}{2} = \overline{c(\omega)},$$

hence the integral in (5) takes the form

$$\int_0^\infty [a(\omega)\cos\omega x + b(\omega)\sin\omega x]d\omega = \int_0^\infty [c(\omega)e^{i\omega s} + c(-\omega)e^{-i\omega s}]d\omega = \qquad ...(7)$$

Changing the variable $\omega \to -\omega$ in the second term we have

$$\int_0^\infty c(-\omega)e^{-i\omega s}d\omega = \int_{-\infty}^0 c(\omega)e^{i\omega s}d\omega$$

and from (7) it follows that

$$\int_0^\infty [a(\omega)\cos\omega x + b(\omega)\sin\omega x]d\omega = \int_{-\infty}^\infty c(\omega)e^{i\omega s}d\omega. \qquad ...(8)$$

If $\omega \geq 0$ then using the formulae (6), (3) and (4) we have

$$c(\omega)=\frac{1}{2\pi}\int_{-\infty}^{\infty}f(t)\cos dt-\frac{i}{2\pi}\int_{-\infty}^{\infty}f(t)\sin\omega t dt$$

$$c(\omega)=\frac{1}{2\pi}\int_{-\infty}^{\infty}f(t)(\cos dt-i\sin\omega t)dt=\frac{1}{2\pi}\int_{-\infty}^{\infty}f(t)e^{-i\omega t}dt$$

If $\omega<0$ then $-\omega>0$ and

$$c(\omega)=\overline{c(-\omega)}=\frac{1}{2\pi}\int_{-\infty}^{\infty}f(t)\overline{e^{i\omega t}}dt$$

or

$$c(\omega)=\frac{1}{2\pi}\int_{-\infty}^{\infty}f(t)e^{-i\omega t}dt.$$

Thus, (5) obtains the form

$$f(x)=\frac{1}{2\pi}\int_{-\infty}^{\infty}f(t)e^{-i\omega t}dt \qquad ...(9)$$

where

$$c(\omega)=\frac{1}{2\pi}\int_{-\infty}^{\infty}f(t)e^{-i\omega t}dt \qquad ...(10)$$

Combining (9) and (10) gives

$$f(x)=\int_{-\infty}^{\infty}\left[\frac{1}{2\pi}\int_{-\infty}^{\infty}f(t)e^{-i\omega t}dt\right]e^{i\omega x}d\omega \qquad ...(11)$$

or

$$f(x)=\frac{1}{2\pi}\int_{-\infty}^{\infty}d\omega\int_{-\infty}^{\infty}f(t)e^{i\omega(x-t)}dt \qquad ...(12)$$

The result (12) is called the complex form of the Fourier integral representation. Now we re-write (11) in the form

$$f(x)=\frac{1}{2\pi}\int_{-\infty}^{\infty}\left[\int_{-\infty}^{\infty}f(t)e^{-i\omega t}dt\right]e^{-i\omega t}d\omega$$

$$=\frac{1}{2\pi}\int_{-\infty}^{\infty}F(\omega)e^{i\omega x}d\omega$$

where

$$F(\omega)=\int_{-\infty}^{\infty}f(t)e^{-i\omega t}dt.$$

Denoting in the last integral the integration variable by x again we have that the complex form of the Fourier integral representation may be written in the form of the pair of equations

$$F(\omega) = \int_{-\infty}^{\infty} f(x)e^{-i\omega x}\,dx \qquad ...(13)$$

and

$$f(x) = \frac{1}{2\pi}\int_{-\infty}^{\infty} F(\omega)e^{i\omega x}\,d\omega. \qquad ...(14)$$

$f(\omega)$ as defined by (13) is called the *Fourier transform* of $f(x)$, and it provides a frequency-domain representation of the non-periodic function $f(x)$ whenever the integral in (13) exists. Equation (14) provides us with a way of reconstructing $f(x)$ if we know its Fourier transform and the transform defined by (14) is called the *inverse Fourier transform*. The transforms (13) and (14) together constitute the Fourier transform pair, and they provide a pathway between the time- and frequency-domain of a function. In contrast with a Fourier series expansion of a periodic function, where the resolved frequencies take discrete values, the expression of a non-periodic function $f(x)$ in the frequency domain is a function of a continuously varying frequency.

Example 1. Does the function $f(x) = 1 - \infty < x < \infty$ have a Fourier integral representation?

Solution:

Clearly

$$\int_{-\infty}^{\infty} |\, f(x)|\, dx$$

is unbounded and the Theorems 1 and 2 do not guarantee the existing of a Fourier integral representation . We can confirm directly from (13) that a Fourier transform does not exist. We have

$$\int_{-\infty}^{\infty} 1e^{-i\omega x}\,dx = \lim_{N\to\infty}\int_{-N}^{N} e^{-i\omega x}\,dx$$

$$= \lim_{N\to\infty}\left[-\frac{1}{i\omega}(e^{-i\omega N} - e^{-i\omega N}) = \right]$$

$$= \lim_{N\to\infty}\frac{2\sin\omega N}{\omega}.$$

Since this limit does not exist, the given function does not have a Fourier integral representation.

Example 2. Find the Fourier transform of the function $f(x) = H(x)e^{-x}$ where $H(x)$ is the *Heaviside unit step function*

$$H(x) = \begin{cases} 1 \text{ if } x \ge 0 \\ 0 \text{ if } x < 0 \end{cases}$$

Solution:

The graph of $f(x)$ is shown in Figure 4.1

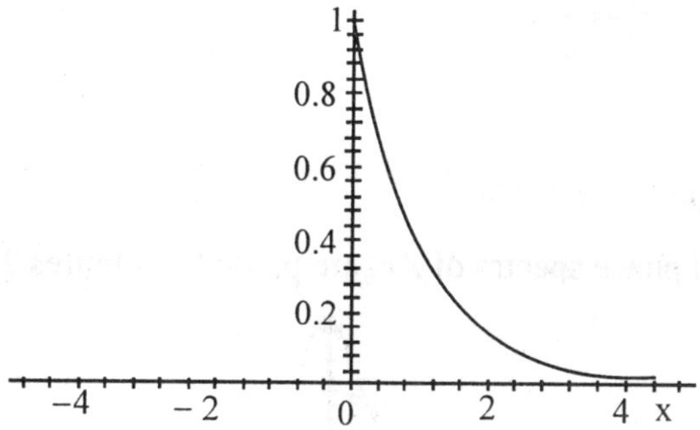

Fig. 1

Since

$$\int_{-\infty}^{\infty} H(x)e^{-x}dx = \lim_{N\to\infty}\int_0^N e^{-x}dx = 1$$

the given function is absolutely integrable on the interval $-\infty < x < \infty$ and by Theorem 2 the Fourier transform exists. Using (4.13), we have

$$F(\omega) = \int_{-\infty}^{\infty} H(x)e^{-x}e^{i\omega x}dx =$$

$$= \int_0^{\infty} e^{-(1+i\omega)x}dx = \left[-\frac{e^{-(1+i\omega)x}}{1+i\omega}\right]_0^{\infty} = \frac{1}{1+i\omega} \qquad ...(15)$$

The Fourier transforms are generally complex-valued functions of the real frequency variable ω. If $F(\omega)$ is the Fourier transform of the signal $f(x)$ then $F(\omega)$ is also called the *complex frequency spectrum* of $f(x)$. If we express $F(\omega)$ in the exponential form

$$F(\omega) = H(x)e^{-x}.$$

then real-valued functions $|F(\omega)|$ and $F(\omega)$ are called the *amplitude and phase spectra* respectively of the signal $f(x)$.

Example 3. Determine the amplitude and phase spectra of the signal

$$f(x) = H(x)\, e^{-x}.$$

Solution:

From (15) we have

$$F(\omega) = \frac{1}{1+i\omega} = \frac{1}{1+\omega^2} - i\frac{\omega}{1+\omega^2}.$$

Thus the amplitude and phase spectra of $F(\omega)$ are

$$|F(\omega)| = \frac{1}{\sqrt{1+\omega^2}}$$

and

$$\arg F(\omega) = -\arctan\omega.$$

The amplitude and phase spectra of $f(x)$ are plotted in Figures 2 and 3 respectively.

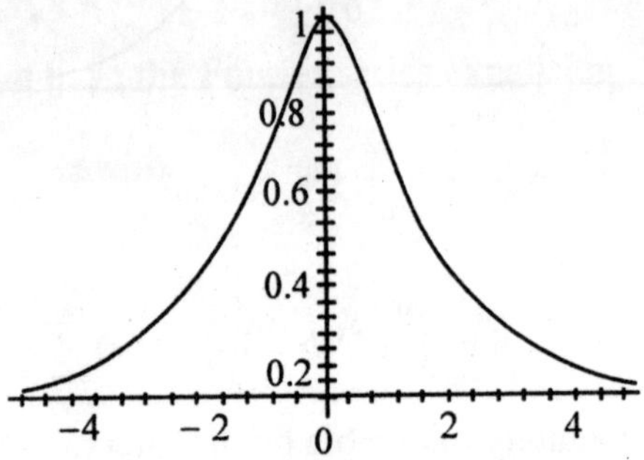

Fig. 2

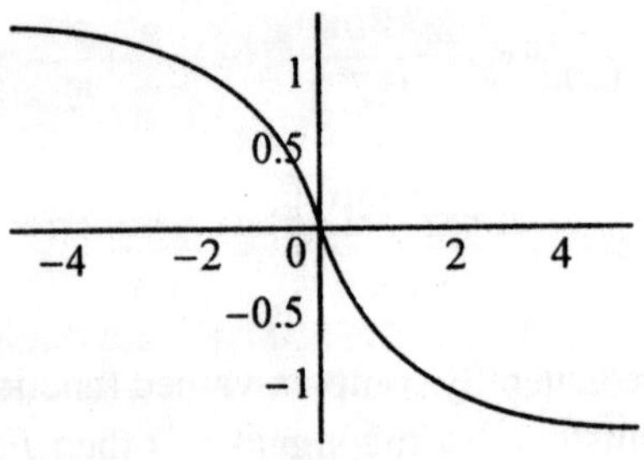

Fig. 2

FOURIER COSINE AND SINE TRANSFORMS

Consider the Fourier integral representation in the form

$$f(x) = \frac{1}{\pi}\int_0^\infty d\omega \int_{-\infty}^\infty f(t)\cos\omega t\cos\omega x dt + \frac{1}{\pi}\int_0^\infty f(t)\omega t \sin\omega x dt =$$

$$= \frac{1}{\pi}\int_0^\infty \cos\omega x d\omega \int_{-\infty}^\infty f(t)\cos\omega t dt + \frac{1}{\pi}\int_0^\infty \sin\omega x d\omega \int_{-\infty}^\infty f(t)\sin\omega t dt. \qquad ...(16)$$

If $f(x)$ is an even function then

$$\int_{-\infty}^\infty f(t)\cos\omega t dt = 2\int_0^\infty f(t)\cos\omega t dt$$

and

$$\int_{-\infty}^{\infty} f(t)\sin\omega t dt = 0.$$

In this case (4.16) reduces to

$$f(x) = \frac{2}{\pi}\int_0^{\infty}\cos\omega x d\omega \int_0^{\infty} f(t)\cos\omega t dt \quad ...(17)$$

which is called the *Fourier cosine integral*. If $f(x)$ is an odd function then (4.16) reduces to

$$f(x) = \frac{2}{\pi}\int_0^{\infty}\sin\omega x d\omega \int_0^{\infty} f(t)\sin\omega t dt \quad ...(18)$$

which is called the *Fourier sine integral*. The Fourier cosine integral may be expressed in the form of the pair of equations

$$F_1(\omega) = \int_0^{\infty} f(x)\cos\omega x dx \quad ...(19)$$

which is referred to as *Fourier cosine transform* and

$$F(x) = \frac{2}{\pi}\int_0^{\infty} F_1(\omega)\cos\omega x d\omega.$$

The Fourier sine integral may be expressed in the form of the pair of equations

$$F_2(\omega) = \int_0^{\infty} f(x)\sin\omega x dx \quad ...(20)$$

which is referred to as *Fourier sine transform* and

$$f(x) = \frac{2}{\pi}\int_0^{\infty} F_2(\omega)\sin\omega x d\omega.$$

If the function $f(x)$ is defined only over the interval $0 < x < \infty$ then it can be extended to the whole number scale so that the result is even function for which (17) holds. Also it can be extended to the whole number scale so that the result is odd function for which (18) holds. It follows from (17) and (18) that if the function $f(x)$ satisfies the conditions of Theorem 1 or Theorem 2 then applying to $f(x)$ the Fourier cosine transform (19) or Fourier sine transform (20) twice and scaling by factor $\frac{2}{\pi}$ we obtain the parent function again.

Example 4. Find the Fourier integral representation of the function

$$f(x) = \begin{cases} 1 \text{ if } |x| < 1, \\ \frac{1}{2} \text{ if } |x| = 1, \\ 0 \text{ if } |x| > 1 \end{cases}$$

Solution:

Clearly $f(x)$ is an even function, thus

$$f(x)=\frac{2}{\pi}\int_0^\infty F_1(\omega)\cos\omega x d\omega$$

where

$$F_1(\omega)=\int_0^\infty f(x)\cos\omega x dx=$$

$$=\int_0^1\cos\omega x dx=\frac{\sin\omega}{\omega}.$$

Hence

$$F(x)=\frac{2}{\pi}\int_0^\infty\frac{\sin\omega}{\omega}\cos\omega x d\omega$$

and using Theorem 2 we obtain that

$$\frac{2}{\pi}\int_0^\infty\frac{\sin\omega}{\omega}\cos\omega x d\omega=\begin{cases}1 \text{ if } |x|<1,\\ \frac{1}{2} \text{ if } |x|=1,\\ 0 \text{ if } |x|>1.\end{cases}$$

Taking in the last equation $x = 0$ we obtain

$$\frac{2}{\pi}\int_0^\infty\frac{\sin\omega}{\omega}d\omega=1$$

that is

$$\int_0^\infty\frac{\sin\omega}{\omega}d\omega=\frac{\pi}{2}.$$

PROPERTIES OF THE FOURIER TRANSFORM

The Linearity Property

If the functions f(x) and g(x) have the Fourier transforms $F(\omega)$ and $G(\omega)$ respectively, α and β are constants, then the Fourier transform of the function $\alpha f(x)+\beta(x)$ is $\alpha F(\omega)+\beta G(\omega)$.

The proof of this property follows immediately from the definition (5.13), since

$$\int_{-\infty}^\infty[\alpha f(x)+\beta g(x)e^{-i\omega x}dx=a\int_{-\infty}^\infty f(x)e^{-i\omega x}dx+\beta\int_{-\infty}^\infty g(x)e^{-i\omega x}dx=$$

$$=\alpha F(\omega)+\beta G(\omega).$$

Clearly the inverse Fourier transform (4.14) is also linear. The linearity property can be extended to any number of terms in the sum, i.e. if $f_k(x)$ have the Fourier transforms $F_k(\omega)$ and α_k are the constants $1 \leq k \leq a$ then

$$\sum_{k=1}^{n} a_k f_k(x)$$

has the Fourier transform

$$\sum_{k=1}^{n} a_k F_k(\omega).$$

The linearity property can be extended also to an infinite number of terms if the series concerned are convergent.

The Similarity Property

If the function f(x) has the Fourier transform $f(\omega)$ then the function f(ax) for a > 0 has the Fourier transform $\frac{1}{a}F\left(\frac{\omega}{a}\right)$.

The proof follows from the definition (5.13) making in the integral

$$\int_{-\infty}^{\infty} f(ax)e^{-i\omega x}dx$$

the substitution $t = ax$. We have

$$\int_{-\infty}^{\infty} f(ax)e^{-i\omega x}dx = \int_{-\infty}^{\infty} f(t)e^{-\frac{i\omega I}{a}} d\left(\frac{t}{a}\right) = \frac{1}{a}F\left(\frac{\omega}{a}\right).$$

Example 5. Determine the Fourier transform of the signal $H(x)e^{-ax}$ $(a > 0)$.

Solution:

By (15) the signal $f(x) = H(x)e^{-x}$ has the Fourier transform $\frac{1}{1+i\omega}$. Hence $H(x)e^{-ax} = f(ax)$ has by the similarity property the Fourier transform

$$\frac{1}{a}\frac{1}{1+i\frac{\omega}{a}} = \frac{1}{a+i\omega} \qquad ...(21)$$

The Shift Property

If the function f(x) has the Fourier transform $F(\omega)$ then the function $e^{i\omega_0 x}f(x)$ has the Fourier transform $F(\omega - \omega_0)$.

The proof follows immediately from the definition (4.13) again, since

$$\int_{-\infty}^{\infty} e^{-i\omega_0 x} f(x) e^{-i\omega x} dx = \int_{-\infty}^{\infty} f(x) e^{-i(\omega-\omega_0)x} = F(\omega - \omega_0).$$

The shift property or the frequency-shift property indicates that multiplication by $e^{-i\omega_0 x}$ shifts the spectrum of $f(x)$ so that it is centered on the point $\omega = \omega_0$ in the frequency domain.

Example 6. Let $f(x)$ have the frequency spectrum $F(\omega)$. Determine the frequency spectrum $G(\omega)$ of the signal $g(x) = f(x) \cos ax$.

Solution:

By Euler's formula

$$\cos ax = \frac{1}{2}(e^{iax} + e^{-iax})$$

thus,

$$g(x) = \frac{1}{2} e^{iax} f(x) + e^{-iax} f(x).$$

Using first the linearity property and then the shift property, we obtain

$$G(\omega) = \frac{1}{2} F(\omega - a) + \frac{1}{2} F(\omega + a).$$

The Delay Property .

If the function $f(x)$ has the Fourier transform $F(\omega)$ then the delayed version of f(x), defined as $g(x) = f(x - a)$ has the Fourier transform $G(\omega) = e^{i\omega a} F(\omega)$.

The proof follows from the definition (5.13) again. We have

$$G(\omega) = \int_{-\infty}^{\infty} f(x-a) e^{-i\omega x} dx.$$

Making the substitution $t = x - a$, we have

$$G(\omega) = \int_{-\infty}^{\infty} f(t) e^{-i\omega(t+a)} dt = e^{-i\omega a} \int_{-\infty}^{\infty} f(t) e^{-i\omega t} dt = e^{-i\omega a} F(\omega).$$

This result, known as delay or time-shift property, implies that delaying a signal by a time a causes its Fourier transform to be multiplied by $e^{-i\omega a} F(\omega)$.

Example 7. Let $\varphi(x)$ have the Fourier transform $\Phi(\omega)$. Determine the Fourier transform $\psi(\omega)$ of the function $\psi(x) = \varphi(2x) - \varphi(2x - 1)$.

Solution:

By the similarity property $\varphi(2x)$ has the Fourier transform $\frac{1}{2}\Phi\left(\frac{\omega}{2}\right)$. By the delay property $\varphi(2x-1)$ has the Fourier transform $\frac{1}{2}e^{-1\omega}\Phi\left(\frac{\omega}{2}\right)$ and by the linearity property $\psi(x)$ has the Fourier transform

$$\psi(\omega)=\frac{1}{2}\Phi\left(\frac{\omega}{2}\right)-\frac{1}{2}e^{-i\omega}\Phi\left(\frac{\omega}{2}\right)=\frac{1}{2}(1-e^{-i\omega})\Phi\left(\frac{\omega}{2}\right).$$

Time-differentiation property

If the function f(x) has a Fourier transform $F(\omega)$ then its derivative f'(x) has the Fourier transform $i\omega F(\omega)$.

By the definition of the inverse Fourier transform (14) we have

$$f(x)=\frac{1}{2\pi}\int_{-\infty}^{\infty}F(\omega)e^{i\omega x}d\omega.$$

Differentiating with respect to *x* gives

$$f'(x)=\frac{1}{2\pi}\int_{-\infty}^{\infty}\frac{\partial}{\partial x}[F(\omega)e^{i\omega x}]d\omega=\frac{1}{2\pi}\int_{-\infty}^{\infty}i\omega F(\omega)e^{i\omega x}d\omega$$

which means, by the definition (14), that $f'(x)$ has the Fourier transform $i\omega F(\omega)$ i.e. the result of differentiation in the time domain is algebraic multiplication in the frequency domain.

Repeating the differentiation with respect to *x n* times gives that *f*(*n*)(*x*) has the Fourier transform $(i\omega)^n F(\omega)$.

Example 8. Determine the Fourier transform $Y(\omega)$ of the time signal $y(x)$ satisfying the differential equation

$$\frac{d^2y(x)}{dx^2}+3\frac{dy(x)}{dx}+7y(x)=H(x)e^{-2x}.$$

Solution:

Using first the linearity property throughout in the differential equation, the time-differentiation property and (21) with $a=2$ we obtain

$$(i\omega)^2Y(\omega)+3(i\omega)Y(\omega)+7Y(\omega)=\frac{1}{2+i\omega}$$

that is, $$(-\omega^2+3i\omega+7)Y(\omega)=\frac{1}{2+i\omega}$$

giving $$Y(\omega)=\frac{1}{(7-\omega^2+3i\omega)(2+i\omega)}.$$

The Symmetry Property

Changing in (14)

$$f(x)=\frac{1}{2\pi}\int_{-\infty}^{\infty}F(\omega)e^{i\omega x}d\omega$$

the variable in the integration to *y* we have

$$f(x)=\frac{1}{2\pi}\int_{-\infty}^{\infty}F(y)e^{iyx}dy$$

so that

$$2\pi f(-x)=\int_{-\infty}^{\infty}F(y)e^{iyx}dy. \qquad ...(22)$$

The right-hand side of (22) is by the definition (13) the Fourier transform of the function *F*(*y*). Replacing *x* by ω and then *y* by *x*, (22) gives the form

$$2\pi f(-\omega)=\int_{-\infty}^{\infty}F(x)e^{-i\omega x}dx.$$

We conclude that if f(x) has the Fourier transform $F(\omega)$ then F(x) has the Fourier transform $2\pi f(-\omega)$ or if f(x) and $F(\omega)$ form a Fourier transform pair then F(x) and $2\pi f(-\omega)$ form another Fourier transform pair. This is the symmetry property.

Example 9. Determine the Fourier transform of the signal

$$g(x)=\begin{cases}\dfrac{\sin x}{x} & \text{if } x\neq 0,\\ & \text{if } x=0.\end{cases}$$

Solution:

From Example 4.6 we know that the function

$$f(x)=\begin{cases}1 \text{ if } |x|<1,\\ \dfrac{1}{2} \text{ if } |x|=1,\\ 0 \text{ if } |x|>1.\end{cases}$$

has the Fourier transform $\frac{\sin\omega}{\omega}$. Thus, by the symmetry property $g(x)=\frac{\sin\omega}{\omega}$ has the Fourier transform

$$2\pi f(-\omega)=\begin{cases}2\pi \text{ if } |\omega|<1,\\ \pi \text{ if } |\omega|=1,\\ 0 \text{ if } |\omega|>1.\end{cases}$$

CONVOLUTION THEOREM

The *convolution* of two given piecewise continuous functions $u(x)$ and $v(x)$, denoted by, $u(x)\neq v(x)$, is defined as

$$u(x)\neq v(x)=\int_{-\infty}^{\infty}u(\xi)v(x-\xi)d\xi.$$

Let $U(\omega)$ and $V(\omega)$ denote the Fourier transforms of $u(x)$ and $v(x)$ respectively i.e.

$$U(\omega)=\int_{-\infty}^{\infty}u(x)e^{-i\omega x}dx$$

and

$$V(\omega)=\int_{-\infty}^{\infty}v(x)e^{-l\omega x}dx.$$

Using (4.13) we find the Fourier transform $Y(\omega)$ of the convolution $y(x)=u(x)\neq v(x)$

$$\begin{aligned}Y(\omega)&=\int_{-\infty}^{\infty}[u(x)\neq v(x)]e^{-i\omega x}dx\\ &=\int_{-\infty}^{\infty}e^{-i\omega x}\left[\int_{-\infty}^{\infty}u(\xi)v(x-\xi)d\xi\right]dx\\ &=\int_{-\infty}^{\infty}u(\xi)d\xi\int_{-\infty}^{\infty}v(x-\xi)e^{-i\omega x}dx.\end{aligned}$$

Changing the variable $y=x-\xi$ that is $x=y+\xi$ the transform can be expressed as

$$Y(\omega)=\int_{-\infty}^{\infty}u(\xi)e^{-i\omega\xi}d\xi\int_{-\infty}^{\infty}v(y)e^{-i\omega y}dy=U(\omega)V(\omega).$$

Consequently, the Fourier transform of the convolution $u(x)\neq v(x)$ is the product of the transforms $U(\omega)\,V(\omega)$.

Next we find the inverse Fourier transform of the convolution

$$Y(\omega)=U(\omega)\neq V(\omega)=\int_{-\infty}^{\infty}U(y)V(\omega-y)dy.$$

By (14) we have

$$Y(x)=\frac{1}{2\pi}\int_{-\infty}^{\infty}e^{i\omega x}\left[\int_{-\infty}^{\infty}U(y)V(\omega-y)dy\right]d\omega.$$

A change of variable $z = \omega - y$ that is $\omega = z + y$ leads to

$$Y(x) = \frac{1}{2\pi}\int_{-\infty}^{\infty}\left[\int_{-\infty}^{\infty} U(y)V(z)e^{i(y+z)x}dy\right]dz =$$

$$= \frac{1}{2\pi}\int_{-\infty}^{\infty} U(y)e^{iyx}dy\int_{-\infty}^{\infty} V(z)e^{izx}dz = 2\pi u(x)v(x).$$

Thus, the convolution of the transforms $U(\omega) \neq V(\omega)$ is the Fourier transform of of the product $2\pi u(x)v(x)$ and consequently, the product $u(x)v(x)$ has the Fourier transform $\frac{1}{2\pi}U(\omega) \neq V(\omega)$.

GENERALIZED FOURIER TRANSFORM

The unit impulse function or Dirac delta function is denoted by $\delta(x)$ and is defined as

$$\delta(x) = 0 \quad (x \neq 0)$$

$$\int_{-\infty}^{\infty} \delta(x)dx = 1.$$

The Dirac delta function is not a function in an ordinary sense, it is a so-called generalized function. An important property of Dirac delta function is the so-called sifting property of Dirac delta function, which states that if f(x) is continuous at $x = a$ then

$$\int_{-\infty}^{\infty} f(x)\delta(x-a)dx = f(a) \qquad ...(23)$$

Using this sifting property we obtain Fourier transforms (13) of Dirac delta functions $\delta(x)$ and $\delta(x-a)$ respectively

$$F_x(\omega) = \int_{-\infty}^{\infty} \delta(x)e^{-i\omega x}dx = e^0 = 1 \qquad ...(24)$$

and

$$F_{x-a}(\omega) = \int_{-\infty}^{\infty} \delta(x-a)e^{-i\omega x}dx = e^{-i\omega a}. \qquad ...(25)$$

Since by (23) $\delta(x)$ has the Fourier transform 1 then using the symmetry property, we deduce that 1 has has the Fourier transform

$$2\pi\delta(-\omega) = 2\pi\delta(\omega). \qquad ...(26)$$

Likewise, since by (24) $\delta(x-a)$ has the Fourier transform $e^{-i\omega a}$ then by symmetry property e^{-iax} has the Fourier transform $2\pi\delta(-\omega - a)$ or, substituting $a = \omega_0$, the function $e^{i\omega_0 x}$ has the Fourier transform

$$2\pi\delta(\omega_0 - \omega) = 2\pi\delta(\omega - \omega_0). \qquad ...(27)$$

Therefore, despite that the functions 1 and $e^{i\omega_0 x}$ do not have a Fourier transform in the ordinary sense, they have generalized Fourier transforms. The generalized Fourier transform is an extension of the Fourier transform. Indeed, if we apply the inverse Fourier transform to $2\pi\delta(\omega)$ then using the sifting property (23) we obtain

$$\frac{1}{2\pi}\int_{-\infty}^{\infty} 2\pi\delta(\omega)e^{i\omega x}d\omega = 1,$$

so the initial function 1 is recovered. Similarily, if we apply the inverse Fourier transform to $2\pi\delta(\omega-\omega_0)$ then using the sifting property again, we obtain

$$\frac{1}{2\pi}\int_{-\infty}^{\infty} 2\pi\delta(\omega-\omega_0)e^{i\omega x}d\omega = e^{i\omega_0 x}$$

so that $e^{i\omega_0 x}$ is also recovered.

Suppose now that a periodic function *f*(*x*) defined on $-\infty < x < \infty$, may be expanded in a (convergent) Fourier series having exponential form

$$f(x) = \sum_{k=-\infty}^{\infty} c_k e^{ik\omega_0 x}$$

and let $F(\omega)$ denotes its Fourier transform. Since $e^{ik\omega_0 x}$ has the Fourier transform $2\pi\delta(\omega-k\omega_0)$ then, using the linearity property, we obtain that

$$F(\omega) = \sum_{k=-\infty}^{\infty} c_k 2\pi\delta(\omega-k\omega_0) = 2\pi\sum_{k=-\infty}^{\infty} c_k\delta(\omega-k\omega_0) \qquad ...(28)$$

where c_k are the coefficients of the exponential form of the Fourier series expansion of *f*(*x*).

Example 10. Determine the generalized Fourier transform of the unit impulse train

$$F(x) = \sum_{k=-\infty}^{\infty} \delta(x-kT).$$

Solution:

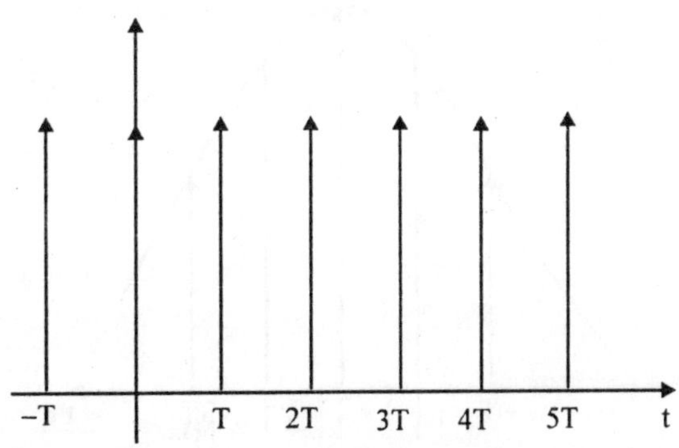

Fig. 3 Unit Impulse Train

Although $f(x)$ is a generalized function, it has the period T, since

$$f(x+T) = \sum_{k=-\infty}^{\infty} \delta(x+T-kT) = \sum_{k=-\infty}^{\infty} \delta(x-(k-1)T)$$

$$= \sum_{k=-\infty}^{\infty} \delta(x-kT) = f(x)$$

and we can formally expand the periodic function $f(x)$ in a Fourier series

$$f(x) = \sum_{k=-\infty}^{\infty} c_k e^{ik\omega_0 x} \left(\omega_0 = \frac{2\pi}{T} \right)$$

where $$c_k = \frac{1}{T} \int_{-\frac{T}{2}}^{\frac{T}{2}} f(x) e^{-ik\omega_0 x} dx = \frac{1}{T} \int_{-\frac{T}{2}}^{\frac{T}{2}} \delta(x) e^{-ik\omega_0 x} dx = \frac{1}{T}.$$

By (28) it follows that the generalized Fourier transform $G(\omega)$ of $f(x)$ is

$$G(\omega) = 2\pi \sum_{k=-\infty}^{\infty} \frac{1}{T} \delta(\omega - k\omega_0)$$

$$= \omega_0 \sum_{k=-\infty}^{\infty} \delta(\omega - k\omega_0)$$

that is, the time-domain impulse train has another impulse train as its transform.

SHANNON-KOTELNIKOV SAMPLING THEOREM

Let us sample the continuous-time non-periodic signal $f(x)$ at equal intervals T, to generate the sequence

$$\{..., f(-kT), ..., f(-T), f(0), f(T), ..., f(kT), ...\}.$$

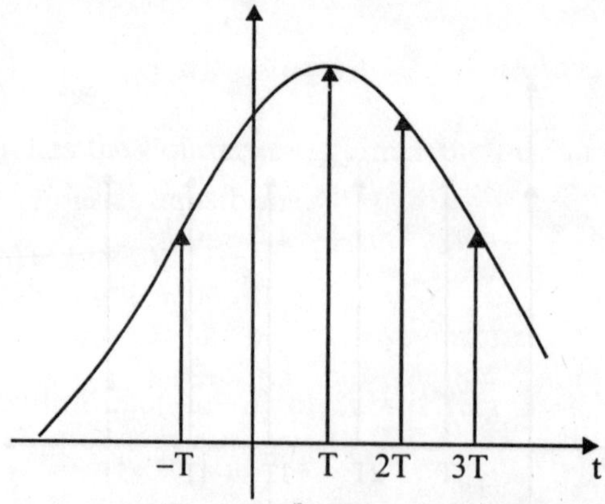

Fig. 4 Sampling of the Continuous-time Signal

We can represent the sampled version of $f(x)$ as

$$f_1(x) = \sum_{k=-\infty}^{\infty} f(kT)\delta(x-kT)$$

$$= \sum_{k=-\infty}^{\infty} f(x)\delta(x-kT) \qquad \text{...(29)}$$

$$= f(x)\sum_{k=-\infty}^{\infty} \delta(x-kT)$$

which is called the continuous-time representation of the sampled version of f(x). This is the product of two functions

$$f(x) \text{ and } = f(x)\sum_{k=-\infty}^{\infty} \delta(x-kT).$$

The Fourier transform of this product is $\frac{1}{2\pi}$ times the convolution of the Fourier transforms $f(\omega)$ of $f(x)$ and $G(\omega)$ from Example 12. Thus the Fourier transform $F_1(\omega)$ of the sampled version $f_s(x)$ is

$$F_1(\omega) = \frac{1}{2\pi}F(\omega) \neq \omega_0 \sum_{k=-\infty}^{\infty} \delta(\omega - k\omega_0)$$

$$= \frac{1}{2\pi}F(\omega) \neq \frac{2\pi}{T}\sum_{k=-\infty}^{\infty} \delta(\omega - k\omega_0)$$

$$= \frac{1}{T}\int_{-\infty}^{\infty} F(\omega - z)\sum_{k=-\infty}^{\infty} \delta\left(z - \frac{2k\pi}{T}\right)dz$$

$$= \frac{1}{T}\sum_{k=-\infty}^{\infty}\int_{-\infty}^{\infty} F(\omega - z)\delta\left(z - \frac{2k\pi}{T}\right)dz$$

$$= \frac{1}{T}\sum_{k=-\infty}^{\infty}\left(\omega - \frac{2k\pi}{T}\right).$$

Thus $$F_1(\omega) = \frac{1}{T}\sum_{k=-\infty}^{\infty} F(\omega - k\omega_0). \qquad \text{...(30)}$$

We see that the spectrum $F_1(\omega)$ of the sampled version $f_s(x)$ consists of repetition of the spectrum $F(\omega)$ of $f(x)$ scaled by a factor $\frac{1}{T}$, these repetitions being spaced at intervals ω_0 apart. Figure 6 shows the amplitude spectrum $|F(\omega)|$ of a *band-limited signal* $f(x)$, that is, a signal whose spectrum is zero for $|\omega| > \omega_m$. The length of the interval $\lfloor -\omega_m ; \omega_m \rfloor$ where $F(\omega)$ is non-zero, i.e. $2\omega_m$ is said to be the *bandwidth* of the signal $f(x)$.

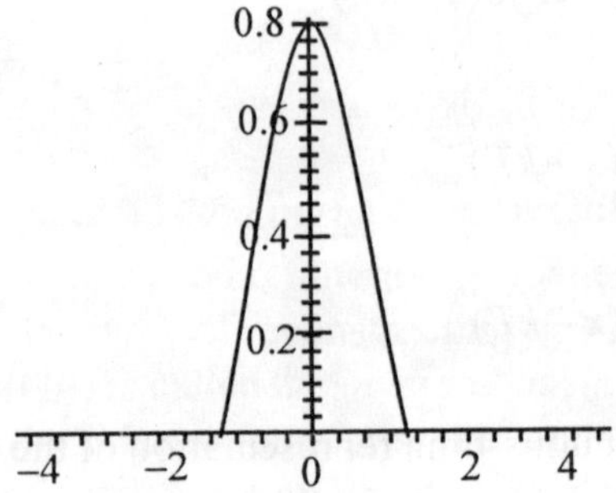

Fig. 5

Figure 7 Shows the amplitude spectrum $|F_1(\omega)|$ of the sampled version i.e. the repeats of the amplitude spectrum $|F(\omega)|$ being spaced at intervals ω_0 apart.

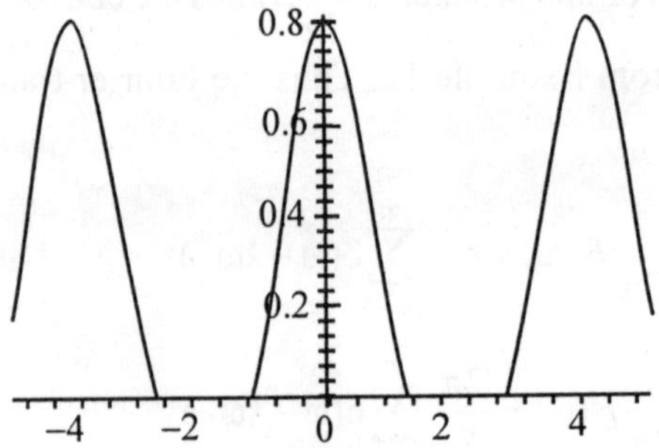

Fig. 6

It is clear that as the sampling interval T increases, ω_0 decreases and starting from some moment two neighboring copies will interact and the amplitude spectrum of the sampled version $|F_1(\omega)|$ does not represent in $-\omega_m < \omega < \omega_m$ the amplitude spectrum $|F(\omega)|$. For a band-limited signal $f(x)$ this interaction between the neighboring copies is avoided if

$$\omega_0 \geq 2\omega_m. \qquad ...(31)$$

Since $\omega_0 = \dfrac{2\pi}{T}$, this implies that

$$T \leq \frac{\pi}{\omega_m}. \qquad ...(32)$$

Thus, the minimum time interval allowed between the samples is

$$T_{\min} = \frac{\pi}{\omega_m}.$$

Let us suppose now that $f(x)$ is a band-limited function with bandwidth $2\omega_m$ and that $f(x)$ has been sampled with sampling interval satisfying (32). Then (31) is also satisfied and in the interval $-\frac{\omega_0}{2} < \omega < \frac{\omega_0}{2}$ the spectrum $F_1(\omega)$ represents the spectrum $F(\omega)$. Moreover, $F_1(\omega)$ is a periodic function with period $\omega_0 = \frac{2\pi}{T}$, that is, with frequency T and has the Fourier series expansion

$$F_1(\omega) = \sum_{k=-\infty}^{\infty} c_k e^{ikT\omega}$$

where

$$c_k = \frac{1}{\omega_0} \int_{-\frac{\omega_0}{2}}^{\frac{\omega_0}{2}} F_1(\omega) e^{ikT\omega} d\omega$$

$$= \frac{1}{\omega_0} \frac{1}{T} \int_{-\frac{\omega_0}{2}}^{\frac{\omega_0}{2}} F(\omega) e^{ikT\omega} d\omega$$

$$= \frac{1}{2\pi} \int_{-\infty}^{\infty} F(\omega) e^{ikT\omega} d\omega$$

or, by the definition of the inverse Fourier transform (4.14) $c_k = f(kT)$. Thus, the Fourier series expansion of $F_1(\omega)$ has the form

$$F_1(\omega) = \sum_{k=-\infty}^{\infty} f(kT) e^{-ikT\omega}. \qquad \text{...(33)}$$

Using the inverse Fourier transform (5.14) again and (5.33) we find

$$F(x) = \frac{1}{2\pi} \int_{-\infty}^{\infty} F(\omega) e^{i\omega x} d\omega = \frac{1}{2\pi} \int_{-\frac{\omega_0}{2}}^{\frac{\omega_0}{2}} TF_1(\omega) e^{i\omega x} d\omega$$

$$= \frac{1}{\omega_0} \int_{-\frac{\omega_0}{2}}^{\frac{\omega_0}{2}} \sum_{k=-\infty}^{\infty} f(kT) e^{-ikT\omega} e^{i\omega x} d\omega$$

$$= \frac{1}{\omega_0} \sum_{k=-\infty}^{\infty} f(kT) \int_{-\frac{\omega_0}{2}}^{\frac{\omega_0}{2}} e^{i\omega(x-kT)} d\omega$$

$$= \frac{1}{\omega_0} \sum_{k=-\infty}^{\infty} f(kT) \frac{1}{i(x-kT)} \left[e^{e\frac{\omega_0}{2}(x-kT)} - e^{-i\frac{\omega_0}{2}(x-kT)} \right]$$

$$= \frac{1}{\omega_0} \sum_{k=-\infty}^{\infty} f(kT) \frac{2}{x-kT} \sin \frac{\omega_0 (x-kT)}{2}$$

that is $$f(x) = \sum_{k=-\infty}^{\infty} f(kT) \frac{\sin \frac{\omega_0 (x-kT)}{2}}{\frac{\omega_0 (x-kT)}{2}}. \qquad ...(34)$$

We have proved the following theorem referred as Shannon-Kotelnikov Sampling Theorem.

Theorem 3. A band-limited signal $f(x)$ with bandwidth $2\omega_m$ is completely determined by the countable set of samples $f(kT)$ of the signal $f(x)$, if the sampling interval satisfies (32). Furthermore, it may be obtained from these values by the equality (34).

For a periodic signal $f(x)$ this theorem can be expressed in the following way.

Theorem 4. If $f(x)$ is a periodic signal of period T, such that the Fourier series expansion has no harmonics above $\frac{N}{2}$ harmonic, then f(x) is completely determined by the set of values $f\left(\frac{kT}{N+1}\right)$, $k = 0, 1, 2, ..., N$. Furthermore, it may be obtained from these values as

$$f(x) = \sum_{k=0}^{N} f\left(\frac{kT}{N+1}\right) \frac{\sin \frac{\omega_0 [x(N+1)-kT]}{2}}{\frac{\omega_0 [x(N+1)-kT]}{2}}.$$

EXERCISE

1. Show that the Fourier trasform of the odd function

$$f(x) = \begin{cases} e^{-\alpha x} & x > 0 \\ -e^{\alpha x} & x < 0 \end{cases}$$

is given by $F(k) = F[f(x)] = \frac{1}{\sqrt{2\pi}} \frac{-2ik}{\alpha^2 + k^2}$

Thus show that $\int_{-\infty}^{+\infty} \frac{k e^{ikx}}{\alpha^2 + k^2} dk = i\pi e^{-\alpha x} \quad x > 0$

and $\int_0^{+\infty} \frac{k \sin kx}{\alpha^2 + k^2} dk = \frac{\pi}{2} e^{-\alpha x} \quad x > 0$

(Notice that since the function $f(x)$ is an odd function of x its Fourier transform is imaginary.)

2. Show that Fourier trasform of the function

$$f(x) = e^{-\alpha|x|} e^{i\beta x}$$

is given by

$$F(k) = F[f(x)] = \frac{1}{\sqrt{2\pi}} \frac{2\alpha}{\alpha^2 + (\beta^2 - k^2)}$$

Hence prove that

$$\int_{-\infty}^{+\infty} \frac{\cos kx}{\alpha^2 + (\beta - k)^2} dk = \frac{\pi}{\alpha} e^{-\alpha|x|} \cos \beta x$$

and $\displaystyle\int_{-\infty}^{+\infty} \frac{\sin kx}{\alpha^2 + (\beta - k)^2} dk = \frac{\pi}{\alpha} e^{-\alpha|x|} \sin \beta x$

3. Prove that

$$F[\delta(x-a)] = \frac{1}{\sqrt{2\pi}} e^{-ika}$$

4. Consider the rectangle function defined by the equation

$$\text{rect}\left(\frac{x}{a}\right) = \begin{cases} 1 & |x| < \frac{1}{2}a \\ 0 & |x| > \frac{1}{2}a \end{cases}$$

Show that

(a) $F\left[\text{rect}\left(\dfrac{x}{a}\right)\right] = \dfrac{a}{\sqrt{2\pi}} \text{sinc}\left(\dfrac{ka}{2\pi}\right)$

where

$$\text{sinc } x \equiv \frac{\sin \pi x}{\pi x}$$

is known as the sinc function.

(b) Prove that

$$\text{rect}\left(\frac{x}{a}\right) * \text{rect}\left(\frac{x}{a}\right) = \begin{cases} 0 & \text{for } |x| > a \\ a - x & \text{for } 0 < x < a \\ a + x & \text{for } -a < x < 0 \end{cases}$$

which is the triangle function.

(c) Find the Fourier transform of the triangle function and verify Eq. (145).

5. Assuming

$f(t) = e^{-t^2/\alpha^2}$ and $g(t) = e^{-t^2/\beta^2}$

determine $F(\omega)$ and $G(\omega)$

Chapter 10

Application of Fourier Transform

INTRODUCTION

In the earlier chapter we have learnt about Fourier transform and their properties. In this chapter we will discuss about the application of Fourier transform.

The Fourier transform has many applications, in fact any field of physical science that uses sinusoidal signals, such as engineering, physics, applied mathematics, and chemistry, will make use of Fourier series and Fourier transforms. It would be impossible to give examples of all the areas where the Fourier transform is involved, but here are some examples from physics, engineering, and signal processing.

Communications

In communications theory the signal is usually a voltage, and Fourier theory is essential to understanding how a signal behaves when it passes through filters, amplifiers and communications channels. Even discrete digital communications which use 0's or 1's to send information still have frequency contents. This is perhaps easiest to grasp in the case of trying to send a single square pulse down a channel.

The field of communications spans a range of applications from high-level network management down to sending individual bits down a channel. The Fourier transform is usually associated with these low level aspects of communications.

If we take simple digital pulse that is to be sent down a telephone line, it will ideally look like this:

If we take the Fourier transform of this to show what frequencies make up this signal we get something like:

This means that the square pulse is a sum of infinite frequencies. However if the telephone line only has a bandwidth of 10MHz then only the frequencies below 10MHz will get through the channel. This will cause the digital pulse to be distorted e.g.

This fact has to be considered when trying to send large amounts of data down a channel, because if too much is sent then the data will be corrupted by channel and will be unusable.

Extending the example of the telephone line, whenever you dial a number on a "touch-tone" phone you hear a series of different tones. Each of these tones is composed of two different frequencies that add together to produce the sound you hear.

The Fourier transform is an ideal method to illustrate this, as it shows these two frequencies e.g.

Astronomy

The Fourier transform is not just limited to simple lab examples. When used in real situations it can have far reaching implications about the world around us.

Take for example the field of astronomy. Some times it isn't possible to get all the information you need from a normal telescope and you need to use radio waves or radar instead of light. These radar signals are treated just like any other ordinary time varying voltage signal and can be processed digitally.

Magellan

One recent example of this technique was the NASA *Magellan* satellite which was released from the space shuttle Atlantis on 4th May 1989 and sent to Venus on a 15 month journey that took it one and a half times around the sun.

Venus is Earth's closest planetary companion, and is comparable in its size and diameter. It is difficult to study the surface of Venus because it is perpetually covered with a cloud layer which normal optical telescopes can't penetrate, so *Magellan* had radar and advanced Digital Signal Processing that was designed to see through this cloud layer. It's mission was map the planet with radar and to reveal surface features as small as 250 meters across.

The black and white pictures that it sent back were strips of the planets surface, about 20 km wide, from the north pole to the south pole. One example image of a surface feature called "Pandora Corona" is shown next. If you look at the image, you will see a two black lines through the picture. This is just a mismatch between the strips sent back by Magellan. It also gives you an idea of the scale of the image as each strip is 20 km wide.

From the radar signal received from the planet *Magellan* was also able to gather other information about the planets surface. The next image is a global map of venus showing emissitivity of the Venus's surface.

By the end of the mission *Magellan* had far exceeded the mission objectives by mapping 99% of the planet's surface, and by giving unpresidented detail. By piecing all the information together from this mission and from earlier Pioneer-Venus orbiter data, as well as the Soviet Venera 13 and 14 spacecraft, they were able to produce the following mosaic image which is a global view of Venus using simulated color to enhance small scale structures.

Magellan was also carrying another smaller antenna that pointed straight down and measured the time taken for a signal to reflect off the surface. This was used to build up a topographic map of the surface beneath the spacecraft. Combining this data with the earlier radar images they were able to construct a 3D model of the surface of Venus. The model was animated to provide an MPEG movie that flies you across the Maxwell Mountains on Venus. Any of this would have been impossible if it had not been for Digital Signal Processing and the Fast Fourier Transform.

Geology

The FFT and Nuclear Explosions

Seismic research has always been a common user for the Discrete Fourier Transform (and the FFT). If you look at the history of the FFT you will find that one of the original uses for the FFT was to distinguish between natural seismic events and nuclear test explosions because they generate different frequency spectra.

Seismic Reflection Method

The major exploration tool used by the oil and gas industry for locating new hydrocarbon reserves is the SEISMIC REFLECTION METHOD. This technique enables sub-surface sedimentary rock layers to be mapped, from measurements of the amplitude and reflection times of events obtained by the reflection of seismic (sound) waves from the layers.

The basic features of the technique are:

1. A source of seismic energy which is activated at the Earth's surface.
2. The downward propagation of seismic wavges into the earth.
3. The reflection of seismic waves from the interfaces seperating geological media having dissimilar acoustic properties.
4. The upward propagation of reflected seismic waves.
5. Detection of the reflected seismic waves at the Earth's surface using receivers located at various distances from the source.
6. Recording of the signals at each receiver.

The data recorded at the receivers under goes extensive DSP to produce a seismic section which is a representation of a cross section of the Earth. This is then interpreted in terms of what the survey was looking for.

One of the many problems in extracting the information from the data is that the receivers that record the reflected seismic waves also pick up the signature of the source. (The signature of a source is the wave-form that it emits upon activation.) The signature can be removed from the seismic data by deconvolution as long as you also have a recording of the signature, as well as information about the ghost reflections, absorption, and instrument response. These can all be recorded at the time of the survey or can be deduced later from the data.

The (fast) Fourier transform is used to convert the signal into the frequency domain, where convolution between two signals is obtained simply multiplying them together. Once again the Fourier Transform is used as a tool in Digital Signal Processing to perform other operations, rather than producing an end result itself.

An example seismic section. This is a very interesting piece of data as it shows various geological features such as the continuous layers at the top. More importantly, at the top of the picture, to the right of the middle, there are sedimentary layers that have cut off by more recent layers. This could possible be an area of trapped hydrocarbons. The data is composed of 1000 traces each of 400 samples, and is of an area in the North Sea mapped by Shell.

In another seismic section the data contains two salt domes. The one on the left is unusual in that its size is far greater then normally encountered. The data is of an area in the North Sea mapped by BP.

Optics

In electromagnetic theory, the intensity of light is proportional to the square of the oscillating electric field which exists at any point in space. The Fourier transform of this signal is the equivalent of breaking the light into it's component parts of the spectrum, a mathematical spectrometer.

One simple example application of the fourier transform in optics is the diffraction of light when it passes through narrow slits. The ideas represented here can be equally applied to acoustic, x-ray, and microwave diffraction, or any other form of wave diffraction.

Suppose we have the experiment shown in the picture below where we have a slit in an opaque screen.

This square transmission function is an equivalent to the to a filter response such as an ideal band-pass filter. Light is only transmitted between the edges of the slit.

The Fourier transform of this gives us it's frequency content, which in turn is used to calculate the diffraction intensity at different angles from the normal. The resultant intensity is given by the square of the Fourier transform:

In other words, on the screen there is a bright region based around the slit which rapidly falls off to either side of the slit. Roughly what you would expect if you shone a light through a hole in a wall.

If the screen is altered to give us two parallel slits, with width W and distance D apart, then the transmission function is:

If we the use this, and the *Fourier transform* to calculate the diffracted intensity we get

This represents closely spaced maxima and minima (light and dark bands) that gradually decrease the further from the center of the two slits you go.

This is the *Fourier transform* used to calculate the result of a Young's double slit experiment.

APPLICATION OF FOURIER TRANSFORM TO DIFFERENTIAL EQUATIONS

Since the beginning Fourier himself was interested to find a powerful tool to be used in solving differential equations. Therefore, it is of no surprise that we discuss in this page, the application of Fourier series differential equations. We will only discuss the equations of the form

$$y^{(n)} + a_{n-1}y^{(n-1)} + \ldots\ldots + a_1 y' + a_0\, y = f(x),$$

where $f(x)$ is a 2π -periodic function.

Note that we will need the complex form of Fourier series of a periodic function. Let us define this object first:

Definition 1. Let $f(x)$ be 2π-periodic. The complex Fourier series of $f(x)$ is

$$\sum_{-\infty}^{\infty} d_n e^{in\pi x}$$

where

$$d_n = \frac{1}{2\pi}\int_{-\pi}^{\pi} f(x)\, e^{-in\pi x}\, dx$$

We will use the notation

$$f(x) \sim \sum_{-\infty}^{\infty} d_n e^{in\pi x}$$

If you wonder about the existence of a relationship between the real Fourier coefficients and the complex ones, the next theorem answers that worry.

Theoreme 1. Let $f(x)$ be 2π-periodic. Consider the real Fourier coefficients $\{a_n\}$ and $\{b_n\}$ of $f(x)$, as well as the complex Fourier coefficients $\{d_n\}$. We have

$$\begin{cases} d_0 = a_0, \\ d_n = \dfrac{1}{2}(a_n - ib_n), \text{ for } n \geq 1 \\ d_{-n} = \dfrac{1}{2}(a_n + ib_n), \text{ for } n \geq 1 \end{cases}$$

The proof is based on Euler's formula for the complex exponential function.

Remark. When $f(x)$ is $2L$-periodic, then the complex Fourier series will be defined as before where

$$d_n = \frac{1}{2c}\int_{-c}^{c} f(x)\, e^{-i(n\pi x/c)}\, dx$$

for any $n = 0, \pm 1, \pm 2 \ldots$

Example 1. Let $f(x) = x$, for $x \in]-1,1[$

and $f(x+2) = f(x)$. Find its complex Fourier coefficients $\{d_n\}$.

Solution. We have $d_0 = 0$ and

$$d_n = \frac{1}{2}\int_{-1}^{1} xe^{-in\pi x}dx$$

Easy calculations give

$$d_n = \frac{-1}{2}\left[\frac{e^{in\pi}}{in\pi} + \frac{e^{-in\pi}}{in\pi} - \frac{e^{in\pi}}{(in\pi)^2} + \frac{e^{-in\pi}}{(in\pi)^2}\right]$$

Since $e^{i\pi} = e{-}i\pi = -1$, we get $d_n = \frac{(-1)^n}{(in\pi)}$

Consequently

$$f(x) \sim \sum_{n=1}^{\infty} \frac{(-1)^{n+1}}{in\pi}\left[e^{in\pi x} - e^{-in\pi x}\right]$$

Back to our original problem. In order to apply the Fourier technique to differential equations, we will need to have a result linking the complex coefficients of a function with its derivative. We have:

Theorem 2. Let $f(x)$ be $2L$-periodic. Assume that $f(x)$ is differentiable. If

$$f(x) \sim \sum_{-\infty}^{\infty} d_n e^{i(n\pi x)/L}$$

then
$$f'(x) \sim \sum_{-\infty}^{\infty} \frac{in\pi}{c} d_n e^{i(n\pi x)/L}$$

Example 2. Find the periodic solutions of the differential equation

where $f(x)$ is a 2π-periodic function.

Solution. Set

$$f(x) \sim \sum_{-\infty}^{\infty} d_n e^{in\pi x}$$

Let y be any 2π-periodic solution of the differential equation. Assume

Then, from the differential equation, we get

$$iny_n + 2y_n = f_n, \text{ for any } n.$$

Hence
$$y_n = \frac{1}{2+in} f_n, \qquad \text{for any } n.$$

Therefore, we have

$$y = \sum_{-\infty}^{\infty} \frac{1}{2+in} f_n e^{in\pi x}$$

Example 3. Find the periodic solutions of the differential equation

$$y'' + 2y' + y = \sin(x)$$

Solution. Because

$$\sin(x) = \frac{e^{ix} - e^{-ix}}{2i}$$

we get $\begin{cases} f_1 = 1/2i \\ f_{-1} = -1/2i \\ f_n = 0 \quad \text{for } n \neq \mp 1 \end{cases}$ with

Let y be a periodic solution of the differential equation. If

$$y \sim \sum_{-\infty}^{\infty} y_n e^{in\pi x}$$

then $[(in)^2 + 2(in) + 1]\, y_n = f_n$. Hence

$$y_1 = y_{-1} = -\frac{1}{4}, \text{and } y_n = 0 \text{ otherwise.}$$

Therefore, the differential equation has only one periodic solution

$$y = -\frac{1}{4}e^{ix} - \frac{1}{4}e^{-ix} = -\frac{1}{2}\cos(x)$$

The most important result may be stated as:

Theorem 3. Consider the differential equation

$$y^{(m)} + a_{m-1}y^{(m-1)} + \ldots\ldots + a_1 y' + a_0 y = f(x)$$

where $f(x)$ is a $2c$-periodic function. Assume

$$p\left((in)/c\right) = \left(\frac{in}{c}\right)^m + a_{m-1}\left(\frac{in}{c}\right)^{m-1} + \ldots + a_1\left(\frac{in}{c}\right) + a_0 \neq 0,$$

for $n = 0, \pm 1, \pm 2 \ldots.$ Then the differential equations has one $2c$-periodic solution given by

$$y \sim \sum_{-\infty}^{\infty} \frac{1}{p\left((in)/c\right)} f_n e^{i(n\pi x)/c}$$

where $$f(x) \sim \sum_{-\infty}^{\infty} f_n e^{i(n\pi x)/c}$$

FOURIER TRANSFORM, COSINE AND SINE TRANSFORMS

If the function $f(t)$ one seeks to find the transform of is even then the exponent in the expression for the integral can be replaced with a cos. Also since the function is even then the integral for -ve t is the same as the integral for +ve t and one only needs to integrate for +ve t and multiply the result by 2. With this change the Fourier transform becomes the Cosine transform and is shown in the following two formulas.

$$g_c(\omega) = 2\int_0^{\infty} f(t)\cos\omega t dt$$

$$f(t) = \frac{1}{\pi}\int_0^{\infty} g_c(\omega)\cos\omega t d\omega$$

When the angular frequency w variable is replaced by the cyclic frequency f, the the Cosine transform is represented by the following 2 formulas.

$$g_c(f) = 2\int_0^{\infty} f(t)\cos 2\pi f t dt$$

$$f(t) = 2\int_0^{\infty} g_c(f)\cos 2\pi f t df$$

Sometimes the following definition with the same factors in front is used.

$$g_c(\omega) = \sqrt{\frac{2}{\pi}}\int_0^{\infty} f(t)\cos\omega t dt$$

$$f(t) = \sqrt{\frac{2}{\pi}}\int_0^{\infty} g_c(\omega)\cos\omega t d\omega$$

If the function f(t) is odd then a similar argument to the above produces the Sine transform.

$$g_s(\omega) = 2\int_0^{\infty} f(t)\sin\omega t dt$$

$$f(t) = \frac{1}{\pi}\int_0^{\infty} g_s(\omega)\sin\omega t d\omega$$

Expressed in terms of the variable f this becomes.

$$g_s(f) = 2\int_0^{\infty} f(t)\sin 2\pi f t dt$$

$$f(t) = 2\int_0^{\infty} g_s(f)\sin 2\pi f t df$$

Even Function

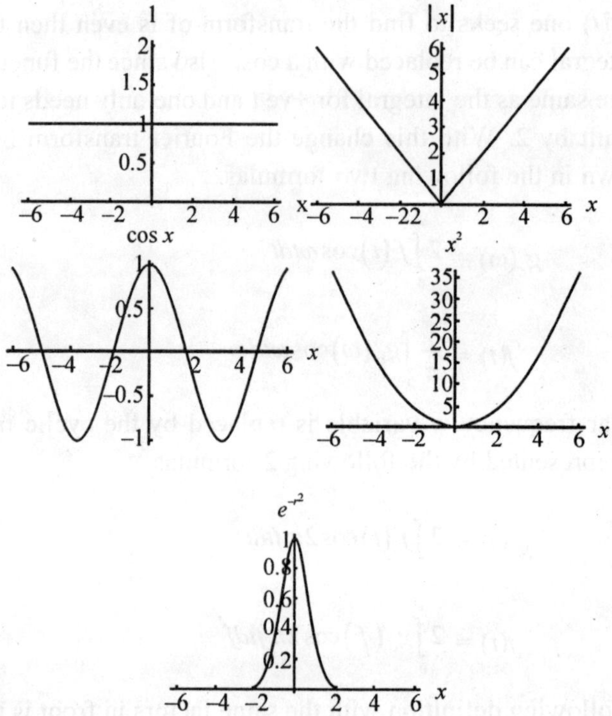

A function $f(x)$ such that $f(x)=f(-x)$. Examples of even functions include 1 (or, in general, any constant function), $|x|$, $\cos x$, x^2, and e^{-x^2}.

An even function times an odd function is odd. If an even function is differentiable, then its derivative is an odd function. Similarly, if an odd function is differentiable, then its derivative is an even function. The Maclaurin series of an even function contains only even powers.

Odd Function

Earlier we have discuss about even function. Here we discuss about odd function, which play an important role in solving various definite integral.

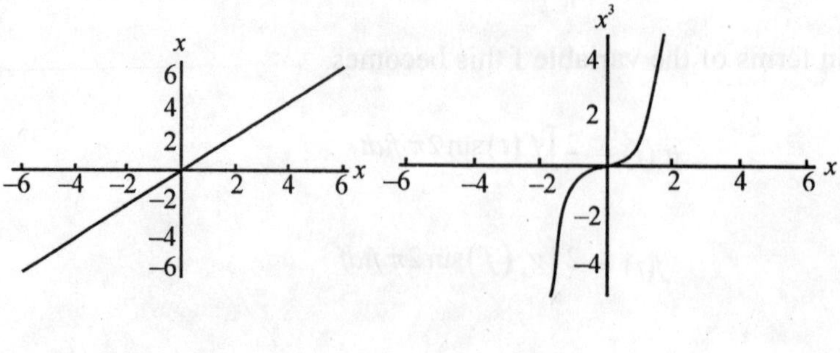

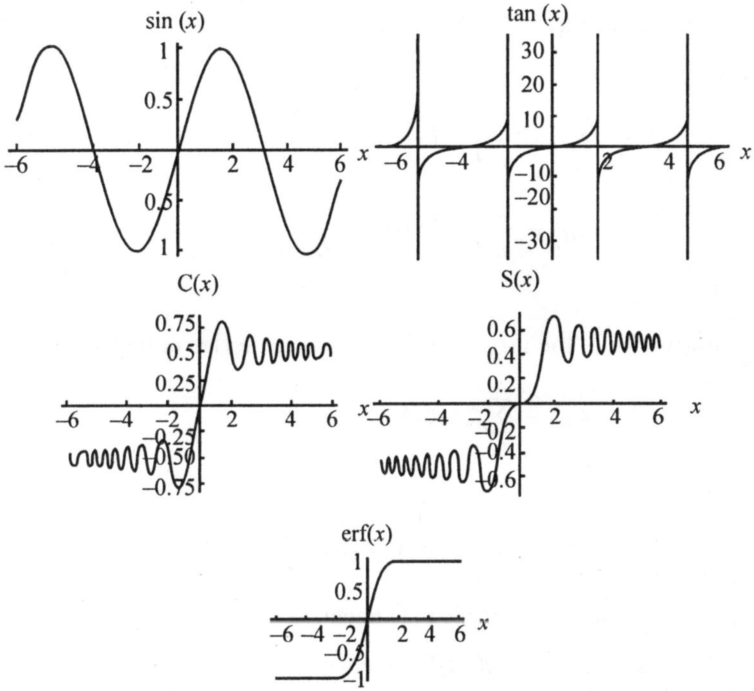

An odd function is a function for which $f(x) = -f(-x)$. Examples of odd functions include x, x^3, $\sin x$, $\tan x$, erf (x), and the Fresnel integrals $C(x)$, and $S(x)$.

An even function times an odd function is odd, and the product of two odd functions is even. If an even function is differentiable, then its derivative is an odd function. Similarly, if an odd function is differentiable, then its derivative is an even function.

Since an odd function is zero at the origin, it follows that the Maclaurin series of an odd function contains only odd powers.

Fourier Cosine Series

If $f(x)$ is an even function, then $b_n = 0$ and the Fourier series collapses to

$$f(x) = \frac{1}{2}a_0 + \sum_{x=1}^{\infty} a_n \cos(nx), \quad \text{...(1)}$$

where

$$a_0 = \frac{1}{\pi}\int_{-\pi}^{\pi} f(x)\,dx = \frac{2}{\pi}\int_0^{\pi} f(x)\,dx \quad \text{...(2)}$$

$$a_n = \frac{1}{\pi}\int_{-\pi}^{\pi} f(x)\cos(nx)\,dx \quad \text{...(3)}$$

$$= \frac{2}{\pi}\int_0^{\pi} f(x)\cos(nx)\,dx, \quad \text{...(4)}$$

where the last equality is true because

$$f(x)\cos(nx) = f(-x)\cos(-nx) \qquad ...(5)$$

Letting the range go to L,

$$a_0 = \frac{2}{L}\int_0^L f(x)\,dx \qquad ...(6)$$

$$a_n = \frac{2}{L}\int_0^L f(x)\cos\left(\frac{n\pi x}{L}\right)dx \qquad ...(7)$$

Fourier Sine Series

If $f(x)$ is an odd function, then $a_n = 0$ and the Fourier series collapses to

$$f(x) = \sum_{n=1}^{\infty} b_n \sin(nx), \qquad ...(8)$$

]where

$$b_n = \frac{1}{\pi}\int_{-\pi}^{\pi} f(x)\sin(nx)\,dx = \frac{1}{\pi}\int_0^{\pi} f(x)\sin(nx)\,dx \qquad ...(9)$$

for $n = 1, 2, 3, ...$ The last equality is true because

$$f(x)\sin(nx) = [-f(-x)]\,[-\sin(-nx)] \qquad ...(10)$$

$$= f(-x)\sin(-nx) \qquad ...(11)$$

Letting the range go to L,

$$b_n = \frac{2}{L}\int_0^L f(x)\sin\left(\frac{n\pi x}{L}\right)dx \qquad ...(12)$$

Fourier Sine Transform

In mathematics, the Fourier sine transform is a special case of the continuous Fourier transform, arising naturally when attempting to transform an odd function. Consider the general Fourier transform:

$$F(\omega) = F(f)(\omega) = \frac{1}{\sqrt{2\pi}}\int_{-\infty}^{\infty} f(t)e^{-iwt}\,dt$$

We may expand the integrand by means of Euler's formula:

$$F(\omega) = \frac{1}{\sqrt{2\pi}}\int_{-\infty}^{\infty} f(t)(\cos\omega t - i\sin\omega t)\,dt$$

or, written as the sum of two integrals:

$$F(\omega) = \frac{1}{\sqrt{2\pi}}\int_{-\infty}^{\infty} f(t)\cos\omega t\,dt - \frac{i}{\sqrt{2\pi}}\int_{-\infty}^{\infty} f(t)\,s\,\omega\,fc$$

Now notice that if we assume *f(t)* is an odd function, the product *f(t)*cosù*t* is also odd whilst the product *f(t)sinùt* is an even function. Since we are integrating over an interval symmetric about the origin (i.e. $-\infty$ to $+\infty$), the first integral must vanish to zero, and the second may be simplified to give:

$$F(\omega) = -i\sqrt{\frac{2}{\pi}}\int_0^\infty f(t)\sin\omega t\,dt,$$

which is the Fourier sine transform for odd *f(t)*. It is clear that the transformed function *F(w)* is also an odd function, and a similar analysis of the general Inverse Fourier transform yields a second sine transform, namely:

$$f(t) = i\sqrt{\frac{2}{\pi}}\int_0^\infty F(\omega)\sin\omega t\,d\omega$$

Note that the numerical factors in the transforms are defined uniquely only by their product, as discussed for general continuous Fourier transforms. For this reason the imaginary units *i* and -*i* can be omitted, with the more commonly seen forms of the Fourier sine transforms being:

$$F(\omega) = \sqrt{\frac{2}{\pi}}\int_0^\infty f(t)\sin\omega t\,dt,$$

and

$$f(t) = \sqrt{\frac{2}{\pi}}\int_0^\infty F(\omega)\sin\omega t\,d\omega$$

Fourier Cosine Transform

In mathematics, the Fourier cosine transform is a special case of the continuous Fourier transform, arising naturally when attempting to transform an even function. Consider the general Fourier transform:

$$F(\omega) = F(f)\,(t) = \frac{1}{\sqrt{2\pi}}\int_{-\infty}^{\infty} f(t)e^{-iwt}\,dt$$

$$F(\omega) = F(f)(t) = \frac{1}{\sqrt{2\pi}}\int_{-\infty}^{\infty} f(t)e^{-iwt}\,dt$$

We may expand the integrand by means of Euler's formula:

$$F(\omega) = \frac{1}{\sqrt{2\pi}}\int_{-\infty}^{\infty} f(t)(\cos\omega r - i\sin\omega t)\,dt$$

or, written as the sum of two integrals:

$$F(\omega) = \frac{1}{\sqrt{2\pi}}\int_{-\infty}^{\infty} f(f)\cos\omega t\,dt - \frac{i}{\sqrt{2\pi}}\int_{-\infty}^{\infty} f(t)\sin\omega t\,dt$$

Now notice that if we assume *f(t)* is an even function, the product *f(t)cosὺt* is also even whilst the product *f(t)sinὺt* is an odd function. Since we are integrating over an interval symmetric about the origin (i.e. $-\infty$ to $+\infty$), the second integral must vanish to zero, and the first may be simplified to give:

$$F(\omega) = \sqrt{\frac{2}{\pi}} \int_0^\infty f(t) \cos \omega t \, dt$$

which is the Fourier cosine transform for even *f(t)*. It is clear that the transformed function *F(ω)* is also an even function, and a similar analysis of the general inverse Fourier transform yields a second cosine transform, namely:

$$f(t) = \sqrt{\frac{2}{\pi}} \int_0^\infty F(\omega) \cos \omega t \, d\omega$$

Note that the numerical factors in the transforms are defined uniquely only by their product, as discussed for general continuous Fourier transforms.

EXERCISE

1. Using the definitions, prove that

   (a) $F[e^{-2\pi\alpha|x|}] = \frac{\alpha}{\pi} \frac{1}{\alpha^2 + u^2}$

   (b) $F[e^{-i\pi x^2}] = \exp\left[+i\pi u^2 - \frac{i\pi}{4}\right]$

2. Using the results of the Problem 6(b) show that

   $F[\cos \pi x^2] = \cos\left[\pi u^2 - \frac{\pi}{4}\right]$.

   and

   $F[\sin \pi x^2] = -\sin\left[\pi u^2 - \frac{\pi}{4}\right]$

3. Consider the different equation

   $\ddot{x} - \frac{\gamma}{\omega_0}\ddot{x} + \omega_0^2 x = \alpha E(t)$

   where γ, ω_0 and α are constants and dots represents differentiation with respect to time. Assuming that $\ddot{x}$, $\dot{x}$ and x vanish as $t \to \pm\infty$, show that the solution of the above equation is given by

$$x(t) = \frac{\alpha}{(2\pi)^{1/2}} \int_{-\infty}^{+\infty} \frac{g(\omega)e^{i\omega t}}{[(\omega_0^2 - \omega^2) + i\gamma\omega^3 / \omega_0]} d\omega$$

where

$$g(\omega) = \frac{1}{(2\pi)^{1/2}} \int_{-\infty}^{+\infty} E(t)e^{-i\omega t} dt$$

[Above Equation appears in the theory of the scattering of a light beam (with field strength $E(t)$ by a bound electron. The bound electron is assumed to be an oscillator such that if the electron performs only free vabrations, its motion will be periodic with frequency ω_0. The quantity $\gamma = 2e^2 \omega_0/3mc^3$, where c is the speed of light in free space, e and m are the charge and mass of the electron respectively and $\alpha = e/m$.].

4. Using the Fourier transform techniques, show that the solution of the integro-differential equation

$$\mu\frac{\partial\Psi}{\partial x} + \sum\Psi(x, \mu) = \frac{1}{2}c\sum\int_{-1}^{+1}\Psi(x,\mu)d\mu + \frac{\delta(x)}{4\pi}$$

subject to the boundary condition that

$$\Psi(x,\mu) \to 0 \text{ and } x \to \pm\infty$$

is given by

$$\Psi(x,\mu) = \frac{1}{8\pi^2}\int_{-\infty}^{+\infty}\frac{e^{i\tau x}}{\sum + i\tau\mu}\left[1 - \frac{c\Sigma}{2i\tau}\ln\frac{\Sigma + i\tau}{\Sigma - i\tau}\right]^{-1} d\tau$$

Evaluate

$$\Phi(x) = 2\pi\int_{-1}^{+1}\Psi(x,\mu)d\mu$$

In above Eq. c and Σ are constants and μ lies between -1 and $+1$.

5. Using the Fourier transform techniques show that the solution of the inhomogeneous part of the quation

$$(\nabla^2 + k^2)\Psi(r) = F(r)$$

is given by

$$\Psi(r) = \iiint G(r,r')F(r')dr'$$

where $G(r,r') = -\frac{1}{4\pi^2 i\rho}\int_{-\infty}^{+\infty} dk' \frac{k' e^{ik'\rho}}{k'^2 - k^2}; \rho = |r - r'|$

is known as the *Green's function.*

6. Using the definition

$$F[f(x)] = F(k) = \frac{1}{\sqrt{2\pi}}\int_{-\infty}^{+\infty} f(x)e^{-ikx}dx$$

Show that

$$\int_{-\infty}^{+\infty} f_1^*(k)F_2(k)dk = \int_{-\infty}^{+\infty} f_1^*(x)f_2(x)dx$$

and

$$\int_{-\infty}^{+\infty} F_1(-k)F_2(k)dk = \int_{-\infty}^{+\infty} f_1(x)f_2(x)dx$$

7. Show that

$$\frac{d^m F(k)}{dk^m} = (-i)^m[x^m f(x)]$$

Thus

$$F\left[x^m e^{-\frac{1}{2}x^2}\right] = i^m \frac{d^m}{dk^m}\left(e^{-\frac{1}{2}k^2}\right)$$

Using the above equation, prove for $n = 0, 1, 2$ and 3.
[$H_0(x) = 1$, $H_1(x) = 2x$, $H_2(x) = 4x^2 - 2$ and $H_3(x) = 8x^3 - 12x$].

Chapter 11

Introduction to Laplace Transform

Introduction

The Laplace transform is used by engineers and mathematicians as a method of solving linear differential and integral equations. The transform itself has a very close relation to fundamental economic ideas, and is a natural tool to treat a variety of economic problems involving stocks, flows, and time, including depreciation patterns, lagged adjustments, adaptive expectations, and dynamic production issues. The Laplace transform plays a role in the theory of probability distributions, where its derivatives can be interpreted as the moments of a probability distribution over the non-negative real. In statistical mechanics the Laplace transform of the indicator function for the state space is the "partition function", which summarizes the properties of the maximum entropy configuration of the system.

There are many kinds of transforms out there in the world. Laplace transforms and Fourier transforms are probably the main two kinds of transforms that are used. As we will see in later sections we can use Laplace transforms to reduce a differential equation to an algebra problem. The algebra can be messy on occasion, but it will be simpler than actually solving the differential equation directly in many cases. Laplace transforms can also be used to solve Initial-value problems that we can't use any previous method on.

For "simple" differential equations such as those in the first few sections of the last chapter Laplace transforms will be more complicated than we need. In fact, for most homogeneous differential equations such as those in the last chapter Laplace transforms is significantly longer and not so useful. Also, many of the "simple" non-homogeneous differential equations that we saw in the Undetermined Coefficients and Variation of Parameters are still simpler (or at the least no more difficult than Laplace transforms) to do as we did them there. However, at this point, the amount of work required for Laplace transforms is starting to equal the amount of work we did in those sections.

Laplace transforms comes into its own when the forcing function in the differential equation starts getting more complicated. In the previous chapter we looked only at non-homogeneous differential equations in which *g(t)* was a fairly simple continuous function. In this chapter we will start looking at $g(t)$'s that are not continuous. It is these problems were the reasons for using Laplace transforms starts to become clear.

We will also see that, for some of the more complicated non-homogeneous differential equations from the last chapter, Laplace transforms are actually easier on those problems as well.

You know, it's always a little scary when we devote a whole section just to the definition of something. Laplace transforms (or just transforms) can seem scary when we first start looking at them. However, as we will see, they aren't as bad as they may appear at first.

Before we start with the definition of the Laplace transform we need to get another definition out of the way.

The Definition

A function is called piecewise continuous on an interval if the interval can be broken into a finite number of subintervals on which the function is continuous on each open subinterval (*i.e.* the subinterval without it's endpoints) and has a finite limit at the endpoints of each subinterval. Below is a sketch of a piecewise continuous function

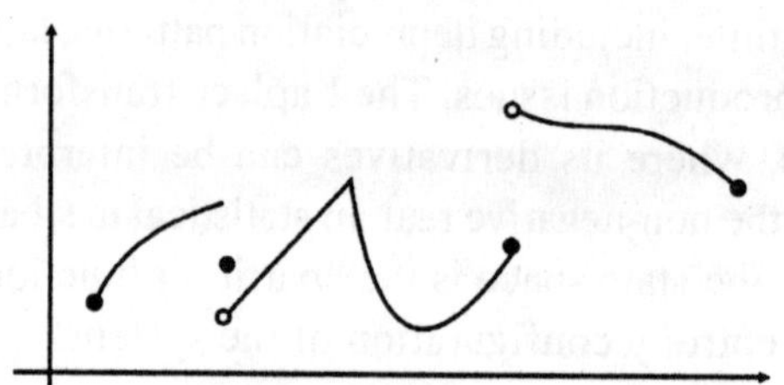

In other words, a piecewise continuous function is a function that has a finite number of breaks in it and doesn't blow up to infinity anywhere.

Now, let's take a look at the definition of the Laplace transform.

The Laplace Transform of a Function

The Laplace Transform of a function $y(t)$ is defined by

$$Y(s) = L[y(t)](s) = \int_0^\infty e^{-st} y(t)\,dt$$

if the integral exists. The notation $L[f(t)](s)$ means take the Laplace transform of $y(t)$. The functions $f(t)$ and $Y(s)$ are partner functions. Note that $F(s)$ is indeed only a function of s since the *definite* integral is with respect to t.

Suppose that $f(t)$ is a piecewise continuous function of t defined for all positive values of t. Then the Laplace transform of $f(t)$, denoted by $L\{f(t)\}$ is defined by

$$L\{f(t)\} = \int_0^\infty e^{-st} f(t)\,dt \qquad ...(1)$$

provided that the integral exists, s is a parameter which may be real or complex number.

$L\{f(t)\}$ being clearly a function of s is briefly written as $F(s)$

i.e., $L\{f(t)\}= F(s)$ ...(2)

which can be also written as

$$f(t) = L^{-1}\{F(s)\} \quad ...(3)$$

Then $f(t)$ is called as inverse Laplace transform. of $F(s)$. The symbol L, which transform $f(t)$ into $F(s)$ is called the Laplace transformation operator

Now, the integral in the definition of the transform is called an improper integral and it would probably be best to recall how these kinds of integrals work before we actually jump into computing some transforms

The so called back transformation or inverse Laplace transformation, i.e. the determination of the original function from the mapped function, is described by the *inverse integral* shown in section . For this inverse Laplace transform an operator notation in the form $f(t) = L^{-1}\{F(s)\}$ can be used.

Examples.

Let $y(t) = \exp(t)$. We have

$$Y(s) = \int_0^\infty e^{-st}e^t dt = \int_0^\infty e^{-(s-1)t}dt = \frac{1}{s-1}$$

The integral converges if s >1. The functions exp(t) and 1/(s-1) are partner functions.

Let $y(t) = \cos(3t)$. We have

$$Y(s) = \int_0^\infty e^{-st}\cos(3t)\,dt = \frac{s}{s^2+9}$$

The integral converges for s>0. The integral can be computed by doing integration by parts twice or by looking in an integration table.

Existence of the Laplace Transform

If y(t) is piecewise continuous for t>=0 and of exponential order, then the Laplace Transform exists for some values of s. A function y(t) is of exponential order c if there is exist constants M and T such that

$$|y(t)| \le Me^{ct} \qquad t \ge T$$

All polynomials, simple exponentials (exp(at), where a is a constant), sine and cosine functions, and products of these functions are of exponential order.

An example of a function not of exponential order is exp(t^2). This function grows too rapidly. The integral

$$\int_0^\infty e^{-st} e^{t^2} dt$$

does not converge for any value of s.

Rules for Computing Laplace Transforms of Functions

There are several formulas and properties of the Laplace transform which can greatly simplify calculation of the Laplace transform of functions. We summarize them below. The properties can verified using the integral formula for the Laplace Transform.

Linearity

Like differentiation and integration the Laplace transformation is a *linear* operation. What does this mean? In words, it means that the Laplace transform of a constant times a function is the constant times the Laplace transform of the function. In addition the Laplace transform of a sum of functions is the sum of the Laplace transforms.

Let us restate the above statement. Let Y_1(s) and $Y_2(s)$ denote the Laplace transforms of $y_1(t)$ and $y_2(t)$, respectively, and let c_1 be a constant. Recall that $L[f(t)](s)$ denotes the Laplace transform off(t). We have

$$L[c_1 y_1(t)](s) = c_1 L[y_1(t)](s) = c_1 Y_1(s)$$

$$L[y_1(t) + y_2(t)](s) = L[y_1(t)](s) + L[y_2(t)](s) = Y_1\,(s) + Y_2(s)$$

As a corollary, we have the third formula:

$$L[c_1 y_1(t)] + c_2 y_2(t)(s) = c_1\ Y_1(s) + c_2 Y_2(s)$$

Here are several examples:

$$L[3e^t](s) = 3L[e^t](s) = 3 + \frac{1}{s-1} = \frac{3}{s-1}$$

Here we have used the results in the table for the Laplace transform of the exponential. Here are a couple of more examples:

$$L[e^t + \cos 2t](s) = L[e^t](s) + L[\cos 2t]\,(s) = \frac{1}{s-1} + \frac{s}{s^2+4}$$

$$L[3e^t + 4\cos 2t](s) = 3L[e^t](s) + 2L\,[\cos 2t](s) = \frac{3}{s-1} + \frac{4s}{s^2+4}$$

Translation Property

The translation formula states that $Y(s)$ is the Laplace transform of $y(t)$, then

$$L[e^{at}y(t)](s) = Y(s-a),$$

where a is a constant. Here is an example. The Laplace transform of the

$$y(t) = t \text{ is } Y(s) = \frac{1}{s^2}. \text{ Hence}$$

$$L[e^{2t}t](s) = Y(s-2)\ \frac{1}{(s-2)^2}$$

Laplace Transform of the Derivative

Suppose that the Laplace transform of $y(t)$ is $Y(s)$. Then the Laplace Transform of $y'(t)$ is

$$L[y(t)](s) = sY(s) - y(0)$$

For the second derivative we have

$$L[y''(t)](s) = s^2Y(s) - sy(0) - y'(0)$$

For the n'th derivative we have

$$L[y(n)(t)](s) = s^nY(s) - s^{n-1}y(0) - s^{n-2}y'(0) - \ldots - y(n-1)(0)$$

Derivatives of the Laplace Transform

Let Y(s) be the Laplace Transform of y(t). Then

$$L[t^n y(t)](s) = (-1)^n \frac{d^nY}{ds^n}(s)$$

Here is an example. Suppose we wish to compute the Laplace transform of tsin(t). The Laplace transform of sin(t) is 1/(s^2+1).

Hence, we have

$$L[t \sin t] = -\frac{d}{ds}\left[\frac{1}{s^2+1}\right] = \frac{2s}{(s^2+1)^2}$$

Laplace Transform of the Integral

If $L\{f(t)\} = F(s)$,

Then $$L\left\{\int_0^t f(u)\,du\right\} = \frac{F(s)}{s} \text{ or } L^{-1}\left\{\frac{F(s)}{s}\right\} = \int_0^t f(u)\,du$$

Correspondences of the Laplace Transform

The Laplace transformation is an *unambiguously reversible* mapping between the original function and the mapped function. $f(t)$ and $F(s)$ are referred to as transform pairs and have a unique correspondence. This is the reason why in most cases one does not need to use the inverse integral. A *Correspondence table*, as shown in Table 2.1, will suffice. For the inverse transformation case ones goes from the right column to the left column. In addition some theorems on the Laplace transform given in the next section may be useful.

Table 1. Corresponding elements of the Laplace transform

Nr.	time response $f(t), f(t) = 0$ for $t < 0$	Laplace transformed $F(s)$
1.	δ pulse $\delta(t)$	1
2.	unit step $\sigma(t)$	$\frac{1}{s}$
3.	t	$\frac{1}{s^2}$
4.	t^2	$\frac{2}{s^3}$
5.	$\frac{t^n}{n!}$	$\frac{1}{s^{n+1}}$
6.	e^{-at}	$\frac{1}{s+a}$
7.	te^{-at}	$\frac{1}{(s+a)^2}$
8.	t^2e^{-at}	$\frac{1}{(s+a)^3}$
9.	t^ne^{-at}	$\frac{n!}{(s+a)^{n+1}}$
10.	$1 - e^{-at}$	$\frac{a}{s(s+a)}$
11.	$\frac{1}{a^2}\left(e^{-at} - 1 + at\right)$	$\frac{1}{s^2(s+a)}$
12.	$(1 - at)e^{-at}$	$\frac{s}{(s+a)^2}$
13.	$\sin w_0 t$	$\frac{w_0}{s^2 + w_0^2}$
14.	$\cos w_0 t$	$\frac{s}{s^2 + w_0^2}$
15.	$e^{-at} \sin w_0 t$	$\frac{w_0}{(s+a)^2 + w_0^2}$
16.	$e^{-at} \cos w_0 t$	$\frac{s+a}{(s+a)^2 + w_0^2}$

17.	$\frac{1}{a}f\left(\frac{t}{a}\right)$	$F(as)\ (a > 0)$
18.	$e^{at}f(t)$	$F(s-a)$
19.	$f(t-a)$ for $t>a\geq 0$ 0 for $t<a$	$e^{-as}F(s)$
20.	$-t\,f(t)$	$\frac{dF(s)}{ds}$
21.	$(-t)^n f(t)$	$\frac{d^nF(s)}{ds^n}$
22.	$f_1(t)f_2(t)$	$\frac{1}{2\pi j}\int_{c-j\infty}^{c+j\infty}F_1(p)F_2(s-p)\,dp$

Example 1. If $c \neq 0$, evaluate the following integral

$$\int_0^\infty e^{ct}\,dt$$

Solution: Remember that you need to convert improper integrals to limits as follows,

$$\int_0^\infty e^{ct}\,dt = \lim_{x\to\infty}\int_0^n e^{a}\,dt$$

Now, do the integral, then evaluate the limit.

$$\int_0^\infty e^{ct}\,dt = \lim_{x\to\infty}\int_0^n e^{ct}\,dt$$

$$= \lim_{n\to\infty}\left(\frac{1}{c}e^{ct}\right)\Bigg|_0^n$$

$$= \lim_{n\to\infty}\left(\frac{1}{c}e^{cn}-\frac{1}{c}\right) \qquad ...(4)$$

Now, at this point, we've got to be careful. The value of c will affect our answer. We've already assumed that c was non-zero, now we need to worry about the sign of c. If c is positive the exponential will go to infinity. On the other hand, if c is negative the exponential will go to zero.

So, the integral is only convergent (*i.e.* the limit exists and is finite) provided $c<0$. In this case we get,

$$\int_0^\infty e^{ct}\,dt = -\frac{1}{c} \text{ provided } c<0 \qquad ...(5)$$

Example 2. Calculate $L\{1\}$.

Solution: There's not really a whole lot do here other than plug the function $f(t) = 1$ into (1)

$$£(1) = \int_0^\infty e^{-st}\, dt$$

Now, at this point notice that this is nothing more than the integral in the previous example with $c = -s$. Therefore, all we need to do is reuse (2) with the appropriate substitution. Doing this gives,

$$£(1) = \int_0^\infty e^{-st}\, dt = -\frac{1}{-s} \qquad \text{provided } -s < 0$$

Or, with some simplification we have,

$$£(1) = \frac{1}{s} \text{ provided } s > 0$$

Notice that we had to put a restriction on s in order to actually compute the transform. All Laplace transforms will have restrictions on s. At this stage of the game, this restriction is something that we tend to ignore, but we really shouldn't ever forget that it's there.

Let's do another example

Example 3. Compute $£(e^{at})$

Solution: Plug the function into the definition of the transform and do a little simplification.

$$£(e^{at}) = \int_0^\infty e^{-st} e^{at}\, dt = \int_0^\infty e^{(a-s)t}\, dt$$

Once again, notice that we can use (2) provided $c = a - s$.

So let's do this.

$$£(e^{at}) = \int_0^\infty e^{(a-s)t}\, dt$$

$$= -\frac{1}{a-s} \qquad \text{provided } a - s < 0$$

$$= \frac{1}{s-a} \qquad \text{provided } s > a \qquad \text{...(6)}$$

Let's do one more example that doesn't come down to an application of (2).

Example 4. Compute $L\{\sin(at)\}$.

Solution: Note that we're going to leave it to you to check most of the integration here. Plug the function into the definition. This time let's also use the alternate notation.

$$£(\sin(at)) = F(s)$$

$$= \int_0^\infty e^{-st} \sin(at)\, dt$$

$$= \lim_{x\to\infty} \int_0^x e^{-st} \sin(at)\, dt$$

Now, if we integrate by parts we will arrive at,

$$F(s) = \lim_{x\to\infty}\left(-\left(\frac{1}{a}e^{-st}\cos(at)\right)\Big|_0^n - \frac{s}{a}\int_0^n e^{-st}\cos(at)\,dt\right)$$

Now, evaluate the first term to simplify it a little and integrate by parts again on the integral. Doing this arrives at,

$$F(s) = \lim_{x\to\infty}\left(\frac{1}{a}\left(1-e^{-sn}\cos(an)\right) - \frac{s}{a}\left(\left(\frac{1}{a}e^{-st}\sin(at)\right)\Big|_0^n + \right.\right.$$

$$\frac{s}{a}\int_0^n e^{-st}\sin(at)\,dt'$$

Now, evaluate the second term, take the limit and simplify.

$$F(s) = \lim_{n\to\infty}\left(\frac{1}{a}\left(1-e^{sn}\cos(an)\right) - \frac{s}{a}\left(\frac{1}{a}e^{-sn}\sin(an) + \right.\right.$$

$$\left.\left.\frac{s}{a}\int_0^n e^{-st}\sin(at)\,dt\right)\right)$$

$$= \frac{1}{a} - \frac{s}{a}\left(\frac{s}{a}\int_0^\infty e^{-st}\sin(at)\,dt\right)$$

$$= \frac{1}{a} - \frac{s^2}{a^2}\int_0^\infty e^{-st}\sin(at)\,dt$$

Now, notice that in the limits we had to assume that $s>0$ in order to do the following two limits.

$$\lim_{n\to\infty} e^{-sn}\cos(an) = 0$$

$$\lim_{n\to\infty} e^{-sn}\sin(an) = 0$$

Without this assumption, we get a divergent integral again. Also, note that when we got back to the integral we just converted the upper limit back to infinity. The reason for this is that, if you think about it, this integral is nothing more than the integral that we started with. Therefore, we now get,

$$F(s) = \frac{1}{a} - \frac{s^2}{a^2}F(s)$$

Now, simply solve for $F(s)$ to get,

$$\pounds\{\sin(at)\} = F(s) = \frac{a}{s^2+a^2} \qquad \text{provides } s > 0$$

As this example shows, computing Laplace transforms is often messy.

Before moving on to the next section, we need to do a little side note. On occasion you will see the following as the definition of the Laplace transform.

$$\pounds(f(t)) = \int_{-\infty}^{\infty} e^{-st} f(t)\,dt$$

Note the change in the lower limit from zero to negative infinity. In these cases there is almost always the assumption that the function *f(t)* is in fact defined as follows,

$$f(t) = \begin{cases} 0 & \text{if } t<0 \\ f(t) & \text{if } t\geq 0 \end{cases}$$

In other words, it is assumed that the function is zero if $t<0$. In this case the first half of the integral will drop out since the function is zero and we will get back to the definition given in (1). A Heaviside function is usually used to make the function zero for $t < 0$. We will be looking at these in a later section.

The Inverse Laplace Transform

The inverse Laplace transform of the function *Y*(*s*) is the unique

function y(t) that is continuous on [0, infty) and satisfies

L[*y*(*t*)](*s*)=*Y*(*s*). If all possible functions *y*(*t*) are discontinuous one can select a piecewise continuous function to be the inverse transform.

We will use the notation

$$L^{-1}[Y(s)](t)$$

or *Li*[*Y*(*s*)](*t*) to denote the inverse Laplace transform of *Y*(*s*).

Determining the Inverse Laplace Transform

The Laplace transform *Y*(*s*) of a function *y*(*t*) defined on [0,infty) is defined by an integral. It turns out that formula for determining *y*(*t*) given *Y*(*s*) also involves an integral. The integral is complex valued integral, and its evaluation is beyond the scope of this course. For this reason we take a more pedestrian approach in computing the inverse transform. We will use tables and a few tricks.

Table 2. Inverse Laplace Transform

Function y(t)	*Transform Y(s)*	*s*
1	1/*s*	$s > 0$
t	1/*s*^2	$s > 0$
t^*n*, *n*=integer	*n*!*s*^(*n*+1)	$s > 0$
exp(*at*), *a*=constant	1/(*s*-*a*)	$s > 0$
cos(*bt*), *b*=constant	*s*/(*s*^2+*b*^2)	$s > 0$
sin(*bt*), *b*=constant	*b*/(*s*^2+*b*^2)	$s > 0$
exp(*at*)cos(*bt*)	(*s*-*a*)/[*s*-*a*]^2+*b*^2	$s > a$

There are only small number of functions listed in the table. The set of invertible functions can be increased by exploiting linearity.

Linearity of the Inverse Transform

Let y_1(t) and y_2(t) be the inverse Laplace transforms of Y_1(s)

and Y_2(s), respectively, and let c be a constant. We have

$$L^{-1}[cY_1(s)](t) = cL^{-1}[Y_1(s)](t) = cy_1(t)$$

$$L^{-1}[Y_1(s) + Y_2(s)](t) = L^{-1}[Y_1(s)](t) + L^{-1}[Y_2(s)](t) = y_1(t) + y_2(t)$$

As a corollary, we have a third formula:

$$L^{-1}[c_1Y_1(s) + c_2\, Y_2(s)](t) = c_1L^{-1}[Y_1(s)](t) + c_2L^{-1}[Y_2(s)](t) = c_1y_1(t) + c_2y_2(t)$$

Here are several examples. What is the inverse transform of 7/s^3?

From the table above we know that the inverse of 2/s^3 is y(t)=t^2.

Hence,

$$L^{-1}[7/s^3](t) = \frac{7}{2}L^{-1}[2/s^3](t) = \frac{7}{2}t^2$$

Consider the Laplace transform Y(s)=(3s+4)/(s^2+4). Using the table above, we know that the inverse of s/(s^2+4) is cos(2t) and that the inverse of 2/(s^2+4) is sin(2t).

Hence,

$$L^{-1}\left[\frac{3s+4}{s^2+4}\right](t) = 3L^{-1}\left[\frac{s}{s^2+4}\right](t) + 2L^{-1}\left[\frac{2}{s^2+4}\right](t)$$
$$= 3\cos 2t + 2\sin 2t.$$

Corollary:

(i) $L^{-1}\left\{\frac{F(s)}{s^2}\right\} = \int_0^t\left(\int_0^t f(t)dt\right)dt$

(ii) $L^{-1}\left\{\frac{F(s)}{s^3}\right\} = \int_0^t\left(\int_0^t\left(\int_0^t f(t)dt\right)dt\right)dt$

(iii) $L^{-1}\left\{\frac{F(s)}{s^n}\right\} = \left(\int_0^t \cdots \int_0^t\left(\int_0^t\left(\int_0^t f(t)dt\right)dt\right)dt\right)dt$

The Method of Partial Fractions

Consider the Laplace transform:

$$Y(s) = \frac{2s-8}{s^2-5s+6}$$

Some manipulations must be done before Y(s) can be inverted since it does not appear directly in our table of Laplace transforms.

As we will show below:

$$Y(s) = \frac{4}{s-2} - \frac{2}{s-3}$$

Now, we can invert $Y(s)$. From the table, we see that the inverse of 1/(s-2) is exp(2t) and that inverse of 1/(s-3) is exp(3t). Using the linearity of the inverse transform, we have

$$y(t) = L^{-1}[Y(s)](t) = 4e^{2y} - 2e^{3t}$$

The method of partial fractions is a technique for decomposing functions like Y(s) above so that the inverse transform can be determined in a straightforward manner. It is applicable to functions of the form

$$Y(s) = \frac{Q(s)}{P(s)}$$

where Q(s) and P(s) are polynomials and *the degree of Q is less than the degree of P*. For simplicity we assume that Q and P have real coefficients. We consider the following cases:

- $P(s)$ is a quadratic with 2 real roots
- $P(s)$ is a quadratic with a double root
- $P(s)$ is a quadratic with complex conjugate roots
- $P(s)$ is of degree 3 and has 1 real root and 2 complex conjugate roots

P(s) is a Quadratic with 2 Real Roots

Consider the initial example on this page:

$$Y(s) = \frac{2s-8}{s^2-5s+6}$$

The denominator can be factored: s^2-5s+6=(s-2)(s-3). The denominator has 2 real roots. In this case we can write

$$\frac{2s-8}{s^2-5s+6} = \frac{A}{s-2} + \frac{B}{s-3}$$

Here A and B are numbers. This is always possible if the numerator is of degree 1 or is a constant. The goal now is to find A and B. Once we find A and B, we can invert the Laplace transform:

$$y(t) = L^{-1}[Y(s)](t) = Ae^{2t} + Be^{3t}$$

If we add the two terms on the RHS, we get

$$\frac{A}{s-2} + \frac{B}{s-3} = \frac{A(s-3)+B(s-2)}{(s-2)(s-3)} = \frac{(A+B)s+(-3A-2B)}{s^2-5s+6}$$

Comparing the LHS and RHS, we have:

$$\frac{2s-8}{s^2-5s+6} = \frac{(A+B)s+(-3A-2B)}{s^2-5s+6}$$

The denominators are equal. The two sides will be equal if the numerators are equal. The numerators are equal for all s if the coefficients of s and the constant term are equal. Hence, we get the following equations:

$$2 = A + B - 8 = -3A - 2B$$

This is a system of 2 linear equations with 2 unknowns. It can be solved by standard methods. We obtain A=4 and B= –2. Hence

$$Y(s) = \frac{4}{s-2} - \frac{2}{s-3}$$

If the denominator has n distinct roots:

$$P(s) = (s - a_1)\,(s - a_2) \ldots (s - a_n),$$

then the decomposition has the form

$$\frac{Q(s)}{P(s)} = \frac{A_1}{s-a_1} + \frac{A_2}{s-a_2} + \ldots + \frac{A_n}{s-a_n}$$

where A_1, A_2, ..., A_n are unknown numbers. The inverse transform is

$$y(t) = L^{-1}\left[\frac{Q(s)}{P(s)}\right](t) = A_1 e^{a_1 t} + A_2 e^{a_2 t} + \ldots + A_n e^{a_n t}$$

The unknown coefficients can be determined using the same technique as in the case of only 2 factors, as shown above.

P(s) is a Quadratic with a Double Root

Consider the example:

$$Y(s) = \frac{s+1}{(s+2)^2}$$

The denominator has double root –2. The appropriate decomposition in this case is

$$Y(s) = \frac{s+1}{(s+2)^2} = \frac{A}{s+2} + \frac{B}{(s+2)^2}$$

Here A and B are numbers. From the table the inverse transform of 1/(s + 2) is exp(–2t) and the inverse transform of $1/(s+2)^2$ is texp(–2t). Hence, the inverse transform of Y(s) is

$$y(t) = L^{-1}[Y(s)](t) = Ae^{-2t} + Bte^{-2t}$$

Using the same technique as in the case of distinct roots above, it can be shown that A=1 and B=–2.

P(s) is a Quadratic and has Complex Conjugate Roots

Consider the example:

$$Y(s) = \frac{3s+9}{s^2+4s+5}$$

The roots of the denominator are –2+/–i. We can complete the square for the denominator. We have

$$s^2 + 4s + 5 = s^2 + 4s + 4 + 1 = (s+2)^2 + 1$$

Hence, we have

$$Y(s) = \frac{3s+9}{(s+2)^2+1}$$

Note the denominator (s+2)^2+1 is similar to that for Laplace transforms of exp(–2t)cos(t) and exp(–2t)sin(t). We need to manipulate the numerator. Note that in the formula in the table, we have a=–2 and b=1. We look for a decomposition of the form

$$\frac{3s+9}{(s+2)^2+1} = \frac{A(s+2)+B}{(s+2)^2+1}$$

If we can find A and B, then:

$$\frac{3s+9}{(s+2)^2+1} = A\frac{s+2}{(s+2)^2+1} + B\frac{1}{(s+2)^2+1}$$

The inverse transform is

$$y(t) = L^{-1}[Y(s)](t) = Ae^{-2t}\cos t + Be^{-2t}\sin t$$

We can determine A and B by equating numerators in the expression

$$\frac{3s+9}{(s+2)^2+1} = \frac{A(s+2)+B}{(s+2)^2} = \frac{As+2A+B}{(s+2)^2+1}$$

Comparing coefficients of s in the numerator we conclude 3=*A*.

Comparing the constant terms we conclude 2*A*+*B*=9.

Hence,

$$A = 3 \quad \text{and} \quad B = 3.$$

P(s) is of Degree 3 and has 1 Real Root and 2 Complex Conjugate Roots

Consider the example:

$$Y(s) = \frac{s^2 + 9s + 3}{(s-1)(s^2 + 4s + 5)}$$

The roots of the denominator are 1 and –2+/–i. In this case we look for a decomposition of the form:

$$\frac{s^2 + 9s + 3}{(s-1)(s^2 + 4s + 5)} = \frac{A}{s-1} + \frac{Bs + C}{s^2 + 4s + 5}$$

where A, B and C are unknowns. The inverse transform of A/(s-1) is Aexp(t). Once B and C are determined the second term can be manipulated as in the previous example.

Solving Linear ODE Using Laplace Transforms

How can we use Laplace transforms to solve ode? The procedure is best illustrated with an example. Consider the ode

$$y - 5y + 6y' = 0$$

$$y(0) = 2, y'(0) = 2$$

This is a linear homogeneous ode and can be solved using standard methods.

Let $Y(s)=L[y(t)](s)$. Instead of solving directly for $y(t)$, we derive a new equation for $Y(s)$. Once we find $Y(s)$, we inverse transform to determine $y(t)$.

The first step is to take the Laplace transform of both sides of the original differential equation. We have

$$L[y'' - 5y' + y](s) = L[0](s)$$

Obviously, the Laplace transform of the function 0 is 0. If we look at the left-hand side, we have

$$L[y'' - 5y' + 6y](s) = L[y''](s) - 5L[y'](s) \;\; 6L[y](s)$$

Now use the formulas for the L[y''] and L[y']:

$$L[y'] = sL[y] - y(0) = sY(s) - 2$$

Here we have used the fact that y(0)=2. And,

$$L[y''] = s^2 L[y] - sy(0) - y'(0) = s^2Y(s) - 2s - 2$$

Hence, we have

$$L[y''](s) - 5L[y'](s) + 6L[y](s) = s^2Y(s) - 2s - 2 - 5sY(s) + 10\; 6Y(s)$$

The Laplace-transformed differential equation is

$$(s^2 - 5s + 6)Y(s) - 2s - 2 + 10 = 0$$

This is a linear algebraic equation for Y(s)! We have converted a differential equation into a algebraic equation! Solving for Y(s), we have

$$Y(s) = \frac{2s-8}{s^2-5s+6}$$

We can simplify this expression using the method of partial fractions:

$$Y(s) = \frac{4}{s-2} + \frac{-2}{s-3}$$

Recall the inverse transforms:

$$L^{-1}[1/(s-2)](t) = e^{2t}$$
$$L^{-1}[1/(s-3)](t) = e^{3t}$$

Using linearity of the inverse transform, we have

$$y(t) = L^{-1}\left[\frac{4}{s-2} + \frac{-2}{s-3}\right](t) = 4e^{2t} - 2e^{3t}$$

Another Example

Consider the ode:

$$y'' + 4y' + 5y = e^t$$
$$y(0) = 1 y'(0) = 2$$

This problem has an inhomogeneous term. In the direct approach one solves for the homogeneous solution and the particular solution separately. For this problem the particular solution can be determined using variation of parameters or the method of undetermined coefficients. Using the Laplace transform technique we can solve for the homogeneous and particular solutions at the same time.

Let $Y(s)$ be the Laplace transform of $y(t)$. Taking the Laplace transform of the differential equation we have:

$L[y + 4y + 5y] = L[10e^t]$

The Laplace transform of the LHS $L[y'''+4y'+5y]$ is

$L[y'' + 4y' + 5y] = L[y''] + 4L[y'] + 5L[y] = s2Y(s) - sy(0) - y'(0) + 4sY(s) - 4y(0) + 5Y(s)$

The Laplace transform of the RHS is

$$L[10e^t] = \frac{10}{s-1}$$

Equating the LHS and RHS and using the fact that $y(0) = 1$ $y'(0) = 2$, we obtain

$$(s^2 + 4s + 5)Y(s) - s - 6 = \frac{10}{s-1}$$

Solving for Y(s), we obtain:

$$Y(s) = \frac{s+6}{s^2+4s+5} + \frac{10}{(s-1)(s^2+4s+5)}$$

Using the method of partial fractions it can be shown that

$$Y(s) = \frac{1}{(s+2)^2+1} + \frac{1}{s-1}$$

Using the fact that the inverse of 1/(s-1) is e^t and that the inverse of 1/[(s+2)^2+1] is exp(2t)sin(t), it follows that

$$y(t) = e^{2t}\sin t + e^t.$$

EXERCISE

Find the Laplace Transform of:

1. $e^{-3.t}.\sin(5.t-4)$
2. $(t+2)^2.\cos(k.t)$

Find the Partial Fractions for the following:

3. $\dfrac{s+5}{s^3+7.s^2+16.s+12}$
4. $\dfrac{s^2+2.s+5}{s^3+7.s^2+17.s+15}$

Find the inverse Laplace Transform of:

5. $\dfrac{2.s+1}{(s+2).(s+3)\,(s^2+4.s+8)}$

Chapter 12

Application of Laplace Transform

INTRODUCTION

In this chapter we discuss some theorem on Laplace transform. Also, we discuss the application of Laplace transform.

MAIN THEOREMS OF THE LAPLACE TRANSFORM

The Laplace transform of the function f(t) is denoted by

$$L(f(t)) = \int_0^{\omega} e^{-st} f(t)\,dt \qquad ...(1)$$

(a) Superposition Theorem:

For arbitrary constants a_1 and a_2 it follows that

$$\{a_1 f_1(t) + a_2 f_2(t)\} = a_1 F_1(s) + a_2 F_2(s)\,. \qquad ...(2)$$

The Laplace transformation is a linear integral transformation.

(b) Similarity Theorem:

For an arbitrary constant a > 0

$$\{f(at)\} = \frac{1}{a} F\left(\frac{s}{a}\right) \qquad ...(3)$$

is valid. This follows from Eq. (8.1) by the substitution of $\tau = at$.

(c) Real Shifting Theorem:

For an arbitrary constant a > 0

$$\{f(t-a)\} = e^{-as} F(s) \qquad ...(4)$$

is valid. This follows directly from Eq. (1) by the substitution of $\tau = t - a$.

(d) Complex Shifting Theorem:

For an arbitrary constant a > 0

$$\{e^{-at} f(t)\} = F(s+a) \qquad ...(5)$$

is valid. This follows directly from Eq. (1).

(e) Derivative Theorem:

For a causal function of time, $f(t)$, for which the derivative for $t > 0$ exists.

$$\left\{\frac{\mathrm{d}f(t)}{\mathrm{d}t}\right\} = s\, F(s) - f(0+), \qquad ...(6)$$

and in the case of multiple differentiation

$$\left\{\frac{\mathrm{d}^n f(t)}{\mathrm{d}t^n}\right\} = s^n\ F(s) - \sum_{i=1}^{n} s^{n-i} \left.\frac{\mathrm{d}^{(i-1)} f(t)}{\mathrm{d}t^{(i-1)}}\right|_{t=0+} \qquad ...(7)$$

(f) Complex Differentiation Theorem:

This theorem shows that a differentiation of the mapped function F(s) corresponds to a multiplication with the time t in the time domain:

$$\{t^k = f(t)\} = (-1)^k \frac{d^k F(s)}{ds^k}. \qquad ...(8)$$

(g) Integral Theorem:

The integral of a function is mapped by

$$\left\{\int_o^t = f(\tau)\mathrm{d}\tau\right\} = \frac{1}{s} F(s). \qquad ...(9)$$

(h) Convolution in the Time Domain:

The convolution of two functions of time $f_1(t)$ and $f_2(t)$, presented by the symbolic notation $f_1(t) * f_2(t)$, is defined as

$$f_1(t) * f_2(t) = \int_o^t f_1(\tau)\ f_2(t-\tau)\mathrm{d}\tau. \qquad ...(10)$$

In the previous section it is shown that the convolution of the two original functions corresponds to the multiplication of the related mapped functions, that is

$$\{f_1(t) * f_2(t)\} = F_1(s)\ F_2(s) \qquad ...(11)$$

(i) Convolution in the Frequency Domain:

Whereas in (h) the convolution of two functions of time was given, a similar result for the convolution of two functions in the frequency domain exists and is given by

$$\{f_1(t)\ f_2(t)\} = \frac{1}{2\pi j} \int_{c-joo}^{c+joo} F_1(p)\ F_2(s-p)\,dp \qquad ...(12)$$

Here $F_1(s)$ •–o $f_1(t)$ and $F_2(s)$ •–o $f_2(t)$ is valid. Furthermore, p is the complex variable of integration. According to this theorem the Laplace transform of the product of two functions of time is equal to the convolution of $F_1(s)$ and $F_2(s)$ in the mapped domain.

(j) Initial and Final Value Theorems:

The *theorem of the initial condition* allows the direct calculation of the function value $f(0+)$ of a causal function of time $f(t)$ from the Laplace transform $F(s)$. If the Laplace transform of $f(t)$ and $f(t)$ exist, then

$$f(0+) = \lim_{t\to 0+} f(t) = \lim_{s\to\infty} s\ F(s) \qquad ...(13)$$

is valid if the $\lim_{t\to 0} f(t)$ exists.

Using the *theorem of the final value* the value of $f(t)$ for $t\to\infty$ can be determined from $F(s)$, if the Laplace transform of $f(t)$ and $f(t)$ exist and the limit $\lim_{t\to\infty} f(t)$ also exists. Then it follows from previous section that

$$f(\infty) = \lim_{t\to\infty} f(t) = \lim_{t\to 0} s\ F(s) \qquad ...(14)$$

One has to observe that

$$\lim_{t\to\infty} f(t) \text{ or } \lim_{t\to 0} f(t)$$

can be calculated only from the corresponding Laplace transform $\{f(t)\}$ by application of the theorems of the initial or final value, if the existence of the related limit in the time domain is a priori assured. The following two examples should explain this:

Example 1.

$$f(t) = e^{at}\ (\alpha > 0) \bullet\!\!-\!\!\circ F(s) = \frac{1}{s-\alpha}$$

The limit $\lim_{t\to\infty} e^{\alpha t}$ does not exist so that the final value theorem may not be applied.

Example 2.

$$f(t) = \cos w_0 t \bullet\!\!-\!\!\circ F(s) = \frac{s}{s^2 + w_0^2}$$

The limit $\lim_{t \to \infty} \cos w_0 t$ does not exist and therefore the final value theorem may not be applied.

It can be concluded from the last two examples that the following general statement is valid: If the Laplace transform $F(s)$ has, apart from a single pole at the origin $s = 0$, poles on the imaginary axis or in the right-half plane, then the initial or final value theorems cannot be applied.

THE INVERSE LAPLACE TRANSFORM

As already mentioned in previous section in many cases a direct evaluation of the complex inverse integral is not necessary. A complicated function, $F(s)$, must be decomposed into a sum of simple functions of s that is

$$F(s) = F_1(s) + F_2(s) + ... + F_n(s), \qquad ...(15)$$

which have a known inverse Laplace transform:

$$\begin{aligned} L^{-1}\{F(s)\} &= L^{-1}\{F_1(s)\} + L^{-1}\{F_2(s)\} + ... + L^{-1}\{F_3(s)\} \\ &= f_1(t) + f_2(t)_n(t) = f(t). \end{aligned} \qquad ...(16)$$

For many problems in control the function $F(s)$ is a ratio of polynomials in s, known as *rational fraction,* that is

$$\mathrm{F}(s) = \frac{n_0 + n_1 s + ... n_m s^m}{d_0 + d_1 s + ... + s^n} = \frac{N(s)}{D(s)}, \qquad ...(17)$$

where $N(s)$ and $D(s)$ are the numerator and the denominator, respectively.

If $m > n$, then $N(s)$ is divided by $D(s)$, where a polynomial in s and a ratio of polynomials are obtained. The numerator of the fraction $N_1(s)$ has a lower order than n. E.g, if $m = n + 2$, then

$$\frac{N(s)}{D(s)} = k_2 s^2 + k_1 s + k_0 + \frac{N_1(s)}{D(s)}, \qquad ...(18)$$

where by degree $N_1(s) < n$ and k_0, k_1 and k_2 are constants.

A rational fraction $F(s)$ given in Eq. (17) can be decomposed into more simple functions by application of *partial fraction decomposition*, as shown in Eq. (15). In order to perform this decomposition the denominator polynomial $D(s)$ must be factorised into the form

$$F(s) = \frac{N(s)}{(s-s-1)\,(s-s_2)...(s-s_n)}. \qquad ...(19)$$

For a denominator polynomial of n -th order one obtains roots or zeros $s = s_1, s_2, ..., s_n$. The zeros of $D(s)$ are also known as the *poles* of $F(s)$, since they define where $F(^s)$ is infinite. The partial fraction decomposition for different types of poles is shown in the following.

Case 1: $F(s)$ has only *single poles.*

Here $F(s)$ can be expanded into the form

$$F(s) = \sum_{k=1}^{n} \frac{c_k}{s - s_k}, \qquad ...(20)$$

where the *residuals* c_k are real or complex constants. Using the table of correspondences one immediately can obtain the corresponding function of time

$$F(t) = \sum_{k=1}^{n} c_k e^{s_k t} \quad \text{for } t > 0. \qquad ...(21)$$

The values c_k can be determined either by comparing the coefficients or by using the theorem of residuals from the theory of functions according to

$$c_k = \frac{N(s_k)}{D'(s_k)} = (s - s_k)\frac{N(s)}{D(s)}\bigg|_{s=s_k}. \qquad ...(22)$$

for $k = 1, 2, ..., n$ with $D'(s_k) = \mathrm{d}\,D/\mathrm{d}s\,|_{s=s_k}$

Case 2: $F(s)$ has *multiple poles.*

For multiple poles of $F(s)$ each with multiplicity $r_k(k = 1, 2, ..., l)$ the corresponding partial fraction decomposition is

$$F(s) = \sum_{k=1}^{l} \sum_{v=1}^{rk} \frac{c_{kv}}{(s - s_k)^v} \quad \text{with } n = \sum_{k=1}^{l} r_k \qquad ...(23)$$

The back transformation of Eq. (8.23) into the time domain is

$$f(t) = \sum_{k=1}^{l} e^{s_k t} \sum_{v=1}^{rk} \frac{c_{kv} t^{v-1}}{(v-1)!} \text{ for } t > 0 \qquad ...(24)$$

The real or complex coefficients $c_k v$ for $v = 1, 2, ... , r_k$ determined by the theorem of residuals are

$$c_{kv} = \frac{1}{(r_k - v)!}\left\{\frac{d(r_k - v)}{ds(r_k - v)!}[F(s)\,(s - s_k)^{r_k}]\right\}_{s=s_k} \qquad ...(25)$$

This general relation also contains the case of single poles of $F(s)$. The poles may be real or complex.

Case 3: *F*(*s*) has also *conjugate complex poles.*

As both, the numerator $N(s)$ and the denominator $D(s)$ of the function $F(s)$ are rational algebraic functions, complex factors always arise as conjugate complex pairs. If $F(s)$ has a conjugate complex pair of poles $s_1, 2 = \sigma_1 \pm jw_1$, then for the function $F_1,2(s)$ in the partial fraction decomposition of

$$F(s) = \frac{N(s)}{D(s)} = F_{1,2}(s) + F_3(s) + ... + F_n(s)$$

Eq. (20) can be applied to give

$$F_{1,2}(s) = \frac{c_1}{s - (\sigma + jw_1)} + \frac{c_2}{s - (\sigma_1 - jw_1)}, \qquad ...(26)$$

where the residuals

$$c_{1,2} = \delta_1 \pm j\varepsilon_1$$

are also a conjugate complex pair. Therefore, both fractions of $F_{1,2}(s)$ can be combined, and one obtains

$$F_{1,2}(s) = \frac{\beta_0 + \beta_1 s}{\alpha_0 + \alpha_1 s + s^2} \qquad ...(27)$$

with the real coefficients

$$\left.\begin{array}{ll} \alpha_0 = \alpha_1^2 + w_1^2 & ; \alpha_1 = -2\sigma_1 \\ \beta_0 = -2(\sigma_2\beta_1 + w_1\varepsilon_1) & ; \beta_1 = 2\delta_1 \end{array}\right\}. \qquad ...(28)$$

The determination of the coefficients β_0 and β_1 is performed again using the theorem of residuals by

$$(\beta_0 + \beta_1 s)\,|_{s=s_1} = (s - s_1)\,(s - s_2)\left.\frac{N(s)}{D(s)}\right|_{s=s_1} \qquad ...(29)$$

As s_1 is complex, both sides of this equation are complex. Comparing the real and imaginary parts of both sides one gets two equations for the calculation of β_0 and β_1. This procedure is demonstrated now using the following example.

Example 3. Find the inverse Laplace transform $f(t)$ of

$$F(s)=\frac{1}{(s^2+2s+2)(s+2)}$$

The partial fraction decomposition of $F(s)$ is

$$F(s)=F_{1,2}(s)+F_3(s)=\frac{\beta_0+\beta_1 s}{s^2+2s+2}+\frac{c_3}{s+2},$$

where the function $F_1,2(s)$ contains the conjugate pair of poles

$s_1,2=\pm$ j.

In addition the third pole of $F(s)$ is

$$s_3=-2$$

For the coefficients β_0 and β_1 it follows from Eq. (29)

$$(\beta_0+\beta_1 s)\,|_{s=s_1}=\left.\frac{1}{s+2}\right|_{s=s_1}$$

$$(\beta_0+\beta_1)+\mathrm{j}\beta_1=\frac{1}{-1+\mathrm{j}+2}=\frac{1}{2}-\mathrm{j}\frac{1}{2}.$$

Comparing the real and imaginary parts on both sides one obtains

$$\beta_0-\beta_1=\frac{1}{2}\ \text{and}\ \beta_1=-\frac{1}{2}$$

and from this finally $\beta_0=0$.

Using Eq. (22) the residual is

$$c_3=(s+2)\left.\frac{1}{(s^2+2s+2)}\right|_{s=s_3}=\frac{1}{2}.$$

The partial fraction decomposition of $F(s)$ is thus

$$F(s)=-\frac{1}{2}\left[\frac{s}{s^2+2s+2}\right]+\frac{1}{2}\frac{1}{s+2},$$

which can be rearranged in the form

$$F(s)=-\frac{1}{2}\left[\frac{s+1}{(s+1)^2+1}-\frac{1}{(s+1)^2+1}-\frac{1}{s+2}\right]$$

which can be directly applied to find the inverse transformation.

$$f(t) = -\frac{1}{2}[e^{-t}\cos t - e^{-t}\sin t - e^{-2t}] \text{ for } t > 0,$$

which can be rearranged as

$$f(t) = \frac{1}{2}e^{-t}[e^{-t} + \sin t - \cos t] \text{ for } t > 0.$$

The graphical representation of $f(t)$ is shown in Figure 2.1a. Figure 8.1b shows the corresponding poles, marked by a x, for this $F(s)$ in the complex s plane.

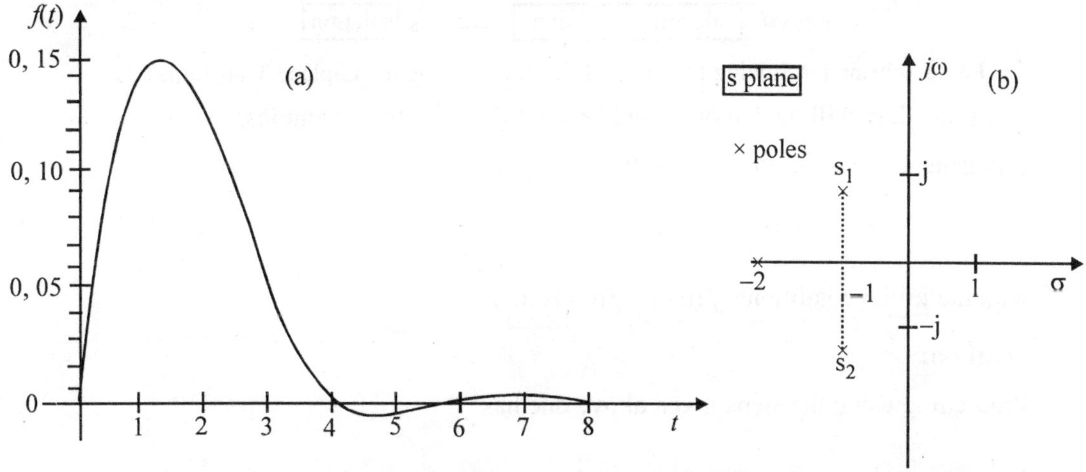

Fig. 1 (a) Graph of the Original Function $f(t)$ (Function in the Time Domain) and (b) Position of the Poles of $F(s)$ in the s Plane

It can be seen from this example that the position of the poles s_1, s_2 and s_3 affects the shape of the graph of $f(t)$. In this case all poles of $F(s)$ have negative real parts, therefore the graph of $f(t)$ shows a damped behaviour, i.e. it decreases to zero for $t \to \infty$. If the real part of one pole be positive, then the graph of $f(t)$ would be infinitely large for $t \to \infty$.

Since in control problems the original function $f(t)$ always represents the time behaviour of a system variable, the behaviour of this system variable $f(t)$ can be judged to a large extent by investigation of the positions of the poles of the corresponding mapped function $F(s)$.

SOLVING LINEAR DIFFERENTIAL EQUATIONS USING THE LAPLACE TRANSFORM

The Laplace transform, the basics of which have been introduced in the sections above, is an elegant way for fast and schematic solving of linear differential equations with constant coefficients. In the following the importance of this approach is demonstrated. Instead of solving the differential equation with the initial conditions directly in the original domain, the detour via a mapping into the frequency domain is taken, where only an algebraic equation has to be solved. Thus solving differential equations is performed according to Figure 2 in the following three steps:

Transformation of the differential equation into the mapped space,
Solving the algebraic equation in the mapped space,
Back transformation of the solution into the original space.

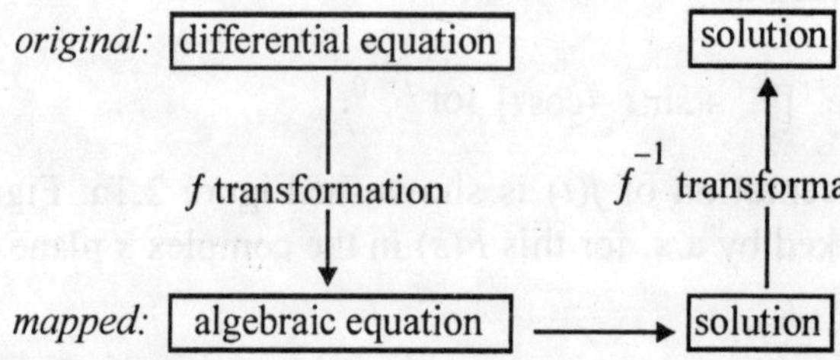

Fig. 2 Schema for Solving Differential Equations Using the Laplace Transformation

The procedure will be demonstrated by the following two examples.

Example 4. Consider the differential equation

$$\ddot{f}(t)+3\dot{f}(t)+2f(t)=e^{-t}$$

with the initial conditions $f(0+)=\dot{f}(0+)=0$

Solution:

Proceeding using the steps given above one has

Step 1: $s^2F(s)+3sF(s)+2F(s)=\dfrac{1}{s+1}$

Step 2: $F(s)=\dfrac{1}{s+1}\,\dfrac{1}{s^2+3s+2}$

Step 3: The complex function $F(s)$ must be decomposed into partial fractions in order to use the tables of correspondences. This gives

$$F(s)=\frac{1}{s+1}-\frac{1}{s+1}+\frac{1}{(s+1)^2}.$$

By using the inverse Laplace transformation that the solution of the given differential equation is

$$f(t)=e^{-2t}-e^{-t}+te^{-t}.$$

Example 5. Given the differential equation

$$\ddot{x}+a_1\dot{x}+a_0x=0\,, \qquad ...(30)$$

where a_0 and a_1 are constants and the initial conditions $\dot{x}(0+)$ and $x(0+)$ are known. Then

Solution:

Step 1: $s^2 X(s) - s\, x(0+) - \dot{x}(0+) + a_2[s\, X(s) - x(0+)] + a_0\, X(s) = 0$

Step 2: $X(s) = \dfrac{s + a_1}{s^2 + a_1 s + a_0} x(0+) + \dfrac{1}{s^2 + a_1 s + a_0} \dot{x}(0+)$, ...(31)

$$X(s) = L_0(s)\, x(0+) + L(s)\, \dot{x}(0+)$$

with the abbreviation

$$L_0(s) = \frac{N_0(s)}{D(s)} = \frac{s + a_1}{s^2 + a_1 s + a_0} \text{ and } L(s) = \frac{N(s)}{D(s)} = \frac{1}{s^2 + a_1 s + a_0}.$$

Step 3: Case (a): two *single real zeros of the denominator:*

This means

$$D(s) = s^2 + a_1 s + a_0 = (s - \alpha_1)(s - \alpha_2).$$

For both rational expressions $L_0(s)$ and $L(s)$ it follows by partial fraction decomposition that

$$L_0(s) = \frac{A_1}{s - \alpha_1} + \frac{A_2}{s - \alpha_2} \text{ and } L(s) = \frac{B_1}{s - \alpha_1} + \frac{B_2}{s - \alpha_2}.$$

The coefficients A_i and B_i can now be determined by comparing coefficients or by applying Eq. (22):

$$A_i = \frac{N_0(\alpha_i)}{D'(\alpha_i)};\ B_i = \frac{N(\alpha_i)}{D'(\alpha_i)} \text{ for } i = 1, 2.$$

Thus for Eq. (8.31) follows

$$X(s) = \left[\frac{A_1}{s - \alpha_1} + \frac{A_2}{s - \alpha_2}\right] x(0+) + \left[\frac{B_1}{s - \alpha_1} + \frac{B_2}{s - \alpha_2}\right] \dot{x}(0+),$$

and by applying the formula the solution of the differential equation is

$$x(t) = [A_1 e^{\alpha_1 t} + A_2 e^{\alpha_2 t}]\, x(0+) + [B_1 e^{\alpha_1 t} + B_2 e^{\alpha_2 t}]\, \dot{x}(0+)$$

$$= [A_1 x(0+) + B_i \dot{x}(0+)] e^{\alpha_1 t} + [A_2\, x(0+) + B_2 \dot{x}(0+)] e^{\alpha_2 t}. \quad ...(32)$$

Case (b): One *double real zero of the denominator:*

Here is $D(s) = (s - \alpha)^2$

For the two rational expressions $L0(s)$ and $L(s)$ of Eq. (31) the partial fraction decomposition is

$$L_0(s) = \frac{A_1}{s-\alpha} + \frac{A_2}{(s-\alpha)^2} \text{ and } L(s) = \frac{B_1}{s-\alpha} + \frac{B_2}{(s-\alpha)^2}.$$

The coefficients A_i and B_i are determined by comparing both sides or by evaluation of Eq. (25):

$$A_1 = \left\{\frac{d}{ds}\left[\frac{N_0(s)}{D(s)}(s-\alpha)^2\right]\right\}_{s=\alpha} = 1, \; A_2 = \left[\frac{N_0(s)}{D(s)}(s-\alpha)^2\right]_{s=\alpha} = \alpha + \alpha_1$$

and $$B_1 = \left\{\frac{d}{ds}\left[\frac{N(s)}{D(s)}(s-\alpha)^2\right]\right\}_{s=\alpha} = 0, \; B_2 = \left[\frac{N(s)}{D(s)}(s-\alpha)^2\right]_{s=\alpha} = 1.$$

From these results one obtains the solution

$$X(s) = \frac{x(0+)}{s-\alpha} + \frac{(\alpha + a_1)\, x(0+) + \dot{x}(0+)}{(s-\alpha)^2}$$

in the mapped space. By applying the inverse Laplace transformation the required solution of the differential equation is

$$x(t) = x(0+)e^{\alpha t} + [(\alpha + \alpha_1)\, x(0+) + \dot{x}(0+)]\, te^{\alpha t}. \qquad ...(33)$$

Case (c): Two *conjugate complex zeroes of the denominator:*
Here

$$D(s) = (s-\alpha_1)(s-\alpha_2) \text{ with } \alpha_{1,2} = \sigma_1 \pm jw_1.$$

Introducing the values of α_1 and α_2 and after multiplication of this expression one obtains

$$D(s) = (s - \sigma_1)^2 + w_1^2.$$

Comparison with the denominator of the original relation, Eq. (31), gives according to Eq. (28)

$$a_0 = \sigma_1^2 + w_1^2 \text{ and } a_1 = -2\sigma_1.$$

With these coefficients Eq. (31) is in the form

$$X(s) = \frac{s - 2\sigma_1}{(\dot{s}-\sigma_1)^2 + w_1^2} x(0+) + \frac{1}{(s-\sigma_1)^2 + w_1^2} \dot{x}(0+)$$

$$= \left[\frac{s-\sigma_1}{(s-\sigma_1)^2 + w_1^2} - \frac{\sigma_1}{w_1}\frac{w_1}{(s-\sigma_1)^2 + w_1^2}\right] x(0+) + \frac{1}{w_1}\frac{w_1}{(s-\sigma_1)^2 + w_1^2} \dot{x}(0+),$$

and from this one gets $X(s)$ the corresponding time function

$$x(t) = e^{\sigma_1 t}\left[\cos w_1 t - \frac{\sigma_1}{w_1}\sin w_1 t\right]x(0+) + \frac{1}{w_1}e^{\sigma_1 t}\dot{x}(0+)\sin w_1 t \text{ or rearranged}$$

$$x(t) = e^{\sigma_1 t}\left\{x(0+)\cos w_1 t + \left[\frac{1}{w_1}\dot{x}(0+) - \frac{\sigma_1}{w_1}x(0+)\right]\sin w_1 t\right\}. \qquad ...(34)$$

Also from Eq. (31) of this example the importance of the position of the zeros of $D(s)$, the poles of $X(s)$, on the solution is clear. For all three cases the solution of the differential equation according to Eqs. (32), (33) and (34) is mainly influenced by the position of the poles of $X(s)$. These poles of $X(s)$ are - as one can see from the two examples - only depend on the left side of the corresponding differential equation, i.e. the homogeneous part of it. As is generally known the solution of the homogeneous differential equation describes the *modes* of the system, that is the behaviour, which depends only on the initial conditions. Therefore, consider for the general case only the homogeneous part of an nth-order ordinary homogeneous linear differential equation with constant coefficients that is

$$\sum_{i=0}^{n} a_i \frac{\mathrm{d}^i x_a(t)}{\mathrm{d}t^i} = 0 \qquad (35)$$

with all $n$initial conditions $(\mathrm{d}^i x_a(t)/\mathrm{d}t^i)|_{t=0+}$ for $i = 0, 1, ..., n - 1$.

One obtains by Laplace transformation

$$X_a(s)\sum_{i=0}^{n} a_i s^i - \left[\sum_{i=1}^{n} a_i \sum_{\nu=1}^{i} s^{i-\nu}\frac{\mathrm{d}^{\nu-1}}{\mathrm{d}t^{\nu-1}}x_a(t)|_{t=0+}\right] = 0$$

and a form according to Eq. (8.31)

$$X_a(s) = \frac{\sum_{i=1}^{n} a_i \sum_{\nu=1}^{i} s^{i-\nu}\frac{\mathrm{d}^{\nu-1}}{\mathrm{d}t^{\nu-1}}x_a(t)|_{t=0+}}{\sum_{i=0}^{n} a_i s^i} = \frac{N(s)}{D(s)}, \qquad ...(36)$$

where $N(s)$ and $D(s)$ are polynomials in s and the initial conditions are only in the numerator polynomial $N(s)$. The poles $s_k(k = 1, 2, ... , n)$ of $X_a(s)$ can be determined directly from the solution of the equation

$$\sum_{i=1}^{n} a_i s^i = 0 \qquad ...(37)$$

After factorisation of this equation one obtains

$$a_n(s - s_1)(s - s_2) \dots (s - s_n) = 0. \qquad \dots(38)$$

The poles s_k of $X_a(s)$ make it possible to perform a partial fraction decomposition of $X_a(s)$, e.g. for the case of single poles according to Eq. (20). For this case one obtains following Eq. (21) the solution of the homogeneous differential equation, Eq. (35), in the form

$$x_a(t) = \sum_{k=1}^{n} c_k e^{s_k t} \quad \text{for } t > 0.$$

From this one can realise that the position of the poles s_k of $X_a(s)$ in the plane completely characterises the modes or inherent behaviour of the system described by Eq. (35). Thus one obtains for Re $s_k < 0$ (left-half s plane) a decreasing and for Re $s_k > 0$ (right-half s plane) an increasing behaviour of $x_a(t)$, while for pairs of poles with Re $s_k = 0$ permanent oscillations occur. Therefore, Eq. (37) or equivalently Eq. (38), is called the *characteristic equation* and the poles s_k of $X_a(s)$ are often called *eigenvalues* of the equation. Therefore investigation of the characteristic equation provides the most important information about the oscillating behaviour of a system.

Laplace transform of the impulse $\delta(t)$ and step $\sigma(t)$

The impulse function $\delta(t)$ is *not a function* in the sense of classical analysis, but a distribution (pseudo-function). Therefore without entering the theory of distributions the integral

$$\{\delta(t)\} = \int_0^{\infty} \delta(t) e^{-st} \, dt \qquad \dots(39)$$

is not defined. The singularity exactly matches with the lower integration limit. The impulse function can be approximately described by the limit

$$\delta(t) = \lim_{e \to 0} r_e(t)$$

with the rectangular impulse function

$$r_e = \begin{cases} 1/\varepsilon & \text{for } 0 \le t \le \varepsilon \\ 0 & \text{otherwise.} \end{cases} \qquad \dots(40)$$

Strictly speaking this representation of $\delta(t)$ is not a distribution, as $r_e(t)$ for $0 \le t \le \infty$ is not arbitrarily often differentiable. Because of the simple description compared with other functions (e.g. Gaussian functions) this approach is preferred here. From Eq. (8.39) it follows that

$$\{\delta(t)\} = \left[\int_0^{\infty} \lim_{e \to 0} r_e(t) \right] e^{-st} \, dt. \qquad \dots(41)$$

As Eq. (8.40) can also be represented in the form

$$r_e(t) = \frac{1}{\varepsilon}[\sigma(t) - \sigma(t - \varepsilon)], \qquad ...(42)$$

where $\sigma(t)$ is the unit step. Since the integration is independent of ε , the limit and integration can be permuted so that

$$\{\delta(t)\} = \lim_{e \to 0} \left\{ \frac{1}{\varepsilon} \int_0^{\infty} [\sigma(t) - \sigma(t - \varepsilon)]\, e^{-st} dt \right\}$$

$$\{\delta(t)\} = \lim_{e \to 0} \left\{ \frac{1}{\varepsilon} \frac{1}{s} (1 - e^{-\varepsilon s}) \right\}.$$

By applying l' Hospital's rule one obtains

$$\{\delta(t)\} = \lim_{e \to 0} \frac{se^{-\varepsilon s}}{s} = 1. \qquad ...(43)$$

As the impulse $\delta(t)$ has an area of unity it is also called unit impulse.

The solution can be determined in the following three steps:

Step 1: The Laplace transform of the given differential equation is:

$$sY(s) - y(0-) = 1 \text{ with } y(0-) = 0.$$

Step 2: The solution of the algebraic equation is:

$$Y(s) = \frac{1}{s}$$

Step 3: From the back transformation the solution follows as

$$y(t) = \sigma(t)$$

where $\sigma(t)$ is the unit step function.

TRANSFER FUNCTIONS

Transfer functions simplify working with linear differential equations, which describe the transition behavior of linear dynamical systems. As transfer functions contain at least a rational function of a complex variable, the physical interpretation of poles and zeros is discussed. Finally, the most popular operations of combining systems described by transfer functions are presented. The properties of transfer functions are shown by examples and interactive questions allow the readers to test their knowledge of transfer functions.

Definition

Linear, continuous-time time-invariant systems with lumped parameters - as initially a dead time is not being taken into account - will be described by the ordinary differential equation

$$\sum_{i=0}^{n} a_i \frac{d^i x_{3.}(t)}{dt^i} = \sum_{j=0}^{m} bj \frac{djx_e(t)}{dt^j}. \quad ...(44)$$

If all *initial conditions* are set to *zero* and the Laplace transformation is applied to both sides of this equation, one obtains

$$X_{3.}(s)\sum_{i=0}^{n} a_i s^i = X_e(s)\sum_{j=0}^{m} b_j s^j,$$

or reordering

$$\frac{X_{3.}(s)}{X_e(s)} = \frac{b_0 + b_1 s + ... + b_m s^m}{a_0 + a_1 a + ... + a_n s^n} = G(s) = \frac{N(s)}{D(s)}, \quad ...(45)$$

where $N(s)$ and $D(s)$ describe the numerator and denominator polynomials, respectively. The quotient of the Laplace-transformed output and input of such a type of system is a rational fraction. The coefficients of this fraction depend only on the structure and parameters of the system. Such a type of function $G(s)$, which describes completely the transfer behaviour of a system, is called the *transfer function* of the system. With such a transfer function the output

$$X_{3.}(s) = G(s)X_e(s) \quad ...(46)$$

INTERPRETATION OF THE TRANSFER FUNCTION

Comparing Eq. (46) with the convolution theorem, Eqs. (10) and (11), it then follows for the representation of Eq. (46) in the time domain that

$$x_{3.}(t) = \int_0^t g(t-r)x_e(\tau)\, d\tau, \quad ...(47)$$

where obviously the inverse Laplace transform of $G(s)$ is the function $g(t)$. This function is generally known as the *weighting function* of the system. In other words, the transfer function is the Laplace-transformed weighting function according to

$$G(s) = \{g(t)\}. \quad ...(48)$$

If the unit impulse $\delta(t)$ is taken as the input signal $x_3.(t)$ for a system described by the transfer function $G(s)$, one obtains according to Eq. (46)

$$X_{3.}(s) = G(s)\ \{\delta(t)\}$$

and after using the Laplace transform of the unit impulse $\delta(t)$ from Eq. (46)

$$X_{3.}(s) = G(s)$$

or from Eq. (8.48)

$$x_{3.}(t) = g(t).$$

This shows that the response to an unit impulse $\delta(t)$ is the weighting function. Therefore the weighting function is also called the *impulse response.*

Another interpretation is when the system is excited by the input signal

$$x_e(t) = \bar{x}_e e^{\sigma t} \sin(wt).$$

For the steady-state case one obtains for the output signal

$$x_{35}(t) = |G(\sigma + jw)| \bar{x}_e e^{\sigma t} \sin(wt + \phi(\sigma + jw)). \qquad ...(49)$$

This shows that the modulus $|G(\sigma + jw)|$ of the transfer function describes the gain, and that $\phi(\sigma + jw) = \arg|G(\sigma + jw)$ describes the phase shift of a sinusoidal function with the frequency w and with increasing or decreasing amplitude according to $e^{\sigma t}$ can be immediately calculated for a known input signal $x_e(t)$, and therefore $X_e(s)$

REALISABILITY AND PROPERNESS OF TRANSFER FUNCTIONS

It must be mentioned that a transfer function with $m > n$ is physically not realisable. The transfer function of an ideal differentiator is described by $G(s) = s$ according to Eq. (8.6). Any transfer function with $m > n$ can be decomposed into

$$G(s) = \frac{N(s)}{D(s)} = \frac{N_1(s)}{D(s)} + k_0 + k_1 s + ... + k_{m-m} s^{n-n},$$

where degree $N_1(s) = n - 1$ and terms in s with positive powers also occur. Such derivative elements would deliver for input signals of arbitrary high frequency corresponding output signals of arbitrary high amplitude, which are physically not realisable. The *condition of realisability* of the transfer function according to Eq. (8.45) is

$$\text{degree } N(s) \leq \text{degree } D(s) \text{ or } m \leq n. \qquad ...(50)$$

This condition is also called as the *condition of properness*. If a transfer function does not follow this condition, it is called an *improper* transfer function. It has the property that $G(s) \to \infty$ as $s \to \infty$. A realisable transfer function is called as *proper* and it always follows $G(s) \to k_0$ as $s \to \infty$. A transfer function with $k_0 = 0$ is called as *strictly proper*.

Transfer functions with dead time

If a time delay or *dead time* T_t is introduced in the input signal $x_e(t)$, one obtains instead of Eq. (8.44) the differential equation

$$\sum_{i=0}^{n} a_i \frac{d^i x_{3.}(t)}{dt^i} = \sum_{j=0}^{m} bj \frac{djx_e(t - T_t)}{dt^j}. \qquad ...(51)$$

In this case taking the Laplace transformation gives the *transcendental* transfer function

$$G(s) = \frac{N(s)}{D(s)} e^{-sT_t}. \qquad ...(52)$$

Poles and zeros of the transfer function

In some cases (e.g. stability analysis) it is expedient to represent the rational transfer function $G(s)$ according to Eq. (45) in the factorised form

$$G(s) = \frac{N(s)}{D(s)} = k_0 \frac{(s - s_{z_1})(s - s_{z_2})...(s - s_{z_m})}{(s - s_{p_1})(s - s_{p_2})...(s - s_{p_n})}. \qquad ...(53)$$

For physical reasons only real coefficients a_i, b_j occur. Therefore the poles sp_i and the *zeros* s_{z_j} of $G(s)$, respectively, can be *real or complex conjugate* pairs. The terms zeros and poles are chosen, because the transfer function is zero at s_{z_j} and infinite at s_{p_i}. Zeros and poles can be graphically represented in the complex plane as shown in Figure 3. A linear time-invariant system *without* dead time is described completely by the distribution of its poles and zeros and the gain factor k_0.

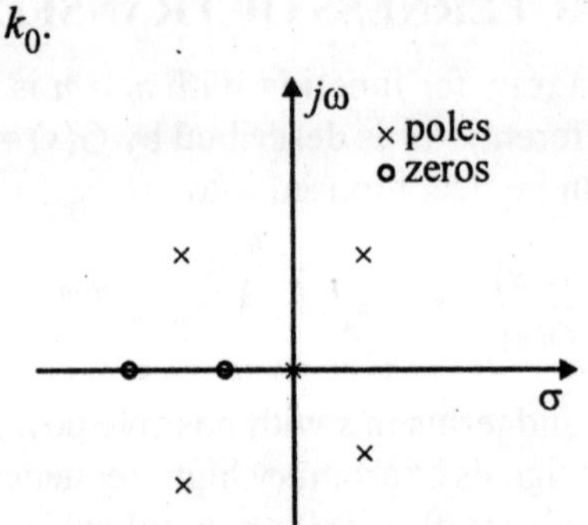

Fig. 3 Example of the Pole and Zero Distribution of a Rational Transfer Function in the Complex Plane

Moreover, the poles and zeros of a transfer function have a further significance. Observing a system without input ($x_e(t) \equiv 0$) according to Eq. (44) and determining the time response $x_3(t)$ for the given *n*initial conditions, one has to solve the associated homogeneous differential equation

$$\sum_{i=0}^{n} a_i \frac{d^i x_3(t)}{dt^i} = 0, \qquad ...(54)$$

which corresponds exactly to Eq. (35). For the approach $x_3(t) = e^{st}$ of Eq. (54) one obtains for the solution in s the *characteristic equation*

$$P(s) = \sum_{i=0}^{n} a_i s^i = 0 \qquad ...(55)$$

which was already mentioned in Eq. (37). This relation can be directly determined by setting the denominator of $G(s)$ to zero ($D(s) = 0$), as long as $D(s)$ and $N(s)$ have no common factor. The zeros s_k of the characteristic equation are the poles sp_i of the transfer function. As already shown in previous section the modes (i.e. $x_e(t) \equiv 0$) are described by the characteristic equation, so that the poles s_{p_i} of a transfer function contain all of this information.

The zeros of a transfer function are those values $s = s_{z_j}$ for which $|G(s_{z_j})| = 0$. This means that the output signal $X_3(s)$ does not contain any components which depend on s_{z_j}. In order to explain this in more detail a stable system with a transfer function according to Eq. (53) is excited by the input signal

$$x_e(t) = e^{s_{zj}t}$$

First for simplification the zero $s_{z_j} = \sigma_{z_j}$ is assumed to be real. For this case the input signal is $x_e(t) = e^{\sigma_{zj}t}$. Because $|G(s_{z_j})| = 0$ one obtains from Eq. (8.49) the steady-state output signal as

$$I_{35}(t) = 0$$

In the case of complex conjugate pairs of zeros s_{z_j}, $s_{z_j+1} = s^*_{zj}$ both zeros have to be taken into consideration in the input signal

$$\begin{aligned} x_e(t) &= 2e^{sz_it} + e^{s^*_{zj}t} \\ &= 2e^{\sigma z_it} \cos \omega t \\ &= 2e^{\sigma z_it} \sin\left(\omega t + \frac{\pi}{2}\right). \end{aligned}$$

Eq. (49) leads also to the result $x_{as}(t) = 0$. This shows that a zero sz_i of a system blocks the transmission of the input signal $e^{\sigma z_it}$.

Example The mass-spring-damper mechanical system in Figure 8.45 with the mechanical constants $c_1 = 1$, $c_2 = 2$, $d = 1.5$, $m_1 = 1$ and $m_2 = 4$ is excited by the force x_e. The transfer function between the force and the position can be shown to be

$$G(s) = \frac{s^2 + 1}{s^1 + 0.5s^3 + 1.75s^2 + 0.5s + 0.5}$$

which has the zeros $sz_{1,2} = \pm j$. If this system is excited by the sinusoidal input signal

$$X_e(s) = \frac{1}{(s - sz_1)(s - sz_2)} = \frac{1}{s^2 + 1}$$

$$^{-1}\{X_e(s)\} = \sin t,$$

which is derived from this pair of zeros, the output signal $x_a(t)$ decays to zero as shown in Figure 4 even though the input signal is a sinusoidal signal and the mass m_1 shows an undamped oscillation (see Figure 3c). The system does not pass this oscillation to the mass when the frequency matches the zeros.

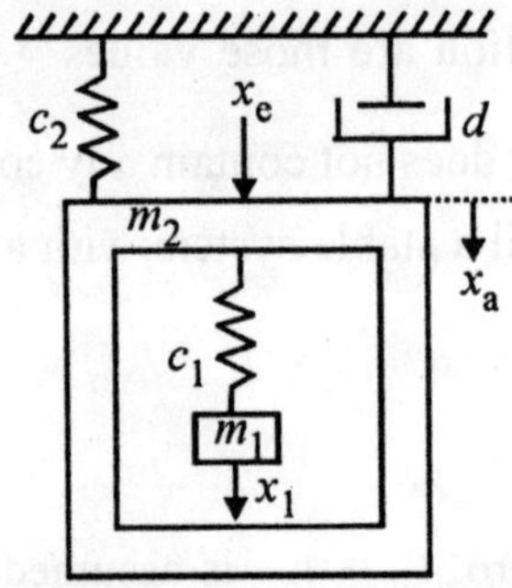

Fig. 4 Mass-spring-damper Mechanical System Used for the Interpretation of Zeros

Using transfer functions for calculations

For combinations of transfer functions simple rules for determining the resulting transfer function can be derived. The combinations are of the type that transfer function blocks are connected. In making any transfer function block connection it is assumed that the connection does not load the block to which the connection is being made.

(a) *Series connection*

From the diagram in Figure 8.4 it follows that

$$Y(s) = G_2(s)X_{e2}(s)$$

$$X_{e2}(s) = X_{a1}(s) = G_1(s)\,U(s)$$

$$Y(s) = G_2(s)G_1(s)\,U(s).$$

The total transfer function of this series connection is

$$G(s) = \frac{Y(s)}{U(s)} = G_1(s)G_2(s) \qquad \text{...(56)}$$

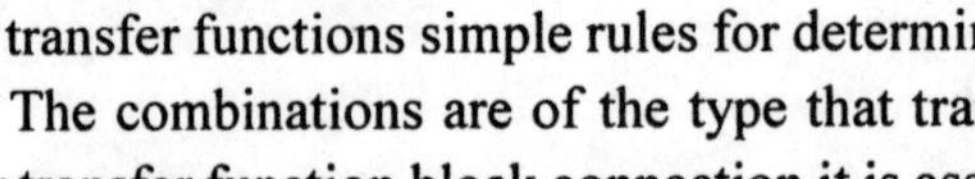

Fig. 5 Series Connection of two Transfer Functions

(b) ***Parallel connection***

For the output of both functions it follows from Figure 8.48 that

$$X_{a1}(s) = G_1(s)\ U(s)$$

$$X_{a2}(s) = G_2(s)\ U(s).$$

The output of the total system is

$$Y(s) = X_a(s) = X_{a_2}(s) + X_{a_2}(s) = [G_1(s) + G_2(s)]U(s),$$

and from this the transfer function of a paralle! connection is

$$G(s) = \frac{Y(s)}{U(s)} = G_1(s) + G_2(s) \qquad ...(8.57)$$

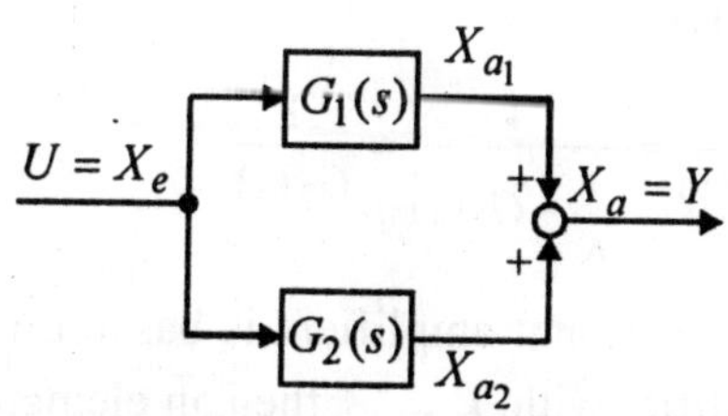

Fig. 6 Parallel Connection of Two Transfer Functions

(c) ***Feedback loop***

From Figure 6 the output is

$$Y(s) = X_a(s) = [U(s) \underset{(+)}{-} X_{a_2}(s) + G_1(s)$$

Since

$$X_{a_2} = G_2(s)Y(s)$$

one obtains

$$Y(s) = [U(s) \underset{(+)}{-} G_2(s)Y(s)]G_1(s),$$

and from this

$$Y(s) = \frac{G_1(s)}{1 \underset{(-)}{+} G_1(s)G_2(s)} U(s)$$

The total transfer function of the feedback loop is

$$G(s) = \frac{Y(s)}{U(s)} = \frac{G_1(s)}{1 \underset{(-)}{+} G_1(s)G_2(s)} \qquad ...(57)$$

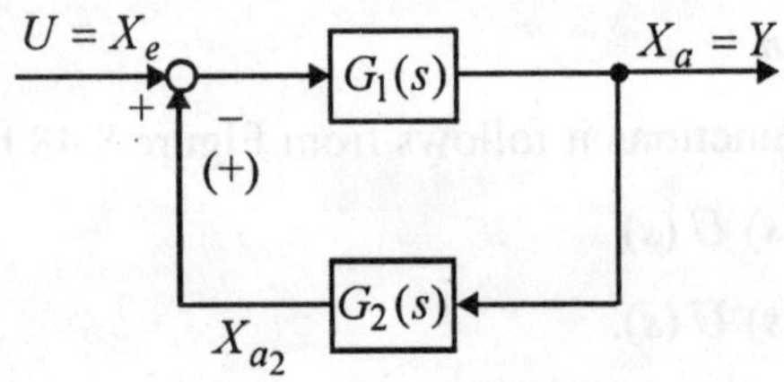

Fig. 7 Feedback Using Two Elements

As the output of $G_1(s)$ is fed back via $G_2(s)$ to the input, this is called *feedback*. One has to distinguish between a positive feedback for positive adding of $X_{a_2}(s)$ and a negative feedback for negative adding of $X_{a_2}(s)$.

Example 1. For the special case of $G_1(s)$ being a pure amplifier with a high gain $K \to \infty$, one obtains for negative feedback

$$G(s) = \frac{K}{1 + KG_2(s)} = \frac{1}{\frac{1}{K} + G_2(s)} \approx \frac{1}{G_2(s)}.$$

The entire technique of operational amplifiers is based on this principle. In a feedback loop for which $G_1(s)$ is an amplifier with $K \to \infty$ then an element $G_2(s)$ can be used to realise any transfer function within certain limits.

EXERCISE

1. A rectangular voltage pulse of height E and duration T is applied to a series RL combination at $t = 0$ (see Figure given below). Determine the current as a function of time. The initial current is zero.

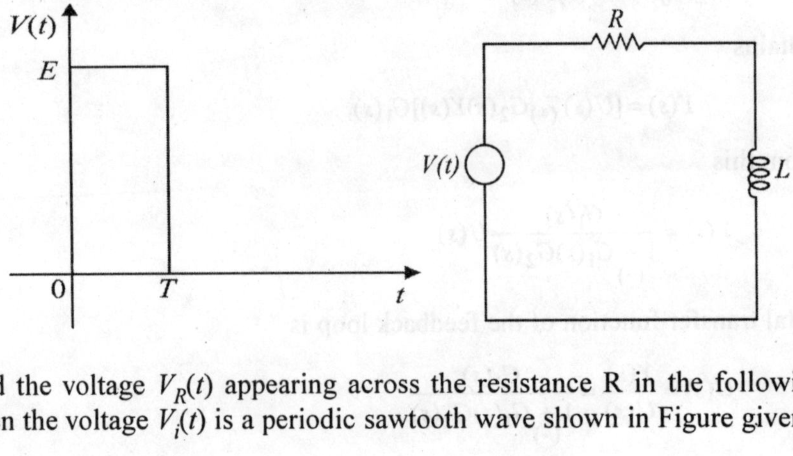

2. Find the voltage $V_R(t)$ appearing across the resistance R in the following circuit when the voltage $V_i(t)$ is a periodic sawtooth wave shown in Figure given below.

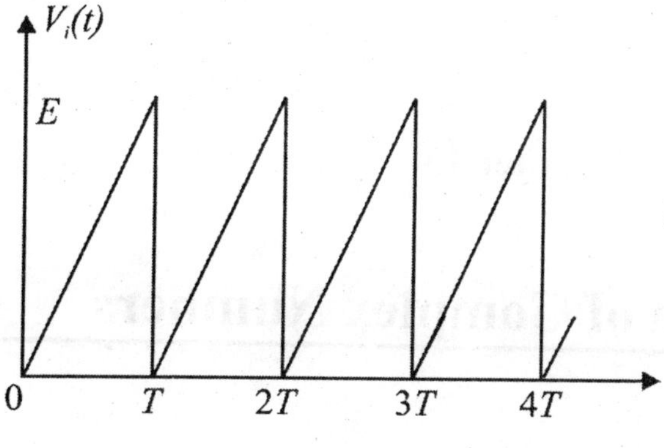

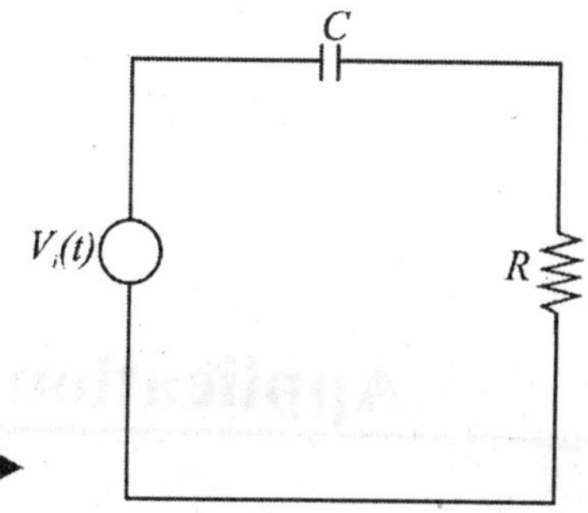

Determine the steady state expression for V_R. Initial condition $V_c(0^+) = \frac{E}{2}$.

3. Solve the following differential equation

$$X''(t) - 4X'(t) + 4X(t) = 2e^{2t}$$

$$X'(0) = 0, X(1) = 0$$

4. Solve the following system of differential equations

$$X''(t) - 3X'(t) - Y'(t) + 2Y(t) = 14t + 3$$

$$X'(t) - 3X(t) + Y'(t) = 0$$

$$X(0) = 0, Y(0) = 6.5, X'(0) = 0.$$

5. A mass m is attached to a spring of stiffness k^2 and rests in equilibrium on a frictionless horizontal plane. The mass is subjected to a impulsive blow P at $t = 0$ to extend the spring (show in Figure given below). Find an expression for the displacement at time t.

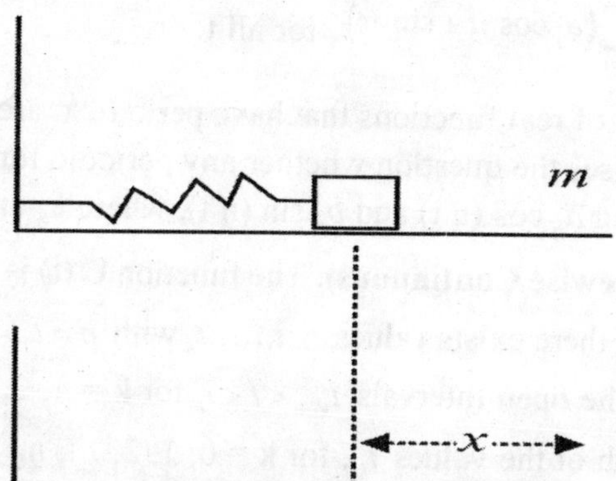

Chapter 13

Application of Complex Numbers

INTRODUCTION

We have already studied more about complex number, their properties and opertions on them. After studying applications of various topics, now in this chapter we are going to study the application of complex numbers. There is a wide application of Complex numbers in various field such as Medical engineering, Electrical engineering, Electnonics, Physics, Chemistry and so on.

INTEGRAL TRANSFORM

In this chapter we show how Integral transform are related to the study of complex analysis. We develop the Fourier Series representation for a real-valued function U(t) of the real variable t. Complex Fourier Series and Fourier Transforms are then discussed. Finally, we develop the Laplace Transform and the complex variable technique for finding its inverse. This chapter focuses on applying these ideas to solving problems involving real-valued functions, so many of the theorems throughout are stated without proof.

Let U(t) be a real-valued function that is periodic with period 2π, that is

$$S(t) = \frac{a_0}{2} + \sum_{j=1}^{n}\left(a_j \cos jt + \sin jt\right);\ \text{for all t.} \qquad ...(1)$$

Familiar examples of real functions that have period 2π, are cos (nt) and sin (nt), where n is an integer. This raises the question whether any periodic function can be represented by a sum of terms involving a_n cos (n t) and b_n sin (n t), where a_n and b_n are real constants.

***Definition. 1* (Piecewise Continuous).** The function U (t) is piecewise continuous on the closed interval [a, b], if there exists values $t_0, t_1, \ldots, t_n$ with $a = t_0 < t_1 < \ldots < t_n = b$ such that U is continuous in each of the open intervals $t_{k-1} < t < t_k$ for k = 1, 2, ..., n and has left-hand and right-hand limits at each of the values t_k, for k = 0, 1, 2, ..., n.

Definition. 2 **(Fourier Series).** If U (t) is periodic with period 2π and is piecewise continuous on (–π, π), then the Fourier Series S (t) for U(t) is

$$S(t) = \frac{a_0}{2} + \sum_{j=1}^{n} \left(a_j \cos jt + b_j \sin jt\right) \qquad ...(2)$$

where the coefficients a_j and b_j are given by the so-called

Euler's Formulae

$$a_j = \frac{1}{\pi} \int_{-\pi}^{\pi} U(t)\cos(jt)\,dt \text{, for j = 0, 1, ..} \qquad ...(3)$$

and

$$b_j = \frac{1}{\pi} \int_{-\pi}^{\pi} U(t)\sin(jt)\,dt \text{, for j = 1, 2, ...} \qquad ...(4)$$

Theorem 1. **(Fourier Expansion).** Assume that S(t) is the Fourier Series for U(t). If U(t) and $U'(t)$ are piecewise continuous on (–π, π), then S(t) is convergent for all $t \in (-\pi, \pi)$.

The relation U(t) = S(t), holds for all $t \in (-\pi, \pi)$, where U is continuous. If $t = a$, is a point of discontinuity of U, then

$$S(a) = \frac{U(a^-) + U(a^+)}{2}, \qquad ...(5)$$

where $U(a^+)$ and $U(a^+)$ denote the left-hand and right-hand limits, respectively. With this understanding, we have the Fourier Series expansion:

$$U(t) = \frac{a_0}{2} + \sum_{j=1}^{n} \left(a_j \cos jt + b_j \sin jt\right) \qquad ...(6)$$

Method I. **Fourier Series - Trigonometric Polynomials.** Calculate the following series to find S[n, t] for finding the Fourier Series - Trigonometric Polynomial of degree n.

$$a_n = \frac{\int_{-\pi}^{\pi} \cos(jt)\, U(t)\,dt}{\pi} \qquad ...(7)$$

$$b_n = \frac{\int_{-\pi}^{\pi} \sin(jt)\, U(t)\,dt}{\pi} \qquad ...(8)$$

$$S_n(t) = \frac{a(0)}{2} + \sum_{j=1}^{n} \{a(j)\cos(jt) + b(j)\sin(jt)\} \qquad ...(9)$$

Example 1. Find the Fourier Series expansion of $U(t)=\frac{t}{2}$ for $-\pi \le t \le \pi$.

Solution. Take the function u(t), and for illustration construct the Fourier Series - Trigonometric Polynomial of degree $n = 9$.

$$U(t)=\frac{t}{2}$$

$$S_9[t]=\sin[t]-\frac{1}{2}\sin[2t]+\frac{1}{3}\sin[3t]-\frac{1}{4}\sin[4t]$$

$$+\frac{1}{5}\sin[5t]-\frac{1}{6}\sin[6t]+\frac{1}{7}\sin[7t]-\frac{1}{8}\sin[8t]+\frac{1}{9}\sin[9t]$$

We can also get this result using Fourier Trig Series.

$$S_9[t]=\sin[t]-\frac{1}{2}\sin[2t]+\frac{1}{3}\sin[3t]-\frac{1}{4}\sin[4t]$$

$$+\frac{1}{5}\sin[5t]-\frac{1}{6}\sin[6t]+\frac{1}{7}\sin[7t]-\frac{1}{8}\sin[8t]+\frac{1}{9}\sin[9t]$$

A graph of U[t], S_1 [t], S_2 [t] and S_3 [t] is given below.

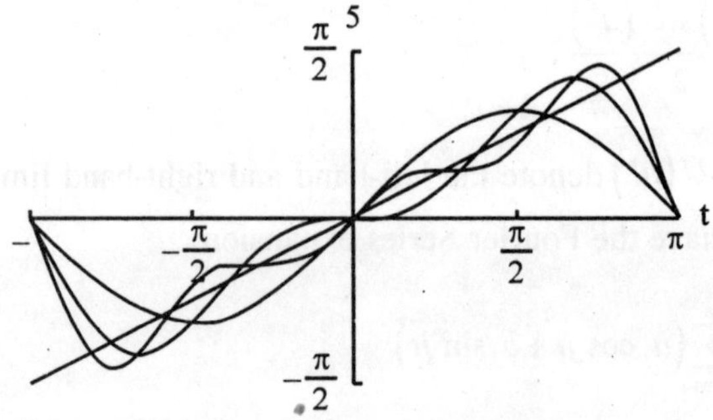

$$U(t)=\frac{t}{2}$$

$$S_1[t]=\sin[t]$$

$$S_1[t]=\sin[t]-\frac{1}{2}\sin[2t]$$

$$S_2[t]=\sin[t]-\frac{1}{2}\sin[2t]+\frac{1}{3}\sin[3t]$$

The general term for the series is $\frac{(-1)^{n-1}}{n}\sin[nt]$. We can sum up 9 terms and see that it agrees with $S_g(t)$.

Then we can sum the infinite series $S[t]=\sum_{j=1}^{\infty}\frac{(-1)^{n-1}}{j}\sin[jt]$ and see if it agrees with $U(t)=\frac{t}{2}$.

$$S[t]=\sum_{j=1}^{\infty}\frac{(-1)^{n-1}}{j}\sin[jt]$$

$$S[t]=\frac{1}{2}i\left(\log\left[1+e^{-it}\right]+\log\left[1+e^{it}\right]\right)$$

$$U(t)=\frac{t}{2}$$

We can plot S(t) to see what it looks like.

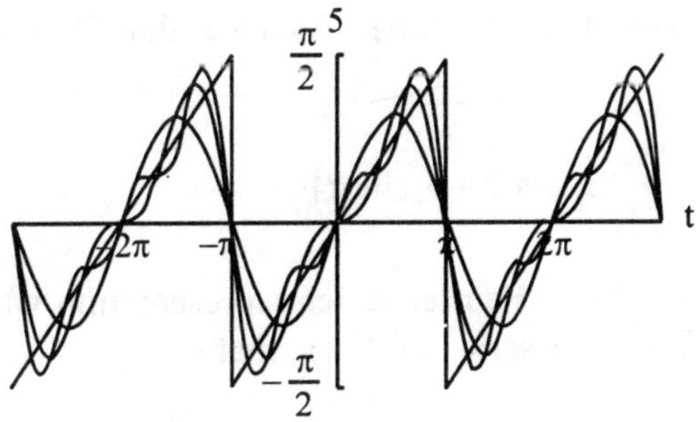

$$S[t]=\frac{1}{2}i\left(\log\left[1+e^{-it}\right]-\log\left[1+e^{it}\right]\right)$$

$$S_1[t]=\sin[t]$$

$$S_1[t]=\sin[t]-\frac{1}{2}\sin[2t]$$

$$S_2[t]=\sin[t]-\frac{1}{2}\sin[2t]+\frac{1}{3}\sin[3t]$$

Therefore, $S[t]=\frac{1}{2}i\left(\log\left[1+e^{-it}\right]-\log\left[1+e^{it}\right]\right)$ is the periodic extension of $U(t)=\frac{t}{2}$.

Theorem 2. If U(t) and V(t) have Fourier series representations, then their sum W(t) = U(t) + V(t), has a Fourier series representation, and the Fourier coefficients of W are obtained by adding the corresponding coefficients of U and V.

***Theorem 3* (Fourier Cosine Series).** Assume that U(t) is an even function and has period 2π. If U(t) and U¢(t) are piecewise continuous, the Fourier series or U(t) involves only the cosine terms, (i.e. $b_j = 0$ for all j), and we write

$$U(t)=\frac{a_0}{2}+\sum_{j=1}^{n}a_j \cos jt \qquad ...(10)$$

where $a_j=\frac{2}{\pi}\int_0^{\pi}U(t)\cos(jt)dt$, for j = 0, 1, 2, ...

***Theorem 4* (Fourier Sine Series).** Assume that U(t) is an odd function and has period 2π. If U(t) and U′(t) are piecewise continuous, the Fourier series for U(t) involves only the sine terms, (i.e. a_j = 0 for all j), and we write

$$U(t)=\sum_{j=1}^{n}b_j \sin(jt), \qquad ...(11)$$

where $b_j=\frac{2}{\pi}\int_0^{\pi}U(t)\sin(jt)dt$, for j = 1, 2, ...

***Theorem 5* (Term wise Integration).** Assume that U(t) has the Fourier series representation

$$U(t)=\frac{a_0}{2}+\sum_{j=1}^{\infty}\left(a_j \cos jt+b_j \sin jt\right). \qquad ...(12)$$

Then the integral of U(t) has a Fourier series representation which can be obtained by term wise integration of the Fourier series of U(t), that is

$$\int_0 U(t)dt = \sum_{j=1}^{\infty}\left[\frac{\left(a_j+(-1)^{j+1}a_0\right)}{j}\sin(jt)-\frac{b_j}{j}\cos(jt)\right]$$

$$\frac{a_0 t}{2} = \sum_{j=1}^{\infty}\frac{(-1)^{j+1}a_0}{j}\sin(jt) \qquad ...(13)$$

where we have used the expansion $\frac{a_0 t}{2} = \sum_{j=1}^{\infty}\frac{(-1)^{j+1}a_0}{j}\sin(jt)$

***Theorem 6* (Term wise Differentiation).** Assume that both U(t) and U¢(t) have Fourier series representation and that

$$U(t)=\frac{a_0}{2}+\sum_{j=1}^{\infty}\left(a_j \cos jt+b_j \sin jt\right) \qquad ...(14)$$

Then U¢(t) can be obtained by term wise differentiation of U(t), that is

$$U'(t)=\sum_{j=1}^{\infty}\left(-ja_j \sin jt+jb_j \cos jt\right)$$

Method II. **Fourier Series - Trigonometric Polynomials.** Modify procedure S[n, t] for finding the Fourier Series - Trigonometric Polynomial of degree n to the case of a piecewise continuous function on the subintervals $[-\pi,0)$ and $(0,\pi]$.

$$a_n = \frac{\int_{-\pi}^{0}\cos(jt)\,U_2(t)dt + \int_{0}^{\pi}\cos(jt)\,U_1(t)dt}{\pi}$$

$$b_n = \frac{\int_{-\pi}^{0}\sin(jt)\,U_2(t)dt + \int_{0}^{\pi}\sin(jt)\,U_1(t)dt}{\pi}$$

$$b_n = \frac{\int_{-\pi}^{0}\sin(jt)\,U_2(t)dt + \int_{0}^{\pi}\sin(jt)\,U_1(t)dt}{\pi}$$

***Theorem** 7* **(Term wise Integration).** Assume that U(t) has the Fourier series representation.

$$U(t) = \frac{a_0}{2} + \sum_{j=1}^{\infty}\left(a_j\cos jt + b_j\sin jt\right)$$

Then the integral of U(t) has a Fourier series representation which can be obtained by term wise integration of the Fourier series of U(t), that is

$$\int_0^t U(t)dt = \sum_{j=1}^{\infty}\left[\frac{\left(a_j+(-1)^{j+1}a_0\right)}{j}\sin(jt) - \frac{b_j}{j}\cos(jt)\right]$$

where we have used the expansion $\frac{a_0 t}{2} = \sum_{j=1}^{\infty}\frac{(-1)^{j+1}a_0}{j}\sin(jt)$.

Theorem 8 **(Term wise Differentiation).** Assume that both U(t) and U′(t) have Fourier series representation and that

$$U(t) = \frac{a_0}{2} + \sum_{j=1}^{\infty}\left(a_j\cos jt + b_j\sin jt\right) \qquad \text{...(15)}$$

Then U′(t) can be obtained by term-wise differentiation of U(t), that is

$$U'(t) = \sum_{j=1}^{\infty}\left(-ja_j\sin jt + jb_j\cos jt\right). \qquad \text{...(16)}$$

Example 2. The function $U(t) = |t|$ for $-\pi \le t \le \pi$ has the Fourier series representation

$$T(t) = \frac{\pi}{2} - \frac{4}{\pi}\sum_{j=1}^{\infty}\frac{\cos\left[(2j-1)t\right]}{(2j-1)^2}$$

Solution. Take the function U[t], and construct the Fourier Series - Trigonometric Polynomial of degree n = 9.

$$U(t) = t \text{ for } 0 \le t \le \pi$$

$$U(t) = -t \text{ for } -\pi \le t \le 0$$

$$T_9(t) = \frac{\pi}{2} - \frac{4\cos(t)}{\pi} - \frac{4\cos(3t)}{9\pi} - \frac{4\cos(5t)}{25\pi} - \frac{4\cos(7t)}{49\pi} - \frac{4\cos(9t)}{81\pi}$$

We can also get this result using Fourier Trig Series.

$$T_9(t) = \frac{\pi}{2} - \frac{4\cos(t)}{\pi} - \frac{4\cos(3t)}{9\pi} - \frac{4\cos(5t)}{25\pi} - \frac{4\cos(7t)}{49\pi} - \frac{4\cos(9t)}{81\pi}$$

A graph of U(t), T_1(t) and T_3 (t) is given below.

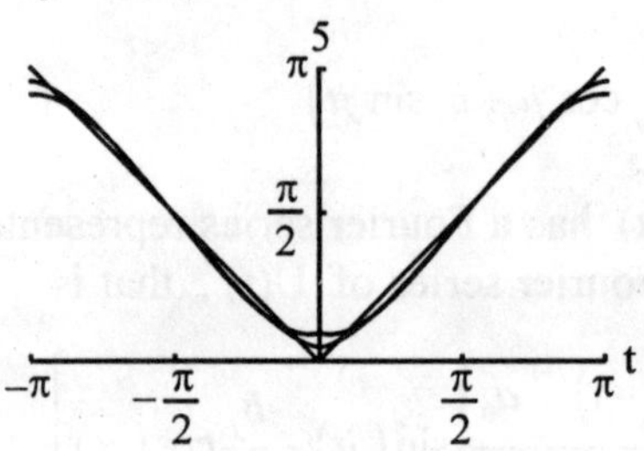

$$U(t) = |t|$$

$$T_1(t) = \frac{\pi}{2} - \frac{4\cos(t)}{\pi}$$

$$T_2(t) = \frac{\pi}{2} - \frac{4\cos(t)}{\pi} - \frac{4\cos(3t)}{9\pi}$$

The general term for the series is $\frac{\cos[(2j-1)t]}{(2j-1)^2}$. We can sum up 5 terms and see that it agrees with T_9 (t).

Then we can sum the infinite series and see if it agrees with U(t) = $|t|$.

$$T(t) = \frac{\pi}{2} - \frac{e^{-it}\left(LerchPhi\left[e^{-2it}, 2, \frac{1}{t}\right]\right) + e^{2it}\left(LerchPhi\left[e^{-2it}, 2, \frac{1}{t}\right]\right)}{2\pi}$$

$$D: |z| \le 1$$

$$u(\cos\theta, \sin\theta) = u(\theta) = \frac{\theta}{2} - \pi < \theta \le \pi C = |z| = 1$$

$$u(\theta) = \sum_{k=1}^{\infty}\frac{(-1)^{k-1}}{k}\sin(k\theta)$$

$$U(r,\theta) = \sin\theta - \frac{1}{2}(\sin 2\theta) + \frac{1}{3}(\sin 3\theta) - \frac{1}{4}(\sin 4\theta) + \frac{1}{5}(\sin 5\theta) - \frac{1}{6}(\sin 6\theta) + \frac{1}{7}(\sin 7\theta)$$

$U(t) = |t|$.

$$T(t) = \frac{\pi}{2} - \frac{4}{\pi}\sum_{j=1}^{5}\frac{\cos[(2j-1)t]}{(2j-1)^2}$$

$$T(t) = \frac{\pi}{2} - \frac{e^{-it}\left(LerchPhi\left[e^{-2it},2,\frac{1}{t}\right]\right) + e^{2it}\left(LerchPhi\left[e^{-2it},2,\frac{1}{t}\right]\right)}{2\pi}$$

$U(t) = |t|$

We can plot $T(t)$ to see what it looks like.

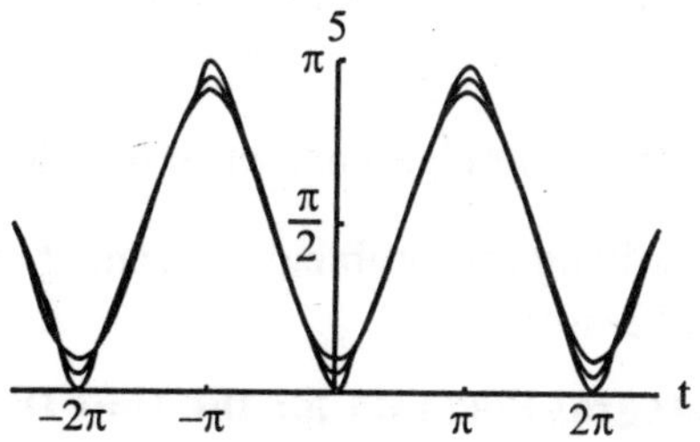

$$T(t) = \frac{\pi}{2} - \frac{e^{-it}\left(LerchPhi\left[e^{-2it},2,\frac{1}{t}\right]\right) + e^{2it}\left(LerchPhi\left[e^{-2it},2,\frac{1}{t}\right]\right)}{2\pi}$$

$$T_1(t) = \frac{\pi}{2} - \frac{4\cos(t)}{\pi}$$

$$T_2(t) = \frac{\pi}{2} - \frac{4\cos(t)}{\pi} - \frac{4\cos(3t)}{9\pi}$$

Therefore,

$$T(t) = \frac{\pi}{2} - \frac{e^{-it}\left(LerchPhi\left[e^{-2it},2,\frac{1}{t}\right]\right) + e^{2it}\left(LerchPhi\left[e^{-2it},2,\frac{1}{t}\right]\right)}{2\pi}$$ is the periodic extension

of $U(t) = |t|$.

APPLICATION OF DIRICHLET THEORY

The Dirichlet problem for the closed unit disk $D:|z|\leq 1$ is to find a real-valued function u(*x* ,*y*) that is harmonic in the unit disk D and that takes on the boundary values

$u(\cos\theta,\sin\theta)=u(\theta)$ for $-\pi<\theta\leq\pi$,

at points z= $(\cos\theta,\sin\theta)$ on the unit circle $C:|z|=1$.

***Theorem 9.* (The Dirichlet problem in D).** If U(t) has period 2π and has the Fourier series representation

$$U(t) = \frac{a_0}{2}+\sum_{j=1}^{\infty}\left(a_j\cos jt+b_j\sin jt\right),$$

then the solution to the Dirichlet problem in the unit disk D is

$$U(r\cos\theta,r\sin\theta) = \int_{-\infty}^{\infty}\frac{(1-r^2)U(t)}{1+r^2-2r\cos(t-\theta)}dt\frac{a_0}{2}$$

$$+\sum_{j=1}^{\infty}\left(a_j r^j\cos(j\theta)+b_j r^j\sin(j\theta)\right)e^{i\theta} \quad ,(17)$$

where $x + i\,y = r\ e^{i\theta}$ denotes a complex number in the closed disk $|z|\leq 1$.

An important method for solving this problem is our next result which is attributed to the French mathematician Siméon Poisson.

***Theorem 10.* (Poisson Integral Formula for the Unit Disk).** Let u (*x*, *y*) be a function that is harmonic in a simply connected domain that contains the closed unit disk $D:|z|\leq 1$. If *u*(*x*, *y*) takes on the boundary values

$u(\cos\theta,\sin\theta)=u(\theta)$ for $-\pi<\theta\leq\pi$,

then u(*x*, *y*) has the integral representation

$$U(r\cos\theta,r\sin\theta)=\int_{-\infty}^{\infty}\frac{(1-r^2)U(t)}{1+r^2-2r\cos(t-\theta)}dt, \quad ...(18)$$

which is valid for $|z|\leq 1$.

Example 3. Find the function u(*x*, *y*) that is harmonic in the unit disk $D:|z|\leq 1$ and takes on the boundary values $u(\cos\theta,\sin\theta)=u(\theta)=\frac{\theta}{2}$ for $-\pi<\theta\leq\pi$.

Solution. We have the Fourier series expansion

$$u(\theta) = \sum_{k=1}^{\infty} \frac{(-1)^{k-1}}{k} \sin(k\theta).$$

Using formula (3) for the solution of the Dirichlet problem, to obtain the boundary value approximation $U_7(r,\theta)$.

$$U_7(r,\theta) = r\sin\theta - \frac{1}{2}r^2(\sin 2\theta) + \frac{1}{3}r^3(\sin 3\theta) - \frac{1}{4}r^4(\sin 4\theta) + \frac{1}{5}r^5(\sin 5\theta) - \frac{1}{6}r^6(\sin 6\theta) + \frac{1}{7}r^7(\sin 7\theta)$$

$$U_7(r,\theta) = r\sin\theta - \frac{1}{2}r^2(\sin 2\theta) + \frac{1}{3}r^3(\sin 3\theta) - \frac{1}{4}r^4(\sin 4\theta) + \frac{1}{5}r^5(\sin 5\theta) - \frac{1}{6}r^6(\sin 6\theta) + \frac{1}{7}r^7(\sin 7\theta)$$

Plot the boundary curve $U_7(1,\theta)$.

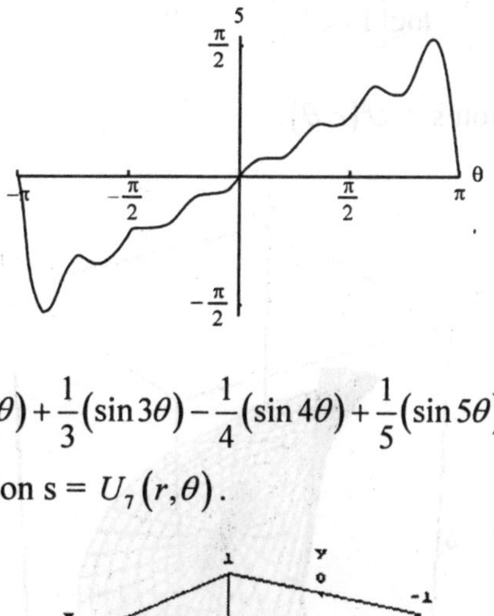

$$U_7(1,\theta) = \sin\theta - \frac{1}{2}(\sin 2\theta) + \frac{1}{3}(\sin 3\theta) - \frac{1}{4}(\sin 4\theta) + \frac{1}{5}(\sin 5\theta) - \frac{1}{6}(\sin 6\theta) + \frac{1}{7}(\sin 7\theta).$$

Plot the harmonic function s = $U_7(r,\theta)$.

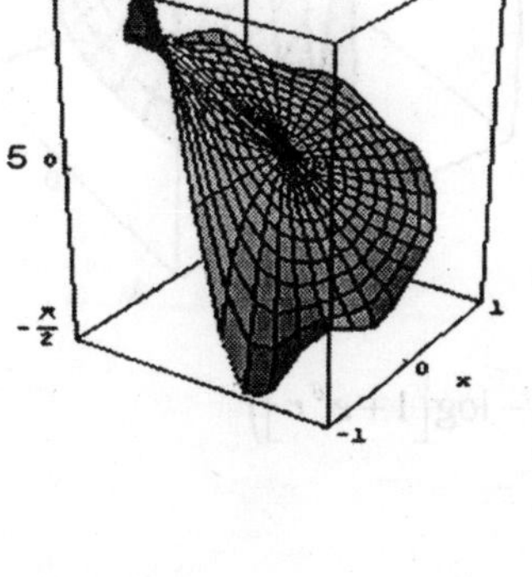

$$U_7(r,\theta) = r\sin\theta - \frac{1}{2}r^2(\sin 2\theta) + \frac{1}{3}r^3(\sin 3\theta) - \frac{1}{4}r^4(\sin 4\theta) + \frac{1}{5}r^5(\sin 5\theta) - \frac{1}{6}r^6(\sin 6\theta) + \frac{1}{7}r^7(\sin 7\theta)$$

Second, investigate the infinite sum $U(r,\theta)$.

$$U(r,\theta) = \frac{1}{2}i\left(\log\left[1+e^{-i\theta}r\right] - \log\left[1+e^{i\theta}r\right]\right)$$

Plot the boundary curve U (1, θ).

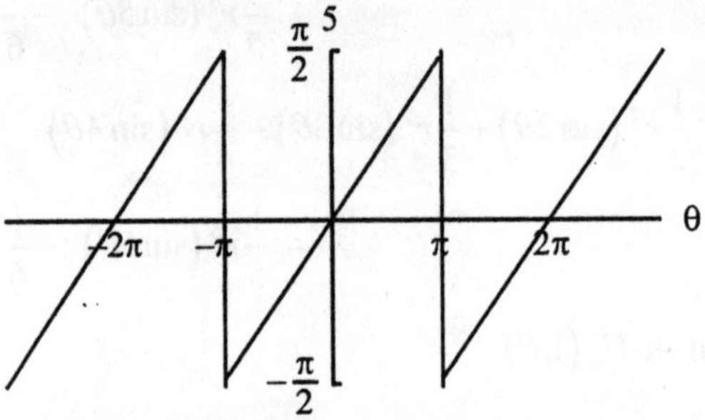

$$U(1,\theta) = \frac{1}{2}i\left(\log\left[1+e^{-i\theta}\right] - \log\left[1+e^{i\theta}\right]\right)$$

Plot the harmonic function s = $U(r,\theta)$.

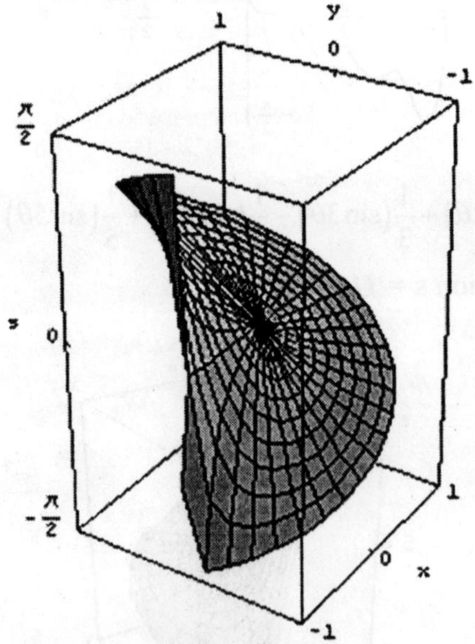

$$U(r,\theta) = \frac{1}{2}i\left(\log\left[1+e^{-i\theta}r\right] - \log\left[1+e^{i\theta}r\right]\right)$$

APPLICATION OF VIBRATIONS

Consider a spring that resists compression as well as extension, that is suspended vertically from a fixed support, and a body of mass m that is attached at the lower end of the spring. We make the assumption that the mass m is much larger than the mass of the spring so that we can neglect the mass of the spring. If there is no motion then the system is in static equilibrium. If the mass is pulled down further and released, then it will undergo an oscillatory motion.

Suppose that there is no friction to slow down the motion of the mass, then we say that the system is undamped. We will determine the Simple Harmonic Motion of this mechanical system by considering the forces acting on the mass during the motion. This will lead to a differential equation relating the displacement as a function of time. For details see page 412 in the textbook. The undamped mechanical system is governed by the linear differential equation

$$mU''(t)+kU(t)=0, \qquad ...(19)$$

and the general solution to this D. E. is known to

be $U(t)=A\cos(\omega t)+B\sin(\omega t)$, where $\omega=\sqrt{\dfrac{k}{m}}$ ...(20)

If we consider frictional forces that slow down the motion of the mass, then we say that the system is damped. This is visualized by connecting a dashpot (or shock adsorber) to the mass. For small velocities it is assumed that the frictional force is proportional to velocity. The damping constant c must be positive. For details see page 414 in the textbook. The Damped Mechanical Motion is governed by the linear differential equation

$$mU''(t)+cU'(t)+kU(t)=0. \qquad ...(21)$$

The two cases mentioned above are called free vibrations because all the forces that affect the motion of the system are internal to the system. We extend our analysis to cover the case in which an external force F(t) acts on the mass. Such a force might occur from vibrations of the support to which the top of the spring is attached, or from the effect of a magnetic field on a mass made of iron. The forced motion of the mechanical system satisfies the non-homogeneous linear differential equation

$$mU''(t)+cU'(t)+kU(t)=F(t). \qquad ...(22)$$

The function F(t) is called the input or driving force and the solution U(t) is called the output or response. Of particular interest are periodic inputs F(t) that can be represented by Fourier Series.

For damped mechanical systems that are driven by a periodic input F(t), the general solution involves a transient part that vanishes as $t\to\infty$, and a steady state part that is periodic.

The transient part of the solution $U_h(t)$ is found by solving the homogeneous differential equation

$$mU_h''(t)+cU_h'(t)+kU_h(t)=0.$$

This D. E. leads to the characteristic equation $m\lambda^2+c\lambda+k=0$, which has roots $\lambda=\frac{-c\pm\sqrt{c^2-4mk}}{2m}$. The constants *m*, *c*, and *k* are all positive, and there are three cases to consider.

If $c^2-4mk>0$, then the roots are real and distinct, and since $\sqrt{c^2-4mk}<c$, it follows that the roots l_1 and l_2 are negative real numbers. Thus for this case we have

$$U_h(t)=A_1e^{\lambda_1 t}+A_2e^{\lambda_2 t} \text{ and } \lim_{t\to\infty}U_h(t)=0$$

If $c^2-4mk=0$, then the roots are real and equal, $\lambda_1=\lambda_2=\lambda$ where λ is a negative real number, and for this case we have

$$U_h(t)=A_1e^{\lambda_1 t}+A_2te^{\lambda_1 t} \text{ and } \lim_{t\to\infty}U_h(t)=0.$$

If $c^2-4mk<0$, then the roots are complex conjugates, $\lambda=-\alpha\pm i\beta$, where a and b are positive real number, and it follows that

$$U_h(t)=A_1e^{-\alpha t}\cos(\beta t)+A_2e^{-\alpha t}\sin(\beta t) \text{ and } \lim_{t\to\infty}U_h(t)=0.$$

In all three cases, we see that the homogeneous solution $U_h(t)$ decays to 0 as $t\to\infty$. The steady state solution $U_p(t)$ can be obtained by representing $U_p(t)$ by its Fourier series and substituting $U_p''(t)$, $U_p'(t)$ and $U_p(t)$ into the non-homogeneous differential equation and solving the resulting system for the Fourier coefficients of $U_p(t)$. The general solution to the non-homogeneous linear differential equation is

$$U(t)=U_h(t)+U_p(t)$$

Example 4. **Find the general solution to $U''(t)+2U'(t)+U(t)=F(t)$,**

where F(t) is given by the Fourier series $F(t)=\sum_{k=1}^{\infty}\frac{\cos[(2k-1)t]}{(2k-1)^2}$.

Solution: **First, find the transient solution to the differential equation.**

$$U(t)+2U'(t)+U''(t)=0$$

$$\{U(t)=e^{-t}C(1)+e^{-t}tC(2)\}$$

The transient solution of D.E is

$$U_h(t)=e^{-t}C_1+e^{-t}tC_2$$

Next, use Fourier series to construct the steady state solution to the differential equation. Consider the formula for the nth terms T(n, t) and F(n, t) of the Fourier Series for U(t) and F(t), respectively. Then form the equation relating the n-terms and the equations involving Cos(n t) and Sin(n t). Then solve these equations for the Fourier coefficients of U(t).

$$T(n,t) = A_n \cos(nt) + B_n \sin(nt)$$

$$F(n,t) = \frac{\left[1-(-1)^n\right]\cos(nt)}{2n^2}$$

$$\cos(nt)A_n - n^2\cos(nt)A_n + \sin(nt)B_n - n^2\sin(nt)B_n + 2\{-n\sin(nt)A_n + n\cos(nt)B_n\}$$

$$= \frac{\left[1-(-1)^n\cos(nt)\right]}{2n^2}$$

$$A_n - n^2A_n + 2nB_n = \frac{\left[1-(-1)^n\right]}{2n^2}$$

$$-2nA_n + B_n - n^2B_n = 0$$

$$a_0 = 0$$

$$a_n = \frac{\left[\{-1+(-1)^n\}\{-1+n^2\}\right]}{2n^2\left(1+n^2\right)^2}$$

$$b_n = \frac{-1+(-1)^n}{n\left(1+n^2\right)^2}$$

Use the Fourier coefficients of U(t) and form the nth term of the series U(n, t), and verify that it satisfies the given D.E.

$$U(0,t) = 0$$

$$U(n,t) = \frac{\{-1+(-1)^n\}\{-1+n^2\}\cos(nt)}{2n^2\left(1+n^2\right)^2} - \frac{\{-1+(-1)^n\}\sin(nt)}{n\left(1+n^2\right)^2}$$

$$\frac{\partial^2 U(n,t)}{\partial t^2} + 2\frac{\partial U(n,t)}{\partial t} + U(n,t) = \frac{\left[1-(-1)^n\cos(nt)\right]}{2n^2}$$

$$-\frac{\{-1+(-1)^n\}\{-1+n^2\}\cos(nt)}{2\left(1+n^2\right)^2} + \frac{\{-1+(-1)^n\}\{-1+n^2\}\cos(nt)}{2n^2\left(1+n^2\right)^2}$$

$$-\frac{\left\{-1+(-1)^n\right\}\sin(nt)}{n\left(1+n^2\right)^2}+\frac{\left\{-1+(-1)^n\right\}\sin(nt)}{\left(1+n^2\right)^2}+$$

$$2\left[-\frac{\left\{-1+(-1)^n\right\}\cos(nt)}{\left(1+n^2\right)^2}-\frac{\left\{-1+(-1)^n\right\}\left\{-1+n^2\right\}\sin(nt)}{2n\left(1+n^2\right)^2}\right]$$

$$=\frac{\left\{-1+(-1)^n\right\}\cos(nt)}{2n^2\left(1+n^2\right)^2}$$

$$\Rightarrow -\frac{\left[-1+(-1)^n\right]\cos(nt)}{2n^2}=\frac{\left[1-(-1)^n\cos(nt)\right]}{2n^2}$$

Now we derive the Trigonometric Polynomial $S_5(t)$ of degree n = 5 and verify that it satisfies the D.E.

$$S_5(t)=\sum_{n=0}^{5}U(n,t)=-\frac{2}{225}\cos(3t)-\frac{6}{4225}\cos(5t)+\frac{1}{2}\sin t$$

$$+\frac{1}{150}\sin(3t)+\frac{1}{1690}\sin 5t$$

$$S_5'(t)=\frac{2}{75}\sin(3t)+\frac{6}{845}\sin(5t)+\frac{1}{2}\cos t+\frac{1}{50}\cos(3t)+\frac{1}{338}\cos(5t)$$

$$S_5''(t)=\frac{2}{25}\cos(3t)+\frac{6}{169}\cos(5t)-\frac{1}{2}\sin t-\frac{3}{50}\sin(3t)-\frac{5}{338}\sin(5t)$$

The left and right side of the D.E is

$$S_5''(t)+2S_5'(t)+S_5(t)=\frac{2}{25}\cos(3t)+\frac{6}{169}\cos(5t)-\frac{1}{2}\sin t$$

$$-\frac{3}{50}\sin(3t)-\frac{5}{338}\sin(5t)+2\left[\frac{2}{75}\sin(3t)+\frac{6}{845}\sin(5t)\right.$$

$$\left.+\frac{1}{2}\cos t+\frac{1}{50}\cos(3t)+\frac{1}{338}\cos(5t)\right]$$

$$+-\frac{2}{225}\cos(3t)-\frac{6}{4225}\cos(5t)+\frac{1}{2}\sin t$$

$$+\frac{1}{150}\sin(3t)+\frac{1}{1690}\sin 5t$$

$$= \cos t+\frac{1}{9}\cos(3t)+\frac{1}{25}\cos(5t)$$

$$f_5(t)=\sum_{n=0}^{5}F(n,t) = \cos t+\frac{1}{9}\cos(3t)+\frac{1}{25}\cos(5t)$$

Hence, $S_5''(t)+2S_5'(t)+S_5(t) = f_5(t)$ $f_5(t)=\sum_{n=0}^{5}F(n,t) = \cos t+\frac{1}{9}\cos(3t)+\frac{1}{25}\cos(5t)$

Now we check the maximum value of coefficients a_n and b_n, we observe that

$$\lim_{n\to\infty} a_n = \lim_{n\to\infty}\frac{\left[\{1-(-1)^n\}\{-1+n^2\}\right]}{2n^2(1+n^2)^2}=0$$

$$\lim_{n\to\infty} b_n - \lim_{n\to\infty}\frac{\{1-(-1)^n\}}{n(1+n^2)^2} = 0$$

We can check these sequences for convergent.

$$\{a_n\}=\left\{0,0,0,-\frac{2}{225},0,-\frac{6}{4225},\ldots\right\}$$

$$\{b_n\}=\left\{\frac{1}{2},0,\frac{1}{150},0,\frac{1}{1690},\ldots\right\}$$

$\{a_n\}$ = {0, 0, 0, – 0.0089, 0, – 0.00142, ...}

$\{b_n\}$ = {0.5, 0, 0.0067, 0, 0.000592, ...}

Maybe S_5 (t) is a close approximation to the solution. A graph of S_5 (t) is given below.

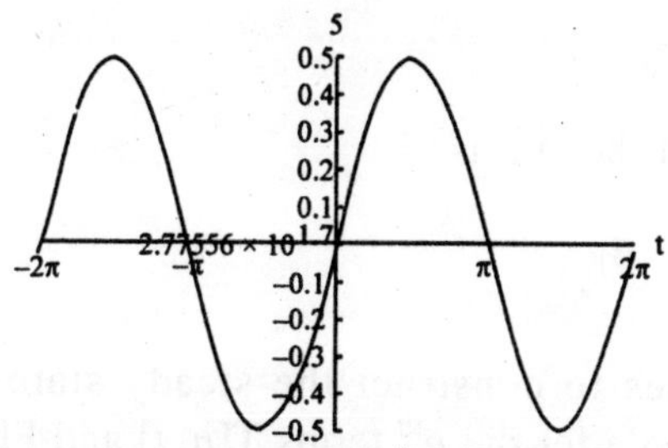

The Trignometric polynomial S_5 (t) is an approximation to the steady state solution is U_p (t) of the D.E.

$$S_5(t)=\sum_{n=0}^{5}U(n,t)=-\frac{2}{225}\cos(3t)-\frac{6}{4225}\cos(5t)+\frac{1}{2}\sin t+\frac{1}{150}\sin(3t)+\frac{1}{1690}\sin 5t$$

For illustration purposes, use the approximate solution $U(t) = U_h(t) + S_5(t)$ and use it so solve U(0) = 1 and U¢(0) = 0 and plot the result.

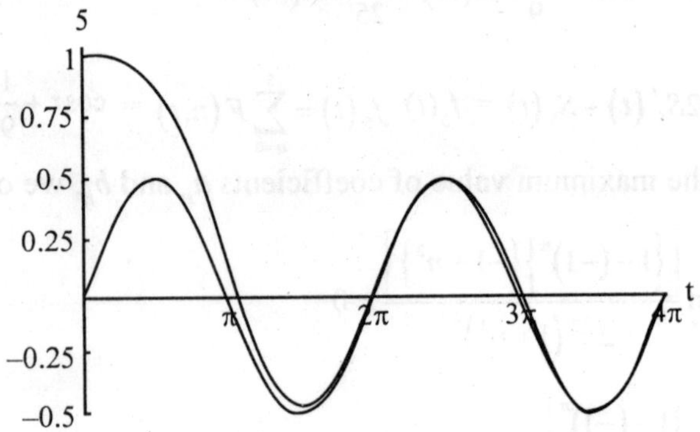

The Trigonometric polynomial S_5 (t) is an approximation to the steady state solution is U_p (t) of the D.E with initial condition is U(0) = 1 and U′(0) = 0

$$U(t)=\frac{38417e^{-t}}{38025}+\frac{2851te^{-t}}{5850}-\frac{2}{225}\cos(3t)$$

$$-\frac{6}{4225}\cos(5t)+\frac{1}{2}\sin t+\frac{1}{150}\sin(3t)+\frac{1}{1690}\sin 5t$$

Exercise 5. Find the general solution to $U(t)+3U'(t)+U''(t)=F(t)$, where F(t) is given by the Fourier series $F(t)=\sum_{n=1}^{\infty}\frac{(-1)^{n-1}}{n}\sin(nt)$.

Solution: First, find the transient solution to the differential equation.

$$U(t)+3U'(t)+U''(t)=0$$

The transient solution of the D.E is

$$U_h(t)=e^{\frac{1}{2}(-3-\sqrt{5})t}C_1+e^{\frac{1}{2}(-3+\sqrt{5})t}C_2$$

Next, use Fourier series to construct the steady state solution to the differential equation. Consider the formula for the nth terms T[*n*, *t*] and F[*n*, *t*] of the Fourier Series for U[t] and F[t], respectively. Then form the equation relating the n-terms and the equations involving Cos(*n t*) and Sin[*n t*]. Then solve these equations for the Fourier coefficients of U[t].

$$T(n,t)=A_n\cos(nt)+B_n\sin(nt)$$

$$F(n,t)=\frac{(-1)^{-1+n}\sin(nt)}{n}$$

$$\cos(nt)A_n - n^2\cos(nt)A_n + \sin(nt)B_n - n^2\sin(nt)B_n + 3\{-n\sin(nt)A_n + n\cos(nt)B_n\}$$
$$=\frac{(-1)^{-1+n}\sin(nt)}{n}$$

$$A_n - n^2A_n + 3nB_n = 0$$

$$-3nA_n + B_n - n^2B_n = \frac{(-1)^{-1+n}}{n}$$

$$a_0 = 0$$

$$a_n = \frac{3(-1)^n}{1+7n^2+n^4}$$

$$b_n = \frac{(-1)^n(-1+n^2)}{n(1+7n^2+n^4)}$$

Use the Fourier coefficients of U(t) and form the nth term of the series U(n, t), and verify that it satisfies the given D.E.

$$U(0,t)=0$$

$$U(n,t)=\frac{3(-1)^n\cos(nt)}{1+7n^2+n^4}+\frac{(-1)^n(-1+n^2)\sin(nt)}{n(1+7n^2+n^4)}$$

$$\frac{\partial^2U(n,t)}{\partial t^2}+3\frac{\partial U(n,t)}{\partial t}+U(n,t) = \frac{(-1)^{-1+n}\sin(nt)}{n}$$

$$\frac{3(-1)^n\cos(nt)}{1+7n^2+n^4}-\frac{3(-1)^n n^2\cos(nt)}{1+7n^2+n^4}$$

$$+\frac{(-1)^n(-1+n^2)\sin(nt)}{n(1+7n^2+n^4)}-\frac{(-1)^n n(-1+n^2)\sin(nt)}{n(1+7n^2+n^4)}$$

$$+3\left\{\frac{(-1)^n(-1+n^2)\cos(nt)}{1+7n^2+n^4}-\frac{3(-1)^n n\sin(nt)}{1+7n^2+n^4}\right\}=\frac{(-1)^{-1+n}\sin(nt)}{n}-\frac{(-1)^n\sin(nt)}{n}=$$

$$\frac{(-1)^{-1+n}\sin(nt)}{n}$$

Now we derive the Trigonometric Polynomial $S_5(t)$ of degree n = 5 and verify that it satisfies the D.E.

$$S_5(t)=\sum_{n=0}^{5}U(n,t)=-\frac{1}{2}\cos t+\frac{1}{15}\cos 2t-\frac{3}{145}\cos 3t$$

$$+\frac{1}{123}\cos 4t-\frac{1}{267}\cos 5t+\frac{1}{30}\sin 2t$$

$$-\frac{8}{435}\sin 3t+\frac{5}{492}\sin 4t-\frac{8}{1335}\sin 5t$$

$$S_5'(t)=\frac{1}{2}\sin t-\frac{2}{15}\sin 2t+\frac{9}{145}\sin 3t-\frac{4}{123}\sin 4t+\frac{5}{267}\sin 5t$$

$$+\frac{1}{15}\cos 2t-\frac{8}{145}\cos 3t+\frac{5}{123}\cos 4t-\frac{8}{267}\cos 5t$$

$$S_5''(t)=\frac{1}{2}\cos t-\frac{4}{15}\cos 2t+\frac{27}{145}\cos 3t-\frac{16}{123}\cos 4t+$$

$$\frac{25}{267}\cos 5t-\frac{2}{15}\sin 2t+\frac{24}{145}\sin 3t-\frac{20}{123}\sin 4t+\frac{40}{267}\sin 5t$$

The Left and Right side becomes,

$$S_5''(t)+3S_5'(t)+S_5(t)=\left[\frac{1}{2}\cos t-\frac{4}{15}\cos 2t+\right.$$

$$\frac{27}{145}\cos 3t-\frac{16}{123}\cos 4t+\frac{25}{267}\cos 5t-\frac{2}{15}\sin 2t+$$

$$\left.\frac{24}{145}\sin 3t-\frac{20}{123}\sin 4t+\frac{40}{267}\sin 5t\right]+3\left[\frac{1}{2}\sin t-\frac{2}{15}\sin 2t\right.$$

$$+\frac{9}{145}\sin 3t-\frac{4}{123}\sin 4t+\frac{5}{267}\sin 5t+\frac{1}{15}\cos 2t-\frac{8}{145}\cos 3t$$

$$\left.+\frac{5}{123}\cos 4t-\frac{8}{267}\cos 5t\right]+\left[-\frac{1}{2}\cos t+\frac{1}{15}\cos 2t\right.$$

$$-\frac{3}{145}\cos 3t+\frac{1}{123}\cos 4t-\frac{1}{267}\cos 5t+\frac{1}{30}\sin 2t-\frac{8}{435}\sin 3t$$

$$\left.+\frac{5}{492}\sin 4t-\frac{8}{1335}\sin 5t\right]$$

$$= \sin t - \frac{1}{2}\sin 2t + \frac{1}{3}\sin 3t - \frac{1}{4}\sin 4t + \frac{1}{5}\sin 5t$$

$$f_5(t) = \sum_{n=0}^{5} F(n,t) = \sin t - \frac{1}{2}\sin 2t + \frac{1}{3}\sin 3t - \frac{1}{4}\sin 4t + \frac{1}{5}\sin 5t$$

$$S_5''(t) + 3S_5'(t) + S_5(t) = f_5(t)$$

Hence, $S_5''(t) + 3S_5'(t) + S_5(t) = f_5(t) = \sin t - \frac{1}{2}\sin 2t + \frac{1}{3}\sin 3t - \frac{1}{4}\sin 4t + \frac{1}{5}\sin 5t$

Look at the size of the coefficients a_n and b_n and observe that

$$\lim_{n\to\infty} a_n = \lim_{n\to\infty} \frac{3(-1)^n}{1+7n^2+n^4} = 0$$

$$\lim_{n\to\infty} b_n = \lim_{n\to\infty} \frac{3(-1)^n(-1+n^2)}{n(1+7n^2+n^4)} = 0$$

We can check these sequences for convergent.

$$\{a_n\} = \left\{0, -\frac{1}{3}, \frac{1}{15}, -\frac{1}{145}, \frac{1}{123}, -\frac{1}{267}, \ldots\right\}$$

$$\{b_n\} = \left\{0, \frac{1}{30}, -\frac{8}{435}, \frac{5}{492}, -\frac{8}{1335}, \ldots\right\}$$

$\{a_n\}$ = {0, – 0.333, 0.0667, – 0.0207, 0.00813, – 0.00375, ...}

$\{b_n\}$ = {0, 0.333, – 0.0184, 0.0102, – 0.00599, ...}

Maybe $S_5(t)$ is a close approximation to the solution. A graph of $S_5(t)$ is given below.

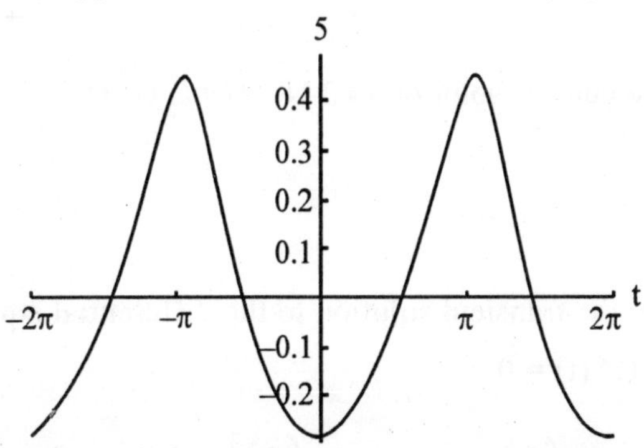

The Trigonometric Polynomial $S_5(t)$ is an approximation to the steady state solution $U_p(t)$ Of the D.E

$$S_5(t)=\sum_{n=0}^{5}U(n,t)=-\frac{1}{2}\cos t+\frac{1}{15}\cos 2t-\frac{3}{145}\cos 3t$$

$$+\frac{1}{123}\cos 4t-\frac{1}{267}\cos 5t+\frac{1}{30}\sin 2t$$

$$-\frac{8}{435}\sin 3t+\frac{5}{492}\sin 4t-\frac{8}{1335}\sin 5t$$

For illustration purposes, use the approximate solution $U(t) = U_h(t) + S_5(t)$ and use it so solve $U(0) = 1$, $U'(0) = 0$ and plot the result.

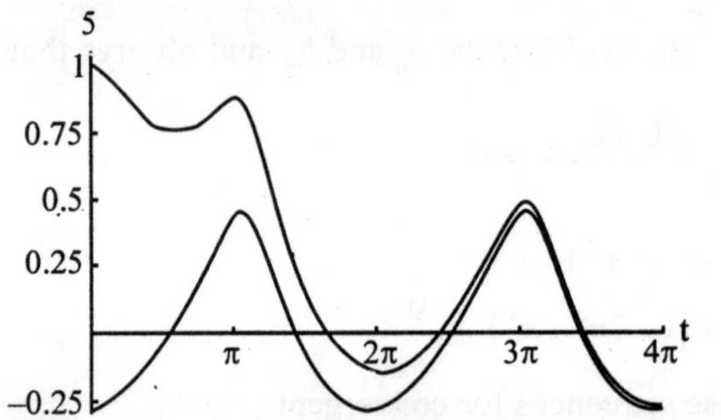

The Trigonometric Polynomial $S_5(t)$ is an approximation to the solution $U(t) = U_h(t) + S_5(t)$ with the initial conditions $U(0) = 1$, $U¢(0) = 0$.

$$U(t) = -0.209238e^{\frac{1}{2}(-3-\sqrt{5})t}+1.49221e^{\frac{1}{2}(-3+\sqrt{5})t}$$

$$-\frac{1}{3}\cos t+\frac{1}{15}\cos 2t-\frac{3}{145}\cos 3t+\frac{1}{123}\cos 4t$$

$$-\frac{1}{267}\cos 5t+\frac{1}{30}\sin 2t-\frac{8}{435}\sin 3t$$

$$+\frac{5}{492}\sin 4t-\frac{8}{1335}\sin 5t$$

Exercise 6. Find the general solution to 2 U" (t) + 2 U" (t) + U (t) = F (t), where F(t) is given by he Fourier series

$$F(t)=\sum_{n=1}^{\infty}\frac{(-1)^{n-1}}{n}\sin(nt)$$

Solution: First, find the transient solution to the differential equation.

U (t) + 2 U' (t) + 2 U" (t) = 0

$$\left[\{(U(t))\}\rightarrow e^{-t/2}C(2)\cos\left(\frac{t}{2}\right)+e^{-t/2}C(1)\sin\left(\frac{t}{2}\right)\right]$$

The transient solution to the D.E. is

$$U_h(t) = e^{-t/2}\sin\left(\frac{t}{2}\right)c_1 + e^{-t/2}\cos\left(\frac{t}{2}\right)c_2$$

Next, use Fourier series to find the steady state solution to the differential equation. Enter the formula for the n-th terms $T(n,t)$ and $F(n,t]$ of the Fourier Series for $U(t)$ and $F(t)$, respectively. Then form the equation relating the n-terms and the equations involving Cos(n,t) and Sin(n,t). Then solve these equations for the Fourier coefficients of $U(t)$.

$$T(n,t) = \cos(n\,t)\,A_n + \sin(n\,t)\,B_n$$

$$F(n,t) = \frac{(-1)^{-1+n}\sin(n\,t)}{n}$$

$$\cos(n\,t)\,A_n + \sin(n\,t)\,B_n + 2(-n\sin(n\,t)\,A_n + n\cos(n\,t)\,B_n) + 2(-n^2\cos(n\,t)\,A_n - n^2\sin(n\,t)\,B_n) = \frac{(-1)^{-1+n}\sin(n\,t)}{n}$$

$$A_n - 2n^2 A_n + 2\,n\,B_n = 0$$

$$-2\,n\,A_n + B_n - 2\,n^2\,B_n = \frac{(-1)^{-1+n}}{n}$$

$$A_0 = 0$$

$$a_n = \frac{2(-1)^n}{1+4n^4}$$

$$b_n = \frac{(-1)^n(-1+2n^2)}{n(1+4n^4)}$$

Use the Fourier coefficients of U[t] and form the n-th term of the series U[n,t], and verify that it satisfies the given D.E.

$$U(0,t) = 0$$

$$U(n,t) = \frac{2(-1)^n\cos(n\,t)}{1+4n^4} + \frac{(-1)^n(-1+2n^2)\sin(n\,t)}{n(1+4n^4)}$$

consider the differential equation

$$2\,\partial_{t,t}U(n,t) + 2\partial_t U(n,t) + U(u,t) = \frac{(-)^{-1+n}\sin(n\,t)}{n}$$

$$\frac{2(-1)^n\cos(n\,t)}{1+4n^4} + \frac{(-1)^n(-1+2n^2)\sin(n\,t)}{n(1+4n^4)} +$$

$$2\left[\frac{(-1)^n\left(-1+2n^2\right)\cos(n\,t)}{1+4n^4}\right]-\frac{2(-1)^n\sin(n\,t)}{1+4n^4}+$$

$$2\left(\frac{2(-1)^n\cos(n\,t)}{1+4n^4}-\frac{(-1)^n n\left(-1+2n^2\right)\sin(n\,t)}{1+4n^4}\right)=$$

$$\frac{(-1)^{-1+n}\sin(n\,t)}{n}-\frac{(-1)^n\sin(n\,t)}{n}=\frac{(-1)^{-1+n}\sin(n\,t)}{n}$$

Construct the Trigonometric Polynomial $S_5(t)$ of degree $n = 5$ and verify that it satisfies the D.E.

$$S_5(t)=\sum_{n=0}^{5}U(n,t)-\frac{2\cos(t)}{5}+\frac{2}{65}\cos(2t)-\frac{2}{325}\cos(3t)+\frac{2\cos(4t)}{1025}$$

$$\frac{2\cos(5t)}{2501}-\frac{\sin(t)}{5}+\frac{7}{130}\sin(2t)-\frac{17}{975}\sin(3t)+\frac{31\sin(4t)}{4100}-\frac{49\sin(5t)}{12505}$$

$$S_5'(t)=-\frac{\cos(t)}{5}+\frac{7}{65}\cos(2t)-\frac{17}{325}\cos(3t)+$$

$$\frac{31\cos(4t)}{1025}-\frac{49\cos(5t)}{2501}+\frac{2\sin(t)}{5}-\frac{4}{65}\sin(2t)+$$

$$\frac{6}{325}\sin(3t)-\frac{8\sin(4t)}{1025}+\frac{10\sin(5t)}{2501}$$

$$S_5''(t)=\frac{2\cos(t)}{5}-\frac{8}{65}\cos(2t)+\frac{18}{325}\cos(3t)-$$

$$\frac{32\cos(4t)}{1025}+\frac{50\cos(5t)}{2501}+\frac{\sin(t)}{5}-\frac{14}{65}\sin(2t)+$$

$$\frac{51}{325}\sin(3t)-\frac{32\cos(4t)}{1025}+\frac{50\cos(5t)}{2501}+\frac{\sin(t)}{5}-$$

$$\frac{14}{65}\sin(2t)+\frac{51}{325}\sin(3t)-\frac{124\sin(4t)}{1025}+\frac{245\sin(5t)}{2501}$$

The left and right side of the D.E. are

$$2\,S_5''(t)+2\,S_5'(t)+S_5(t)=\sin(t)-\frac{1}{2}\sin(2t)+\frac{1}{3}\sin(3t)-\frac{1}{4}\sin(4t)+\frac{1}{5}\sin(5t)$$

$$f_5(t)=\sum_{n=0}^{5}F(n,t)=\sin(t)-\frac{1}{2}\sin(2\,t)+\frac{1}{3}\sin(3t)-\frac{1}{4}\sin(4t)+\frac{1}{5}\sin(5t)$$

$$F_5(t) = \sum_{n=0}^{5} F(n,t) = \sin(t) - \frac{1}{2}\sin(2t) + \frac{1}{3}\sin(3t) - \frac{1}{4}\sin(4t) + \frac{1}{5}\sin(5t)$$

$$2\,S_5''(t) + 2\,S_5'(t) + S_5(t) = f_5(t)$$

$$\sin(t) - \frac{1}{2}\sin(2t) + \frac{1}{3}\sin(3t) - \frac{1}{4}\sin(4t) + \frac{1}{5}\sin(5t)$$

$$= \sin(t) - \frac{1}{2}\sin(2t) + \frac{1}{3}\sin(3t) - \frac{1}{4}\sin(4t) + \frac{1}{5}\sin(5t)$$

True

Look at the size of the coefficients a_n and b_n and observe that

$$\lim_{n\to\infty} a_n = \lim_{n\to\infty} \frac{2(-1)^n}{1+4n^4} = 0$$

$$\lim_{n\to\infty} b_n = \lim_{\delta x\to 0} \frac{(-1)^n\left(-1+2n^2\right)}{n\left(1+4n^4\right)} = 0$$

We can check the above for convergent

$$\{a_n\} = \left\{0, -\frac{2}{5}, \frac{2}{65}, -\frac{2}{325}, \frac{2}{1025}, -\frac{2}{2501}, \ldots\right\}$$

$$\{b_n\} = \left\{-\frac{1}{5}, \frac{7}{130}, -\frac{7}{975}, \frac{31}{4100}, -\frac{49}{12505}, \ldots\right\}$$

$$\{a_n\} = \{0., -0.4, 0.0308, -0.00615, 0.00195, -0.0008, \ldots\}$$

$$\{b_n\} = \{-0.2, 0.0538, -0.0174, 0.00756, -0.00392, \ldots\}$$

Maybe $S_5(t)$ is a close approximation to the solution. A graph of $S_5(t)$ is given below.

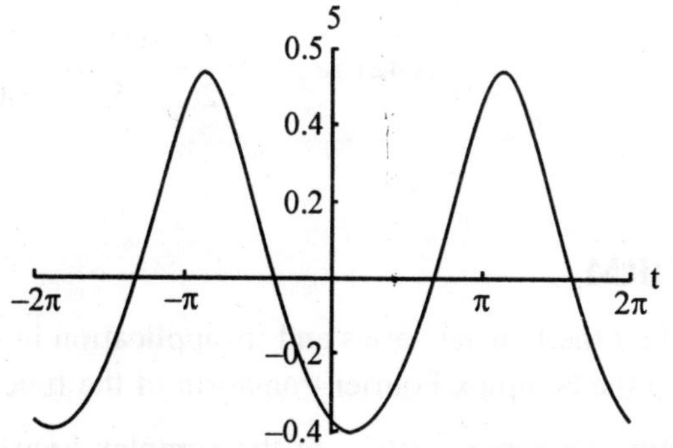

The Trigonometric Polynomical $S_5(t)$ is an approximation to the steady state solution $U_p(t)$ of the D.E.

$$S_5(t) = -\frac{2\cos(t)}{5} + \frac{2}{65}\cos(2t) - \frac{2}{325}\cos(3t) + \frac{2\cos(4t)}{1025}$$

$$\frac{2\cos(5t)}{2501}-\frac{\sin(t)}{5}+\frac{7}{130}\sin(2t)-\frac{17}{975}\sin(3t)+$$
$$\frac{31\sin(4t)}{4100}-\frac{49\sin(5t)}{12505}$$

For illustration purposes, use the approximate solution $U(t) = U_h(t) + S_5(t)$ and use it so solve $U(0) = 1$, $U'(0) = 0$, and plot the result.

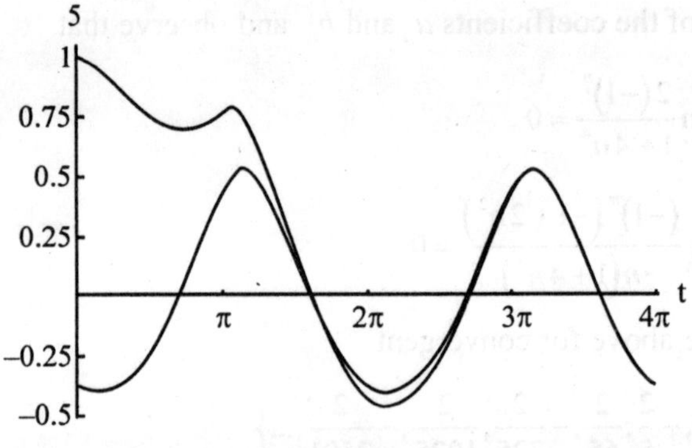

The Trigonometric Polynomial $S_5(t)$ is used to form an approximation to the solution $U(t) = U_h(t) + S_5(t)$ with the initial conditions $U(0) = 1$, and $U'(0) = 0$

$$U(t) = 1.37423e^{-t/t}\cos\left(\frac{t}{2}\right)-\frac{2\cos(t)}{5}+\frac{5}{65}\cos(2t)-$$
$$\frac{2}{325}\cos(3t)+\frac{2\cos(4t)}{1025}-\frac{2\cos(5t)}{2501}+$$
$$1.64216e^{-t/1}\sin\frac{t}{2}-\frac{\sin(t)}{5}+\frac{7}{130}\sin(2t)-\frac{17}{975}\sin(3t)+$$
$$\frac{31\sin(4t)}{4100}-\frac{49\sin(5t)}{12505}$$

FOURIER TRANSFORM

Earlier we have studied the Fourier series and its application in complex numbers. Now in this section we develop the complex Fourier transform of the function $U(t)$.

The Fourier transform is a generalization of the complex Fourier series in the limit as $L \to \infty$. Replace the discrete A_x with the continuous $F(k)d\,k$ while letting $n/L \to k$. Then change the sum to an integral, and the equations become

$$f(x) = \int_{-\infty}^{\infty} F(k)e^{2\pi ikx}dk \qquad ...(23)$$

$$F(k) = \int_{-\infty}^{\infty} f(x)e^{-2\pi ikx}dx \qquad ...(24)$$

Here,

$$F(x) = F_x[f(x)](k) = \int_{-\infty}^{\infty} f(x)e^{-2\pi ikx}dx \qquad ...(25)$$

is called the *forward* $(-i)$ Fourier transform, and

$$f(x) = F_x^{-1}[F(k)](x) = \int_{-\infty}^{\infty} F(k)e^{2\pi ikx}dk \qquad ...(26)$$

is called the *inverse* $(+i)$ Fourier transform. The notation $F_x[f(x)](k)$ is $\hat{f}(k)$ and $\hat{f}(x)$ are sometimes also used to denote the Fourier transform and inverse Fourier transform, respectively (Krantz 1999, p. 202). Note that some authors (especially physicists) prefer to write the transform in terms of angular frequency $\omega = 2\pi v$ instead of the oscillation frequency v. However, this destroys the symmetry, resulting in the transform pair

$$H(\omega) = F(h(t)) = \int_{-\infty}^{\infty} h(t)e^{-i\omega t}dt \qquad ...(27)$$

$$h(t) = F^{-1}[H(\omega)] \frac{1}{2\pi}\int_{-\infty}^{\infty} H(\omega)e^{-i\omega t}d\omega \qquad ...(28)$$

To restore the symmetry of the transforms, the convention

$$g(y) = F[f(t)] = \frac{1}{\sqrt{2\pi}}\int_{-\infty}^{\infty} f(t)e^{-iyt}dt \qquad ...(29)$$

$$f(t) = F^{-1}[g(y)] = \frac{1}{\sqrt{2\pi}}\int_{-\infty}^{\infty} g(y)e^{iyt}dy \qquad ...(30)$$

is sometimes used. In general, the Fourier transform pair may be defined using two arbitrary constants a and b as

$$F(\omega) = F(\omega) = \sqrt{\frac{|b|}{(2\pi)^{1-a}}}\int_{-\infty}^{\infty} f(t)e^{ib\omega t}dt \qquad ...(31)$$

$$f(t) = \sqrt{\frac{|b|}{(2\pi)^{1+a}}}\int_{-\infty}^{\infty}(\omega)e^{-ib\omega t}d\omega \qquad ...(32)$$

The Fourier transform $F(k)$ of a function $f(x)$ is implemented as Fourier Transform $[f, x, k]$ Since any function can be split up into even and odd portions $E(x)$ and $O(x)$, $f(x) = \frac{1}{2}[f(x)+f(-x)] + \frac{1}{2}[f(x)-f(-x)=]$

$$= E(x) + O(x) \qquad ...(33)$$

a Fourier transform can always be expressed in terms of the Fourier cosine transform and Fourier sine transform as

$$F_x[f(x)]\,(k) = \int_{-\infty}^{\infty} E(x)\cos(2\pi kx)\,dx - i\int_{-\infty}^{\infty} O(x)\sin(2\pi kx)\,dx \qquad ...(34)$$

A function $f(x)$ has a forward and inverse Fourier transform such

$$\text{that } f(x) = \begin{cases} \int_{-\infty}^{\infty} e^{2\pi ikx}\left[\int_{-\infty}^{\infty} f(x)e^{-\pi ikx}\,dx\right]dk & \textit{for } f(x) \text{ continuous at } x \\ \frac{1}{2}\left[f(x_+ + f(x_-))\right] & \textit{for } f(x) \text{ discontinuous at } x, \end{cases} \qquad ...(35)$$

provided that:

1. $\int_{-\infty}^{\infty} \left|f(x)\,dx\right|$ exists.
2. There are a finite number of discontinuities.
3. The function has bounded variation.

A sufficient weaker condition is fulfillment of the Lipschitz condition The smoother a function (i.e., the larger the number of continuous derivatives), the more compact its Fourier transform. The Fourier transform is linear, since if $f(x)$ and $f(x)$ have Fourier transforms $F(k)$ and $F(k)$, then

$$\int\left[a\,f(x) + b\,g(x)\right]e^{-2\pi ikx}\,dx = a\int_{-\infty}^{\infty} f(x)e^{-2\pi ikx}\,dx + b\int_{-\infty}^{\infty} g(x)e^{-2\pi ikx}\,dx \qquad ...(36)$$

$$= aF(k) + bG(k,) \qquad ...(37)$$

Therefore, $\quad F[af(x) + b\,g(x)] = aF[f(x)] + nF[g(x)] = \mathrm{a}F(\mathrm{k}) + b\,G(k) \qquad ...(38)$

The Fourier transform is also symmetric since $F(k) = F_x[f(x)]\,(k)$ implies $F(-k) = F_x\,[f(-x)]\,(k)$. Let $f * g$ denote the convolution, then the transforms of convolutions of functions have particularly nice transforms,

$$F[f * g] = F[f]\,F[g] \qquad ...(39)$$

$$F[f\,g] = F[f] * F[g] \qquad ...(40)$$

$$F^{-1}[F(f)\,F(g)] = f * g \qquad ...(41)$$

$$F^{-1}[F(f) * F(g)] = \mathrm{f\,g} \qquad ...(42)$$

The first of these is derived as follows:

$$F[f * g] = \int_{-\infty}^{\infty}\int_{-\infty}^{\infty} e^{-2\pi ikx} f(x')g(x - x')\,dX'dx \qquad ...(43)$$

$$= \int_{-\infty}^{\infty}\int_{-\infty}^{\infty}\left[e^{-2\pi ikx'} f(x')\,dx\right]\left[e^{-2\pi ik(x-x')} g(x - x')\,dx\right] \qquad ...(44)$$

$$= \left[\int_{-\infty}^{\infty} e^{-2\pi ikx'} f(x')\,dx'\right]\left[\int_{-\infty}^{\infty} e^{-2\pi ikx''} g(x'')\,dx''\right] \qquad ...(45)$$

$$= F[f]\, F[g], \qquad ...(46)$$

where $\qquad x'' = x - x'.$

There is also a somewhat surprising and extremely important relationship between the autocorrelation and the Fourier transform known as the Wiener-Khinchin theorem. Let Fx $[f(x)\,(k)] = F(k)$, and $\bar{f}$ denote the complex conjugate of f, then the Fourier transform of the absolute square of $F(k)$ is given by

$$F_k[|F(k)|^2]\,(x) = \int_{-\infty}^{\infty} \bar{f}(\tau) f(\tau + x)\,d\tau \qquad ...(47)$$

The Fourier transform of a derivative $f'(x)$ of a function $f(x)$ is simply related to the transform of the function $f(x)$ itself. Consider

$$F_k[f'(x)]\,(k) = \int_{-\infty}^{\infty} f'(x) e^{-2\pi ikx}\,dx \qquad ...(48)$$

Now use integration by parts

$$\int v\,du - [u\,v] - \int v\,du \qquad ...(49)$$

$$\text{with } du = f'(x)\,dx \qquad ...(50)$$

$$V = e^{-2\pi ikx} \qquad ...(51)$$

and

$$u = f(x) \qquad ...(52)$$

$$dv = -2\pi ike^{-2\pi ikx}\,dx, \qquad ...(53)$$

then

$F_x\,[f'](x)(k) = \left[f(x)e^{-\pi ikx}\right]_{-\infty}^{\infty} - \int_{-\infty}^{\infty} f(x)\left(-2\pi ike^{-2\pi ikx}\,dx\right)$ (54) The first term consists of an oscillating function times $f(x)$. But if the function is bounded so that

$$\lim_{x\to\pm\infty} f(x) = 0 \qquad ...(55)$$

(as any physically significant signal must be), then the term vanishes, leaving

$$F_x\,[f'(x)]\,(k) = 2\pi ik \int_{-\infty}^{\infty} f(x)e^{-2\pi ikx}\,dx \qquad ...(56)$$

$$= 2\pi ik\, F_x\,[f(x)]\,(k) \qquad ...(57)$$

This process can be iterated for the nth derivative to yield

$$F_x\,[f^{(n)}(x)](k) = (2\pi ik)^n\, F_x[f(x)]\,(k) \qquad ...(58)$$

The important modulation theorem of Fourier transforms allows

$F_x\,[\cos(2\pi k_0\,x) f(x)]\,(k)$ to be expressed in terms of

$F_x\,[f(x)]\,(k) = F(k)$ as follows,

$$F_x\,[\cos(2\,\pi\,k_0\,x)\,f(x)]\,(k)$$

$$= \int_{-\infty}^{\infty} f(x)\cos(2\pi k_0\,x)e^{-2\pi ikx}dx \quad ...(59)$$

$$= \frac{1}{2}\int_{-\infty}^{\infty} f(x)e^{2\pi ik_0 x}e^{-2\pi ikx}d\,x + \frac{1}{2}\int_{-\infty}^{\infty} f(x)e^{-2\pi ik_0 x}e^{-2\pi ikx}dx \quad ...(60)$$

$$= \frac{1}{2}\int_{-\infty}^{\infty} f(x)e^{-2\pi i(k-k_0)x}dx + \int_{-\infty}^{\infty} f(x)e^{-2\pi i(k+k_0)x}dx \quad ...(61)$$

$$= \frac{1}{2}\left[F(k-k_0)+F(k+k_0)\right] \quad ...(62)$$

Since the derivative of the Fourier transform is given by

$$F'(k) = \frac{d}{d\,k}F_x\left[f(x)\right](k) = \int_{-\infty}^{\infty}(-2\pi ix)f(x)e^{-2\pi ikx}dx \quad ...(63)$$

it follows that

$$F'(0) = -2\pi i\int_{-\infty}^{\infty} xf(x)dx \quad ...(64)$$

Iterating gives the general formula

$$\mu n = \int_{-\infty}^{\infty} x^n f(x)dx = \frac{F^{(x)}(0)}{(-2\pi i)^n} \quad ...(65)$$

The variance of a Fourier transform is

$$\sigma_f^2 = ((x\,f - (xf))^2), \quad ...(66)$$

and it is true that

$$\sigma_{f+g} = \sigma f + \sigma_g \quad ...(67)$$

If $f(x)$ has the Fourier transform $F_x\,[f(x)]\,(k) = F(k)$,then the Fourier transform has the shift property

$$\int_{-\infty}^{\infty} f(x-x_0)e^{-2\pi ikx}dx$$

$$= \int_{-\infty}^{\infty} f(x-x_0)e^{-2\pi i(x-x_0)k}e^{-2\pi i(kx_0)}d(x-x_0) \quad ...(68)$$

$$= e^{-2\pi ikx_0}F(k) \quad ...(69)$$

so $f(x - x_0)$ has the Fourier transform

$$F_x\,[f(x-x_0)]\,(k) = e^{-2\pi ikx_0}F(k) \qquad ...(70)$$

If $f(x)$ has a Fourier transform $F_x\,[f(x)]\,(k) = F(k)$, then the Fourier transform obeys a similarity theorem.

$$\int_{-\infty}^{\infty} f(ax)e^{-2\pi ikx}dx = \frac{1}{|a|}\int_{-\infty}^{\infty} f(ax)e^{-2\pi i(ax)(k/a)}d(ax) = \frac{1}{|a|}F\left(\frac{k}{a}\right), \qquad ...(71)$$

so $f(a\,x)$ has the Fourier transform

$$F_x\,[f(a\,x)]\,(k) = |a|^{-1}F\left(\frac{k}{a}\right) \qquad ...(72)$$

The "equivalent width" of a Fourier transform is

$$w_e = \frac{\int_{-\infty}^{\infty} f(x)dx}{f(0)} = \frac{F(0)}{\int_{-\infty}^{\infty} F(k)dk} \qquad ...(73)$$

The "autocorrelation width" is

$$w_a = \frac{\int_{-\infty}^{\infty} f * \bar{f}dx}{\left[f * \bar{f}\right]_0} = \frac{\int_{-\infty}^{\infty} f\,dx\int_{-\infty}^{\infty} \bar{f}dx}{\int_{-\infty}^{\infty} f\bar{f}dx} \qquad ...(74)$$

where $f * g$ denotes the cross-correlation of f and g and $\bar{f}$ is the complex conjugate. Any operation on $f(x)$ which leaves its area unchanged leaves $F(0)$ unchanged, since

$$\int_{-\infty}^{\infty} f(x)dx = F_x[f(x)](0) = F(0) \qquad ...(75)$$

The following table summarized some common Fourier transform pairs.

Function	$f(x)$	$F(k) = F_x\,[f(x)]\,(k)$
Fourier transform—1	1	$\delta(k)$
Fourier transform—Cosine	$\cos(2\pi k_0 x)$	$\frac{1}{2}[\delta(k-k_0) + \delta(k+k_0)]$
Fourier transform—Delta function	$\delta(x-x_0)$	$e^{-2\pi ikx_0}$
Fourier transform—Exponential function	$e^{-2\pi k_0}\|x\|$	$\frac{1}{\pi}\frac{k_0}{k^2+k_0^2}$
Fourier transform—Gaussian	e^{-ax^2}	$\sqrt{\frac{\pi}{a}}e^{-\pi^2k^2/a}$
Fourier transform—Heaviside step function	$H(x)$	$\frac{1}{2}\left[\delta(k) - \frac{i}{\pi k}\right]$
Fourier transform—Inverse function	$-\text{PV}\frac{1}{\pi x}$	$i[1-2H(-k)]$

Fourier transform—Lorentzian function	$\frac{1}{\pi}\frac{\frac{1}{2}\Gamma}{(x-x_0)^2+\left(\frac{1}{2}\Gamma\right)^2}$	$e^{-2\pi ikx_0-\Gamma\pi\lvert k\rvert}$
Fourier transform—Ramp function	$R(x)$	$\pi i\delta'(2\pi k)-\frac{1}{4\pi^2k^2}$
Fourier transform—Sine	$\sin(2\pi k_0x)$	$\frac{1}{2}i\left[\delta(k+k_0)-\delta(k-k_0)\right]$

In two dimensions, the Fourier transform becomes

$$F(x,y) = \int_{-\infty}^{\infty}\int_{-\infty}^{\infty} f(k_x,k_y)\, e^{-2\pi i(k_xx+k_yy)}dk_xdk_y \qquad ...(76)$$

$$f(k_x,k_y) = \int_{-\infty}^{\infty}\int_{-\infty}^{\infty} F(x,y)\, e^{2\pi i(k_xx+k_yy)}dx_xdy \qquad ...(77)$$

Similarly, the n-dimensional Fourier transform can be defined for $k,x\in R^x$ by

$$F(x)=\underbrace{\int_{-\infty}^{\infty}\int_{-\infty}^{\infty}}_{x} f(k)e^{-2\pi ikx}d^xk \qquad ...(78)$$

$$f(k)=\underbrace{\int_{-\infty}^{\infty}\int_{-\infty}^{\infty}}_{x} f(x)e^{2\pi ikx}d^xx.$$

***Theorem 11* (Fourier Transform).** Let $U(t)$ and $U'(t)$ be piecewise continuous, and $\int_{-\infty}^{\infty}|U(t)|\,dt<M$, for some positive constant M. The Fourier transform $F(w)$ of $U(t)$ is defined as

$$F(w) = \frac{1}{2\pi}\int_{-\infty}^{\infty} U(t)e^{-iwt}dt.$$

At points of continuity, $U(t)$ has the integral representation

$$U(t) = \int_{-\infty}^{\infty} F(w)e^{-iwt}dw,$$

and at a point $t=a$ of discontinuity of $U(t)$, this integral converges to $\frac{U(a^-)+U(a^+)}{2}$.

Example 7. Find the Fourier transform of $U(t) = e^{-|t|}$.

Solution: Enter the integrand $U(t) = e^{-|t|}$ and perform integration.

$$U[t] = e^{-Abs[t]}$$

$$U_1[t] = e^t$$

$$U_2[t] = e^{-t}$$

$$W_1[t] = \int e^te^{-twt}dt = \frac{ie^{t(1-iw)}}{i+w}$$

$$W_2[t] = \int e^{-t}e^{-twt}dt = \frac{ie^{-it(-i+w)}}{-i+w}$$

Because of the exponential order $w_1[-\infty] = 0$ and $w_2[\infty] = 0$ and the Fourier transform is given by the following computation:

$$U[t] = e^{-Abs[t]}$$

$$U[w] = \frac{1}{2\pi}(w_1[0] - w_2[0])$$

$$w_1[0] = \frac{i}{i+w}$$

$$w_2[0] = \frac{i}{-i+w}$$

$$F[w] = \frac{1}{2\pi}\left(\left(\frac{i}{i+w}\right) - \left(\frac{i}{-i+w}\right)\right)$$

$$F[w] = \frac{1}{\pi(-i+w)(i+w)}$$

$$U(t) = e^{Abs[t]}$$

$$V_1 = \int_{-\infty}^{0} e^{t}e^{-iwt}dt = \frac{i}{i+w}$$

$$V_2 = \int_{-\infty}^{0} e^{-t}e^{-twt}dt = -\frac{i}{-i+w}$$

$$F[w] = \frac{1}{2\pi}(v_1 + v_2) = \frac{1}{2\pi}\left(\left(\frac{i}{i+w}\right) - \left(-\frac{i}{-i+w}\right)\right)$$

$$F[w] = \frac{-\frac{i}{-i+\infty} + \frac{i}{i+\infty}}{2\pi} = \frac{1}{\pi(-i+w)(i+w)} = \frac{1}{\pi + \pi w^2}$$

$$F[w] = \frac{1}{\pi(1+w^2)}$$

LAPLACE TRANSFORM

Although a function f(t) may be defined for all values of t, its Laplace transform is not influenced by values of f(t), when T < 0. The Laplace transform of f(t) is actually defined for the function U(t) given by

$$U(t) = \begin{cases} f(t) \; for \, t \geq 0, \\ 0 \quad for \, t < 0. \end{cases} \qquad \text{...(79)}$$

A sufficient condition for the existence of the Laplace transform is that | f (t) | does not grow too rapidly as $t \to +\infty$. We say that the function f (t) is of exponential order if there exists real constants M > 0 such that

$| f(t) | \leq Me^{Kt}$ holds for all $t \geq 0$.

All functions in this chapter are assumed to be of exponential order. The next theorem shows that their Laplace transform F(s) = $F(\sigma + i\tau)$ exists for values of s = $\sigma + i\tau$ in a domain that includes the right half-plane Re (s) > K.

LAPLACE TRANSFORM OF FUNCTION

The Laplace transform is an integral transform perhaps second only to the Fourier transform in its utility in solving physical problems. The Laplace transform is particularly useful in solving linear ordinary differential equations such as those arising in the analysis of electronic circuits. The (unilateral) Laplace transform £ (not to be confused with the Lie derivative, also commonly denoted

£) is defined by $£t[f(t)](s) \equiv \int_0^\infty f(t)e^{-st}\,dt$, ...(80)

where f (t) is defined for $t \geq 0$. The unilateral Laplace transform is almost always what is meant by "the" Laplace transform, although a bilateral Laplace transform is sometimes also defined as

$$£t^{(2)}[f(t)](s) = \int_0^\infty f(t)e^{-st}\,dt \qquad ...(81)$$

The inverse Laplace transform is known as the Bromwich integral, sometimes known as the Fourier-Mellin integral (see also the related Duhamel's convolution principle). A table of several important one-sided Laplace transforms is given below.

f	$£[f(t)](s) = F(s)$	Range
1	$\frac{1}{s}$	s > 0
t	$\frac{1}{s^2}$	s > 0
t^n	$\frac{x}{s^{n+1}}$	$n \in Z > 0$
t^a	$\frac{\Gamma(a+1)}{s^{a+1}}$	a > 0
e^{at}	$\frac{1}{s-a}$	s > a
Cos(at)	$\frac{s}{s^2+a^2}$	s > 0
Sin(at)	$\frac{a}{s^2+a^2}$	s > 0

Cosh(at)	$\frac{s}{s^2 - a^2}$	s > | a |
Sinh(at)	$\frac{a}{s^2 - a^2}$	s > | a |
$e^{at} sin(bt)$	$\frac{b}{(s-a)^2 + b^2}$	s > a
$e^{at} cos(bt)$	$\frac{s-a}{(s-a)^2 + b^2}$	s > a
$\delta(t-c)$	e^{-cs}	c > 0
$H_c(t)$	$\frac{e^{-cs}}{s}$	s > 0
$J_0(t)$	$\frac{1}{\sqrt{s^2+1}}$	
$J^n(at)$	$\frac{\left(\sqrt{s^2+a^2} - s\right)^n}{a^n \sqrt{s^2+a^2}}$	s > 0, n > –1

In the above table, $J_0(t)$ is the zeroth-order Bessel function of the first kind, $J_0(t)$ is the delta function, and $H_c(t)$ is the Heaviside step function.

The Laplace transform has many important properties. The Laplace transform existence theorem states that, if f(t) is piecewise continuous on every finite interval in $(0,\infty)$ satisfying

$$| f(t)| \leq Me^{at} \quad ...(82)$$

for all $t \in (0,\infty)$, then $£[f(t)](s)$ exists for all s > a. The Laplace transform is also unique, in the sense that, given two functions F_1(t)and F_2(t) with the same transform so that

$$£[f_1(t)](s) = £[f_2(t)](s) \equiv f(s), \quad ...(83)$$

then Lerch's theorem guarantees that the integral

$$\int_0^a N(t)dt = 0 \quad ...(84)$$

vanishes for all a > 0 for a null function defined by

$$N(t) \equiv F_1(t), \quad ...(85)$$

The Laplace transform is linear since

$$£[af(t) + bg(t)] = \int_0^\infty [af(t) + bg(t)e^{-st}dt] \quad ...(86)$$

$$= a\int_0^\infty fe^{-st}dt + b\int_0^\infty ge^{-st}dt \quad ...(87)$$

$$= a\pounds[f(t)] + b\pounds[g(t)] \quad ...(88)$$

The Laplace transform of a convolution is given by

$$\pounds[f(t) * +g(t)] = \pounds[f(t)]\pounds[g(t)] \quad ...(89)$$

$$\pounds^{-1}[F * G] = \pounds^{-1}[F] * \pounds^{-1}[G] \quad ...(90)$$

Now consider differentiation. Let $f(t)$be continuously differentiable n–1 times in $[0,\infty]$. If $|f(t)| \leq Me^{at}$, then

$$\pounds[f^n(t)](s) = S^n\pounds[f(t)] - S^{n-1}f(0) - S^{n-2}f(0) - ... - f^{(n-1)} \quad ...(91)$$

This can be proved by integration by parts,

$$\pounds[f'(t)](s) = \lim_{a\to\infty} \int_0^a e^{-st} f'(t)\,dt \quad ...(92)$$

$$= \lim_{a\to\infty} \left\{ \left[e^{-st} f(t)\right]_0^a + s\int_0^a e^{-st} f(t)\,dt \right\} \quad ...(93)$$

$$= \lim_{a\to\infty} \left[e^{-sa} f(a) - f(0) + s\int_0^a e^{-st} f(t)\,dt \right] \quad ...(94)$$

$$= s\pounds[f(t)] - (0) \quad ...(95)$$

Continuing for higher-order derivatives then gives

$$\pounds[f''(t)](s) = s^2\pounds[f(t)](s) - sf(0) - f'(0) \quad ...(96)$$

This property can be used to transform differential equations into algebraic equations, a procedure known as the Heaviside calculus, which can then be inverse transformed to obtain the solution. For example, applying the Laplace transform to the equation

$$f''(t) + a_1 f'(t) + a_0 f(t) = 0 \quad (97)$$

gives

$$\left\{s^2\pounds[f(t)](s)\right\} - sf(0) - f'(0) + a_1\left\{s\pounds[f(t)](s) - f(0)\right\} + a_0\pounds[f(t)](s) = 0 \quad ...(98)$$

$$\pounds[f(t)](s)\left(s^2 + a_1 s + a_0\right) - sf(0) - f'(0) - a_1 f(0)s = 0 \quad ...(99)$$

which can be rearranged to

$$\pounds[f(t)](s) = \frac{sf(0) + f'(0) + a_1 f(0)}{s^2 + a_1 s + a_0} \quad ...(100)$$

If this equation can be inverse Laplace transformed, then the original differential equation is solved.

The Laplace transform satisfied a number of useful properties. Consider exponentiation. If $£\left[f(t)\right](s) = F(s)$ for s > a (i.e., *F*(s)is the Laplace transform of *f*), then $£\left[e^{at} f\right](s) = f(s - a)$ for s > a + a. This follows from

$$F(s-a) = \int_0^\infty fe^{-(s-a)t}\,dt = \int_0^\infty \left[f(t)e^{at}\right]e^{-st}\,dt \qquad ...(101)$$

$$= £\left[e^{at} f(t)\right](s) \qquad ...(102)$$

The Laplace transform also has nice properties when applied to integrals of functions. If *f*(t)is piecewise continuous and $|\, f(t) \,| \leq Me^{at}$ then

$$£\left[\int_0^t f(f')\,dt'\right] = \frac{1}{s}£t\left[f(t)\right](s) \qquad ...(103)$$

***Theorem 12* (Laplace Transform).** If *f*(t) is of exponential order, then its Laplace Transform $£t(f(t))$ = F(s) exists and is given by

$$F(s) = \int_{-\infty}^{\infty} f(t)e^{-st}\,dt \textit{ where } s = \sigma + i\tau.$$

The defining integral for F(s) exists at points $s = \sigma + i\tau$ in the right half plane $\tau > k$.

Remark: The domain of definition of the defining integral for the Laplace transform $£(f(t))$ seems to be restricted to a half plane. However, the resulting formula F(s) might have a domain much larger than this half plane. Later we will show that F(s) is an analytic function of the complex variables. For most applications involving Laplace transforms that we will study, the Laplace transforms are rational functions that have the form $\frac{P(s)}{Q(s)}$, where P and Q are polynomials, and some other important ones will have the form $\frac{e^{is}P(s)}{Q(s)}$.

***Theorem 13:* (Linearity of the Laplace Transform).** Let f(t) have Laplace transforms F(s) and G(s), respectively. If a and b are constants, then

$$£(af(t) + bg(t)) = aF(s) + bG(s) \qquad ...(104)$$

***Theorem 14* (Uniqueness of the Laplace Transform).** Let f(t) and g(t) have Laplace transforms F(s) and G(s), respectively.

If F(s) ≡ G(s) than f(t) ≡ g(t)

Example 8. Find the Laplace transform of the step function

$f(t) = 1$ for $0 \leq t < c$,

$f(t) = 0$ for $c < t$.

Solution: Consider the function f(t) and calculate the integral defining $\pounds(f(t))$.

$$f[t] = 1$$

$$g[t] = \int 1\, e^{-st} dt = -\frac{e^{-st}}{s}$$

$$F[s] = g[c] - g[0] = \frac{1}{s} - \frac{e^{-cs}}{s}$$

$$F(s) = \int_0^c e^{-st} dt = \frac{1}{s} - \frac{e^{-cs}}{s}$$

$$F(s) = \frac{1}{s} - \frac{e^{-cs}}{s}$$

$$f[t] = 1$$

$$f(s) = \frac{1}{s}$$

Example 9. Find the Laplace transform of $f(t) = e^{it}$.

Solution: Take the function $f(t)$ and calculate the integral defining $\pounds(f(t))$.

$$f[t] = e^{it}$$

$$g[t] = \int e^{it} e^{-st} dt = \frac{e^{(1-s)t}}{a-s}$$

$$F[s] = \int e^{it} e^{-st} dt = g[\infty] - g[0]$$

$$F[s] = -\frac{1}{a-s} + \frac{e^{(1-s)\infty}}{a-s} = \frac{1}{-a+s}$$

Assuming that f(t) is of exponential order, we have $g[t] \to 0$ as $t \to \infty$. Then $g[\infty] = 0$, and the Laplace transform can be computed by the calculation $g[\infty] - g[0] = -g[0]$.

$$f[t] = e^{it}$$

$$g[t] = \int e^{it} e^{-st} dt = \frac{e^{(1-s)t}}{a-s}$$

$$F[s] = -g[0] = \frac{1}{-a+s}$$

$$f[t] = e^{it}$$

$$F(s) = \frac{1}{-a+s}$$

Example 10. Find the Laplace transform of f[t] = Sinh[at].

Solution:

Since $f[t] = \text{Sinh}[at] = \dfrac{e^{it} - e^{-it}}{2} = \dfrac{1}{2}e^{it} - \dfrac{1}{2}e^{-it}$, we can apply the results of Example 11.8;

$£\left(e^{it}\right) = \dfrac{1}{s-a}$ and $£\left(e^{-it}\right) = \dfrac{1}{s+a}$.

$$f_1[t] = e^{it}$$

$$f_2[t] = e^{-it}$$

$$f_{12}[t] = \frac{1}{2}\left(-e^{it} + e^{it}\right)$$

$$f_{12}[t] = \text{Sinh}[at]$$

$$f[t] = \text{Sinh}[at]$$

$$F_1[s] = \frac{1}{-a+s}$$

$$F_2[s] = \frac{1}{a+s}$$

$$F_{12}[s] = \frac{F_1[s] - F_2[s]}{2} = \frac{1}{2}\left(\frac{1}{-a+s} - \frac{1}{a+s}\right)$$

$$F_{12}[s] = \frac{F_1[s] - F_2[s]}{2} = -\frac{a}{(a-s)(a+s)}$$

$$F(s) = £\left(\sin h[at]\right) = \frac{a}{-a^2+s^2}$$

$$f[t] = \sin h[at]$$

$$F[s] = \frac{a}{-a^2+s^2}$$

We can use the property of linearity to find new Laplace transforms from known transforms.

Example 11. Find the Laplace transform of f[t] = t.

Solution: Consider the function f (t) and calculate the integral defining £ (f(t)).

f [t] = t

$$g[t] = \int te^{-st}dt = e^{-st}\left(-\frac{1}{s^2} - \frac{t}{s}\right)$$

$$F[s] = \int_0^{\infty} t\,e^{-st}dt = g[\infty] - g[0]$$

$$F[s] = \frac{1}{s^2} + e^{s(-\infty)}\left(-\frac{1}{s^2} + \frac{-\infty}{Sign[s]}\right) = \frac{1}{s^2}$$

Assuming that f(t) is of exponential order, we have $f[t] \to 0$ as $t \to \infty$. Then $g(\infty) = 0$, and the Laplace transform can be computed by the calculation $g[\infty] - g[0] = -g[0]$.

f [t] = t

$$g[t] = \int te^{-st}dt = e^{-st}\left(-\frac{1}{s^2} - \frac{t}{s}\right)$$

$$F[s] = g[0] = \frac{1}{s^2}$$

f [t] = t

$$F[s] = \frac{1}{s^2}$$

Example 12. Find the Laplace transform of f (t) = t^3

Solution: Consider the function f(t) and calculate the integral defining £ (f(t)).

f (t)= t^3

$$g[t] = \int t^3e^{-st}dt = e^{-st}\left(-\frac{6}{s^4} - \frac{6t}{s^3} - \frac{3t^2}{s^2} - \frac{t^3}{s}\right)$$

$$g[s] = \int_0^{\infty} t^3e^{-st}dt = g[\infty] - g[0]$$

$$F[s] = +e^{s(-\infty)}\left(-\frac{6}{s^4}+\frac{-\infty}{Sign[s]^3}+\frac{-\infty}{Sign[s]^2}+\frac{-\infty}{Sign[s]}\right) = \frac{6}{s^4}$$

Assuming that f(t) is of exponential order, we have $g[t]\to 0$ as $t\to 0$. Then $g[\infty]\to 0$ and the Laplace transform can be computed by the calculation $g[\infty]-g[0]=-g[0]$.

$$f[t] = t^3$$

$$g[t] = \int t^3 e^{-st}dt = e^{-st}\left(-\frac{6}{s^4}-\frac{6t}{s^3}-\frac{3t^2}{s^2}-\frac{t^3}{s}\right)$$

$$F[s] = g[0] = \frac{6}{s^4}$$

Example 13. Find the Laplace transform of f [t] = Cos [bt].

Solution:

Method 1. Consider the function f(t) and calculate the integral defining £(f(t)), were

$$f[t] = Cos(bt)$$

$$g[t] = \int Cos[bt]e^{-st}dt = \frac{e^{-st}(-s\,Cos[bt]+b\,Sin[bt])}{b^2+s^2}$$

$$F[s] = \int_0^{\infty} Cos[bt]e^{-st}dt = g[\infty]-g[0]$$

$$F[s] = \frac{s-e^{s(-\infty)}s\,Cos[b\infty]+be^{s(-\infty)}Sin[b\infty]}{b^2+s^2} = \frac{s}{b^2+s^2}$$

Assuming that f(t) is of exponential order, we have $g[t]\to 0$ as $t\to 0$. Then $g[\infty]\to 0$. Then $g(\infty)=0$, and the Laplace transform can be computed by the calculation $g[\infty]-g[0]=-g[0]$.

$$f[t] = Cos[bt]$$

$$g[t] = \int Cos[bt]e^{-st}dt = \frac{e^{-st}(-s\,Cos[bt]+b\,Sin[bt])}{b^2+s^2}$$

$$F[s] = g[0] = \frac{s}{b^2 + s^2}$$

Method 2. Since $f[t] = \quad \text{Cos}[bt] = \frac{e^{ibt} + e^{-ibt}}{2} = \frac{1}{2}e^{ibt} + \frac{1}{2}e^{-ibt}$, we can use $£\left(e^{ibt}\right)$

$$= \frac{1}{s - ib} \text{ and } £\left(e^{-ibt}\right) = \frac{1}{s + ib}$$

$$f_1[t] = e^{ibt}$$

$$f_2[t] = e^{-ibt}$$

$$f_{12}[t] = \frac{1}{2}\left(e^{-ibt} + e^{ibt}\right)$$

$$f_{12}[t] = \text{Cos}[bt]$$

$$f_1[s] = \frac{1}{-ib + s}$$

$$F_2[s] = \frac{1}{ib + s}$$

$$F_{12}[s] = \frac{F_1[s] + F_2[s]}{2} = \frac{1}{2}\left(\frac{1}{-ib + s} + \frac{1}{ib + s}\right)$$

$$F_{12}[s] = \frac{F_1[s] + F_2[s]}{2} = \frac{s}{(b - is)(b + is)}$$

$$F[s] = £(\text{Cos}[b\,t]) = \frac{s}{b^2 + s^2}$$

$$f[t] = \text{Cos}[bt]$$

$$F[s] = \frac{s}{b^2 + s^2}$$

Example 14. Find the inverse Laplace transform of $F[s] = \frac{3s + 6}{s^2 + 9}$.

Solution: Enter the function F[s] and find its partial fraction expansion.

$$F[s] = \frac{6 + 3s}{9 + s^2} = \frac{6}{9 + s^2} = \frac{3s}{9 + s^2}$$

The transform F [s] is a linear combination.

$$f_1[t] = \text{Cos}[3t]$$

$$f_2\,[t] = \text{Sin}\,[3t]$$

$$L_1\,[t] = \frac{s}{9+s^2}$$

$$L_2\,[t] = \frac{3s}{9+s^2}$$

$$F\,[s] = \frac{6+3s}{9+s^2}$$

Is $F\,[s] = 3L_1\,[s] + 2\,L_2\,[s]$?

$$\frac{6+3s}{9+s^2} = \frac{6}{9+s^2} + \frac{3s}{9+s^2}\text{, True}$$

The inverse of L_1 [s] is f_1 [t]= Cos[3t] and the inverse of L_2[s] is f_2[t] = Sin [3t].
Hence $f\,[t] = 3f_1\,[t] + 2f_2\,[t]$.

$$f\,[t] = 3f_1\,[t] + 2f_2\,[t]$$

$$f\,[t] = 3\,\text{Cos}[3t] + 2\,\text{Sin}\,[3t]$$

$$F[s] = \frac{6+3s}{9+s^2}$$

$$f\,[t] = 3\,\text{Cos}\,[3t] + 2\,\text{Sin}\,[3t]$$

LAPLACE TRANSFORMS OF DERIVATIVES AND INTEGRALS

Theorem 15. **(Differentiation of f(t)).** Let f(t) and f'(t) be continuous for $t \geq 0$, and of exponential order. Then

$$£(f'(t)) = sF(s) = f(s), \qquad ...(104)$$

where $F(s) = £(f(t))$.

Corollary (Differentiation of f(t)). If f(t), f'(t) and f''(t) are of exponential order, then

$$£(f''(t)) = s^2F(s) - s\,f\,(0) - f\,'(0). \qquad ...(105)$$

Example 15. Using Theorem 15 and find $£\left(cos^2(t)\right)$

Solution: Consider f [t] and calculate f [0] and f '[t].

$$f\,[t] = \text{Cos}\,[t]^2$$

$$f[0] = 1$$
$$f'[t] = -2\text{Cos}[t]\,\text{Sin}[t]$$
$$f'[t] = 4\text{-Sin}[2t]$$

Use the fact that $£(\text{Sin}[2t]) = \dfrac{2}{s^2+4}$

$$£(f'(t)) = sF(s) = f(s), \tag{104}$$

where $F(s) = £(f(t)).$

Corollary (Differentiation of f(t)). If f(t), f'(t) and f''(t) are of exponential order, then

$$£(f''(t)) = s^2F(s) - s\,f(0) - f'(0). \tag{105}$$

Example 16. Using Theorem 15 and find $£(cos^2(t))$

Solution: Consider f [*t*] and calculate f [0] and f '[*t*].

$$f[t] = \text{Cos}[t]^2$$
$$f[0] = 1$$
$$f'[t] = 2\text{Cos}[t]\,\text{Sin}[t]$$
$$f'[t] = -\text{Sin}[2t]$$

Use the fact that $£(\text{Sin}[2t]) = \dfrac{2}{s^2+4}$

$\dfrac{2}{s^2}$ £(sin[2*t*]) $= \dfrac{2}{s^2+4}$ to help solve for £ (f[t]).

$$£(f'[t]) = -\frac{2}{4+s^2}$$
$$f[0] = 1$$
$$£(f'[t]) = s\,F[s] - f[0]$$
$$-\frac{2}{4+s^2} = -1 + s\,F[s]$$
$$1 - \frac{2}{4+s^2} = s\,F[s]$$
$$\frac{2+s^2}{4+s^2} = s\,F[s]$$

$$\left\{\left\{F[s]\to\frac{2+s^2}{4s+s^3}\right\}\right\}$$

$$£\,(f[t]) = \frac{2+s^2}{4s+s^3}$$

$$f[t] = \text{Cos}[t]^2$$

$$£\,(f[t]) = \frac{2+s^2}{4s+s^3}$$

Theorem 16. **(Integration of f(t)).** Let f (t) be continuous for $0 \le t$, and of exponential order and £ (f(t)) = F(s), then

$$£\left(\int f(t)\,dt\right) = \frac{F(s)}{s}. \tag{106}$$

Example 17. Using Theorem 16 find £(f[t]). Find (a) £(t^2) (b) £(t^3).

Solution: (a) Consider the function $f[t] = t^2$.

$$f[t] = t^2$$

Use the fact that $f'[t] = 2t$ and $£(3t^2)=\frac{2}{s^2}$.

$$f'[t] = 2t$$

$$£(f'[t]) = \frac{2}{s^2}$$

$$£\left(\int f(t)\,dt\right) = \frac{£(f'[t])}{s}$$

$$£\left(\int f(t)\,dt\right) = \left(\frac{2}{s^2}\right)/s = \frac{2}{s^3}$$

$$£(f\,[t]) = \frac{2}{s^3}$$

$$f\,[t] = t^2$$

$$£(f\,[t]) = \frac{2}{s^3}$$

(b) Take the function $f\,[t] = t^3$

$$f\,[t] = t^3$$

Use the fact that $f'[t] = 3t^3$ and $£(3t^2) = \frac{2}{s^3}$

$$f'[t] = 3t^2$$

$$£(f'[t]) = \frac{6}{s^3}$$

$$£\left(\int f(t)\,dt\right) = \frac{£(f'[t])}{s}$$

$$£\left(\int f(t)\,dt\right) = \left(\frac{6}{s^3}\right)/s = \frac{6}{s^4}$$

$$£(f[t]) = \frac{6}{s^4}$$

$$f[t] = t^3$$

$$£(f[t]) = \frac{6}{s^4}$$

For many physical problems involving mechanical systems and electric circuits, the transform F(s) is known, and the inverse of Y(s) can easily be computed. This process is referred to as the operational calculus and has the advantage of changing problems in differential equations into problems in algebra. Then the solution obtained will satisfy the specific initial conditions.

Example 18. Solve the initial value problem

$$Y''[t] + Y[t] = 0$$

with

$$Y[0] = 2, \text{ and } Y'[0] = 3$$

Solution: Take the initial conditions and the functions f[t] and F[s]. Then Laplace transform the D.E. and solve for Y[s].

$$Y[t] + Y''[t] = f[t]$$

$$-sy0-yl + Y[s] + s^2\, Y[s] = F[s]$$

$$Y[0] = 2$$

$$Y'[0] = 3$$

$$f[t] = 0$$

$$f[s] = 0$$

$$Y[t] + Y''[t] = 0$$

$$-3 - 2s + Y[s] + s^2 Y[s] = 0$$

$$\left\{\left\{Y[s] \to \frac{3+2s}{1+s^2}\right\}\right\}$$

$$Y[s] = \frac{3+2s}{1+s^2}$$

$$Y[s] = \frac{3}{1+s^2} + \frac{2s}{1+s^2}$$

Using the known Laplace transforms £(Sin[t]) = $\frac{1}{1+s^2}$ and £(Cos[t]) = $\frac{s}{1+s^2}$ the solution f[t] is the linear combination Y[t] = 2 Cos[t] + 3Sin[t].

$$Y[s] = \frac{3}{1+s^2} + \frac{2s}{1+s^2}$$

$$Y[t] = 3Sin[t] + 2\ Cos[t]$$

We can check this result using the Laplace Transform

$$Y[s] = \frac{3}{1+s^2} + \frac{2s}{1+s^2}$$

$$Y[s] = 2\ Cos[t] + 3Sin[t]$$

We can substitute y[t] into the D.E. and verify it is a solution.

$$\begin{aligned} Y[t] &= 2\ Cos[t] + 3Sin[t] \\ Y''[t] &= -2\ Cos[t] - 3Sin[t] \\ Y[t] + Y''[t] &= 0 \\ Y[0] &= 2 \\ Y'[0] &= 3 \\ Y[t] + Y''[t] &= 0 \end{aligned}$$

{Y[0] = = 2, Y'[0] = = 3}

{{Y[t] → 2 Cos [t] + 3 Sin [t]}}

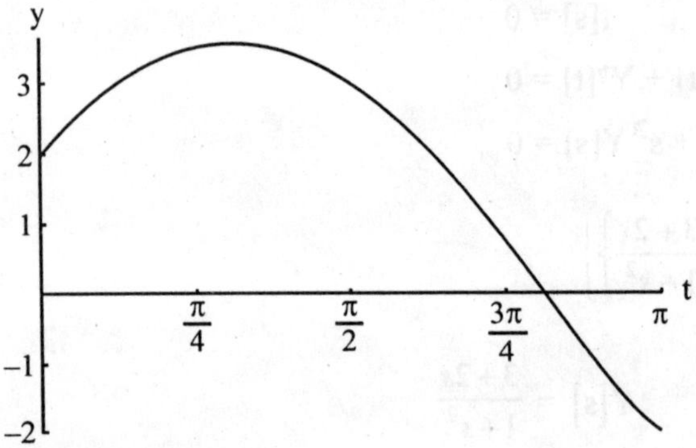

$Y[t] = 2\,Cos[t] + 3Sin[t]$

Example 19. Solve the initial value problem

$Y''[t] + Y'[t] - 2Y[t] = 0$

and

$Y[0] = 1$, and $Y'[0] = 4$

Solution: Take the initial conditions and the functions f[t] and F[s]. Then Laplace transform the D.E. and solve for Y[s].

$-2Y[t] + Y'[t] + Y''[t] == f[t]$

$-Y0 - sY0 - Y1 - 2Y[s] + sY[s] + s^2Y[s] == F[s]$

$$\begin{aligned} Y[0] &= 0 \\ Y'[0] &= 4 \\ f[t] &= 0 \\ F[s] &= 0 \\ -2Y[t] + Y'[t] + Y''[t] &== 0 \\ -5 - s - 2Y[s] + sY[s] + s^2Y[s] &== 0 \end{aligned}$$

$$\left\{\left\{Y[s] \to \frac{5+s}{-2+s+s^2}\right\}\right\}$$

$$Y[s] = \frac{5+s}{-2+s+s^2}$$

$$Y[s] = \frac{2}{-2+s} - \frac{1}{2+s}$$

Using the known Laplace transforms $\pounds\left(e^t\right) = \frac{1}{s-1}$ and $\pounds\left(e^{-2t}\right) = \frac{1}{s+2}$ the solution f [t] is the linear combination $Y[t] = 2e^t - e^{-2t}$.

$$Y[s] = \frac{2}{-1+s} + \left(-\frac{1}{2+s}\right)$$

$$Y[t] = 2e^{t} - e^{-2t}$$

We can check this with result using the Laplace Transform package.

$$Y[s] = \frac{2}{-1+s} - \frac{1}{2+s}$$

$$Y[t] = -e^{-2t} - 2e^{t}$$

We can substitute y[t] into the D.E. and verify it is a solution.

$$Y[t] = -e^{-2t} - 2e^{t}$$

$$Y''[t] = -4e^{-2t} - 2e^{t}$$

$$Y''[t] + Y' - 2Y = -2e^{-2t} + 4e^{t} - 2(-e^{-2t} - 2e^{t})$$

$$Y'' + Y' - 2Y = 0$$

$$Y[0] = 1$$

$$Y'[0] = 4$$

We solve the D.E. and plot the solution.

$$-2Y[t] + Y'[t] + Y''[t] == 0$$

$$\{Y[0] == 1, Y'[0] == 4\}$$

$$\{\{Y[t] \to -e^{-2t} - 2e^{t}\}\}$$

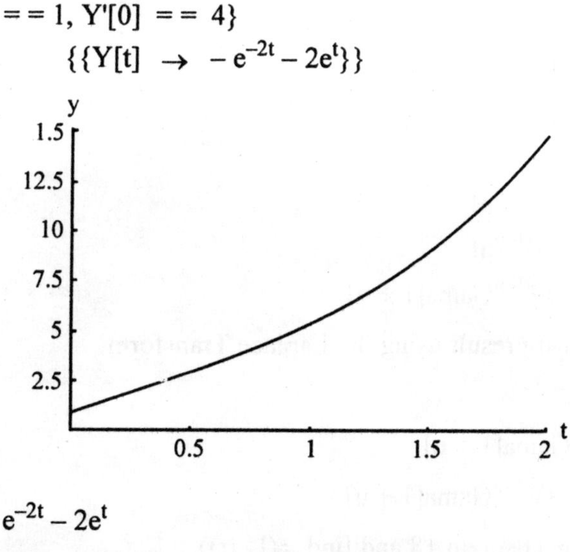

$$Y[t] \to -e^{-2t} - 2e^{t}$$

SHIFTING THEOREMS AND THE STEP FUNCTION

We have seen how the Laplace transform can be used to solve linear differential equations. Familiar functions that arise in solutions to differential equations are $e^{lt}\cos(bt)$ and $e^{lt}\sin(bt)$. The first shifting theorem will show how their transforms are related to those of cos (bt) and sin (bt) by shifting the variable s in F(s). A companion result, called thee second shifting theorem, will show how the transform of f(t-a) can be obtained by multiplying F(s) by e^{-is}. Loosely speaking, these results show that multiplication of f(t) by e^{it} corresponds to shifting F(s-a), and shifting f(t-a) corresponds to multiplication of the transform $F(s)e^{-is}$ by e^{-is}.

***Theorem 17.* (Shifting the Variable s)** If F(s) is the Laplace transform of f(t), then

$£(e^{it}f(t)) = F(s-a).$ ...(107)

***Definition:* (The Unit Step Function).** Let $a \geq 0$. Then, the **unit step function $U_1(t)$** is

$$\mathbf{U_1(t)} = \begin{cases} 0, \text{ for } < a, \\ 1, \text{ for } t > a. \end{cases} \qquad ...(108)$$

***Theorem 18.* (Shifting the Variable t).** If F(s) is the Laplace transform of f (f), and $a \geq 0$, then

$£(U_1 (t) f (t-a)) = e^{-1s} F(s).$ ...(109)

Example 20. Use Theorem 11.15 and find $£(t^n e^{at})$.

Solution:

Use the fact that $£(t^n) = \dfrac{n!}{s^{n+1}}$.

$$f[t] = t^n$$
$$£(t^n) = s^{-1-n}n!$$
$$F[s] = s^{-1-n}n!$$
$$£(f[t]\, e^{1t}) = F\,[s-a]$$
$$F\,[s-a] = (-a+s)^{-1-n}n!$$
$$£(t^n e^{1t}) = (-a+s)^{-1-n}n!$$
$$£(t^n e^{1t}) = (-a+s)^{-1-n}\text{Gama}[1+n]$$

We can check this with result using the Laplace Transform

$$f[t] = t^n$$
$$£(t^n) = s^{-1-n}\text{Gama}[1+n]$$
$$£(t^n e^{1t}) = (-a+s)^{-1-n}\text{Gama}[1+n]$$

Example 21. Using Theorem 18 and find $£(U_c(t))$.

Solution:

Consider the function f [t] = 1 and $F[s] = \dfrac{1}{s}$, and apply Theorem 11.16.

$$f[t] = 1$$

$$£(1)= \frac{1}{s}$$

$$F(s)= \frac{1}{s}$$

$$£(U_c(t)f[t-c]) = e^{-cs}F[s]$$

$$f[t-c] = 1$$

$$£(U_c(t)(1)) = \frac{e^{cs}}{s}$$

We can check this with result using the Laplace Transform.

$$g[t] = \text{UnitStep }[-c+t]\ (t)$$

$$g[t] = \text{UnitStep }[-c+t]$$

$$£(\text{UnitStep }[-c+t]) = \frac{e^{cs}}{s}$$

$$£(U_c(t)(1)) = \frac{e^{cs}}{s}$$

Example 22. Using Theorem 18 and find $£(U_c(t)\cos[t-c])$.

Solution:

Consider the function f [t] = cos [t] and $F[s] = \frac{s}{s^2+1}$, and applying Theorem 118, we get

$$f[t] = \cos[t]$$

$$£(\cos[t]) = \frac{s.}{1+s^2}$$

$$F[s] = \frac{s}{1+s^2}$$

$$£(U_c(t)f[t-c]) = e^{-cs}F[s]$$

$$f[t-c] = \cos[c-t]$$

$$£(U_c(t)\cos[t-c]) = \frac{e^{-cs}s}{1+s^2}$$

We can check this with result using the Laplace Transform

$$g[t] = \text{UnitStep}[-c+t]\ (\cos[c-t])$$

$$g[t] = \cos[c-t]\ \text{UnitStep }[-c+t]$$

$$£\ (\cos\,[c - t]\ \text{UnitStep}\ [-c + t]) = \frac{e^{-cs}s}{1+s^2}$$

$$£(U_c(t)\cos\,[c - t]) = \frac{e^{-cs}s}{1+s^2}$$

Example 23. Find £ ([t]) for f[t] given in Figure 11.24.

Solution:

Consider the function $f(t) = 1 - U_1(t) + U_2(t) - U_3(t) + U_4(t) - U5(t)$.

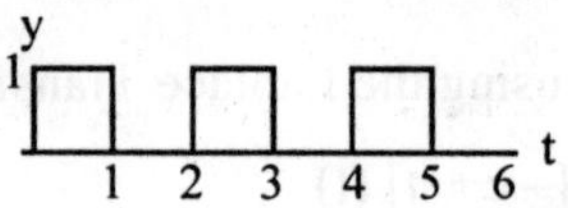

f [t] = 1 – UnitStep[–5 + t] + UnitStep[–4 + t] – UnitStep[–3 + t] + UnitStep[–2 + t] – UnitStep[–1 + t].

Use Laplace Transform package to find £(f [t]).

f [t] = 1 – UnitStep[–5 + t] + UnitStep[–4 + t] – UnitStep[–3 + t] + UnitStep[–2 + t] – UnitStep[–1 + t].

$$F[s] = \frac{1}{s} - \frac{e^{-5s}}{s} + \frac{e^{-4s}}{s} - \frac{e^{-3s}}{s} + \frac{e^{-2s}}{s} - \frac{e^{-s}}{s}$$

Example 24. Solve the initial value problem

$Y''(t) + Y(t) = U\pi\,(t)$

with

$Y(0) = 0$, and $Y'(0) = 0$

Solution:

Take the initial conditions and the functions $f[t]$ and $F[s]$. Then Laplace transform the D.E. and solve for $Y[s]$.

$Y(t) + Y''(t) = f(t)$

$-sY_0 - Y_1 + Y(s) + s^2\,Y(s) = F(s)$

$Y(0) = 0$

$Y'(0) = 0$

$f(t) = \text{UnitStep}\ (-\pi + t)$

$$F(s) = \frac{e^{-\pi s}}{s}$$

$Y(t) + Y''(t) = \text{UnitStep}\ (-\pi + t)$

$$Y(s) + s^2\, Y(s) = \frac{e^{-\pi s}}{s}$$

$$\left\{\left\{Y(s) \to \frac{e^{-\pi s}}{s\left(1+s^2\right)}\right\}\right\}$$

$$Y(s) = \frac{e^{-\pi s}}{s\left(1+s^2\right)}$$

$$Y(s) = e^{-\pi s}\left(\frac{1}{s} - \frac{s}{1+s'}\right)$$

$$Y(s) = \frac{e^{-\pi s}}{s} - \frac{e^{-\pi s}}{1+s^2}$$

Use the facts

that

$$£(Up\ (t)) = \frac{e^{-\pi s}}{s} \text{ and } £\ (U_\pi\ (t) \cos\ (t - \pi)) = \frac{s\, e^{-\pi s}}{s^2+1}$$

and construct $y[t]$.

$Y(t)$ = UnitStep $(-\pi + t)$ + cos (t) UnitStep $(-\pi + t)$

We can check this with result using the Laplace Transform

$$Y(s) = \frac{e^{-\pi s}}{s} - \frac{e^{-\pi s}}{1+s^2}$$

$Y(t)$ = (1 + cos (t)) UnitStep $(-\pi + t)$

We can substitute y[t] into the D.E. and verify it is a solution.

$Y(t)$ = (1 + cos (t)) UnitStep $(-\pi + t)$

$Y''(t)$ = –2 DiracDelta $(\pi - t)$ sin (t) – cos (t) UnitStep $(-\pi + t)$ – (1 + cos (t) DiracDelta'$(\pi - t)$)

$Y(t)$ + Y'' (t) = –2 DiracDelta $(\pi - t)$ sin (t) –

cos (t) UnitStep $(-\pi + t)$ + (1 + cos (t) UnitStep $(-\pi + t)$ – (1 + cos (t))

$Y(0) = 0$

$Y'(0) = 0$

Notice that the derivative of the unit step function involves the Dirac delta function.

We use D Solve package to solve the D.E. and plot the solution.

$Y''(t) + Y''(t)$ = UnitStep $(-\pi + t)$

$\{Y(0) = 0, Y'(0) = 0\}$

$\{(Y(t) \to \cos(t)\ \text{UnitStep}(-\pi + t) + \cos(t)^2\ \text{UnitStep}(-\pi + t) + \sin(t)^2\ \text{UnitStep}(-\pi + t))\}$

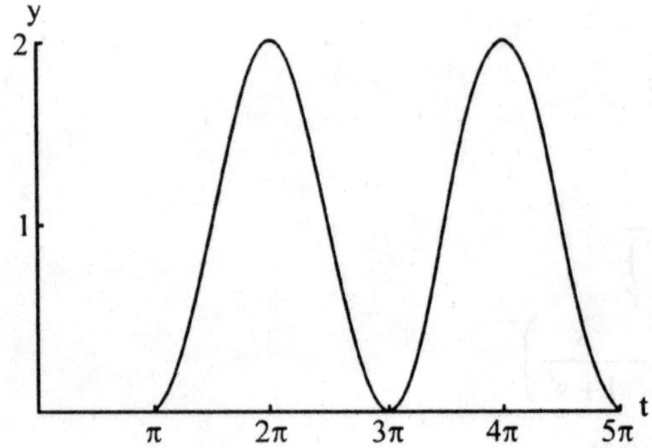

Y(t) = cos (t) UnitStep $(-\pi + t)$ + cos $(t)^2$ UnitedStep $(-\pi + t)^2$ UnitStep $(-\pi + t)$

Multiplication and division by t

Sometimes the solutions to nonhomogeneous linear differential equations with constant coefficients involve the functions t cos(b t), t sin(b t), or $t^n e^t$ as part of the solution. We now show how the Laplace transforms of $tf(t)$ and $\frac{f(t)}{t}$ are related to the Laplace transform of f(t). The transform of $tf(t)$ will be obtained via differentiation and the transform of $\frac{f(t)}{t}$ will be obtained via integration. To be precise, we state the following theorems.

Theorem 19. **(Multiplication by t)** If F(S) is the Laplace transform of f(t), then

$$£(tf(t)) = f'(s). \tag{110}$$

Theorem 20. **(Division by t)** Let both f(t) and $\frac{f(t)}{t}$ have Laplace transforms and let F(s) denote the Laplace transform of f(t). If $\lim_{t \to 0^+} \frac{f(t)}{t}$ exists then

$$£\left(\frac{f(t)}{t}\right) = \int_s^{\infty} F(\sigma)\,d\sigma \tag{111}$$

Example 25. Use Theorem 11.17 and find £(t cos (b t)).

Solution:

Use the fact that £(cos (b t)) = $\frac{s}{s^2 + b^2}$

f(t) = cos (b t)

$$£(f(t)) = \frac{s}{b^2 + s^2}$$

$$F(s) = \frac{s}{b^2 + s^2}$$

$$t\,f(t) = t \cos(b\,t)$$

$$£(t\,f(t)) = -F'(s)$$

$$£(t \cos(b\,t)) = \frac{2s^2}{\left(b^2 + s^2\right)^2} - \frac{1}{b^2 + s^2}$$

$$£(t \cos(b\,t)) = \frac{-b^2 + s^2}{\left(b^2 + s^2\right)^2}$$

We can check this with result using the Laplace Transform package.

$$t\,f(t) = t \cos(b\,t)$$

$$£\,(t\,f(t)) = \frac{-b^2 + s^2}{\left(b^2 + s^2\right)^2}$$

Example 26. Use Theorem 11.18 and find $£\left(\frac{\sin[t]}{t}\right)$.

Solution:

Use the fact that $£(\sin(t)) = \frac{1}{s^2 + 1}$

$$f(t) = \sin(t)$$

$$£\,(f(t)) = \frac{1}{1 + s^2}$$

$$F(s) = \frac{1}{1 + s^2}$$

$$\frac{f(t)}{t} = \frac{\sin(t)}{t}$$

$$£\left(\frac{f(t)}{t}\right) = \int_s^{\infty} F(r)\,dr$$

$$£\left(\frac{\sin(t)}{t}\right) = \int_s^{\infty} \frac{1}{1 + r^2}\,dr$$

$$£\left(\frac{\sin(t)}{t}\right) = \text{ArcTan}\left(\frac{1}{s}\right)$$

We can check this with result using the Laplace Transform package.

$$\frac{f(t)}{t} = \frac{\sin(t)}{t}$$

$$£\left(\frac{f(t)}{t}\right) = \text{ArcTan}\left(\frac{1}{s}\right)$$

Example 27. Solve the initial value problem

$ty''(t) - ty''(t) - y(t) = 0$

with

$y(0) = 0$, and $y'(0) = 0$

Solution:

Take the initial conditions and the functions f[t] and F[s]. Then Laplace transform the D.E. and solve for Y[s]. Observe that a D.E. involving Y'[s] will arise in the computations.

$-y(t) - ty'(t) + ty''(t) = f(t)$

$y_0 - 2s\, Y(s) + s\, Y''(s) - s_2\, Y'(s) = F(s)$

$Y(0) = 0$

$f(t) = 0$

$F(s) = 0$

$-2\, sY(s) + s\, Y'(s) - s_2\, Y'(s) = 0$

$$\left\{\left\{Y(s) \to \frac{c(1)}{(-1+s)^2}\right\}\right\}$$

$$Y(s) = \frac{c}{(-1+s)^2}$$

$$Y(t) = ce^2t$$

We can substitute y[t] into the D.E. and verify it is a solution.

$$Y(t) = ce^2\, t$$

$$Y'(t) = ce^2 + ce^2\, t$$

$$Y'' = 2\, ce^2 + ce^2\, t$$

$t\, y''(t) - t\, y'(t) - y(t) = -\, ce^t\, t - t(ce^t + ce^t\, t) + t\,(2\, ce^t + ce^t\, t)$

$t\, y''(t) - t\, y'(t) - y(t) = 0$

For this problem, the solution involves a family of functions.

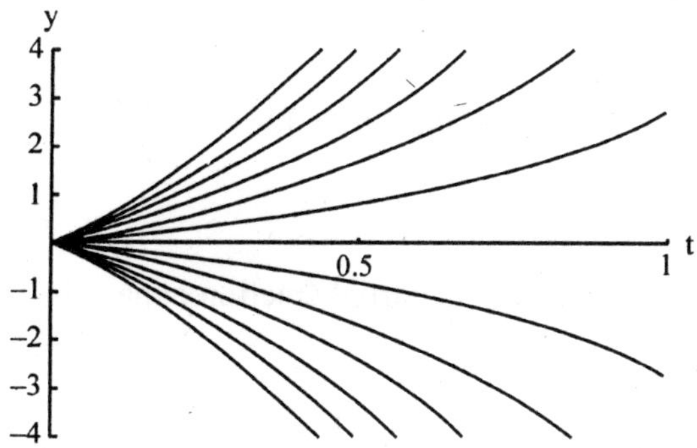

y(t = {$-6e^t$ t, – 5 e^t t, $-e^t$ t, – 3e^t t, $-2e^t$ t, $-e^t$ t, 0, e^t t, 2e^t t, 3e^t t, 4e^t t, 5e^t t, 6e^t t}

INVERSE LAPLACE TRANSFORM

So far, most of the applications involving the Laplace transform involves a transform (or part of a transform) that is expressed by

$$Y(s) = \frac{P(s)}{Q(s)}$$

where P and Q are polynomials that have no common factors. The inverse of Y(s) is found by using its partial fraction representation and referring to Table of Laplace Transforms. We now show how the theory of complex variables can be used to systematically find the partial fraction representation.

Theorem 21. **(Nonrepeated Linear Factors)** Let P(s) be a polynomial of degree at most n – 1. If Q(s) has degree n, and has distinct complex roots a_1, a_2, ...a_n,then

$$Y(s)=\frac{P(s)}{Q(s)}=\frac{P(s)}{(s-a_1)(s-a_2)...(s-a_n)}=\sum_{k=1}^{n}\frac{\text{Res}[Y,a_k]}{(s-a_k)}$$

Theorem 22. **(A Repeated Linear Factors)** If P(s) and Q (s) are polynomials of degree μ and *v*, respectively and μ < v + n and Q (a) $\neq$ 0, them

$$Y(s) = \frac{P(s)}{(s-a)^n A(s)}=\sum_{k=1}^{n}\frac{A_k}{(s-a)^k}+R(s),$$

where R(s) is the sum of all partial fractions that do not involve factors of the form $(s - a)^j$. Furthermore, the coefficients A_k can be computed with the formula

$$A_k = \frac{1}{(n-k)!}\lim_{s\to a}\frac{d^{n-k}}{d\,s^{n-k}}\frac{P(s)}{Q(s)} \text{ for } k = 1,2,\ldots,n$$

Example 28. Find the inverse Laplace transform of $Y(s) = \frac{s^3 - 4s + 1}{s(s-1)^3}$

Solution:

Take the function Y[s] and find its partial fraction expansion.

$$Y(s) = \frac{1-4s+s^3}{(-1+s)^3 s}$$

$$Y(s) = -\frac{2}{(-1+s)^2} + \frac{1}{(-1+s)^2} + \frac{2}{-1+s} - \frac{1}{s}$$

The partial fraction expansion can be obtained with the residue calculus.

$$Y(s) = \frac{a_3}{(s-1)^3} + \frac{a_2}{(s-1)^2} + \frac{a_1}{s-1} + \frac{h_1}{s}$$

At s = 1, we compute a_1, a_2, and a_3 as follows.

$$a_3 = \lim_{s\to 1}\frac{1-4s+s^3}{s} = -2$$

$$a_2 = \lim_{s\to 1}\frac{-1+2s^3}{s^2} = 1$$

$$a_1 = \lim_{s\to 1}\frac{2(1+s^3)}{s^3} = 2$$

At s = 0, we compute b_1 as follows.

$$b_1 = \lim_{s\to 0}\frac{1-4s+s^3}{(-1+s)^3} = -$$

Using the residue calculus, we construct the partial fraction expansion.

$$Y(s) = -\frac{2}{(-1+s)^3} + \frac{1}{(-1+s)^2} + \frac{2}{-1+s} - \frac{1}{s}$$

Use a table of Laplace transforms to solve for Y(t) = $£^{-1}$(Y[s])

f[t]	÷(f[t])
$-e^t t^2$	$-\frac{2}{(-1+s)^3}$

$e^t t$	$\frac{2}{(-1+s)^3}$
2et	$\frac{2}{-1+s}$
−1	$-\frac{1}{2}$

$$Y(s) = -\frac{2}{(-1+s)^3}+\frac{1}{(-1+s)^2}+\frac{2}{-1+s}-\frac{1}{s}$$

$$Y(t) = -1 + 2e^t + e^t t - e^t t^2$$

We can check this with result using the Laplace Transform.

$$Y(s) = -\frac{2}{(-1+s)^3}+\frac{1}{(-1+s)^2}+\frac{2}{-1+s}-\frac{1}{s}$$

$$Y(t) = -1 + 2e^t + e^t t - e^t t^2$$

Theorem 23. **(Irreducible Quadratic Factors)** Let P(s) and Q(s) be polynomials with real coefficients such that the degree of P(s) is at most 1 larger than the degree of Q(s). If T(s) does not have a factor of the form $(s - a)^2 + b^2$, then

$$Y(s) = \frac{P(s)}{Q(s)}=\frac{P(s)}{((s\text{-}a)^2+b^2)T(s)} \quad ...(112)$$

$$= \frac{2A(s - a) - 2Bb}{(s - a)^2+b^2}+R(s),$$

where $A + iB = \frac{P(a+ib)}{Q'(a+ib)}$.

Example 29. Find the inverse Laplace transform of Y(s) =

$$\frac{5s}{(s^2+4)(s^2+9)}.$$

Solution:

Consider the function Y[s] and find its partial fraction expansion.

$$Y(s) = \frac{5s}{(4+s^2)(9+s^2)}$$

$$Y(s) = \frac{s}{4+s^2}-\frac{s}{9+s^2}$$

The partial fraction expansion can be obtained with the residue calculus.

The simple poles in the upper half plane occur at $z_1 = 2\,i$ and $z_t = 3\,i$. First determine their residues.

$$Y(s) = \frac{5s}{(4+s^2)(9+s^2)}$$

$$\text{Res}[Y(s), 2i] = \frac{1}{2}$$

$$\text{Res}\,[Y(s), 3\,i] = -\frac{1}{2}$$

Using the residue calculus, we construct the partial fraction expansion.

$$Y_1(s) = \frac{s}{4+s^2}$$

$$Y_2(s) = -\frac{s}{9+s^2}$$

$$Y(s) = \frac{s}{4+s^2} - \frac{s}{9+s^2}$$

Use a table of Laplace transforms to solve for $Y(t) = £^{-1}\,(Y(s))$

f[t]	÷{(f [t])
cos (2 t)	$\frac{s}{4+s^2}$
–cos (3 t)	$-\frac{s}{9+s^2}$

$$Y(s) = \frac{s}{4+s^2} - \frac{s}{9+s^2}$$

$$Y(t) = \cos(2\,t) - \cos(3\,t)$$

We can check this result using the Laplace Transform.

$$Y(s) = \frac{s}{4+s^2} - \frac{s}{9+s^2}$$

$$Y(t) = \cos(2\,t) - \cos(3\,t)$$

Example 30. Find the inverse Laplace transform of $Y(s) = \dfrac{s^3+3s^2-s+1}{s(s+1)^2(s^2+1)}$

Solution: Consider the function $Y[s]$ and find its partial fraction expansion.

$$Y(s) = \frac{5s}{(4+s^2)(9+s^2)}$$

$$Y(s) = \frac{s}{4+s^2} - \frac{s}{9+s^2}$$

The partial fraction expansion can be obtained with the residue calculus.

The simple poles in the upper half plane occur at $z_1 = 2i$ and $z_2 = 3i$. First determine their residues.

$$Y(s) = \frac{5s}{(4+s^2)(9+s^2)}$$

$$\text{Res}\,(Y(s), 2\,i) = \frac{1}{2}$$

$$\text{Res}\,(Y(s), 3\,i) = -\frac{1}{2}$$

Using the residue calculus, we construct the partial fraction expansion.

$$Y_1(s) = \frac{s}{4+s^2}$$

$$Y_2(s) = -\frac{s}{9+s^2}$$

$$Y(s) = \frac{s}{4+s^2} - \frac{s}{9+s^2}$$

Use a table of Laplace transforms to solve for $Y(t) = £^{-1}(Y(s))$

f[t]	÷{(f[t])
cos(2 *t*)	$\frac{s}{4+s^2}$
–cos (3 *t*)	$-\frac{s}{9+s^2}$

$$Y(s) = \frac{s}{4+s^2} - \frac{5}{9+s^2}$$

$$Y(t) = -\cos(2\,t) - \cos(3\,t)$$

We can check this with result using the Laplace Transform .

$$Y(s) = \frac{s}{4+s^2} - \frac{s}{9+s^2}$$

$$Y(s) = \cos(2\,t) - \cos(3\,t)$$

Example 31. Solve the initial value problem

$Y'(t) = Y(t) - x(t)$ with $Y(0) = 1$,

$X'(t) = 5\,Y(t) - 3x(t)$ and $x(0) = 2$.

Solution: Take the initial conditions and the system of D.E.s. Then Laplace transform the D.E.s and solve for the transforms $Y[s]$ any $X[s]$ and the functions $y[t]$ and $x[t]$.

$$Y'(t) = -x(t) + Y(t)$$

$$X'(t) = -3\,x\,(t) + 5\,Y(t)$$

$$Y(0) = 1$$

$$X(0) = 2$$

$$-1\,m+ sY\,(s) = -X(s) + Y(s)$$

$$-2 + sX(s) = -\,3\,X\,(s) + 5\,Y(s)$$

$$\left\{\left\{X(s) \to -\frac{-3-2s}{2+2s+s^2}, Y(s) \to -\frac{-1-s}{2+2s+s^2}\right\}\right\}$$

$$Y(s) = -\frac{-1-s}{2+2s+s^2}$$

$$X(s) = -\frac{-3-2s}{2+2s+s^2}$$

$$Y(t) = e^{-t}\cos\,(t)$$

$$x(t) = E^{-t}\,(2\cos\,(t) + \sin\,(t))$$

We can substitute $y[t]$ and $x[t]$ into the D.E.'s and verify that they are the solution.

$$Y'\,(t) = -e^{-t}\cos\,(t) - e^{-t}\sin\,(t)$$

$$y(t) - x(t) = -e^{-t}\cos\,(t) - e^{-t}\sin\,(t)$$

Does $Y'(t) = Y(t) - X(t)$? True

$$X'(t) = e^{-t}\,(\cos\,(t) - 2\sin\,(t)) - e^{-t}\,(2\cos\,(t) + \sin\,(t))$$

$$X'(t) = -e^{-t}\cos\,(t) - 3\,e^{-t}\sin\,(t)$$

$$5Y(t) - 3X(t) = -e^{-t}\cos\,(t) - 3\,e^{-e}\sin(t)$$

Does $X'(t) = 5Y(t) - 3X(t)$? True

We solve the system of D.E.'s and plot the solution.

$$Y'(t) = -X(t) + Y(t)$$

$$X'(t) = -3\,X(t) + 5Y(t)$$

$$\left\{\left\{Y(t) \to \frac{1}{2}e^{(-1-i)t}\left(1+e^{2it}\right), X(t) \to \frac{1}{2}e^{(-1-i)t}\left((2+i)+(2-i)e^{2it}\right)\right\}\right\}$$

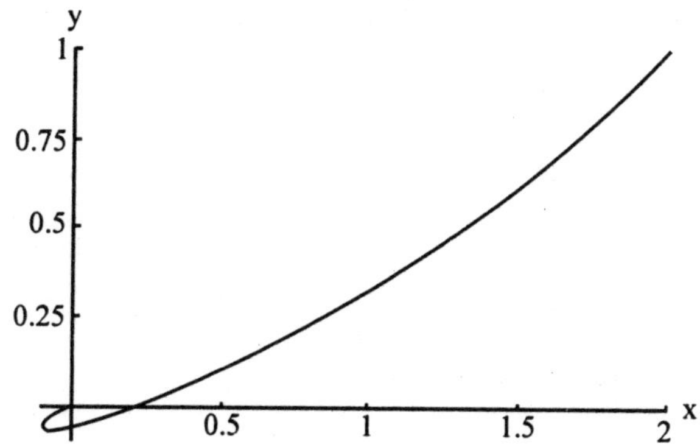

$Y(t) = e^{-t} \cos (t) \cos(t)$

$X(t) = e^{-t} (2 \cos (t) + \sin (t))$

$Y(0) = 1$

$X(0) =$

***Theorem 24.* (Inverse Laplace Transform)** Let $Y(s) = \dfrac{P(s)}{Q(s)}$, where P(s) and Q(s) are polynomials of degree m and n, respectively, and n > m. The inverse Laplace transform of Y(s) is given by

$$Y(t) = £^{-1} (Y(s)) = \Sigma \text{ Res}[F(s) e^{st}, s_k]$$

where the sum is taken over all the residues of the complex function $F(s) e^{st}$.

Using the residue calculus, the English mathematician Oliver Heaviside discovered a method for inverting the Laplace transform.

***Theorem 25.* (Heaviside Expansion Theorem)** Let P(s) and Q(s) be polynomials of degree m and n, respectively, where n > m. If Q(s) has n distinct simple zeros at the points $s_1, s_2, \ldots, s_n$, then $\dfrac{P(s)}{Q(s)}$ is the Laplace transform of the function Y(t) given by

$$Y(t) = £^{-1} \frac{P(s)}{Q(s)} = \sum_{k=1}^{n} \frac{P(s_k)}{Q'(s_k)} e^{s} k^{t} \tag{113}$$

Example 32. Find the inverse Laplace transform of

$$Y(s) = \frac{4s+3}{s^3 + 2s^2 + s + 2}$$

Solution: Consider the function $Y[s]$ and use the Heaviside Expansion Theorem to find y[t].

$P(s) = 3 + 4s$

$$Q(s) = 2 + s + 2\,s^2 + s^2$$

$$Y(s) = \frac{3+4s}{2+s+2s^2+s^3}$$

Solve the equation Q(s) = 0

$2 + s + 2\,s^2 + s^3 = 0$

$(2 + s)\,(1 + s^2) = 0$

$\{(s \to -2)\}, \{2 \to -i\}, \{s \to i\}$

$s_1 = -2$

$s_2 = -i$

$s_3 = i$

$Q[s]$ has simple zeros at the points $s_1 = -2$, $s_t = -i$, and $s_3 = i$. Construct the three terms $\frac{P[s_k]}{Q'[s_k]} e^{s_k t}$ for $k = 1, 2, 3$.

$P(-2)/Q'(-2) = (-5)/(5) = -1$

$P(-i)/Q'(-i) = (3 - 4i)/(-2 - 4i) = \frac{1}{2} + i$

$P(i)/Q'(i) = (3 + 4i)/(-2 + 4i) = \frac{1}{2} + i$

$P(-2)/Q'(-2)\,e^{-2t} = -e^{-2t}$

$P(-i)/Q'(-i)\,e^{-it} = \left(\frac{1}{2} + i\right) e^{-it}$

$P(-i)/Q'(-i)\,e^{-it} = \frac{\cos(t)}{2} + i\left(\cos(t) - \frac{\sin(t)}{2}\right) + \sin(t)$

$P(i)/Q'(i)\,e^{it} = \left(\frac{1}{2} - i\right) e^{it}$

$P(i)/Q'(i)\,e^{it} = \frac{\cos(t)}{2} + i\left(-\cos(t) + \frac{\sin(t)}{2}\right) + \sin(t)$

Use the Heaviside Expansion Theorem to construct y[t].

$Y(t) = P[-2]/Q'[-2]\,e^{-2t} + P[-i]/Q'[-i]\,e^{-it} + P[i]/Q'[i]e^{it}$

$Y[t] = (-e^{-it}) + \left(\left[\frac{1}{2} + i\right] e^{-it}\right) + \left(\left[\frac{1}{2} - i\right] e^{it}\right)$

$$Y[t] = (-e^{-2t}) + \left(\frac{\cos[t]}{2} + i\left(\cos[t] - \frac{\sin[t]}{2}\right) + \sin[t]\right)$$

$$+\left(\frac{\cos[t]}{2} + i\left(-\cos[t] + \frac{\sin[t]}{2}\right) + \sin[t]\right)$$

$$Y[t] = -e^{-2t} + \cos[t] + 2\sin[t]$$

We can check this with result using the Laplace Transform

$$F[s] = -\frac{1}{2+s} + \frac{2}{1+s^2} + \frac{s}{1+s^2}$$

$$Y[t] = -e^{-2t} + \cos[t] + 2\sin[t]$$

Convolution for the Laplace Transform

Let F(s) and G(s) denote the transforms of f(t) and g(t), respectively. Then the inverse of the product F(s) G(s) is given by the function h(t) = (f + g) (t) and is called the convolution of f(t) and g(t) and can be regarded as s generalized product of f(t) and g(t). Convolution will assist us in solving integral equations.

***Theorem 26.* (Convolution Theorem)** Let F(s) and G(s) denote the Laplace transforms of f(t) and g(t), respectively. Then the product H(s) = F(s) G(s) is the Laplace transform of the convolution of f(f) and g (t) and is denoted by h(t) = (f + g) (t), and has the integral representation

$$h(t) = (f+g)[t] = \int_0^t f[\tau]g[t-\tau]d\tau \tag{114}$$

or $$h(t) = (g+f)(t) = \int_0^t g[\tau]f[t-\tau]d\tau \tag{115}$$

Example 33. Use convolution theorem to find the inverse Laplace transform of H(s) = $\frac{2s}{(s^2+1)^2}$

Solution: The transform H[s] is the product of F(s) = $\frac{1}{s^2+1}$, and G(s) = $\frac{2s}{s^2+1}$, which are known to be the Laplace transforms of f(t) = sin(t), and g(t) = 2 cos(t), respectively. Consider the function H[s].

$$F(s) = \frac{1}{1+s^2}$$

$$G(s) = \frac{2s}{1+s^2}$$

$$F(s)\,G(s) = \frac{2s}{\left(1+s^2\right)^2}$$

$$H(s) = \frac{2s}{\left(1+s^2\right)^2}$$

Use the convolution formula to construct h(t) = (f + g) (t) = $\int_0^t f(\tau)g(t-\tau)d\tau$

$$f(\tau) = \sin(\tau)$$

$$g(t) = 2\cos(t)$$

$$g(t-\tau) = 2\cos(t-\tau)$$

$$f(\tau)\,g(t-\tau) = 2\cos(t-\tau)\sin(\tau)$$

$$\int f(\tau)\,g(t-\tau)d\tau = 2\left(\frac{1}{2}\tau\sin(t)+\frac{1}{2}\left(-\frac{1}{2}\cos(t)\cos(2\tau)-\frac{1}{2}\sin(t)\sin(2\tau)\right)\right)$$

$$h(t) = \int_0^t f(\tau)g(t-\tau)d\tau$$

$$h(t) = 2\left(\frac{\cos(t)}{4}-\frac{1}{4}\cos(t)\cos(2t)+\frac{1}{2}t\sin(t)-\frac{1}{4}\sin(t)\sin(2t)\right)$$

$$h(t) = t\sin(t)$$

We can use the other convolution formula to construct h(t) = (g * f) (t) = $\int_0^t g(\tau)f(t-\tau)d\tau$

$$f(t) = \sin(t)$$

$$f(t-\tau) = \sin(t-\tau)$$

$$g(\tau) = 2\cos(\tau)$$

$$g(\tau)\,f(t-\tau) = 2\cos(\tau)\sin(t-\tau)$$

$$\int g(\tau)\,f(t-\tau) = d\tau - 2\left(\frac{1}{2}\tau\sin(t)+\frac{1}{2}\left(\frac{1}{2}\cos(t)\cos(2\tau)+\frac{1}{2}\sin(t)\sin(2\tau)\right)\right)$$

$$\int g(\tau)\,f(t-\tau) = 2\left(\frac{1}{2}\tau\sin(t)+\frac{1}{2}\left(\frac{1}{2}\cos(t)\cos(2\tau)+\frac{1}{2}\sin(t)\sin(2\tau)\right)\right)$$

$$\int g(\tau)f(t-\tau)d\tau = \frac{1}{2}\cos(t)\cos(2\tau)+\tau\sin(t)+\frac{1}{2}\sin(t)\sin(2\tau)$$

$$h(t) = \int_0^t g(\tau)g(t-\tau)d\tau$$

$$h(t) = 2\left(-\frac{\cos(t)}{4}+\frac{1}{4}\cos(t)\cos(2t)+\frac{1}{2}t\sin(t)+\frac{1}{2}t\sin(t)+\frac{1}{4}\sin(t)\sin(2t)\right)$$

h(t) = t sin (t)

We can check this with result using the Laplace Transform package.

$$H(s) = \frac{2s}{\left(1+s^2\right)^2}$$

h(t) = t sin (t)

Example 34. Use the convolution theorem to solve the integral equation f (t) = 2 cos (t) $- \int_0 g(t-\tau)f(\tau)d\tau$

Solution: Use the facts that £(t) $= \frac{1}{s^2}$ and £ (cos (t)) = $\frac{s}{s^2+1}$.

Set up the integral equation and Laplace transform each term.

$$f(t) = 2\cos(t) - \int_0 g(t-\tau)f(\tau)d\tau$$

$$F(s) = \frac{2s}{s^2+1} - \frac{1}{s^2}F(s)$$

$$F(s) = \frac{2s}{1+s^2} - \frac{F(s)}{s^2} \quad \left\{\left\{F(s) \to \frac{2s}{\left(1+\frac{1}{s^2}\right)\left(1+s^2\right)}\right\}\right\}$$

$$F(s) = -\frac{2s}{\left(1+s^2\right)^2} + \frac{2s}{1+s^2}$$

Use known Laplace transforms to find the solution.

f1(t) = t sin (t)

$$£(f_1(t)) = \frac{2s}{\left(1+s^2\right)^2}$$

$f_2(t) = \cos(t)$

$$£(f_2(t)) = \frac{s}{1+s^2}$$

$$F(s) = -\frac{2s}{\left(1+s^2\right)^2} + \frac{2s}{1+s^2} \; -£(f_1(t)) + 2\,£(f_1(t)) = -\frac{2s}{\left(1+s^2\right)^2} + \frac{2s}{1+s^2}$$

$F(T) = -F_1(t) + 2\,f_2(t) = 2\cos(t) - t\sin(t)$

We can check this with result using the Laplace Transform.

$$F(s) = -\frac{2s}{\left(1+s^2\right)^2} + \frac{2s}{1+s^2}$$

$$f(t) = 2\cos(t) - t\sin(t)$$

Engineers and physicists sometimes consider forces that produce large effects that are applied over a very short time interval. The force acting at the time an earthquake starts is an example. This leads to the idea of a *unit impulse function* δ(t). Consider the small positive constant a, then the function $\delta_a(t)$ is defined by

$$\delta_a(t) = \begin{cases} \frac{1}{a} & \text{for } 0<t<a \\ 0 & \text{otherwise} \end{cases} \qquad \text{...(118)}$$

The unit impulse function is obtained by letting the interval width go to zero, i.e.,

$$\delta(t) = \lim_{1\to 0} \delta_a(t) \qquad \text{...(119)}$$

Although δ(t) is called the Dirac Delta Function, it is not an ordinary function. To be precise it is a distribution, and the theory of distributions permits manipulations of δ(t) as though it were a function. For our work, we will treat δ(t) as a function and investigate its properties.

Example 35. Find the Laplace transform of the Dirac delta function δ(t).

Solution: For any $a > 0$ we define $\delta_a(t)$ on the interval [0, a] as follows.

$$\delta(a, t) = \frac{1}{a} \text{ for } 0 \le t \le a$$

$$F(a, s) = \int_0^a \delta(a,t)e^{-st}dt$$

$$F(a, s) = \int_0^a \frac{1}{a}e^{-st}dt$$

$$F(a, s) = \frac{\frac{1}{s} - \frac{e^{-1s}}{s}}{a}$$

$$F(a, s) = \frac{e^{-1s}\left(-1+e^{1s}\right)}{as}$$

Plot $\delta_a(t)$ for $a = \frac{1}{10}, \frac{1}{20}, \frac{1}{50, 100}$.

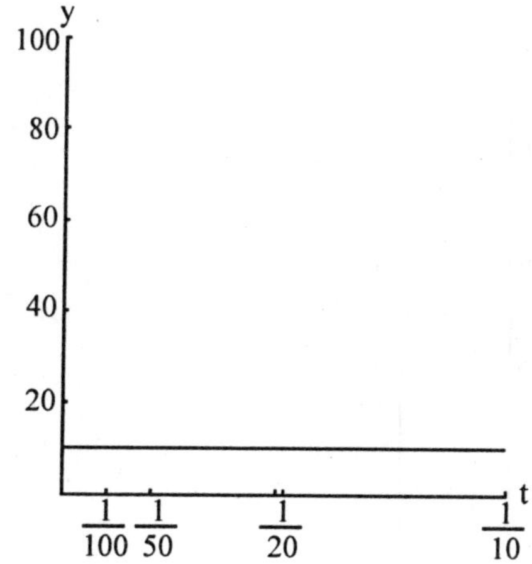

$$\delta\left(\frac{1}{10}, t\right) = 10 \text{ for } 0 \le t \le \frac{1}{10}$$

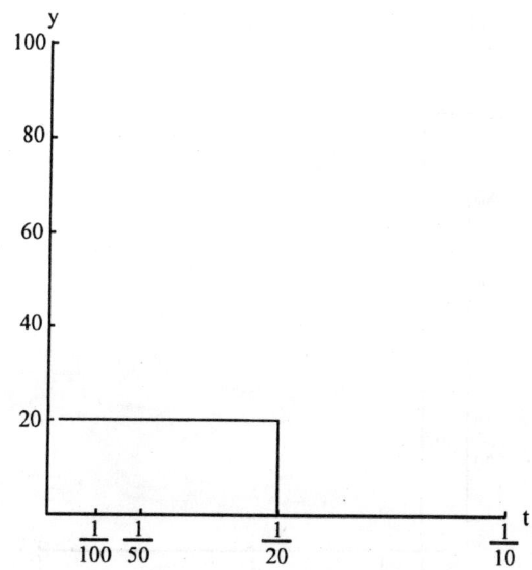

$$\delta\left(\frac{1}{20}, t\right) = 20 \text{ for } 0 \le t \le \frac{1}{20}$$

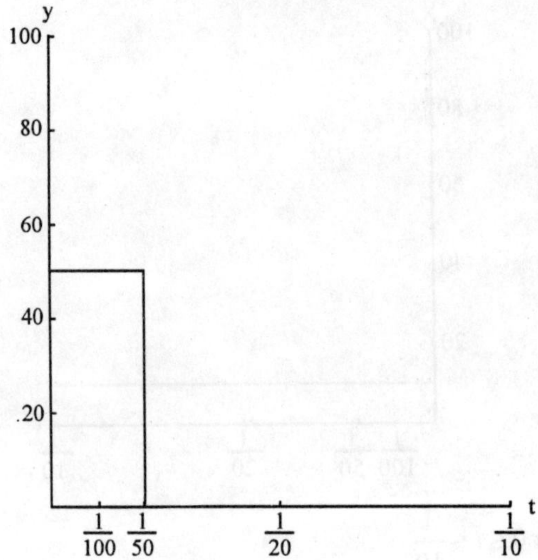

$$\delta\left(\frac{1}{50},t\right) = 50 \text{ for } 0 \le t \le \frac{1}{50}$$

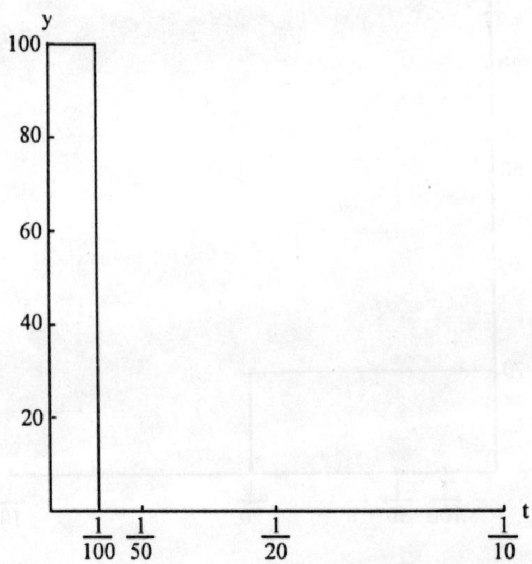

$$\delta\left(\frac{1}{100},t\right) = 100 \text{ for } 0 \le t \le \frac{1}{100}$$

Find

$\lim\limits_{1\to 0}$ £(δ_a(t)) to obtain the Laplace transform of δ(t).

$$\delta(a,t) = \frac{1}{a} \text{ for } 0 \le t \le a$$

$$£(\delta(a,t)) = \frac{e^{-1s}\left(-1+e^{1s}\right)}{as}$$

$$\delta(t) = \text{DiracDelta (t)}$$

$$\delta(t) = \lim_{1\to 0}\frac{1}{a} \text{ for } 0 \le t \le a$$

$$£(\delta(t)) = \lim_{1\to 0}\frac{e^{-1s}\left(-1+e^{1s}\right)}{as}$$

$$£(\delta(t)) = 1$$

We can check this with result using the Laplace Transform.

$$\delta(t) = \text{DiracDelta(t)}$$

$$f(S) = 1$$

Example 36. Solve the initial value problem Y″(t) + 4Y′(t) + 13 Y(t) = 3δ(t) with Y(0) = 0 and Y′(0) = 0.

Solution: Taking transforms results in the following equation.

13Y(t) + 4Y′(t) + Y″(t) = 3 DiracDelta(t)

Y(0) = 0

Y′(0) = 0

$$13Y(s) + 4sY(s) + s^2Y(s) = 3$$

$$\left\{\left\{Y(s) \to \frac{3}{13+4s+s^2}\right\}\right\}$$

$$Y(s) = \frac{3}{13+4s+s^2}$$

From a table of Laplace transforms we get $£\,(e^{-2}t \sin(3t)) = \frac{3}{(s+2)^2+3^2} = \frac{3}{13+4s+s^2}$, and the solution is $Y(t) = e^{-2}t \sin(3\,t)$

We can check this with *Mathematica*'s result using the LaplaceTransform package.

$$Y(s) = \frac{3}{13+4s+s^2}$$

$$Y(t) = -\frac{1}{2} i e^{(-2-3i)t} \left(-1+e^{6it}\right)$$

$$Y(t) = \frac{1}{2}e^{-2t}\sin(3t) - \frac{1}{2}e^{-2t}\cos(6t)\sin(3t) + \frac{1}{2}e^{-2t}\cos(3t)\sin(6t) +$$

$$i\left(\frac{1}{2}e^{-2t}\cos(3t) - \frac{1}{2}e^{-2t}\cos(3t)\cos(6t) - \frac{1}{2}e^{-2t}\sin(3t)\sin(6t)\right)$$

$Y(t) = e^{-2}t \sin(3t)$

The initial condition for must be changed to $Y'(0) = 3$ This is required because $Y'(0^-) = 0$ and $Y'(0^+) = 3$.

13 Y(t) + 4Y′ (t) + Y′(t) = 3 DiracDelta (t)

{Y(0) = 0, Y′(0) = 3}

{{Y(t) → e^{-2}t sin (3 t) UnitStep (t)}}

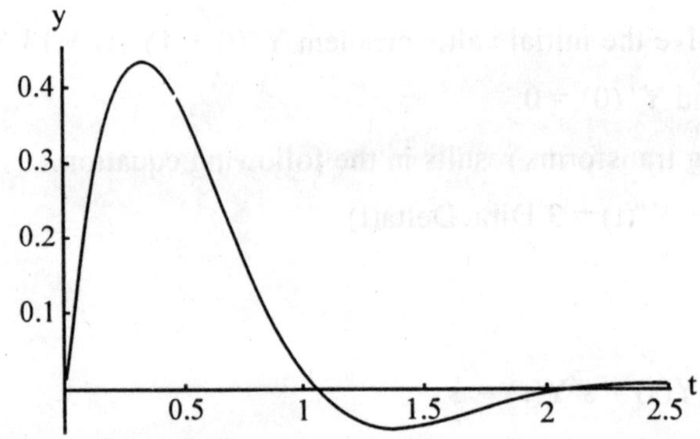

Y(t) = e^{-2}t sin(3 t) UnitStep (t)

Theorem 27. **(Initial value Problem - IVP convolution method)** The unique solution to the initial value problem

$aY''(t) + by'(t) + cy(t) = g(t)$ with $Y(0) = Y0$, and $Y'(0) = Y_1$ is given by

$Y(t) = u(t) + (h * g)(t)$,

where u(t) is the solution to the homogeneous equation $au''(t) + hu'(t) = 0$ with $u(0) = Y0$, and $u'(0) = Y_1$, and h(t) has the Laplace transform given by $H(s) = \frac{1}{as^2 + bs + c}$.

Example 37. Use convolution to solve the initial value problem $Y''(t) + Y(t) = \text{Tan}(t)$ with $Y(0) = 1$ and $Y'(0) = 2$.

Solution:

The solution is accomplished in two steps.

(a) First solve u″(t) + u(t) = 0 with u(0) = 1 and u′(0) = 2.

(b) Second solve for v″(t) + v(t) = Tan (t) with v(0) = 0 and v′ (0) = 0 Then the desired solution is Y(t) = u(t) + v(t).

(a) First solve u″ (t) + u(t) = 0 with u (0) = 1 and u′ (0) = 2.

$$u(t) + u''(t) = 0$$

$$u(0) = 1$$

$$u' = 2$$

$$-2-s + U(s) + s^2U(s) = 0$$

$$\left\{\left\{U(s) \to \frac{2+3}{1+s^2}\right\}\right\}$$

$$U(s) = \frac{2+s}{1+s^2}$$

$$U(s) = \frac{2}{1+s^2} + \frac{3}{1+s^2}$$

$$u(t) = \cos(t) + 2\sin(t)$$

$$u(0) = 1$$

$$u' = 2$$

(b) Second solve for $v''[t] + v[t] = \tan[t]$ with v[0] and v′[0] = 0. The convolution theorem is used to make the construction of $v[t] = (h*g)[t]$ where $H[s] = \frac{1}{s^2+1}$ and g[t] = tan[t]

$$h[t] = \sin[t]$$

$$g[t] = \tan[t]$$

$$h[t-\tau]g[\tau] = \sin[t-\tau]\tan[\tau]$$

$$\int h[t-\tau]g[\tau] = \cos[t]\log\left[\cos\left[\frac{\tau}{2}\right] - \sin\left[\frac{\tau}{2}\right]\right] - \cos[t]\log\left[\cos\left[\frac{t}{2}\right] + \sin\left[\frac{t}{2}\right]\right] - \cos[t]\sin[t] + \cos[t]\sin[t]$$

$$v[t] = \int_0^t h[t-\tau]g[\tau]d\tau$$

$$v[t] = \cos[t]\log\left[\cos\left[\frac{\tau}{2}\right] - \sin\left[\frac{\tau}{2}\right]\right] - \cos[t]\log\left[\cos\left[\frac{t}{2}\right] + \sin\left[\frac{t}{2}\right]\right] + \sin[t]$$

v[0] = 0

v'[0] = 0

Now using trigonometric substitutions to simplify the expression for v[t].

u[t] = cos [t] + 2 sin [t]

u[0] = 1

u'[0] = 2

$$v[t] = \cos[t] \log\left[\frac{cos[t]}{1+sin[t]}\right] + \sin[t]$$

Y[t] = u[t] + v[t]

$$Y[t] = cps[t] + \cos[t] \log\left[\frac{cos[t]}{1+sin[t]}\right] + 3\sin[t]$$

Y[0] = 1

Y'[0] = 2

We can substitute y[t] into the D.E.'s and verify that it is the solution.

$$Y''[t] + Y[t] = -2(1 + \sin[t])\left(-\frac{cos[t]^2}{(1+sin[t])^2} - \frac{sin[t]}{1+sin[t]}\right) \tan[t]$$

$$+ \cos[t]\left(-\frac{cos[t]^2}{(1+sin[t])^2} - \frac{sin[t]}{1+sin[t]} + \sec[t]\,(1+\sin[t])\left(\frac{2cos[t]^3}{(1+sin[t])^3}\right.\right.$$

$$\left.\left. + \frac{3cos[t]sin[t]}{(1+sin[t])^2} - \frac{cos[t]}{1+sin[t]}\right) + \sec[t]\,(1+\sin[t])\right.$$

$$\left(-\frac{cos[t]^2}{(1+sin[t])^2} - \frac{sin[t]}{1+sin[t]}\right) \tan[t]$$

Y''[t] + Y[t] = tan[t]

We use D Solve package to solve the D.E. and plot the solution.

Y''[t] + Y[t] = tan[t]

{Y[0] = = 1, Y'[0] = = 2}

$$\{\{Y[t]\to \cos[t]+\cos[t]\ \log\left[\cos\left[\frac{t}{2}\right]-\sin\left[\frac{t}{2}\right]\right]-\cos[t]\ \log\left[\cos\left[\frac{t}{2}\right]+\sin\left[\frac{t}{2}\right]\right]+3\sin[t]\}\}$$

Remark. This involves the version un-simplified of v[t]. So we will plot our previously simplified function y[t].

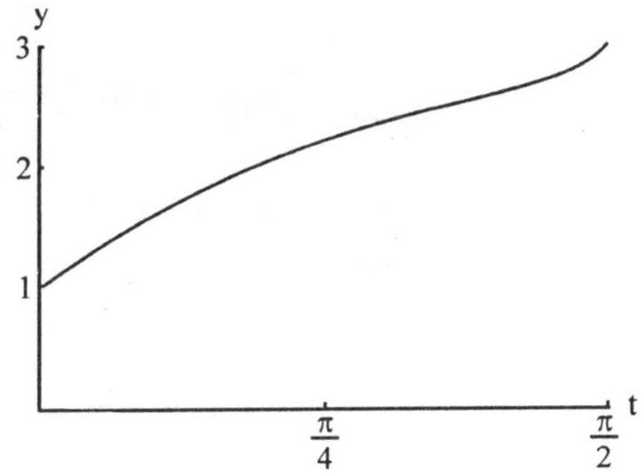

$$Y[t] = \cos[t] + \cos[t]\log\left[\frac{\cos[t]}{1+\sin[t]}\right] + 3\sin[t].$$

EXERCISE

1. Find the Laplace trasnsform of the following functions:

   (i) $t^2\sin^2 t$ (ii) $t\sinh at$ (iii) $t^2e^{-2t}\cos t$

   (iv) $\frac{e^{-at}-e^{-bt}}{2}$ (v) $\frac{1-\cos t}{t^2}$ (v) $f(t)=\begin{cases}\cos\left(t-\frac{2\pi}{3}\right), & t>\frac{2\pi}{3}\\ 0, & t<\frac{2\pi}{3}\end{cases}$

2. Evaluate the following:

   (i) $\int_0^\infty te^{-3t}\sin t\,dt$ (ii) $\int_0^\infty te^{-2t}\cos t\,dt$

3. Prove that $\int_0^\infty \frac{e^{-t}-e^{-3t}}{t}dt = \log 3$

4. Evaluate

   (i) $L\left(\int_0^t \frac{\sin t}{t}dt\right)$ (ii) $L\left(\int_0^t e^{-t}\cos t\,dt\right)$

5. Find the Inverse Laplace trasnsform of the following functions:

   (i) $\frac{3\left(s^2-2\right)^2}{2s^5}$ (ii) $\frac{3s+7}{s^2-2s-3}$ (iii) $\frac{s^3}{s^4-a^4}$

(iv) $\dfrac{s^2+6}{(s^2+1)(s^2+4)}$ (v) $\dfrac{s^2+s}{(s^2+1)(s^2+2s+2)}$ (vi) $\dfrac{1}{2}\log\left(\dfrac{s^2+b^2}{s^2+a^2}\right)$.

6. Using convolution theorem evaluate the following:

(i) $L^{-1}\left\{\dfrac{1}{(s+a)(s+b)}\right\}$ (ii) $L^{-1}\left\{\dfrac{1}{s^2(s^2+a^2)}\right\}$ (iii) $L^{-1}\left\{\dfrac{1}{(s-2)(s+2)^2}\right\}$.

7. Show that $L^{-1}\left(\dfrac{1}{s}\sin\dfrac{1}{s}\right)=t-\dfrac{t^3}{(3!)^2}+\dfrac{t^5}{(5!)^2}-\dfrac{t^7}{(7!)^2}+\ldots$

8. Express $f(x)=1$ for $0\le x\le\pi,$
$= 0 \text{ for } x>\pi$

as a Fourier sine integral and hence evaluate $\displaystyle\int_0^\infty \frac{1-\cos(\pi\lambda)}{\lambda}\sin(x\lambda)\,d\lambda$.

9. Find the Fourier transform of $f(x)=\begin{cases}1 \text{ for } |x|<1\\ 0 \text{ for } |x|>1\end{cases}$

Hence evaluate $\displaystyle\int_0^\infty \frac{\sin x}{x}dx$.

10. Find the Fourier transform of $e^{-|x|}$.

Hence show that $\displaystyle\int_0^\infty \frac{x\sin mx}{1+x^2}dx=\frac{\pi e^{-m}}{2}$, $m>0$.

11. Find the Fourier sine transform of $e^{\frac{-ax}{x}}$

12. Express the function $f(x)=\begin{cases}1, \text{ for } |x|\le 1\\ 0, \text{ for } |x|>1\end{cases}$ as Fourier integral

Hence evaluate $\displaystyle\int_0^\infty \frac{\sin\lambda\cos\lambda x}{\lambda}d\lambda$

13. Use the method of residues, evaluate the inverse Laplace transform of each of the following:

(i) $\dfrac{4s+5}{(s+1)(s-2)^2}$ (ii) $\dfrac{1}{(s-2)(s^2+1)}$ (iii) $\dfrac{1}{s^2(s^2-a^2)}$

(iv) $\dfrac{4s+5}{(s+1)(s-1)^2}$